Drug Resistance in Bacteria, Fungi, Malaria, and Cancer

Gunjan Arora • Andaleeb Sajid •
Vipin Chandra Kalia
Editors

Drug Resistance in Bacteria, Fungi, Malaria, and Cancer

Springer

Editors
Gunjan Arora
CSIR - Institute of Genomics
and Integrative Biology
Delhi, India

Andaleeb Sajid
National Institute of Allergy and
Infectious Diseases
National Institutes of Health
Bethesda, Maryland
USA

Vipin Chandra Kalia
CSIR - Institute of Genomics
and Integrative Biology
Delhi, India

ISBN 978-3-319-48682-6 ISBN 978-3-319-48683-3 (eBook)
DOI 10.1007/978-3-319-48683-3

Library of Congress Control Number: 2017935491

Printed on acid-free paper

This Springer imprint is published by Springer Nature
The registered company is Springer International Publishing AG
The registered company address is: Gewerbestrasse 11, 6330 Cham, Switzerland

Dedicated to my Parents

Preface

Human beings interact with almost all kinds of living beings. The interaction is complexed by the influence of environmental factors. These interactions assume various relationships ranging from symbiotic at one end of the spectrum to pathogenic/parasitic at the other end. Human body—cells, tissues, and organs—has a complete network of metabolic activities and systems, which are well organized. Human systems are tuned to recognize and sense almost all kinds of living and dead molecules and organisms. Taking advantage of these human senses, vaccines are developed to protect humans against pathogenic organisms. Vaccine development being a very complex phenomenon has been pursued for a limited number of diseases. Interestingly, in spite of the presence of a mechanism to prevent entry of foreign entities, human bodies house a large number of microbes on the skin surface, within the gut all the times. A third set of microbes infect the human body and are labeled as pathogenic. Here, bacteria start multiplying at different rates. The bacteria choose to either stay back and cause infectious diseases such as tuberculosis—*Mycobacterium tuberculosis*, *Pseudomonas aeruginosa*, and *Burkholderia cepacia*—or multiply rapidly and hurriedly leave the human body such as in the cases of diarrhea, cholera, and septicemia. However, in all cases, microbes cause enormous damage to human health. Discovery of antibiotics provided mankind with much needed relief from diseases, which had been assuming epidemic dimensions. Incidentally, this discovery could not sustain for long since microbes started reacting to this environmental stress. Microbes developed resistance to antibiotics. This provoked scientists to discover newer antibiotics. This game of success and failure has reached a virtual dead end. From single antibiotic-resistant microbes, we have now seen multiple drug-resistant and extremely drug-resistant microbes. The pathogens have evolved resistance to almost all the antibiotics, which thus are proving ineffective. Scientific morale is going down as pharmaceutical companies are not showing any interest in investing money in this area. Thus, the need is to look for novel drug targets and drugs. We have to develop innovative systems to beat microbial intelligence. The question which is being raised repeatedly is: Shall we mutely witness this onslaught or prepare novel antibacterial?

Research for human welfare needs a strong team of dedicated scientific minds. The young researchers and brilliant scientists can make innovative and significant contributions. This book provides recent developments in the area of Drug Resistance, which can provide leads to the young ignited minds and enable them to translate them into novel drugs for the benefits of the society. In this book, scientists with extremely strong academic knowledge and practical experience have agreed to share their recent research works. This book is a compilation of the research works of dedicated teams of scientific workers who are committed to human welfare. I am sincerely humbled by the contributions made by each of the respected authors. I am really thankful to them for their efforts. Words are not sufficient enough to truly and adequately acknowledge the worthiness of their efforts. My true inspirations to write this book were bestowed on me by the constant faith and support of my mother (Late Mrs. Kanta Kalia), who passed away, during the preparation of this book; my father (Mr. R.B. Kalia); Amita (wife), Sunita and Sangeeta (sisters); Ravi, Vinod, and Satyendra (brothers); Daksh and Bhrigu (sons), my teachers, especially Dr. A.P. Joshi; and my friends Rup, Hemant, Yogendra, Rakesh, Atya, Jyoti, Malabika, Neeru, and Ritusree. I must also acknowledge the support of my young friends Asha, Sadhana, Sanjay, Mamtesh, Subhasree, Shikha, Jyotsana, Ravi, Priyanka, and Rahul.

Delhi, India Vipin Chandra Kalia

Contents

Author Biographies

Andaleeb Sajid completed her Ph.D. in Biotechnology from the CSIR-Institute of Genomics and Integrative Biology (CSIR-IGIB, Delhi, India) in 2011. As a postgraduate student, she was awarded National Eligibility fellowship to support her research on regulation of bacterial physiology through signal transduction events. In 2013, Dr. Sajid joined the postdoctoral position at Laboratory of Clinical Infectious Diseases at the National Institute of Allergy and Infectious Diseases (National Institutes of Health) and worked on antimycobacterial drug discovery with elucidation of novel drug targets. She has more than 11 years of experience of working on biology of human pathogens and their eradication. In this tenure, Dr. Sajid has authored 17 peer-reviewed publications with more than 200 citations and an h-index of 8. Currently, Dr. Sajid is working as Research Fellow at National Cancer Institute (NIH).

Gunjan Arora obtained his Ph.D. in Zoology in 2011 from CSIR-Institute of Genomics and Integrative Biology (CSIR-IGIB, Delhi, India) and the University of Delhi (Delhi, India). After his postgraduation in Marine Biotechnology, Dr. Arora was awarded Junior Research Fellowship and Senior Research fellowship from CSIR (India) to pursue doctoral studies. He examined the role of Ser/Thr/Tyr protein phosphorylation and lysine protein acetylation in physiology of *B. anthracis*, *M. tuberculosis*, and *M. ulcerans* during 7 years of research. He was awarded Vaccine Research Innovation award (Department of Biotechnology, India) in 2013. In 2016, he received Young Investigator Award (Honorable mention) from the American Society of Tropical medicine and Hygiene. He is also a recipient of Global Health Award from Bill and Melinda Gates foundation. His current research interests focus on systems level approaches to understand tropical infectious diseases. During this time, Dr. Arora has authored 18 peer-reviewed publications with 206 citations and an h-index of 9. He is currently working as postdoctoral fellow in NIH Intramural research program.

Vipin Chandra Kalia is presently working as Emeritus Scientist. He has been the Chief Scientist, and the Deputy Director, at Microbial Biotechnology and Genomics, CSIR-Institute of Genomics and Integrative Biology, Delhi. He is a Professor, AcSIR, who obtained his M.Sc. and Ph.D. in Genetics from Indian Agricultural Research Institute, New Delhi. He has been elected as (1) Fellow of the National Academy of Sciences (FNASc), (2) Fellow of the National Academy of Agricultural Sciences (FNAAS), and (3) Fellow of the Association of Microbiologists of India (FAMI). His main areas of research are Microbial biodiversity, Bioenergy, Biopolymers, Genomics, Microbial evolution, Quorum sensing, Quorum quenching, Drug discovery, and Antimicrobials. He has published more than 100 papers in Scientific journals such as (1) Nature Biotechnology, (2) Biotechnology Advances, (3) Trends in Biotechnology, (4) Annual Review of Microbiology, (5) Critical Reviews in Microbiology, (6) Bioresource Technology, (7) PLoS ONE, (8) BMC Genomics, (9) International Journal of Hydrogen Energy, and (10) Gene. He has authored 14 book chapters. His works have been cited 3702 times with an h-index of 32 and an i10 index of 71 (http://scholar.google.co.in/citations?hl=en&user=XaUw-VIAAAAJ). He has edited three books: (1) Quorum Sensing vs Quorum Quenching: A Battle with No End in Sight (2015), http://link.springer.com/book/10.1007/978-81-322-1982-8; (2) Microbial Factories Vol. 1: Biofuels, Waste treatment (2015) http://link.springer.com/book/10.1007%2F978-81-322-2598-0; and (3) Microbial Factories Vol. 2: Biodiversity, Biopolymers, Bioactive Molecules (2015) http://link.springer.com/book/10.1007%2F978-81-322-2595-9. He is presently the Editor-in-Chief of the Indian Journal of Microbiology and editor of (1) Journal of Microbiology & Biotechnology (Korea), (2) International Scholarly Res. Network Renewable Energy, (3) Dataset Papers in Microbiology, and (4) PLoS ONE. He is a life member of the following Scientific societies: (1) Society of Biological Chemists of India; (2) Society for Plant Biochemistry and Biotechnology, India; (3) Association of Microbiologists of India; (4) Indian Science Congress Association; (5) BioEnergy Society of India; and (6) the Biotech Research Society of India (BRSI). He is also a member of the American Society for Microbiology. He can be contacted at vckalia@igib.res.in; vc_kalia@yahoo.co.in

Emerging Themes in Drug Resistance

Gunjan Arora, Ankur Kulshreshtha, Kriti Arora, Puneet Talwar, Rishi Raj, Gurpreet Grewal, Andaleeb Sajid, and Ritushree Kukreti

Abstract In the last century, advances in biomedical sciences led to improvements in quality of human life mainly by decreasing the disease burden and providing various healthcare innovations. Effective medications were developed for most infectious diseases, lifestyle disorders, and other diseases. For once, we all believed that we can create a disease-free world; however, the excessive use of medications generated drug-resistant human pathogens leading to untreatable forms of many diseases. In the last few decades, the focus has been to understand the evolution of drug resistance, tackle the current drug-resistant disease agents, and develop new drugs that are potent and reliable for long-term usage. In this chapter, we will discuss the emerging themes in drug resistance research. As biologists are gaining deeper understanding of cellular complexity and disease agents, numerous modern themes have been pursued. The focus is to identify novel drug targets and develop specific drug molecules detrimental for the disease causing pathogens, and to harness host immunity. For the scope of this chapter, we will primarily discuss

G. Arora (✉)
National Institute of Allergy and Infectious Diseases, National Institutes of Health, Rockville 20852, MD, USA
e-mail: arorag1983@gmail.com

A. Kulshreshtha
CSIR-Institute of Genomics and Integrative Biology, Mall Road, Delhi 110007, India

International Centre for Genetic Engineering and Biotechnology (ICGEB), Aruna Asaf Ali Marg, New Delhi, Delhi 110067, India

Jawaharlal Nehru University, New Delhi, Delhi 110067, India

K. Arora • A. Sajid
Tuberculosis Research Section, National Institute of Allergy and Infectious Diseases, National Institutes of Health, Bethesda 20892, MD, USA

P. Talwar • G. Grewal • R. Kukreti (✉)
CSIR-Institute of Genomics and Integrative Biology, Mall Road, Delhi 110007, India
e-mail: ritus@igib.res.in

R. Raj
Molecular Biophysics Unit, Indian Institute of Science, Bangalore, Karnataka 560012, India

G. Arora et al. (eds.), *Drug Resistance in Bacteria, Fungi, Malaria, and Cancer*,
DOI 10.1007/978-3-319-48683-3_1

pharmacogenomics, therapeutic antibodies, cell-to-cell communication, exosomes, structuromics, and posttranslational modifications.

1 Pharmacogenomics

1.1 Introduction

The completion of human genome project in 2003 followed by emergence of high-throughput sequencing and genotyping technologies has led to the initiation of a new era in the research field with central focus on genomics. This has changed the way of looking at conventional study approaches leading to merging of different domains, which resulted in the opening of new research frontiers such as pharmacogenomics, network medicine, and systems pharmacology to name a few. In this section, we aim to provide a broad overview of the pharmacogenomics, including major developments that have occurred in this field and finally highlighting the significant achievements and future perspective.

Pharmacogenomics (PGx) and pharmacogenetics (PGt) involve individual's genetic signatures to study drug response and/or drug toxicity behavior often referred to as phenotype. PGx differs from PGt in terms of the involvement of entire genome instead of few selected genes with no prior hypothesis. Genetic signature refers to the markers such as single-nucleotide polymorphisms (SNPs) and copy number variations (CNVs). Drug response phenotype implies to individual who responds well to the treatment (the good responder) and to one who doesn't (the poor responder) based on drug metabolism profile categorized into extensive metabolizer, intermediate metabolizer, and poor metabolizer. Individuals with extensive-metabolizer phenotype excrete the drug from the body completely earlier than expected, failing to achieve its desired therapeutic effect. These individuals require higher doses than the normal individuals which are basically intermediate metabolizers. Poor metabolizers accumulate the drug in the body and may experience adverse drug reactions (ADR). Some people also show drug toxicity due to hypersensitivity reactions that may also be due to the individual specific genetic signatures such as genes in the human leukocyte antigen (HLA) system. So, these individuals may require a lower dose or switch to another drug.

Several factors affect the outcome of a PGx/PGt study including ethnicity, disease (brain disorders-serum drug levels are not useful), genetics (host susceptibility genes) and environment (diet, smoking, alcohol); and therefore these are considered an important part of study design based on the study hypothesis. The steps involved in a PGx/PGt study are depicted in Fig. 1.

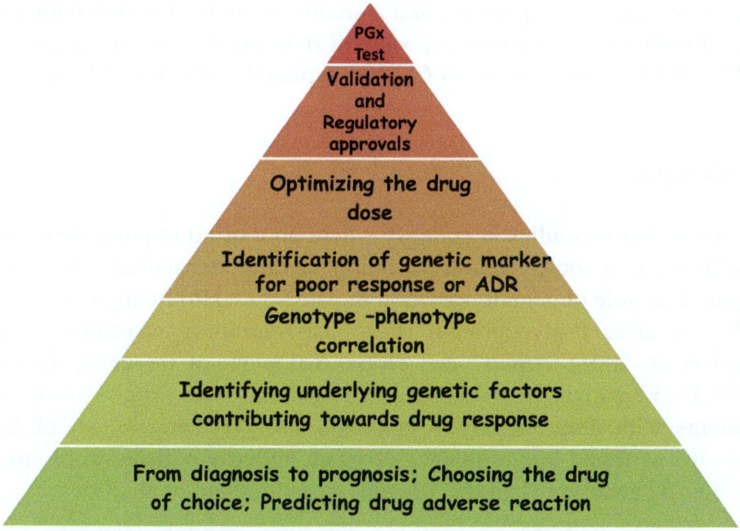

Fig. 1 Steps toward personalized medicine: To date around 136 FDA-approved drugs are available with pharmacogenetics information in their labeling

1.2 Role of Pharmacogenomics in Drug Resistance

The ultimate goal of pharmacogenomics is the development of optimized drug therapy based on the genetic makeup of an individual with maximum efficacy. The drug effects are determined by number of drug-metabolizing genes categorized into three groups—phase I (functionalization by cytochrome P450 superfamily), phase II (conjugation by conjugating enzymes such as sulfotransferases and UDP-glucuronosyltransferases), and phase III (excretion by drug transporters). The genetic variation in these genes may lead to differential response to drug treatment which can have important clinical implications. Drug transporters (phase III) have pivotal role in regulating the absorption, distribution, and excretion of many medications. There are number of studies where inherited variation in these transporters has been associated with differential response in different diseases and development of drug resistance in many diseases. Few such studies are discussed below.

1.2.1 Tuberculosis

Rifampicin is a first-line drug used to treat tuberculosis. Weiner et al. conducted a pharmacokinetics study in pulmonary tuberculosis patients from Africa, North America, and Spain compared with North American healthy controls to investigate the reasons for the interindividual variations in rifampicin levels. The study found that polymorphisms in the *SLCO1B1* gene (encodes drug transporter OATP1B1—

organic anion transporter peptide) had a significant influence on rifampicin exposure. Approximately 36 % lower exposure of rifampicin was found in patients with *SLCO1B1* 463 CA genotypes than CC genotypes (Weiner et al. 2010).

1.2.2 Epilepsy

There are number of studies in epilepsy where poor or no response to antiepileptic drugs (AEDs) is associated with variation in drug transporters. Siddiqui et al. investigated the role of genetic variation in *ABCB1* (ATP-binding cassette subfamily, B1), encoding P-glycoprotein (P-gp) for multidrug resistance in epilepsy (Siddiqui et al. 2003). *ABCB1* can efflux AEDs out of the cells. They studied *ABCB1* C3435T variant (rs1045642) in 200 patients with drug-resistant epilepsy, 115 patients with drug-responsive epilepsy, and 200 controls, all of European ancestry. It was found that multidrug-resistant epilepsy patients were more likely to have the CC genotype than the TT genotype (Siddiqui et al. 2003).

1.2.3 HIV

The pharmacokinetics profile of antiretroviral drugs used in HIV therapy is also dependent on genetic differences in drug transporters. Anderson et al. conducted a study to understand the relationship of antiretroviral drug pharmacokinetics and pharmacodynamics with polymorphisms in drug-metabolizing genes. They observed that "indinavir" oral clearance was 24 % faster in individuals having multidrug resistance-associated protein 2 (MRP2)-24C/T variant which can interfere with drug efficacy as indinavir is a substrate of MRP2 and thus can limit drug absorption and accelerate drug clearance (Anderson et al. 2006).

1.3 Role of Pharmacogenomics in Drug Discovery

The information from PGx can help in drug discovery and development. The example of Herceptin (trastuzumab), a humanized monoclonal antibody against ErbB2 receptor used in breast cancer treatment (Vogel and Franco 2003; Noble et al. 2004), explains how well PGt tests can help in progressing drug discovery and development. It has been observed that more positive response was observed in patients with tumors overexpressing ErbB2. With PGt testing, individuals having overexpression of ErB2 and appropriate for Herceptin treatment can be identified (McCarthy et al. 2005). Similar approach can be used with drug transporters. One approach can be the use of specific inhibitors against transporters to stop efflux of drugs to overcome drug resistance in subgroup where overexpression of specific transporter is observed. The other approach can be identifying the molecular players like nuclear receptors regulating expression of transporters as their genetic variability is responsible for nonresponsiveness to drugs and drugs can be designed against those molecular players to achieve therapeutic efficacy. More research in PGx and development of specific technologies can really help in progress of medicinal sciences.

2 Therapeutic Antibodies for Bacterial Diseases

2.1 The Growing Problem of Antimicrobial Resistance

Antibiotics are the most commonly prescribed class of therapeutic agents both prophylactically and therapeutically. However, the use of antibiotics is not just limited to clinical settings. In an attempt to make food safer to eat, antibiotics have been introduced in poultry and dairy farms. In addition, several antibiotics are also being exploited in water sources for irrigation to prevent crop from diseases. This indiscriminate use of antibiotics has contributed heavily to the spread of antimicrobial resistance. Resistance to most antibiotics has been observed within 4–7 years of their introduction (Clatworthy et al. 2007). This has led to reemergence of infectious diseases, which had previously been effectively controlled by chemotherapy. This has in turn prompted surveillance programs by the World Health Organization (WHO) and Centers of Disease Control and Prevention (CDC) and European Center for Disease Prevention and Control (ECDC) to monitor the rise and spread of antibiotic resistance (http://www.who.int/drugresistance/global_action_plan/en/, http://www.cdc.gov/drugresistance/cdc_role.html, http://ecdc.europa.eu/en/healthtopics/antimicrobial_resistance/database/Pages/database.aspx).

According to current estimates by the CDC, there were two million cases of infections with 23,000 deaths from antibiotic-resistant organisms every year in the United States alone (2013 Threat Report, CDC http://www.cdc.gov/drugresistance/threat-report-2013/index.html). Several bacterial and fungal pathogens have made a comeback and have once again become an unmet clinical need of infectious disease (Fauci and Morens 2012). The ESKAPE pathogens, namely, *Enterococcus faecium*, *Staphylococcus aureus*, *Klebsiella pneumoniae*, *Acinetobacter baumannii*, *Pseudomonas aeruginosa, and Enterobacter* species, are posing a serious challenge in the nosocomial environment and are rapidly becoming pan resistant to all known classes of antibiotics used against them (Rice 2008).

2.2 The Need for New Therapeutic Interventions

Owing to the growing problem of antimicrobial resistance, there is an urgent need for development of new therapeutic agents against infectious pathogens. Antimicrobial resistance has surfaced against all known classes of drugs including colistin which has been considered the last-resort antibiotic against several multidrug-resistant pathogens (Liu et al. 2016). With the exhaustion of the clinically useful antibiotic repertoire, attention is now being focused on other biologics that can serve to incapacitate pathogens. The post-antibiotic era has begun, and with that drug discovery programs are shifting focus toward large-molecule therapies which are based on an understanding of the host's own immune defenses. Antibodies are one such class of agents that have historic precedence for use in successful treatment of certain infections. In the pre-antibiotic era, as early as the 1890s, antibodies were being used for the treatment of select infections, but with the discovery of antibiotics, antibody therapies took a backseat. As a result, their potential was neither fully realized nor exploited.

2.3 The Foundations of Antibody Therapy

The discovery of passive antibody immunotherapy for infectious diseases is credited to Emil von Behring and Shibasaburo Kitasato. They were the first to show that serum from guinea pigs infected with a sublethal dose of *Corynebacterium diphtheriae* could be used to protect healthy guinea pigs from a lethal challenge of the organism. Similar protection was also shown by them against another bacterial pathogen, *Clostridium tetani* (Von Behring and Kitasato 1965, 1991). For this discovery in 1901, von Behring secured the first physiology/medicine Nobel Prize. As is evident, the protection was afforded by antibodies directed against the diphtheria or tetanus toxins. At the same time, Paul Ehrlich made his contribution to the field by standardizing the therapeutic dose by titrating the antiserum in infected animals and establishing methods for generation of high-titer antiserum using large animals such as horses (Bosch and Rosich 2008). This was an important advancement in dose determination for immune-based therapies for which he was awarded the Nobel Prize in 1908 (along with Élie Metchnikoff for his discovery of phagocytosis). In the United States, the mass production of diphtheria antiserum for therapeutic use began in 1895 by the H. K. Mulford Company in Glenolden, Philadelphia (later Merck Sharp & Dohme Corp. in 1929). Subsequently, antiserum was successfully used in the treatment of some bacterial diseases such as diphtheria, tetanus, pneumococcal pneumonia, and meningitis (Casadevall 1996). These early advances not only helped lay the foundations of vaccinology but also established the therapeutic potential of antibodies. However, despite these early achievements, interest in serum-based therapies rapidly declined in the 1930s. The reasons for this decline included the problem of serum sickness (due to the presence of foreign animal proteins), unpredictable therapeutic outcomes (which was due to instability of preparations, differences in batches, and a lack of understanding of disease stage at the time of administration), and most importantly perhaps the rise of antibiotics. While antibodies continued to be exploited for vaccine discovery, their application to treatment of infections quickly decreased. After the discovery of penicillin by Alexander Fleming in 1928, antibiotics quickly became the magic bullets for treatment of life-threatening and nonlife-threatening microbial infections.

2.4 *Antibody Structure*

The typical immunoglobulin G (IgG) molecule is a Y-shaped molecule which has two longer heavy chains each of which is in turn linked to two shorter light chains by disulfide bonds. The heavy chains are also linked to each other by disulfide bonds. The light chains and their associated heavy-chain regions which form the fork of the "Y," known as the Fab fragment, are involved in antigen binding. The stem of the "Y" which is composed of the two heavy chains is known as the Fc portion and is involved in binding to receptors on the surface of phagocytic cells. The Fab and Fc fragments are separated from each other by a hinge region which allows flexibility of movement to the Fab fragment. Further, the heavy and light chains are divided into distinct domains known as the variable and constant regions. Each heavy chain has three constant regions (C_H1, C_H2, and C_H3) and one variable (V_H) region. Each light chain has one constant (C_L) and one variable (V_L) region. Each Fab fragment is therefore composed of the V_H and V_L regions (which form the V region) and the C_H1 and C_L regions (which form the C region). The amino acid residues in the variable region define the specificity of the Ab. The V region of the Fab fragment which makes contact with the antigen is known as the paratope, and it is complementary to the residues on the antigen which form the antigen's epitope (Putnam et al. 1979; Litman et al. 1993; Mattu et al. 1998; Maverakis et al. 2015).

2.5 *Types of Therapeutic Antibody Molecules*

Therapeutic antibodies come in many flavors (Chames et al. 2009). From the conventional polyclonal sera to antibodies with multiple specificities and unconventional structure, all forms are currently being explored in therapy (Fig. 2).

1. While polyclonal antibodies are not as common in development anymore, *monoclonal* antibodies are still at the forefront of antibody therapies.
2. *Chimeric* antibodies combine the portions of mouse and human antibodies where the human portion makes up to 70 %, allowing better specificity of the Fc region and reduced toxicity.
3. Further advancement of technology has made possible the generation of *humanized* antibodies which are 85–50 % human, and these therefore have even better safety profiles.
4. Antibody fragments comprising solely of the VL and VH domains, termed *scFv* (single-chain variable fragment), are used to provide specificity in the absence of Fc receptor-mediated opsonophagocytosis.

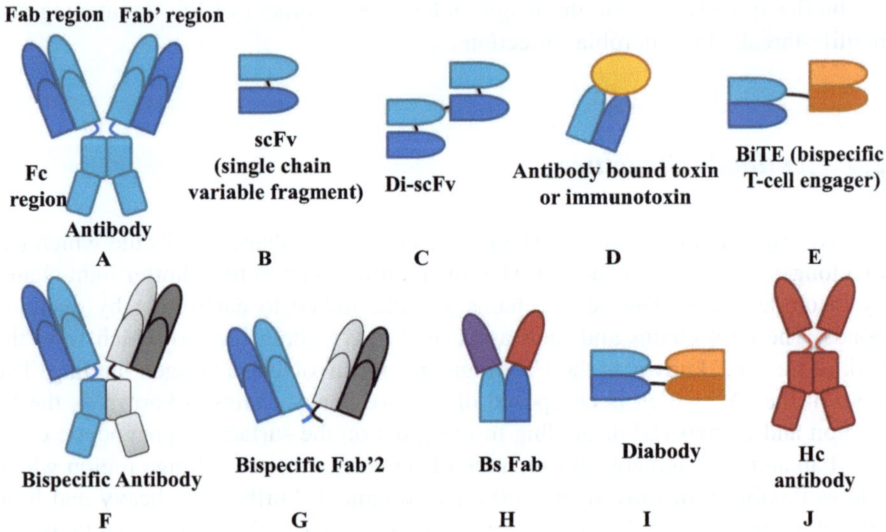

Fig. 2 Therapeutic antibody structural representation: (**a**) basic antibody structure, (**b**) single-chain variable fragment, (**c**) dimer of single-chain variable fragment, (**d**) toxin/therapeutic compound bound to single-chain variable fragment, (**e**) bispecific T-cell engager, (**f**) bispecific antibody, (**g**) bispecific F(ab)2 region, (**h**) bispecific Fab region, (**i**) diabody, and (**j**) heavy-chain only antibody

5. Two scFv linked together with a very short linker make a structure known as the *diabody*.

2.6 Clinical Pipeline

The first monoclonal antibody to be approved for clinical use against an infectious agent was Palivizumab/Synagis for respiratory syncytial virus (RSV) in 1998. Palivizumab has been developed by MedImmune against the F protein of RSV (Johnson et al. 1997), for treatment of severe respiratory infection in high-risk infants. In recent years, antibody therapies have been in development against many infectious pathogens. Antibody therapies are being developed to enable pathogen clearance by diverse approaches.

2.6.1 Neutralization of Viruses and Toxins

Direct neutralization by formation of antigen-antibody complexes which can be cleared by the immune system.

2.6.2 Opsonophagocytic Killing

Antibodies specific for proteins on the pathogen's surface. The Fc region is in turn able to bind the Fc receptors on surface of phagocytic cells. This allows the phagocyte to directly engulf and mediate killing of the pathogen.

2.6.3 Complement Activation

Binding of antibodies can also lead to complement activation which in turn can lead to pathogen lysis by formation of the membrane attack complex (particularly in Gram-negative bacteria and fungi).

2.6.4 Antibody-Drug Conjugates

Antibody-drug conjugate is a novel class of biologics where an antibody provides a means of delivery of a small-molecule chemotherapeutic agent.

There are several candidates currently in clinical development that are based on one or more of the abovementioned approaches (ter Meulen 2007, 2011; Bebbington and Yarranton 2008; Nagy et al. 2008; Saylor et al. 2009). These therapies are frequently directed against surface structures of the pathogens which are often critical for attachment, toxin injection, and immune evasion. Also, many candidates in development are directed against bacterial toxins which continue to cause injury to the host even after elimination of the pathogen. Some of them are discussed below:

Bacillus One such target is the protective antigen (PA) of *Bacillus anthracis*. PA is a pore-forming protein of *B. anthracis* which heptamerizes, forming a pore in the endosomal membrane through which it delivers lethal factor (LF) and edema factor (EF) into the cell cytosol of host. All the antibodies in development against *B. anthracis* PA act by blocking the binding of PA to its receptor. Raxibacumab (Human Genome Sciences, NCT02016963), AVP-21D9 (Emergent BioSolutions, NCT01202695), Anthim (Elusys Therapeutics, NCT01453907), and Valortim (PharmAthene, NCT00964561) are antibodies in development for anthrax. Raxibacumab has recently been approved for treatment of inhalation anthrax in combination with antibiotics (Tsai and Morris 2015).

Clostridium Toxin neutralization is also an effective strategy for *Clostridium botulinum*. There are seven toxins elaborated by *C. botulinum* BoNT/A-G. Of these BoNT/A is the most devastating toxin causing paralysis. A mixture of three IgG monoclonal antibodies with humanized regions, XOMA 3AB, is currently under development against BoNT/A (Nayak et al. 2014) (NCT01357213). The antibody targets different regions of the BoNT/A toxin. *Clostridium difficile*, which causes pseudomembranous colitis or diarrhea associated with antibiotics, exerts its effects by production of toxins A (TcdA) and B (TcdB), which mediate damage to epithelial cells (Voth and Ballard 2005). A combination of human monoclonal antibodies to TcdA (GS-CDA1) and TcdB (MDX-1388) is currently in development by Medarex (NCT00350298). Similarly, Merck Sharp & Dohme Corp. is also developing actoxumab-bezlotoxumab (a mixture of monoclonal antibodies to TcdA and TcdB which has completed phase III clinical trial (NCT01513239, Yang et al. 2015)).

Staphylococcus *Staphylococcus aureus* produces many surface-associated as well as secreted virulence factors and toxins (such as proteases, adhesins, superantigens, leukocidins, autolysins) that serve as good targets for antibody therapy (Sause et al. 2016). Altastaph is a polyclonal IgG preparation directed against capsular polysaccharides 5 and 8 which has completed phase II trial (NCT00063089). Another polyclonal antibody in development (from Bristol-Myers Squibb) is Veronate which is directed against fibrinogen-binding protein, clumping factor A (ClfA) (DeJonge et al. 2007), but it failed to show any significant protection in the phase III trial (NCT00113191). Similarly a monoclonal Ab, Aurexis/tefibazumab (also developed by Bristol-Myers Squibb) directed against *ClfA* and aimed to be used in combination with antibiotic therapy has also completed phase IIa dose-escalation study (NCT00198289). Pagibaximab (Biosynexus Inc.) is a monoclonal antibody directed against lipoteichoic acids of staphylococcal cell wall and is in development for sepsis in neonates (NCT00646399). Aurograb (NeuTec Pharma.) has completed phase III trial for use in combination with vancomycin for treatment of methicillin-resistant *S. aureus* infections. Sanofi is developing SAR279356 (NCT01389700) a monoclonal antibody against the surface polysaccharide poly-N-acetylated glucosamine (PNAG) (Kelly-Quintos et al. 2006).

Escherichia coli Chimeric monoclonal antibodies to Shiga toxins 1 and 2 (Stx 1 and 2) are in development by Thallion Pharmaceuticals (NCT01252199) against bacteria that produce Shiga toxin such as "Shiga toxin-producing *E. coli* (STEC)." This organism can cause hemolytic uremic syndrome and acute renal failure.

Pseudomonas *Pseudomonas aeruginosa* is one of the most recalcitrant nosocomial pathogens. *Pseudomonas* pathogenesis relies on production of cytotoxins, surface adhesion molecules, and biofilms. Panobacumab (also known as KBPA-101 and Aerumab11) is a pentameric *IgM* monoclonal antibody targeting the O11 lipopolysaccharide of *P. aeruginosa* which is being developed by Kenta Biotech Ltd. and has completed phase IIa safety and pharmacokinetics trial (NCT00851435, Lazar et al. 2009; Que et al. 2014). *Pseudomonas* infections are particularly problematic in cystic fibrosis patients. KB001-A is a humanized monoclonal antibody targeting the pcrV structural component of the *Pseudomonas* type III secretion system, essential for translocation of *Pseudomonas* cytotoxins, and is being developed by KaloBios Pharmaceuticals (NCT01695343) for treatment of cystic fibrosis patients. MedImmune is developing MEDI3902 which is a bispecific antibody against *P. aeruginosa*. The antibody simultaneously targets the PcrV protein as well as Psl (a capsular lipopolysaccharide) and has shown promise for use independently as well as an adjunct to existing chemotherapeutic regimens (DiGiandomenico et al. 2014). It has recently completed a phase I clinical trial (NCT02255760).

2.7 Advantages and Disadvantages of Antibody Therapy

Antibody therapeutics offer several advantages over antibiotic therapy. First and foremost, antibodies have a good safety profile (especially monoclonals) and hence are better tolerated. Owing to their specificity, they suffer less from off-target and nonspecific effects. Also due to their specificity, antibodies do not have adverse effects on normal microflora of the body. However, the extreme specificity of antibodies makes them very narrow-spectrum therapeutics which in turn makes them less attractive for pharmaceutical companies. Nevertheless, antibodies have multifunctional capabilities. They don't just exert their effects by directly blocking epitopes but also enhance immune function by mediating opsonization, agglutination, complement activation, and engagement of toxic cells by antibody-dependent cellular cytotoxicity (ADCC). Thus, they not only block pathogen attack but also aid in pathogen clearance from tissue spaces and bloodstream. When used in combination with antibiotics, they can work in conjunction with curtail pathogens and in some instances have even been shown to work synergistically to enhance the efficiency of antibiotics to which resistance has developed (DiGiandomenico et al. 2014). Another important advantage of antibodies is that they can be used not only therapeutically but also prophylactically.

Antibody therapies do suffer from several challenges as well. Besides being extremely narrow in their spectrum, they also pose the risk of being rendered ineffective if epitopes that they are directed against undergo a change. Bacterial and viral pathogens are known to show antigenic variation which can make antibodies ineffective. However in such a case, the paratope on the antibody may be modified to allow targeting of the new epitope. Further, since these antibodies are proteins, there is a chance that the immune system could mount a response against the antibody itself. Such anti-idiotypic antibodies can neutralize the therapeutic molecule, and they have been documented in treatment with therapeutic antibodies at least in case of rheumatoid arthritis (Isaacs et al. 1992). Additionally, certain types of antigens are only elaborated by select strains of the pathogen making such antigens less attractive for clinical development. Knowledge of the antigenic epitope is also crucial for developing an antibody therapeutic.

3 Development of Drug Resistance Through Inter- and Intracellular Communication

Cell-to-cell communication is the key to cellular adaptation and survival. Different types of cells have developed ways to communicate with each other. Initially, the intercellular communication was believed to be maintained by soluble paracrine and endocrine factors and by physical cell-to-cell contacts. The first of such contacts discovered were cell-cell synapses, gap junctions, plasmodesmata, and cellular projections (Abounit and Zurzolo 2012). In eukaryotes, one of these ways is through formation of nanotubular networks that provide valuable communicative means. Tunneling nanotubes, the long extensions of membrane between two cells placed distantly, and their networks were first discovered in plant and animal cells (Tarakanov and Goncharova 2009). In 2011, Dubey et al. showed how bacteria communicate with each other by nanotubes (Dubey and Ben-Yehuda 2011). Interestingly, bacteria can form nanotubes not only between same species but between different species as well (Dubey and Ben-Yehuda 2011; Pande et al. 2015). These nanotubes have been shown to be involved in cargo transport and antibiotic resistance (Dubey and Ben-Yehuda 2011; Pande et al. 2015). It is not clear if these bacterial nanotubes are just a communication mechanism or a survival strategy to counteract drug pressure.

Further, the molecular mechanisms of nanotube formation are still unknown. The study by Pande et al. has shown that bacteria form nanotubes during stress conditions such as nutrient starvation (Pande et al. 2015). These bacterial nanotubes reflect complex social behavior as bacteria from different species can form nanotubes to communicate (Dubey and Ben-Yehuda 2011; Pande et al. 2015). Hence, such nanotube formation may be a mechanism to exchange nutrients and other metabolites in bacterial communities such as biofilms. Biofilm-forming bacteria are known to survive drug pressure, and it is possible that in community some bacteria transfer metabolites to protect others from drug-induced pressure. The fitness advantage can answer how some bacteria survive drug pressure and acquire antibiotic resistance by staying metabolically quiescent for very long time. Further, mechanism of this vectorial transport also needs to be elaborated. One of the ways to maintain directionality is quorum sensing, the other could be electrical coupling. Also, concentration-dependent gradient between the two bacteria may help in determining the direction of flow.

Structurally, bacterial nanotubes resemble vectorial structure. Temperature may affect membrane motility, permeability, and membrane potential that in turn may affect nanotube composition, structure, and number. Another mode of intracellular communication is synapse, a link between nerve cells that are separated by a small gap across which neurotransmitters can be diffused through to generate an impulse. Synapse has also been proposed in plants and between immune and target cells, long after the corresponding notion of neuronal synapse in animals. Tunneling nanotubes have been shown to regulate neuronal synapses. These F-actin-based structures are involved in organelle transfer and electrical coupling between cells

(Wang and Gerdes 2015). These nanotubes may help drug-sensitive cancer cell to connect with drug-resistant mutant cells and survive drug pressure (Wang and Gerdes 2015). Although tunneling nanotubes are absent in bacteria, similar structural forms are involved in antibiotic resistance indicating divergent evolution.

Nanotubes have been shown to transport proteins, plasmids, and other macromolecules between the cells. This mechanism may confer high levels of resistance markers being transferred between the cells. Consequently, nanotubular structures will be an important target for designing new drugs to overcome development of drug resistance. These drugs can be given in addition to other drugs that kill the target cells. Thus, combination therapies can be developed targeting bacterial physiology as well as nanotube formation. These therapies would be sensitive as well as specific for a given disease, considering the fact that nanotubular structures are specific for a given cell type.

4 Exosome-Mediated Drug Resistance

4.1 Extracellular Vesicles as Carriers of Drug-Resistant Traits Across Pathologies

Drugs impart an evolutionary selection pressure on the target cells. The phenomenon of transmission of drug resistance across population involves multiple mechanisms and helps in selecting for resistant cells from heterogeneous population. One recently discovered mechanism that has garnered much attention is the transfer of these traits by means of secreted vesicles, also termed as extracellular vesicles (EVs) (Yanez-Mo et al. 2015). These vesicles are further subclassified depending upon their size and density as either large oncosomes derived from cancerous cells (ranging in size from 1 μm to 10 μm), microvesicles (100 nm to 1 μm), and exosomes (30–100 nm) (Minciacchi et al. 2015). While large oncosomes and microvesicles are the recent focus of attention, exosomes have long been studied, and sufficient evidence exist that implicate their role in pathologies (Robbins and Morelli 2014; Braun and Moeller 2015; Campos et al. 2015; Coleman and Hill 2015; Fujita et al. 2015; Mahmoudi et al. 2015; Schorey et al. 2015).

4.2 Vesicle-Mediated Drug Resistance in Noncommunicable Disorders

A major impediment in managing the epidemic of noncommunicable disorders (NCDs) is the emergence of resistance to the existing line of drugs. Drug resistance is a phenomenon that comes across disease boundaries including emergence of steroid-resistant asthma (Luhadia 2014), insulin resistance leading to non-insulin-dependent diabetes mellitus, obesity, hypertension, dyslipidemia, atherosclerosis (DeFronzo 1997), and most importantly chemoresistance in cancers (Singh and Settleman 2010). While the direct role for exosomes has not been established for other conditions, their role in cancer drug resistance has been documented in several studies. Acquired and de novo resistance to existing therapies including chemotherapy, radiation therapy, and other targeted therapies has become a huge concern in the treatment of cancer (Meads et al. 2009), and exosomes have been implicated in several of these mechanisms (Azmi et al. 2013).

Radiation therapy employs high-energy waves, such as X-rays, electron beams, gamma rays, or protons for eliminating cancer cells, and theoretically this presents an attractive therapeutic choice; however, few cancerous cells were found to survive these high-energy radiations, the detailed cause for which is still being investigated. It has emerged recently that one of the mechanisms by which cancerous cells counters radiations is by secreting "survivin" which is a member of the inhibitor of apoptosis protein (IAP) family in exosomes (Khan et al. 2009). This creates a microenvironment that promotes further metastasis and drug resistance. Cancer cells have been shown to use exosomal pathway to physically efflux drugs, cisplatin, and doxorubicin that are one of the most widely used drugs (Shedden et al. 2003; Safaei et al. 2005). Similarly, in a breast cancer cell line that expresses Her2, the resistance to trastuzumab (monoclonal antibody-targeting Her2 receptor) arises due to exosomes that overexpress and secrete Her2. These exosomes are released by cancer-associated fibroblasts that increase cancer stem cells and induce several anti-apoptotic pathways (Ciravolo et al. 2012). Exosomes transferred from stroma to breast cancer cells lead to resistance against chemotherapy and radiation through antiviral and NOTCH3 pathways (Boelens et al. 2014). Tumor microenvironment has classically been associated with chronic hypoxia that can affect the DNA damage repair pathways and thereby induce DNA replication errors. This leads to genetic instability contributing to radiation resistance (Bristow and Hill 2008). Horizontal transfer of miRNAs and phosphorylated glycoproteins by drug-resistant breast cancer cell lines has also been suggested to be a novel mechanism of transmission of chemoresistance (Chen et al. 2014). Substantiating this principle, it was found that exosomes from docetaxel (DOC /exo)-resistant MCF-7 breast cancer cells can confer drug resistance in drug-sensitive variant MCF-7 cell line (MCF-7/S).

Exosomes also play a role in thwarting body's own defense mechanisms against cancerous cells as was demonstrated in several studies. It was reported that cancerous cells avoid complement-mediated lysis by exosomal secretion of protein mortalin (Pilzer and Fishelson 2005; Pilzer et al. 2005). Another study reported that TNF-α, which is secreted in association with exosomes, prevents cell death induced by activation of cytotoxic T cells (Zhang et al. 2006). Separately, lymphoma exosomes were found to protect target cells from attack of antibodies by releasing CD20 (Aung et al. 2011).

Exosomes carry several proteins that are unique to cancer cells. While some promote drug resistance, others have been used for cancer detection as biomarkers; for instance, exosome-associated glypican-1 (GPC-1) has been shown to sensitively differentiate between healthy individuals and patients with pancreatic cancer (Melo et al. 2015). Apart from this exosome, encapsulated DNA has also been used for cancer detection as well as to determine mutational status of parental tumor cells (Skog et al. 2008; Guescini et al. 2010; Thakur et al. 2014). Thus, exosomes play multifaceted roles in several noncommunicable disorders.

4.3 Vesicle-Mediated Drug Resistance in Pathogenic Disorders

Most human pathogens are bacterial species that lack typical machinery required for exosome production; however, they are known to produce outer membrane vesicles (OMVs) that bud from their plasma membranes and are known to carry proteins, phospholipids, lipopolysaccharides, and nucleic acids. OMVs of bacteria play a variety of roles on extracellular activities within intra- and interspecies microbes. Experimental data suggests that these vesicles often act as bacterial virulence factor and play significant role in pathogenesis. They've also been implicated in transmission of drug resistance by bacterial species such as *S. aureus* and *P. aeruginosa* to entire polymicrobial community by transfer of vesicle enclosed β-lactamase enzyme that allows recipient gram-positive and gram-negative bacteria to survive in the presence of antibiotic penicillin (Ciofu et al. 2000; Lee et al. 2013).

Transfer of plasmid containing genes for antibiotic resistance has been a classic route by which drug resistance can rapidly spread across population. Several of the gram-negative, but not gram-positive, eubacteria-derived vesicles were shown to carry nuclease-protected linear or supercoiled DNAs (Dorward and Garon 1990). In a laboratory demonstration of similar phenomenon in malaria parasite *Plasmodium falciparum*, Neta Regev-Rudzki et al. demonstrated that strains that were separately transfected with plasmids conferring resistance to WR99210 or blasticidin S upon co-culture allowed parasite to grow in the presence of both the drugs, while culturing them separately didn't allowed growth of parasite (Regev-Rudzki et al. 2013).

Though much is known about the role of extracellular vesicles from bacteria and fungi in pathogenesis, the details of the role played by them in drug resistance is still an active area of research. Future research in extracellular vesicles is expected to shed new lights on the mechanisms of emergence and transmission of drug resistance across various physiological and pathological processes.

5 Structuromics

Transcription is one of the most important and basic steps for a cellular system. During transcription, RNA is formed using DNA as a template along with RNA polymerase and different kinds of modulator proteins. After transcription, primary transcript forms and additional modifications make mature RNAs, which are subsequently used for their respective functions. Mature RNA molecules retain some conserved sequences responsible for the formation of signature secondary structures for the interaction with different set of proteins. These secondary structures remain the same at a given time in the cell. But, these structures may change when RNA interacts with different set of proteins to maintain the physiology of the cell. Level of a given protein varies with time and requirement. It is intriguing to understand how the protein concentration varies for a given protein with time, keeping the sequence and hence the signature secondary structures the same. It suggests that RNA sequence or secondary structure is being modulated in response to specific signal. Thus, variations in different protein concentrations keep the cells alive and healthy in temporal environment (biotic and abiotic stress).

A highly informative concept came as "structuromics" by Howard Y. Chang's group, which gave the comprehensive analysis of in vivo RNA secondary structure. The approach which they named "in vivo click selective 2′-hydroxyl acylation and profiling experiment (icSHAPE)" utilizes biochemical modifications of unpaired RNA bases and determines the structural dynamics of whole-cell RNA modules (Spitale et al. 2015). Understanding of this machinery will provide information such as protein-binding sites within specific secondary structures.

5.1 Role of Structuromics in Deciphering the Drug Resistance Mechanism

Exploring the structure of RNAs and analyzing their interaction with protein molecules in vivo will shed light on intracellular regulatory networks. These mechanisms resemble hidden weapons of bacteria and they are activated by the cell when needed. If we know the way and time of their activation, we can target those modules by making them nonfunctional. For example, RNAIII of *S. aureus*, which is 514 nucleotides long RNA molecule, regulates many genes encoding exportins and cell wall-associated proteins (Waters and Storz 2009). RNAIII interacts with different sets of proteins with its 5′ and 3′ domains and regulate their expression. It also controls expression of many virulence factors; these factors keep them viable and give strength to fight against host (Morfeldt et al. 1995; Boisset et al. 2007; Novick and Geisinger 2008; Chevalier et al. 2010; Liu et al. 2011). Targeting the modules formed by RNAIII under specific conditions will help in reducing the virulence of *S. aureus*. Though most of these structures arise during infection of the host and, hence, are difficult to target technically. In such cases,

RNA structuromics might play a significant role, giving the vision of RNAIII complexes.

5.2 *Role of Structuromics in Drug Designing*

Till now, siRNA-mediated drugs have been used for some of the diseases like age-related macular degeneration (AMD), pachyonychia congenita (PC, a rare form of hereditary keratoderma that can affect the skin, mouth, hair, and eyes), and coronary artery disease and are also being used in HIV (Rossi 2006; Kleinman et al. 2008; Smith et al. 2008). Deciphering the structure of RNA-protein complexes (for some specific cases) either by structuromics or by cryo-electron microscopy will explore the mechanism of action of specific targets of siRNA molecules. This may play a crucial role in understanding the possible mechanisms of cells becoming resistant to an siRNA-based drug molecule. Subsequently, with the available information, chemical modifications within an siRNA molecule might provide additional information in its target inhibition. For example, addition of a drug (e.g., cross-linker which will inhibit the dissociation of protein and RNA molecule at its specific time of action) with siRNA will improve its stability as well as its action. Thus, RNA structuromics is a revolutionary aspect of RNA-based drug discovery.

6 Posttranslationally Modified Protein Networks as Drug Targets

In cellular information transfer, the most important transducers are proteins that define vital nature of the cell. For long, the central dogma of molecular information was thought to be DNA to RNA to protein. However, the translated proteins are often only the preprocessed forms which need further modifications to make them the elegant biological devices. The most common post-transnational modifications are phosphorylation, acetylation, glycosylation, and ubiquitinylation. Interestingly, these conserved modifications modify the protein activity, localization, and stability in both prokaryotes and eukaryotes. Therefore, these modifications and the enzymes responsible for carrying out these modifications are considered one of the most promising drug targets against many diseases such as cancer, malaria, and tuberculosis, among others. Specific posttranslationally modified (PTM) events are crucial in clinical manifestations and successful infections in most infectious disease. Therefore, targeting these modifications can be more effective than aiming a single drug target.

In *M. tuberculosis*, there are 11 Ser/Thr protein kinases, 2 Tyr phosphatases, and 1 Ser/Thr phosphatase. The genome also encodes at least one protein lysine acetyl transferase and one deacetylase (Nambi et al. 2010, 2013; Singhal et al. 2015). These enzymes together modify at least 400 important proteins reversibly to help the bacteria handle the stress and adapt by physiological changes (Prisic et al. 2010; Singhal et al. 2015). Another related pathogen *M. ulcerans* that causes Buruli ulcer has a kinase present in pathogenicity-related plasmid pMUM001 (Arora et al. 2014). This kinase phosphorylates some other structural proteins, also coded by virulence-associated plasmid, and helps the pathogen in successful infection (Arora et al. 2014). Interestingly, while proteins such as PrkC or PknB are conserved in most gram positive and actinomycetes, there are some other kinases that are exclusive and possess dual specificity (Arora et al. 2012, 2013). Two such kinases present in *B. anthracis*: one is the first DYRK-like kinase of prokaryotes PrkD and another is Ser-/Thr-/Tyr-specific enzyme PrkG (Arora et al. 2012). *B. anthracis* has evolved for its pathogenic lifestyle needs, lost tyrosine kinases, and expressed dual specificity protein kinases that are important for growth and possibly pathogenicity of the organism (Arora et al. 2012). Interestingly, all Ser/Thr protein kinases, dual specificity protein kinases, Tyr kinases, and acetyltransferases are shown to modify and regulate key physiological pathway proteins such as glycolysis, protein translation, and one-carbon metabolism (Arora et al. 2010, 2012, 2013; Singhal et al. 2013, 2015; Maji et al. 2015; Pereira et al. 2015; Prisic et al. 2015; Sajid et al. 2015). Several of these modifications regulated the key metabolic events in overlapping manner. The network of such signaling events forms nodes that fine-tune cellular response and help in adaptation and thriving under different stress conditions. Therefore, identification and targeting such nodes will be key to novel drug discovery.

7 Conclusions

The human genome has 46 chromosomes and around 20,000–25,000 protein-coding genes. Interestingly, the total number of human diseases known is estimated to be 10,000–30,000. In the years before the discovery of antibiotics, the general perception was humankind will never be able to successfully combat the pathogens such as tuberculosis, malaria, and cholera. In the next hundred years, we have achieved an unparallel understanding of life at the molecular level. Medical science has provided limited but effective solutions to combat major diseases. However, drug-resistant disease forms have constantly challenged our success. The emergence of drug-resistant forms has resulted in prolonged illness and increased rate of death even for common infections. According to WHO estimate, people infected with drug-resistant form of MRSA are 64 % more likely to die than compared to nonresistant forms. To end our misery from palpable drug-resistant diseases, in the future we will have to adapt cooperative drug discovery strategies and benefit from emerging themes such as immunotherapy, pharmacogenomics, and structuromics to annihilate these enemies of mankind. It is a necessity to use new and powerful beacons that target different disease forms. To achieve this, discovery of effective therapeutic agents that can provide protection against both acute and chronic forms along with better disease management is urgently required. With the experience of combating these diseases in last century, we must not rely on conserved drug target strategies and may have to develop multiple subordinate strategies each potent enough to strike pestilence effectively.

References

Abounit S, Zurzolo C (2012) Wiring through tunneling nanotubes–from electrical signals to organelle transfer. J Cell Sci 125:1089–1098. doi:10.1242/jcs.083279

Anderson PL, Lamba J, Aquilante CL, Schuetz E, Fletcher CV (2006) Pharmacogenetic characteristics of indinavir, zidovudine, and lamivudine therapy in HIV-infected adults: a pilot study. J Acquir Immune Defic Syndr 42:441–449. doi:10.1097/01.qai.0000225013.53568.69

Arora G, Sajid A, Gupta M, Bhaduri A, Kumar P, Basu-Modak S, Singh Y (2010) Understanding the role of PknJ in *Mycobacterium tuberculosis*: biochemical characterization and identification of novel substrate pyruvate kinase A. PLoS One 5, e10772. doi:10.1371/journal.pone.0010772

Arora G, Sajid A, Arulanandh MD, Singhal A, Mattoo AR, Pomerantsev AP, Leppla SH, Maiti S, Singh Y (2012) Unveiling the novel dual specificity protein kinases in *Bacillus anthracis*: identification of the first prokaryotic dual specificity tyrosine phosphorylation-regulated kinase (DYRK)-like kinase. J Biol Chem 287:26749–26763. doi:10.1074/jbc.M112.351304

Arora G, Sajid A, Arulanandh MD, Misra R, Singhal A, Kumar S, Singh LK, Mattoo AR, Raj R, Maiti S, Basu-Modak S, Singh Y (2013) Zinc regulates the activity of kinase-phosphatase pair (BasPrkC/BasPrpC) in *Bacillus anthracis*. Biometals 26:715–730. doi:10.1007/s10534-013-9646-y

Arora G, Sajid A, Singhal A, Joshi J, Virmani R, Gupta M, Verma N, Maji A, Misra R, Baronian G, Pandey AK, Molle V, Singh Y (2014) Identification of Ser/Thr kinase and forkhead associated domains in *Mycobacterium ulcerans*: characterization of novel association between protein kinase Q and MupFHA. PLoS Negl Trop Dis 8, e3315. doi:10.1371/journal.pntd.0003315

Aung T, Chapuy B, Vogel D, Wenzel D, Oppermann M, Lahmann M, Weinhage T, Menck K, Hupfeld T, Koch R, Trumper L, Wulf GG (2011) Exosomal evasion of humoral immunotherapy in aggressive B-cell lymphoma modulated by ATP-binding cassette transporter A3. Proc Natl Acad Sci U S A 108:15336–15341. doi:10.1073/pnas.1102855108

Azmi AS, Bao B, Sarkar FH (2013) Exosomes in cancer development, metastasis, and drug resistance: a comprehensive review. Cancer Metastasis Rev 32:623–642. doi:10.1007/s10555-013-9441-9

Bebbington C, Yarranton G (2008) Antibodies for the treatment of bacterial infections: current experience and future prospects. Curr Opin Biotechnol 19:613–619. doi:10.1016/j.copbio.2008.10.002

Boelens MC, Wu TJ, Nabet BY, Xu B, Qiu Y, Yoon T, Azzam DJ, Twyman-Saint VC, Wiemann BZ, Ishwaran H, Ter Brugge PJ, Jonkers J, Slingerland J, Minn AJ (2014) Exosome transfer from stromal to breast cancer cells regulates therapy resistance pathways. Cell 159:499–513. doi:10.1016/j.cell.2014.09.051

Boisset S, Geissmann T, Huntzinger E, Fechter P, Bendridi N, Possedko M, Chevalier C, Helfer AC, Benito Y, Jacquier A, Gaspin C, Vandenesch F, Romby P (2007) *Staphylococcus aureus* RNAIII coordinately represses the synthesis of virulence factors and the transcription regulator Rot by an antisense mechanism. Genes Dev 21:1353–1366. doi:10.1101/gad.423507

Bosch F, Rosich L (2008) The contributions of Paul Ehrlich to pharmacology: a tribute on the occasion of the centenary of his Nobel Prize. Pharmacology 82:171–179. doi:10.1159/000149583

Braun GS, Moeller MJ (2015) Progenitor cell-derived extracellular vesicles: an emerging diagnostic and therapeutic tool for renal disease. Nephrol Dial Transplant 30:339–341. doi:10.1093/ndt/gfv027

Bristow RG, Hill RP (2008) Hypoxia and metabolism. Hypoxia, DNA repair and genetic instability. Nat Rev Cancer 8:180–192. doi:10.1038/nrc2344

Campos JH, Soares RP, Ribeiro K, Andrade AC, Batista WL, Torrecilhas AC (2015) Extracellular vesicles: role in inflammatory responses and potential uses in vaccination in cancer and infectious diseases. J Immunol Res 2015:832057. doi:10.1155/2015/832057

Casadevall A (1996) Antibody-based therapies for emerging infectious diseases. Emerg Infect Dis 2:200–208. doi:10.3201/eid0203.960306

Chames P, Van RM, Weiss E, Baty D (2009) Therapeutic antibodies: successes, limitations and hopes for the future. Br J Pharmacol 157:220–233. doi:10.1111/j.1476-5381.2009.00190.x

Chen WX, Liu XM, Lv MM, Chen L, Zhao JH, Zhong SL, Ji MH, Hu Q, Luo Z, Wu JZ, Tang JH (2014) Exosomes from drug-resistant breast cancer cells transmit chemoresistance by a horizontal transfer of microRNAs. PLoS One 9, e95240. doi:10.1371/journal.pone.0095240

Chevalier C, Boisset S, Romilly C, Masquida B, Fechter P, Geissmann T, Vandenesch F, Romby P (2010) *Staphylococcus aureus* RNAIII binds to two distant regions of coa mRNA to arrest translation and promote mRNA degradation. PLoS Pathog 6, e1000809. doi:10.1371/journal.ppat.1000809

Ciofu O, Beveridge TJ, Kadurugamuwa J, Walther-Rasmussen J, Hoiby N (2000) Chromosomal beta-lactamase is packaged into membrane vesicles and secreted from *Pseudomonas aeruginosa*. J Antimicrob Chemother 45:9–13. doi:10.1093/jac/45.1.9

Ciravolo V, Huber V, Ghedini GC, Venturelli E, Bianchi F, Campiglio M, Morelli D, Villa A, Della MP, Menard S, Filipazzi P, Rivoltini L, Tagliabue E, Pupa SM (2012) Potential role of HER2-overexpressing exosomes in countering trastuzumab-based therapy. J Cell Physiol 227:658–667. doi:10.1002/jcp.22773

Clatworthy AE, Pierson E, Hung DT (2007) Targeting virulence: a new paradigm for antimicrobial therapy. Nat Chem Biol 3:541–548. doi:10.1038/nchembio.2007.24

Coleman BM, Hill AF (2015) Extracellular vesicles–their role in the packaging and spread of misfolded proteins associated with neurodegenerative diseases. Semin Cell Dev Biol 40:89–96. doi:10.1016/j.semcdb.2015.02.007

DeFronzo RA (1997) Insulin resistance: a multifaceted syndrome responsible for NIDDM, obesity, hypertension, dyslipidaemia and atherosclerosis. Neth J Med 50:191–197

DeJonge M, Burchfield D, Bloom B, Duenas M, Walker W, Polak M, Jung E, Millard D, Schelonka R, Eyal F, Morris A, Kapik B, Roberson D, Kesler K, Patti J, Hetherington S (2007) Clinical trial of safety and efficacy of INH-A21 for the prevention of nosocomial staphylococcal bloodstream infection in premature infants. J Pediatr 151:260–265. doi:10.1016/j.jpeds.2007.04.060

DiGiandomenico A, Keller AE, Gao C, Rainey GJ, Warrener P, Camara MM, Bonnell J, Fleming R, Bezabeh B, Dimasi N, Sellman BR, Hilliard J, Guenther CM, Datta V, Zhao W, Gao C, Yu XQ, Suzich JA, Stover CK (2014) A multifunctional bispecific antibody protects against *Pseudomonas aeruginosa*. Sci Transl Med 6:262ra155. doi:10.1126/scitranslmed.3009655

Dorward DW, Garon CF (1990) DNA is packaged within membrane-derived vesicles of gram-negative but not gram-positive bacteria. Appl Environ Microbiol 56:1960–1962

Dubey GP, Ben-Yehuda S (2011) Intercellular nanotubes mediate bacterial communication. Cell 144:590–600. doi:10.1016/j.cell.2011.01.015

Fauci AS, Morens DM (2012) The perpetual challenge of infectious diseases. N Engl J Med 366:454–461. doi:10.1056/NEJMra1108296

Fujita Y, Kosaka N, Araya J, Kuwano K, Ochiya T (2015) Extracellular vesicles in lung microenvironment and pathogenesis. Trends Mol Med 21:533–542. doi:10.1016/j.molmed.2015.07.004

Guescini M, Genedani S, Stocchi V, Agnati LF (2010) Astrocytes and Glioblastoma cells release exosomes carrying mtDNA. J Neural Transm (Vienna) 117:1–4. doi:10.1007/s00702-009-0288-8

Isaacs JD, Watts RA, Hazleman BL, Hale G, Keogan MT, Cobbold SP, Waldmann H (1992) Humanised monoclonal antibody therapy for rheumatoid arthritis. Lancet 340:748–752. doi:10.1016/0140-6736(92)92294-P

Johnson S, Oliver C, Prince GA, Hemming VG, Pfarr DS, Wang SC, Dormitzer M, O'Grady J, Koenig S, Tamura JK, Woods R, Bansal G, Couchenour D, Tsao E, Hall WC, Young JF (1997) Development of a humanized monoclonal antibody (MEDI-493) with potent in vitro and in vivo activity against respiratory syncytial virus. J Infect Dis 176:1215–1224. doi:10.1086/514115

Kelly-Quintos C, Cavacini LA, Posner MR, Goldmann D, Pier GB (2006) Characterization of the opsonic and protective activity against *Staphylococcus aureus* of fully human monoclonal antibodies specific for the bacterial surface polysaccharide poly-N-acetylglucosamine. Infect Immun 74:2742–2750. doi:10.1128/IAI.74.5.2742-2750.2006

Khan S, Aspe JR, Asumen MG, Almaguel F, Odumosu O, Acevedo-Martinez S, De Leon M, Langridge WH, Wall NR (2009) Extracellular, cell-permeable survivin inhibits apoptosis while promoting proliferative and metastatic potential. Br J Cancer 100:1073–1086. doi:10.1038/sj.bjc.6604978

Kleinman ME, Yamada K, Takeda A, Chandrasekaran V, Nozaki M, Baffi JZ, Albuquerque RJ, Yamasaki S, Itaya M, Pan Y, Appukuttan B, Gibbs D, Yang Z, Kariko K, Ambati BK, Wilgus TA, DiPietro LA, Sakurai E, Zhang K, Smith JR, Taylor EW, Ambati J (2008) Sequence- and target-independent angiogenesis suppression by siRNA via TLR3. Nature 452:591–597. doi:10.1038/nature06765

Lazar H, Horn MP, Zuercher AW, Imboden MA, Durrer P, Seiberling M, Pokorny R, Hammer C, Lang AB (2009) Pharmacokinetics and safety profile of the human anti-*Pseudomonas aeruginosa* serotype O11 immunoglobulin M monoclonal antibody KBPA-101 in healthy volunteers. Antimicrob Agents Chemother 53:3442–3446. doi:10.1128/AAC.01699-08

Lee J, Lee EY, Kim SH, Kim DK, Park KS, Kim KP, Kim YK, Roh TY, Gho YS (2013) *Staphylococcus aureus* extracellular vesicles carry biologically active beta-lactamase. Antimicrob Agents Chemother 57:2589–2595. doi:10.1128/AAC.00522-12

Litman GW, Rast JP, Shamblott MJ, Haire RN, Hulst M, Roess W, Litman RT, Hinds-Frey KR, Zilch A, Amemiya CT (1993) Phylogenetic diversification of immunoglobulin genes and the antibody repertoire. Mol Biol Evol 10:60–72

Liu Y, Mu C, Ying X, Li W, Wu N, Dong J, Gao Y, Shao N, Fan M, Yang G (2011) RNAIII activates map expression by forming an RNA-RNA complex in *Staphylococcus aureus*. FEBS Lett 585:899–905. doi:10.1128/AAC.00522-12

Liu YY, Wang Y, Walsh TR, Yi LX, Zhang R, Spencer J, Doi Y, Tian G, Dong B, Huang X, Yu LF, Gu D, Ren H, Chen X, Lv L, He D, Zhou H, Liang Z, Liu JH, Shen J (2016) Emergence of plasmid-mediated colistin resistance mechanism MCR-1 in animals and human beings in China: a microbiological and molecular biological study. Lancet Infect Dis 16:161–168. doi:10.1016/S1473-3099(15)00424-7

Luhadia SK (2014) Steroid resistant asthma. J Assoc Physicians India 62:38–40

Mahmoudi K, Ezrin A, Hadjipanayis C (2015) Small extracellular vesicles as tumor biomarkers for glioblastoma. Mol Aspects Med 45:97–102. doi:10.1016/j.mam.2015.06.008

Maji A, Misra R, Kumar MA, Kumar D, Bajaj D, Singhal A, Arora G, Bhaduri A, Sajid A, Bhatia S, Singh S, Singh H, Rao V, Dash D, Baby SE, Sarojini MJ, Chaudhary A, Gokhale RS, Singh Y (2015) Expression profiling of lymph nodes in tuberculosis patients reveal inflammatory milieu at site of infection. Sci Rep 5:15214. doi:10.1038/srep15214

Mattu TS, Pleass RJ, Willis AC, Kilian M, Wormald MR, Lellouch AC, Rudd PM, Woof JM, Dwek RA (1998) The glycosylation and structure of human serum IgA1, Fab, and Fc regions and the role of N-glycosylation on Fc alpha receptor interactions. J Biol Chem 273:2260–2272. doi:10.1074/jbc.273.4.2260

Maverakis E, Kim K, Shimoda M, Gershwin ME, Patel F, Wilken R, Raychaudhuri S, Ruhaak LR, Lebrilla CB (2015) Glycans in the immune system and The Altered Glycan Theory of Autoimmunity: a critical review. J Autoimmun 57:1–13. doi:10.1016/j.jaut.2014.12.002

McCarthy AD, Kennedy JL, Middleton LT (2005) Pharmacogenetics in drug development. Philos Trans R Soc Lond B Biol Sci 360:1579–1588. doi:10.1098/rstb.2005.1688

Meads MB, Gatenby RA, Dalton WS (2009) Environment-mediated drug resistance: a major contributor to minimal residual disease. Nat Rev Cancer 9:665–674. doi:10.1038/nrc2714

Melo SA, Luecke LB, Kahlert C, Fernandez AF, Gammon ST, Kaye J, LeBleu VS, Mittendorf EA, Weitz J, Rahbari N, Reissfelder C, Pilarsky C, Fraga MF, Piwnica-Worms D, Kalluri R (2015) Glypican-1 identifies cancer exosomes and detects early pancreatic cancer. Nature 523:177–182. doi:10.1038/nature14581

Minciacchi VR, Freeman MR, Di VD (2015) Extracellular vesicles in cancer: exosomes, microvesicles and the emerging role of large oncosomes. Semin Cell Dev Biol 40:41–51. doi:10.1016/j.semcdb.2015.02.010

Morfeldt E, Taylor D, von Gabain A, Arvidson S (1995) Activation of alpha-toxin translation in *Staphylococcus aureus* by the trans-encoded antisense RNA, RNAIII. EMBO J 14:4569–4577

Nagy E, Giefing C, von Gabain A (2008) Anti-infective antibodies: a novel tool to prevent and treat nosocomial diseases. Expert Rev Anti Infect Ther 6:21–30. doi:10.1586/14787210.6.1.21

Nambi S, Basu N, Visweswariah SS (2010) cAMP-regulated protein lysine acetylases in mycobacteria. J Biol Chem 285:24313–24323. doi:10.1074/jbc.M110.118398

Nambi S, Gupta K, Bhattacharyya M, Ramakrishnan P, Ravikumar V, Siddiqui N, Thomas AT, Visweswariah SS (2013) Cyclic AMP-dependent protein lysine acylation in mycobacteria regulates fatty acid and propionate metabolism. J Biol Chem 288:14114–14124. doi:10.1074/jbc.M113.463992

Nayak SU, Griffiss JM, McKenzie R, Fuchs EJ, Jurao RA, An AT, Ahene A, Tomic M, Hendrix CW, Zenilman JM (2014) Safety and pharmacokinetics of XOMA 3AB, a novel mixture of three monoclonal antibodies against botulinum toxin A. Antimicrob Agents Chemother 58:5047–5053. doi:10.1128/AAC.02830-14

Noble ME, Endicott JA, Johnson LN (2004) Protein kinase inhibitors: insights into drug design from structure. Science 303:1800–1805. doi:10.1126/science.1095920

Novick RP, Geisinger E (2008) Quorum sensing in staphylococci. Annu Rev Genet 42:541–564. doi:10.1146/annurev.genet.42.110807.091640

Pande S, Shitut S, Freund L, Westermann M, Bertels F, Colesie C, Bischofs IB, Kost C (2015) Metabolic cross-feeding via intercellular nanotubes among bacteria. Nat Commun 6:6238. doi:10.1038/ncomms7238

Pereira SF, Gonzalez RL Jr, Dworkin J (2015) Protein synthesis during cellular quiescence is inhibited by phosphorylation of a translational elongation factor. Proc Natl Acad Sci U S A 112:E3274–E3281. doi:10.1073/pnas.1505297112

Pilzer D, Fishelson Z (2005) Mortalin/GRP75 promotes release of membrane vesicles from immune attacked cells and protection from complement-mediated lysis. Int Immunol 17:1239–1248. doi:10.1093/intimm/dxh300

Pilzer D, Gasser O, Moskovich O, Schifferli JA, Fishelson Z (2005) Emission of membrane vesicles: roles in complement resistance, immunity and cancer. Springer Semin Immunopathol 27:375–387. doi:10.1007/s00281-005-0004-1

Prisic S, Dankwa S, Schwartz D, Chou MF, Locasale JW, Kang CM, Bemis G, Church GM, Steen H, Husson RN (2010) Extensive phosphorylation with overlapping specificity by *Mycobacterium tuberculosis* serine/threonine protein kinases. Proc Natl Acad Sci U S A 107:7521–7526. doi:10.1073/pnas.0913482107

Prisic S, Hwang H, Dow A, Barnaby O, Pan TS, Lonzanida JA, Chazin WJ, Steen H, Husson RN (2015) Zinc regulates a switch between primary and alternative S18 ribosomal proteins in *Mycobacterium tuberculosis*. Mol Microbiol 97:263–280. doi:10.1111/mmi.13022

Putnam FW, Liu YS, Low TL (1979) Primary structure of a human IgA1 immunoglobulin. IV. Streptococcal IgA1 protease, digestion, Fab and Fc fragments, and the complete amino acid sequence of the alpha 1 heavy chain. J Biol Chem 254:2865–2874

Que YA, Lazar H, Wolff M, Francois B, Laterre PF, Mercier E, Garbino J, Pagani JL, Revelly JP, Mus E, Perez A, Tamm M, Rouby JJ, Lu Q, Chastre J, Eggimann P (2014) Assessment of panobacumab as adjunctive immunotherapy for the treatment of nosocomial *Pseudomonas aeruginosa* pneumonia. Eur J Clin Microbiol Infect Dis 33:1861–1867. doi:10.1007/s10096-014-2156-1

Regev-Rudzki N, Wilson DW, Carvalho TG, Sisquella X, Coleman BM, Rug M, Bursac D, Angrisano F, Gee M, Hill AF, Baum J, Cowman AF (2013) Cell-cell communication between malaria-infected red blood cells via exosome-like vesicles. Cell 153:1120–1133. doi:10.1016/j.cell.2013.04.029

Rice LB (2008) Federal funding for the study of antimicrobial resistance in nosocomial pathogens: no ESKAPE. J Infect Dis 197:1079–1081. doi:10.1086/533452

Robbins PD, Morelli AE (2014) Regulation of immune responses by extracellular vesicles. Nat Rev Immunol 14:195–208. doi:10.1038/nri3622

Rossi JJ (2006) RNAi as a treatment for HIV-1 infection. Biotechniques Suppl:25–29

Safaei R, Larson BJ, Cheng TC, Gibson MA, Otani S, Naerdemann W, Howell SB (2005) Abnormal lysosomal trafficking and enhanced exosomal export of cisplatin in drug-resistant human ovarian carcinoma cells. Mol Cancer Ther 4:1595–1604. doi:10.1158/1535-7163.MCT-05-0102

Sajid A, Arora G, Singhal A, Kalia VC, Singh Y (2015) Protein phosphatases of pathogenic bacteria: role in physiology and virulence. Annu Rev Microbiol 69:527–547. doi:10.1146/annurev-micro-020415-111342

Sause WE, Buckley PT, Strohl WR, Lynch AS, Torres VJ (2016) Antibody-based biologics and their promise to combat *Staphylococcus aureus* infections. Trends Pharmacol Sci 37:231–241. doi:10.1016/j.tips.2015.11.008

Saylor C, Dadachova E, Casadevall A (2009) Monoclonal antibody-based therapies for microbial diseases. Vaccine 27(Suppl 6):G38–G46. doi:10.1016/j.vaccine.2009.09.105

Schorey JS, Cheng Y, Singh PP, Smith VL (2015) Exosomes and other extracellular vesicles in host-pathogen interactions. EMBO Rep 16:24–43. doi:10.15252/embr.201439363

Shedden K, Xie XT, Chandaroy P, Chang YT, Rosania GR (2003) Expulsion of small molecules in vesicles shed by cancer cells: association with gene expression and chemosensitivity profiles. Cancer Res 63:4331–4337

Siddiqui A, Kerb R, Weale ME, Brinkmann U, Smith A, Goldstein DB, Wood NW, Sisodiya SM (2003) Association of multidrug resistance in epilepsy with a polymorphism in the drug-transporter gene ABCB1. N Engl J Med 348:1442–1448. doi:10.1056/NEJMoa021986

Singh A, Settleman J (2010) EMT, cancer stem cells and drug resistance: an emerging axis of evil in the war on cancer. Oncogene 29:4741–4751. doi:10.1038/onc.2010.215

Singhal A, Arora G, Sajid A, Maji A, Bhat A, Virmani R, Upadhyay S, Nandicoori VK, Sengupta S, Singh Y (2013) Regulation of homocysteine metabolism by *Mycobacterium tuberculosis* S-adenosylhomocysteine hydrolase. Sci Rep 3:2264. doi:10.1038/srep02264

Singhal A, Arora G, Virmani R, Kundu P, Khanna T, Sajid A, Misra R, Joshi J, Yadav V, Samanta S, Saini N, Pandey AK, Visweswariah SS, Hentschker C, Becher D, Gerth U, Singh Y (2015) Systematic analysis of mycobacterial acylation reveals first example of acylation-mediated regulation of enzyme activity of a bacterial phosphatase. J Biol Chem 290:26218–26234. doi:10.1074/jbc.M115.687269

Skog J, Wurdinger T, van Rijn S, Meijer DH, Gainche L, Sena-Esteves M, Curry WT Jr, Carter BS, Krichevsky AM, Breakefield XO (2008) Glioblastoma microvesicles transport RNA and proteins that promote tumour growth and provide diagnostic biomarkers. Nat Cell Biol 10:1470–1476. doi:10.1038/ncb1800

Smith FJ, Hickerson RP, Sayers JM, Reeves RE, Contag CH, Leake D, Kaspar RL, McLean WH (2008) Development of therapeutic siRNAs for pachyonychia congenita. J Invest Dermatol 128:50–58. doi:10.1038/sj.jid.5701040

Spitale RC, Flynn RA, Zhang QC, Crisalli P, Lee B, Jung JW, Kuchelmeister HY, Batista PJ, Torre EA, Kool ET, Chang HY (2015) Structural imprints in vivo decode RNA regulatory mechanisms. Nature 519:486–490. doi:10.1038/nature14263

Tarakanov AO, Goncharova LB (2009) Cell-cell nanotubes: tunneling through several types of synapses. Commun Integr Biol 2:359–361

ter Meulen J (2007) Monoclonal antibodies for prophylaxis and therapy of infectious diseases. Expert Opin Emerg Drugs 12:525–540. doi:10.1517/14728214.12.4.525

ter Meulen J (2011) Monoclonal antibodies in infectious diseases: clinical pipeline in 2011. Infect Dis Clin North Am 25:789–802. doi:10.1016/j.idc.2011.07.006

Thakur BK, Zhang H, Becker A, Matei I, Huang Y, Costa-Silva B, Zheng Y, Hoshino A, Brazier H, Xiang J, Williams C, Rodriguez-Barrueco R, Silva JM, Zhang W, Hearn S, Elemento O, Paknejad N, Manova-Todorova K, Welte K, Bromberg J, Peinado H, Lyden D (2014) Double-stranded DNA in exosomes: a novel biomarker in cancer detection. Cell Res 24:766–769. doi:10.1038/cr.2014.44

Tsai CW, Morris S (2015) Approval of raxibacumab for the treatment of inhalation anthrax under the US food and drug administration "Animal Rule". Front Microbiol 6:1320. doi:10.3389/fmicb.2015.01320

Vogel CL, Franco SX (2003) Clinical experience with trastuzumab (herceptin). Breast J 9:452–462. doi:10.1046/j.1524-4741.2003.09602.x

Von Behring E, Kitasato S (1965) On the development of immunity to diphtheria and tetanus in animals. Dtsch Med Wochenschr 90:2183

Von Behring E, Kitasato S (1991) The mechanism of diphtheria immunity and tetanus immunity in animals. 1890. Mol Immunol 28:1317, 1319–1320

Voth DE, Ballard JD (2005) *Clostridium difficile* toxins: mechanism of action and role in disease. Clin Microbiol Rev 18:247–263. doi:10.1128/CMR.18.2.247-263.2005

Wang X, Gerdes HH (2015) Transfer of mitochondria via tunneling nanotubes rescues apoptotic PC12 cells. Cell Death Differ 22:1181–1191. doi:10.1038/cdd.2014.211

Waters LS, Storz G (2009) Regulatory RNAs in bacteria. Cell 136:615–628. doi:10.1016/j.cell.2009.01.043

Weiner M, Peloquin C, Burman W, Luo CC, Engle M, Prihoda TJ, Mac Kenzie WR, Bliven-Sizemore E, Johnson JL, Vernon A (2010) Effects of tuberculosis, race, and human gene SLCO1B1 polymorphisms on rifampin concentrations. Antimicrob Agents Chemother 54:4192–4200. doi:10.1128/AAC.00353-10

Yanez-Mo M, Siljander PR, Andreu Z, Zavec AB, Borras FE, Buzas EI, Buzas K, Casal E, Cappello F, Carvalho J, Colas E, Cordeiro-da SA, Fais S, Falcon-Perez JM, Ghobrial IM, Giebel B, Gimona M, Graner M, Gursel I, Gursel M, Heegaard NH, Hendrix A, Kierulf P, Kokubun K, Kosanovic M, Kralj-Iglic V, Kramer-Albers EM, Laitinen S, Lasser C, Lener T, Ligeti E, Line A, Lipps G, Llorente A, Lotvall J, Mancek-Keber M, Marcilla A, Mittelbrunn M, Nazarenko I, Nolte-'t Hoen EN, Nyman TA, O'Driscoll L, Olivan M, Oliveira C, Pallinger E, Del Portillo HA, Reventos J, Rigau M, Rohde E, Sammar M, Sanchez-Madrid F, Santarem N, Schallmoser K, Ostenfeld MS, Stoorvogel W, Stukelj R, Van der Grein SG, Vasconcelos MH, Wauben MH, De Wever O (2015) Biological properties of extracellular vesicles and their physiological functions. J Extracell Vesicles 4:27066. doi:10.3402/jev.v4.27066

Yang Z, Ramsey J, Hamza T, Zhang Y, Li S, Yfantis HG, Lee D, Hernandez LD, Seghezzi W, Furneisen JM, Davis NM, Therien AG, Feng H (2015) Mechanisms of protection against *Clostridium difficile* infection by the monoclonal antitoxin antibodies actoxumab and bezlotoxumab. Infect Immun 83:822–831. doi:10.1128/IAI.02897-14

Zhang HG, Liu C, Su K, Yu S, Zhang L, Zhang S, Wang J, Cao X, Grizzle W, Kimberly RP (2006) A membrane form of TNF-alpha presented by exosomes delays T cell activation-induced cell death. J Immunol 176:7385–7393. doi:10.4049/jimmunol.176.12.7385

Molecular Mechanism of Drug Resistance: Common Themes

C.M. Santosh Kumar

Abstract Antibiotics are the chemical or biochemical moieties that specifically and effectively inhibit the growth of a pathogen but not the host organism and thus are employed in the treatment of the infection by the pathogen. However, the survival pressure in the presence of antimicrobial agents forces a minor fraction of the bacterial population to evolve mechanisms that evade the inhibitory effects of the administered antibiological agent, thereby emerging as a drug-resistant variety and consequently challenging the treatment regime. Additionally, the emergence of multidrug-resistant (MDR) variants of the pathogens that are capable of resisting several structurally dissimilar drugs is becoming very common and resulting in the infections that are difficult or impossible to treat. To counter the vast array of chemically and structurally dissimilar antibiotics within their lifestyle boundaries, different bacteria have developed different antibiotic- and pathogen-specific resistance mechanisms. Although distinct and involving several molecular events, these mechanisms can be broadly classified into two modes, the intrinsic and the acquired modes of resistance, which are further classified into subclasses. This chapter reviews recent advances and current understanding of the molecular details of these mechanisms.

1 Introduction

The discovery of antibiotics strongly supported modern lifestyle with their variety of applications, such as insecticides, herbicides and chemotherapeutic agents in treatment of bacterial, viral and parasitic infections and even cancer (Hayes and Wolf 1990; Blair et al. 2015). Therefore, research aimed at identifying new antibiotics had been the top priority once, and the resulted broad-spectrum drugs provided extraordinary clinical efficacy. Ironically, however, for any effective

C.M.S. Kumar (✉)
Laboratory of Structural Biology (LSB), National Centre for Cell Science (NCCS), Pune 411007, MH, India
e-mail: cms@nccs.res.in

G. Arora et al. (eds.), *Drug Resistance in Bacteria, Fungi, Malaria, and Cancer*,
DOI 10.1007/978-3-319-48683-3_2

antibiotic that is employed in human therapies, clinically significant resistance appears typically in the period of a few months to years (Davies 1996). For example, resistance to the popular drug penicillin has been observed as soon as 2 years of its discovery during mid-1940s (Abraham and Chain 1988). Although the vulnerability of antibiotic resistance is implicated in several medical conditions including cancer, scope of this chapter is limited to the hazards of antibiotic resistance in human pathogens. Incidentally, the list of microbes that have emerged resistance to various classes of drugs is alarmingly expanding with increasing instances of the multidrug-resistant bacterial infections that agonizingly include several untreatable infections. Understandably, the incidence of drug resistance is of major economic importance owing to its grave consequences to health, and therefore, the World Economic Forum listed antibiotic resistance as the second greatest threats to human health owing to a sharp increase in the infections through the past years (Blair et al. 2015). Antimicrobial Resistance Global Report 2014 on Surveillance of the World Health Organization estimates annual deaths due to multidrug-resistant bacteria in Europe and the United States at 25,000 and 23,000, respectively, while costs estimated to 1.5 billion euros annually. Understandably, these figures are much higher in countries with limited medical facilities. As explained in the earlier chapters, the bacterial infectious diseases pose a serious public health and economic threat, thereby signifying the need to comprehend the molecular mechanisms that confer the resistance to the antibiotic drugs. During the past four decades, studies towards empathizing the mechanisms of drug resistance have been increasingly popular, and the resulting research involving the classical genetic tools and the modern recombinant DNA technologies dissected the intricate details of the variety of mechanisms involved. Principally the drug resistance in the pathogens can be classified into two modes, the intrinsic and acquired modes of resistance (Hayes and Wolf 1990). While the intrinsic resistance arises due to inherent structural or functional features of the bacteria, the acquired resistance is attained largely either through mutations or horizontal transfer of resistance conferring genes or several other means. Understanding the numbers, bacteria dwell in large numbers in an infection cycle (about 10^{10} bacteria) with rapid generation time (about 20 min for *E. coli* to 23 h to *M. tuberculosis*) that is supplemented with high mutation rate (one mutation in 10^7 bases replicated). So, in one generation time, the population of bacteria gets mutated at around thousand loci. If any of these mutations result in antibiotic resistance, the bacteria survive the drug load and multiply into dominant form in the population (Walsh 2000).

2 Mechanisms of Drug Resistance

The modes of resistance could be classified for the practical purpose as following. While the case of pathogen being initially drug resistant represents intrinsic (natural or de novo) resistance, the pathogen becoming resistant only after treatment represents acquired resistance. Thus, the aim of this chapter is to highlight the molecular mechanisms and biochemical events that are responsible for the antibiotic resistance and to discuss the possible modes of evolution and contradiction of this phenomenon.

2.1 Intrinsic Resistance

Intrinsic resistance is described as a characteristic feature of an organism, which allows a population of the organism to tolerate the encountered antibiotic. Since every organism evolves to adopt to its habitat, it is reasonable to presume that organisms living in an antibiotic-rich environment might have evolved the said resistance features to combat the antibiotic stress. Therefore, during the course of evolution, some microorganisms have developed intrinsic resistance to antibiotics via the pathways that are probably guided by the exposure levels and chemical nature of the encountered drug. Consequently, the intrinsic resistance is largely genetic and chromosome encoded. Let us understand the intrinsic resistance with the following examples. Considering a simplest example of the *Streptomyces*, which produce several clinically useful natural antibiotics (Kieser et al. 2000; Bibb 2013), has evolved mechanisms to resist the adverse effects of their own product antibiotics prior to the evolution of the antibiotic production pathways (Hayes and Wolf 1990). Another example relates to the membrane targeting lipopeptide, daptomycin, which acts on the membranes by inserting into the anionic phospholipid membranes and thus ineffective against Gram-negative bacteria owing to lower phospholipid content in their cytoplasmic membrane (Zhu et al. 2010). Likewise, membrane impermeability in Gram-negative bacteria resulted in the intrinsic resistance to many compounds, for example, the glycopeptide antibiotic vancomycin acts by binding to D-Ala-D-Ala peptides and inhibiting peptidoglycan cross-linking (Randall et al. 2013). Another interesting example is of a popular antimicrobial drug, triclosan, which effectively acts against the Gram-positive and several Gram-negative bacteria but ineffective against the members of the genus *Pseudomonas*, since these bacteria lack the target enzyme, the lipid anabolic enoyl-ACP reductase (Zhu et al. 2010; Blair et al. 2015). Interestingly, this resistance has been earlier attributed to increased efflux of the antibiotic (Chuanchuen et al. 2003).

2.2 Acquired Resistance

Apart from the intrinsic resistance, bacteria gain resistance, termed acquired resistance, through either genetic (largely nonchromosomal) routes that include mutations in existing DNA and acquisition of new DNA or nongenetic processes (Walsh 2000). Unlike the intrinsic mode of resistance, the acquired resistance is broadcasted faster since the latter relies on the spread of the antibiotic resistance genes between different bacterial cells and species (Davies 1994) through the mobile genetic elements such as plasmids (Arthur and Courvalin 1993; Walsh et al. 1996) and transposons (Schentag et al. 1998). A simple example of acquired resistance is observed in surgical hospitals, wherein methicillin is administered to counter the enterobacteria, *Staphylococcus aureus* and *Enterococcus faecalis*, that inhabit abdominal cavity following a surgical procedure. However, these bacteria rapidly acquire resistance to methicillin (Barber 1961) and vancomycin (Schentag et al. 1998) within 7 and 14 days after administering the drugs, respectively. Methicillin resistance, which results in the popular MRSA phonotype, is acquired from resistant bacteria through horizontal transfer of a mobile genetic element: the staphylococcal cassette chromosome *mec* (sccMec) element, which encodes a mutant penicillin-binding protein (PBP2) that enables cell wall biosynthesis even in the presence of the antibiotic (Shore et al. 2011). Additionally, resistance to vancomycin emerges in MRSA via the acquisition of genetic elements contributing to the vancomycin-resistant enterococci (VRE) phenotype. This emphasizes the speed and the frequency with which resistant phenotypes can emerge by acquisition of genetic elements from neighbouring bacteria. The mechanisms of the acquired resistance are broadly classified into three modes: (a) diminution of the intracellular antibiotic load, either by preventing the entry of the antibiotic (Baroud et al. 2013) or by increasing antibiotic efflux (Ogawa et al. 2014); (b) modification of the target molecules, either by mutating the gene encoding the target molecule (Gao et al. 2010) or by post-translationally modifying the target molecule (Cannatelli et al. 2013); and (c) inactivation of the antibiotic, either by hydrolytically cleaving the drug (Woodford et al. 2011; Johnson and Woodford 2013) or by modifying the drug by adding functional groups (Qin et al. 2012; Romanowska et al. 2013). All these mechanisms have been well established with support of the vast documentation obtained through the past decade (Nikaido and Takatsuka 2009; Wright 2011; Fernández and Hancock 2012), as summarized in the following sections.

2.2.1 Diminution of Intracellular Antibiotic Load: The External Affairs

Reaching the target molecule and rapidly mounting to therapeutic levels are essential for any antibiotic to be effective. Therefore, the primary defence of any pathogen to counter an antibiotic is to deplete it from the cell by blocking/minimizing its entry into the pathogen and/or quickly expelling it out by enhancing the efflux mechanisms (Fig. 1a). Different bacteria employ different mechanisms and players in reducing the intracellular levels of antibiotics as described in the following sections.

2.2.1.1 Reducing Permeability: The Consular Affairs

One of the early understood mechanisms of drug resistance involves the prevention of the entry of the antibiotic into the cell (Kumar and Schweizer 2005). For

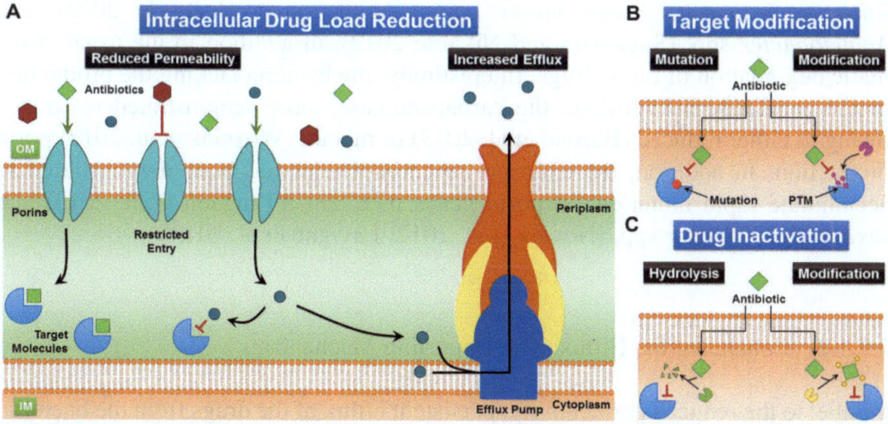

Fig. 1 Mechanisms of drug resistance. (**a**) Intracellular drug load is reduced by either limited permeability to certain antibiotics or increased efflux. Bulkier or hydrophobic antibiotics (*red hexagon*) are prevented from entering the cell, while the antibiotics (*green diamond*) that can enter the cell via the porin channels are efficiently removed by efflux. A few antibiotics (*blue circle*), however, reach their target and inhibit the target activity. (**b**) Modification of the target molecules by mutations (*red circle*) that result in a functional target that are unable to bind the drug efficiently. Alternatively, target biomolecules are modified by post-translational (or transcriptional) modifications (PTM), which limit the access for the antibiotic and thus prevent antibiotic binding. (**c**) Antibiotics are inactivated by enzymatic hydrolysis that prevent their binding to the target molecules and thereby conferring resistance. Antibiotics that resist hydrolysis are modified by enzymatic addition of different functional groups, which alter antibiotic structure, sterically prevent target binding and thereby confer resistance

example, owing to their cell wall's lipopolysaccharide (LPS), several Gram-negative bacteria like *Escherichia coli* (Kojima and Nikaido 2013), *P. aeruginosa* (Strateva and Yordanov 2009), *Vibrio cholera* (Kitaoka et al. 2011) and *Salmonella enterica* (Gunn 2008) are intrinsically less permeable and thus resistant to many hydrophobic antibiotics (Wiese et al. 1999; Delcour 2009), such as the erythromycin, roxithromycin, clarithromycin and azithromycin. Besides, the outer membrane porin channels confer reduced permeability by imposing a size threshold for the antibiotic molecules (Kumar and Varela 2013) and thus allowing passage to only small-sized antimicrobial agents (Pages et al. 2009) although the major porins, the OmpF and OmpC, function as non-specific channels (Kojima and Nikaido 2013; Tran et al. 2013a). Therefore, to limit toxic antibiotic entry into the cell, the drug-resistant bacteria either downregulate the porins or replace the porins with more selective channels (Lin et al. 2002). Using this well-established mechanism, several clinically dreadful bacterial pathogens such as *Serratia marcescens*, *E. cloacae*, *S. enterica*, *E. aerogenes*, *Klebsiella pneumoniae* (Papagiannitsis et al. 2013; Poulou et al. 2013) and *P. aeruginosa* (Page 2012) acquire resistance to several antimicrobial agents such as the aminoglycosides, β-lactams, chloramphenicols and fluoroquinolones. Surprisingly, similar mechanism has been effective in resisting the newer drugs such as carbapenems and cephalosporins by several

Enterobacteriaceae, *Pseudomonas* spp. (Tamber and Hancock 2003) and *Acinetobacter* spp. (Sugawara and Nikaido 2012), in addition to the usual enzymatic degradation of these drugs. Interestingly, the bacteria lacking the production of the carbapenem hydrolase, the carbapenemase, have demonstrated resistance owing to either reduced (Baroud et al. 2013) or mutant (Wozniak et al. 2012) porin production. In addition, the exposure to carbapenems has been demonstrated to accumulate rapid mutations in porin genes in *E. coli* (Tängdén et al. 2013) and several *Enterobacter* spp. (Novais et al. 2012; Lavigne et al. 2013).

2.2.1.2 Enhancing the Efflux: The Deporting Mechanisms

Parallel to the reduced permeability, persistent efflux of the drugs from the bacterial cells is another common drug-resistant mechanism (Webber and Piddock 2003) that is mediated by a family of proteins called energy-driven drug efflux pumps (Fig. 1a), which pump out the drugs faster than they can diffuse in, thereby maintaining the 'intra-bacterial' concentrations of the drug at sub- or non-inhibitory levels (Levy 1992; Paulsen et al. 1995). These ubiquitous efflux pumps are variants of the lipophilic/amphipathic metabolite pumps (Ross et al. 1995; Paulsen et al. 1995; Zgurskaya and Nikaido 2000) and can be mechanically classified into two types: (1) the primary active transporters—ATP-binding cassette (ABC) (Higgins 1992), also known as the permeability glycoprotein (Pgp) transporters (Davidson and Maloney 2007)—that utilize energy from ATP hydrolysis to efflux the drugs out of the cell and (2) the secondary active transporters that are major players in conferring drug resistance (Kumar and Varela 2012) and utilize ion gradient for drug efflux (Mitchell et al. 1998; Henderson and Strauss 1991). The classical example for efflux-induced drug resistance is the resistance to tetracycline by several Gram-positive and Gram-negative bacteria, which is conferred by overproducing a set of plasmid-encoded proteins, the Tet A, B and C proteins that are membrane-located secondary active transport pumps (McMurry and Levy 1978; McMurry et al. 1980; Chopra 1984, 1986). However, unlike the tetracycline pumps, several efflux pumps transport a plethora of structurally unrelated chemical moieties and thus are known as multidrug-resistant (MDR) efflux pumps, examples of which are alarmingly increasing in several bacteria such as MdeA in *Streptococcus mutans* (Huang et al. 2004), FuaABC in *Stenotrophomonas maltophilia* (Hu et al. 2012), KexD in *K. pneumoniae* (Ogawa et al. 2012) and QacA (Tennent et al. 1989; Brown and Skurray 2001) and LmrS (Floyd et al. 2010) in *S. aureus*. In addition, examples of secondary active multidrug efflux pumps have been identified in several prokaryotes and eukaryotes such as the multidrug and toxic compound extrusion (MATE) efflux pump family (Kuroda and Tsuchiya 2009), resistance-nodulation-division (RND) superfamily pumps (Nikaido and Takatsuka 2009; Nikaido 2011) and major facilitator superfamily (MFS) (Marger and Saier 1993; Paulsen et al. 1996; Saidijam et al. 2006). In summary, understanding the molecular mechanisms of the bacterial efflux pumps might lead to the novel chemical modulators that could regulate the efflux and facilitate the antibiotic action in the novel treatment regimens.

2.2.2 Modification of the Target: The Blacking Out

In addition to the membrane-borne resistance mechanisms above, bacteria defend the antibiotics by structurally altering the target molecules to reduce drug-binding affinity. This type of drug resistance often results either from a mutation in the target molecule by genetic means or modification of the target by post-translational modifications (Fig. 1b). The major cellular targets of different antibiotics and their possible modes of action have been understood over the past few decades (Table 1).

Table 1 Cellular processes affected by the major classes of antibacterial drugs

Cellular process	Antibiotic	Target	Mode of action	Resistance mechanism
Membrane biogenesis	β-Lactams	Transpeptidases/ transglycosylases (PBPs)	Blocking of cross-linking enzymes in cell wall peptidoglycan layer	Hydrolysis by β-lactamases, PBP mutants
	Vancomycin	D-Ala-D-Ala termini of peptidoglycan and lipid II	Sequestrating the cross-linking substrate	Employing altered building blocks: D-Ala-D-Lac or D-Ala-D-Serin place of D-Ala-D-Ala
Folic acid metabolism	Sulphonamide	Dihydropteroate synthase (DHPS)	Blocking folate synthesis	Mutations in DHPS Overproduction of PABA
	Trimethoprim	Dihydrofolate reductase (DHFR)	Blocking of the formation of THFA	Reduced permeability Mutant DHFR production
Nucleic acid metabolism	Fluoroquinolones	DNA gyrase	Inhibiting DNA replication	Gyrase mutants to drug resistance
	Rifampacin	RNA polymerase	Blocking DNA transcription	Mutations in RNA polymerase gene
Protein synthesis	Macrolides (erythromycin class)	Peptidyl transferase, centre of the ribosome	Protein synthesis Inhibition	rRNA methylation, drug efflux
	Tetracyclines	Peptidyl transferase		Drug efflux
	Aminoglycosides	Peptidyl transferase		Enzymatic modification of drug
	Oxazolidinones	Peptidyl transferase		Unknown

2.2.2.1 Mutation of the Target

Generally, antibiotics bind their specific targets with high affinity either competing with the substrate for the active site or binding to an allosteric site and preventing access for the substrate and consequently blocking the normal activity of the target (Spratt 1994; Chu et al. 1996). Therefore, changes to the target structure, often by point mutations, that prevent antibiotic binding without affecting the function of target can confer resistance (Fig. 1b). In the large population of pathogens at the site of infection, a single point mutation in the target gene in a single bacterium that bestows resistance can result in the strains, which proliferate to form a population of the resistant bacteria. The classical example of drug target modification is the penicillin resistance mechanism in staphylococci, which, in addition to the expression of the hydrolase β-lactamase (Fisher et al. 2005), arises due to various alterations of the penicillin-binding proteins (PBPs) and converting them to lower-affinity forms (Katayama et al. 2000). *S. aureus*, which causes several infectious diseases, has been resistant to penicillin and its derivative methicillin by different mechanisms including (a) mutation in PBP (Hackbarth et al. 1995; Gao et al. 2010), (b) overexpression of PBP (Boyle-Vavra et al. 2000) and, principally, (c) horizontal acquisition of staphylococcal cassette chromosome *mec* (sccMec) element, which harbours the *mecA* gene that codes for the β-lactam-insensitive PBP2a protein, which takes over the cell wall biogenesis from the β-lactam-inhibited wild-type PBP (Dowson et al. 1989; Katayama et al. 2000). These mechanisms therefore present the molecular basis for the methicillin-resistant *S. aureus* (MRSA) phenotype (Schentag et al. 1998). Another example of target protein mutation is exhibited by the vancomycin-resistant enterococci (VRE), wherein the resistance to the antibiotic vancomycin, which inhibits cell wall biogenesis by binding to the D-Ala-D-Ala dipeptides and preventing the peptidoglycan cross-linking, is conferred by a three-gene cassette, *vanHAX*, that encodes a new pathway enzyme (D'Costa et al. 2011). The product of the first gene, *vanH*, reduces pyruvate to D-lactate, and the product of the second gene, *vanA*, ligates D-alanine to produce D-Ala-D-Lac, followed by the product of the third gene, *vanX*, which selectively hydrolyses the normal dipeptide D-Ala-D-Ala but not D-Ala-D-Lac (Walsh et al. 1996). Therefore, D-Ala-D-Lac serves as the building block for peptidoglycan cross-linking rather than D-Ala-D-Ala (Reynolds and Courvalin 2005), thereby lowering the affinity to vancomycin by about 1000-fold (Bugg et al. 1991) and enabling the bug to dwell at 1000-fold higher levels of the antibiotic. On the other hand, resistance to streptomycin in *Mycobacterium tuberculosis* is due to the mutations in the *rrs* genes encoding ribosomal subunits that lead to altered ribosomal protein targets and thereby prevent binding of this aminoglycoside (Finken et al. 1993).Other examples for target mutation include the amino acid substitution in DNA gyrase that confers resistance to quinolone antimicrobials (Robicsek et al. 2006; Fàbrega et al. 2009), mutations leading to structural changes in the β-subunit of bacterial RNA polymerase that confer resistance to rifampin (Wehrli 1983) and rifampacin (Goldstein 2014) and mutations in the folate metabolic enzymes DHPS and DHFR that confer resistance to sulphonamides (Sköld 2000) and trimethoprim (Huovinen and Kotilainen 1987), respectively.

2.2.2.2 Protection or Modification of the Target

In addition to the mutation-borne alterations, bacteria have developed ways to resist the antibiotic by specifically modifying the target biomolecules such that the antibiotic affinity is considerably reduced (Fig. 1b).Recent studies have demonstrated that several clinically relevant bacteria have employed modification of the target molecules, notably the ribosomal components (Table 1), as the principal mechanism of drug resistance. For example, site-specific mono- or di-methylation of 16S (Jana and Deb 2006) and 23S rRNA (Xiong et al. 2000) by the erythromycin ribosomal methylases (ERM) irreversibly alters the drug-binding sites of macrolide, lincosamine and streptogramin classes of antibiotics (Bussiere et al. 1998; Lambert 2005). Such modifications do not, however, impair protein biosynthesis function of these ribosomal components but lower affinity for the erythromycin and pristinamycin classes of antibiotics. This plasmid-encoded ERM mechanism is the self-defence mechanism in erythromycin-producing organisms and the main resistance mechanism in resistant *S. aureus* and other bacteria that acquire this plasmid (Leclercq and Courvalin 2002). Another example of target modification is the methylation of 23S rRNA at A2503 by the chloramphenicol-florfenicol resistance (*cfr*) methyltransferase (Zhang et al. 2014). The plasmid-borne *cfr* is widely distributed among Gram-positive and Gram-negative bacteria (Shen et al. 2013) and confers resistance to at least six different classes of antibiotics (Atkinson et al. 2013) including the phenicols, pleuromutilins, streptogramins, lincosamides and oxazolidonones (Long et al. 2006). Moreover, resistance to the aminoglycoside protein synthesis inhibitors is facilitated by methylation of the 16S rRNA by the rRNA methyltransferases (Jana and Deb 2006) that are widely distributed in resistant *Enterobacteriaceae* strains worldwide and encoded by *armA* (Fritsche et al. 2008) and *rmt* genes (Gutierrez et al. 2012). Furthermore, resistance to tetracyclines has also been demonstrated to be due to the mutations and substitutions in the peptidyl transferase loop of the 23S rRNA (Roberts 1996; Schnappinger and Hillen 1996; Widdowson and Klugman 1998).

Apart from the ribosomal RNA, modification of several protein and lipid targets has been reported to confer resistance. The *qnr* families of quinolone resistance genes encode pentapeptide repeat proteins (PRPs) that bind the DNA gyrase and topoisomerase IV and protect from quinolones, which otherwise result in double-strand DNA breaks (Vetting et al. 2011). Another example includes the resistance to polymyxin antibiotics, the cyclic antimicrobial peptides with long hydrophobic tail, which bind to lipopolysaccharide (LPS) and disrupt cell membranes (Lim et al. 2010; Cai et al. 2012). Resistance to polymyxins is generally linked to the expression levels of the regulators that affect the levels of LPS. Mutations which bring about over expression of *pmrC* (Beceiro et al. 2011) result in the increased addition of phosphoethanolamine to lipid A and reduced negative charge on the LPS, which consequently reduces colistin binding. In addition, mutations in the *K. pneumoniae* PhoPQ that increases expression of the PmrAB system have been demonstrated to induce polymyxin resistance (Miller et al. 2011; Cannatelli et al. 2013, 2014, 2015). Furthermore, daptomycin is an antibiotic secreted by Gram-positive bacteria to eradicate surrounding bacteria by disrupting multiple aspects of bacterial cell membrane function, including leakage of intracellular contents and loss of membrane potential (Blake and O'Neill 2013; Randall et al. 2013). Therefore, daptomycin has been employed in the treatment of several infections by multiple drug resistance strains (Davlieva et al. 2013). Similar to polymyxin, the resistance to daptomycin occurs due to the remodelling of the phospholipid content of the membrane upon the addition of L-Lys to the phosphatidylglycerol by the action of a mutant peptide resistance factor (Mishra and Bayer 2013). In addition, a novel pathway of resistance to daptomycin wherein relocation of the target away from the primary binding site has been reported in enterococci (Tran et al. 2013b).

2.2.3 Neutralization of the Antibiotic: Warhead Destruction

In addition to keeping the intracellular levels low and protecting the target biomolecules, bacteria have also developed ways to directly inactivate the antibiotics by either hydrolytic cleavage or addition of functional groups to the antibiotics (Fig. 1c). We will discuss the examples for these two modes of drug neutralization in the following sections.

2.2.3.1 Hydrolytic Cleavage of the Antibiotic

The enzyme-catalysed fragmentation of antibiotics has been the principal mechanism of antibiotic resistance since the introduction of the first antibiotic, penicillin, and its hydrolase, the β-lactamase (Philippon 1985). Incidentally, the discovery of the β-lactamase has actually preceded the discovery of penicillin itself, and the enzyme was then believed to play an important role in the peptidoglycan assembly (Abraham and Chain 1988). Discovery of thousands of enzymes that can either degrade or modify different classes of antibiotics, such as the aminoglycosides, β-lactams, macrolides and phenicols, has followed in the later years (Livermore et al. 2008; Nordmann et al. 2011; Voulgari et al. 2013). Moreover, the enzymes that specifically degrade different antibiotics within the same class have been classified into subclasses, for example, different β-lactamases that hydrolyse different β-lactams; for example, the penicillins, cephalosporins, clavams, carbapenems and monobactams have been classified accordingly (Livermore et al. 2008; Nordmann et al. 2011; Voulgari et al. 2013). The mechanism of action of β-lactam class of antibiotics includes the acylation and irreversible modification of the cell wall cross-linking PBPs, using a characteristic strained four-membered lactam ring. The β-lactamases, on the other hand, hydrolyse the β-lactam ring and convert the antibiotic into the inactive penicilloic acid product, which no longer binds PBP and thus is ineffective as a drug. In terms of the numbers, since one β-lactamase molecule can hydrolytically inactivate a thousand penicillin molecules per second (Walsh 2000), a pool of about 10^5 β-lactamase molecules secreted per cell can hydrolyse 100 million penicillin molecules every second, creating an alarmingly difficult task for the clinicians. Moreover, the less-threatening early β-lactamases, such as AmpC β-lactamase, were active against the first-generation β-lactams and were less mobilized owing to the chromosomal localization of their *bla* gene (Jacoby 2009), albeit rare instances of plasmid-borne versions (Coudron et al. 2003). However, several β-lactamases identified later in Gram-negative bacteria are plasmid-borne, such as the TEM-1 β-lactamase of *E. coli* (Datta and Kontomichalou 1965), SHV-1 (sulfhydryl variable active site) (Livermore 1995), IMP (imipenemase) (Lynch et al. 2013), VIM (Verona integron-encoded metallo-β-lactamase), carbapenemase (KPC) from *Klebsiella pneumoniae* (Voulgari et al. 2013), OXA (oxacillinase) (Gniadkowski 2001) and NDM (Johnson and Woodford 2013). Wider dissemination of these enzymes among Gram-negative bacteria, however, resulted in evolution of β-lactamase variants that efficiently hydrolyse a broad range of β-lactam derivatives (Philippon 1985) and thus led to the development of the extended-spectrum β-lactamases (ESBLs) (Johnson and Woodford 2013). In other words, the expansion of antibiotic classes for improved derivatives has been countered by the emergence of novel hydrolytic enzymes, which thereby reinforced the rapid emergence of resistance in several bacterial pathogens to these novel antibiotics, exposing a serious clinical concern (Lynch et al. 2013). In addition, a novel and extremely diverse class of ESBLs, the cefotaxime degrading enzyme (CTX-M) class, has been identified in Gram-negative bacteria (Bauernfeind et al. 1990). The CTX-M class ESBLs are believed to have followed a complex diversification pathway, beginning with the escape of the chromosomal genes from soil bacteria, followed by *ISEcp1*-mediated insertion and conjugation-mediated transfer (Rossolini et al. 2008) that finally resulted in hundreds of variants of these enzymes (Bonnet 2004). These instances, therefore, indicate that different evolutionary paths, (a) the expansion of the resistant bacterial clones and (b) plasmid-meditated mobilization of resistant genes, have operated towards the emergence of the β-lactamase variants resulting in the rapid development and dissemination of resistance. As a consequence, several pathogenic bacteria that are capable of producing multiple ESBLs have emerged and have been compounding the medical threat.

In addition to the β-lactamases, the macrolide esterases from several *Enterobacteria* cease that hydrolyse the lactone ring of erythromycin A and oleandomycin (Barthélémy et al. 1984), and fosfomycin epoxidases that cleave the functional epoxide ring fosfomycin (Rigsby et al. 2005) have been characterized. Besides, a novel mechanism of fosfomycin inactivation has been demonstrated, wherein the carbon-phosphorus lyase enzyme complex hydrolyses the carbon-phosphorus bond in the antibiotic (McGrath et al. 1998).

2.2.3.2 Addition of Functional Group

Although the inactivation of the antibiotic by hydrolysis is the prevalent mechanism leading to resistance, for the antibiotics that lack hydrolysable groups, bacteria have developed enzymes that can decorate the surface of the antibiotic with different chemical substituents, thereby sterically preventing the binding of the antibiotic to its target molecule (Fig. 1c). The list of functional groups that are transferred currently includes acyl, nucleotidyl, phosphoryl, ribitoyl and thiol moieties and is continuously increasing with the discovery of several new inactivating enzymes and understanding their mechanisms (Wright 2005).

The aminoglycoside antibiotics, the rRNA-binding protein synthesis inhibitors, lack hydrolysable bonds and resist to hydrolysis but susceptible to modification owing to several exposed amide and hydroxyl groups on these large molecules that are attacked majorly by three classes of enzymes, acetyltransferases, phosphotransferases and nucleotidyltransferases, which add acetyl, phosphoryl and adenyl groups, respectively (Shaw et al. 1993), thereby reducing the affinity of the antibiotic to the ribosome and rendering the drug ineffective (Jana and Deb 2006). Although these three classes of enzymes bind to similar class of antibiotics by mimicking the rRNA-binding cleft (Norris and Serpersu 2013), they are evolutionarily and mechanistically diverse (Romanowska et al. 2013). Surprisingly, structural studies on the phosphotransferases indicated a close structural relation with the eukaryotic protein kinases (Hon et al. 1997), suggesting horizontal acquisition of these enzymes by the resistant bacteria. Moreover, the genome of the food-borne *Campylobacter coli* harbours a genomic island of 14 genes, which includes six genes for the aminoglycoside-modifying enzymes that span the three classes (Qin et al. 2012) and paradoxically pose an immense challenge for the treatment of the campylobacter infectious intestinal disease (IID) (Tam et al. 2003). In addition, other examples for enzyme-catalysed modification of the antibiotics include chloramphenicol acetyltransferase (CAT)-mediated acetylation of chloramphenicol (Schwarz et al. 2004); streptogramin acetyltransferase (SAT)- and virginiamycin acetyltransferase (VAT)-mediated acetylation of streptogramins and virginamycins, respectively (Allignet and El Solh 1997); macrolide phosphotransferase (MPH)- and glycosyltransferase (MGH)-mediated phosphorylation and glycosylation of macrolide antibiotics, respectively (Wright 2005); glutathione-induced fosfomycin inactivation by FosA and FosB (Arca et al. 1997); ADP-ribosyltransferase (ARR)-mediated ADP-ribosylation of rifampin (Spanogiannopoulos et al. 2014); and TetX (a flavin-dependent monooxygenase)-mediated reduction of tetracycline (Yang et al. 2004).

3 Multi-, Extreme- and Pan-Drug Resistance: The Current Threats

Over and above the resistance to single antibiotics as explained in the above sections, several bacteria have developed simultaneous resistance to various drugs leading to the multiple, extensive or pan-drug-resistant isolates worldwide, the MDR, XDR or PDR 'superbugs' (Magiorakos et al. 2012), which pose threatening clinical consequences and public health crisis (Watanabe 1963; Alekshun and Levy 2007; Nikaido 2009; Nikaido and Takatsuka 2009). Rapid advent of new resistance mechanisms blended with the decrease in treatment efficacy leads to the emergence of MDR and XDR phenotypes in several dreadful bacteria: *E. coli* against cephalosporin and fluoroquinolones, *K. pneumoniae* against cephalosporin and carbapenems, *S. aureus* against methicillin, *S. pneumoniae* against different β-lactams, non-typhoidal *Salmonella* and different species of *Shigella* against fluoroquinolones (Nikaido 2009), *Neisseria gonorrhoeae* against cephalosporin and *M. tuberculosis* against rifampicin, isoniazid and fluoroquinolone (Sharma and Mohan 2006). Additionally different infecting agents such as the fungi (Loeffler and Stevens 2003), virus (Margeridon-Thermet and Shafer 2010; Strasfeld et al. 2010) and protozoan parasites (Ullman 1995; Bansal et al. 2006) have demonstrated DR phenotypes. Since the MDR strains could simply employ the combination of the mechanisms that resist individual drugs (Tanwar et al. 2014), emergence of MDR has become a rapid and unavoidable natural hazard. In addition, inappropriate use of antimicrobial drugs, inadequate sanitation and meagre infection control practices promote the further spread of MDR, therefore persuading for an urgent need to understand mechanisms and means of combating the microbial infections by the superbug variants.

4 Conclusions and Future Perspectives

Bacteria isolated from several pristine sites, including isolated caves (Bhullar et al. 2012) and permafrost (Hernández et al. 2012), have exhibited antibiotic resistance indicating a natural course of evolution for billions of years has been instrumental in shaping the antibiotic resistance even in the absence of human activity (D'Costa et al. 2011). Although antibiotic resistance is a natural phenomenon, wide and especially indiscriminate use of antibiotics in human medicine, veterinary medicine, different forms of agriculture and genetic manipulation studies has ironically hurried the evolution of antibiotic-resistant bacteria. Fortunately, system level understanding of the biology of the pathogens employing genomic, transcriptomic and proteomic approaches and advances in structural tools in understanding the molecular details of antibiotic action has uncovered and will continue to uncover the precise events underpinning resistance, which could boost the discovery and development of novel drugs and advanced therapeutic strategies to counter the current resistance mechanisms. Therefore, rapid evaluation of the novel antibiotics for potential resistance emanation and early identification of the pathways involved in the emanation of resistance will aid in the development of agents that are unlikely either to develop resistance or become ineffective for microbiological reasons. In addition, strategies needed to be developed that can enhance the life span of the antibiotics. However, since resistance is a continuously evolving process, development of next-generation antibacterial drugs using the available technologies, information and expertise is the current and continuous challenge in the field. Moreover, early understanding of the mechanisms of drug resistance with the continuous cross talk between the academic institutions and industry might counter this perennial problem.

Acknowledgments This work was supported by grants from the Department of Biotechnology, India (BT/PR3260/BRB/10/967/2011). We thank Kalyani PV for assistance with proof reading the manuscript.

References

Abraham EP, Chain E (1988) An enzyme from bacteria able to destroy penicillin. 1940. Rev Infect Dis 10:677–678. doi:10.1038/146837a0

Alekshun MN, Levy SB (2007) Molecular mechanisms of antibacterial multidrug resistance. Cell 128:1037–1050. doi:10.1016/j.cell.2007.03.004

Allignet J, El Solh N (1997) Characterization of a new staphylococcal gene, *vgaB*, encoding a putative ABC transporter conferring resistance to streptogramin A and related compounds. Gene 202:133–138. doi:10.1016/S0378-1119(97)00464-2

Arca P, Reguera G, Hardisson C (1997) Plasmid-encoded fosfomycin resistance in bacteria isolated from the urinary tract in a multicentre survey. J Antimicrob Chemother 40:393–399. doi:10.1093/jac/40.3.393

Arthur M, Courvalin P (1993) Genetics and mechanisms of glycopeptide resistance in enterococci. Antimicrob Agents Chemother 37:1563–1571. doi:10.1128/AAC.37.8.1563

Atkinson GC, Hansen LH, Tenson T, Rasmussen A, Kirpekar F, Vester B (2013) Distinction between the Cfr methyltransferase conferring antibiotic resistance and the housekeeping RlmN methyltransferase. Antimicrob Agents Chemother 57:4019–4026. doi:10.1128/AAC.00448-13

Bansal D, Sehgal R, Chawla Y, Malla N, Mahajan RC (2006) Multidrug resistance in amoebiasis patients. Indian J Med Res 124:189–194

Barber M (1961) Methicillin-resistant staphylococci. J Clin Pathol 14:385–393. doi:10.1136/jcp.14.4.385

Baroud M, Dandache I, Araj GF, Wakim R, Kanj S, Kanafani Z, Khairallah M, Sabra A, Shehab M, Dbaibo G, Matar GM (2013) Underlying mechanisms of carbapenem resistance in extended-spectrum β-lactamase-producing *Klebsiella pneumoniae* and *Escherichia coli* isolates at a tertiary care centre in Lebanon: role of OXA-48 and NDM-1 carbapenemases. Int J Antimicrob Agents 41:75–79. doi:10.1016/j.ijantimicag.2012.08.010

Barthélémy P, Autissier D, Gerbaud G, Courvalin P (1984) Enzymic hydrolysis of erythromycin by a strain of *Escherichia coli*. A new mechanism of resistance. J Antibiot (Tokyo) 37:1692–1696. doi:10.7164/antibiotics.37.1692

Bauernfeind A, Grimm H, Schweighart S (1990) A new plasmidiccefotaximase in a clinical isolate of *Escherichia coli*. Infection 18:294–298. doi:10.1007/BF01647010

Beceiro A, Llobet E, Aranda J, Bengoechea JA, Doumith M, Hornsey M, Dhanji H, Chart H, Bou G, Livermore DM, Woodford N (2011) Phosphoethanolamine modification of lipid A in colistin-resistant variants of *Acinetobacter baumannii* mediated by the pmrAB two-component regulatory system. Antimicrob Agents Chemother 55:3370–3379. doi:10.1128/AAC.00079-11

Bhullar K, Waglechner N, Pawlowski A, Koteva K, Banks ED, Johnston MD, Barton HA, Wright GD (2012) Antibiotic resistance is prevalent in an isolated cave microbiome. PLoS One 7, e34953. doi:10.1371/journal.pone.0034953

Bibb MJ (2013) Understanding and manipulating antibiotic production in actinomycetes. Biochem Soc Trans 41:1355–1364. doi:10.1042/BST20130214

Blair JM, Webber MA, Baylay AJ, Ogbolu DO, Piddock LJ (2015) Molecular mechanisms of antibiotic resistance. Nat Rev Microbiol 13:42–51. doi:10.1038/nrmicro3380

Blake KL, O'Neill AJ (2013) Transposon library screening for identification of genetic loci participating in intrinsic susceptibility and acquired resistance to antistaphylococcal agents. J Antimicrob Chemother 68:12–16. doi:10.1093/jac/dks373

Bonnet R (2004) Growing group of extended-spectrum beta-lactamases: the CTX-M enzymes. Antimicrob Agents Chemother 48:1–14. doi:10.1128/AAC.48.1.1-14.2004

Boyle-Vavra S, Berke SK, Lee JC, Daum RS (2000) Reversion of the glycopeptide resistance phenotype in *Staphylococcus aureus* clinical isolates. Antimicrob Agents Chemother 44: 272–277. doi:10.1128/AAC.44.2.272-277.2000

Brown MH, Skurray RA (2001) Staphylococcal multidrug efflux protein QacA. J Mol Microbiol Biotechnol 3:163–170

Bugg TD, Dutka-Malen S, Arthur M, Courvalin P, Walsh CT (1991) Identification of vancomycin resistance protein VanA as a D-alanine: D-alanine ligase of altered substrate specificity. Biochemistry 30:2017–2021. doi:10.1021/bi00222a002

Bussiere DE, Muchmore SW, Dealwis CG, Schluckebier G, Nienaber VL, Edalji RP, Walter KA, Ladror US, Holzman TF, Abad-Zapatero C (1998) Crystal structure of ErmC', an rRNA methyltransferase which mediates antibiotic resistance in bacteria. Biochemistry 37: 7103–7112. doi:10.1021/bi973113c

Cai Y, Chai D, Wang R, Liang B, Bai N (2012) Colistin resistance of Acinetobacter baumannii: clinical reports, mechanisms and antimicrobial strategies. J Antimicrob Chemother 67: 1607–1615. doi:10.1093/jac/dks084

Cannatelli A, D'Andrea MM, Giani T, Di Pilato V, Arena F, Ambretti S, Gaibani P, Rossolini GM (2013) *In vivo* emergence of colistin resistance in *Klebsiella pneumoniae* producing KPC-type carbapenemases mediated by insertional inactivation of the PhoQ/PhoPmgrB regulator. Antimicrob Agents Chemother 57:5521–5526. doi:10.1128/AAC.01480-13

Cannatelli A, Giani T, D'Andrea MM, Di Pilato V, Arena F, Conte V, Tryfinopoulou K, Vatopoulos A, Rossolini GM, Group CS (2014) *MgrB* inactivation is a common mechanism of colistin resistance in KPC-producing *Klebsiella pneumoniae* of clinical origin. Antimicrob Agents Chemother 58:5696–5703. doi:10.1128/AAC.03110-14

Cannatelli A, Santos-Lopez A, Giani T, Gonzalez-Zorn B, Rossolini GM (2015) Polymyxin resistance caused by *mgrB* inactivation is not associated with significant biological cost in *Klebsiella pneumoniae*. Antimicrob Agents Chemother 59:2898–2900; 40:11–17. doi:10.1128/AAC.04998-14

Chopra I (1984) Antibiotic resistance resulting from decreased drug accumulation. Br Med Bull 40:11–17

Chopra I (1986) Genetic and biochemical basis of tetracycline resistance. J Antimicrob Chemother 18:51–56. doi:10.1093/jac/18.Supplement_C.51

Chu DT, Plattner JJ, Katz L (1996) New directions in antibacterial research. J Med Chem 39: 3853–3874. doi:10.1021/jm960294s

Chuanchuen R, Karkhoff-Schweizer RR, Schweizer HP (2003) High-level triclosan resistance in *Pseudomonas aeruginosa* is solely a result of efflux. Am J Infect Control 31:124–127. doi:10.1067/mic.2003.11

Coudron PE, Hanson ND, Climo MW (2003) Occurrence of extended-spectrum and AmpC beta-lactamases in bloodstream isolates of *Klebsiella pneumoniae*: isolates harbor plasmid-mediated FOX-5 and ACT-1 Amp. beta-lactamases. J Clin Microbiol 41:772–777. doi:10.1128/JCM.41.2.772-777.2003

Datta N, Kontomichalou P (1965) Penicillinase synthesis controlled by infectious R factors in Enterobacteriaceae. Nature 208:239–241. doi:10.1038/208239a0

Davidson AL, Maloney PC (2007) ABC transporters: how small machines do a big job. Trends Microbiol 15:448–455. doi:10.1016/j.tim.2007.09.005

Davies J (1994) Inactivation of antibiotics and the dissemination of resistance genes. Science 264: 375–382. doi:10.1126/science.8153624

Davies J (1996) Origins and evolution of antibiotic resistance. Microbiologia 12:9–16

Davlieva M, Zhang W, Arias CA, Shamoo Y (2013) Biochemical characterization of cardiolipin synthase mutations associated with daptomycin resistance in enterococci. Antimicrob Agents Chemother 57:289–296. doi:10.1128/AAC.01743-12

D'Costa VM, King CE, Kalan L, Morar M, Sung WW, Schwarz C, Froese D, Zazula G, Calmels F, Debruyne R, Golding GB, Poinar HN, Wright GD (2011) Antibiotic resistance is ancient. Nature 477:457–461. doi:10.1038/nature10388

Delcour AH (2009) Outer membrane permeability and antibiotic resistance. Biochim Biophys Acta 1794:808–816. doi:10.1016/j.bbapap.2008.11.005

Dowson CG, Hutchison A, Brannigan JA, George RC, Hansman D, Liñares J, Tomasz A, Smith JM, Spratt BG (1989) Horizontal transfer of penicillin-binding protein genes in penicillin-resistant clinical isolates of *Streptococcus pneumoniae*. Proc Natl Acad Sci U S A 86: 8842–8846

Fàbrega A, Madurga S, Giralt E, Vila J (2009) Mechanism of action of and resistance to quinolones. Microb Biotechnol 2:40–61. doi:10.1111/j.1751-7915.2008.00063.x

Fernández L, Hancock RE (2012) Adaptive and mutational resistance: role of porins and efflux pumps in drug resistance. Clin Microbiol Rev 25:661–681. doi:10.1128/CMR.00043-12

Finken M, Kirschner P, Meier A, Wrede A, Böttger EC (1993) Molecular basis of streptomycin resistance in Mycobacterium tuberculosis: alterations of the ribosomal protein S12 gene and point mutations within a functional 16S ribosomal RNA pseudoknot. Mol Microbiol 9: 1239–1246. doi:10.1111/j.1365-2958.1993.tb01253.x

Fisher JF, Meroueh SO, Mobashery S (2005) Bacterial resistance to beta-lactam antibiotics: compelling opportunism, compelling opportunity. Chem Rev 105:395–424. doi:10.1021/cr030102i

Floyd JL, Smith KP, Kumar SH, Floyd JT, Varela MF (2010) LmrS is a multidrug efflux pump of the major facilitator superfamily from *Staphylococcus aureus*. Antimicrob Agents Chemother 54:5406–5412. doi:10.1128/AAC.00580-10

Fritsche TR, Castanheira M, Miller GH, Jones RN, Armstrong ES (2008) Detection of methyltransferases conferring high-level resistance to aminoglycosides in enterobacteriaceae from Europe, North America, and Latin America. Antimicrob Agents Chemother 52: 1843–1845. doi:10.1128/AAC.01477-07

Gao W, Chua K, Davies JK, Newton HJ, Seemann T, Harrison PF, Holmes NE, Rhee HW, Hong JI, Hartland EL, Stinear TP, Howden BP (2010) Two novel point mutations in clinical *Staphylococcus aureus* reduce linezolid susceptibility and switch on the stringent response to promote persistent infection. PLoS Pathog 6, e1000944. doi:10.1371/journal.ppat.1000944

Gniadkowski M (2001) Evolution and epidemiology of extended-spectrum beta-lactamases (ESBLs) and ESBL-producing microorganisms. Clin Microbiol Infect 7:597–608. doi:10.1046/j.1198-743x.2001.00330.x

Goldstein BP (2014) Resistance to rifampicin: a review. J Antibiot (Tokyo) 67:625–630. doi:10.1038/ja.2014.107

Gunn JS (2008) The *Salmonella PmrAB* regulon: lipopolysaccharide modifications, antimicrobial peptide resistance and more. Trends Microbiol 16:284–290. doi:10.1016/j.tim.2008.03.007

Gutierrez B, Escudero JA, San Millan A, Hidalgo L, Carrilero L, Ovejero CM, Santos-Lopez A, Thomas-Lopez D, Gonzalez-Zorn B (2012) Fitness cost and interference of Arm/Rmt aminoglycoside resistance with the RsmF housekeeping methyltransferases. Antimicrob Agents Chemother 56:2335–2341. doi:10.1128/AAC.06066-11

Hackbarth CJ, Kocagoz T, Kocagoz S, Chambers HF (1995) Point mutations in *Staphylococcus aureus PBP2* gene affect penicillin-binding kinetics and are associated with resistance. Antimicrob Agents Chemother 39:103–106. doi:10.1128/AAC.39.1.103

Hayes JD, Wolf CR (1990) Molecular mechanisms of drug resistance. Biochem J 272:281–295. doi:10.1042/bj2720281

Henderson GB, Strauss BP (1991) Evidence for cAMP and cholate extrusion in C6 rat glioma cells by a common anion efflux pump. J Biol Chem 266:1641–1645

Hernández J, Stedt J, Bonnedahl J, Molin Y, Drobni M, Calisto-Ulloa N, Gomez-Fuentes C, Astorga-España MS, González-Acuña D, Waldenström J, Blomqvist M, Olsen B (2012) Human-associated extended-spectrum β-lactamase in the Antarctic. Appl Environ Microbiol 78:2056–2058. doi:10.1128/AEM.07320-11

Higgins CF (1992) ABC transporters: from microorganisms to man. Annu Rev Cell Biol 8:67–113. doi:10.1146/annurev.cb.08.110192.000435

Hon WC, McKay GA, Thompson PR, Sweet RM, Yang DS, Wright GD, Berghuis AM (1997) Structure of an enzyme required for aminoglycoside antibiotic resistance reveals homology to eukaryotic protein kinases. Cell 89:887–895. doi:10.1016/S0092-8674(00)80274-3

Hu RM, Liao ST, Huang CC, Huang YW, Yang TC (2012) An inducible fusaric acid tripartite efflux pump contributes to the fusaric acid resistance in *Stenotrophomonas maltophilia*. PLoS One 7, e51053. doi:10.1371/journal.pone.0051053

Huang J, O'Toole PW, Shen W, Amrine-Madsen H, Jiang X, Lobo N, Palmer LM, Voelker L, Fan F, Gwynn MN, McDevitt D (2004) Novel chromosomally encoded multidrug efflux transporter MdeA in *Staphylococcus aureus*. Antimicrob Agents Chemother 48:909–917. doi:10.1128/AAC.48.3.909-917.2004

Huovinen P, Kotilainen P (1987) In vitro activity of a new cyclic lipopeptide antibiotic, LY146032, against gram-positive clinical bacteria. Antimicrob Agents Chemother 31: 455–457. doi:10.1128/aac.31.3.455

Jacoby GA (2009) AmpC beta-lactamases. Clin Microbiol Rev 22:161–182. doi:10.1128/CMR. 00036-08

Jana S, Deb JK (2006) Molecular understanding of aminoglycoside action and resistance. Appl Microbiol Biotechnol 70:140–150. doi:10.1007/s00253-005-0279-0

Johnson AP, Woodford N (2013) Global spread of antibiotic resistance: the example of New Delhi metallo-β-lactamase (NDM)-mediated carbapenem resistance. J Med Microbiol 62:499–513. doi:10.1099/jmm.0.052555-0

Katayama Y, Ito T, Hiramatsu K (2000) A new class of genetic element, staphylococcus cassette chromosome mec, encodes methicillin resistance in *Staphylococcus aureus*. Antimicrob Agents Chemother 44:1549–1555. doi:10.1128/AAC.44.6.1549-1555.2000

Kieser T, Bibb MJ, Buttner MJ, Chater KF, Hopwood DA (2000) General introduction to actinomycete biology. In: Hopwood DA (ed) Practical *Streptomyces* genetics, 1st edn. John Innes Foundation, Norwich, pp 1–21

Kitaoka M, Miyata ST, Unterweger D, Pukatzki S (2011) Antibiotic resistance mechanisms of *Vibrio cholerae*. J Med Microbiol 60:397–407. doi:10.1099/jmm.0.023051-0

Kojima S, Nikaido H (2013) Permeation rates of penicillins indicate that *Escherichia coli* porins function principally as nonspecific channels. Proc Natl Acad Sci U S A 110:E2629–E2634. doi:10.1073/pnas.1310333110

Kumar A, Schweizer HP (2005) Bacterial resistance to antibiotics: active efflux and reduced uptake. Adv Drug Deliv Rev 57:1486–1513. doi:10.1016/j.addr.2005.04.004

Kumar S, Varela MF (2012) Biochemistry of bacterial multidrug efflux pumps. Int J Mol Sci 13: 4484–4495. doi:10.3390/ijms13044484

Kumar S, Varela MF (2013) Molecular mechanisms of bacterial resistance to antimicrobial agents. In: Méndez-Vilas A (ed) Microbial pathogens and strategies for combating them: science, technology and education, 1st edn. Formatex Research Center, Badajoz, pp 522–534

Kuroda T, Tsuchiya T (2009) Multidrug efflux transporters in the MATE family. Biochim Biophys Acta 1794:763–768. doi:10.1016/j.bbapap.2008.11.012

Lambert PA (2005) Bacterial resistance to antibiotics: modified target sites. Adv Drug Deliv Rev 57:1471–1485. doi:10.1016/j.addr.2005.04.003

Lavigne JP, Sotto A, Nicolas-Chanoine MH, Bouziges N, Pagès JM, Davin-Regli A (2013) An adaptive response of *Enterobacter aerogenes* to imipenem: regulation of porin balance in clinical isolates. Int J Antimicrob Agents 41:130–136. doi:10.1016/j.ijantimicag.2012.10.010

Leclercq R, Courvalin P (2002) Resistance to macrolides and related antibiotics in *Streptococcus pneumoniae*. Antimicrob Agents Chemother 46:2727–2734. doi:10.1128/AAC.46.9.2727-2734.2002

Levy SB (1992) Active efflux mechanisms for antimicrobial resistance. Antimicrob Agents Chemother 36:695–703. doi:10.1128/AAC.36.4.695

Lim LM, Ly N, Anderson D, Yang JC, Macander L, Jarkowski A, Forrest A, Bulitta JB, Tsuji BT (2010) Resurgence of colistin: a review of resistance, toxicity, pharmacodynamics, and dosing. Pharmacotherapy 30:1279–1291. doi:10.1592/phco.30.12.1279

Lin J, Huang S, Zhang Q (2002) Outer membrane proteins: key players for bacterial adaptation in host niches. Microbes Infect 4:325–331. doi:10.1016/S1286-4579(02)01545-9

Livermore DM (1995) beta-Lactamases in laboratory and clinical resistance. Clin Microbiol Rev 8:557–584

Livermore DM, Hope R, Mushtaq S, Warner M (2008) Orthodox and unorthodox clavulanate combinations against extended-spectrum beta-lactamase producers. Clin Microbiol Infect 14: 189–193. doi:10.1111/j.1469-0691.2007.01858.x

Loeffler J, Stevens DA (2003) Antifungal drug resistance. Clin Infect Dis 36:S31–S41. doi:10.1086/344658

Long KS, Poehlsgaard J, Kehrenberg C, Schwarz S, Vester B (2006) The CfrrRNA methyltransferase confers resistance to Phenicols, Lincosamides, Oxazolidinones, Pleuromutilins, and Streptogramin A antibiotics. Antimicrob Agents Chemother 50:2500–2505. doi:10.1128/AAC.00131-06

Lynch JP, Clark NM, Zhanel GG (2013) Evolution of antimicrobial resistance among Enterobacteriaceae (focus on extended spectrum β-lactamases and carbapenemases). Expert Opin Pharmacother 14:199–210. doi:10.1517/14656566.2013.763030

Magiorakos AP, Srinivasan A, Carey RB, Carmeli Y, Falagas ME, Giske CG, Harbarth S, Hindler JF, Kahlmeter G, Olsson-Liljequist B, Paterson DL, Rice LB, Stelling J, Struelens MJ, Vatopoulos A, Weber JT, Monnet DL (2012) Multidrug-resistant, extensively drug-resistant and pandrug-resistant bacteria: an international expert proposal for interim standard definitions for acquired resistance. Clin Microbiol Infect 18:268–281. doi:10.1111/j.1469-0691.2011.03570.x

Marger MD, Saier MH (1993) A major superfamily of transmembrane facilitators that catalyse uniport, symport and antiport. Trends Biochem Sci 18:13–20. doi:10.1016/0968-0004(93)90081-W

Margeridon-Thermet S, Shafer RW (2010) Comparison of the mechanisms of drug resistance among HIV, Hepatitis B, and Hepatitis C. Viruses 2:2696–2739. doi:10.3390/v2122696

McGrath JW, Hammerschmidt F, Quinn JP (1998) Biodegradation of phosphonomycin by *Rhizobium huakuii* PMY1. Appl Environ Microbiol 64:356–358

McMurry L, Levy SB (1978) Two transport systems for tetracycline in sensitive *Escherichia coli*: critical role for an initial rapid uptake system insensitive to energy inhibitors. Antimicrob Agents Chemother 14:201–209. doi:10.1128/AAC.14.2.201

McMurry L, Petrucci RE, Levy SB (1980) Active efflux of tetracycline encoded by four genetically different tetracycline resistance determinants in *Escherichia coli*. Proc Natl Acad Sci U S A 77:3974–3977

Miller AK, Brannon MK, Stevens L, Johansen HK, Selgrade SE, Miller SI, Høiby N, Moskowitz SM (2011) PhoQ mutations promote lipid A modification and polymyxin resistance of *Pseudomonas aeruginosa* found in colistin-treated cystic fibrosis patients. Antimicrob Agents Chemother 55:5761–5769. doi:10.1128/AAC.05391-11

Mishra NN, Bayer AS (2013) Correlation of cell membrane lipid profiles with daptomycin resistance in methicillin-resistant *Staphylococcus aureus*. Antimicrob Agents Chemother 57: 1082–1085. doi:10.1128/AAC.02182-12

Mitchell BA, Brown MH, Skurray RA (1998) QacA multidrug efflux pump from *Staphylococcus aureus*: comparative analysis of resistance to diamidines, biguanidines, and guanylhydrazones. Antimicrob Agents Chemother 42:475–477

Nikaido H (2009) Multidrug resistance in bacteria. Annu Rev Biochem 78:119–146. doi:10.1146/annurev.biochem.78.082907.145923

Nikaido H (2011) Structure and mechanism of RND-type multidrug efflux pumps. Adv Enzymol Relat Areas Mol Biol 77:1–60. doi:10.1002/9780470920541.ch1

Nikaido H, Takatsuka Y (2009) Mechanisms of RND multidrug efflux pumps. Biochim Biophys Acta 1794:769–781. doi:10.1016/j.bbapap.2008.10.004

Nordmann P, Poirel L, Toleman MA, Walsh TR (2011) Does broad-spectrum beta-lactam resistance due to NDM-1 herald the end of the antibiotic era for treatment of infections caused by Gram-negative bacteria? J Antimicrob Chemother 66:689–692. doi:10.1093/jac/dkq520

Norris AL, Serpersu EH (2013) Ligand promiscuity through the eyes of the aminoglycoside N3 acetyltransferase IIa. Protein Sci 22:916–928. doi:10.1002/pro.2273

Novais A, Rodrigues C, Branquinho R, Antunes P, Grosso F, Boaventura L, Ribeiro G, Peixe L (2012) Spread of an OmpK36-modified ST15 *Klebsiella pneumoniae* variant during an outbreak involving multiple carbapenem-resistant Enterobacteriaceae species and clones. Eur J Clin Microbiol Infect Dis 31:3057–3063. doi:10.1007/s10096-012-1665-z

Ogawa W, Onishi M, Ni R, Tsuchiya T, Kuroda T (2012) Functional study of the novel multidrug efflux pump KexD from *Klebsiella pneumoniae*. Gene 498:177–182. doi:10.1016/j.gene.2012.02.008

Ogawa Y, Saraya T, Koide T, Kikuchi K, Ohkuma K, Araki K, Makino H, Yonetani S, Takizawa H, Goto H (2014) Methicillin-resistant *Staphylococcus aureus* enterocolitis sequentially complicated with septic arthritis: a case report and review of the literature. BMC Res Notes 7:21. doi:10.1186/1756-0500-7-21

Page MG (2012) The role of the outer membrane of Gram-negative bacteria in antibiotic resistance: Ajax' shield or Achilles' heel? Handb Exp Pharmacol 211:67–86. doi:10.1007/978-3-642-28951-4_5

Pages G, Torres AM, Ju P, Bansal PS, Alewood PF, Kuchel PW, Vandenberg JI (2009) Structure of the pore-helix of the hERGK(+) channel. Eur Biophys J 39:111–120. doi:10.1007/s00249-009-0433-1

Papagiannitsis CC, Giakkoupi P, Kotsakis SD, Tzelepi E, Tzouvelekis LS, Vatopoulos AC, Miriagou V (2013) OmpK35 and OmpK36 porin variants associated with specific sequence types of *Klebsiella pneumoniae*. J Chemother 25:250–254. doi:10.1179/1973947813Y.0000000075

Paulsen IT, Brown MH, Dunstan SJ, Skurray RA (1995) Molecular characterization of the staphylococcal multidrug resistance export protein QacC. J Bacteriol 177:2827–2833

Paulsen IT, Brown MH, Skurray RA (1996) Proton-dependent multidrug efflux systems. Microbiol Rev 60:575–608

Philippon A (1985) Susceptibility of *Pseudomonas aeruginosa* to beta-lactam antibiotics. Chemioterapia 4:424–428

Poulou A, Voulgari E, Vrioni G, Koumaki V, Xidopoulos G, Chatzipantazi V, Markou F, Tsakris A (2013) Outbreak caused by an ertapenem-resistant, CTX-M-15-producing *Klebsiella pneumoniae* sequence type 101 clone carrying an OmpK36 porin variant. J Clin Microbiol 51:3176–3182. doi:10.1128/JCM.01244-13

Qin S, Wang Y, Zhang Q, Chen X, Shen Z, Deng F, Wu C, Shen J (2012) Identification of a novel genomic island conferring resistance to multiple aminoglycoside antibiotics in *Campylobacter coli*. Antimicrob Agents Chemother 56:5332–5339. doi:10.1128/AAC.00809-12

Randall CP, Mariner KR, Chopra I, O'Neill AJ (2013) The target of daptomycin is absent from *Escherichia coli* and other gram-negative pathogens. Antimicrob Agents Chemother 57: 637–639. doi:10.1128/AAC.02005-12

Reynolds PE, Courvalin P (2005) Vancomycin resistance in enterococci due to synthesis of precursors terminating in D-alanyl-D-serine. Antimicrob Agents Chemother 49:21–25. doi:10.1128/AAC.49.1.21-25.2005

Rigsby RE, Fillgrove KL, Beihoffer LA, Armstrong RN (2005) Fosfomycin resistance proteins: a nexus of glutathione transferases and epoxide hydrolases in a metalloenzyme superfamily. Methods Enzymol 401:367–379. doi:10.1016/S0076-6879(05)01023-2

Roberts MC (1996) Tetracycline resistance determinants: mechanisms of action, regulation of expression, genetic mobility, and distribution. FEMS Microbiol Rev 19:1–24. doi:10.1111/j.1574-6976.1996.tb00251.x

Robicsek A, Jacoby GA, Hooper DC (2006) The worldwide emergence of plasmid-mediated quinolone resistance. Lancet Infect Dis 6:629–640. doi:10.1016/S1473-3099(06)70599-0

Romanowska J, Reuter N, Trylska J (2013) Comparing aminoglycoside binding sites in bacterial ribosomal RNA and aminoglycoside modifying enzymes. Proteins 81:63–80. doi:10.1002/prot.24163

Ross JI, Eady EA, Cove JH, Baumberg S (1995) Identification of a chromosomally encoded ABC-transport system with which the staphylococcal erythromycin exporter MsrA may interact. Gene 153:93–98

Rossolini GM, D'Andrea MM, Mugnaioli C (2008) The spread of CTX-M-type extended-spectrum beta-lactamases. Clin Microbiol Infect 14(Suppl 1):33–41. doi:10.1111/j.1469-0691.2007.01867.x

Saidijam M, Benedetti G, Ren Q, Xu Z, Hoyle CJ, Palmer SL, Ward A, Bettaney KE, Szakonyi G, Meuller J, Morrison S, Pos MK, Butaye P, Walravens K, Langton K, Herbert RB, Skurray RA, Paulsen IT, O'Reilly J, Rutherford NG, Brown MH, Bill RM, Henderson PJ (2006) Microbial drug efflux proteins of the major facilitator superfamily. Curr Drug Targets 7:793–811. doi:10.2174/138945006777709575

Schentag JJ, Hyatt JM, Carr JR, Paladino JA, Birmingham MC, Zimmer GS, Cumbo TJ (1998) Genesis of methicillin-resistant *Staphylococcus aureus* (MRSA), how treatment of MRSA infections has selected for vancomycin-resistant *Enterococcus faecium*, and the importance of antibiotic management and infection control. Clin Infect Dis 26:1204–1214. doi:10.1086/520287

Schnappinger D, Hillen W (1996) Tetracyclines: antibiotic action, uptake, and resistance mechanisms. Arch Microbiol 165:359–369. doi:10.1007/s002030050339

Schwarz S, Kehrenberg C, Doublet B, Cloeckaert A (2004) Molecular basis of bacterial resistance to chloramphenicol and florfenicol. FEMS Microbiol Rev 28:519–542. doi:10.1016/j.femsre.2004.04.001

Sharma SK, Mohan A (2006) Multidrug-resistant tuberculosis: a menace that threatens to destabilize tuberculosis control. Chest 130:261–272. doi:10.1378/chest.130.1.261

Shaw KJ, Rather PN, Hare RS, Miller GH (1993) Molecular genetics of aminoglycoside resistance genes and familial relationships of the aminoglycoside-modifying enzymes. Microbiol Rev 57: 138–163

Shen J, Wang Y, Schwarz S (2013) Presence and dissemination of the multiresistance gene *cfr* in Gram-positive and Gram-negative bacteria. J Antimicrob Chemother 68:1697–1706. doi:10.1093/jac/dkt092

Shore AC, Deasy EC, Slickers P, Brennan G, O'Connell B, Monecke S, Ehricht R, Coleman DC (2011) Detection of staphylococcal cassette chromosome mec type XI carrying highly divergent *mecA*, *mecI*, *mecR1*, *blaZ*, and *ccr* genes in human clinical isolates of clonal complex

130 methicillin-resistant *Staphylococcus aureus*. Antimicrob Agents Chemother 55: 3765–3773. doi:10.1128/AAC.00187-11
Sköld O (2000) Evolution and mechanisms for spread of antimicrobial resistance. Acta Vet Scand Suppl 93:23–27, discussion 28–36
Spanogiannopoulos P, Waglechner N, Koteva K, Wright GD (2014) A rifamycin inactivating phosphotransferase family shared by environmental and pathogenic bacteria. Proc Natl Acad Sci U S A 111:7102–7107. doi:10.1073/pnas.1402358111
Spratt BG (1994) Resistance to antibiotics mediated by target alterations. Science 264:388–393. doi:10.1126/science.8153626
Strasfeld L, Lee I, Tatarowicz W, Villano S, Chou S (2010) Virologic characterization of multidrug-resistant cytomegalovirus infection in 2 transplant recipients treated with maribavir. J Infect Dis 202:104–108. doi:10.1086/653122
Strateva T, Yordanov D (2009) *Pseudomonas aeruginosa* – a phenomenon of bacterial resistance. J Med Microbiol 58:1133–1148. doi:10.1099/jmm.0.009142-0
Sugawara E, Nikaido H (2012) OmpA is the principal nonspecific slow porin of *Acinetobacter baumannii*. J Bacteriol 194:4089–4096. doi:10.1128/JB.00435-12
Tam CC, O'Brien SJ, Adak GK, Meakins SM, Frost JA (2003) *Campylobacter coli* – an important foodborne pathogen. J Infect 47:28–32. doi:10.1016/S0163-4453(03)00042-2
Tamber S, Hancock RE (2003) On the mechanism of solute uptake in *Pseudomonas*. Front Biosci 8:s472–s483. doi:10.2741/1075
Tängdén T, Adler M, Cars O, Sandegren L, Löwdin E (2013) Frequent emergence of porin-deficient subpopulations with reduced carbapenem susceptibility in ESBL-producing *Escherichia coli* during exposure to ertapenem in an in vitro pharmacokinetic model. J Antimicrob Chemother 68:1319–1326. doi:10.1093/jac/dkt044
Tanwar J, Das S, Fatima Z, Hameed S (2014) Multidrug resistance: an emerging crisis. Interdiscip Perspect Infect Dis 2014:541340. doi:10.1155/2014/541340
Tennent JM, Lyon BR, Midgley M, Jones IG, Purewal AS, Skurray RA (1989) Physical and biochemical characterization of the *qacA* gene encoding antiseptic and disinfectant resistance in *Staphylococcus aureus*. J Gen Microbiol 135:1–10. doi:10.1099/00221287-135-1-1
Tran QT, Williams S, Farid R, Erdemli G, Pearlstein R (2013a) The translocation kinetics of antibiotics through porin OmpC: insights from structure-based solvation mapping using WaterMap. Proteins 81:291–299. doi:10.1002/prot.24185
Tran TT, Panesso D, Gao H, Roh JH, Munita JM, Reyes J, Diaz L, Lobos EA, Shamoo Y, Mishra NN, Bayer AS, Murray BE, Weinstock GM, Arias CA (2013b) Whole-genome analysis of a daptomycin-susceptible *Enterococcus faecium* strain and its daptomycin-resistant variant arising during therapy. Antimicrob Agents Chemother 57:261–268. doi:10.1128/AAC.01454-12
Ullman B (1995) Multidrug resistance and P-glycoproteins in parasitic protozoa. J Bioenerg Biomembr 27:77–84. doi:10.1007/BF02110334
Vetting MW, Hegde SS, Zhang Y, Blanchard JS (2011) Pentapeptide-repeat proteins that act as topoisomerase poison resistance factors have a common dimer interface. Acta Crystallogr Sect F Struct Biol Cryst Commun 67:296–302. doi:10.1107/S1744309110053315
Voulgari E, Poulou A, Koumaki V, Tsakris A (2013) Carbapenemase-producing Enterobacteriaceae: now that the storm is finally here, how will timely detection help us fight back? Future Microbiol 8:27–39. doi:10.2217/fmb.12.130
Walsh C (2000) Molecular mechanisms that confer antibacterial drug resistance. Nature 406: 775–781. doi:10.1038/35021219
Walsh CT, Fisher SL, Park IS, Prahalad M, Wu Z (1996) Bacterial resistance to vancomycin: five genes and one missing hydrogen bond tell the story. Chem Biol 3:21–28. doi:10.1016/S1074-5521(96)90079-4
Watanabe T (1963) Infective heredity of multiple drug resistance in bacteria. Bacteriol Rev 27: 87–115

Webber MA, Piddock LJ (2003) The importance of efflux pumps in bacterial antibiotic resistance. J Antimicrob Chemother 51:9–11. doi:10.1093/jac/dkg050

Wehrli W (1983) Rifampin: mechanisms of action and resistance. Rev Infect Dis 5(Suppl 3): S407–S411. doi:10.1093/clinids/5.Supplement_3.S407

Widdowson CA, Klugman KP (1998) Emergence of the M phenotype of erythromycin-resistant pneumococci in South Africa. Emerg Infect Dis 4:277–281. doi:10.3201/eid0402.980216

Wiese A, Brandenburg K, Ulmer AJ, Seydel U, Müller-Loennies S (1999) The dual role of lipopolysaccharide as effector and target molecule. Biol Chem 380:767–784. doi:10.1515/BC.1999.097

Woodford N, Turton JF, Livermore DM (2011) Multiresistant Gram-negative bacteria: the role of high-risk clones in the dissemination of antibiotic resistance. FEMS Microbiol Rev 35: 736–755. doi:10.1111/j.1574-6976.2011.00268.x

Wozniak A, Villagra NA, Undabarrena A, Gallardo N, Keller N, Moraga M, Román JC, Mora GC, García P (2012) Porin alterations present in non-carbapenemase-producing Enterobacteriaceae with high and intermediate levels of carbapenem resistance in Chile. J Med Microbiol 61: 1270–1279. doi:10.1099/jmm.0.045799-0

Wright GD (2005) Bacterial resistance to antibiotics: enzymatic degradation and modification. Adv Drug Deliv Rev 57:1451–1470. doi:10.1016/j.addr.2005.04.002

Wright GD (2011) Molecular mechanisms of antibiotic resistance. Chem Commun (Camb) 47: 4055–4061. doi:10.1039/c0cc05111j

Xiong L, Kloss P, Douthwaite S, Andersen NM, Swaney S, Shinabarger DL, Mankin AS (2000) Oxazolidinone resistance mutations in 23S rRNA of *Escherichia coli* reveal the central region of domain V as the primary site of drug action. J Bacteriol 182:5325–5331. doi:10.1128/JB.182.19.5325-5331.2000

Yang W, Moore IF, Koteva KP, Bareich DC, Hughes DW, Wright GD (2004) TetX is a flavin-dependent monooxygenase conferring resistance to tetracycline antibiotics. J Biol Chem 279: 52346–52352. doi:10.1074/jbc.M409573200

Zgurskaya HI, Nikaido H (2000) Cross-linked complex between oligomeric periplasmic lipoprotein AcrA and the inner-membrane-associated multidrug efflux pump AcrB from *Escherichia coli*. J Bacteriol 182:4264–4267. doi:10.1128/JB.182.15.4264-4267.2000

Zhang WJ, Xu XR, Schwarz S, Wang XM, Dai L, Zheng HJ, Liu S (2014) Characterization of the *IncA/C* plasmid pSCEC2 from *Escherichia coli* of swine origin that harbours the multi-resistance gene *cfr*. J Antimicrob Chemother 69:385–389. doi:10.1093/jac/dkt355

Zhu L, Lin J, Ma J, Cronan JE, Wang H (2010) Triclosan resistance of *Pseudomonas aeruginosa* PAO1 is due to FabV, a triclosan-resistant enoyl-acyl carrier protein reductase. Antimicrob Agents Chemother 54:689–698. doi:10.1128/AAC.01152-09

Molecular Mechanism of Drug Resistance

Shilpa Ray, Susmita Das, and Mrutyunjay Suar

Abstract The treatment of microbial infections has suffered greatly in this present century of pathogen dominance. Inspite of extensive research efforts and scientific advancements, the worldwide emergence of microbial tolerance continues to plague survivability. The innate property of microbe to resist any antibiotic due to evolution is the virtue of intrinsic resistance. However, the classical genetic mutations and extrachromosomal segments causing gene exchange attribute to acquired tolerance development. Rampant use of antimicrobials causes certain selection pressure which increases the resistance frequency. Genomic duplication, enzymatic site modification, target alteration, modulation in membrane permeability, and the efflux pump mechanism are the major contributors of multidrug resistance (MDR), specifically antibiotic tolerance development. MDRs will lead to clinical failures for treatment and pose health crisis. The molecular mechanisms of antimicrobial resistance are diverse as well as complex and still are exploited for new discoveries in order to prevent the surfacing of "superbugs." Antimicrobial chemotherapy has diminished the threat of infectious diseases to some extent. To avoid the indiscriminate use of antibiotics, the new ones licensed for use have decreased with time. Additionally, in vitro assays and genomics for anti-infectives are novel approaches used in resolving the issues of microbial resistance. Proper use of drugs can keep it under check and minimize the risk of MDR spread.

1 Introduction

With the advent of technological advancements, the rising scientific era witnessed the emergence of infectious diseases. This led to a sharp increase in global mortality and morbidity rate. Hence the research community held up the war against these pathogens by investigating deep into their molecular mechanisms, their host–pathogen interaction, and their epidemiology for the discovery of fine effective antimicrobial measures for host survival and safety. The researchers treated the

S. Ray • S. Das • M. Suar (✉)
School of Biotechnology, KIIT University, Bhubaneswar, Odisha, India
e-mail: msuar@kiitbiotech.ac.in

G. Arora et al. (eds.), *Drug Resistance in Bacteria, Fungi, Malaria, and Cancer*,
DOI 10.1007/978-3-319-48683-3_3

pathogenic ailments with useful inventions for long-term medication. Drug generally implies to foreign elements or agents that have some medicinal properties for common therapeutic usage. They can be used for bacterial infections, even as antifungal or antiparasitic agents, for cancer treatments, etc. The discovery of antibiotics was the greatest medical intervention affecting human survivability and health regime (Chain 1979; Fleming 1944; Dougherty and Pucci 2011; Ligon 2004). However indiscriminate usage had dramatically introduced new biological problems that are hard to confront with the present-day scientific solutions. Hence failure of medications did set the dawn of a post-antimicrobial era. The time of the Second World War had limited access to these expensive, rare, systemic medications (sulfonamides, penicillin, etc.). With time, simplified production of formulations eased the use of such treatments. Gradually, these antimicrobial agents mostly antibiotics became the elixir for the ailments from time then (Dougherty and Pucci 2011). Moreover the discoverer of penicillin Sir Alexander Fleming warned the surfacing of resistant forms of *Staphylococcus aureus* due to improper penicillin usage which would cause serious host complications (Hartman and Tomasz 1984). Few years later resistant forms emerged with 50 % of susceptible strains becoming resistant to the drug. Similar trend was observed in many other microbial species switching their drug sensitivity approach to a severe resistance mechanism thereby affecting healthy non-vulnerable population. This section will discuss in detail the emergence of drug-resistant microbial populations and the factors that govern their drug-resistant feature. The major focus of this segment will highlight the molecular, cellular, clinical, and genetic factors that bring about this severe cause of drug resistance. Beginning from the natural microbial resistance to the evolutionary alteration in the pathogen's genome, this chapter will cover the idea of how dealing with the conventional drug resistance mechanisms in the twenty-first century will create new frontiers for innovative therapeutic development. The problems and the complex challenge of dealing the multidrug resistance (MDR) mechanism at the molecular level will enable strategies for futuristic drug development for combating fungal, bacterial, and viral resistance mechanism.

2 Emergence of Drug Resistance: The Road So Far

Adaption is a very essential condition for survival as well as sustenance. All living organisms nurture themselves with crucial components from their living system. In addition to fundamental requirements, adaption against the toxic agents also requires armors of endurance. The adage "survival of the fittest" also applies to the environmental sustenance of microbes. This microbial tolerance has enabled the mechanism of resistance as one of the means to combat the harmful environmental effects. This results in conferring multiple drug resistance within pathogens against idle treatments. The first drug resistance occurred against penicillin and sulfonamides against *S. aureus* (Rammelkamp and Maxon 1942; Sabath et al. 1977). The discovery of antibiotics led to the emergence of antibiotic resistance in the following two or more decades. The pathogens in the hospitals were not only reported to be resistant to the therapeutics but also remained viable for further infecting the vulnerable individuals with weakened immune system. The nineteenth century had an impressive pattern of increased tolerance mechanism among the pathogens from sulfonamide and penicillin-resistant *S. aureus* to multidrug-resistant *M. tuberculosis*. Some gastroenteric pathogens like *Shigella*, *Salmonella*, *V. cholera*, *E. coli*, *P. aeruginosa*, etc., also developed resistance against many antimicrobials during the course of time. Some strains also enabled community-dependent infection spread like *Streptococcus* developing resistance to penicillin and *S. aureus* and *Enterococcus* developing resistance to vancomycin.

2.1 History of Antibiotic Resistance Development: A Problem Getting Worse

The emergence of drug resistance has always been a major concern worldwide right after the introduction of drugs for common use. The crucial role of microbes in causing diseases led to the discovery of antimicrobial drugs (Davies and Davies 2010). Penicillin was the first of its kind as mentioned before to be introduced by Alexander Fleming in 1928 (Fleming 1944). Its effectiveness was against the Gram-positive bacteria, especially *Staphylococcus aureus* followed by few more antibiotics including streptomycin, tetracycline, chloramphenicol, vancomycin, macrolides, nalidixic acid, etc. (J. T. Park and Stromistger 1957). However, different drug-resistant microorganisms also started to show up with due time course. In the 1950s, penicillinase-producing *S. aureus* found its way to the society resulting in the gradual emergence and spread of multidrug-resistant *S. aureus*. To combat the harmful effects of penicillinase-producing *S. aureus*, methicillin was developed, but to utter disappointment, methicillin-resistant *Staphylococcus aureus* (MRSA) thwarted mankind in the UK (Brumfitt and Hamilton-Miller 1989; Klevens et al. 2007). Meanwhile, ampicillin and piperacillin were produced as broad-spectrum antibiotics, which also proved to be effective against Gram-negative *Enterobacteriaceae* and *Pseudomonas aeruginosa*, respectively. In the 1960s, a new genre of drugs named cephems was designed and widely used (Bryskier 2000). With time, different generations of cephems were developed according to their antimicrobial spectra. But simultaneously, there was emergence of penicillin-intermediate *S. pneumoniae* (PISP) in the latter half of 1967 and penicillin-resistant *S. pneumoniae* (PRSP) in the 1970s. Frequent use of cephems was believed to be responsible for the increase of PRSP. Ampicillin, which was earlier effective against *Haemophilus influenzae*, failed in the 1980s, when the strains gained resistance against antibiotic by producing β-lactamase (Rubin et al. 1981). In the 1990s, β-lactamase-producing strains reduced in Japan, but highly resistant β-lactam strains increased in number through mutations in their PBP (penicillin-binding protein) genes. Such were named β-lactam-negative ampicillin-resistant

(BLNAR) strains which were more common in Japan than in other parts of the world. In the latter half of the 1990s, vancomycin-resistant *S. aureus* (VRSA) was reported in the USA, which was thought to acquire the antibiotic resistance gene from vancomycin-resistant enterococci (VRE) via horizontal gene transfer (Goldrick 2002; Sung and Lindsay 2007). In the early twenty-first century, the increased use of third-generation cephems and carbapenem, quinolones gave rise to increased risk of resistant *Gonococci*, multidrug-resistant *P. aeruginosa* (MDRP), and quinolone-resistant *E. coli.* A relatively new concern in this field is the multidrug-resistant (MDR) and extensively drug-resistant (XDR) tuberculosis. According to 2013 reports, out of all TB cases worldwide, approximately 5 % were estimated to be MDR-TB where the bacterium is resistant to minimum two of the most powerful first-line anti-TB drugs, isoniazid and rifampicin (Klopper et al. 2013). When MDR-TB becomes resistant to at least one drug from each group of second-line anti-TB drugs like fluoroquinolones and other injectable drugs, it's defined as XDR-TB. Apart from resistance in microbes, resistance to chemotherapy by cancer cells emerged to be a major concern in cancer research.

3 Factors for Drug Resistance Development: The Ongoing Phenomena

Antibiotic resistance is a serious global issue that has seized the roots of development. Antimicrobial resistance affects host immune profile, modulates with pathogen's fitness cost, and influences the genetic co-selection of resistant species with their frequency of reversibility potential (Andersson and Hughes 2010). The biologic mechanisms of the microbe are mostly responsible for such a resistant feature to fight the environmental toxic conditions. The inherent property of the pathogen, i.e., the natural resistance of the microbe, is a reason of resistance emergence. The major causative factor of resistance development is also the frequency of appearance of resistant bacteria due to genetic mutations or evolutionary horizontal gene transfer (Dzidic and Bedeković 2003; Thomas and Nielsen 2005). These mutations can modulate with the pathogen's drug uptake and efflux ability along with target alteration. Secondly the exposure of the pathogen to the drug/therapeutic influences the screening of resistant strains. The drug pharmacokinetic properties which affect pathogen's sustenance and clearance directly measure the degree of resistance mechanism. Human microflora is a hub of microbes and the release or exposure of wide-spectrum antibiotics can trigger resistant microbes to flourish and spread their tentacles of tolerance. Antibacterial agents mostly target bacterial cell wall synthesis, protein synthetic machinery, DNA duplication, and repair processes or transcriptional regulatory processes. Thereby the resistant mutants lack biological fitness. Hence, the natural selection theory suggests less prevalence of antibiotics to decrease the emergence of resistant species. Even the frequency with which the genetic alterations occur within the microbe affecting varying degree of resistance is another factor. The antibiotic *selection pressure* is another factor of modifying the host's microbiota. The host system is subjected to a phenomenon of "selective pressure" when treated with antibiotics during infection. With a greater activity scale, the resistance frequency increases. This results in the resistant species surviving in the host population as compared to the susceptible strain with the harsh effect of the antibiotic. So being a reservoir increases the chances of infection spread to a greater extent. Antibiotics like cephalosporins, azithromycin, and fluoroquinolones bring about this effect of "selective pressure" in eukaryotic hosts. A careful and considerate use of antimicrobials is highly recommended for human safety. This high-end technical and materialistic world should concern the appropriate prescription of a much selective and narrow-range antibiotic thereby minimizing the risk of resistance development. Other factors include the drug exposure properties and concentration dosage dealing with their pharmacokinetic profile, the drug-pattern usage and its distribution scale, the immune system of the host (immunocompromised individuals are more prone to infection), pathogen's fitness cost of tolerance, and influence of nonresistant therapeutics, thereby affecting the pathogen's transmission intensity. However the poor and developing nations are reeling under the burden of drug resistance which directly affects their economic turnover. As a consequence, one has to raise the standards of high-end sophisticated health services and prevent the prevalence of community-associated infections.

Different factors acts as fulcrum in the process of drug resistance development (Wright and Poinar 2012). But which factors determine the most influential parameter for developing resistance and which factors remain insignificant are still under investigation. It's yet to be defined that how adaptation enable the microbe to culminate their survival strategy with resistance development. The understanding of the molecular and clinical mechanisms that ignite the "trigger responses" for microbial acclimatization has just begun. The literature of therapeutic invention has led to the conclusion that the once susceptible strains have developed weapons of resistance against the sensitive drug. For instance, a span of five decades made *Streptococcus* species resistant to penicillin and *S. aureus* resistant to vancomycin. This will enable one to formulate new mechanisms for restructuring new designs for drug development against many life-threatening diseases. Among the risk factors responsible for the emergence of drug resistance, illogical and rampant use of antibiotics is one of the crucial reasons for such an issue (Alanis 2005). The uncontrolled application of drugs in agricultural industry and in animals is becoming another rising factor for not only food production but also resistance development. Moreover treating immunocompromised individuals with life-saving drugs, increasing the survival chances of patients with unrestrained drug usage, and greater usage of invasive processes are other factors adding to the therapeutic tolerances. Greater medical advancements have led to control of infection and increased life expectancy of many patients. Even minor causes like lack of effective methods of hygiene and restrictions while handling infected patients can minimize the risk level of resistance to some extent. Not only widespread use of antibiotics but also harsh chemicals like herbicides, organic pesticides, and other toxic agents along with chemotherapies for cancer and viral, parasitic, and fungal treatments have raised the concern for resistance emergence. The resistance mechanism in microbes, plants, and humans portrays certain homology in the proteins conferring such tolerance mechanism. Microbial genetics play a central role in shaping up the molecular structure for centralizing the mechanistic studies of drug resistance. The horizontal gene transfer influencing resistance linkage enables the co-selection of resistant species with tolerance to more than drug (Alonso et al. 2001; Baker-Austin et al. 2006). Reduced use of one drug can't revert back the resistance if the genetic alterations already support other resistance genes. Moreover co-selection can also lead to clonal substitution of the resistance-linked allele. Similar structural framework of drugs can even confer co-selection. This again challenges the regulatory checks. Drug resistance can be categorized into intrinsic and acquired resistance. The detailed mechanistic approach will be discussed in the sections ahead. Intrinsic mechanism deals with the natural ability of the microbe as an innate immunity mechanism, whereas acquired mechanism deals with the environmental influence bringing about genetic modification, thereby giving a new dimension to resistance aspect.

4 General Mechanism of Drug Resistance

Drug development still forms the top headed research enterprise globally due to unsuccessful therapeutic reign of potent drugs over microbial weapons. The term "drug" is generally applied to all foreign chemicals including antibiotics, herbicides, and therapeutic agents against virus, parasites, cancer, etc. The host–microbe warfare has led to the compromise of clinical interventions and rise of multidrug-resistant species (*Streptomyces*). Resistance to seven or more antibiotics has even led to a resistance phenotype for around 20 drugs. Such mechanisms have made the environment emerge into a reservoir of pathogen tolerance. The emergence of new infectious agents causing AIDS, SARS, etc., has modulated the resistance standards with raised clinical challenges. The fast-growing drug resistance mechanism will become the signature of potent microbes inhabiting the environment with new emerging diseases and higher tolerance level causing mortality and morbidity. Understanding of microbial genetics and gene manipulation modes will give a greater insight and provide a new dimension into fighting the resistance mechanisms (Hayes and Wolf 1990). The molecular mechanism of resistance can be categorized into *intrinsic* and *acquired* mode of tolerance. Intrinsic relates to the inherent and integral property of the microbe that has built up evolutionarily for resistance characteristics (Cox and Wright 2013). Additionally the procedure of mutations and selective characterization forms the major genetic changes for resistance emergence. Methods of gene transfer, gene alterations in stress-regulating genes causing altered protein expression, and gene amplification can bring about the change in the pathogen's genetic constitution. Both the molecular approaches of resistance mechanisms will be discussed in detail in the following sections.

4.1 Intrinsic Resistance

Intrinsic resistance is defined as the ability of an organism to resist the antimicrobial/chemical compounds using a characteristic feature, which is an inherent or integral property developed by virtue of evolution. This can also be referred to as "insensitivity" due to the invulnerable nature of the organism toward that particular drug. The natural resistance feature, though less prevalent, sometimes undergoes spontaneous genomic alterations due to the absence of antibiotic-based selective pressure. However mostly the antibacterial-based microecological pressure triggers the stimulus for pathogen adaptation by the development of drug resistance. Mutations or evolutionary competition enables drug resistance gene uptake. It can arise due to certain events as outlined in Fig. 1 and mentioned below:

Absence/Modification of Target Site Microbial uptake of an antimicrobial drug is essential for a target-oriented action. Porins serve as the passageways for the drugs to cross the outer membrane of the bacterial cell. Some bacteria have the ability to manipulate their cell wall or membrane in order to protect themselves

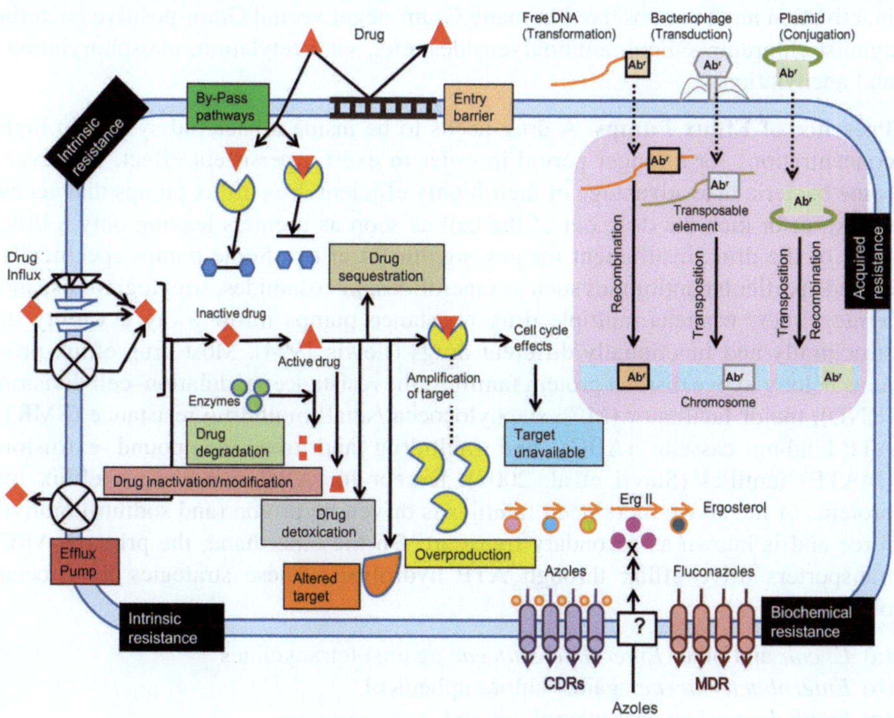

Fig. 1 Schematic presentation of multiple diverse molecular mechanism of microbial resistance

from foreign drugs. For example, certain Gram-negative bacteria can significantly lessen the uptake of certain antibiotics like aminoglycosides by altering the membrane porin frequency, size, and selectivity. On the other hand, the modification in the PBP (penicillin-binding protein) site led to the insensitivity toward the β-lactam antibiotics (Malouin and Bryan 1986).

Species-Specific Structure of Target Site Although the mode of action of antibiotics is almost similar across the same community of bacteria, species specificity has been detected in some cases. This is due to the lack of affinity of the drug to its target site. Different species under a single genus of a bacterium can alter the binding site of the drug by presenting various structural motifs for the same target, thus developing resistance. For example, the crystal structures of the large ribosomal subunit in *Staphylococcus aureus* showed specific structural motifs and binding modes for different antibiotics of same function as well as for a particular drug against different species of the bacteria.

Inactivation of Antimicrobial Agents via Modification/Degradation Destroying or manipulating the active component of the antimicrobial drug has always been considered as one of the effective techniques adopted by microbes for protection. For example, in penicillins and cephalosporins, the bacterial enzyme beta-lactamase hydrolyzes and deactivates the beta-lactam ring producing inactive penicilloic acid. It is then unable to bind to the PBPs, thereby maintaining the cell wall synthesis of the bacteria (Waxman and Strominger 1983). This kind of inactivation has been observed in many Gram-negative and Gram-positive bacteria against chloramphenicol, aminoglycosides, etc., via acetylation, phosphorylation, and adenylation.

Presence of Efflux Pumps A drug needs to be inside a bacterial system in high concentrations for a longer period in order to exert a persistent effect. However, some bacteria take advantage of their highly efficient drug efflux pumps that act as an export or kick the drug out of the cell as soon as it enters leaving only a little trace of the drug, insufficient for any significant effect. Some pumps specifically extrude particular antibiotics such as macrolides, lincosamides, streptogramins, and tetracyclines, whereas multiple drug resistance pumps throw away a variety of structurally and functionally different drugs (Lewis 1994). Most drug efflux proteins belong to five distinct protein families: the resistance–nodulation–cell division (RND), major facilitator (MF), staphylococcal/small multidrug resistance (SMR), ATP-binding cassette (ABC), and multidrug and toxic compound extrusion (MATE) families (Stavri et al. 2007). Except for ABC transporters, efflux by proteins of the above mentioned families is driven by proton (and sodium) motive force and is known as secondary transport. On the other hand, the primary ABC transporters drive efflux through ATP hydrolysis. These strategies have been observed in:

(a) *E. coli* and other *Enterobacteriaceae* against tetracyclines
(b) *Enterobacteriaceae* against chloramphenicol
(c) *Staphylococci* against macrolides and streptogramins
(d) *Staphylococcus aureus* and *Streptococcus pneumonia* against fluoroquinolones

High Detoxication Capacity Many bacteria secrete toxic compounds to protect themselves from their predators and other competitors. But they also need to evade the harmful effects of those noxious chemicals they produce. This is seen in the case of antibiotic-producing bacteria such as *Streptomyces* spp. In their defense, they develop resistance involving inactivation of their own antibiotic products streptomycin and neomycin by phosphotransferases and acetyltransferases and also by protecting the target site, i.e., rRNA by methylation in erythromycin-producing *S. erythraeus*. Additionally, in higher organisms, protein expression related to protection against chemicals is highly tissue specific. For example, the mammalian lung withstands the damage due to oxygen-induced free radicals. So this tissue has developed a large number of defense mechanisms including glucose-6-phosphate dehydrogenase, α-tocopherol, glutathione, glutathione peroxidase, glutathione reductase, superoxide dismutase, catalase, etc. Also, the bronchiolar epithelium contains high levels of detoxication enzymes due to its direct exposure to the environment. These features make sure that it has a natural resistance to many drugs that work through the generation of free radicals or perform as alkylating agents.

Low Drug Delivery This is due to low bioavailability and stability, fast metabolism, and less time for circulation of the drug inside the host system. All these things contribute to low drug delivery into the target site, but since the drug is exposed to the environment, resistance can be developed by the microbes or harmful cells such as tumor cells.

Cell Cycle Effects In mammalian cells, the rate of cell division is a major cause of intrinsic drug resistance. This is due to the fact that the main dose-limiting determinant in the cancer chemotherapy is the toxicity to the rapidly diving normal cells. Mostly, the anticancer drugs are effective against the quickly proliferating malignant cells. So, solid tumors that are slow growing develop resistance as most of the cells are in G_0 resting state.

Chemically Induced Adaptive Change Drugs or other types of toxic agents, upon entry into the cell, evoke many biochemical changes inside that lead to adaptation of the cells against the same or other compounds. The difference between intrinsic resistance by adaptive changes and other forms of intrinsic resistance is that the former is temporary and reversible in the absence of the toxic agent. This fact is observed in clinical practices, especially in cancer therapy where it protects the normal host cells but not the tumor cells from the adverse effects of the chemotherapeutic agent.

Stress Response Environmental factors other than drugs, such as pH, osmotic shock, UV irradiation, heat, trauma, viral infection, anoxia, and oxidative stress, can contribute to stimulate a genetic reflex in the cells that provide resistance not only to the stress factors involved but also against drugs. Prokaryotes have mainly four stress-induced regulons, namely, the SOS response, the oxyR network, the heat-shock response, and the adaptive response to alkylating agents. Like in *E. coli*, the *gro*EL and *dna*K heat-shock proteins are not only induced by hyperthermia but also by UV irradiation or nalidixic acid. Both of them affect the SOS response. Similarly, in *Salmonella* Typhimurium, the ability of a cell to adapt to H_2O_2-induced oxidative stress also gives resistance to heat killing. Many biochemical events influence resistance development which mainly includes decreased drug delivery and uptake, high efflux as mentioned above, greater inhibition of metabolic drug activity, and drug sequestration mechanisms.

4.2 Acquired Resistance

Microbial drug resistance development is related to the organization of their genetic material that becomes tolerant and the ease of uptake of exogenous DNA to alter their inherent genetic makeup. The continued selective pressure has thereby led to different modes of pathogen survival against the harsh medications. The emergence of resistance mostly involves two categories of pathogen: one involving the susceptible group and the other heterogeneous group comprising at least one microbe with drug-resistant determinant. The resistant group emerges fit with renewed genetic composition coding for the resistance which further assists in its propagation. Efflux mechanisms, drug modulation, membrane permeability alteration, etc., form the bullets of superbug evolution as depicted in Fig. 1. Thus the MDRs like *Pseudomonas*, *Klebsiella*, methicillin-resistant *Staphylococcus aureus* (MRSA), and XDRs like *Mycobacterium tuberculosis* have evaded the clinician's arsenal with their remarkable virulence potential. For an insight into drug invention, an understanding of molecular mechanism of drug resistance will help to sort out the therapeutic trade-offs with novel approaches. As mentioned before, pathogen drug resistance mechanism can be either *intrinsic* or *acquired* (Fig. 1). The "biological" aspect of resistance development is either absent from a majority of microbial population or is underexpressed before drug exposure. Microbial resistance has its basis at the genetic level which is modified either by gene knockout or introduction. This alters the genetic composition and cellular gene expression forming myriads of biological resistance forms (Mazodier and Davies 1991). Where intrinsic mechanism is solely due to the inherent microbial property of natural chromosomal genes and efflux system, acquired mechanism involves genetic mutations or gene transfer/exchange methods through the process of *transformation*, *transduction*, or *conjugation* (Fig. 1) (Flintoff 1989). Conjugation, the most general mode of drug resistance transmission, is facilitated by plasmids by forming a temporary "pilus" between two adjacent bacteria for genetic material exchange. Transformation is the process of uptake of exogenous DNA from the surrounding due to microbial degradation/lysis for further incorporation into any recipient organism's genetic cassette. Transduction essentially requires a vector specifically viruses that carry up the drug resistance genes for further introduction into bacterial host (bacteriophage mode of resistance transmission). Gene transfer is not genus specific, so the divergence of genetic exchange has led to the evolutionary buildup of the resistance reservoir. The independently replicating plasmids distinguishable with their origin of replication mostly contain the genes of antibiotic resistance. Transposons encoding the resistance determinants are the "jumping genes" present either on plasmids or host chromosome (Frost et al. 2005). The DNA terminal sequences enable recombination and encode proteins which facilitate their stable integration into host genome. Conjugative transposons bear unique plasmid-like properties which add on the advantage of funneling up many endogenous extra-chromosomal elements. Integrons comprise of gene cassettes that bear stable recombination features of undergoing multiple gene exchange within a single crossover. One super-integron was reported in *Vibrio cholerae* which comprised of about 3 % of host's genetic makeup. Numerous plasmids existing within a single microbe frame the genetic composition of the organism. They even comprise of "R factors" annotated as the resistance units forming the means of resistance spread among microbes. *Shigella* strains bearing self-replicating as well as self-transferable elements were identified to exhibit sulfonamide tolerance. Streptomycin, chloramphenicol, and tetracycline were used as optional medications; however the susceptible strains started developing tolerance against all three antibiotics with due time course.

After resistance gene transfer, the gene overexpression or mechanistic biological activity modulates the drug treatment in a way to neutralize its effect. The biological mechanism of resistance generally involves chemical/enzymatic degradation or modification of the therapeutic agent rendering it inactive against the bug. Such is the mode of resistance development in case of β-lactam antibiotics. Secondly, the active drug efflux mechanisms, much intensified than influx modes, promote effective resistance development. Efflux mode of microbial tolerance was evident in tetracycline and fluoroquinolones. Thirdly, target modification involves the microbe to alter the substrate binding affinity of the drug thereby hampering its activity. Similar mechanistic methods involve structural conformational changes in PBPs which renders penicillin resistance and DNA gyrase modulation which leads to fluoroquinolone tolerance (Wolfson and Hooper 1985). The frequency of mutations within the wild-type microbial population that has emerged irrespective of any selective pressure or drug exposure is attributed to the natural selection of emergence of acquired resistance. This differentiates from the intrinsic mechanism where the genetic alteration becomes a part of the biological variation. Selection of the microbial units that can withstand and sustain the chemical insult proves superior. Examples include the PBPs in *E. coli* causing cephalosporin resistance and alterations in acetylcholinesterase conferring tolerance to Rabon (Tripathi and O'Brien 1973). Drugs or antimicrobials aren't mutagenic. However, certain tumor treatments involve mutagens in chemotherapy that can evolve the selection pressure for resistance thereby causing genetic instability with high mutations or amplifications. In tumors specifically the differentiation between natural selection and acquired tolerance is a highly debatable topic. Such phenomenon leads to the constitutive expression of certain phenotypic changes instrumental for adaptive response. In *E. coli*, the altered LexA repressor gene has an impact in regulation of SOS signals (Little and Mount 1982). Even *S.* Typhimurium when resistant to hydrogen peroxide modulates the expression of certain stress-regulating genes including catalase, SOD, glutathione peroxidase, etc. Changes at the transcriptional level or early mutational occurrence influence significant mechanistic cascades that lead to overexpression of proteins modulating the microbial genotype to display resistance phenotype. The modes of development of acquired resistance involve mutations, efflux systems, gene amplifications, drug modification, or target alterations.

Chromosomal-Based Genetic Alteration Mutation in drug targets is basically the most common mechanism of microbial resistance emergence. The fluoroquinolone resistance mechanism can be attributed to genetic alterations as well as efflux pump machinery. The drug targets DNA gyrase as well as topoisomerase IV which when altered confer fluoroquinolone resistance. These multi-subunit targets play a pivotal role during DNA duplication each comprising of two subunits: GyrA and GyrB for DNA gyrase and ParC/GrlA and ParE/GrlB for topoisomerase IV. One subunit of these complexes functions for the DNA-binding role, whereas the other carries up the ATP-binding and hydrolysis role. The quinolone-resistance determining-region in DNA-biding domain bears the mutational changes that confer antibiotic resistance. Innumerable mutations impose additive effects to build up the bacterium's resistant trait. Similarly rifamycins form the front-line therapeutic against *tuberculosis* infection either individually or in combination with isoniazid, streptomycin, etc. However, RpoB point mutations prevent the drug-binding affinity at the RNA polymerase subunit conferring combinatorial drug resistance (Mariam et al. 2004). Sulfonamide targets dihydropteroate synthase whose alteration results in decreased enzymatic activity for the drug. Trimethoprim blocks dihydrofolate reductase enzyme whose mutation causes protein over-induction with reduced drug affinity. Point mutations at 16S rRNA and 23S rRNA operons confer tetracycline and MLS antibiotic resistance, respectively (Ross et al. 1998).

Genomic Duplication Gene alterations mainly include gene mutational events along with gene amplification or overexpression. The method of genomic doubling is quite prevalent in conferring drug resistance among eukaryotic cells, the sole reason in tumors. This genetic induction leads to the modulation at the protein level for augmented biosynthetic machinery leading to the overexpression of many transporters. In *E. coli*, the genomic amplification of *acrAB* locus in tetracycline exposure led to induction of the AcrAB efflux pump systems contributing to multiple drug-resistant phenotype (Nikaido and Zgurskaya 2001). Such duplication phenomenon has also been observed in *S. aureus* with respect to methicillin resistance. Genome amplification forms one of the mechanisms of resistance avoiding the boundaries of mutational aspects. However the absence of drug made the microbes revert back to their normal phenotypes. So the tolerance mechanism is basically unstable.

Enzymatic Approach of Drug Modification General mechanism of drug modification involves two specific classes of enzymes; one group which causes drug degradation and another catalyzes chemical modifications. The β-lactamases encoded by plasmids and transposons confer adaptive resistance as compared to the chromosomal chunk which attributes intrinsic property. The structural and functional characterization classifies β-lactamases into two groups, one having serine at the active catalytic site (classes A, C, D) and another with zinc-dependent metalloenzyme (class B). Being zinc dependent, class B enzymes are susceptible to EDTA and hydrolyze carbapenems. AmpC, a prototypic class C-type enzyme being plasmid borne, is easily transferred among many Gram-negative strains like *Salmonella* spp., *Klebsiella* spp., etc. (Jacoby 2009). Even the acetyltransferases, phosphotransferases, and adenylates that modify the aminoglycosides exist on certain mobile genetic units or integrons where they enable resistance transmission. The MLS antibiotics are inhibited by groups of esterases as well as phosphotransferases that modulate 14- and 15-membered macrolides (Nakajima 1999). The acetyltransferases and hydrolases affect streptogramin A and B drugs, respectively. The transferase enzyme concerned with nucleotidyl moiety transfer bestows resistance to lincosamides antibiotics. Even chloramphenicol, which targets the protein biosynthetic machinery, is a bacteriostatic drug which is inactivated by certain groups of acetyltransferases. Prevalent mostly in *Enterococcus* and *Staphylococcus*, the enzyme's translational attenuation depends on the regulation of their protein expression and induction.

Modulated Drug Targets β-Lactamase-producing *S. aureus* was the first penicillin- and methicillin-resistant strain. The mechanism involved genes contributing to changes in PBPs that confers β-lactam resistance in *Streptococcus* as well as *Staphylococcus*. This gene is encoded by *mecA* on a mobile genetic unit in resistant *Staphylococcus aureus* (Wielders et al. 2001). The "staphylococcal cassette chromosome" is the mobile element comprising of regulatory segments and enzymes responsible for site-specific recombination. The cell wall synthetic process requires a number of PBPs in *Staphylococcus aureus*. PBP2 enzyme plays dual role in resistant *S. aureus* where the transpeptidase and transglycosylase activities switch in accordance to drug exposure for imparting susceptibility or tolerance features in the microbe (Brown and Reynolds 1980). The plasmid-borne *qnr* determinants found widely in Gram-negative species of *E. coli*, *Shigella*, non-typhoidal *Salmonella*, etc., affect fluoroquinolone sensitivity (Piekarska et al. 2011). The pentapeptide repeat protein family includes Qnr as well as MfpA which regulate fluoroquinolone resistance by shielding DNA gyrase and topoisomerase II, respectively. Qnr additionally protects topoisomerase from drug effect. Moreover MfpA in *Mycobacterium* forms an identical structural outlook of B-DNA-inhibiting ciprofloxacin activity by interaction with DNA gyrase (Montero et al. 2001). Coupled with other modes, Qnr and MfpA can amplify the resistance profile to higher extents. Glycopeptides form the major example of bearing the drug resistance feature due to drug target modifications in Gram-positive spherical bacteria. Multiple clustered proteins contribute to resistance by modulating the peptidoglycan production. The complex elucidation of unveiling the glycopeptide resistance concerns gene clusters like racemases or dehydrogenases forming serine and lactate, respectively, which alter the peptidoglycan framework. The two-component unit regulates the cellular biosynthetic mechanisms. With intact

D-Ala available, the reduced interaction efficiency with serine and lactate substrates nullifies. The dipeptidase and carboxypeptidase act in accordance to their respective function of cleaving and removing the terminal D-alanine. The vanA gene cluster conferring vancomycin resistance in *Enterococcus* has transferred its resistance determinants even in *S. aureus* encoded on Tn*1546* transposon. However, plasmid from *E. faecalis* forms the initial mode of *vanA* transfer into *Staphylococcus*. The erythromycin resistance methylase (*erm*) influences the macrolide drug-binding affinity by methylation of adenine residues of 23S rRNA (Maravic 2004); Vester and Douthwaite 2001. The resistance markers are usually constitutively expressed or in certain cases MLS drug exposure induces expression.

Efflux Mechanisms and Membrane Permeability Channel Rather than restricting drug uptake and internalization, resistance is mostly due to the failure of undergoing drug interaction with cellular targets due to drug effusion (Fig. 1). Efflux pump mechanism ejects the drug out of the cell and was initially observed during tetracycline resistance development. The five protein transporter families involved in efflux machinery are ATP-binding cassette (ABC) transporters, resistance–nodulation–cell division (RND) protein superfamily, major facilitator (MF) protein groups, small multidrug resistance (SMR) units, and multidrug and toxic compound extrusion (MATE) protein superfamily (Poole 2005). ABC group of primary transporters employ ATP hydrolysis for efflux mechanism, whereas the rest secondary groups involve proton-motive gradient force for conferring drug expulsion (Kobayashi et al. 2001). The drug discharge proteins are either categorized into single protein component systems having narrow substrate range or bear two proteins to facilitate the binding of variable structural compounds conferring wide spectrum of resistance phenotypes. RND transporters enable efflux of cytosolic proteins through the inner and outer membrane barriers. The 20 tetracycline efflux transporters comprise of transmembrane spanning regions where the protein expression is regulated under the transcriptional repressor. The inactivation of the repressor by the drug promotes the expression of the tetracycline efflux machinery. TetA-E are the efflux proteins in Gram-negative bacteria, with TetK and L in Gram-positive microbes, and TetR is the repressor (Schnappinger and Hillen 1996). Certain proteins are also involved in electroneutral chemical reactions. Additionally, macrolide and streptogramin tolerance is attributed by a group of ABC efflux protein family called MsrA. Its homologues VgaA and VgaB in *Staphylococcus* bring about streptogramin A and pristinamycin resistance, respectively (Lina et al. 1999). Mef efflux systems also confer macrolide resistance in *Streptococcus*. Removal of efflux machinery can revert back the genetic profile to antibiotic susceptibility even with the persistence of chromosomal alterations. E1–E8 are the groups of efflux protein superfamily that confer phenicol resistance. CmlA is an efflux system aiding in chloramphenicol resistance, which promotes an induced attenuation-based resistance mechanism (Bischoff et al. 2005). Porin proteins enable the smooth flow of molecules across the cell membrane barrier in Gram-negative bacteria (Delcour 2009). OmpF in *Escherichia coli* and OprD in *Pseudomonas* act as checkpoints to monitor the nonspecific entry of many compounds. OprD mutational changes lead to imipenem tolerance. Simultaneous expression and regulation of OprD and MexEF-OprM efflux complex will confer carbapenem resistance (Lee and Ko 2012; Quale et al. 2006). The bacterial cell envelop restricts and permits selective entry of many hydrophobic and hydrophilic components. Resistance to polymyxin B however doesn't involve porins but cell envelope alterations. The PmrAB system regulates the LPS and lipid A moiety changes which confers polymyxin B resistance in *Salmonella* Typhimurium (Gunn 2008).

5 Drug Resistance Mechanism (Disease Specific)

5.1 Tumor Drug Resistance: An Evolving Paradigm

After all the advancements in cancer research, emergence of drug resistance restricted the efficacy of the therapies. Resistance to cancer chemotherapy results from a range of factors, starting from individual variations in patients and somatic cell genetic differences in tumor, even those from the same tissue of origin (Dean et al. 2005). Resistance can be intrinsic as well as acquired. Chemotherapy resistance occurs when cancers that have been responding to a drug suddenly stop reacting. There are several possible reasons responsible which mainly include:

1. Some cancer cells that escape the harmful effects of the drug mutate and develop resistance toward the drug. Later, upon multiplication, they become more resistant.
2. Gene amplification: A cancer cell has the ability to produce hundreds of copies of genes of a particular gene. This leads to the overexpression of the corresponding protein, which in turn makes the anticancer drug ineffective.
3. With the help of a molecule called p-glycoprotein (P-gp), cancer cells pump out the drug entering the system using their proficient drug efflux pumps.
4. Highly efficient DNA damage repair machinery, one of the survival secrets of cancer cells, also plays vital role in contributing resistance against anticancer drugs (Holohan et al. 2013).
5. Cancer cells may also develop strategies to inactivate the drugs (Holohan et al. 2013).

The molecular mechanisms of drug resistance in tumors are discussed in details in the following segments:

1. *Altered membrane transport*: One of the most promising drug resistance mechanisms against antineoplastic agents is the method by which the cell flushes out the cytotoxic compound with the help of some membrane proteins that helps to reduce the inside drug concentration below the cell-killing threshold. These proteins modulate absorption, distribution, and excretion of many pharmacological compounds. ABC transporters are encoded by as many as 48 genes. In the

clinical transport-associated MDR, the most commonly involved gene is the MDR1 that encodes for the P-glycoprotein (P-gp; MDR1, ABCB1) which is a phosphorylated and glycosylated 170 kDa protein. Other well-known ABC transporters are the MDR-associated protein 1 (MRP1, ABCC1), the mitoxantrone resistance protein (MXR1/BCRP, ABCG2), and the ABCB4 (MDR3) and ABCB11 (sister P-gp or BSEP) proteins involved in the secretion of hepatic phosphatidylcholine and bile acids, respectively, as well as transport of certain drugs. The most interesting feature of differentiating MDR proteins from other mammalian transporters is their high substrate specificity. Unlike classical transporters, MDR transporters translocate a variety of structurally different hydrophobic compounds along with other unique compounds, and this forms the platform for the cross-resistance to many chemically unrelated compounds. Overexpression of MDR proteins in tumors like hepatomas and lung or colon carcinomas often shows intrinsic resistance (Gottesman 2002; Gottesman et al. 2002). The role of P-gp, in the absence of any therapeutic agent or toxin, is thought to protect the cell from xenobiotics. However, several reports suggest P-gp to have prognostic significance in certain types of neoplasms as well as to play an important role in CNS penetration of drugs (Begley 2004). But all these have failed to correlate with the clinical evidence. So, its mode of action has always been controversial. According to some reports, MDR proteins are not responsible for transporting drugs, but they alter ion transport or signal transduction, thus later on affecting drug distribution. In tumor cells, anticancer drugs and cytotoxic cytokines like TNF/Fas ligand family play an important role in induction of apoptosis and tumor therapy (Reed 2003). Drug-resistant tumor cell lines show resistance to Fas-induced caspase-3 activation and apoptosis which is reported to be mediated by P-gp. The cells expressing P-gp are resistant to a lot of stimuli responsible for the activation of caspase apoptotic cascade, whereas it is not the case in caspase-independent cell death where cell dies by the action of pore-forming proteins GzB.

2. *Genetic responses*: Drugs such as methotrexate inhibit key enzymes involved in the proliferation pathway of mammalian cells. When transcription of the gene encoding for the enzyme increases, large amount of enzyme is produced that leads to faster proliferation. But the concentration of the drug is limited which cannot block the additional enzyme that is produced. Thus the cells develop resistance against the drug. One way to overexpress the enzyme is by the amplification of the gene encoding the enzyme, which is achieved by replication of a region from the chromosome that results in multiple copies of the same gene. Several drug-resistant cancer cell lines and DNA from two drug-resistant leukemic patients have shown gene rearrangements in their chromosome resulting in the initial activation or enhanced expression of MDR1 gene. Moreover, therapy with rifampicin has also shown to induce MDR1 expression in healthy individuals. Hence, MDR1 overexpression can be affected by gene amplification/rearrangement, rifampicin induction, etc. Another important factor responsible in drug-resistant cancer is the mutation in the apoptotic gene p53. p53 usually induces apoptosis in cells which have undergone DNA damage

(Chen et al. 1996). Thus, the target DNA won't be affected and will continue replicating in the presence of mutated p53. So, the drugs that increase DNA damage will come to no use in certain cancers. In many cancer cases, deletion of p53 was reported to be linked to MDR. Also, reduced expression of p53 in human breast cancer cells altered response to paclitaxel and 5-FU. Other genes involved in apoptotic pathway, like h-ras and bcl2/bax, have also been observed to contribute to drug resistance (Davis et al. 2003). Thus drug resistance arising from genetic responses affects a large variety of anticancer drugs and increases the percentage of surviving mutant cells, which in turn leads to greater tumor heterogeneity.

3. *Enhanced DNA repair*: Cancer cells develop resistance to drugs such as cisplatin by an enhanced ability to remove cisplatin-DNA adducts and to repair the cisplatin-induced lesions with the help of certain DNA repair proteins like XPE-BF (xeroderma pigmentosum group E binding factor). ERCC1 (excision repair cross-complementing protein), a DNA-binding protein, has the ability to recognize cisplatin-induced DNA damage and thus its level increases in cisplatin-resistant cells (Siddik 2003). The level of ERCC1 is also found to increase in carboplatin-resistant tumors.
4. *Alterations in target molecules*: Modifications in the target of a drug is a common way to develop resistance against the same. As seen in antiestrogen (e.g., tamoxifen) therapy for breast cancer, patients undergo a transition from a responsive state to an endocrine-resistant state due to an apparent loss of estrogen receptors in the resistant cancer cells (Ring and Dowsett 2004). So they finally stop responding to tamoxifen treatment, while the growth of their tumors can still be inhibited for a short span by estrogen synthesis inhibitors like aromatase inhibitors followed by complete unresponsiveness to any endocrine modification. Hence, the surviving cancer cells no longer depend on estrogen for growth and the original drug that targets estrogen receptors becomes useless. Another example is a tyrosine kinase inhibitor, imatinib, which induces apoptosis in cancer cells by disabling the damaged bcr-abl receptors, preventing ATP binding (Capdeville et al. 2002). But reports suggest that during clinical trials, the chronic myeloid leukemic patients in remission had reactivated bcr-abl activity, few patients had amplified copies of the bcr-abl gene, and others had a single point mutation within the ATP-binding site of the gene. Hence, this gene shows to play an important role in initiation and maintenance of cancer and thus related to anticancer drug resistance (Dean et al. 2005). Mutation in the topoisomerase gene is another cause of drug resistance due to its vital role in DNA replication process. Chemotherapeutic drug like etoposide that targets topoisomerase II suffers from resistance when cancer cells mutate the latter in a way to alter its nuclear localization. Chromosomal losses are very common in cancer, and due to its aneuploid nature, there has been the emergence of MDR. While undergoing repeated cell division for a number of times, a cancer cell has the chance of losing the drug-sensitive gene from the chromosome, and also chromosomal rearrangement during mitosis can contribute to the activation or inactivation of different biochemical pathways that can affect the mode of action

of the drug. The size of the tumor also matters as the center part of most tumors has limited blood supply. So, the larger the tumor, the lower the drug efficacy. Apart from this, some metabolic enzymes, either alone or together with transporters like P-gps, can alter the drug absorption, distribution, metabolism, and excretion. For example, enzymes like cytochrome p450s (cyp450) in combination with P-gp greatly affect the drug absorption and bio-distribution to the tissues preventing intestinal transcellular permeability, biliary disposition in the liver, urinary elimination through the kidney, and placental transport.

5. *Metabolic effects*: Effective clearance of drugs can often be achieved by some xenobiotics that have the ability to modify high-density apolipoprotein or by overexpression of the drug-metabolizing enzymes and/or carrier molecules. The increased production of glutathione or ubiquitin leads to drug inactivity by the formation of conjugates that are excreted, for example, cisplatin that becomes resistant to ovarian carcinomas after the overexpression of dihydrodiol dehydrogenase. In some cases, the underexpression of few drug-metabolizing enzymes (e.g., deoxycytidine kinase) can also lessen drug (e.g., arabinosidase) activity in a situation where the drug needs to be catalytically cleaved to be in its active form. Additionally, protein kinase C has been found to have increased activity in the drug-resistant breast carcinoma cells because of its role in both drug exclusion and apoptosis (Caponigro et al. 1997). Breast cancer cells have shown resistance against paclitaxel and vincristine due to the involvement of the extracellular matrix as well. The ligation of the b1 integrins by the extracellular matrix inhibits apoptosis mediated by these two drugs.
6. *Growth factors*: High levels of serum interleukin-6 (IL-6) have been observed in different drug-resistant cancer cells, whereas cells sensitive to the chemotherapy did not produce any detectable IL-6. The reason behind this resistance was attributed to the activation of the CCAAT enhancer-binding protein family of transcription factors and induction of MDR1 gene expression (Okamura et al. 2004). Reports have also suggested that extracellular factors can contribute to drug resistance against a particular cancer. Like increased levels of acidic and basic fibroblast, growth factors in the media of solid and metastatic tumors can affect the broad-spectrum drug (paclitaxel, doxorubicin, and 5-FU) efficacy and lead to develop resistance. When applied in combination, these two growth factors can give rise to a tenfold increase in drug resistance.

5.2 Antibiotic Resistance: The Bacterial Weapons

The era of the twentieth century witnessed the discovery of essential antibacterial drugs to control bacterial proliferation for limiting infectious agents. Even though vaccines and other public health agendas were instrumental, still antimicrobial therapy checked the further transmission of infectious pathogens (Donadio et al. 2010). Antibiotics literally implies "against life" but scientifically are compounds that hinder with the normal functioning of the bacterium without interfering with the biological processes of the eukaryotic host harboring the microbe (Fischbach and Walsh 2009). The present scientific world now struggles to combat the issue of antimicrobial resistance. The frequency in resistance has constrained the mob to question the efficacy of these conventional medications. The discovery of every drug is followed by the bacterial strategic mechanisms to overcome the stringency by developing tolerance. This has also led the researchers to investigate into the pathogen's clinical, molecular, and cellular factors that make them the superbugs of resistance (Arnold 2007; Neu 1992). Hence the new drugs have raised question on their proficiency to avoid the emergence of multidrug resistance microbes. The MDR pathogenic strains of *M. tuberculosis*, *S. pneumonia*, *S. aureus*, etc., have posed innumerable challenges for further antibiotic development (Wright and Poinar 2012). With few antimicrobials in hand, the post-antimicrobial era seems to be approaching soon. Antibiotics can be either bacteriocidal (bacterial death) or bacteriostatic (bacterial growth inhibition). The last five decades had enabled the discovery of many natural antibiotics like fungal penicillins and cephalosporins that kill the bacteria. Even streptomycin, tetracycline, etc., are known microbial targets from *Streptomyces*. Certain semisynthetic alterations led to the production of second- and third-generation antibiotics like β-lactams of penicillin and azithromycin from erythromycin. However, a complete synthetic antibiotic like ciprofloxacin came later into existence.

The targets of antibacterial drugs and their mechanism of action will help one to understand the emerging resistance profile among these bugs. The antibacterial drugs normally target the genes responsible for *bacterial cell wall synthesis* (Green 2002), genes involved in *protein biosynthetic pathway*, and the ones modulating the *microbial DNA replication and repair*. The rigid, flexible peptidoglycan lining of the bacterial cell is a meshwork of peptide and glycan cross-links which provides integrity and osmotic balance to the microbe. The transpeptidases and transglycosylases act on the amide and glycan links, respectively, to strengthen the osmotic rigidity of the cell. Both these enzymes are the antibacterial targets of β-lactam in penicillin and cephalosporin. These pseudosubstrates enable acetylation at the enzymatic active site which leads to weak cross-linkage of peptide bonds in the glycan lining, thereby making the cell susceptible to lysis. Even vancomycin targets the peptidoglycan layer, but not the cross-linking factor rather alters the substrate interaction with the enzyme. Target alteration weakens the cell integrity subjecting the bacteria to lysis. The high reactivity of vancomycin is due to the hydrogen bonds with the D-alanine dipeptide of peptidoglycan side chain. Both β-lactam and vancomycin work synergistically on the substrate and the enzyme when used in a combined recipe. Secondly, with distinct prokaryotic RNA and protein synthetic machinery, certain classes of antibiotics like erythromycin, tetracycline, and aminoglycosides selectively target the microbial survival by hindering with essential steps of ribosomal functioning. The new class of protein synthesis inhibitor have a huge spectrum of antibacterial target as the ribosomal machinery involves the protein synthesis steps of initiation, elongation, and termination of codons to build up the peptide chain. So the protein synthesis inhibitors target such large supramolecular machinery with essential biosynthetic process that further alters the binding and catalysis of many enzyme-catalyzed reactions. The third group of antibacterial compounds disrupt with the DNA doubling and repair mechanism. For instance, DNA gyrase that helps in DNA strand uncoiling is targeted by the fluoroquinolones like ciprofloxacin. DNA topoisomerases are categorized as Type I and Type II in accordance to the single- or double-stranded breaks. Moreover, DNA gyrase is a Type II topoisomerases, and ciprofloxacin acts by forming a complex between the transient double-stranded break and inactive enzyme. Since the cleaved DNA mounts up, an SOS repair mechanism ultimately leads to the death of the microbe. Similarly topoisomerase IV is a major target in *Staphylococcus aureus*. The antibacterial drugs alter and modulate with the cell wall, protein, and DNA synthetic pathway. They act in a selective biochemical manner to target the microbial machinery as compared to that of host. The present generation of antibiotics not only require better efficacy and less toxicity use but also new unaltered targets for universal and regulatory acceptance.

The epidemiological studies have implicated the feature of selective advantage for pathogens. For instance, prior antibiotic exposure is a crucial factor causing salmonellosis. Secondly, antibiotic-resistant microbes are known to be more virulent with an overexpression of adhesins and toxins on R-plasmids thereby increasing the scale of virulence. Resistance can also lead to increased frequency of the disease. An infected individual with MDRs will be the carrier of infection causing transmission risk as compared to the susceptible strains. So reservoirs are quite important in the persistence of infectious agents. They also enable genetic elements swap and selective pressure exposure adds on to the evolution of resistance mechanism. A study hereby reported the clinical isolates of *S. aureus* and *S. epidermidis* to have similar drug-resistant profile obtained from the same hospital source during *S. aureus* epidemic (Cohen et al. 1982). Hence *S. epidermidis* served an important reservoir affecting the emergence and chances of resistance occurrence. Treating drug-resistant microbes requires more effort and expenditure for long-term antibacterial medications. Literature has suggested variant microbial properties, environmental factors, and frequent use of antimicrobial agents to be responsible for the incidence of drug-resistant microbes. A case occurred where a woman under the medication of chloramphenicol died due to chloramphenicol-resistant *S.* Typhimurium infection (Tacket et al. 1985). So chloramphenicol was no longer the drug option against salmonellosis. This further leads either to the emergence, persistence, or transmission of resistant ability. A case illustrates that the medications with cephalosporins lead to the augmentation of resistance in enterococci (Dahms et al. 1998). Where reservoirs are concerned, infants in nurseries serving as a depot for producing greater number of *staphylococci* were termed as "cloud babies." Even certain microbes are posing health problems due to high-end technological and societal advancements which are influencing their resistant transmission. Economic changes are also a major factor destabilizing the health infrastructure in many advanced countries like the USA. These enable the application of certain control programs for the curtailment of pathogenic infections.

5.2.1 Survival Strategies

The invention of an antibiotic and its widespread acceptance for clinical use have limited time period for their medications. From months to few years, these antibiotics have proven to emerge resistant to their targeted bacterial strain (Bush et al. 2011). For instance, penicillin developed resistance among the bacterial populations within a span of 2 years after its global use. Similarly, vancomycin resistance against enterococci did spread at an alarming rate within 5 years of its introduction during the late 1980s. The resistance involves a collective action of five genes, the major reason why vancomycin resistance took a greater time to evolve (Walsh et al. 1996). The clinically important resistance is dependent bacterial doubling time, where the intrinsic resistance can lead to genetic alterations in 1:100 ratio finally amassing a pool of resistant superbugs. In this population, if the resistant species proves superior and tolerates the antibiotic, then the pool of sensitive population are killed where the resistant species fills up the numbers of the susceptible ones with their own dominant resistant type. A subtherapeutic medication assures practical resistant species outgrowth. Antibacterial resistance generally involves four basic mechanisms of target modification, drug inactivation, decreased drug uptake, and increased efflux systems (Alekshun and Levy 2007). Bacterial drug resistance thereby can be intrinsic due to the inherent property of the microorganism or acquired due to the evolutionary process of spontaneous mutations or gene uptake. Intrinsic mechanism makes bacterium naturally resistant to antibiotics like all mycoplasma withstand β-lactam action due to lack of peptidoglycan wall. A primary reason for resistance development is due to the spread of the antibiotic resistance genes on the plasmids that get multiplied independently and carried over to the next-generation doublets thereby conferring resistance (Yoneyama and Katsumata 2006). In certain cases, such genes can be evenly segregated onto the mobile genetic elements like transposons that can jump from one DNA locus into another. Similar occurrence happens in *Enterococcus*-resistant vancomycin strain where the five genes of microbial cargo hop into variant genomic locales. In medical environments, selective antibiotic pressures on pathogens like *S. aureus* and *Enterococcus faecalis* enable them to switch their antibiotic-susceptible profile to resistant ones. For instance, the methicillin-susceptible strain of *S. aureus* (MSSA) becomes resistant (MRSA) in patients with surgical treatments within a span of 5 days in hospitals. These isolates become resistant to vancomycin treatment. Hence, the antibiotic pressure alters the sensitivity genomic contour of the pathogen triggering the antibiotic resistance switch, for which new discoveries should target the molecular approach of resistance that makes these pathogens superbugs (Walsh 2000).

The major antibacterial drug resistance strategy involved is mostly the *efflux pump mechanism* where the microbe pumps out the antibiotic preventing their disruption of cellular processes (Poole 2005). The intrinsic mechanism of resistance is known to induce the expression of efflux pump systems. In *Pseudomonas aeruginosa*, the disruption of the functioning of MexB pump raises the sensitivity to antibiotics like tetracyclines, chloramphenicol, β-lactams, etc. (Lomovskaya et al. 2001). Hence antibiotics can regulate the efflux pump expression in microbes at the transcription level of gene function. Transporters can be expressed due to the modulation of these regulators. In *Neisseria gonorrhoeae*, the efflux machinery encoded by the *mtr* operon eases the transport of antimicrobial components. However, mutation in the mtrR gene increases the bug's resistance to penicillins, macrolides, as well as rifamycins (Veal et al. 2002). Secondly in *P. aeruginosa*, exposure to fluoroquinolones leads to mutations in the genes encoding for efflux proteins as well as topoisomerases. Again mutation in efflux genes augments the resistance level to fluoroquinolones in *P. aeruginosa*. *Streptococcus pneumoniae* is known to be resistant mostly due to the drug efflux pumping out bacterial strategy. The resistance in *Streptococcus* occurs in a stepwise regulated way initiating in *parC* mutations, followed by *gyrA* and then PmrA pump (Jones et al. 2000; Piddock et al. 2002). Each level of mutation confers to a greater degree of resistance. In the 1980s, active efflux became major players in antibiotic resistance that led to the emergence of multiple drug-resistant strains (MDRs). Efflux transporters are known to be polyspecific where they eject out a huge diversity of structurally nonidentical components (Piddock 2006b). Additionally the MDR efflux systems eject out toxic compounds enabling the microbe to escape the classic antibiotic therapy (Paulsen 2003). There are five basic protein families that form the category of bacterial efflux pump system. Two well-known protein families are the ABC superfamily (ATP-binding cassette) along with the MFS (major facilitator superfamily). Other three small units include the SMR (small multidrug resistance), the RND (resistance–nodulation–cell division), and the MATE (multidrug and toxic compounds extrusion) family (Piddock 2006a). Other than ABC family pumps, the rest transporters are termed as secondary transporters conjugated to proton influx, hence called as H1-drug antiporters. Conversely, the primary transporters, i.e., ABC, utilize ATP for their activity.

5.2.2 Antibiotic Efflux Pump System

Antibiotics require a particular site of action and minimal level of concentration for targeting microbial cellular processes. For instance, if the antibiotics have to target the protein biosynthetic pathway, then the membrane barrier has to be surpassed for disrupting the protein assembly at high accumulated concentrations. The Gram (+ve) and the Gram (−ve) bacterial strains have overexpressed the active efflux pumps to expel out tetracycline to gain antibiotic tolerance (Nikaido 1996). A huge variety and diversity of export pumps are employed for a large spectrum of lipophilic or amphipathic compounds to keep the diffusion level low for unhindered bacterial sustenance. Some examples show the export pumps to be expressed by the antibiotic-producing cells for expelling out the antibacterial compounds as a protective shield to prevent the suicidal death of the microbe by its intrinsic weapon.

Efflux Transporters The most important factor in causing bacterial death is the accumulation of harsh antibiotics within the cytoplasmic fraction for inhibitory effects or growth attenuation. With Gram-positive bacteria lacking an outer plasma membrane barrier, the production of transporters enables the efflux mechanism to confer resistance within such species. Additionally the tough exterior of Gram-negative bacteria avoids drug buildup. As mentioned above, the Gram-positive bacteria comprise of three protein transporter superfamilies, namely, the ABC family of transporters which harnesses cytoplasmic ATP for antibiotic ejection and MF and SMR transporter protein groups which exploit an electrochemical proton gradient for efflux mechanism (Markham and Neyfakh 2001). The molecular mechanism of relating the ATP and proton exchange with drug efflux is still under investigation. Failures in X-ray structural characterization of transporter molecules have also raised major scientific concern. The examples of substrate-specific efflux pumps are either macrolide specific or tetracycline targeted (TetKLZ) which confer an additional function of immunity for the microbe. The issue of drug recognition can be resolved by getting an insight into the structural aspect of proteins regulating the efflux protein expression. The TetR protein binds to tetracycline leading to the overexpression of tetracycline substrate-specific transporter. This drug DNA-binding repressor undergoes a conventional van der Waals mode of interaction where the H bonds facilitate the linkage of tetracycline–Mg^{2+} complex with the polar amino acids bound with water molecules. The ligand binding occurs proper with a defined chemical architecture. So tetracycline transporters undergo substrate-specific interaction that governs their efflux mechanism. Another level of interaction occurs in the transcriptional regulator of the well-known bacterial species of *Bacillus subtilis*. The BmrR regulator induces the production of Bmr multidrug efflux protein which expels out intracellular hydrophobic cations (Paulsen 2003). In such case of regulators, the ligand binding occurs by means of both electrostatic and hydrophobic interactions rather than hydrogen bonds. So the dependency on structural outlook of substrate specificity becomes limited. The substrate specificity of the transporters is linked to its multidrug abilities. For instance, the evolutionarily distinct transporters of *B. subtilis* (Bmr, Blt) and *S. aureus* (NorA) share functional and sequence homology with transporters of *Lactococcus lactis* (LmrP) and *Staphylococcus* (QacA) itself as they belong to the identical MF family. They are much more homologous to the tetracycline transporters of Gram-negative bacteria. Even there is very insignificant identity among the same group of transporters like tetracycline, efflux proteins TetK and TetL bear very less similarity to TetZ of *C. glutamicum*. Both TetK and TetL share much similarity rather than the multidrug efflux units of *Staphylococcus* QacA. Another example to illustrate the mechanism of substrate recognition and binding is the class of lipocalins (Gunn 2008). These proteins share a similar tertiary structural outlook and bear high binding affinity for diversified substrates (ligands). Some lipocalins are even identical to multidrug efflux units enabling the organic compounds to interact with the hydrophobic protein core. The diversity of substrate binding by lipocalins was well demonstrated through mutagenesis of the hydrophobic binding residues. This resulted in switching the affinity partially toward the hydrophilic fluorescein moieties. This led to the modulation of substrate recognition. Hereby, Gram-positive bacterial species bear different modulated transporters under the ABC, MF, and SMR family to force out the harsh antimicrobial agents.

There are around a dozen or two efflux membrane transporters in Gram-positive bacteria. One or more efflux transporters can also be present on the same microbe forming multiple drug transporters with an array of substrates. In *B. subtilis*, around four multidrug transporters have been reported, namely, Bmr, Bmr3, and Blt MF efflux systems (Ahmed et al. 1995), and one SMR family protein EbrAB. The sequence homologies screened many other hypothetical transporter genes in *Bacillus*. Efflux mechanism can be an opportunistic attempt by these multidrug transporters. For instance, tetracycline transporters have an additional function of transporting monovalent ions like Na^+ and K^+; other than tetracycline and genetic alterations of such transporters confers antibiotic and saline susceptibility. Examples of such transporters are TetL and TetK from *B. subtilis* and *S. aureus*, respectively. Being translationally regulated by tetracycline, these transporters channel the transport of monovalent ions as a side effect of drug transport. Similarly in *B. subtilis*, the two specific multidrug transporters Bmr and Blt are distinctly and differentially regulated at the transcriptional level (Ahmed et al. 1995). Blt and Bmr transporters extrude polyamines from the bacterium. However the Bmr–BmrR combination functions to assure toxin protection. Specific transporters when studied, their functional relevance for multiple purposes comes into significance.

The MLS class of antibiotics hinder protein synthesis by targeting the 50S ribosomal subunit. Other than target modulation and enzymatic blockage, the efflux pump mechanism in these antibiotic classes showcases tolerance features in Gram-positive strains. The efflux mechanism was due to *msrA* and *msrB* genes discovered in *Staphylococcus* and *msrC* gene in *Enterococcus* provided 200-fold and eightfold increase in resistance, respectively (Schmitz et al. 2000). In macrolide-resistant *Streptococcus*, the *mefA* or *mefE* genes augment tolerance level to 60-fold. These genes code for MF family transporters. The *mef* genes account for macrolide resistance in many Gram-positive as well as Gram-negative pathogens. It is present within mobile transposons; thereby, the tolerance mechanism will spread rapidly among other pathogenic species. Similarly, tetracycline transporters TetK and TetL predominantly account for antibiotic tolerance in *Staphylococcus* as well as *Enterococcus*. For greater antibiotic efficiency, decreased affinity of tetracycline derivatives via efflux pump mechanism has been approached, for example, glycyclines. The second approach encourages the use of tetracycline analogs in conjugation with the antibiotic blocking efflux transporters. Fluoroquinolone resistance occurs due to topoisomerase and DNA gyrase modulation by multidrug efflux transporters, for instance, NorA of *S. aureus* belonging to MF family. Greater NorA expression leads to acquired ciprofloxacin tolerance in addition to intrinsic fluoroquinolone resistance. Targeting efflux pump mechanisms increases the response of pathogen to antimicrobials as well as promotes drug accumulation within the bacterium. Cationic peptides target different resistance mechanism of *Pseudomonas* against antimicrobials, namely, efflux method and target site modification to highlight the role of membrane barrier as a target for overcoming pathogen tolerance (Lin et al. 2010).

5.2.3 Degradation of Antibiotic

The second strategy involves degradation of the chemically active component of the antibiotic weapon. Accumulation or expelling out antibacterial agents from cells doesn't modulate with the structure of the antibiotic. Like in penicillins and cephalosporins, the inactivation of β-lactam ring will cripple the efficacy of the antibiotic itself. B-lactamase is the enzyme that catalyzes this modulation. The active ring enables acetylation and modulation of the peptidoglycan cross-links, whose disruption renders the antibiotic nonfunctional. The lactamase enzyme is produced in the bacterial periplasm to inactivate the cytoplasmic antibiotic targets. A single enzyme can cripple about hundred penicillin particles, so the greater the enzyme, the higher the intensity of antibiotic destruction and the more efficient the strategy. However, other antibacterial compounds like aminoglycosides aren't prone to such hydrolytic cleavage. Aminoglycosides target the protein synthetic machinery and bear three specific chemical alternates that bring modulation in the ribosomal RNA binding. The aminoglycoside resistance can be due to the adenylyl, phosphoryl, or acetyl transferases which insert either an AMP moiety or phosphate group or bring about amino acid acetylation. These modifications decrease the RNA binding affinity and disrupt with the protein synthesis. The structural elucidation of phosphotransferase shows a direct evidence of evolutionary link to kinase enzyme thereby facilitating the recruitment of bacterial resistance strategies.

5.2.4 Alteration of Bacterial Target

The third resistance approach employed by the microbes involves the modulation or reprogramming of the target enzyme in the resistant pathogen. This camouflage mechanism can occur in conjugation with the efflux mechanism thereby adding up to the resistance strategy. For example, the erythromycin-resistant strains alter the adenine moieties by methylation in the peptidyl transferase loop of ribosomal RNA unit (McCusker and Fujimori 2012). The erythromycin ribosomal methylase gene targets the decreased RNA affinity for erythromycin as well as pristinamycin drugs without blocking the protein synthesis. This method of methylation is the prime machinery of resistance in the virulent species of *S. aureus* and acts as immunity armor against erythromycin-expressing strains. Other than erythromycin, target modulation is also observed in vancomycin-resistant enterococci (VRE) to escape the harsh antibiotic effects. In the resistant *Enterococcus*, the *vanHAX* gene encodes a pathway where these three genes play different modulatory roles for providing a survival advantage to the bacterium (Sood et al. 2008). *vanH* gene enables pyruvate reduction to D-lactate followed by *vanA* forming D-Ala-D-Lac and *vanX* hydrolyzing the D-alanine dipeptide rather than D-Ala-D-Lac linkage. Overall, the accumulation of D-Ala-D-Lac substrate becomes the point of elongation and extension at the peptidoglycan terminal strands. This remodulation of the D-alanine dipeptide to D-Ala-D-Lac affects the degree of vancomycin binding by 1000-folds without impairing the glycan and peptide cross-linking efficiency. This tolerance confers a greater profile of vancomycin resistance. Consequently not only does β-lactamase production affect the resistance mechanism, but also the penicillin-binding proteins bring about lower antibiotic affinity. In *Staphylococcus aureus*, the characterization of a penicillin-binding protein encoded by the *mecA* gene will help to elucidate the molecular mechanism of MRSA phenotype.

5.3 Tuberculosis: The Unsolved Puzzle

The last millennium has witnessed the generations of antibiotic development with the startling increase in resistance against those antibacterial drugs. Among the life-threatening species, *Mycobacterium tuberculosis* has become a global threat for mortality with time. The major survival advantage of *M. tuberculosis* is its dormant stage which sustains in asymptomatic hosts that later on leads to disease. *Mycobacterium tuberculosis* makes use of all the efflux transporters for their survival. The major two mechanisms thought to play a pivotal role in mycobacterial drug resistance are the cell wall barrier and the efflux pump machinery (Silva and Palomino 2011). The genes that encode for efflux transporters have been extensively studied as they encode for proteins that channelize compounds like tetracycline, fluoroquinolones, aminoglycosides, and drugs like isoniazid used for tuberculosis treatment itself. Hence, the balance between pumping out antibiotics and enabling cellular drug intake is yet to be explored further for new inventions (De Rossi et al. 2006). In spite of the BCG vaccine, resistance has also been observed against many anti-TB compounds. With the course of time, the *M. tuberculosis* strains have developed mutations that have actively targeted the drug stimulation giving rise to MDR-TB strains. For instance, the streptomycin resistance develops due to changes in the genes like *rrs* and *rpsL* which alters the ribosomal binding site for streptomycin. Even the *pncA* gene alteration leads to pyrazinamidase resistance in this notorious bug. Isoniazid has drug targets that are involved in the cell wall biosynthesis (mycolic acid), but they don't entirely confer resistance. The prime reason of natural resistance in *M. tuberculosis* is decreased cell wall permeability due to high lipid content which limits cellular drug intake. Another factor that comes into play is the efflux system that forces out antimicrobial agents. The microbial efflux and influx balance thereby contributes to the microbial sustenance.

5.4 *Antifungal Drug Resistance*

The evolutionary problem of resistance has become well documented with time. Antifungal drug resistance also isn't exempted from such threat, though antibacterial resistance is a greater concern. The microbial world encompasses various carriers of infections among which fungal pathogens flourish in adaptable population. Under the administration of antifungal agents, the sensitive fungal lot evolves resistance mechanism against the drugs. The era of the 1980s had a very thorough study on the biochemical, genetic, and clinical aspect of antifungal resistance, but presently the elucidation of cellular and molecular mechanisms is under investigation (Anderson 2005). The drug targets are mostly designed for the fungus and less for the host, so a depth of understanding into the molecular mechanism of antifungal resistance can promote a divergent long-term host survivability. However the pathogen fitness on environmental impact and evolution of potential mutations causing divergent resistance can be explored further with experimentation. The strategy implementation for combating drug resistance would require hindering with the pathogen's evolutionary sustenance approach. A study of mutation, pathogen fitness, and multiresistant factor interacting in combination for a collective phenotype can undoubtedly modulate with the pathogen's gene expression and help us reduce the chances of increased drug resistance. Greater incidence of resistant fungal pathogens has increased the risk factor of mortality in patients bearing severe immunosuppression. Though novel drugs have come up, still patients under long-term antifungal medication undergo a microflora transition during the course of time which further leads to the development of an apparent resistance mechanism.

The nineteenth century had witnessed drug resistance problem pertaining to a range of infectious diseases like tuberculosis, salmonellosis, HIV, etc. The scientific world also came across the problem of fungal infections during that time which posed a threat to health and life. This was mainly due to a change in the immune profile of the patients who were inflicted with AIDS or cancer or had undergone any sort of transplantation. That time demanded the urgent requirement of new invention of antifungal drugs as compared to the conventional ones with least side effects and with more impact on combating infections rather than being resistant to the new emerging pathogens. One study reported 33 % of patients with AIDS did bear resistance against *Candida albicans* (Sanglard et al. 1995). About 200 out of the 1.5 million species within the fungal kingdom are associated with human diseases. Some are commensals, whereas others like *Candida* are opportunistic species which infect when the host's immune system cripples down. Though skin infections are initial symptoms, systemic fungal infections causing dissemination are difficult for diagnosis and cause greater incidence of mortality. The epidemiological survey lists *Candida*, *Aspergillus*, and *Cryptococcus* species to be the causative agents of infection-related mortality. Mostly azole antifungals are used to treat infections and fluconazole usage decreases *Aspergillus* infection from 10 % to 20 %. However azole resistance has emerged due to acquired mechanism in the opportunistic species or due to selection pressure in the innate resistant strains.

The drug resistance mechanism among the fungal pathogens is mostly due to the reason of increased efflux where there happens to be an overexpression of certain transporters of cell membrane (Cannon et al. 2009). This mutational change in the transcriptional regulator confers a resistance characteristic. Secondly, an alteration in the protein target causes either a change in antifungal drug binding or allosteric inactivation of the enzyme. Some minor changes in amino acid sequences bestow the pathogen with such resistant phenotype by altering the drug activity. However, higher extent of amino acid alteration leads to functional loss of protein and accumulation of unwanted products called toxin with no significance of the drug. Thereby, altering metabolism also confers resistance mechanism. The standardized measurement of antifungal drug resistance is with the protocol of minimum inhibitory concentration (MIC) where the terms "drug sensitive" and "drug resistant" are outlined. However when MIC is not clear during growth transition, parameters of fitness analysis can be used to quantify resistance. At times, tolerance assays can be used for lethal drug measurements though it's not positively related to drug resistance.

The resistance in microbial population is due to evolutionary processes. However in a mixture of resistant and sensitive strains, the phenotype of resistance is not clearly defined. The population size and mutational effects confer drug resistance within the eukaryotic environment. With high incidence of opportunistic fungal pathogens, the immunocompromised individuals fall into the trap of mortality of these invasive species. *Candida* and *Aspergillus* being the most threatening species of concern are the cause of death rate of about 40–50 %. The antifungal resistance has mainly risen up due to triazole drugs that have conferred both primary as well as secondary resistance with apparent shift of colonization markers in the susceptible strains. Triazoles like fluconazole, posaconazole, etc., are mostly used to treat *Candida* infections. A study done over a decade showed 140,000 *Candida* strains to be resistant to fluconazole and voriconazole by 6 % and 3 %, respectively (Pfaller et al. 2010). The Netherlands and UK reported triazole resistance to shoot up to sixfolds over a period of 14 years. This section will outline the details of antifungal drug resistance mechanism and the strategies to combat such problems in the future.

5.4.1 Antifungal Agents and Their Mechanism of Action

The mechanism of action of different antifungal drugs is an essential prerequisite for getting an insight into their resistance mechanism. The choice of the antifungal drug should be based on factors concerning the host specificities and drug properties like its absorption and toxicity features (Odds et al. 2003). The host immune profile, the pathogen specificities (i.e., fungal species and its response to drugs) affecting the site of infection, and the pharmacokinetic properties of the antifungal agent should be taken into consideration. Very less antimycotic agents have been used to treat systemic infections. According to the action mechanism, the antifungal drugs are categorized into four different classes, namely, polyenes, azoles, nucleic acid synthesis inhibitors, and inhibitors of glucan synthesis. Their mode of action enabled a clear understanding and elucidation of their resistance mechanism. Out of these, three antimycotic agents, namely, polyenes, azoles, and allylamines, have their antifungal effect due to their inhibitory property on synthesis/interaction of ergosterol, a major fungal membrane component.

5.4.2 Ergosterol Biosynthesis Inhibitors (Azoles and Triazoles)

The 1970s witnessed the discovery of azoles, clotrimazole being the first azole-based drugs for systemic infections. These N-substituted imidazoles are compounds ranging from miconazole to ketoconazole and fluconazole. Due to certain limitations of miconazole, ketoconazole became the first commercialized oral antifungal medication against chronic candidiasis (Petersen et al. 1980), with an exception to *C. glabrata*. For human use, itraconazole and fluconazole were the triazoles for oral as well as intravenous administration. The safety and efficiency of fluconazole has been clinically approved for global use. Itraconazole was used against *Candida* spp. as well as *Aspergillus* spp. (Pfaller et al. 2005). Similarly, fluconazole intake could decrease invasive candidiasis in patients undergoing chemotherapy or transplantations. The frequent use of fluconazole has led to a resistant host microflora against the medication. The twentieth century witnessed the clinical approval of voriconazole for global use in the USA. Additionally, two other triazoles, namely, posaconazole and ravuconazole, were scrutinized for their action and efficacy against *Candida* species. Voriconazole, structurally and functionally similar to fluconazole and itraconazole, was used in conjugation with liposomal AmB for medication. Posaconazole was effective against *Candida* spp., *Aspergillus*, as well as *Cryptococcus*. From experimentation, posaconazole was found to be the most efficient triazole against itraconazole-sensitive strains. Ravuconazole also showed additional effect on *Fusarium*, histoplasma, *Blastomyces*, etc. The triazole drugs differ in their mechanism of action, resistance, as well as cross-resistance pattern among microbes. For instance, a voriconazole-resistant *Aspergillus* isolate has slight cross-resistance among itraconazole, posaconazole, as well as voriconazole. Similarly for *Scedosporium* strains, no cross-resistance was observed among triazole drugs like miconazole, itraconazole, or voriconazole. Neither was any sort of resistance reported against posaconazole medication.

The plasma membrane bioregulator ergosterol maintains the fungal cell integrity. The demethylation of lanosterol is catalyzed by 14a-demethylase in a cytochrome P-450-dependent manner. Alteration of this target enzyme results in structural and functional modulation of fungal membrane. Azoles inhibit the synthesis of ergosterol, an essential fungal membrane component by blocking the activity of the enzyme lanosterol demethylase which catalyzes the reaction of ergosterol biosynthetic pathway (Bossche 1985). The heme domain of the enzyme is bound with the nitrogen atom of azole ring to prevent lanosterol's demethylation. Azoles also target methylsterol synthetic pathway. Azole resistance is prevalent in patients with HIV infections which undergo long-term treatment procedures to combat mucosal or oral *Candida* colonization (Lupetti et al. 2002). This frequency of azole resistance has markedly increased with the course of time. With the amount of CD4 cells, pathogen load, and therapy dosage, the incidence of resistance varies. A study reported the presence of resistant *C. neoformans* from a healthy patient without having any previous fluconazole medication (Orni-Wasserlauf et al. 1999). Another study reported the HIV-infected patients bearing *C. albicans* infection to be resistant to clotrimazole (Pelletier et al. 2000). Certain cases also witness a profile of cross-resistance (Müller et al. 2000). Many species of *Candida* like *C. krusei* have an intrinsic resistant characteristic to fluconazole which is prevalent in patients infected with HIV, cancer, or undergoing transplantation. Azole antimycotic drugs have an array of heterogeneous functions ranging from acting as inhibitors of membrane-bound enzymes to the blockage of lipid biosynthetic pathway. Furthermore, azoles like fluconazole and itraconazole bring about the aggregation of sterol precursors in *Cryptococcus* by the reducing obtusifolione. Even the demethylation affects the cholesterol synthesis in mammals by a greater dosage of azoles. A study done by Hitchcock et al. reported that 50 % inhibitory concentration of voriconazole had 250-fold more activity against the mammalian demethylase as compared to the fungal enzyme (Martin et al. 1997). So azoles have their action to be genus based.

Azole resistance mechanisms are mostly similar to antibacterial mechanism like the target enzyme modification, the efflux pump resistance mechanism, and the aminoglycoside tolerance with membrane alterations. The specificity against azoles is still a question as cross-resistance among this class of drugs is quite common. Bacterial strains have best evolved with efflux pumps for resistance mechanism like the ***mar*** (multiple antibiotic resistance) genes in *E. coli* (Cohen et al. 1993; Alekshun and Levy 1999). These genes are also associated for chloramphenicol and tetracycline resistance. These multidrug efflux pumps have also conferred resistance in *S. aureus* and *P. aeruginosa* against fluoroquinolones and β-lactams, respectively. The phospholipid and fatty acid content influence the membrane permeability as well as miconazole resistance in *C. albicans*. Similarly, in *P. aeruginosa* diminished membrane D2 porin expression as well as enhanced amphotericin β-lactamase expression enables imipenem resistance. No mutation in the gene; just the membrane composition can alter the microbe tolerance levels. Vancomycin's size exclusion by the bacterial membrane is another factor for its resistance.

Alteration of Drug Efflux Efflux pump machinery is mostly responsible for the dominance of resistance mechanism among the fungal pathogens. This is a common mechanism of antibiotic resistance in *S. pneumoniae*. The two efflux pumps in pathogens conferring azole resistance comprise of proteins belonging to the major facilitator superfamily (**MFS**) and ATP-binding cassette (**ABC**) superfamily. Moreover, ABC (ATP-binding cassette) transporters are the major culprits of drug resistance. MFS protein pumps involve the passage of structurally diverse components. MDR1 in fluconazole-resistant *Candida* strains were known to encode resistance for benomyl as well as methotrexate. The ABC transporters require ATP for substrate channelization for which they bear two ATP-binding cytoplasmic moieties. Other than that, there are four core integral domains that span the membrane a couple of times. In *S. cerevisiae*, the ABC transporters recognized are classified into MDR, CFTR, YEF, and PDR families. Five CDR (*Candida* drug resistance) genes in *Candida* are responsible for azole resistance (White et al. 2002). CDR1 present in *Cryptococcus* and *Candida* is structurally similar to human P-glycoprotein. When *C. albicans* was experimented for its ability to mount up fluconazole, an overexpression of CDR1 levels was found with reduced drug concentrations. Secondly, reduced fluconazole accumulation could be due to the ATP-dependent drug efflux mechanism. A study by Sanglard et al. carried out experiments involving 16 *C. albicans* clinical isolates from five individuals with HIV infections (Sanglard et al. 1995). The aim was to observe the accumulation of fluconazole with treatment. Fewer amounts of fluconazole levels did correspond to tenfold higher CDR1 mRNA profile. However, with overexpressed CaMDR1 mRNA level, normal CDR1 transcriptomic levels were observed. This concludes that CDR1 is mostly involved in the transport of azole antifungals, whereas CaMDR1 gene specifically enables fluconazole resistance development. A CDR1 mutation enabled greater susceptibility of *C. albicans* to triazoles. Similarly, CDR2 induction resulted in azole resistance. A double knockout mutant (CDR1 and CDR2 deletion) strain showed greater susceptibility than a single gene disruption. Moreover using membrane potential as the driving force, MDR1 overexpression also leads to azole resistance. A study reported the accumulation of a fluorescent rhodamine 123 dye in *C. albicans* and *C. glabrata* (Clark et al. 1996). This dye is specifically transported by the MDR machinery. This mechanism also leads to a phenomenon of efflux competition. Additionally, ABC transporters overexpression is a crucial factor in promoting azole resistance in *C. glabrata* isolates. However, exposure to azoles can contribute to a transcriptional alteration in CDR profiling. The sterol composition doesn't influence resistance; rather the genes like ERG16, MDR1, and CDR1 are involved in microbial tolerance mechanism. Secondly continuous azole exposure can lead to an induced expression of *ERG16* as well as *CDR1* genes, thereby leading to cross-resistance among other azole medications.

Alteration of the Target Enzyme The most common example with respect to enzyme alteration is lanosterol demethylase whose overexpression confers azole resistance in *C. glabrata*. The modification of this enzyme plays a pivotal role in conferring azole resistance mechanism. No difference in sterol distribution was found in two fluconazole resistant and susceptible *C. krusei* strains. The inhibitory effect of fluconazole was 20–40 % higher in *C. krusei* than *C. albicans*. This is due to greater active efflux mechanisms in *C. krusei* isolates. With respect to 14a-demethylase enzyme modification, the azole-susceptible units were found to have an altered peak in the carbon monoxide spectra of the cytochrome. Moreover, this enzyme has low affinity for azole drugs. Whether alteration of the enzyme is the sole factor remains still a question. Thereby, certain cases have reported the overexpression as major criteria for resistance mechanism. The overexpression leads to enhanced copies of the enzyme which promotes ergosterol synthesis and enables resistance development against fluconazole as well as itraconazole. Here the ergosterol increase could be attributed to the overexpressed enzyme as well as to the less susceptibility of both azoles and amphotericin B. Some studies have focused on P-450 levels for conferring cross-resistance in azoles. But in *C. albicans*, the increased expression doesn't have much of an impact. Azoles are inhibitors of ergosterol synthesis by modulating the binding with the demethylase enzyme. Further the *ERG11* gene was investigated for its role in drug selection pressure. Sequence analysis in *C. albicans* targeted amino acid substitutions at the active site of the enzyme (i.e., the heme domain) to be responsible for resistance development. *ERG11* (*ERG16*) is the gene encoding for the protein, also termed as CYP51A1 in *C. albicans* (Marichal et al. 1999). Any alteration or mutation in this gene conferred azole tolerance. Lysine substitution with arginine at 467th residue near the heme domain brought about functional alteration to the enzymatic activity. Even a 464th residue substitution in *C. albicans* caused heme domain alteration contributing to fluconazole resistance with reduced activity. A study reported a T315A substitution in *C. albicans* showed twofold reduction in lanosterol demethylase catalytic activity and diminished affinity for fluconazole (Lamb et al. 1997). These substitutions also lead to decreased accessibility of the enzyme active site as was observed when the 105th phenylalanine residue was replaced with leucine in *C. albicans*. The antifungal agent fluconazole has a channelized entry into the active site where certain point mutations alter the accessibility of the substrates. These mutations are huddled up around three specific regions that are linked with resistance.

Alteration of the *ERG3* Genes *ERG3* gene encodes for sterol D(5,6) desaturase and an alteration or mutation in this gene confers azole resistance. In sensitive species, the azole exposure enables 14-methyl-3,6-diol accumulation causing a fungistatic property, however the mutation in *ERG3* gene together with the accumulation of the precursor (14-methylfecosterol) promotes fungal growth. *S. cerevisiae* has shown to exhibit azole resistance due to *ERG3* gene mutation (Martel et al. 2010). Some clinical isolates of HIV patients bearing resistant *C. albicans* showed an amassing of 14-methylfecosterol due to an impaired sterol D(5,6) desaturase. This impairment has also led to azole resistance in the fungi *U. maydis* (Joseph-home et al. 1995).

Alteration of the Drug Influx The drug uptake can be influenced by the plasma membrane composition and fluidity; for instance, the sterol content can affect the cellular drug influx. An alteration in the cell membrane constituents of *C. albicans*, like the phospholipids as well as the fatty acid content, can confer resistance to miconazole. A study by Hitchcock and Whittle reported a greater lipid profile and less polar to neutral lipid content in *C. albicans*-resistant strain as compared to wild-type one (Hitchcock and Whittle 1993). This hints reduced membrane permeability and decreased azole intake. Some reports demonstrated the fluconazole-resistant *Candida albicans* species to have reduced amount of ergosterol as well as low phosphatidylcholine: phosphatidylethanolamine profile (Löffler et al. 2000). Such phenomenon modulates with the fluconazole uptake and intracellular accumulation. *C. krusei* also had decreased itraconazole accumulation rather than modulations in drug efflux mechanism or alteration in ergosterol amounts.

A study observed 20 isolates from HIV patients with oropharyngeal candidiasis (both susceptible and resistant to fluconazole) previously treated with azoles for the resistance frequency (Perea et al. 2001). About 85 % of those resistant strains had overexpression of drug efflux pumps with similar expression profile of CDRs and *MDR1* gene. Around 60 % isolates showed alteration in the lanosterol demethylase enzyme with 35 % having an overexpression of the gene encoding the enzyme. Around 75 % of these strains showed multiple mechanisms of resistance. Another study reported only overexpression of CDR genes in resistant strains without any correlation of tolerance to amino acid substitution or *MDR1*, *ERG11* expression profile (Marr et al. 1998). CDR-encoded pumps (*CDR1* and *CDR2*) play a regulatory role in conferring azole resistance mechanism.

5.4.3 Polyenes

Action and Resistance Mechanism From the 1950s, polyenes (AmB) have been known to be the standard treatment for systemic fungal infections. Amphotericin B, the broad-spectrum antifungal agent, targets ergosterol and is active against *Candida* species, *Cryptococcus neoformans*, strains of *Aspergillus*, *Zygomycetes*, etc. (Brajtburg et al. 1990). The polyene-susceptible pathogens are known to bear sterols in their cell membrane as compared to the resistant ones. The literature suggests sterol addition for counterbalancing the fungal inhibitory effect of polyenes. The physicochemical interplay between membrane sterols and polyene antifungals restricts the binding affinity of drugs. Their interaction can be directly quantified with UV absorbance. The mode of action of amphotericin B involves the aggregation of 8–10 polyene molecules to form a porin channel within bilayer membrane for disrupting fungal ionic gradient by loss of potassium ions. This pore enables polyene hydroxyl moieties to protrude inward thereby causing altered permeability and loss of essential cytosolic components. The fatty acyl components also render polyene antifungal susceptibility in yeast. Additionally, amphotericin B also acts as an oxidative load on fungal membranes thereby causing the death of *C. albicans*. At higher doses, polyenes also hamper the functioning of fungal chitin synthase enzyme. For the unpleasant activity of AmB, the research developed conjugated AmB with liposomes for better functioning in host and least toxicity. Some examples include Amphotec, AmBisome, etc. During liposome and AmB formulation, selective transfer method enables the AmB's transfer from donor liposome to target the membranal ergosterol facilitated by phospholipases (either pathogen or host) (Boswell et al. 1998). A formulation of liposomal nystatin and polyene Nyotran is under clinical trials for evaluation.

Primary resistance to amphotericin B has been reported in the isolates of *Candida* like *C. lipolytica*, *Candida lusitaniae*, etc. Intrinsic resistance has been observed on *Trichosporon beigelii*, *Pseudallescheria* spp., as well as *Scopulariopsis* spp. Secondary resistance against polyenes has been described in *C. neoformans* and many species of *Candida*. Some hypotheses have led to the conclusion that initial medications with azoles can lead to amphotericin B resistance. The resistance mechanism can be either due to the category of sterol or the fungal membrane composition. In one study, cross-resistance was observed to amphotericin B in fluconazole-resistant *Candida* strains from HIV-infected individuals due to modulation in ergosterol synthesis (Heinic et al. 1993). The same isolates also showed cross-resistance to nystatin. Such polyene (AmB as well as nystatin) resistance has also been observed in trauma patients with compromised immune profile. The lipid complexes with AmB have the active amphotericin moiety released from tissue lipases in vivo. This confers resistance to AmB contributing to less drug efficacy but with lipid formulations; the edge shifts to increased efficacy and uptake during drug administration. An alteration in the membrane's lipid composition due to deficit ergosterol amount lowers the affinity of amphotericin B for binding to plasma membrane. However, the resistance is conferred majorly due to ergosterol and not due to altered sterol composition. Another factor for resistance mechanism is deposition of β-1,3 glucans in the pathogen's cell wall which enables greater access of larger molecules to membrane due to high cell wall stability.

Polyene tolerance develops by selection pressure when certain resistant isolates multiply naturally within small number of population. These naturally developed resistant strains produce altered sterols for binding nystatin at decreased affinity. The binding of nystatin to sterol influences cell membrane damage. With greater affinity, the membrane damage increases. Generally, the resistant strain is expected to grow slow as compared to the susceptible one. Polyene tolerance is lost after certain passages in media devoid of nystatin. However, sterol contents result in increased affinity for nystatin. So it's mutation that plays a role in tolerance development rather than selection. Polyene resistance is mostly studied using cells grown in increasing or gradient concentration of antifungals that generates mutants. The biochemical hypothesis outlines resistance to be a qualitative as well as quantitative factor of sterol content in cells. Altered sterol content decreases the binding of polyenes in resistant cells either due to complete lack of cellular ergosterol amount without associated change in overall sterol composition or by replacing the higher affinity polyene-binding sterols with less binding ones. At times steric modification or thermodynamic alteration in polyene binding can also lead to the resistance mechanism. However the major reason of tolerance due to decreased ergosterol content is not caused due to enzymatic degradation but blockage on synthetic pathway which leads to decreased polyene susceptibility. However it was reported that certain *Candida* strains possessed certain key sterols that enabled the resistance to polyenes (Hamilton-Miller 1972). D8 sterols possessing strains are more resistant to polyenes than the ones bearing D7 sterols.

In one study, AmB resistance in *Leishmania* species was attributed to ergosterol substitution in the membrane which alters the membrane fluidity and binding of amphotericin B (Mbongo et al. 1998). The interaction of polyenes with the membrane components have been extensively studied and explored by researchers. Stationary phase cells are more prone to polyene resistance as compared to their exponential counterparts because of less active cells involved in synthesis of cell wall components during static phase. So the access to membrane slows down. Thereby drug modulation can't influence polyene tolerance mechanism. Efflux pump method can't be involved in polyene resistance development. The genetic basis of polyene resistance in *S. cerevisiae* is related to the mutations in *pol* genes (Molzahn and Woods 1972). The mutants had a decreased amount or complete lack of ergosterol which establishes the much talked about antifungal resistance mechanism.

Allylamines The functional and chemical aspect of allylamines makes them distinct from the group of ergosterol biosynthetic inhibitors. These antifungal agents like terbinafine are effective against dermatophytes and azole-resistant *Candida* as well as *Cryptococcus* species. Squalene accumulation during ergosterol biosynthetic pathway at the step of squalene epoxidation is the direct target of allylamine inhibition (Ryder 1992). Squalene epoxidase is essential for allylamine activity, thereby hinting fungal death due to higher levels of squalene rather than ergosterol deficiency. Accumulation of squalene leads to greater membrane permeability and obstruction to cellular organization. However, extensive usage of terbinafine and naftifine can confer cross-tolerance to fluconazole-resistant *Candida* strains. Azoles, polyenes, and allylamines play similar mechanism of action targeting cell wall synthesis like penicillin, vancomycin, and other antibacterial agents.

5.4.4 5-Fluorocytosine: Nucleic Acid Synthesis Inhibitors

Action and Resistance Mechanism A fluorinated pyrimidine fluorocytosine has been prevalent for use against fungal infections since the era of 1960. With an efficient penetrating capability into body fluids, 5-FC was targeted against *Candida* as well as *C. neoformans*. However, the incidence of primary resistance among fungal pathogens led to the combinational therapy of 5-FC along with other antifungal drugs like amphotericin B along with fluconazole. The cytosine permease enzyme aids the absorption of 5-FC into the fungal cells (Andriole 1999). Immediate to its entry, the compound undergoes modification to 5-fluorouracil by deaminase enzyme. This compound undergoes conversion to fluorouridine triphosphate by UMP pyrophosphorylase which after further phosphorylation is incorporated into the fungal RNA chain, thereby blocking the protein synthesis. Fluorouracil also undergoes conversion to 5-fluorodeoxyuridine monophosphate which disrupts with the functioning of thymidylate synthase, which aids in DNA synthesis. Thereby, 5-FC disrupts with DNA, RNA, as well as protein synthesis of pathogen cell. The resistance mechanism have been widely investigated and hence have related the disrupted functioning of the enzyme cytosine permease or deaminase (which enables two consecutive absorption and conversion steps of 5-FC) to diminished drug uptake by fungal cells. This is due to a mutational change that not only promotes primary but also intrinsic resistance. Secondly, less catalytic activity of uridine monophosphate pyrophosphorylase and uracil phosphoribosyltransferase also confers 5-FC resistance in *C. albicans*. Both cytosine deaminase and uracil phosphoribosyltransferase are involved in the pyrimidine salvage pathway and are unessential for de novo synthesis. A change in the bases due to mutation in either one of these salvage pathway enzymes confers resistance in *C. albicans* and *Cryptococcus neoformans*. In *C. albicans*, less phosphoribosyltransferase activity was linked to 5-FC resistance in a dose-dependent manner. Resistance to 5-FC due to decreased uptake has been well observed in *S. cerevisiae* and *C. glabrata* (Vandevelde et al. 1972). However, this mechanism is not of significance in *C. albicans* or *Cryptococcus neoformans*.

5.4.5 Inhibitors of Glucan Synthesis: Fungal Cell Wall Compounds

Mode of Action and Resistance Mechanism The multilayer fungal cell wall consists of unique components like α- and β-glucans, mannan, chitin, etc., which provide antifungal drug targets for safe use in mammalian hosts. These compounds have selective toxicity benefits in host. The medically important pathogen *C. albicans* has a multilayered cell wall comprising of β-glucan and mannoprotein (80 % of cell mass) along with chitin. Among the glucan synthesis inhibitors, chitin biosynthetic genes are disrupted by compounds named as nikkomycins which have been scientifically investigated for commercial use. The three groups categorized as glucan synthesis inhibitors are aculeacins, echinocandins, and papulacandins which disrupt the functioning of β-(1,3)-glucan synthetase enzyme that leads to synthesis of 1,3-β, D-glucan. However among the inhibitors, the lipoproteins, i.e., echinocandins, are clinically approved for global use due to their safety, efficacy, and tolerance level. Echinocandins are tested to be potently active against *Candida* spp., *Aspergillus* strains, dimorphic molds, and *Pneumocystis carinii*. Three antimycotic compounds under echinocandins have been thoroughly investigated for safe use, namely, caspofungin, FK-463, and VER-002. No cross-resistance was observed against these three compounds in isolates resistant to triazoles. These inhibitors form a noncompetitive inhibition with secondary effects on either the chitin content or ergosterol content of fungal cell. One study reported the in vitro comparative activity analysis of echinocandin in fluconazole-resistant *Candida* strains with itraconazole and amphotericin B as test parameters (Cuenca-Estrella et al. 2000). The results showed echinocandin to have potent activity against a range of *Candida* spp. like *C. albicans*, *C. tropicalis*, *C. krusei*, etc. However, echinocandins showed to have less activity against *C. parapsilosis* and *C. guilliermondii*. Nevertheless, in *C. neoformans*, the reduced activity of echinocandins is due to decreased fungal glucan synthase activity (Maligie and Selitrennikoff 2005). The resistance mechanism of pathogens against echinocandin is very limited. Kurtz and Douglas experimented with the laboratory-resistant mutants of *S. cerevisiae*. Echinocandins target the β-glucan synthase enzyme which is mainly encoded by two genes called *FKS1* and *RHO1* and regulated by a third gene (Sekiya-Kawasaki et al. 2002). FKS2 is another gene in S. *cerevisiae* which is homologous to *FKS1*. Any base alteration in *FKS1* genetically causing mutations leads to the development of in vitro resistance to echinocandins (Balashov et al. 2006). This is specifically due to the alteration of the target enzyme, glucan synthase. Some sort of resistance is also developed due to mutations in *GNS1* gene coding for fatty acid chain synthesis in cell wall. Conversely, changes in *FKS2* gene does not attribute to resistance. The resistance mechanisms for azoles include efflux pumps and fungal membrane composition, which howsoever seems inappropriate for echinocandins as they don't undergo cytosolic pathway of penetration. Since they don't traverse the fungal membrane barrier, the entry mechanisms don't serve as the methods of resistance development in lipopeptides. Even MDR-like gene activation doesn't confer resistance mechanism to echinocandins. These results finally conclude that *FKS1* mutation can alter the protein forming the catalytic target of glucan synthase enzyme, thereby facilitating lipopeptide resistance development in *S. cerevisiae*. Lastly, a study established that both *S. cerevisiae* and *C. albicans* have similar mechanism of resistance to β-glucan synthase inhibitors (Douglas et al. 1994).

5.5 *Antiviral Drug Resistance*

In the late twentieth century, the development of potent antiviral drugs was considered as an important achievement in the field of biomedical science. Highly effective drugs against a wide range of viruses like herpes, HIV, hepatitis B, influenza, human papillomavirus, respiratory viruses, enteroviruses, hepatitis C, etc., have been designed and proved to be of human benefit. But sadly, with time, antiviral drug resistance has emerged at a considerably higher rate. The resistance to antiviral agents is considered to be a natural phenomenon because of the rapid replication ability of the virus under a selective pressure (Richman 2006). On prolonged drug exposure only those mutants survive which can replicate continuously in that environment and thus become resistant. The development of resistance is a major point of concern in the immunocompromised patients too (Strasfeld and Chou 2010). Quick diagnosis of the resistance type can be made by observing the different mutations in the genome of the viruses that made them resistant. A lot of literatures describe drug resistance in influenza virus, retroviruses (HIV), herpes simplex virus, cytomegalovirus, varicella-zoster virus, and hepatitis B virus (discussed in detail below).

5.5.1 Drug-Resistant Influenza Virus

Adamantanes like amantadine and rimantadine are drugs mainly given for treating influenza A viral infections. But clinical studies have shown this virus to be increasingly resistant to both the drugs, especially amantadine in both animal and human isolates (Englund et al. 1998). The drug-resistant strains showed point mutations in a recognized second reading frame of the M segment of the influenza RNA genome. This region, named as M2 encodes for a tetrameric, transmembrane H^+ ion channel, required for pH mediated entry of the viral ribonucleoprotein into the cytoplasm. Amantadines used to block this channel, hence preventing viral replication. Early experiments with viral neuraminidase inhibitory drugs like oseltamivir and zanamivir showed sensitivity toward both influenza A and B viruses, but recent reports show incidence of higher resistance for oseltamivir than zanamivir. A seasonal influenza H1N1 virus found to be mutated in its neuraminidase gene (H274Y) and thus contribute for developing resistance against oseltamivir (Control and Prevention 2009). Some isolates of avian influenza A virus (H5N1) have also been adamantane resistant.

5.5.2 Drug-Resistant Herpes Simplex Virus

Mucocutaneous HSV infections are usually administered with acyclovir, valacyclovir, and famciclovir, whereas the first is always preferred in the treatment of serious invasive disease like encephalitis. Drug-resistant acyclovir was first encountered as early as in the year 1982 when the drug was systemically circulated, but later drug-resistant strains have been isolated from patients without any history of preexposure to the drug (Nugier et al. 1992). Resistance of HSV to acyclovir is often associated with the viral TK or DNA polymerase mutations (Morfin and Thouvenot 2003). This mutation can lead to a loss of TK function or a modification in TK substrate specificity. Mutations in the TK gene are mainly due to addition or deletion of nucleotides in homopolymer runs of guanines and cytosines, resulting in frame shifting and loss of its function. Drug-resistant TK mutants retain susceptibility to drugs like foscarnet and cidofovir that are independent of viral-mediated phosphorylation, unless a viral DNA polymerase mutation is also present. Given the essential role of DNA polymerase in viral replication, mutations in this gene occur less frequently and have been found to cluster in functional domains II and III.

5.5.3 Drug-Resistant Varicella-Zoster Virus

Normally, the same drugs are administered for VZV as in the case of HSV. Acyclovir is less effective in this case, while famciclovir and valacyclovir prove to be more competent. Like in the case of HSV, drug resistance in VZV is also attributed to the TK mutations in the viral genome (Lacey et al. 1991). It mainly results due to a premature stop codon that leads to a TK-deficient virus. Other mutations related to resistance appear to cluster at particular VZV TK gene loci. Cross-resistance is seen for drugs acyclovir and penciclovir in some in vitro studies.

5.5.4 Drug-Resistant Cytomegalovirus

CMV is usually an opportunistic pathogen associated with AIDS patients. The principle drugs currently being administered for CMV infections are ganciclovir and valganciclovir. But shortly after the introduction of ganciclovir in the late 1980s, cases of drug resistance came into picture (Akalin et al. 2003). On prolonged exposure to ganciclovir, 90 % of resistant CMV isolates were found to have characteristic mutations in the viral UL97 kinase gene (Chou 2008). These mutations apparently reduce the ganciclovir phosphorylation without impairing the major functions of the kinase in viral replication. CMV UL97 drug resistance mutations cluster tightly at codons 460, 520, and 590–607. Mutations M460V/I, H520Q, C592G, A594V, L595S, and C603W are among the most frequently encountered in ganciclovir-resistant isolates. Apart from this, CMV UL54 DNA polymerase mutations can lead to resistance against almost all the available drugs for CMV infection. Many ganciclovir resistance mutations are located in the exonuclease domains and typically confer cross-resistance to cidofovir whereas mutations in and between the catalytic domains can contribute to foscarnet resistance as well as cross-resistance in cidofovir and ganciclovir in a low grade. Usually, UL97 mutation occurs first with ganciclovir resistance, followed by one or more UL54 mutations after prolonged therapy.

5.5.5 Drug-Resistant Hepatitis B Virus

There are currently seven FDA-approved agents for the treatment of hepatitis B out of which, three are nucleoside analogs (lamivudine, entecavir, and telbivudine) and two are nucleotide analogs (adefovir and tenofovir). All of these target HBV DNA polymerase, which includes reverse transcriptase activity. These kinds of drugs are phosphorylated by cellular enzymes to active form and then incorporated into growing DNA, resulting in premature chain termination, among other inhibitory functions associated with viral replication. Due to the high viral replication rate and the error-prone nature of HBV reverse transcriptase, there has been the emergence of resistance against the above said classes of drugs. Reports suggest that signature mutations in the reverse transcriptase domains of the viral polymerase gene are the main causes of drug resistance as it changes the interaction between the virus and drugs. High-level lamivudine resistance is mostly caused by M204I/V mutations, which are in the YMDD (tyrosine–methionine–aspartate–aspartate) motif in the C domain of the polymerase gene and infrequently by A181V/T mutations. The M204I mutation confers high-level cross-resistance to telbivudine, either alone or in association with the secondary mutations L80I/V or L180M. The N236T mutation, on the other hand, decreases viral replicative capacity in vitro and provides cross-resistance to tenofovir but not to lamivudine or telbivudine. During continued entecavir treatment, additional mutations at I169T and M250V or T184G and S202I are selected, conferring resistance to the same. Report on tenofovir resistance has been seen in two HBV/HIV co-infected patients with prior lamivudine exposure (L180M/M204V mutations) and an extra A194T mutation (Lacombe et al. 2010).

5.5.6 Drug-Resistant HIV

Resistance of HIV to antiretroviral drugs is one of the most common causes for therapeutic failure in HIV-infected patients. Despite of continuous research and anti-HIV drug development, no combination of drug studied till date has shown to completely block viral replication. Instead, the virus has developed smart mutations in its different survival pathways and continues to be a threat to mankind. Antiretroviral drugs are either nucleoside reverse transcriptase inhibitors (NRTIs), non-nucleoside reverse transcriptase inhibitors (NNRTIs), or protease inhibitors. Two important mechanisms by which NRTIs work involve mutations (e.g., M184V, K65R, Q151M) occurring at or near the drug-binding site of the reverse transcriptase gene, leading to increased drug discrimination by this gene, and another way is to make key mutations that actually work to undo the action of the drugs, even if they bind to their target RT correctly (Clavel and Hance 2004). NRTIs behave like plugs by blocking nonextendable nucleoside analog monophosphate to the 3′ end of the growing proviral DNA chain, thus inhibiting viral replication. But this phenomenon can be reversed by a reverse transcriptase reaction where the chain-terminating residue is removed and an extendable primer is reinstated. This kind of reverse reaction of DNA polymerization is called pyrophosphorolysis, and it enables reverse transcription and DNA synthesis to resume. This mechanism can be enhanced by some mutations, mostly those selected by zidovudine (Retrovir) and stavudine. As these two drugs are thymidine analogs, these mutations are often referred to as thymidine analog mutations (TAMs).

The mechanism of drug resistance that NNRTIs follow is simpler. All the drugs falling under NNRTIs are designed in such a way to bind to the amino acids packed in a hydrophobic binding pocket within the reverse transcriptase. This pocket does not belong to the active site of the enzyme but near to it and is found only in the presence of the drug. The drugs open this pocket, thus blocking some enzymatic movements inhibiting DNA synthesis. Different mutations conferring resistance to NNRTIs like L100I, Y181C, G190S/A, and M230I involve the amino acids that form the hydrophobic binding pocket. However the K103N mutation is slightly different in the aspect that instead of forming the pocket, it is present near the entrance of the same and it creates a hydrogen bond in the unliganded enzyme. This bond makes the pocket entrance closed for the drug to enter.

Now, the HIV protease gene acts as a homodimer; each of the units constitutes two chains composed of 99 amino acids that make gag (p55) and gag-pol (p160) polyprotein products into active core proteins and viral enzymes (Park and Morrow 1991). These proteases cleave the polyproteins immediately after or during the budding process at nine different positions to give rise to various structural proteins (p17, p24, p7, and p6), reverse transcriptase, integrase, and protease. All the available protease inhibitors bind to the active site amino acids at the center of the homodimer and prevent cleavage of gag and gag-pol protein precursors in severely infected cells. Hence it arrests the maturation and infection by nascent virions. The virus particles develop resistance by mutations which force the pocket at the center of the homodimer to widen, resulting the drug to freely float and not able to bind to its target tightly.

5.6 Antiparasitic Drug Resistance

Parasitic infections have continuously dwindled the global health status, most prominently in the tropical regions. The protozoan and helminth-related diseases have led to the invention of many drugs for their specific treatment decades ago. At that time, about 0.1 % of the global financial asset was invested in therapeutic inventions for tropical diseases including malaria, leishmaniasis, etc. The invention of any drug against targeted pathogens or their structural aspects is an iterative process which involves specific strategies followed by target recognition and validation. The assays after development undergo a thorough screening process to detect structural hits for activity inhibition and their further assessment, and analysis will tag the leads for clinical evaluation. In the case of malaria, drug resistance has been well observed in three malarial species, namely, *P. falciparum*, *P. vivax*, and *P. malariae* (White 2004). With time the parasites have been able to flourish well and replicate within the host system in spite of the drug dosage and absorption. With greater access to parasitic system or the infected erythrocyte, the pharmacokinetics of malarial drugs and understanding of host metabolism have reached new insights. It essentially requires pharmacologically active metabolites for therapeutic purposes. Antiparasitic drug discovery has evolved with new impetus with advancements in genome sequencing, international collaborations, and national programs for fund generation toward this significant impact.

Emergence and Spread The emergence of antiparasitic drug resistance has transformed the global epidemiology profile. Drug dosage and pharmacodynamic properties affect the efficiency of any therapeutic regime. The resistance emergence generally initiates with a genetic change followed by a selection process offering the parasite an endurance profile within host for its survival. Parasitic treatment failures depend on the host immune system as well as parasitic factor. Drug dosage, administration frequency, the period of transition from susceptible to resistant form, time point of host infection, and the fitness cost effects influence the resistance transmission. Acquired tolerance, cross-resistance, drug adherence, and absorption features lead to parasite recrudescence. Cross-resistance is a factor which complicates the administration of different classes of antiparasitic compounds with varied modes of action which are rendered futile due to superbug tolerance mechanisms. The aminoquinolines and antifolates are mostly affected. In malaria, for instance, the non-clearance of asexual parasites leads to gametocyte assembly which transmits resistance determinants. The resistance development occurs in two phases of tolerance emergence and spread (Klein 2013). Initially a rare, random, and spontaneous genetic mutation provides the parasite a survival advantage which later gets selective, replicates, and becomes insusceptible to treatment. Other than chromosomal mutations, even the gene copies influence the drug target alteration or efflux of therapeutics. A single de novo mutation can lead to multiple events; however, non-immune individuals acting as parasite reservoirs can even contribute to de novo tolerance. Selection of the resistant parasites occurs due to subtherapeutic drug concentration exposure where the sensitive lot succumbs and leads to the drug resistance spread. The time frame for selection and resistance transmission of parasites portrays the "fitness cost" which enables subsequent gametocyte production and cessation of drug activity. The major factors for emergence and spread of antiparasitic drug resistance include the intrinsic frequency of the parasitic genetic mutation, the fitness cost of tolerance, the drug selection pressure, drug pharmacokinetic properties and dosage, host immune profile, and transmission profile. There is a list of extrinsic as well as intrinsic factors that contribute to the emergence of resistance. Access and availability of drugs due to economical hindrances, distribution statistics, and reluctance of usage are the general factors of reduced remedial aid. Additionally, complex antimalarial drug regimes are being self-prescribed by individuals with the threat of inadequate adherence. Unregulated pharmaceutical trade can lead to the commercialization and use of counterfeit antiparasitic drugs. Other than the above mentioned extrinsic factors, the intrinsic elements include parasite's cellular, molecular, and clinical features, their species-based innate resistance, drug activity spectrum, and drug response to parasite's stage susceptibility. All these influence drug resistance in the presence of complete therapeutic adherence. Among parasites, the malarial infections are symptomatic and thereby the individuals develop partial immune response (premunition) against these bugs which keeps a check on the resistance spread. Infection-based immune response selectively eliminates the blood parasites including the de novo resistant strains. The immune status of individuals infected with parasite hereby affects the drug efficiency level. Among the drugs, amphotericin B is the most widely used treatment against leishmaniasis. But the developed resistance is species dependent. The vector control and effective case management hence will enable a check on parasitic diseases.

5.6.1 Genetics of Antimalarial Resistance

With increased resistance, the treatment against the parasites slows down causing increased parasitic recrudescence. The fraction of drug-resistant bugs as compared to sensitive parasites drives the spread of resistance determinants. So with increased resistance, the treatment failures augment accelerating the transmission of resistance. The intracellular parasitic drug concentrations are dependent on the genetic composition of the parasite. The rare spontaneous genetic alterations, additions, or mutations in the genes encoding the drug targets or efflux machinery influence the parasite tolerance to drug (Wernsdorfer 1991). Changes in the genes may be linked. Mutations regulate the pathogen's fitness disadvantage independent of the drug exposure. Chromosomal mutations are linked to pathogen fitness which reduces with drug exposure. In *P. falciparum*, chloroquine resistance is developed due to alterations in the gene encoding PfCRT and PfMDR1 transporters (Wellems and Plowe 2001). Cytochrome b (*cytB*) single-base deletions cause atovaquone tolerance. Pyrimethamine resistance is due to alterations in dihydrofolate reductase (dhfr) gene (Cowman et al. 1988). Additionally another factor that modulates spontaneous genetic changes is host immunity. Host immunity takes its own time to reach its peak with all weapons to ward off the immune evasion parasitic strategies. The resistance mechanism is specific to the provision of antiparasitic treatment. For instance, most antimalarial treatments are given in response to the asymptomatic features without dipstick confirmation which reduces the chances of resistance selection. Host defense system restricts the parasitic survival by limiting the gametocyte production due to asexual stage as well as antigametocyte immunity. There are even other mechanisms contributing to the parasitic multigenic tolerance. The var gene encodes the *P. falciparum* erythrocyte membrane protein 1 (PfEMP1) which undergoes alterations during the course of asexual parasitic cycle. This expresses antigenically variant epitopes for immune responses contributing to distinct surface phenotype. These variant subpopulations don't hamper the transmissible densities. The genomic duplication in *Pfmdr* gene is significantly responsible for contributing to *P. falciparum* resistance to mefloquine (Sidhu et al. 2005). *Pfmdr* is also linked to environmental stress responses and codes *Pgh*, an ATP-dependent P-glycoprotein pump. Antifols tolerance is related to stepwise acquisition of chromosomal mutations in *dhfr* gene. The tolerance for the synergistic recipe of sulfonamides and sulfones with antifols can be attributed due to changes in dihydropteroate synthase gene. *PfATPase6* polymorphism confers artemisinin resistance. Limiting spread of resistance can only put a check on global resurgence of parasitic infections. However, a single point mutation isn't the sole contributor of such events, so deep assessment of *Pfdhfr* gene sequences will cater to very limited advantage. A check on the activation of the gametocytogenesis process can help build up therapeutics. Resistance profile augments the gametocyte carriage which if targeted can curb the resistance spread.

Antimalarial Pharmacokinetics The unbound drug in the host plasma contributes to its therapeutic effect. Innumerable factors like parasitic behavioral, metabolic, and molecular attributes influence the drug effects, and even the drug's pharmacokinetic properties affect the pathogen's response to subtherapeutic treatment levels. The degree of absorption and distribution links the drug's bioavailability within host. The therapeutic ratio and oral bioavailability influences the emergence of resistance. The extended half-life during the drug's elimination phase increases the chances of parasitic encounter with selective drug concentrations. For instance, as chloroquine resistance augments, the drug elimination phase gathers insufficient drug amounts, which decreases the selective nature of tolerance. However, prolonged drug exposure extends the selective capacity of parasites in host blood system. The transmission again depends on the host immune profile, drug exposure, parasite duplication profile, and host intracellular processes which influence drug-sensitive pathogens. Repeated drug exposure and parasite subpopulation can contribute to drug resistance. The bar of sensitive parasitic MIC if raised will cause a decrease in the susceptible as well as the selective resistant generations. The balance between de novo resistance and drug elimination phase imposes a discriminatory filter for resistance development and susceptibility inhibition. This selection gives an edge for tolerance in parasites highlighting the elimination phase to be instrumental in conferring antiparasitic resistance. The major challenge is the failure in surpassing phase II clinical trials for antiparasitic drugs. The drug synthesis is economically viable and isn't driven by commercial requirements. Yet the detailed insight into the genomics profile for unraveling the translational formulations for novel drug discovery is still a challenge. Secondly, effective partnership for antiparasitics is lacking mainly because they are basically field driven. Proper resources with high-throughput screening for molecular targets that enable optimization of treatment regime will drive the preclinical setups.

The approaches for novel antiparasitic drug discovery will require different entities for innovative therapeutic formulations. Monitoring the therapeutic resistance by investigating into pathogen's phenotypic susceptibility and cellular and molecular alterations in response to drug and developing molecular probes for limiting tolerance can be one of the strategies for refining the present approaches (Pink et al. 2005). The cost and supply of drugs, the diagnostic methodology, and the design of combinatorial therapies can enhance the efficacy of antiparasitic treatments. The combinations are designed in formulations for increased spectrum of activity with either synergistic or additive effect, reduced toxicity, and dosage requirement and prevention of resistance, for instance, combinations like eflornithine and melarsoprol for trypanosomiasis. The combinational therapy works best for combating parasitic infections particularly malaria. Two or more drugs with identical pharmacokinetic properties and different modes of action can slacken the resistance development. Mostly artemisinin derivatives are potent combinatorial medications against malaria due to their pathogen-killing efficiency, reduced toxic effects, and preventive drug resistance features. In combination with mefloquine, artemisinin leaves the mefloquine "tail" unshielded. This drug "tail" eradication phase edges the sieve for resistance emergence. However, the challenges in combination therapeutics that can cause resistance emergence are insufficient treatment and partial population coverage.

Artemisinins are potent drugs for combating malarial infections. The point of remedial action concerns stage specificity. Additionally, parasite doubling and survival have a greater impact and drug targets can inhibit the further disease progression. Antimalarial drug formulations have also employed resistance reverser mechanism. Another approach will involve refining the conventional drugs for broadening the spectrum of drug action. Such indication for therapeutics had been observed in pneumonia medication, DB289, which is now being clinically tested for malaria as well as African trypanosomiasis (Legros et al. 2002). However, the reluctance of companies to experiment with the risk of testing and toxicity remains an economic challenge. Other strategies involve modulation in drug design for better functioning, like in the case of malarial drugs pyrimethamine whose analogs are being developed to surmount dihydrofolate reductase chromosomal mutations. Antimalarial drug ferroquine bears a chloroquine-like nucleus but with an altered side chain ferrocenic group which lights up its excellent antiparasitic activity against resistant strains (Biot et al. 2005). In vitro tests and molecular approaches ease the parasitic drug monitoring and validation of their therapeutic profile. Chloroquine resistance has been experimented nowadays for producing agents for resistance reversal. Still opportunities for new targets and renewed treatments are being sorted out in the research world today for a ray of anticipation and optimism to combat drug resistance.

6 The Future Ahead

The genesis of drugs/antimicrobial medications gave the world a new hope of survival for fighting the superbugs. Nevertheless, this explosive augmentation of widespread drug usage in the last 20 years has worsened the present global health status by stirring up the tolerance level in microbes. This has also led to the restricted approval of drugs for the present generation of microbial threats. Innumerable complex factors intertwined together contribute to the paucity in the rate of innovative drug development. These multiple forces aren't individually insurmountable, but when combined, their effect imposes a significant and steady proportionate crisis on public welfare with unforeseen consequences. The grave crisis of therapeutic efficacy to combat the microbial resistance mechanisms initiated the introduction of novel strategies and approaches for not only monitoring the drug standards but also limiting the drug tolerance level. This led to the major objective of keeping public health as a priority. Keeping in consideration the past issues, the new generation drugs will employ molecular targets as the line of action against resistance and in vitro tests with molecular markers for drug validation. New drugs specifically new class antibiotics will continue their journey of development for the need of mankind. For instance, medication for multidrug tuberculosis is very much under health requirement. Similarly, *Shigella* outbreaks on a global scale demand cheap antibiotics for oral administration. Generally a huge gap exists between drug development and their clinical approval for worldwide use. This has led to the use of combinational therapies which were found to be effective against many disease outbreaks, malaria being one of them. With a possibility of allergic reactions, these therapies however demanded high-end cost for medications. The problem of generating host-specific "selective pressure" would result in the continued emergence of microbial tolerance. Nevertheless approaches to make sensible usage of drugs along with other essential strategies have decreased the factor of selective pressure. Secondly, early diagnosis might also encourage narrow-range drugs in practice. Two factors are equally important: the duration of drug therapy and the efficacy of the medication. A keen observation on these parameters would open the doors for the blueprints of next-generation drugs keeping the microbial tolerance level in check. Such surveillance systems also consider the frequency of occurrence, persistence, and spread of drug-resistant organisms to make vital public welfare decisions. Major consideration should focus on the prevention of resistance transmission rather than illness cure when time demands. Vaccine development against fatal diseases and infection control strategies can rather maintain a line of limitation for resistance transmission to humans. The greater incidence of drug resistance puts forward a direct link between clinicians and general public health. The present time demands a greater effective approach to combat resistance for the development of ultra-new class of medications against the superbugs of this era. Otherwise there won't be any delay in the commencement of post-microbial era.

6.1 The Major Hurdles: Challenges

The steady pace of time has enabled the surfacing of fatal infectious pathogens like HIV, human metapneumovirus, etc. Such life-threatening diseases have constrained the pharmaceutical industries for the discovery of safe, novel, effective antiviral drugs with not only increased host life but also limitation in pathogen spread. With such emergence, certain new targets have been defined and unfolded lately. This has resulted in the advancement of medications for AIDS patients either individually or in combinational therapies for increased life expectancy. These life-saving drug discoveries have increased the economical pressure on the R&D sections that have directed their investments for new antiviral drugs at the expense of other general antibiotics. The 5-year report in the USA (1998–2003) shows figuratively similar levels of antibacterial and anti-HIV agents to be discovered and approved by pharmaceuticals and FDA, respectively, during that span of time. Hence, antibacterial research has faced unexpectedly decreased productivity due to greater attrition in the discovery of new antibacterial agents. Even hi-tech throughout techniques and molecular modulations into genome research and computational advancements have failed to stand up to the expectations of defined goals of antibacterial development (Bush et al. 2011). This hindrance elevates the barrier of antibacterial requirement and discovery. The design of present era antibiotics requires certain modified criteria that need to be etched onto the discovery panel for withstanding the present-day challenges of multidrug resistance (Walsh 2003). The same problem continues to persist. Ultimately due to other factors of manufacturing defects, efficacy problem, and economic concerns, there happens to be a halt or delay in the pipeline of antibiotic discovery. The world today requires an urgent transformational change in the field of antibiotic development (Wright 2014). The conventional chemical structures can no longer be experimented with for safe and effective drug breakthroughs. A radical change in invention requires an approach from a different perspective targeting new microbial mechanisms and biological aspects for a truly novel platform of antibiotic production. An appropriate novel resource intensive strategy will involve greater time and economy as compared to the approach of alteration to the classical drugs. A partnership among the government, academic, and pharmaceutical units can promote the investments for lucrative antibiotic market discoveries. The major limitation is on the restricted antibiotic use for microbial resistance check. These control measures rationalize drug usage thereby lowering their market value and making investments less striking. The challenge of drug development now concerns demand, market profits, as well as technical superiority with desired potency, apt activity, and necessary safe profile for enduring microbial tolerance.

With an expansion of the pathophysiological mechanisms and molecular targets, the spectrum of drug synthesis has increased the opportunities for business turnover. A defined set of "priorities" can enable better investments for refining public lifestyle. Private organizations pool out better drug discovery as compared to the government units, for instance, a 10-year report enlisted a clean sweep of 93 % antibiotic discovery from private organizations as compared to the 3 % from government and academic bodies. The investments are not only a means for meeting the medical needs but also a profitable turnover payoff to stand up for return investment. The pharmaceutical industries invest around 800 million dollars for drug discovery to approval. This figure has significantly increased around fourfold from the nineteenth century. Conversely the increased manufacture cost with prolonged time of 10–12 years of research for drug development (from formulation to clinical approval), patent span of 20 years from the date of invention, and economic burden on pharmaceutical industries have restrained the companies to channelize their priorities for targeted developments. This has narrowed down the spectrum of anti-infective discoveries and their research programs. Overall the combinatorial effects have led to an alarming diminution in antibiotic development. Some incentives can lure private companies for investment start-up in the anti-infective production field. Nevertheless the problem of drug resistance can only be dealt with if more inputs, collaborations, and research are involved in the academic units and research wings for innovative, effective, and safe medications.

6.1.1 Targeting Resistance with Strategic Approach

A deep understanding of pathogen's survival strategy against the drug will provide insights into the molecular targets for discovery of new class of microbial therapy for combating the problem of resistance. For instance, β-lactam resistance facilitated the target alteration of the warhead at the first shot (Mark et al. 2011). With greater clinical menace, β-lactamase-producing strains were targeted for the hydrolase enzyme by mechanistic-oriented inhibition. Augmentin, the combinational therapy of clavulanate and amoxicillin, formed the façade of antibacterial therapy. Another combinational mix with sulbactam and ampicillin marketed under the name of Unasyn was another essential therapy. Sulbactam was an inhibitor of β-lactamase activity similar to other combinational antibacterial treatments like

Timentin and Zosyn. Modulation of tetracycline and erythromycin structures with alteration in their efflux pump strategies would be appropriate antibacterial therapy. Augmentin is under experimentation with certain additive changes of efflux pump inhibition in combination with macrolides or tetracycline. Even certain semisynthetic analogs of vancomycin have been designed for effective activity against vancomycin-resistant strains. The analogs comprise of a vancosamine biphenyl sugar moiety substitution with higher hydrophobicity and more membrane-oriented and greater activity inhibition between transpeptidases and transglycosylases. Even pristinamycin a combination of two synergistic drugs (quinupristin and dalfopristin) inhibits protein synthetic machinery of vancomycin-resistant pathogens.

Another approach to increase microbial drug susceptibility is to disrupt the outer protective barrier that shields the pathogen from the harsh external environment or to target their efflux pump machinery that expels out undesired products thereby preventing drug accumulation. Genetic alteration in *E. coli* (AcrAB) and *Pseudomonas aeruginosa* (Mex efflux machinery) producing efflux knockout strains compromises with the pathogen's resistance mechanism. The synergy among drug resistance mechanisms enables the efflux systems even to affect non-efflux modes of drug tolerance. The inhibitors of MF and RND transporters in *Pseudomonas* and *E. coli* have been well discussed. With such inhibitors, even *gyrA* and *parC* targets against fluoroquinolone resistance are undermined (Cluzet et al. 2015). Such activity of inhibitors even restricted the pathogen tolerance profile of *H. influenza*, *Klebsiella*, etc. The membrane permeabilization mechanism involves cationic peptides as one of the drug-targeted options. A study in *Pseudomonas* reports the cationic peptides having efflux targets along with β-lactamase directed inhibition to highlight interplay of variant modes of microbial resistance (Gellatly and Hancock 2013).

The contribution of multidrug efflux pumps in conferring intrinsic or acquired resistance mechanism isn't validated. In *E. coli*, about seven genes and nine operons have been studied to regulate the intrinsic efflux mechanism, from which very few contribute to antimicrobial resistance. The roles of transporter pump systems like RND, MF, etc., for antimicrobial resistance in Gram-negative bacteria are still under analysis. Many of the functions of these efflux pumps target the release of internal cellular constituents instead of securing the cell from exogenous toxins. Elucidation of substrate recognition mechanism by efflux transporters is still a challenging aspect. Basically the *DNA array* technique for detection of co-regulated genes with efflux systems in induced strains can explicate the functioning of such pumping machinery (Card et al. 2013, 2014). Secondly, certain *genetic alterations* of mutant generation can result in the accumulation of intracellular metabolites, thereby leading to the functional annotation of many transporters. The genomic profile can provide distinct picture for identification of essential multidrug transporter targets that do contribute to intrinsic resistance or can be a factor for acquired resistance mechanism.

The *structural delineation* of many proteins opens gateways to the understanding of molecular mechanism of action. This is true even for elucidation of multidrug resistance mechanism. The crystal structure of dimeric MarR drug resistance regulator of *E. coli* was solved at 2.3 Å with a DNA-binding domain (Duval et al. 2013). Similar approaches have been employed in deciphering the roles of other transporters that confer resistance mechanism, namely, SmeDEF as well as AcrAB-TolC pump systems. Additionally identification of *tolC* homologue will help to investigate into the mode of antimicrobial resistance in *Vibrio cholerae.*

7 Monitoring Drug Efficacy and Resistance

The four basic methods for monitoring drug efficacy mainly include some in vitro tests, utilization of molecular markers, drug concentration analysis, and detailed drug efficacy studies. Therapeutic effectiveness deals with the direct inspection of drug activity and efficacy over a prolonged period of treatment duration. This clinical standard acts as a blueprint for monitoring subtle alterations consistently for making therapeutic policy outcomes. Following the standard protocol of therapeutic efficacy studies, additional information is also required for the necessary drug characterization and surveillance aspect. Other methods concern the in vitro survival aspect of superbugs and their phenotype depiction, molecular marker experiments regarding genetic knockouts, and drug concentration analysis. The need for a protocol requires a standard test to monitor the in vivo response of the superbugs' resistance causing the disease, for instance, chloroquine resistance against *P. falciparum* (Witkowski et al. 2013). These protocols are modified and revised with the changing drug patterns over time for monitoring the therapeutic proficiency, emergence of resistant strains, and new medications for their treatment. The therapeutic efficacy studies mostly comprise of certain inclusive and exclusive standards, required sample size, assessment parameters, case follow-up strategy, data management and analysis units, ethical committees, and quality check management as the methods of drug evaluation. The treatment is comprised of early medications, clinical failure assessment, and then pathogen response validation. The standard guidelines also involve some national control programs for effective cure policies. However, therapeutic efficacy studies aren't universal methods for monitoring drug remedy parameters. The limitations deal with mostly low sample size of patients for which other surveillance methods and validation techniques (molecular markers/in vitro assessment) can be employed.

7.1 In Vitro Assays

In vitro methods enable to monitor drug efficacy and resistance by evaluating the intrinsic susceptibility of the pathogen to the drugs. The pathogens are given varied drug concentration for generating an optimized and standardized methodology to gauge their sensitivity profile. These studies can complement the epidemiological breakout of diseases. In vitro studies offer a more broad objective perspective to the problem for determining the resistance mechanism. These tests don't consider host factors to puzzle with the outcome which makes them different from therapeutic efficacy protocols. Most importantly, multiple sample tests can be experimented with a range of drug concentrations against the microbes/parasites. Conversely, the innumerable methods and testing approaches question the comparability as well as compatibility of test outcomes. The various tests include certain radioactive-labeled isotope approach followed by antibody-specific ELISA and certain fluorometric assays. Every step presents a new metabolism profile for data quantification. Different laboratories vary with the same set of experimentations. Data presentation in geometric means scales rather percentage profile can solve the issue of standard protocol requirement. In comparison to therapeutic studies, the outcomes are mostly inconsistent due to lack of a stringent protocol and limitation in monitoring the resistance threshold. In vitro studies have huge complexity with a folly in organization of methodologies. These in vitro tests have a greater advantage to study the combinatorial effects of two or more against the conventional ones (Palomino et al. 2014). A synergistic, antagonistic additive profile can enable useful drug combinations for human welfare. Some factors influencing drug trends can be examined through extended time lengths in such studies. With high technical difficulties and significant variability, the present research encourages the invention of other high-throughput assays for detecting drug resistance and efficacy profile.

7.2 *Molecular Markers: Approach for Insights into Drug Resistance*

For determination of drug efficacy and understanding of drug resistance mechanism, molecular markers serve one of the most crucial methods. The major genetic alterations responsible for microbial tolerance once identified can be validated and explained with molecular approach. Serving advantageous over other methods, molecular biology experiments can involve greater sample numbers for detection or analysis within restricted time period and the ease of sample storage and transportation as compared to in vitro tests. The genes potentially involved in conferring resistance to drugs after being detected needs to be screened for identification of the molecular markers. Any point mutation at the targeted gene strengthens the pathogen's tolerance level to therapeutics. The greater the frequency and the higher the number of mutational changes, the greater the degree of resistance. For instance, the parasitic dihydrofolate reductase gene is instrumental in contributing to resistance to antimalarial drugs. Sulfadoxine resistance is conferred due to five specific mutations in dihydrofolate reductase gene (Sharma et al. 2015). Higher degree of resistance is contributed due to genetic alterations at three specific positions of 436, 581, and 613, whereas the other positions 437 and

540 add to some amount of parasitic tolerance. The genetic composition of the microbe and the drug response is controlled by many microbial and host factors including drug pharmacokinetic properties. Cumulative genetic modifications forming variant mutants can also influence pathogen clearance and spread of resistance. Increased copy numbers can even attribute to resistance features in pathogens which hampers medication. Several transporter genes encoding putative sodium hydrogen exchange pumps are associated with low drug response. Molecular detection of drug efficacy provides an early indication for geographical monitoring. Such molecular biology methods are essential for detecting the frequency of mutations with drug introduction or withdrawal. Certain drugs bearing technical difficulties of solubility and suspension can be easily tested with such molecular approach. Molecular markers are a direct predictive approach to test the drug efficacy or therapeutic failures that might have led to the selective emergence of resistant species. However, challenges on method sensitivity still continue to persist due to rising mixed pathogen infections in widespread areas. The tests if sensitive will prevent the camouflage of resistant species and avoid discordant outcomes. Due to high consistency in this approach, greater collaborations between research wings and national programmes are being encouraged.

7.3 Drug Concentration Measurement: A Yardstick for Drug Efficacy

The drug design, invention, and its clinical approval take into consideration various physical and absorptive properties of the therapeutic agent before host administration. Thereby it's very essential for understanding the pharmacological and pharmacokinetic properties of any drug before being metabolized and distributed throughout the host system. Intracellular drug absorption, its interaction, host metabolism of drug, and its elimination from the body will influence the dosage of the medication in accordance to its pharmacokinetic properties. The decision of the drug dosage will be required for enabling proper adaptation to the diversified population. For instance, in antimalarial treatment, the poor host absorption property gives rise to variant blood concentrations as the pharmacodynamic features of the drug vary universally. Majorly monitoring the drug efficacy requires consideration of the success rate of the drug or the treatment failures either due to improper dosage or arising microbial resistance strategy. During treatment failure, the drug amount might be slightly less than the MIC of the proliferating superbugs. A lower therapeutic agent dose implies reappearance of the sensitive pathogen after removal of the administered drug with certain drug modulations. Such situations form the argumentative base of microbial resistance mechanism. The drug dosage thereby becomes a very crucial factor for influencing the pharmacological population kinetics of drug variability. Together with such features, certain software modules can formulate these pharmacokinetic attributes and drug characteristic differences to understand interindividual disparity. For assessing the minimum inhibitory concentrations, research on the drug properties should be intense for further in vivo evaluation. Simple assay method protocols are developed for an approximation of drug exposure. This initiates national control programmes for monitoring the therapeutic agent effectiveness into a routine wide-scale schedule. Additionally, result interpretation would require proper analysis of factors influencing either reduced drug dosage or microbial survival contributing to resistance.

8 Novel Approaches for Drug Development and Resistance Control

The thirst for new drug development with refined microbial targets is under progress. This has led to the development of novel synthetic structures with higher activity range and adequate potency. For instance, the class of oxazolidinones disrupts the protein synthesis by modulating the binding with 23S ribosomal RNA subunit close to the peptidyl transferase junction (Koleva et al. 2015; Jadhavar et al. 2015). Linezolid is an example of clinically approved synthetic oxazolidinones with unmatched potency and selectivity. Another cyclic lipid pentapeptide named ramoplanin equivalent to vancomycin targets substrates of cell wall synthesis. This acts against the vancomycin-resistant *Enterococcus*. The mechanism of drug resistance puts forth two essential queries that require urgent research attention. The first is the approaches that bring up the formulation of new antibiotics and strategies to undermine the resistance development. The first objective necessitates novel approaches for better formulated drugs molecules with efficient delivery and killing mechanism. The second objective requires careful and considerate usage of drug for treating infections.

8.1 Genomics for Anti-Infectives: Quest for Novel Molecular Targets to Circumvent Resistance

The present century now encompasses a huge genomic library database with whole genome sequences of many bacterial pathogens which can further give the researchers a keen insight into the divergent pathogenesis factors (Farhat et al. 2013). Complete sequences of pathogenic strains like *S. aureus*, *Streptococcus* spp., *Salmonella* spp., *M. tuberculosis*, *V. cholerae*, etc., provide a detailed genome profile of the operons regulating the functioning of their antibiotic synthetic mechanism and organization. The gene functional annotation with identical sequences in the database enables identification of targets for mutant library generation. Such functional interruption narrows down the room of choice for targeting the virulence attributes of the pathogen. Such microbial immunomodulatory pathogenic components enable efficient sustenance within host. These defined pooled down targets responsible for forming the pathogen's survival weaponry once optimized can form a database for automated screening and assessment of inhibitors. The hits of the protein targets can be validated for specificity and efficacy. These hits after validation and optimization can undergo in vitro screening process in cell lines followed up by in vivo mechanistic studies. Such strategies offer significant and secured targets for antibiotic generation. Many examples bring up such bioinformatics analysis (Gautam et al. 2016) followed by wet lab validation for greater success rates of drug development. The famous bacterial metallopeptidase, i.e., peptide deformylase, is the sweet target of many potent drugs which mainly blocks the formyl group transfer (Kumari et al. 2013). The major virulence determinants of microbes that are mostly being screened and targeted for drug development involve the secretion system and the signaling cascade mechanism. The bacterial secretion system forms an injection system that punctures into the membrane barrier for transport of virulence proteins modulating with host functioning. The signaling pathway mainly targets either the two-component network system comprising of a sensor kinase and regulatory transcriptional factor or the quorum sensing mechanism involving differential gene expression in conditional responses. Such diverse spectrum of targets opens up the opportunities for development of novel medications with potential antimicrobial activity.

The library development is categorized into more effective synthetic analogs and natural products. The synthetic groups are large, divergent, and modulated structurally with chemical substitutes and functional groups. Some architectural substitutions are a manner to depict the natural formulations. An in situ library approach of combinational therapy is under research which involves the cloning of the entire stretch of the antibiotic biosynthetic operon clustered within 100 kb of the genome especially for polyketides. This gains advantage over nonculturable microbes. In the interim, the multimodular compilation of the polyketide synthase domain facilitates domain programming, module rearrangement, and swap, substitution, and encoding strategies to formulate a new library of transformed polyketides. The same approach has also been for erythromycin derivative generation. The prospective of such process will require defined and targeted objectives of required mutations, active domain substitutions, and domain swap elements that can form a combinational assembly for new formulations. The libraries can involve many tailoring enzymes involved in the multiple steps of antibiotic biosynthesis be it the additive sugars to the macrocyclic lactone scaffold moiety which are active targets of erythromycin or providing alternate deoxy and amino sugars for in vivo studies.

8.2 New Avenues for Increasing Drug Life Span

The emergence of drug resistance mechanism is competing at par in pace with the development of antibiotics. An inevitable interplay between resistance and discovery brings in new challenges either to preserve the efficacy of the present therapeutics or to expand the life of such drugs. This explicit approach had to be central to the present era of resistant bacterial havoc. A check has been imposed for considerate drug prescription for the patients as well as for physicians. Inappropriate and unnecessary antibiotic prescription in developed countries can lead to intermittent therapeutic availability, uncertain effectiveness due to regular use, or chances of self-medication leading to worse consequences. Generally subtherapeutic drug dosage can lead to inhabitation of certain bacterial strains at their dormancy state which cause acute complications during medical crisis. At this juncture, the resistance proves dominant over the infection mechanism. So the major issues are with the carriers of dormant infective strains termed as reservoirs. To conserve the efficacy of drugs as the final resort, a rotation principle of antibiotic usage was implied which led to a devastating scenario in the occurrence of vancomycin-resistant MRSA. Not only swap and switch principle but also combinational primary medications have been approved for life-threatening infections where two units work for neutralizing a specific target, but mixture of variant classes target different activities concomitantly. Examples are Augmentin and Synercid specifically. This combinatorial strategy is a standard approach for anticancer regimes and even for antiretroviral therapy to control cancer and AIDS progression. However, the likelihood of emergence of microbial multidrug tolerance increases. Major cases of drug resistance arise during the extended tenure of treatment therapy (McNairy et al. 2013). Previously there have been reports for widespread use of antibiotics as growth promoters in cattle feed. This results in cross-resistance between different host systems which further reduces the antibiotic life span. A study reported the usage of 1000-fold higher vancomycin for human infections when a vancomycin derivative avoparcin was used for animals in the same year (Walsh 2000). When the *Enterococcus* isolates from those animals were screened, similar five operons encoding for vancomycin resistance were found as in the non-infected human carriers. This led to a complete ban of avoparcin. Similar cases have also been reported due to tolerance mechanism of *Enterococcus* against quinupristin/dalfopristin therapy. This resistance was mostly developed due to the use of virginiamycin in Europe in cattle feed since two decades which led to acetyl transferase effect in animal carriers (Walsh 2000). From then, stringency was employed in the approval and usage of antibacterial compounds in cattle feed. The "waste" in feed would cause "haste" in drug development for deteriorating human health. Overall, the modern molecular and high-throughput approaches can enable screening of microbial genes for candidate targets forming the library of novel synthetic and natural molecules that can be experimented and formulated structurally for better functioning. This will crave new genera of modern antibiotics having effective broad-spectrum activity against the conventional ones. The new age antibiotics will however not hamper the resistance cycle, but haphazard use can affect behavioral changes which would be difficult in achieving with regard to antibiotic value.

9 Conclusion

Drug resistance is one of the greatest concerns of modern science. With grave impacts on survival, the potential biochemical and molecular factors for resistance complexity are still under investigation. The research and understanding of pathogen's fitness cost and tolerance dynamics have evolved new opportunities in clinical field for a greater biological interest. The intrinsic procedures make the reversibility process to be sluggish. The quest and urgency for developing new drugs is in pipeline with strategies to circumvent microbial tolerance. New molecular markers, reduced probability of reversibility, and co-selection of resistance mechanism can be used to exploit the fitness cost for choosing drug targets enabling decent predictions on resistance emergence.

9.1 Key Terms and Definitions

1. A **drug** is a natural or synthetic agent which when ingested into the host system stimulates therapeutic effects for disease treatment or prevention.
2. The term **drug resistance** is the reduction in the drug efficacy that defines the ability of microbes to bear or tolerate the drug (chemical or natural agent) dosage that would otherwise inhibit the growth or kill the pathogen.
3. **Intrinsic resistance** defines the inherent/innate property of the microbe to resist the effect of therapeutics due to evolutionary virtue.
4. **Acquired resistance** is the ability the pathogen obtains to withstand the antimicrobial effect due to exogenous gene transfer/exchange methods or majorly due to genetic mutations.
5. **MDR** abbreviates for multiple drug resistance exhibited by the microbe for its insensitivity to a range of antimicrobials.
6. **Antimicrobials** are agents or drugs (natural or synthetic) that modulate the natural functioning of a microbe either by inhibiting the microbial growth or by killing them. Their primary mode of action against microbes designates their classification as antifungals against fungi or antibiotics against bacteria.
7. **Pharmacokinetics** deals with the fate of the drugs after administration within host system. The kinetics of drugs, their absorption, localization, distribution, metabolism, and elimination are the processes studied in pharmacokinetics.
8. **Antibiotics** are medications used to treat bacterial infections, hence termed as antibacterials, for example, penicillins, cephalosporins, etc.
9. **Antimicrobial susceptibility** defines the sensitivity of a particular bacterium or fungus to the dosage of antimicrobial agent thereby affecting the pathogen's growth or survivability.

References

Ahmed M, Lyass L, Markham PN, Taylor SS, Vazquez-Laslop N, Neyfakh AA (1995) Two highly similar multidrug transporters of Bacillus subtilis whose expression is differentially regulated. J Bacteriol 177:3904–3910

Akalin E, Sehgal V, Ames S, Hossain S, Daly L, Barbara M, Bromberg JS (2003) Cytomegalovirus disease in high-risk transplant recipients despite ganciclovir or valganciclovir prophylaxis. Am J Transplant 3:731–735

Alanis AJ (2005) Resistance to antibiotics: are we in the post-antibiotic era? Arch Med Res 36:697–705. doi:10.1016/j.arcmed.2005.06.009

Alekshun MN, Levy SB (1999) The mar regulon: multiple resistance to antibiotics and other toxic chemicals. Trends Microbiol 7:410–413

Alekshun MN, Levy SB (2007) Molecular mechanisms of antibacterial multidrug resistance. Cell 128:1037–1050. doi:10.1016/j.cell.2007.03.004

Almeida Da Silva PE, Palomino JC (2011) Molecular basis and mechanisms of drug resistance in Mycobacterium tuberculosis: classical and new drugs. J Antimicrob Chemother 66:1417–1430. doi:10.1093/jac/dkr173

Alonso A, Sanchez P, Martinez JL (2001) Environmental selection of antibiotic resistance genes. Environ Microbiol 3:1–9. doi:10.1046/j.1462-2920.2001.00161.x

Anderson JB (2005) Evolution of antifungal-drug resistance: mechanisms and pathogen fitness. Nat Rev Microbiol 3:547–556. doi:10.1038/nrmicro1179

Andersson DI, Hughes D (2010) Antibiotic resistance and its cost: is it possible to reverse resistance? Nat Rev Microbiol 8:260–271. doi:10.1038/nrmicro2319

Andriole VT (1999) Current and future antifungal therapy: new targets for antifungal agents. J Antimicrob Chemother 44:151–162

Arnold SR (2007) Revenge of the killer microbe. CMAJ 177:895–896. doi:10.1503/cmaj.071270

Baker-Austin C, Wright MS, Stepanauskas R, McArthur J (2006) Co-selection of antibiotic and metal resistance. Trends Microbiol 14:176–182. doi:10.1016/j.tim.2006.02.006

Balashov SV, Park S, Perlin DS (2006) Assessing resistance to the echinocandin antifungal drug caspofungin in Candida albicans by profiling mutations in FKS1. Antimicrob Agents Chemother 50:2058–2063. doi:10.1128/AAC.01653-05

Begley DJ (2004) Delivery of therapeutic agents to the central nervous system: the problems and the possibilities. Pharmacol Ther 104:29–45. doi:10.1016/j.pharmthera.2004.08.001

Biot C, Taramelli D, Forfar-Bares I, Maciejewski LA, Boyce M, Nowogrocki G, Egan TJ (2005) Insights into the mechanism of action of ferroquine. Relationship between physicochemical properties and antiplasmodial activity. Mol Pharm 2:185–193. doi:10.1021/mp0500061

Bischoff KM, White DG, Hume ME, Poole TL, Nisbet DJ (2005) The chloramphenicol resistance gene cmlA is disseminated on transferable plasmids that confer multiple-drug resistance in swine Escherichia coli. FEMS Microbiol Lett 243:285–291

Bossche HV (1985) Biochemical targets for antifungal azole derivatives: hypothesis on the mode of action. In: McGinnis MR (ed) Current topics in medical mycology. Springer, New York, pp 313–351. doi:10.1007/978-1-4613-9547-8_12

Boswell G, Buell D, Bekersky I (1998) Am Bisome (liposomal amphotericin B): a comparative review. J Clin Pharmacol 38:583–592. doi:10.1002/j.1552-4604.1998.tb04464.x

Brajtburg J, Powderly WG, Kobayashi GS, Medoff G (1990) Amphotericin B: current understanding of mechanisms of action. Antimicrob Agents Chemother 34:183

Brown DF, Reynolds PE (1980) Intrinsic resistance to β-lactam antibiotics in staphylococcus aureus. FEBS Lett 122:275–278

Brumfitt W, Hamilton-Miller J (1989) Methicillin-resistant Staphylococcus aureus. N Engl J Med 320:1188–1196

Bryskier A (2000) Cephems. Fifty years of continuous research. J Antibiot 53:1028–1037

Bush K, Courvalin P, Dantas G, Davies J, Eisenstein B, Huovinen P, Jacoby GA, Kishony R, Kreiswirth BN, Kutter E, Lerner SA, Levy S, Lewis K, Lomovskaya O, Miller JH,

Mobashery S, Piddock LJ, Projan S, Thomas CM, Tomasz A, Tulkens PM, Walsh TR, Watson JD, Witkowski J, Witte W, Wright G, Yeh P, Zgurskaya HI (2011) Tackling antibiotic resistance. Nat Rev Microbiol 9:894–896. doi:10.1038/nrmicro2693
Cannon RD, Lamping E, Holmes AR, Niimi K, Baret PV, Keniya MV, Tanabe K, Niimi M, Goffeau A, Monk BC (2009) Efflux-mediated antifungal drug resistance. Clin Microbiol Rev 22:291–321. doi:10.1128/CMR.00051-08
Capdeville R, Buchdunger E, Zimmermann J, Matter A (2002) Glivec (STI571, imatinib), a rationally developed, targeted anticancer drug. Nat Rev Drug Discov 1:493–502. doi:10.1038/nrd839
Caponigro F, French RC, Kaye SB (1997) Protein kinase C: a worthwhile target for anticancer drugs? Anticancer Drugs 8:26–33
Card R, Zhang J, Das P, Cook C, Woodford N, Anjum MF (2013) Evaluation of an expanded microarray for detecting antibiotic resistance genes in a broad range of gram-negative bacterial pathogens. Antimicrob Agents Chemother 57:458–465. doi:10.1128/AAC.01223-12
Card RM, Warburton PJ, MacLaren N, Mullany P, Allan E, Anjum MF (2014) Application of microarray and functional-based screening methods for the detection of antimicrobial resistance genes in the microbiomes of healthy humans. PLoS One 9, e86428. doi:10.1371/journal.pone.0086428
Chain E (1979) The early years of the penicillin discovery. Trends Pharmacol Sci 1:6–11. doi:10.1016/0165-6147(79)90004-X
Chen X, Ko LJ, Jayaraman L, Prives C (1996) p53 levels, functional domains, and DNA damage determine the extent of the apoptotic response of tumor cells. Genes Dev 10:2438–2451. doi:10.1101/gad.10.19.2438
Chou S (2008) Cytomegalovirus UL97 mutations in the era of ganciclovir and maribavir. Rev Med Virol 18:233–246. doi:10.1002/rmv.574
Clark FS, Parkinson T, Hitchcock CA, Gow NA (1996) Correlation between rhodamine 123 accumulation and azole sensitivity in Candida species: possible role for drug efflux in drug resistance. Antimicrob Agents Chemother 40:419–425
Clavel F, Hance AJ (2004) HIV drug resistance. N Engl J Med 350:1023–1035. doi:10.1056/NEJMra025195
Cluzet VC, Lautenbach E, Nachamkin I, Cary MS, Fishman NO, Shih NN, Morales KH, Linkin DR (2015) Risk factors for gyrA and parC mutations in Pseudomonas aeruginosa. Infect Control Hosp Epidemiol 36:387–393. doi:10.1017/ice.2014.87
Cohen M, Wong E, Falkow S (1982) Common R-plasmids in Staphylococcus aureus and Staphylococcus epidermidis during a nosocomial Staphylococcus aureus outbreak. Antimicrob Agents Chemother 21:210–215
Cohen SP, Hächler H, Levy S (1993) Genetic and functional analysis of the multiple antibiotic resistance (mar) locus in Escherichia coli. J Bacteriol 175:1484–1492
Control, C. f. D, & Prevention (2009) Oseltamivir-resistant 2009 pandemic influenza A (H1N1) virus infection in two summer campers receiving prophylaxis–North Carolina, 2009. MMWR 58:969
Cowman AF, Morry MJ, Biggs BA, Cross GA, Foote SJ (1988) Amino acid changes linked to pyrimethamine resistance in the dihydrofolate reductase-thymidylate synthase gene of Plasmodium falciparum. Proc Natl Acad Sci U S A 85:9109–9113
Cox G, Wright GD (2013) Intrinsic antibiotic resistance: mechanisms, origins, challenges and solutions. Int J Med Microbiol 303:287–292. doi:10.1016/j.ijmm.2013.02.009
Cuenca-Estrella M, Mellado E, Díaz-Guerra TM, Monzón A, Rodríguez-Tudela JL (2000) Susceptibility of fluconazole-resistant clinical isolates of Candida spp. to echinocandin LY303366, itraconazole and amphotericin B. J Antimicrob Chemother 46:475–477
Dahms RA, Johnson EM, Statz CL, Lee JT, Dunn DL, Beilman GJ (1998) Third-generation cephalosporins and vancomycin as risk factors for postoperative vancomycin-resistant enterococcus infection. Arch Surg 133:1343–1346

Davies J, Davies D (2010) Origins and evolution of antibiotic resistance. [Review]. Microbiol Mol Biol Rev 74:417–433. doi:10.1128/MMBR.00016-10

Davis JM, Navolanic PM, Weinstein-Oppenheimer CR, Steelman LS, Hu W, Konopleva M, Blagosklonny MV, McCubrey JA (2003) Raf-1 and Bcl-2 induce distinct and common pathways that contribute to breast cancer drug resistance. Clin Cancer Res 9:1161–1170

De Rossi E, Ainsa JA, Riccardi G (2006) Role of mycobacterial efflux transporters in drug resistance: an unresolved question. FEMS Microbiol Rev 30:36–52. doi:10.1111/j.1574-6976.2005.00002.x

Dean M, Fojo T, Bates S (2005) Tumour stem cells and drug resistance. Nat Rev Cancer 5:275–284. doi:10.1038/nrc1590

Delcour AH (2009) Outer membrane permeability and antibiotic resistance. Biochim Biophys Acta 1794:808–816. doi:10.1016/j.bbapap.2008.11.005

Donadio S, Maffioli S, Monciardini P, Sosio M, Jabes D (2010) Antibiotic discovery in the twenty-first century: current trends and future perspectives. J Antibiot 63:423–430. doi:10.1038/ja.2010.62

Dougherty TJ, Pucci MJ (2011) Antibiotic discovery and development. Springer Science & Business Media, New York

Douglas C, Marrinan J, Li W, Kurtz M (1994) A Saccharomyces cerevisiae mutant with echinocandin-resistant 1, 3-beta-D-glucan synthase. J Bacteriol 176:5686–5696

Duval V, McMurry LM, Foster K, Head JF, Levy SB (2013) Mutational analysis of the multiple-antibiotic resistance regulator MarR reveals a ligand binding pocket at the interface between the dimerization and DNA binding domains. J Bacteriol 195:3341–3351

Dzidic S, Bedeković V (2003) Horizontal gene transfer-emerging multidrug resistance in hospital bacteria. Acta Pharmacol Sin 24:519–526

Englund JA, Champlin RE, Wyde PR, Kantarjian H, Atmar RL, Tarrand J, Yousuf H, Regnery H, Klimov AI, Cox NJ, Whimbey E (1998) Common emergence of amantadine-and rimantadine-resistant influenza A viruses in symptomatic immunocompromised adults. Clin Infect Dis 26:1418–1424

Farhat MR, Shapiro BJ, Kieser KJ, Sultana R, Jacobson KR, Victor TC, Warren RM, Streicher EM, Calver A, Sloutsky A, Kaur D, Posey JE, Plikaytis B, Oggioni MR, Gardy JL, Johnston JC, Rodrigues M, Tang PKC, Kato-Maeda M, Boroesky M, Muddukrishna B, Kreiswirth BN, Kurepina N, Galagan J, Gagneux S, Birren B, Rubin EJ, Lander ES, Sabeti PC, Murray M (2013) Genomic analysis identifies targets of convergent positive selection in drug-resistant Mycobacterium tuberculosis. Nat Genet 45:1183–1189. doi:10.1038/ng.2747

Fischbach MA, Walsh CT (2009) Antibiotics for emerging pathogens. Science 325:1089–1093. doi:10.1126/science.1176667

Fleming A (1944) The discovery of penicillin. Br Med Bull 2:4–5

Flintoff WF (1989) Acquired drug resistance. Genome 31:447

Frost LS, Leplae R, Summers AO, Toussaint A (2005) Mobile genetic elements: the agents of open source evolution. Nat Rev Microbiol 3:722–732. doi:10.1038/nrmicro1235

Gautam A, Chaudhary K, Kumar R, Gupta S, Singh H, Raghava GP (2016) Managing drug resistance in cancer: role of cancer informatics. Methods Mol Biol 1395:299–312. doi:10.1007/978-1-4939-3347-1_17

Gellatly SL, Hancock RE (2013) Pseudomonas aeruginosa: new insights into pathogenesis and host defenses. Pathog Dis 67:159–173. doi:10.1111/2049-632X.12033

Goldrick B (2002) First reported case of VRSA in the United States: an alarming development in microbial resistance. AJN 102:17

Gottesman MM (2002) Mechanisms of cancer drug resistance. Annu Rev Med 53:615–627

Gottesman MM, Fojo T, Bates SE (2002) Multidrug resistance in cancer: role of ATP–dependent transporters. Nat Rev Cancer 2:48–58

Green DW (2002) The bacterial cell wall as a source of antibacterial targets. Expert Opin Ther Targets 6:1–20. doi:10.1517/14728222.6.1.1

Gunn JS (2008) The Salmonella PmrAB regulon: lipopolysaccharide modifications, antimicrobial peptide resistance and more. Trends Microbiol 16:284–290

Hamilton-Miller J (1972) Sterols from polyene-resistant mutants of Candida albicans. J Gen Microbiol 73:201–203

Hartman BJ, Tomasz A (1984) Low-affinity penicillin-binding protein associated with beta-lactam resistance in Staphylococcus aureus. J Bacteriol 158:513–516

Hayes JD, Wolf CR (1990) Molecular mechanisms of drug resistance. Biochem J 272:281–295. doi:10.1042/bj2720281

Heinic GS, Stevens DA, Greenspan D, MacPhail LA, Dodd CL, Stringari S, Struti WM, Hollander H (1993) Fluconazole-resistant Candida in AIDS patients: report of two cases. Oral Surg Oral Med Oral Pathol 76:711–715. doi:10.1016/0030-4220(93)90039-7

Hitchcock C, Whittle P (1993) Chemistry and mode of action of fluconazole. Clin Dermatol 7:183

Holohan C, Van Schaeybroeck S, Longley DB, Johnston PG (2013) Cancer drug resistance: an evolving paradigm. Nat Rev Cancer 13:714–726. doi:10.1038/nrc3599

Jacoby GA (2009) AmpC β-lactamases. Clin Microbiol Rev 22:161–182

Jadhavar PS, Vaja MD, Dhameliya TM, Chakraborti AK (2015) Oxazolidinones as anti-tubercular agents: discovery, development and future perspectives. Curr Med Chem 22:4379–4397. doi:10.2174/0929867323666151106125759

Jones ME, Sahm DF, Martin N, Scheuring S, Heisig P, Thornsberry C, Köhrer K, Schmitz FJ (2000) Prevalence of gyrA, gyrB, parC, and parE mutations in clinical isolates of Streptococcus pneumoniae with decreased susceptibilities to different fluoroquinolones and originating from worldwide surveillance studies during the 1997–1998 respiratory season. Antimicrob Agents Chemother 44(2):462–466

Joseph-home T, Manning NJ, Hollomon D, Kelly SL (1995) Defective sterol Δ5 (6) desaturase as a cause of azole resistance in Ustilago maydis. FEMS Microbiol Lett 127:29–34

Klein E (2013) Antimalarial drug resistance: a review of the biology and strategies to delay emergence and spread. Int J Antimicrob Agents 41:311–317

Klevens RM, Morrison MA, Nadle J, Petit S, Gershman K, Ray S, Harrison LH, Lynfield R, Dumyati G, Townes JM, Craig AS, Zell ER, Fosheim GE, McDougal LK, Carey RB, Fridkin SK, ABCs MRSA Investigators (2007) Invasive methicillin-resistant Staphylococcus aureus infections in the United States. JAMA 298(15):1763–1771

Klopper M, Warren RM, Hayes C, van Pittius NCG, Streicher EM, Müller B (2013) Emergence and spread of extensively and totally drug-resistant tuberculosis, South Africa. Emerg Infect Dis 19:449–455. doi:10.3201/EID1903.120246

Kobayashi N, Nishino K, Yamaguchi A (2001) Novel macrolide-specific ABC-type efflux transporter in Escherichia coli. J Bacteriol 183:5639–5644

Koleva B, Aggen J, Beuning P (2015) Investigation of the mechanism of action of oxazolidinones. FASEB J 29:575.510

Kumari T, Arora R, Kakkar R (2013) Peptide deformylase inhibitors for addressing the issue of bacterial resistance. Int Rev Biophys Chem 4:19–48

Lacey S, Suzutani T, Powell K, Purifoy D, Honess R (1991) Analysis of mutations in the thymidine kinase genes of drug-resistant varicella-zoster virus populations using the polymerase chain reaction. J Gen Virol 72(3):623–630

Lacombe K, Boyd A, Gozlan J, Lavocat F, Girard PM, Zoulim F (2010) Drug-resistant and immune-escape HBV mutants in HIV-infected hosts. Antivir Ther 15:493–497

Lamb DC, Kelly DE, Schunck W-H, Shyadehi AZ, Akhtar M, Lowe DJ, Baldwin BC, Kelly SL (1997) The mutation T315A in Candida albicans sterol 14α-demethylase causes reduced enzyme activity and fluconazole resistance through reduced affinity. J Biol Chem 272:5682–5688. doi:10.1074/jbc.272.9.5682

Lee JY, Ko KS (2012) OprD mutations and inactivation, expression of efflux pumps and AmpC, and metallo-β-lactamases in carbapenem-resistant Pseudomonas aeruginosa isolates from South Korea. Int J Antimicrob Agents 40:168–172

Legros D, Ollivier G, Gastellu-Etchegorry M, Paquet C, Burri C, Jannin J, Büscher P (2002) Treatment of human African trypanosomiasis—present situation and needs for research and development. Lancet Infect Dis 2:437–440. doi:10.1016/S1473-3099(02)00321-3

Lewis K (1994) Multidrug resistance pumps in bacteria: variations on a theme. Trends Biochem Sci 19:119–123

Ligon BL (2004) Penicillin: its discovery and early development. Paper presented at the seminars in pediatric infectious diseases

Lin YM, Wu SJ, Chang TW, Wang CF, Suen CS, Hwang MJ, Chang MD, Chen YT, Liao YD (2010) Outer membrane protein I of Pseudomonas aeruginosa is a target of cationic antimicrobial peptide/protein. J Biol Chem 285:8985–8994. doi:10.1074/jbc.M109.078725

Lina G, Quaglia A, Reverdy ME, Leclercq R, Vandenesch F, Etienne J (1999) Distribution of genes encoding resistance to macrolides, lincosamides, and streptogramins among staphylococci. Antimicrob Agents Chemother 43:1062–1066

Little JW, Mount DW (1982) The SOS regulatory system of Escherichia coli. Cell 29:11–22. doi:10.1016/0092-8674(82)90085-X

Löffler J, Einsele H, Hebart H, Schumacher U, Hrastnik C, Daum G (2000) Phospholipid and sterol analysis of plasma membranes of azole-resistant *Candida albicans* strains. FEMS Microbiol Lett 185:59–63

Lomovskaya O, Warren MS, Lee A, Galazzo J, Fronko R, Lee M, Blais J, Cho D, Chamberland S, Renau T, Leger R, Hecker S, Watkins W, Hoshino K, Ishida H, Lee VJ (2001) Identification and characterization of inhibitors of multidrug resistance efflux pumps in Pseudomonas aeruginosa: novel agents for combination therapy. Antimicrob Agents Chemother 45:105–116

Lupetti A, Danesi R, Campa M, Del Tacca M, Kelly S (2002) Molecular basis of resistance to azole antifungals. Trends Mol Med 8:76–81

Maligie MA, Selitrennikoff CP (2005) Cryptococcus neoformans resistance to echinocandins: (1, 3) β-glucan synthase activity is sensitive to echinocandins. Antimicrob Agents Chemother 49:2851–2856

Malouin F, Bryan L (1986) Modification of penicillin-binding proteins as mechanisms of beta-lactam resistance. Antimicrob Agents Chemother 30:1–5

Maravic G (2004) Macrolide resistance based on the Erm-mediated rRNA methylation. Curr Drug Targets Infect Disord 4:193–202

Mariam DH, Mengistu Y, Hoffner SE, Andersson DI (2004) Effect of rpoB mutations conferring rifampin resistance on fitness of Mycobacterium tuberculosis. Antimicrob Agents Chemother 48:1289–1294. doi:10.1128/AAC.48.4.1289-1294.2004

Marichal P, Koymans L, Willemsens S, Bellens D, Verhasselt P, Luyten W, Borgers M, Ramaekers FC, Odds FC, Bossche HV (1999) Contribution of mutations in the cytochrome P450 14α-demethylase (Erg11p, Cyp51p) to azole resistance in Candida albicans. Microbiology 145:2701–2713

Mark BL, Vocadlo DJ, Oliver A (2011) Providing β-lactams a helping hand: targeting the AmpC β-lactamase induction pathway. Future Microbiol 6:1415–1427. doi:10.2217/fmb.11.128

Markham PN, Neyfakh AA (2001) Efflux-mediated drug resistance in Gram-positive bacteria. Curr Opin Microbiol 4:509–514. doi:10.1016/S1369-5274(00)00243-5

Marr KA, Lyons CN, Rustad T, Bowden RA, White TC (1998) Rapid, transient fluconazole resistance in Candida albicans is associated with increased mRNA levels of CDR. Antimicrob Agents Chemother 42:2584–2589

Martel CM, Parker JE, Bader O, Weig M, Gross U, Warrilow AG, Rolley N, Kelly DE, Kelly SL (2010) Identification and characterization of four azole-resistant erg3 mutants of Candida albicans. Antimicrob Agents Chemother 54:4527–4533. doi:10.1128/AAC.00348-10

Martin M, Yates J, Hitchcock C (1997) Comparison of voriconazole (UK-109,496) and itraconazole in prevention and treatment of Aspergillus fumigatus endocarditis in guinea pigs. Antimicrob Agents Chemother 41:13–16

Mazodier P, Davies J (1991) Gene transfer between distantly related bacteria. Annu Rev Genet 25:147–171. doi:10.1146/annurev.ge.25.120191.001051

Mbongo N, Loiseau PM, Billion MA, Robert-Gero M (1998) Mechanism of amphotericin B resistance in Leishmania donovani promastigotes. Antimicrob Agents Chemother 42:352–357. doi:10.1128/AAC.00030-11

McCusker KP, Fujimori DG (2012) The chemistry of peptidyl transferase center-targeted antibiotics: enzymatic resistance and approaches to countering resistance. ACS Chem Biol 7:64–72. doi:10.1021/cb200418f

McNairy ML, Cohen M, El-Sadr WM (2013) Antiretroviral therapy for prevention is a combination strategy. Curr HIV/AIDS Rep 10:52–158. doi:10.1007/s11904-013-0152-1

Molzahn S, Woods R (1972) Polyene resistance and the isolation of sterol mutants in Saccharomyces cerevisiae. Microbiology 72:339–348

Montero C, Mateu G, Rodriguez R, Takiff H (2001) Intrinsic resistance of Mycobacterium smegmatis to fluoroquinolones may be influenced by new pentapeptide protein MfpA. Antimicrob Agents Chemother 45:3387–3392. doi:10.1128/AAC.45.12.3387-3392.2001

Morfin F, Thouvenot D (2003) Herpes simplex virus resistance to antiviral drugs. J Clin Virol 26:29–37. doi:10.1016/S1386-6532(02)00263-9

Müller FMC, Weig M, Peter J, Walsh TJ (2000) Azole cross-resistance to ketoconazole, fluconazole, itraconazole and voriconazole in clinical Candida albicans isolates from HIV-infected children with oropharyngeal candidiasis. J Antimicrob Chemother 46:338–341. doi:10.1093/jac/46.2.338

Nakajima Y (1999) Mechanisms of bacterial resistance to macrolide antibiotics. J Infect Chemother 5:61–74. doi:10.1007/s101560050011

Neu HC (1992) The crisis in antibiotic resistance. Science 257:1064–1073. doi:10.1126/science.257.5073.1064

Nikaido H (1996) Multidrug efflux pumps of gram-negative bacteria. J Bacteriol 178:5853–5859

Nikaido H, Zgurskaya HI (2001) AcrAB and related multidrug efflux pumps of Escherichia coli. J Mol Microbiol Biotechnol 3:215–218

Nugier F, Colin J, Aymard M, Langlois M (1992) Occurrence and characterization of acyclovir-resistant herpes simplex virus isolates: report on a two-year sensitivity screening survey. J Med Virol 36:1–12

Odds FC, Brown AJ, Gow NA (2003) Antifungal agents: mechanisms of action. Trends Microbiol 11:272–279. doi:10.1016/S0966-842X(03)00117-3

Okamura H, Yoshida K, Sasaki E, Morimoto H, Haneji T (2004) Transcription factor NF-Y regulates mdr1 expression through binding to inverted CCAAT sequence in drug-resistant human squamous carcinoma cells. Int J Oncol 25:1031–1038. doi:10.3892/ijo.25.4.1031

Orni-Wasserlauf R, Izkhakov E, Siegman-Igra Y, Bash E, Polacheck I, Giladi M (1999) Fluconazole-resistant Cryptococcus neoformans isolated from an immunocompetent patient without prior exposure to fluconazole. Clin Infect Dis 29:1592–1593. doi:10.1086/313546

Palomino JC, Vandamme P, Martin A (2014) Classical and new assays for detecting drug resistance in tuberculosis. Biomark Med 8:1105–1114. doi:10.2217/bmm.14.73

Park J, Morrow CD (1991) Overexpression of the gag-pol precursor from human immunodeficiency virus type 1 proviral genomes results in efficient proteolytic processing in the absence of virion production. J Virol 65:5111–5117

Park JT, Stromistger J (1957) Mode of action of penicillin. Biochemical basis for the mechanism of action of penicillin and for its selective toxicity. American Association for the Advancement of Science. Science 125:99–101

Paulsen IT (2003) Multidrug efflux pumps and resistance: regulation and evolution. Curr Opin Microbiol 6:446–451. doi:10.1016/j.mib.2003.08.005

Pelletier R, Peter J, Antin C, Gonzalez C, Wood L, Walsh TJ (2000) Emergence of resistance of Candida albicans to clotrimazole in human immunodeficiency virus-infected children: in vitro and clinical correlations. J Clin Microbiol 38:1563–1568

Perea S, López-Ribot JL, Kirkpatrick WR, McAtee RK, Santillán RA, Martínez M, Calabrese D, Sanglard D, Patterson TF (2001) Prevalence of molecular mechanisms of resistance to azole antifungal agents in Candida albicans strains displaying high-level fluconazole resistance isolated from human immunodeficiency virus-infected patients. Antimicrob Agents Chemother 45:2676–2684. doi:10.1128/AAC.45.10.2676-2684.2001

Petersen EA, Alling DW, Kirkpatrick CH (1980) Treatment of chronic mucocutaneous candidiasis with ketoconazole: a controlled clinical trial. Ann Intern Med 93:791–795

Pfaller MA, Diekema DJ, Gibbs DL, Newell VA, Ellis D, Tullio V, Rodloff A, Fu W, Ling TA, Global Antifungal Surveillance Group (2010) Results from the ARTEMIS DISK Global Antifungal Surveillance Study, 1997 to 2007: a 10.5-year analysis of susceptibilities of Candida species to fluconazole and voriconazole as determined by CLSI standardized disk diffusion. J Clin Microbiol 48:1366–1377. doi:10.1128/JCM.02117-09

Pfaller M, Boyken L, Hollis R, Messer S, Tendolkar S, Diekema D (2005) In vitro susceptibilities of clinical isolates of Candida species, Cryptococcus neoformans, and Aspergillus species to itraconazole: global survey of 9,359 isolates tested by clinical and laboratory standards institute broth microdilution methods. J Clin Microbiol 43:3807–3810. doi:10.1128/JCM.43.8.3807-3810.2005

Piddock LJ (2006a) Clinically relevant chromosomally encoded multidrug resistance efflux pumps in bacteria. Clin Microbiol Rev 19:382–402. doi:10.1128/CMR.19.2.382-402.2006

Piddock LJ (2006b) Multidrug-resistance efflux pumps? Not just for resistance. Nat Rev Microbiol 4:629–636. doi:10.1038/nrmicro1464

Piddock LJ, Johnson MM, Simjee S, Pumbwe L (2002) Expression of efflux pump gene pmrA in fluoroquinolone-resistant and-susceptible clinical isolates of Streptococcus pneumoniae. Antimicrob Agents Chemother 46:808–812. doi:10.1128/AAC.46.3.808-812.2002

Piekarska K, Rzeczkowska M, Zacharczuk K, Chróst A, Januszkiewicz A, Bareja E, Olak M, Gierczyński R (2011) Prevalence of qnr genes in clinical Enterobacteriaceae non-susceptible to fluoroquinolone in Poland. Med Dosw Mikrobiol 64:211–219

Pink R, Hudson A, Mouriès MA, Bendig M (2005) Opportunities and challenges in antiparasitic drug discovery. Nat Rev Drug Discov 4:727–740. doi:10.1038/nrd1824

Poole K (2005) Efflux-mediated antimicrobial resistance. J Antimicrob Chemother 56:20–51. doi:10.1093/jac/dki171

Quale J, Bratu S, Gupta J, Landman D (2006) Interplay of efflux system, ampC, and oprD expression in carbapenem resistance of Pseudomonas aeruginosa clinical isolates. Antimicrob Agents Chemother 50:1633–1641. doi:10.1128/AAC.50.5.1633-1641.2006

Rammelkamp CH, Maxon T (1942) Resistance of Staphylococcus aureus to the action of penicillin. Exp Biol Med 51:386–389

Reed JC (2003) Apoptosis-targeted therapies for cancer. Cancer Cell 3:17–22. doi:10.1016/S1535-6108(02)00241-6

Richman DD (2006) Antiviral drug resistance. Antiviral Res 71:117–121. doi:10.1016/j.antiviral.2006.03.004

Ring A, Dowsett M (2004) Mechanisms of tamoxifen resistance. Endocr Relat Cancer 11:643–658. doi:10.1677/erc.1.00776

Ross JI, Eady EA, Cove JH, Cunliffe WJ (1998) 16S rRNA mutation associated with tetracycline resistance in a gram-positive bacterium. Antimicrob Agents Chemother 42:1702–1705

Rubin L, Yolken R, Medeiros A, Moxon ER (1981) Ampicillin treatment failure of apparently β-lactamase-negative Haemophilus influenzae type b meningitis due to novel β-lactamase. Lancet 318:1008–1010

Ryder N (1992) Terbinafine: mode of action and properties of the squalene epoxidase inhibition. Br J Dermatol 126:2–7. doi:10.1111/j.1365-2133.1992.tb00001.x

Sabath L, Laverdiere M, Wheeler N, Blazevic D, Wilkinson B (1977) A new type of penicillin resistance of Staphylococcus aureus. Lancet 309:443–447

Sanglard D, Kuchler K, Ischer F, Pagani J, Monod M, Bille J (1995) Mechanisms of resistance to azole antifungal agents in Candida albicans isolates from AIDS patients involve specific multidrug transporters. Antimicrob Agents Chemother 39:2378–2386

Schmitz FJ, Sadurski R, Kray A, Boos M, Geisel R, Köhrer K, Verhoef J, Fluit AC (2000) Prevalence of macrolide-resistance genes in Staphylococcus aureus and Enterococcus faecium

isolates from 24 European university hospitals. J Antimicrob Chemother 45:891–894. doi:10.1093/jac/45.6.891
Schnappinger D, Hillen W (1996) Tetracyclines: antibiotic action, uptake, and resistance mechanisms. Arch Microbiol 165:359–369
Sekiya-Kawasaki M, Abe M, Saka A, Watanabe D, Kono K, Minemura-Asakawa M, Ishihara S, Watanabe T, Ohya Y (2002) Dissection of upstream regulatory components of the Rho1p effector, 1, 3-β-glucan synthase, in Saccharomyces cerevisiae. Genetics 162:663–676
Sharma J, Khan S, Dutta P, Mahanta J (2015) Molecular determination of antifolate resistance associated point mutations in Plasmodium falciparum dihydrofolate reductase (dhfr) and dihydropteroate synthetase (dhps) genes among the field samples in Arunachal Pradesh. J Vector Borne Dis 52:116–121
Siddik ZH (2003) Cisplatin: mode of cytotoxic action and molecular basis of resistance. Oncogene 22:7265–7279. doi:10.1038/sj.onc.1206933
Sidhu ABS, Valderramos SG, Fidock DA (2005) pfmdr1 mutations contribute to quinine resistance and enhance mefloquine and artemisinin sensitivity in Plasmodium falciparum. Mol Microbiol 57:913–926. doi:10.1111/j.1365-2958.2005.04729.x
Sood S, Malhotra M, Das B, Kapil A (2008) Enterococcal infections & antimicrobial resistance. Indian J Med Res 128:111–121
Stavri M, Piddock LJ, Gibbons S (2007) Bacterial efflux pump inhibitors from natural sources. J Antimicrob Chemother 59:1247–1260. doi:10.1093/jac/dkl460
Strasfeld L, Chou S (2010) Antiviral drug resistance: mechanisms and clinical implications. Infect Dis Clin North Am 24:413–437. doi:10.1016/j.idc.2010.01.001
Sung JML, Lindsay JA (2007) Staphylococcus aureus strains that are hypersusceptible to resistance gene transfer from enterococci. Antimicrob Agents Chemother 51:2189–2191
Tacket CO, Dominguez LB, Fisher HJ, Cohen ML (1985) An outbreak of multiple-drug-resistant Salmonella enteritis from raw milk. JAMA 253:2058–2060. doi:10.1001/jama.1985.03350380074024
Thomas CM, Nielsen KM (2005) Mechanisms of, and barriers to, horizontal gene transfer between bacteria. Nat Rev Microbiol 3:711–721. doi:10.1038/nrmicro1234
Tripathi RK, O'Brien RD (1973) Insensitivity of acetylcholinesterase as a factor in resistance of houseflies to the organophosphate Rabon. Pest Biochem Physiol 3:495–498. doi:10.1016/0048-3575(73)90075-8
Vandevelde AG, Mauceri AA, Johnson JE (1972) 5-Fluorocytosine in the treatment of mycotic infections. Ann Inter Med 77:43–50
Veal WL, Nicholas RA, Shafer WM (2002) Overexpression of the MtrC-MtrD-MtrE efflux pump due to an mtrR mutation is required for chromosomally mediated penicillin resistance in Neisseria gonorrhoeae. J Bacteriol 184:5619–5624. doi:10.1128/JB.184.20.5619-5624.2002
Vester B, Douthwaite S (2001) Macrolide resistance conferred by base substitutions in 23S rRNA. Antimicrob Agents Chemother 45:1–12. doi:10.1128/AAC.45.1.1-12.2001
Walsh C (2000) Molecular mechanisms that confer antibacterial drug resistance. Nature 406:775–781. doi:10.1038/35021219
Walsh C (2003) Where will new antibiotics come from? Nat Rev Microbiol 1:65–70. doi:10.1038/nrmicro727
Walsh CT, Fishe SL, Park IS, Prahalad M, Wu PZ (1996) Bacterial resistance to vancomycin: five genes and one missing hydrogen bond tell the story. Chem Biol 3:21–28. doi:10.1016/S1074-5521(96)90079-4
Waxman DJ, Strominger JL (1983) Penicillin-binding proteins and the mechanism of action of beta-lactam antibiotics. Annu Rev Biochem 52:825–869. doi:10.1146/annurev.bi.52.070183.004141
Wellems TE, Plowe CV (2001) Chloroquine-resistant malaria. J Infect Dis 184:770–776. doi:10.1086/322858
Wernsdorfer W (1991) The development and spread of drug-resistant malaria. Parasitol Today 7:297–303. doi:10.1016/0169-4758(91)90262-M

White NJ (2004) Antimalarial drug resistance. J Clin Invest 113:1084–1092. doi:10.1172/jci21682

White TC, Holleman S, Dy F, Mirels LF, Stevens DA (2002) Resistance mechanisms in clinical isolates of Candida albicans. Antimicrob Agents Chemother 46:1704–1713. doi:10.1128/AAC.46.6.1704-1713.2002

Wielders CL, Vriens MR, Brisse S, de Graaf-Miltenburg LA, Troelstra A, Fleer A, Schmitz FJ, Verhoef J, Fluit AC (2001) Evidence for in-vivo transfer of mecA DNA between strains of Staphylococcus aureus. Lancet 357:1674–1675. doi:10.1016/S0140-6736(00)04832-7

Witkowski B, Amaratunga C, Khim N, Sreng S, Chim P, Kim S, Lim P, Mao S, Sopha C, Sam B, Anderson JM, Duong S, Chuor CM, Taylor WR, Suon S, Mercereau-Puijalon O, Fairhurst RM, Menard D (2013) Novel phenotypic assays for the detection of artemisinin-resistant Plasmodium falciparum malaria in Cambodia: in-vitro and ex-vivo drug-response studies. Lancet Infect Dis 13:1043–1049. doi:10.1016/S1473-3099(13)70252-4

Wolfson JS, Hooper DC (1985) The fluoroquinolones: structures, mechanisms of action and resistance, and spectra of activity in vitro. Antimicrob Agents Chemother 28:581–586

Wright GD (2014) Something old, something new: revisiting natural products in antibiotic drug discovery. Can J Microbiol 60:147–154. doi:10.1139/cjm-2014-0063

Wright GD, Poinar H (2012) Antibiotic resistance is ancient: implications for drug discovery. Trends Microbiol 20:157–159. doi:10.1016/j.tim.2012.01.002

Yoneyama H, Katsumata R (2006) Antibiotic resistance in bacteria and its future for novel antibiotic development. Biosci Biotechnol Biochem 70:1060–1075. doi:10.1271/bbb.70.1060

Functional Roles of Highly Conserved Amino Acid Sequence Motifs A and C in Solute Transporters of the Major Facilitator Superfamily

Prathusha Kakarla, Ranjana KC, Ugina Shrestha, Indrika Ranaweera, Mun Mun Mukherjee, T. Mark Willmon, Alberto J. Hernandez, Sharla R. Barr, and Manuel F. Varela

Abstract The biological membrane covers all living cells and provides an effective barrier against the passage of biologically important water-soluble solutes. This natural passage barrier is essentially overcome with the use of integral membrane proteins known as solute transporters. These transport systems translocate solutes across the membrane such as in the case of bacterial drug and multidrug resistance efflux pumps. One of the largest groups of transporters is referred to as the major facilitator superfamily. This group contains secondary active transporters such as symporters and antiporters and passive transporters such as uniporters. The transporters within the major facilitator superfamily share conserved structures and primary amino acid sequences. In particular, several highly conserved amino acid sequence motifs have been discovered and studied extensively, providing substantial evidence for their critical functional roles in the transport of solutes across the membrane.

1 Importance of Solute Transport in Living Organisms

All known living cells are surrounded by a biological membrane that provides an effective barrier against the passage of aqueous-based solutes and ions. Living cells, however, must be able to acquire helpful substances while also extruding harmful ones. Biological membranes solve this barrier problem by using integral membrane proteins that selectively catalyze the acquisition and efflux of helpful

P. Kakarla • R. KC • U. Shrestha • I. Ranaweera • M.M. Mukherjee • T.M. Willmon •
A.J. Hernandez • S.R. Barr • M.F. Varela (✉)
Department of Biology, Eastern New Mexico University, Station 33, Portales, NM 88130, USA
e-mail: Manuel.Varela@enmu.edu

G. Arora et al. (eds.), *Drug Resistance in Bacteria, Fungi, Malaria, and Cancer*,
DOI 10.1007/978-3-319-48683-3_4

and harmful water-soluble molecules, respectively. Therefore, integral membrane solute transporters are important for all life on Earth (Broome-Smith 1999).

When solute transporters are defective, medical disease may occur such as those seen in glucose–galactose malabsorption (Wright et al. 2002), Fanconi–Bickel syndrome (Santer et al. 2002), and De Vivo disease (De Vivo et al. 1991), which are genetic diseases involving impaired transport of glucose across the membranes of cells and develop from inheritable mutations which occur in the genes that encode monosaccharide sugar transporters, thus impairing the uptake of monosaccharides into cells.

Bacteria use solute transporters to efflux multiple antimicrobial agents, often causing loss of chemotherapeutic efficacy during treatment of infectious diseases (Chopra 1992; Kumar and Varela 2013; Li et al. 2015). Solute transporters that multidrug-resistant bacteria use to efflux antimicrobial agents can be grouped into several protein families, such as the ABC (ATP-binding cassette) transporters (Higgins 1992), the resistance-nodulation-cell division (RND) superfamily (Tseng et al. 1999), the small multidrug resistance (SMR) superfamily (Chung and Saier 2001), the multidrug and toxic compound extrusion (MATE) superfamily (Kuroda and Tsuchiya 2009; Kumar et al. 2013), and the major facilitator superfamily (MFS) (Paulsen et al. 1996b; Pao et al. 1998; Saier et al. 1999; Kumar and Varela 2012; Andersen et al. 2015). This review will focus on the antimicrobial agent efflux pumps of the MFS and especially MFS pumps of known structures. Particular attention will be paid to studies which have involved amino acid residues that belong to highly conserved sequence motifs A and C of the MFS (Griffith et al. 1992; Marger and Saier 1993).

2 Acquisition of Helpful Nutrients and Efflux of Harmful Solutes

Substances are routinely transported across biological membranes of living organisms. These substances include an extremely diverse range of water-soluble solutes such as amino acids, Krebs cycle intermediates, sugars, nucleic acids, neurotransmitters, antimicrobial agents, and other small molecules (Henderson et al. 1998). Nutrient uptake via solute transport is a crucial process in which living cells acquire and accumulate molecules from the external environment in order to support metabolism, cell growth, and cell maintenance. On the other hand, living organisms must be able to efflux toxic substances from the inside of their cells into the extracellular milieu in order to maintain growth and survival. Living bacterial cells, for example, have developed integral membrane proteins to facilitate efflux of toxic molecules, a trait that confers antimicrobial resistance (Kumar and Varela 2013).

3 Types of Solute Transporter Systems

Transport systems play important roles in the cellular uptake of helpful molecules such as nutrients, ions, and small molecules and in the exit of harmful or inhibitory molecules. Cellular entry and exit of solutes can occur in two general ways: passive and active transport. Passive transport entails the movement of small molecules across the membrane and does not require biological energy to do so (Mitchell 1967; West and Mitchell 1972). Active transport systems move solutes across the membrane against their own solute concentration gradients (i.e., from low to high concentrations), using integral membrane proteins, called pumps or active transporters. This type of solute transport is referred to as active because of the energy required to conduct transport across the biological membrane (Henderson 1991; Hediger 1994).

3.1 Passive Solute Transport

In passive transport systems, solutes are translocated across the membrane from a side of the membrane with relatively high solute concentration toward the side with relatively low solute concentration, i.e., down the solute concentration gradient (Hediger 1994). The passive solute transport systems generally do not require the expenditure of biological energy. Transport systems use integral membrane carriers to catalyze solute uniport, a facilitative diffusion process that enables a single molecular species to be transported down their concentration gradients (Henderson 1991; Saier 2000).

3.1.1 Facilitated Diffusion

Facilitated diffusion refers to solute transport involving pore- or carrier-forming molecules. In this process, solute reversibly binds to a solute-specific carrier protein that resides integral to the membrane. The complex of solute and carrier oscillates between the inner- and outer-facing surfaces of the biological membrane, thus causing binding and release of the solute to the other side of the same membrane (Henderson 1991).

A special class of integral membrane proteins, called porins, form large nonspecific water-filled channels within the outer membrane to allow the acquisition of nutrients from the periplasm of Gram-negative bacteria. These channels are also associated with the efflux of the waste products (Nikaido 1994). Many so-called classical porins examined so far are OmpC, OmpF, and PhoE from *Escherichia coli* (Nikaido and Vaara 1985; Nikaido 1992). These porins exist as closely associated trimeric complexes that cannot be dissociated even with sodium dodecyl sulfate (SDS), unless heated denatured beforehand (Reid et al. 1988).

These porins show preferences on the basis of solute size and charge. In the case of charge, OmpC and OmpF prefer cations slightly more compared to anions, and PhoE prefers anions. OmpF allows translocation of relatively larger solutes compared to OmpC, showing preferences according to the size of the solute (Nikaido 2003).

3.2 Active Transporter Systems

Two main energy-requiring solute transporter systems, i.e., primary active transport (energized by hydrolysis of ATP) and secondary active transport (energized by ion gradients), are used to efflux biomolecules from bacteria (Mitchell 1966, 1972, 1991, 2011; Harold 2001). Among the dozens of primary and secondary active transporter families, two such superfamilies in particular occur in a ubiquitous manner across all taxonomic categories of living organisms. These systems include a superfamily called the ATP-binding cassette (ABC) transporters and another group called the major facilitator superfamily (MFS) of transporters (Pao et al. 1998; Saier et al. 1999; Davidson and Maloney 2007; Law et al. 2008).

3.2.1 Primary Active Solute Transporters

In primary active transport, the free energy required for solute transport against the electrochemical gradient is provided by the very protein performing the transport. They do so by the hydrolysis of adenosine triphosphate (ATP) (Tarling et al. 2013). Often referred to ABC transporters (Higgins 1992), these primary active transporters represent a large group of integral membrane proteins that couple the transport of a substrate like amino acids, ions, sugars, lipids, and drugs across the membrane (Chang 2003) to the hydrolysis of the phosphate bond between the γ- and the β-phosphate of ATP (ter Beek et al. 2014). It includes both importers and exporters (Locher 2009), bringing nutrients and other molecules into cells or exporting toxins, drugs, and lipids across membranes (Rees et al. 2009). To attain export, ABC transporters use four types of subunits called domains, two transmembrane domains (TMDs) plus two nucleotide binding domains (NBDs). TMDs provide specificity and form the binding sites for ligand, and NBDs undertake ATP hydrolysis to accomplish the translocation across the membrane of its bound solute. However, import requires an additional periplasmic binding domain (PBP) (Linton 2007; Procko et al. 2009). A conformational change in the TMDs occurs once substrate binds, followed by transmission to the NBDs to initiate ATP hydrolysis (Higgins 2001). ABC transporters adopt at least two conformations, i.e., the cis-side or the trans-side. The binding site for the solute is exposed when the transporter is in either one of these two conformations. Alternation between the two conformations allows substrate translocation to occur across the membrane (ter Beek et al. 2014).

3.2.2 Secondary Active Transporters

Secondary active solute transport systems have significant roles in the uptake and efflux of biologically important molecules. Metabolic and bioenergetic systems of organisms convert the energy stored in nutrients during catabolism into an electrochemical energy of protons or sodium ions, generating proton-motive or sodium-motive forces (Mitchell 1967, 1991). These energies are then used to drive biological work such as the translocation of solutes across the membrane against their concentration gradients to accumulate solute on one side of the membrane (Poolman and Konings 1993; Krämer 1994; Wilson and Ding 2001). In the chemiosmosis mode of biological energy generation during respiration and fermentation, light, chemical, or redox energies are converted to electrochemical energies, which in turn are used to drive other biological work. This bioenergetic process takes place by coupling biochemical reactions to the transport of solutes, ions, and other small molecules across the cell and plasma membranes. In bacteria, protons, and sodium are the coupling ions that are used during energy transduction (Krämer 1994).

4 The Major Facilitator Superfamily

The MFS has become an extremely well-studied and important compilation of solute transporters across all taxa of living organisms (Maloney 1994; Paulsen et al. 1996b; Saier et al. 1999; Pao et al. 1998; Law et al. 2008). The substrates or solutes of these MFS transporters are extremely diverse and include structurally distinct small molecules like sugars, amino acids, intermediary metabolites, nucleic acids, antimicrobial agents, and ions. To date, the MFS encompasses thousands of members conveniently stored and organized in a well-maintained database called the Transporter Classification Database (TCD) www.tcdb.org (Saier et al. 2014), which currently includes well over 15,000 proteins of the MFS (Saier et al. 2014).

4.1 Discovery of the MFS

As integral membrane solute transporters were refractory to isolation and purification by traditional biochemical approaches, making their study difficult, molecular biological approaches became available and, thus, quite useful in the cloning of the genes that encoded solute transporters (Teather et al. 1978). Gene cloning, in turn, allowed almost the immediate determination of the nucleotide sequences encoding solute transporters (Büchel et al. 1980). Soon after the cloning and DNA sequence determinations of additional genes that encoded solute transporters became available, a remarkable discovery was made by Henderson and colleagues in which comparison of the sequences between several sugar transporters from prokaryotic and eukaryotic organisms demonstrated that these seemingly distinct proteins were in fact homologous (Maiden et al. 1987), indicating a shared or common evolutionary origin. As many more transporter gene sequences were determined and compared, investigators began to compile these transporters in families and superfamilies, referred to initially as the transporter superfamily (TSF) (Henderson 1993), the uniporter–symporter antiporter (USA) family (Goswitz and Brooker 1995), and the generally accepted term major facilitator superfamily (MFS) (Marger and Saier 1993).

4.2 General Features of the MFS

These transporter members of the MFS include *(a)* uniporters, which catalyze facilitated diffusion of solute across the membrane down their solute concentration gradients; *(b)* symporters, which catalyze ion-driven secondary active transport of solutes in the same directions across the biological membrane; and *(c)* antiporters, which catalyze ion-driven secondary active solute transport across the membrane in opposite directions (Mitchell 1991). These transporters have on average between approximately 400 and 600 amino acids along their polypeptide chains (Pao et al. 1998; Law et al. 2008).

The MFS transporters catalyze the translocation of water-soluble solutes across the membrane using the energy stored in chemiosmotic ion gradients (Marger and Saier 1993). The ions, for instance, are either protons (i.e., H^+) or sodium (i.e., Na^+), and their gradients across the membrane are formed by the respiratory chain during catabolism of nutrients (Mitchell 1991; Harold 2001). The substrate will accumulate extracellularly in an energy-dependent fashion. Thus, these substrate/H^+ antiport (efflux) systems allow all cells, including bacteria, to survive and grow while in the presence of potentially inhibitory molecules. Therefore, these biomolecule efflux systems allow bacteria to tolerate unusually high concentrations of potentially lethal molecules, such as antimicrobial agents, heavy metals, industrial waste molecules, etc. An interesting and unique property of several MFS efflux systems is that they have the ability to transport multiple structurally different substrates (Levy 1992, 2002; Lewis 1994; Piddock 2006). Also known as uniporter–symporter–antiporter superfamily (Goswitz and Brooker 1995), members include both passive and secondary active transport systems.

4.3 Key Secondary Active Transporters of the MFS

The energy of ion gradients drives solute transport across the membrane during secondary active solute transport. Many of the solute transporters that are members of the MFS use these particular types of ion gradient energies for the cellular uptake and efflux of solutes (Poolman and Konings 1993; Krämer 1994; Kumar and Varela 2013). The term symport is used to describe the co-transport movement of solute and ion in the same direction across the cell or plasma membrane; that is, ion translocation down its gradient drives solute transport up its gradient. On the other hand, the term antiport is used to describe the co-transport of solute and its driving ion in the opposite directions across the same types of biological membranes; again, the ion moves down its concentration gradient to mediate solute transport against its own gradient. In both of these symport and antiport systems, the transported solute accumulates on one side of the membrane (Saier 2000).

The lactose permease, LacY, a secondary active transporter from *E. coli*, has been studied in the laboratories of Brooker (Brooker 1990), Kaback (Guan and Kaback 2006), and Wilson (Varela and Wilson 1996) and is considered to be a useful model system for investigation of newer transport systems of the major facilitator superfamily, such as novel multidrug efflux pumps (Floyd et al. 2013). LacY was originally described as an important component of the well-known *lac* operon and is encoded by *lacY*, a regulated structural gene contained within operon itself (Müller-Hill 1996; Varela and Wilson 1996). Using protons, the LacY symporter transports lactose and other related sugars across the inner membrane, and it uses the energy of the electrochemical gradient of protons to couple this movement of sugar and proton symport. This causes sugar to accumulate against a concentration (Mitchell 1967, 1991; Varela and Wilson 1996).

EmrD is a proton-dependent multidrug efflux pump of *E. coli* that belongs to MFS family (Sulavik et al. 2001). EmrD transports detergents, such as benzalkonium chloride and sodium dodecyl sulfate (Nishino and Yamaguchi 2001). Not only does it confer resistance to detergents, the EmrD efflux pump influences the formation of biofilm (Matsumura et al. 2011). The X-ray crystal structure of EmrD exhibits hydrophobic interiors which is a means for transporting various substrates in the drug efflux mechanism. An additional area consisting of two long helical regions that are located on cytoplasmic side can provide additional substrate specificity and transport (Yin et al. 2006).

TetA(B) is the most extensively studied efflux pump of the MFS family, members of which transport sugar, intermediate metabolites, and drugs (Buivydas and Daugelavièius 2006). The gene has been encoded on transposon Tn10 and represents a metal–tetracycline/H^+ antiporter (Tamura et al. 2003). The efflux of tetracycline from bacteria is driven by a proton gradient as the driving force (Kaneko et al. 1985). The presence of TetA(B) in *Bacillus cereus* represents the transfer of the antibiotic resistance genes from other bacteria (Rather et al. 2012). This efflux pump actively expels tetracycline by a membrane-associated protein, resulting in the reduction in the accumulation of tetracycline (Levy 1992; Nelson and Levy 2011).

The bacterial pathogen *S. aureus* harbors many antimicrobial agent efflux pumps that are members of the MFS of transporters, and several are well studied (Hooper 2000; Brown and Skurray 2001; Costa et al. 2013; Andersen et al. 2015). One of the most intensively studied is QacA (Brown and Skurray 2001; Saidijam et al. 2006), a plasmid-encoded multidrug pump that confers resistance to multiple antiseptics, diamidines, and dyes (Tennent et al. 1989). The deduced sequence shows

514 residues, and QacA is the first MFS discovered to have 14 TMS instead of 12 as has previously been observed in other superfamily members. The 14-transmembrane domain topology was supported by fusion studies of QacA with enzymatic reporters (Paulsen et al. 1996a). Presently, many MFS efflux pumps have the 14 TMS motif (Saidijam et al. 2006). QacA transports ethidium bromide using the proton gradient as the driving force (Littlejohn et al. 1992).

Another MFS efflux pump for multiple structurally distinct antimicrobial agents is NorA of *S. aureus* (Ubukata et al. 1989; Yoshida et al. 1990). NorA has 388 amino acid residues and 12 predicted transmembrane segments (Yoshida et al. 1990). Originally discovered in a clinical isolate (Ubukata et al. 1989), NorA was thought to be a single-drug efflux pump for the antimicrobial agent norfloxacin. NorA is now well known to be a multidrug transporter (Neyfakh et al. 1993) which is closely related to Bmr from *Bacillus subtilis* (Neyfakh 1992). Physiological studies show that NorA transports structurally different antimicrobial agents like the fluoroquinolones (e.g., ciprofloxacin and norfloxacin), dyes (e.g., rhodamine and ethidium), and quaternary ammonium compounds (e.g., benzalkonium chloride and tetraphenylphosphonium) (Yoshida et al. 1990; Kaatz et al. 1993; Neyfakh et al. 1993; Kaatz and Seo 1995). Recent primary studies of NorA have emphasized on efflux pump inhibitors of NorA (Holler et al. 2012a, b; Kalia et al. 2012; Roy et al. 2013; Shiu et al. 2013; Thai et al. 2015) and regulation of NorA expression (Fournier et al. 2000, 2001; Truong-Bolduc et al. 2003, 2005; Kosmidis et al. 2010; Deng et al. 2012), both topics of which are beyond the scope of this review but have been reviewed elsewhere (Zhang and Ma 2010; Costa et al. 2013).

The protein MdeA from *S. aureus* is predicted to have 479 amino acids, 14 transmembrane domains (Huang et al. 2004; Yamada et al. 2006), and transport Hoechst 33342 and ethidium bromide (Yamada et al. 2006). Predictions also indicate that MdeA confers resistance to tetraphenylphosphonium chloride, norfloxacin, rhodamine 6G, doxorubicin, and daunorubicin (Yamada et al. 2006; Huang et al. 2004). The MdeA efflux pumps of *S. aureus* N315 (Yamada et al. 2006) and *S. aureus* Buttle (Huang et al. 2004) are 99 % identical, differing at five key residues and likely explaining why MdeA from *S. aureus* Buttle confers resistance to benzalkonium chloride while MdeA from *S. aureus* N315 does not. Additionally, it was shown that piperine inhibits MdeA transport activity and potentiates the effects of the antimicrobial agent mupirocin (Mirza et al. 2011).

A more recently discovered multidrug efflux pump, LmrS, encoded on the chromosome and cloned from a clinical isolate of a methicillin-resistant *S. aureus* (MRSA) strain, actively transports ethidium bromide and confers resistance to structurally dissimilar substrates, such as linezolid, lincomycin, tetraphenylphosphonium chloride, chloramphenicol, erythromycin, florfenicol, fusidic acid, gatifloxacin, kanamycin, oxytetracycline, streptomycin, and trimethoprim (Floyd et al. 2010). The LmrS multidrug efflux pump is predicted to harbor 14 transmembrane domains, which is identical to that predicted for QacA (Paulsen et al. 1996a; Floyd et al. 2010). Furthermore, LmrS shares homology with LmrB of

B. subtilis (Kumano et al. 1997), VceB from *V. cholerae* (Colmer et al. 1998), and EmrB from *E. coli* (Lomovskaya and Lewis 1992).

4.4 Structures of MFS Transporters

Generally, these MFS transporters contain 12 (Fig. 1) or 14 transmembrane-spanning domains (TMS), with an occasional duplication of two 12 TMS to constitute 24 TMS transporters (Moir and Wood 2001; Hirai et al. 2003; Saidijam et al. 2006). Thus far, high-resolution crystal structures have been elucidated for more than a dozen of these MFS transporters. These known MFS protein crystal structures include the multiple drug efflux pump, EmrD, from *E. coli* (Yin et al. 2006); the fucose transporter, FucP, from *E. coli* (Dang et al. 2010); the glucose–H^+ symporter, $GlcP_{Sc}$, from *Staphylococcus epidermidis* (Iancu et al. 2013); the glycerol-3-phosphate transport protein, GlpT, from *E. coli* (Huang et al. 2003); the glucose transporter, GLUT1, from *Homo sapiens* (Sun et al. 2012); the lactose–proton symporter, LacY, from *E. coli* (Abramson et al. 2003); the nitrate/nitrite exchange transporter, NarK, from *E. coli* (Zheng et al. 2013); the nitrate/nitrite antiport protein, NarU, from *E. coli* (Yan et al. 2013); the oligopeptide–H^+ symport protein, $PepT_{So}$, from *Shewanella oneidensis* (Newstead et al. 2011); the phosphate transport protein, PipT, from *Piriformospora indica* (Pedersen et al. 2013); the xylose transporter, XylE, from *E. coli* (Sun et al. 2012); the multidrug transporter, YajR, from *E. coli* (Jiang et al. 2013); the peptide transport protein, YbgH, from *E. coli* (Zhao et al. 2014); the multiple drug efflux pump, MdfA, from *E. coli* (Heng et al. 2015); and, more recently, the mammalian fructose transporter, GLUT5, from *Rattus norvegicus* and *Bos taurus* (Nomura et al. 2015).

Thus far, these high-resolution protein structures support the general notion that the MFS transporters harbor two structurally symmetrical and functionally asymmetrical bundles or domains (Pao et al. 1998; Saier et al. 1999) composed of the first half (N-terminus) 6 TMDs and second half (C-terminus) 6 TMDs, at least for the 12-TMD solute transporters, which is not surprising given the early observation that the two halves of the modern MFS transporter likely arose from an internal sequence duplication and subsequent tandem repeat of a common ancestor with

Fig. 1 Two-dimensional topology model of an MFS transporter

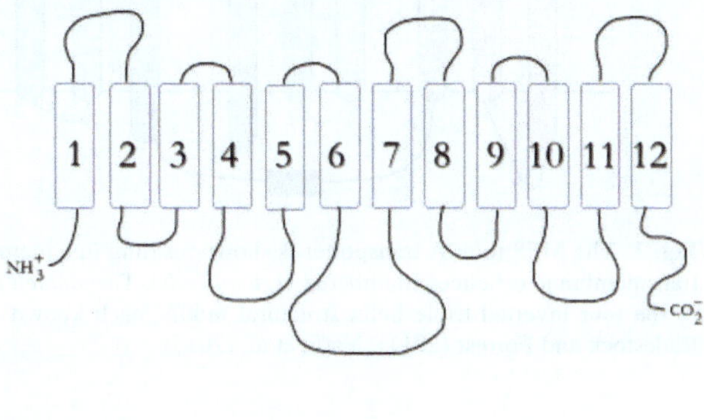

6 TMDs (Griffith et al. 1992). Another feature apparently common to the known crystal structures of the MFS transporters is the presence of a large central aqueous cavity formed by the two halves, supporting previous genetic analyses of the tetracycline efflux pump, TetA(C), where the N- and C-termini bundles or domains interact functionally (McNicholas et al. 1992, 1995), plus low-resolution structural data for the oxalate transporter, OxyT (Heymann et al. 2001, 2003), and Mitchell's notion of a proton gradient as an energy source for driving solute transport across the membrane (Mitchell 1977, 1991). Considering how these structural features related to the mechanism by which solute is translocated across the membrane, the so-called alternating access mechanism has been invoked to explain this important biological process in which the substrate binding site alternately faces one or the other sides of the membrane (Jencks 1980; West 1980, 1997; Tanford 1982). In principle, the substrate binding site of the MFS transporter faces one side of the biological membrane and then upon binding of the substrate orients itself via a conformational change such that the substrate binding site faces the other side to facilitate transport (Henderson 1991; Law et al. 2008), and these MFS transporters, in general, use their flexible gating structures to form inward- or outward-facing states that are occluded in order to prevent unwanted leakage and dissipation of the ion gradients (Stelzl et al. 2014). As shown in Fig. 2, intrinsic in the conserved structure is the so-called MFS fold consisting of inverted triple helices that are repeated four times to form four 3-helix inverted-topology repeats that make up the MFS fold in MFS transporters (Radestock and Forrest 2011).

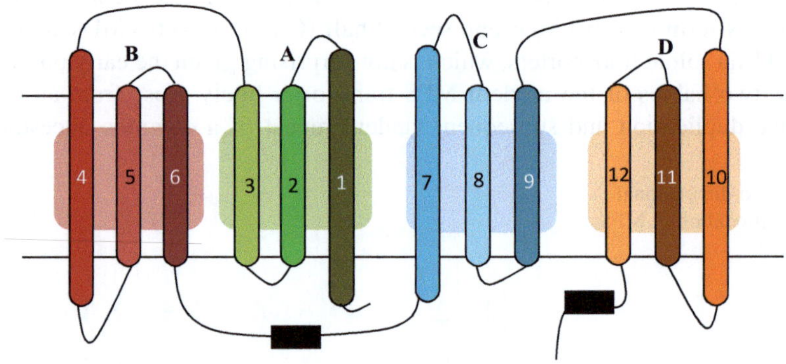

Fig. 2 The MFS fold. A transporter is shown residing in a membrane (*horizontal lines*) with the transmembrane α-helices (numbered *vertical rods*). The *shaded rectangles* A, B, C, and D depict of the four inverted triple helix structural motifs, each known as the MFS fold. Adapted from Radestock and Forrest (2011), Yaffe et al. (2013)

5 Evolutionarily Conserved Sequence Motifs Involving Amino Acid Sequences in Transporters of the MFS

Early studies that discovered the high degree of relatedness between members of the MFS also definitively demonstrated their shared evolutionary conservation of certain amino acid sequences (Fig. 3) (Henderson 1990a, b; Rouch et al. 1990; Griffith et al. 1992; Henderson et al. 1993). These investigators further discovered that members of the MFS shared similar hydrophobicity profiles and similar predicted secondary structures (i.e., 12 or 14 TMDs), suggesting that these family members share conserved three-dimensional structures and, thus, a common ancestral origin. Taken together, these findings suggested that the MFS transporters share a common solute transport mechanism, independent of the transporters' substrate specificities and modes of energy (Henderson and Maiden 1990; Rouch et al. 1990; Griffith et al. 1992; Marger and Saier 1993; Pao et al. 1998; Saier et al. 1998, 1999).

6 Motif A "G X X X D R/K X G R R/K" and Functional Roles

This highly conserved amino acid residue sequence motif from the MFS was discovered by Henderson and coauthors in 1987 (Maiden et al. 1987; Henderson and Maiden 1990). Now known as Motif A, it is widely accepted that elements of this motif reside in a hydrophilic loop between helices 2 and 3 of virtually all transporters of the MFS (Griffith et al. 1992; Pao et al. 1998; Saier et al. 1999; Kumar and Varela 2012; Andersen et al. 2015; see Fig. 3a). Hence, the functional importance of this motif cannot be understated. Perhaps the earliest clues to the importance of residues in Motif A arose well before it was established that elements in this protein region were conserved. First, in a series of studies working with lactose permease, LacY, a key transporter first purified from *E. coli* by Newman and Wilson (Newman and Wilson 1980), truncated LacY protein fragments were later generated by limited proteolysis and deletion mutation analyses by the laboratory of Ehring and colleagues, who found that residues of the N-terminal region where Motif A resides must be important for lactose transport across the membrane (Stochaj et al. 1986, 1988; Stochaj and Ehring 1987). Subsequent follow-up studies were conducted in which co-expression of inactive truncated nonoverlapping LacY fragments functionally complemented each other, restoring active lactose transport, thus further demonstrating the important functional roles of N-terminal residues (Wrubel et al. 1990, 1994).

A

B

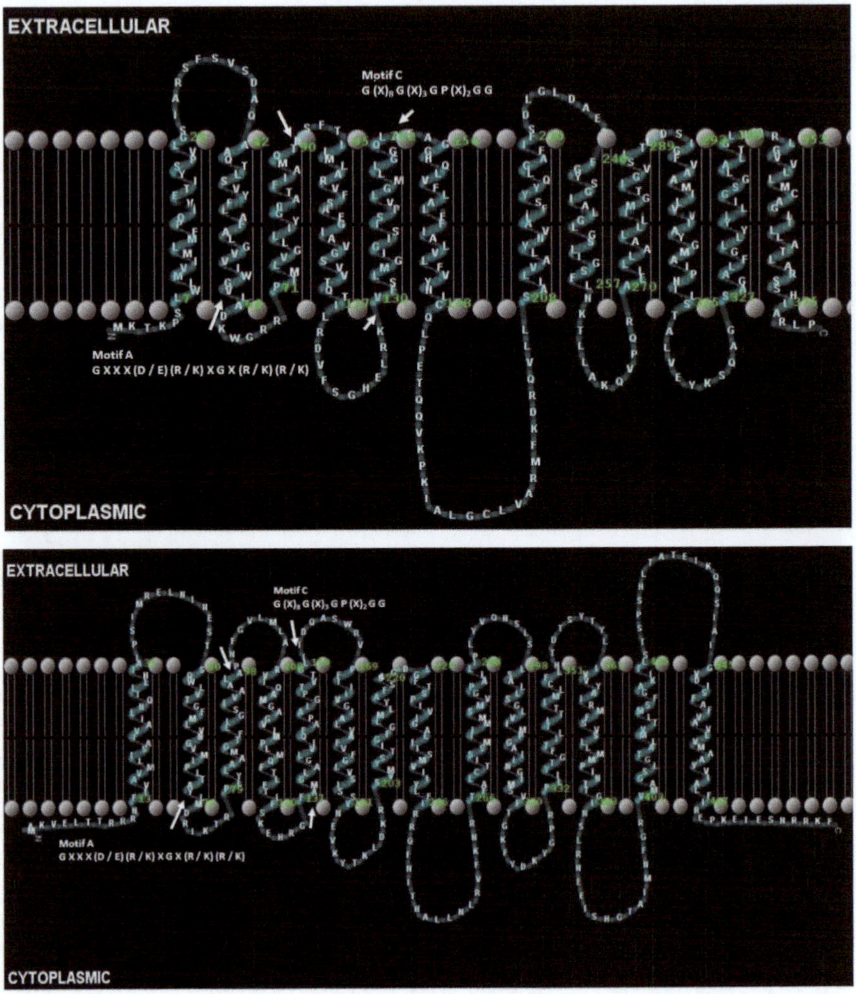

Fig. 3 Highly conserved sequence motifs A and C in 12-TMS and 14-TMS MFS transporters. Figure (**a**) indicates 12 different transmembrane helices joined together by loops. The *white arrows* point to conserved motif A [G X X X (D/E)(R/K) X G X (R/K)(R/K)] and motif C [G $(X)_8$ G $(X)_3$ G P $(X)_2$ G G] of the multidrug efflux pump EmrD-3 (Smith et al. 2009; Floyd et al. 2010) from the microorganism *Vibrio cholerae*, a pathogenic bacterium. Figure (**b**) indicates 14 different transmembrane helices joined together by intra-helical loops. The *white arrows* point to conserved motif A [G X X X (D/E)(R/K) X G X (R/K)(R/K)] and motif C [G $(X)_8$ G $(X)_3$ G P $(X)_2$ G G] in the multidrug efflux pump LmrS from the bacterial pathogen *Staphylococcus aureus*. These figures were generated using TMHMM and Tmpres2D servers

6.1 Early Studies of Motif A

Perhaps, the first site-directed mutational analysis of individual amino acid residues of Motif A in an MFS transporter was conducted by the laboratory of Yamaguchi

(Yamaguchi et al. 1990). The Ser-65–Asp-66 dipeptide of the motif was closely examined (Yamaguchi et al. 1990) in the Tn10 TetA(B) tetracycline efflux pump, which was discovered in the laboratory of Levy (McMurry et al. 1980). Because replacements at position Ser-65 but not at Asp-66 in the Motif A of TetA (B) showed some transport activity, it was concluded that a negative charge and the loop were both necessary for gating but not for substrate binding in the channel (Yamaguchi et al. 1990). The possibility remained, however, that the residues in the loop between helices 2 and 3 did participate in initial substrate binding, as previously postulated (Chopra 1986), as later studies involving Cys-scanning mutagenesis showed that residues in helix 3 (Asp-84) and elements of Motif A (Gly-62, Asp-66, Arg-70, and Ser-77) were also implicated in forming a tetracycline transport pathway and further interpreted as together undergoing conformational changes during transport (Yamaguchi et al. 1993a; Kimura et al. 1998b). The importance of the conserved Asp residue at this locus in TetA(B) was confirmed also in KgtP, an α-ketoglutarate permease (Seol and Shatkin 1992), and TetA(C), a plasmid-encoded tetracycline efflux pump from *E. coli* (McNicholas et al. 1992). Follow-up studies from the Yamaguchi laboratory systematically investigated the rest of the residues in Motif A of TetA(B) and found that only the Asp and Arg residues of the Motif A in the loop 2-3 were essential for tetracycline transport (Yamaguchi et al. 1992a, b), further solidifying the notion that the conserved loop structure participated in a gating function, as previously postulated (Baker and Widdas 1973), while the two Gly residues of the motif were interpreted to function in the formation of a supportive structure in order to stabilize a β-turn in the conserved loop (Yamaguchi et al. 1993b). In a study evaluating the functional roles of Arg residues of TetA(B), Arg-67, Arg-70, and Arg-71, all belong to Motif A, only replacements for Arg-70 lost both tetracycline resistance and transport (Kimura et al. 1998a). Along these lines, a defective primary mutation in TetA (B), in which Asp-66 changed to a Cys, was suppressed by a second-site mutation where Ala-40 was also changed to Asp, supporting the notion that a charged residue is an important requirement for transport (Yamaguchi et al. 1995). Similarly, a defective mutation in which Gly-62 of Motif A was changed to Leu was compensated for by a second-site mutation on the other side of the same membrane in which Leu-30 was changed to a Ser residue, and the authors interpreted this finding as the double mutation providing a "conformational hook" that blocks deleterious conformational changes at a remote location elsewhere in the protein (Kimura et al. 1997). A similar so-called remote conformational suppression effect was observed later when the primary mutation in Motif A in which Gly-62 changed to Leu in TetA(B) was suppressed by the second-site mutation where Ala-354, also on the other side of the cytoplasmic membrane, was changed to Asp (Kawabe and Yamaguchi 1999). This latter effect was interpreted as TetA(B) having a close structural proximity between helices 2 and 11 on the periplasmic side of the cytoplasmic membrane (Kawabe and Yamaguchi 1999). The seminal discovery of salt bridges in the *E. coli* lactose permease, LacY, by the Wilson laboratory, reviewed in ref Varela and Wilson (1996) and see Lee et al. (1996), prompted an evaluation of possible salt bridges in TetA(B) in which Arg-70 of the Motif A was found to interact with Asp-120, which resides at the distal end of helix 4 (Someya et al. 2000). Similarly, using molecular simulation dynamics of the proton-coupled oligopeptide symporters $PepT_{So}$ from *Shewanella oneidensis* and $PepT_{St}$ from *Streptococcus thermophilus*, a salt bridge involving a Motif A residue, Asp-79, was predicted to form with Lys-84 which resides near helix 3 (Fowler et al. 2015). This salt bridge was further predicted to stabilize the outward-facing conformation of $PepT_{So}$, thus potentially participating in the gating topology of symporters in this closely related family (Fowler et al. 2015). In a separate study of the TetA(P) efflux pump for tetracycline from *Clostridium perfringens*, the site-directed mutations at Pro-61 and Arg-71 abolished tetracycline resistance levels (Bannam et al. 2004).

6.2 More Recent Studies of Motif A

Interestingly, a human glucose transporter, GLUT-1, expressed in red blood cells, was studied in patients with GLUT-1 deficiency syndrome, and mutations were found in elements of Motif A: Gly-91 changed to Asp and Arg-93 changed to Gln or Trp (Pascual et al. 2008). These mutations showed reduced glucose transport, and it was concluded from these findings that Gly-91 may be important for substrate docking within the recognition site and that Arg-93 may serve to help anchor GLUT-1 to the membrane (Pascual et al. 2008). Additionally, a study of autosomal dominant missense mutations showed that alteration of the Motif A residue Gly-91 to either Asp or Ala in GLUT1 from *Homo sapiens*, when expressed *Xenopus* oocytes, had severely reduced glucose transport activities (Klepper et al. 2001). In a separate study involving another eukaryotic organism, the fungus *Aspergillus nidulans*, various mutations in the high-affinity nitrate transporter, NrtA, were isolated (Kinghorn et al. 2005). Of this set of mutations, residues of Motif A were altered in which Cys-90 was changed to Phe and Gly-91 was changed to Ser, and both mutants showed reduced nitrate uptake compared to wild-type NrtA (Kinghorn et al. 2005).

The internal duplication event postulated to occur for MFS transporters (Henderson and Maiden 1990; Griffith et al. 1992), particularly the tetracycline efflux pumps (Rubin et al. 1990), prompted the evaluation of the residues of the loop between helices 8 and 9 of TetA(B) (Yamaguchi et al. 1993b). In this analysis, only Gly-273 of TetA(B) in the second loop between helices 8 and 9 was demonstrated to be essential for tetracycline transport (Yamaguchi et al. 1993b).

6.3 Studies of Motif A in Symporters

Prior to the discovery of Motif A, the roles of glycine residues along the LacY protein of *E. coli* (including glycines of Motif A) had been examined in the laboratory of Kaback (Jung et al. 1995), and it had been deemed that no such glycines throughout the symporter were critical for the transport of lactose. The first systematic study using site-directed mutagenesis to specifically address the functional importance of Motif A residues in LacY (Brooker 1990; Varela and Wilson 1996) was conducted in the laboratory of Brooker (Jessen-Marshall et al. 1995). In their first study, most amino acid replacements for Gly-64 and Asp-68 showed dramatic losses of lactose transport activities, while replacements for Lys-69, Gly-72, Arg-73, and Lys-74 showed only moderate to no loss of lactose transport (Jessen-Marshall et al. 1995), and it was concluded that the loop 2-3 structure formed by Motif A facilitates access of lactose entry into the cell by allowing conformational changes to occur upon sugar binding to the symporter (Jessen-Marshall et al. 1995). Using the mutation in which Asp-68 was changed to Thr, second-site revertant mutants were isolated that compensated for the defect conferred by the primary mutation, and it was found that most second sites were located in proximal ends of helices 2, 7, and 11 at the periplasm–membrane juncture (Jessen-Marshall and Brooker 1996). These results were interpreted as the suppressor mutations having altered the protein topology in order to facilitate the interaction between the two bundles of the symporter and helix 2 behaving as an interface between these two symmetrical bundles (Jessen-Marshall and Brooker 1996; Pazdernik et al. 1997a), a finding later supported by extensive molecular physiological analyses (Green et al. 2000; Green and Brooker 2001). In another study, Brooker used second-site suppressor analysis with Gly-64 mutations as the first-site mutation and found second sites dispersed throughout the symporter concluding that Gly-64 allows conformational changes to occur that are necessary for lactose transport across the membrane and that this residue is at the interface between two symmetrical bundles of the LacY protein (Jessen-Marshall et al. 1997; Pazdernik et al. 1997a). As mentioned above, the primary amino acid sequences of the N-terminal halves of the MFS transporters are closely related to their corresponding C-terminal halves. Motif A in the loop between helices 2 and 3 of these transporters is thus duplicated at the cytoplasmic loop between helices 8 and 9 (Griffith et al. 1992). Thus, the functional roles of these conserved amino acids in the loop 8-9 of LacY were evaluated and determined that they, too, serve to facilitate conformational changes that are believed to occur in these transporters during solute and ion transport catalysis (Pazdernik et al. 1997b; Cain et al. 2000).

6.4 Studies of Motif A in Multidrug Efflux Pumps

In the multidrug transporter LmrP from *Lactococcus lactis* (Bolhuis et al. 1995), the functional role of Asp-68, which resides in Motif A, was explored. First, molecular physiological evidence showed that an interaction between Asp-68 and phosphatidylethanolamine, a polar head group of the biological membrane, provides a sensor mechanism for detection of a proton gradient by the cell (Hakizimana et al. 2008). This particular notion that in this position of Motif A, a conserved Asp plays a role in proton gradient sensing, is supported by an apparent lack of conservation of Asp in this location of Motif A within MFS transporters that are not proton driven, such as in the case of the glucose facilitators (Hruz and Mueckler 2001), and the family of organic anion transporters (OATs), which are instead sodium driven (Zhou and You 2007). In another study using a biophysical analysis and molecular simulation dynamics of LmrP, it was found that during substrate transport, protonation of Asp-68 facilitated an outward-facing closed and inward-facing open conformation of the transporter, and deprotonation of Asp-68 to release protons into the cytoplasm favored a resetting back to the resting state conformation (Masureel et al. 2014); that is, Asp-68 plays a functional role in mediating conformational switching of the transporter during the multidrug efflux pump transport cycle. A study of the crystal structure of a proton-dependent oligopeptide transporter, YbgH from *E. coli*, combined with mutagenesis and comparisons with previously elucidated transporter crystal structures, found that a variant of Motif A, called Motif 1, functions as a conformational switch mechanism in order to stabilize YbgH in an outward-facing conformation (Zhao et al. 2014). An interesting development occurred with respect to Motif A and the mechanism of solute transport with the recent crystal structure determination of an *E. coli* outward-facing multidrug efflux pump, YajR, with a clearly defined loop 2-3 structure (Jiang et al. 2013). Based on this YajR crystal structure, the investigators provided structural and functional roles for individual residues of Motif A (Jiang et al. 2013). For instance, Gly-69 of YajR is believed to interact with Gly-337 and Gly-341, which are located on helix 11 of the same protein, thus forming an interface between the two domains (i.e., bundles) and allowing the formation of the outward-facing conformation of the pump (Jiang et al. 2013). Additionally, since Asp-73 was buried deep within the interface between the two bundles adjacent to helix 11 in the YajR structure, it is thus thought that this residue stabilizes both helix 11 and the bundle interface via a dipole-helix interaction; in support of this notion, the mutation Asp-73 changed to Arg decreased the melting temperature, suggesting that Asp-73 becomes solvent accessible (i.e., unburied) during the formation of an inward-facing conformation (Jiang et al. 2013). The Arg-74 residue is believed to interact with membrane phospholipid, thus possibly stabilizing the YajR protein within the membrane (Jiang et al. 2013). Gly-76 may stabilize the interaction within the N-terminal bundle, i.e., Gly-76 may confer an intra-domain stabilization (Jiang et al. 2013). Arg-77, on the other hand, is believed to form salt bridges with both Asp-73 (of Motif A) and Asp-126, the latter residue of which is located at the C-terminal end of helix 4 (Jiang et al. 2013). Incidentally, this same type of salt bridge formation is known to occur in LacY, in which Lys-319 interacts with both Asp-240 and Glu-269 to form alternating ion pairs (Lee et al. 1993). Lastly, Lys-73 of YajR is thought to interact with the C-terminal portion of helix 6 (Jiang et al. 2013). Taken together, the residues of Motif A in the YajR multidrug efflux pump are thought to stabilize the outward-facing conformation of the protein and thus participate in the conformational changes between the outward- and inward-facing stages of the transporter (Jiang et al. 2013). Strikingly, these investigators further found that elements of Motif A of loop 2-3 (called L2-3) are also present to a certain extent in three other loops of YajR, i.e., those loops between helices 5 and 6 (L5-6), 8 and 9 (L8-9), and 11 and

12 (L11-12), suggesting a widespread influence in the solute transport cycle for Motif A and Motif A-like sequences not only throughout a given MFS transporter, but in all transporters of the MFS as well (Jiang et al. 2013). Recently, the three crystal structures were elucidated for the multidrug efflux pump, MdfA, from *E. coli* in which each structure was bound to its substrate chloramphenicol or one of its analogs deoxycholate or n-dodecyl-N,N-dimethyl-amine-N-oxide (Heng et al. 2015). Since Motif A is known to stabilize the outward-facing conformation, as mentioned above for YajR (Jiang et al. 2013), the structural element conferred by this conserved motif is apparently not involved in dictating the inward-facing conformation seen in any of the three MdfA crystal structures (Heng et al. 2015).

7 Motif C "G $(X)_8$ G X X X G P X XG G" and Functional Roles

This conserved sequence motif was discovered by Rouch et al. to reside within the fifth TMD of transporters of the MFS (Rouch et al. 1990; see Fig. 3b). Initially thought to be found only with antiporters of the MFS but not in symporters or uniporters, Motif C was referred to as the "antiporter motif" (Varela et al. 1995; Varela and Griffith 1993). Recently, however, manual adjustments were performed during an extensive multiple sequence comparative analysis to surprisingly discover that sequence elements of the so-called antiporter motif are apparently found in the symporters and uniporters of the MFS as well (Yaffe et al. 2013).

7.1 Early Studies of Motif C in Efflux Pumps for Tetracycline

One of the earliest studies conducted to address the functional importance of Motif C was performed by Varela et al. in which they systematically replaced the most highly conserved residue of the motif, namely, Gly-147 of the tetracycline efflux pump, TetA(C), encoded on plasmid pBR322, with all other 19 amino acid residues (Varela et al. 1995). Interestingly, these investigators found that only Ala and Ser residues were acceptable in place of Gly-147 as tetracycline resistance was reduced to only 26 % and 19 % of the wild-type TetA(C), respectively (Varela et al. 1995). Molecular modeling analysis indicated a slight bend or kink in the fifth helix in the wild-type protein (Varela et al. 1995). Taken together, these investigators concluded that the residues of motif C dictate subtle structural differences inherent in determining substrate specificities and direction of solute transport (Varela et al. 1995). A study by Ginn et al. directly examined the structure–function relationships for all residues of Motif C of the TetA(K) tetracycline efflux pump from *S. aureus* by site-directed mutagenesis and tetracycline efflux assays (Ginn et al. 2000). These investigators found that tetracycline efflux pump activities were moderately to severely reduced for those mutants in which only the conserved residues of the motif were altered by mutation (Ginn et al. 2000). Thus, it was demonstrated in this study that the conserved residues of Motif C confer active tetracycline efflux; furthermore, because of the relative abundance of glycine residues in the motif, it was concluded that such flexible residues mediate conformational changes necessary for the efflux pump to respond to its immediate microenvironment (Ginn et al. 2000). Cysteine-scanning mutagenesis and accessibility of such mutations to the aqueous microenvironment that were studied by the laboratory of Yamaguchi and colleagues who showed that all residues of Motif C within TMD-5 of the Tn10-derived tetracycline efflux pump, TetA(B) from *E. coli*, line a water-filled channel and are thus probably able to bind substrate to facilitate transport (Iwaki et al. 2000). Additionally, these authors concluded that residues of TMD-5 of TetA(B), along with residues of TMD-4, form a permeability barrier that serves to avoid undesirable uncoupling (Iwaki et al. 2000). The laboratory of Levy conducted a second-site suppressor study in which four second-site mutations that complemented a defective mutation at Gly-247 of TetA(B) were found in TMD-5 indicating that residues of Motif C interact with residues of TMD-8 to stabilize their close association to each other (Saraceni-Richards and Levy 2000). These authors further concluded that residues of Motif C that are forming the permeability barrier in TetA(B) mediate conformational switching that occurs during solute transport across the membrane (Saraceni-Richards and Levy 2000).

7.2 Studies of Motif G

As mentioned earlier, bioinformatics evidence indicated an internal tandem repeat of a primordial 6-helix ancestor to form a modern 12-helix structure (Griffith et al. 1992) implying that Motif C is duplicated as well. The duplicated Motif C, denoted Motif G, was found in TMD-11 of the 12-helix MFS transporters (Paulsen et al. 1996b). This notion was confirmed experimentally in a study by Levy and colleagues in which they characterized Mdt(A), a multiple drug efflux pump encoded on a plasmid originating from *Lactococcus lactis*, and found the two Motif C-like sequences, one residing in TMD-5 and the other in TMD-9 (Perreten et al. 2001). Remarkably, these investigators also found an ATPase domain, which is routinely found in primary active transporters (Perreten et al. 2001). In another study involving Mdt(A) from a naturally occurring drug-susceptible variant of *Lactococcus garvieae*, Motif C was found two be altered in two of the canonical residues, thus possibly explaining the observed drug susceptibilities (Walther et al. 2008).

7.3 The Glycine–Proline Dipeptide in Motif C

A molecular mechanics and modeling study showed that a glycine–proline (GP) dipeptide within Motif C specified a bend or kink within the TMD-5 of the MFS efflux pumps (Varela et al. 1995). This particular notion was evaluated by the laboratory of Krulwich in which they closely examined mutations at these two residues, Gly-155 and Pro-156, of the tetracycline efflux pump, TetA(L), from *Bacillus subtilis* and found that the replacements showed, in general, tetracycline binding and a potassium leak, but not transport of tetracycline, suggesting that the GP dipeptide from Motif C is important for tight helix packing and leak proofing of the pump and providing an explanation for observed discrepancies between transport and resistance levels (Jin and Krulwich 2002; De Jesus et al. 2005).

The sole conserved proline residue of Motif C (of the GP dipeptide) was closely studied in QacA, a 14-TMD efflux pump encoded on the chromosome of *S. aureus* in a study focusing mainly on intramembranous Pro residues (Hassan et al. 2006). Replacement of Pro-161 of Motif C with Gly, Ser or Ala residues did not abolish resistance to any QacA substrates, but did show slightly altered drug resistance levels in host cells, suggesting this Pro residue may help form the permeability barrier and allow molecular motions or interactions with substrate to occur during transport of monovalent dyes (Hassan et al. 2006).

7.4 A Conformational Switch and Motif C

An analysis of residues of Motif C in a vesicular acetylcholine transporter, VAChT, from a eukaryote, *Rattus norvegicus*, showed profound loss of acetylcholine transport across the membrane and altered kinetic behavior of transport, indicating that minor and relatively stiff kinks in TMD-5 of VAChT are formed by residues of Motif C and that the motif not only allows conformational flexibility, i.e., switching, but also confers a tight proton seal to prevent dissipation of the membrane potential (Chandrasekaran et al. 2006). Another study of VAChT using homology modeling and molecular dynamics simulations found both kinking and wobbling behavior in structures formed by residues of Motif C and a lowering of the energy barrier for structures in which residues of Motif C were mutated (Luo and Parsons 2010). The authors of this study concluded that the structure formed by Motif C is at the interface between two helical bundles, consisting of TM helices 1–6 and 7–12 of VAChT, and that Motif C forms a complex hinge region between the two helical bundles in order to provide an energy barrier during conformational changes that occur during solute transport (Luo and Parsons 2010). Motif C from another eukaryotic MFS efflux pump, CaMdr1p from *Candida albicans*, which transports antifungal agents, was studied for its functional importance (Pasrija et al. 2007), and the investigators concluded that residues of this motif possibly mediate helix packing. A recent study of VMAT2 from *R. norvegicus* discovered that Motif

C plays a significant role in forming a so-called molecular hinge structure in which helices 5 and 8 interact with helices 2 and 11 to mediate the conformational switching between the two symmetric bundles that is thought to transpire during solute transport (Yaffe et al. 2013).

In a more recent study in which the crystal protein structure was determined for the *E. coli* MdfA multidrug efflux pump, it was shown that the protein was complexed with chloramphenicol or one of two substrate analogs; and it was further demonstrated that elements of Motif C (Ala-150, Ala-153, and Pro-154) (Rouch et al. 1990; Varela et al. 1995) surrounded two critical acidic residues Glu-26 and Asp-34 that reside in helix 1 of MdfA, thus constituting part of a central aqueous substrate binding cavity, a seemingly ubiquitous property of MFS solute transporters (Heng et al. 2015).

Acknowledgments This publication was supported in part by a grant from the National Institute of General Medical Sciences (P20GM103451) of the National Institutes of Health and by an internal research grant, ENMU.

References

Abramson J, Smirnova I, Kasho V, Verner G, Kaback HR, Iwata S (2003) Structure and mechanism of the lactose permease of *Escherichia coli*. Science 301:610–615. doi:10.1126/science.1088196

Andersen JL, He GX, Kakarla P, Crow RK, Kumar S, Lakra WS, Mukherjee MM, Ranaweera I, Shrestha U, Tran T, Varela MF (2015) Multidrug efflux pumps from *Enterobacteriaceae*, *Vibrio cholerae* and *Staphylococcus aureus* bacterial food pathogens. Int J Environ Res Public Health 12:1487–1547. doi:10.3390/ijerph120201487

Baker GF, Widdas WF (1973) The permeation of human red cells by 4,6-O-ethylidene-D-glucopyranose (ethylidene glucose). J Physiol 231:129–142, http://www.ncbi.nlm.nih.gov/pubmed/4715341

Bannam TL, Johanesen PA, Salvado CL, Pidot SJ, Farrow KA, Rood JI (2004) The *Clostridium perfringens* TetA (P) efflux protein contains a functional variant of the Motif A region found in major facilitator superfamily transport proteins. Microbiology 150:127–134. doi:10.1099/mic.0.26614-0

Bolhuis H, Poelarends G, van Veen HW, Poolman B, Driessen AJ, Konings WN (1995) The Lactococcal *lmrP* gene encodes a proton motive force-dependent drug transporter. J Biol Chem 270:26092–26098, http://www.ncbi.nlm.nih.gov/pubmed/7592810

Brooker RJ (1990) The lactose permease of *Escherichia coli*. Res Microbiol 141:309–315, http://www.ncbi.nlm.nih.gov/pubmed/2177910

Broome-Smith JK (1999) Transport of molecules across microbial membranes. Cambridge University Press, Cambridge

Brown MH, Skurray RA (2001) Staphylococcal multidrug efflux protein QacA. J Mol Microbiol Biotechnol 3:163–170, http://www.ncbi.nlm.nih.gov/pubmed/11321569

Büchel DE, Gronenborn B, Müller-Hill B (1980) Sequence of the lactose permease gene. Nature 283:541–545, http://www.ncbi.nlm.nih.gov/pubmed/6444453

Buivydas A, Daugelavièius R (2006) Use of tetraphenylphosphonium ions for studies of activity of tetracycline-extruding pumps. Biologija 2:24–27

Cain SM, Matzke EA, Brooker RJ (2000) The conserved motif in hydrophilic loop 2/3 and loop 8/9 of the lactose permease of *Escherichia coli*. Analysis of suppressor mutations. J Membr Biol 176:159–168, http://www.ncbi.nlm.nih.gov/pubmed/10926681

Chandrasekaran A, Ojeda AM, Kolmakova NG, Parsons SM (2006) Mutational and bioinformatics analysis of proline- and glycine-rich motifs in vesicular acetylcholine transporter. J Neurochem 98:1551–1559. doi:10.1111/j.1471-4159.2006.03975.x

Chang G (2003) Multidrug resistance ABC transporters. FEBS Lett 555:102–105. doi:10.1016/S0014-5793(03)01085-8

Chopra I (1986) Genetic and biochemical basis of tetracycline resistance. J Antimicrob Chemother 18(Suppl C):51–56, http://www.ncbi.nlm.nih.gov/pubmed/3542941

Chopra I (1992) Efflux-based antibiotic resistance mechanisms: the evidence for increasing prevalence. J Antimicrob Chemother 30:737–739, http://www.ncbi.nlm.nih.gov/pubmed/1289349

Chung YJ, Saier MH Jr (2001) SMR-type multidrug resistance pumps. Curr Opin Drug Discov Devel 4:237–245, http://www.ncbi.nlm.nih.gov/pubmed/11378963

Colmer JA, Fralick JA, Hamood AN (1998) Isolation and characterization of a putative multidrug resistance pump from *Vibrio cholerae*. Mol Microbiol 27:63–72, http://www.ncbi.nlm.nih.gov/pubmed/9466256

Costa SS, Viveiros M, Amaral L, Couto I (2013) Multidrug efflux pumps in *Staphylococcus aureus*: an update. Open Microbiol J 7:59–71. doi:10.2174/1874285801307010059

Dang S, Sun L, Huang Y, Lu F, Liu Y, Gong H, Wang J, Yan N (2010) Structure of a fucose transporter in an outward-open conformation. Nature 467:734–738. doi:10.1038/nature09406

Davidson AL, Maloney PC (2007) ABC transporters: how small machines do a big job. Trends Microbiol 15:448–455. doi:10.1016/j.tim.2007.09.005

De Jesus M, Jin J, Guffanti AA, Krulwich TA (2005) Importance of the GP dipeptide of the antiporter motif and other membrane-embedded proline and glycine residues in tetracycline efflux protein Tet(L). Biochemistry 44:12896–128904. doi:10.1021/bi050762c

De Vivo DC, Trifiletti RR, Jacobson RI, Ronen GM, Behmand RA, Harik SI (1991) Defective glucose transport across the blood-brain barrier as a cause of persistent hypoglycorrhachia, seizures, and developmental delay. N Engl J Med 325:703–709. doi:10.1056/NEJM199109053251006

Deng X, Sun F, Ji Q, Liang H, Missiakas D, Lan L, He C (2012) Expression of multidrug resistance efflux pump gene *norA* is iron responsive in *Staphylococcus aureus*. J Bacteriol 194:1753–1762. doi:10.1128/JB.06582-11

Floyd JL, Smith KP, Kumar SH, Floyd JT, Varela MF (2010) LmrS is a multidrug efflux pump of the major facilitator superfamily from *Staphylococcus aureus*. Antimicrob Agents Chemother 54:5406–5412. doi:10.1128/AAC.00580-10

Floyd JT, Kumar S, Mukherjee MM, He G, Varela MF (2013) A review of the molecular mechanisms of drug efflux in pathogenic bacteria: a structure-function perspective. Recent Res Develop Memb Biol 3:55–66

Fournier B, Aras R, Hooper DC (2000) Expression of the multidrug resistance transporter NorA from *Staphylococcus aureus* is modified by a two-component regulatory system. J Bacteriol 182:664–671, http://www.ncbi.nlm.nih.gov/pubmed/10633099

Fournier B, Truong-Bolduc QC, Zhang X, Hooper DC (2001) A mutation in the 5′ untranslated region increases stability of *norA* mRNA, encoding a multidrug resistance transporter of *Staphylococcus aureus*. J Bacteriol 183:2367–2371. doi:10.1128/JB.183.7.2367-2371.2001

Fowler PW, Orwick-Rydmark M, Radestock S, Solcan N, Dijkman PM, Lyons JA, Kwok J, Caffrey M, Watts A, Forrest LR, Newstead S (2015) Gating topology of the proton-coupled oligopeptide symporters. Structure 23:290–301. doi:10.1016/j.str.2014.12.012

Ginn SL, Brown MH, Skurray RA (2000) The TetA(K) tetracycline/H^+ antiporter from *Staphylococcus aureus*: mutagenesis and functional analysis of motif C. J Bacteriol 182:1492–1498, http://www.ncbi.nlm.nih.gov/pubmed/10692352

Goswitz VC, Brooker RJ (1995) Structural features of the uniporter/symporter/antiporter superfamily. Protein Sci 4:534–537. doi:10.1002/pro.5560040319

Green AL, Brooker RJ (2001) A face on transmembrane segment 8 of the lactose permease is important for transport activity. Biochemistry 40:12220–12229, http://www.ncbi.nlm.nih.gov/pubmed/11580298

Green AL, Anderson EJ, Brooker RJ (2000) A revised model for the structure and function of the lactose permease. Evidence that a face on transmembrane segment 2 is important for conformational changes. J Biol Chem 275:23240–23246. doi:10.1074/jbc.M909202199

Griffith JK, Baker ME, Rouch DA, Page MG, Skurray RA, Paulsen IT, Chater KF, Baldwin SA, Henderson PJ (1992) Membrane transport proteins: implications of sequence comparisons. Curr Opin Cell Biol 4:684–695, http://www.ncbi.nlm.nih.gov/pubmed/1419050

Guan L, Kaback HR (2006) Lessons from lactose permease. Annu Rev Biophys Biomol Struct 35:67–91. doi:10.1146/annurev.biophys.35.040405.102005

Hakizimana P, Masureel M, Gbaguidi B, Ruysschaert JM, Govaerts C (2008) Interactions between phosphatidylethanolamine headgroup and LmrP, a multidrug transporter: a conserved mechanism for proton gradient sensing? J Biol Chem 283:9369–9376. doi:10.1074/jbc.M708427200

Harold FM (2001) Gleanings of a chemiosmotic eye. Bioessays 23:848–855. doi:10.1002/bies.1120

Hassan KA, Galea M, Wu J, Mitchell BA, Skurray RA, Brown MH (2006) Functional effects of intramembranous proline substitutions in the staphylococcal multidrug transporter QacA. FEMS Microbiol Lett 263:76–85. doi:10.1111/j.1574-6968.2006.00411.x

Hediger MA (1994) Structure, function and evolution of solute transporters in prokaryotes and eukaryotes. J Exp Biol 196:15–49

Henderson PJ (1990a) The homologous glucose transport proteins of prokaryotes and eukaryotes. Res Microbiol 141:316–328, http://www.ncbi.nlm.nih.gov/pubmed/2177911

Henderson PJ (1990b) Proton-linked sugar transport systems in bacteria. J Bioenerg Biomembr 22:525–569, http://www.ncbi.nlm.nih.gov/pubmed/2172229

Henderson PJ (1991) Studies of translocation catalysis. Biosci Rep 11:477–538, discussion 534–538. http://www.ncbi.nlm.nih.gov/pubmed/1823597

Henderson PJ (1993) The 12-transmembrane helix transporters. Curr Opin Cell Biol 5:708–721, http://www.ncbi.nlm.nih.gov/pubmed/8257611

Henderson PJ, Maiden MC (1990) Homologous sugar transport proteins in *Escherichia coli* and their relatives in both prokaryotes and eukaryotes. Philos Trans R Soc Lond B Biol Sci 326:391–410, http://www.ncbi.nlm.nih.gov/pubmed/1970645

Henderson PJ, Roberts PE, Martin GE, Seamon KB, Walmsley AR, Rutherford NG, Varela MF, Griffith JK (1993) Homologous sugar-transport proteins in microbes and man. Biochem Soc Trans 21:1002–1006, http://www.ncbi.nlm.nih.gov/pubmed/8131886

Henderson P, Griffith J, Sansom C (1998) Function and structure of membrane transport proteins. In: Griffith JK, Sansom C (eds) The transporter factsbook. Academic, San Diego, pp 3–29

Heng J, Zhao Y, Liu M, Liu Y, Fan J, Wang X, Zhao Y, Zhang XC (2015) Substrate-bound structure of the *E. coli* multidrug resistance transporter MdfA. Cell Res 25:1060–1073. doi:10.1038/cr.2015.94

Heymann JA, Sarker R, Hirai T, Shi D, Milne JL, Maloney PC, Subramaniam S (2001) Projection structure and molecular architecture of OxlT, a bacterial membrane transporter. EMBO J 20:4408–4413. doi:10.1093/emboj/20.16.4408

Heymann JA, Hirai T, Shi D, Subramaniam S (2003) Projection structure of the bacterial oxalate transporter OxlT at 3.4Å resolution. J Struct Biol 144:320–326, http://www.ncbi.nlm.nih.gov/pubmed/14643200

Higgins CF (1992) ABC transporters: from microorganisms to man. Annu Rev Cell Biol 8:67–113. doi:10.1146/annurev.cb.08.110192.000435

Higgins CF (2001) ABC transporters: physiology, structure and mechanism–an overview. Res Microbiol 152:205–210. doi:10.1016/S0923-2508(01)01193-7

Hirai T, Heymann JA, Maloney PC, Subramaniam S (2003) Structural model for 12-helix transporters belonging to the major facilitator superfamily. J Bacteriol 185:1712–1718. doi:10.1128/JB.185.5.1712-1718.2003

Holler JG, Christensen SB, Slotved HC, Rasmussen HB, Guzman A, Olsen CE, Petersen B, Molgaard P (2012a) Novel inhibitory activity of the *Staphylococcus aureus* NorA efflux pump by a kaempferol rhamnoside isolated from *Persea lingue* Nees. J Antimicrob Chemother 67:1138–1144. doi:10.1093/jac/dks005

Holler JG, Slotved HC, Molgaard P, Olsen CE, Christensen SB (2012b) Chalcone inhibitors of the NorA efflux pump in *Staphylococcus aureus* whole cells and enriched everted membrane vesicles. Bioorg Med Chem 20:4514–4521. doi:10.1016/j.bmc.2012.05.025

Hooper DC (2000) Mechanisms of action and resistance of older and newer fluoroquinolones. Clin Infect Dis 31(Suppl 2):S24–S28. doi:10.1086/314056

Hruz PW, Mueckler MM (2001) Structural analysis of the GLUT1 facilitative glucose transporter (review). Mol Membr Biol 18:183–193, http://www.ncbi.nlm.nih.gov/pubmed/11681785

Huang Y, Lemieux MJ, Song J, Auer M, Wang DN (2003) Structure and mechanism of the glycerol-3-phosphate transporter from *Escherichia coli*. Science 301:616–620. doi:10.1126/science.1087619

Huang J, O'Toole PW, Shen W, Amrine-Madsen H, Jiang X, Lobo N, Palmer LM, Voelker L, Fan F, Gwynn MN, McDevitt D (2004) Novel chromosomally encoded multidrug efflux transporter MdeA in *Staphylococcus aureus*. Antimicrob Agents Chemother 48:909–917. doi:10.1128/AAC.48.3.909-917.2004

Iancu CV, Zamoon J, Woo SB, Aleshin A, Choe JY (2013) Crystal structure of a glucose/H^+ symporter and its mechanism of action. Proc Natl Acad Sci U S A 110:17862–17867. doi:10.1073/pnas.1311485110

Iwaki S, Tamura N, Kimura-Someya T, Nada S, Yamaguchi A (2000) Cysteine-scanning mutagenesis of transmembrane segments 4 and 5 of the Tn10-encoded metal-tetracycline/H^+ antiporter reveals a permeability barrier in the middle of a transmembrane water-filled channel. J Biol Chem 275:22704–22712, http://www.ncbi.nlm.nih.gov/pubmed/10930423

Jencks WP (1980) The utilization of binding energy in coupled vectorial processes. Adv Enzymol Relat Areas Mol Biol 51:75–106, http://www.ncbi.nlm.nih.gov/pubmed/6255774

Jessen-Marshall AE, Brooker RJ (1996) Evidence that transmembrane segment 2 of the lactose permease is part of a conformationally sensitive interface between the two halves of the protein. J Biol Chem 271:1400–1404, http://www.ncbi.nlm.nih.gov/pubmed/8576130

Jessen-Marshall AE, Paul NJ, Brooker RJ (1995) The conserved motif, GXXX(D/E)(R/K)XG[X](R/K)(R/K), in hydrophilic loop 2/3 of the lactose permease. J Biol Chem 270:16251–16257, http://www.ncbi.nlm.nih.gov/pubmed/7608191

Jessen-Marshall AE, Parker NJ, Brooker RJ (1997) Suppressor analysis of mutations in the loop 2-3 motif of lactose permease: evidence that glycine-64 is an important residue for conformational changes. J Bacteriol 179:2616–2622, http://www.ncbi.nlm.nih.gov/pubmed/9098060

Jiang D, Zhao Y, Wang X, Fan J, Heng J, Liu X, Feng W, Kang X, Huang B, Liu J, Zhang XC (2013) Structure of the YajR transporter suggests a transport mechanism based on the conserved motif A. Proc Natl Acad Sci U S A 110:14664–14669. doi:10.1073/pnas.1308127110

Jin J, Krulwich TA (2002) Site-directed mutagenesis studies of selected motif and charged residues and of cysteines of the multifunctional tetracycline efflux protein Tet(L). J Bacteriol 184:796–1800. doi:10.1128/JB.184.6.1796-1800.2002

Jung K, Jung H, Colacurcio P, Kaback HR (1995) Role of glycine residues in the structure and function of lactose permease, an *Escherichia coli* membrane transport protein. Biochemistry 34:1030–1039, http://www.ncbi.nlm.nih.gov/pubmed/7827019

Kaatz GW, Seo SM (1995) Inducible NorA-mediated multidrug resistance in *Staphylococcus aureus*. Antimicrob Agents Chemother 39:2650–2655, http://www.ncbi.nlm.nih.gov/pubmed/8592996

Kaatz GW, Seo SM, Ruble CA (1993) Efflux-mediated fluoroquinolone resistance in *Staphylococcus aureus*. Antimicrob Agents Chemother 37:1086–1094, http://www.ncbi.nlm.nih.gov/pubmed/8517696

Kalia NP, Mahajan P, Mehra R, Nargotra A, Sharma JP, Koul S, Khan IA (2012) Capsaicin, a novel inhibitor of the NorA efflux pump, reduces the intracellular invasion of *Staphylococcus aureus*. J Antimicrob Chemother 67:2401–2408. doi:10.1093/jac/dks232

Kaneko M, Yamaguchi A, Sawai T (1985) Energetics of tetracycline efflux system encoded by Tn10 in *Escherichia coli*. FEBS Lett 193:194–198. doi:10.1016/0014-5793(85)80149-6

Kawabe T, Yamaguchi A (1999) Transmembrane remote conformational suppression of the Gly-332 mutation of the Tn10-encoded metal-tetracycline/H^+ antiporter. FEBS Lett 457:169–173. doi:10.1016/S0014-5793(99)01032-7

Kimura T, Sawai T, Yamaguchi A (1997) Remote conformational effects of the Gly-62 –> Leu mutation of the Tn10-encoded metal-tetracycline/H^+ antiporter of *Escherichia coli* and its second-site suppressor mutation. Biochemistry 36:6941–6946. doi:10.1021/bi9631879

Kimura T, Nakatani M, Kawabe T, Yamaguchi A (1998a) Roles of conserved arginine residues in the metal-tetracycline/H^+ antiporter of *Escherichia coli*. Biochemistry 37:5475–5780. doi:10.1021/bi973188g

Kimura T, Shiina Y, Sawai T, Yamaguchi A (1998b) Cysteine-scanning mutagenesis around transmembrane segment III of Tn10-encoded metal-tetracycline/H^+ antiporter. J Biol Chem 273:5243–5247, http://www.ncbi.nlm.nih.gov/pubmed/9478980

Kinghorn JR, Sloan J, Kana'n GJ, Dasilva ER, Rouch DA, Unkles SE (2005) Missense mutations that inactivate the *Aspergillus nidulans nrtA* gene encoding a high-affinity nitrate transporter. Genetics 169:1369–1377. doi:10.1534/genetics.104.036590

Klepper J, Willemsen M, Verrips A, Guertsen E, Herrmann R, Kutzick C, Florcken A, Voit T (2001) Autosomal dominant transmission of GLUT1 deficiency. Hum Mol Genet 10:63–68. doi:10.1093/hmg/10.1.63

Kosmidis C, DeMarco CE, Frempong-Manso E, Seo SM, Kaatz GW (2010) *In silico* genetic correlations of multidrug efflux pump gene expression in *Staphylococcus aureus*. Int J Antimicrob Agents 36:222–229. doi:10.1016/j.ijantimicag.2010.05.015

Krämer R (1994) Functional principles of solute transport systems: concepts and perspectives. Biochim Biophys Acta 1185:1–34. doi:10.1016/0005-2728(94)90189-9

Kumano M, Tamakoshi A, Yamane K (1997) A 32 kb nucleotide sequence from the region of the lincomycin-resistance gene (22 degrees-25 degrees) of the *Bacillus subtilis* chromosome and identification of the site of the *lin-2* mutation. Microbiology 143:2775–2782. doi:10.1099/00221287-143-8-2775

Kumar S, Varela MF (2012) Biochemistry of bacterial multidrug efflux pumps. Int J Mol Sci 13:4484–4495. doi:10.3390/ijms13044484

Kumar S, Varela MF (2013) Molecular mechanisms of bacterial resistance to antimicrobial agents. In: Méndez-Vilas A (ed) Microbial pathogens and strategies for combating them: science, technology and education. Formatex Research Center, Badajoz, Spain, pp 522–534. ISBN 978-84-939843-9-7

Kumar S, Floyd JT, He G, Varela MF (2013) Bacterial antimicrobial efflux pumps of the MFS and MATE transporter families: a review. Recent Res Dev Antimicrob Agents Chemother 7:1–21, ISBN: 978-81-308-0465-1

Kuroda T, Tsuchiya T (2009) Multidrug efflux transporters in the MATE family. Biochim Biophys Acta 1794:763–768. doi:10.1016/j.bbapap.2008.11.012

Law CJ, Maloney PC, Wang DN (2008) Ins and outs of major facilitator superfamily antiporters. Annu Rev Microbiol 62:289–305. doi:10.1146/annurev.micro.61.080706.093329

Lee JI, Hwang PP, Wilson TH (1993) Lysine 319 interacts with both glutamic acid 269 and aspartic acid 240 in the lactose carrier of *Escherichia coli*. J Biol Chem 268:20007–20015, http://www.ncbi.nlm.nih.gov/pubmed/8104184

Lee JI, Varela MF, Wilson TH (1996) Physiological evidence for an interaction between Glu-325 and His-322 in the lactose carrier of *Escherichia coli*. Biochim Biophys Acta 1278:111–118. doi:10.1016/0005-2736(95)00209-X

Levy SB (1992) Active efflux mechanisms for antimicrobial resistance. Antimicrob Agents Chemother 36:695–703. doi:10.1128/AAC.36.4.695

Levy SB (2002) Active efflux, a common mechanism for biocide and antibiotic resistance. Symp Ser Soc Appl Microbiol 31:65S–71S. doi:10.1046/j.1365-2672.92.5s1.4.x

Lewis K (1994) Multidrug resistance pumps in bacteria: variations on a theme. Trends Biochem Sci 19:119–123. doi:10.1016/0968-0004(94)90204-6

Li XZ, Plesiat P, Nikaido H (2015) The challenge of efflux-mediated antibiotic resistance in Gram-negative bacteria. Clin Microbiol Rev 28:337–418. doi:10.1128/CMR.00117-14

Linton KJ (2007) Structure and function of ABC transporters. Phys Chem Chem Phys 22:122–130. doi:10.1152/physiol.00046.2006

Littlejohn TG, Paulsen IT, Gillespie MT, Tennent JM, Midgley M, Jones IG, Purewal AS, Skurray RA (1992) Substrate specificity and energetics of antiseptic and disinfectant resistance in *Staphylococcus aureus*. FEMS Microbiol Lett 74:259–265. doi:10.1111/j.1574-6968.1992.tb05376.x

Locher KP (2009) Structure and mechanism of ATP-binding cassette transporters. Philos Trans R Soc Lond B Biol Sci 364:239–245, http://www.jstor.org/stable/40486111

Lomovskaya O, Lewis K (1992) *Emr*, an *Escherichia coli* locus for multidrug resistance. Proc Natl Acad Sci U S A 89:8938–8942, http://www.ncbi.nlm.nih.gov/pubmed/1409590

Luo J, Parsons SM (2010) Conformational propensities of peptides mimicking transmembrane helix 5 and motif c in wild-type and mutant vesicular acetylcholine transporters. ACS Chem Neurosci 1:381–390. doi:10.1021/cn900033s

Maiden MC, Davis EO, Baldwin SA, Moore DC, Henderson PJ (1987) Mammalian and bacterial sugar transport proteins are homologous. Nature 325:641–643. doi:10.1038/325641a0

Maloney PC (1994) Bacterial transporters. Curr Opin Cell Biol 6:571–582. doi:10.1016/0955-0674(94)90079-5

Marger MD, SaierM H Jr (1993) A major superfamily of transmembrane facilitators that catalyse uniport, symport and antiport. Trends Biochem Sci 18:13–20. doi:10.1016/0968-0004(93)90081-W

Masureel M, Martens C, Stein RA, Mishra S, Ruysschaert JM, McHaourab HS, Govaerts C (2014) Protonation drives the conformational switch in the multidrug transporter LmrP. Nat Chem Biol 10:149–155. doi:10.1038/nchembio.1408

Matsumura K, Furukawa S, Ogihara H, Morinaga Y (2011) Roles of multidrug efflux pumps on the biofilm formation of *Escherichia coli* K-12. Biocontrol Sci 16:69–72. doi:10.4265/bio.16.69

McMurry L, Petrucci RE Jr, Levy SB (1980) Active efflux of tetracycline encoded by four genetically different tetracycline resistance determinants in *Escherichia coli*. Proc Natl Acad Sci U S A 77:3974–3977, http://www.ncbi.nlm.nih.gov/pubmed/7001450

McNicholas P, Chopra I, Rothstein DM (1992) Genetic analysis of the *tetA(C)* gene on plasmid pBR322. J Bacteriol 174:7926–7933, http://www.ncbi.nlm.nih.gov/pubmed/1459940

McNicholas P, McGlynn M, Guay GG, Rothstein DM (1995) Genetic analysis suggests functional interactions between the N- and C-terminal domains of the TetA(C) efflux pump encoded by pBR322. J Bacteriol 177:5355–5357, http://www.ncbi.nlm.nih.gov/pubmed/7665527

Mirza ZM, Kumar A, Kalia NP, Zargar A, Khan IA (2011) Piperine as an inhibitor of the MdeA efflux pump of *Staphylococcus aureus*. J Med Microbiol 60:1472–1478. doi:10.1099/jmm.0.033167-0

Mitchell P (1966) Chemiosmotic coupling in oxidative and photosynthetic phosphorylation. Biol Rev Camb Philos Soc 41:445–502. doi:10.1111/j.1469-185X.1966.tb01501.x

Mitchell P (1967) Translocations through natural membranes. Adv Enzymol Relat Areas Mol Biol 29:33–87. doi:10.1002/9780470122747.ch

Mitchell P (1972) Chemiosmotic coupling in energy transduction: a logical development of biochemical knowledge. J Bioenerg 3:5–24. doi:10.1007/BF01515993

Mitchell P (1977) Vectorial chemiosmotic processes. Annu Rev Biochem 46:996–1005. doi:10.1146/annurev.bi.46.070177.005024

Mitchell P (1991) Foundations of vectorial metabolism and osmochemistry. Biosci Rep 11:297–344. doi:10.1007/BF01130212

Mitchell P (2011) Chemiosmotic coupling in oxidative and photosynthetic phosphorylation. Biochim Biophys Acta 1807:1507–1538. doi:10.1016/j.bbabio.2011.09.018

Moir JW, Wood NJ (2001) Nitrate and nitrite transport in bacteria. Cell Mol Life Sci 58:215–224. doi:10.1007/PL00000849

Müller-Hill B (1996) The *Lac* Operon. Walter de Gruyter, Berlin

Nelson ML, Levy SB (2011) The history of the tetracyclines. Ann N Y Acad Sci 1241:17–32. doi:10.1111/j.1749-6632.2011.06354.x

Newman MJ, Wilson TH (1980) Solubilization and reconstitution of the lactose transport system from *Escherichia coli*. J Biol Chem 255:10583–10586, http://www.ncbi.nlm.nih.gov/pubmed/7000781

Newstead S, Drew D, Cameron AD, Postis VL, Xia X, Fowler PW, Ingram JC, Carpenter EP, Sansom MS, McPherson MJ, Baldwin SA, Iwata S (2011) Crystal structure of a prokaryotic homologue of the mammalian oligopeptide-proton symporters, PepT1 and PepT2. EMBO J 30:417–426. doi:10.1038/emboj.2010.309

Neyfakh AA (1992) The multidrug efflux transporter of *Bacillus subtilis* is a structural and functional homolog of the *Staphylococcus* NorA protein. Antimicrob Agents Chemother 36:484–485. doi:10.1128/AAC.36.2.484

Neyfakh AA, Borsch CM, Kaatz GW (1993) Fluoroquinolone resistance protein NorA of *Staphylococcus aureus* is a multidrug efflux transporter. Antimicrob Agents Chemother 37:128–129. doi:10.1128/AAC.37.1.128

Nikaido H (1992) Porins and specific channels of bacterial outer membranes. Mol Microbiol 6:435–442. doi:10.1111/j.1365-2958.1992.tb01487.x

Nikaido H (1994) Porins and specific diffusion channels in bacterial outer membranes. J Biol Chem 269:3905–3908, http://www.jbc.org/content/269/6/3905.full.pdf+html

Nikaido H (2003) Molecular basis of bacterial outer membrane permeability revisited. Microbiol Mol Biol Rev 67:593–656. doi:10.1128/MMBR.67.4.593-656.2003

Nikaido H, Vaara M (1985) Molecular basis of bacterial outer membrane permeability. Microbiol Rev 49:1–32, http://www.ncbi.nlm.nih.gov/pmc/articles/PMC373015/

Nishino K, Yamaguchi A (2001) Analysis of a complete library of putative drug transporter genes in *Escherichia coli*. J Bacteriol 183:5803–5812. doi:10.1128/JB.183.20.5803-5812.2001

Nomura N, Verdon G, Kang HJ, Shimamura T, Nomura Y, Sonoda Y, Hussien SA, Qureshi AA, Coincon M, Sato Y, Abe H, Nakada-Nakura Y, Hino T, Arakawa T, Kusano-Arai O, Iwanari H, Murata T, Kobayashi T, Hamakubo T, Kasahara M, Iwata S, Drew D (2015) Structure and mechanism of the mammalian fructose transporter GLUT5. Nature 526:397–401. doi:10.1038/nature14909

Pao SS, Paulsen IT, Saier MH Jr (1998) Major facilitator superfamily. Microbiol Mol Biol Rev 62:1–34, http://www.ncbi.nlm.nih.gov/pubmed/9529885

Pascual JM, Wang D, Yang R, Shi L, Yang H, De Vivo DC (2008) Structural signatures and membrane helix 4 in GLUT1: inferences from human blood-brain glucose transport mutants. J Biol Chem 283:16732–16742. doi:10.1074/jbc.M801403200

Pasrija R, Banerjee D, Prasad R (2007) Structure and function analysis of CaMdr1p, a major facilitator superfamily antifungal efflux transporter protein of *Candida albicans*: identification of amino acid residues critical for drug/H^+ transport. Eukaryot Cell 6:443–453. doi:10.1128/EC.00315-06

Paulsen IT, Brown MH, Littlejohn TG, Mitchell BA, Skurray RA (1996a) Multidrug resistance proteins QacA and QacB from *Staphylococcus aureus*: membrane topology and identification of residues involved in substrate specificity. Proc Natl Acad Sci U S A 93:3630–3635, http://www.ncbi.nlm.nih.gov/pubmed/8622987

Paulsen IT, Brown MH, Skurray RA (1996b) Proton-dependent multidrug efflux systems. Microbiol Rev 60:575–608, http://www.ncbi.nlm.nih.gov/pubmed/8987357

Pazdernik NJ, Cain SM, Brooker RJ (1997a) An analysis of suppressor mutations suggests that the two halves of the lactose permease function in a symmetrical manner. J Biol Chem 272:26110–26116, http://www.ncbi.nlm.nih.gov/pubmed/9334175

Pazdernik NJ, Jessen-Marshall AE, Brooker RJ (1997b) Role of conserved residues in hydrophilic loop 8-9 of the lactose permease. J Bacteriol 179:735–741, http://www.ncbi.nlm.nih.gov/pubmed/9006028

Pedersen BP, Kumar H, Waight AB, Risenmay AJ, Roe-Zurz Z, Chau BH, Schlessinger A, Bonomi M, Harries W, Sali A, Johri AK, Stroud RM (2013) Crystal structure of a eukaryotic phosphate transporter. Nature 496:533–536. doi:10.1038/nature12042

Perreten V, Schwarz FV, Teuber M, Levy SB (2001) Mdt(A), a new efflux protein conferring multiple antibiotic resistance in *Lactococcus lactis* and *Escherichia coli*. Antimicrob Agents Chemother 45:1109–1114. doi:10.1128/AAC.45.4.1109-1114.2001

Piddock LJ (2006) Multidrug-resistance efflux pumps – not just for resistance. Nat Rev Microbiol 4:629–636. doi:10.1038/nrmicro1464

Poolman B, Konings WN (1993) Secondary solute transport in bacteria. Biochim Biophys Acta 1183:5–39. doi:10.1016/0005-2728(93)90003-X

Procko E, O'Mara ML, Bennett WD, Tieleman DP, Gaudet R (2009) The mechanism of ABC transporters: general lessons from structural and functional studies of an antigenic peptide transporter. FASEB J 23:1287–1302. doi:10.1096/fj.08-121855

Radestock S, Forrest LR (2011) The alternating-access mechanism of MFS transporters arises from inverted-topology repeats. J Mol Biol 407:698–715. doi:10.1016/j.jmb.2011.02.008

Rather MA, Aulakh RS, Gill JPS, Mir AQ, Hassan MN (2012) Detection and sequencing of plasmid encoded tetracycline resistance determinants (*tetA* and *tetB*) from food–borne *Bacillus cereus* isolates. Asian Pac J Trop Med 5:709–712. doi:10.1016/S1995-7645(12)60111-4

Rees DC, Johnson E, Lewinson O (2009) ABC transporters: the power to change. Nat Rev Mol Cell Biol 10:218–227. doi:10.1038/nrm2646

Reid J, Fung H, Gehring K, Klebba P, Nikaido H (1988) Targeting of porin to the outer membrane of *Escherichia coli*. Rate of trimer assembly and identification of a dimer intermediate. J Biol Chem 263:7753–7759, http://www.jbc.org/content/263/16/7753

Rouch DA, Cram DS, DiBerardino D, Littlejohn TG, Skurray RA (1990) Efflux-mediated antiseptic resistance gene *qacA* from *Staphylococcus aureus*: common ancestry with tetracycline- and sugar-transport proteins. Mol Microbiol 4:2051–2062. doi:10.1111/j.1365-2958.1990.tb00565.x

Roy SK, Kumari N, Pahwa S, Agrahari UC, Bhutani KK, Jachak SM, Nandanwar H (2013) NorA efflux pump inhibitory activity of coumarins from *Mesua ferrea*. Fitoterapia 90:140–150. doi:10.1016/j.fitote.2013.07.015

Rubin RA, Levy SB, Heinrikson RL, Kezdy FJ (1990) Gene duplication in the evolution of the two complementing domains of gram-negative bacterial tetracycline efflux proteins. Gene 87:7–13. doi:10.1016/0378-1119(90)90489-E

Saidijam M, Benedetti G, Ren Q, Xu Z, Hoyle CJ, Palmer SL, Ward A, Bettaney KE, Szakonyi G, Meuller J, Morrison S, Pos MK, Butaye P, Walravens K, Langton K, Herbert RB, Skurray RA, Paulsen IT, O'Reilly J, Rutherford NG, Brown MH, Bill RM, Henderson PJ (2006) Microbial drug efflux proteins of the major facilitator superfamily. Curr Drug Targets 7:793–811. doi:10.2174/138945006777709575

Saier MH (2000) A functional-phylogenetic classification system for transmembrane solute transporters. Microbiol Mol Biol Rev 64:354–411. doi:10.1128/MMBR.64.2.354-411.2000

Saier MH Jr, Paulsen IT, Sliwinski MK, Pao SS, Skurray RA, Nikaido H (1998) Evolutionary origins of multidrug and drug-specific efflux pumps in bacteria. FASEB J 12:265–274, http://www.ncbi.nlm.nih.gov/pubmed/9506471

Saier MH Jr, Beatty JT, Goffeau A, Harley KT, Heijne WH, Huang SC, Jack DL, Jahn PS, Lew K, Liu J, Pao SS, Paulsen IT, Tseng TT, Virk PS (1999) The major facilitator superfamily. J Mol Microbiol Biotechnol 1:257–279. doi:10.1111/j.1742-4658.2012.08588.x

Saier MH Jr, Reddy VS, Tamang DG, Vastermark A (2014) The transporter classification database. Nucleic Acids Res 42:D251–D258. doi:10.1093/nar/gkt1097

Santer R, Groth S, Kinner M, Dombrowski A, Berry GT, Brodehl J, Leonard JV, Moses S, Norgren S, Skovby F, Schneppenheim R, Steinmann B, Schaub J (2002) The mutation spectrum of the facilitative glucose transporter gene SLC2A2 (GLUT2) in patients with Fanconi-Bickel syndrome. Hum Genet 110:21–29. doi:10.1007/s00439-001-0638-6

Saraceni-Richards CA, Levy SB (2000) Second-site suppressor mutations of inactivating substitutions at gly247 of the tetracycline efflux protein, Tet(B). J Bacteriol 182:6514–6516. doi:10.1128/JB.182.22.6514-6516.2000

Seol W, Shatkin AJ (1992) Site-directed mutants of *Escherichia coli* alpha-ketoglutarate permease (KgtP). Biochemistry 31:3550–3554. doi:10.1021/bi00128a032

Shiu WK, Malkinson JP, Rahman MM, Curry J, Stapleton P, Gunaratnam M, Neidle S, Mushtaq S, Warner M, Livermore DM, Evangelopoulos D, Basavannacharya C, Bhakta S, Schindler BD, Seo SM, Coleman D, Kaatz GW, Gibbons S (2013) A new plant-derived antibacterial is an inhibitor of efflux pumps in *Staphylococcus aureus*. Int J Antimicrob Agents 42:513–518. doi:10.1016/j.ijantimicag.2013.08.007

Smith KP, Kumar S, Varela MF (2009) Identification, cloning, and functional characterization of EmrD-3, a putative multidrug efflux pump of the major facilitator superfamily from *Vibrio cholerae* O395. Arch Microbiol 191:903–911. doi:10.1007/s00203-009-0521-8

Someya Y, Kimura-Someya T, Yamaguchi A (2000) Role of the charge interaction between Arg (70) and Asp(120) in the Tn10-encoded metal-tetracycline/H^+ antiporter of *Escherichia coli*. J Biol Chem 275:210–214, http://www.ncbi.nlm.nih.gov/pubmed/10617606

Stelzl LS, Fowler PW, Sansom MS, Beckstein O (2014) Flexible gates generate occluded intermediates in the transport cycle of LacY. J Mol Biol 426:735–751. doi:10.1016/j.jmb.2013.10.024

Stochaj U, Ehring R (1987) The N-terminal region of *Escherichia coli* lactose permease mediates membrane contact of the nascent polypeptide chain. Eur J Biochem 163:653–658. doi:10.1111/j.1432-1033.1987.tb10914.x

Stochaj U, Bieseler B, Ehring R (1986) Limited proteolysis of lactose permease from *Escherichia coli*. Eur J Biochem 158:423–428. doi:10.1111/j.1432-1033.1986.tb09770.x

Stochaj U, Fritz HJ, Heibach C, Markgraf M, von Schaewen A, Sonnewald U, Ehring R (1988) Truncated forms of *Escherichia coli* lactose permease: models for study of biosynthesis and membrane insertion. J Bacteriol 170:2639–2645, http://www.ncbi.nlm.nih.gov/pubmed/3286614

Sulavik MC, Houseweart C, Cramer C, Jiwani N, Murgolo N, Greene B, DiDomenico B, Shaw KJ, Miller GH, Hare R (2001) Antibiotic susceptibility profiles of *Escherichia coli* strains lacking multidrug efflux pump genes. Antimicrob Agents Chemother 45:1126–1136. doi:10.1128/AAC.45.4.1126-1136.2001

Sun L, Zeng X, Yan C, Sun X, Gong X, Rao Y, Yan N (2012) Crystal structure of a bacterial homologue of glucose transporters GLUT1-4. Nature 490:361–366. doi:10.1038/nature11524

Tamura N, Konishi S, Yamaguchi A (2003) Mechanisms of drug/H^+ antiport: complete cysteine-scanning mutagenesis and the protein engineering approach. Curr Opin Chem Biol 7:570–579. doi:10.1016/j.cbpa.2003.08.014

Tanford C (1982) Simple model for the chemical potential change of a transported ion in active transport. Proc Natl Acad Sci U S A 79:2882–2884, http://www.ncbi.nlm.nih.gov/pubmed/6283549

Tarling EJ, de Aguiar Vallim TQ, Edwards PA (2013) Role of ABC transporters in lipid transport and human disease. Trends Endocrinol Metab 24:342–350. doi:10.1016/j.tem.2013.01.006

Teather RM, Müller-Hill B, Abrutsch U, Aichele G, Overath P (1978) Amplification of the lactose carrier protein in *Escherichia coli* using a plasmid vector. Mol Gen Genet 159:239–248. doi:10.1007/BF00268260

Tennent JM, Lyon BR, Midgley M, Jones IG, Purewal AS, Skurray RA (1989) Physical and biochemical characterization of the *qacA* gene encoding antiseptic and disinfectant resistance in *Staphylococcus aureus.* J Gen Microbiol 135:1–10, http://www.ncbi.nlm.nih.gov/pubmed/2778425

ter Beek J, Guskov A, Slotboom DJ (2014) Structural diversity of ABC transporters. J Gen Physiol 143:419–435. doi:10.1085/jgp.201411164

Thai KM, Ngo TD, Phan TV, Tran TD, Nguyen NV, Nguyen TH, Le MT (2015) Virtual screening for novel *Staphylococcus Aureus* NorA efflux pump inhibitors from natural products. Med Chem 11:135–155. doi:10.2174/1573406410666140902110903

Truong-Bolduc QC, Zhang X, Hooper DC (2003) Characterization of NorR protein, a multifunctional regulator of *norA* expression in *Staphylococcus aureus.* J Bacteriol 185:3127–3138. doi:10.1128/JB.185.10.3127-3138.2003

Truong-Bolduc QC, Dunman PM, Strahilevitz J, Projan SJ, Hooper DC (2005) MgrA is a multiple regulator of two new efflux pumps in *Staphylococcus aureus.* J Bacteriol 87:2395–2405. doi:10.1128/JB.187.7.2395-2405.2005

Tseng TT, Gratwick KS, Kollman J, Park D, Nies DH, Goffeau A, Saier MH Jr (1999) The RND permease superfamily: an ancient, ubiquitous and diverse family that includes human disease and development proteins. J Mol Microbiol Biotechnol 1:107–125, http://www.ncbi.nlm.nih.gov/pubmed/10941792

Ubukata K, Itoh-Yamashita N, Konno M (1989) Cloning and expression of the *norA* gene for fluoroquinolone resistance in *Staphylococcus aureus.* Antimicrob Agents Chemother 33:1535–1539. doi:10.1128/AAC.33.9.1535

Varela MF, Griffith JK (1993) Nucleotide and deduced protein sequences of the class D tetracycline resistance determinant: relationship to other antimicrobial transport proteins. Antimicrob Agents Chemother 37:1253–1258. doi:10.1128/AAC.37.6.1253

Varela MF, Wilson TH (1996) Molecular biology of the lactose carrier of *Escherichia coli.* Biochim Biophys Acta 1276:21–34. doi:10.1016/0005-2728(96)00030-8

Varela MF, Sansom CE, Griffith JK (1995) Mutational analysis and molecular modelling of an amino acid sequence motif conserved in antiporters but not symporters in a transporter superfamily. Mol Membr Biol 12:313–319. doi:10.3109/09687689509072433

Walther C, Rossano A, Thomann A, Perreten V (2008) Antibiotic resistance in Lactococcus species from bovine milk: presence of a mutated multidrug transporter *mdt(A)* gene in susceptible *Lactococcus garvieae* strains. Vet Microbiol 131:348–357. doi:10.1016/j.vetmic.2008.03.008

West IC (1980) Energy coupling in secondary active transport. Biochim Biophys Acta 604:91–126. doi:10.1016/0005-2736(80)90586-6

West IC (1997) Ligand conduction and the gated-pore mechanism of transmembrane transport. Biochim Biophys Acta 1331:213–234. doi:10.1016/S0304-4157(97)00007-5

West I, Mitchell P (1972) Proton-coupled beta-galactoside translocation in non-metabolizing *Escherichia coli.* J Bioenerg 3:445–462, http://www.ncbi.nlm.nih.gov/pubmed/4570991

Wilson TH, Ding PZ (2001) Sodium-substrate cotransport in bacteria. Biochim Biophys Acta 1505:121–130. doi:10.1016/S0005-2728(00)00282-6

Wright EM, Turk E, Martin MG (2002) Molecular basis for glucose-galactose malabsorption. Cell Biochem Biophys 36:115–121. doi:10.1385/CBB:36:2-3:115

Wrubel W, Stochaj U, Sonnewald U, Theres C, Ehring R (1990) Reconstitution of an active lactose carrier in vivo by simultaneous synthesis of two complementary protein fragments. J Bacteriol 172:5374–5381, http://www.ncbi.nlm.nih.gov/pubmed/2203750

Wrubel W, Stochaj U, Ehring R (1994) Construction and in vivo analysis of new split lactose permeases. FEBS Lett 349:433–438. doi:10.1016/0014-5793(94)00719-5

Yaffe D, Radestock S, Shuster Y, Forrest LR, Schuldiner S (2013) Identification of molecular hinge points mediating alternating access in the vesicular monoamine transporter VMAT2. Proc Natl Acad Sci U S A 110:E1332–E1341. doi:10.1073/pnas.1220497110

Yamada Y, Shiota S, Mizushima T, Kuroda T, Tsuchiya T (2006) Functional gene cloning and characterization of MdeA, a multidrug efflux pump from *Staphylococcus aureus*. Biol Pharm Bull 29:801–804. doi:10.1248/bpb.29.801

Yamaguchi A, Ono N, Akasaka T, Noumi T, Sawai T (1990) Metal-tetracycline/H^+ antiporter of *Escherichia coli* encoded by a transposon, Tn10. The role of the conserved dipeptide, Ser65-Asp66, in tetracycline transport. J Biol Chem 265:15525–15530, http://www.ncbi.nlm.nih.gov/pubmed/2168416

Yamaguchi A, Akasaka T, Ono N, Someya Y, Nakatani M, Sawai T (1992a) Metal-tetracycline/H^+ antiporter of *Escherichia coli* encoded by transposon Tn10. Roles of the aspartyl residues located in the putative transmembrane helices. J Biol Chem 267:7490–7498, http://www.ncbi.nlm.nih.gov/pubmed/1313805

Yamaguchi A, Someya Y, Sawai T (1992b) Metal-tetracycline/H^+ antiporter of *Escherichia coli* encoded by transposon Tn10. The role of a conserved sequence motif, GXXXXRXGRR, in a putative cytoplasmic loop between helices 2 and 3. J Biol Chem 267:19155–19162, http://www.ncbi.nlm.nih.gov/pubmed/1326546

Yamaguchi A, Akasaka T, Kimura T, Sakai T, Adachi Y, Sawai T (1993a) Role of the conserved quartets of residues located in the N- and C-terminal halves of the transposon Tn10-encoded metal-tetracycline/H^+ antiporter of *Escherichia coli*. Biochemistry 32:5698–5704, http://www.ncbi.nlm.nih.gov/pubmed/8389190

Yamaguchi A, Kimura T, Someya Y, Sawai T (1993b) Metal-tetracycline/H^+ antiporter of *Escherichia coli* encoded by transposon Tn10. The structural resemblance and functional difference in the role of the duplicated sequence motif between hydrophobic segments 2 and 3 and segments 8 and 9. J Biol Chem 268:6496–6504, http://www.ncbi.nlm.nih.gov/pubmed/8384213

Yamaguchi A, Inagaki Y, Sawai T (1995) Second-site suppressor mutations for the Asp-66–>Cys mutant of the transposon Tn10-encoded metal-tetracycline/H^+ antiporter of *Escherichia coli*. Biochemistry 34:11800–11806. doi:10.1021/bi00037a018

Yan H, Huang W, Yan C, Gong X, Jiang S, Zhao Y, Wang J, Shi Y (2013) Structure and mechanism of a nitrate transporter. Cell Rep 3:716–723. doi:10.1016/j.celrep.2013.03.007

Yin Y, He X, Szewczyk P, Nguyen T, Chang G (2006) Structure of the multidrug transporter EmrD from *Escherichia coli*. Science 312:741–744. doi:10.1126/science.1125629

Yoshida H, Bogaki M, Nakamura S, Ubukata K, Konno M (1990) Nucleotide sequence and characterization of the *Staphylococcus aureus norA* gene, which confers resistance to quinolones. J Bacteriol 172:6942–6949, http://www.ncbi.nlm.nih.gov/pubmed/2174864

Zhang L, Ma S (2010) Efflux pump inhibitors: a strategy to combat P-glycoprotein and the NorA multidrug resistance pump. Chem Med Chem 5:811–822. doi:10.1002/cmdc.201000006

Zhao Y, Mao G, Liu M, Zhang L, Wang X, Zhang XC (2014) Crystal structure of the *E. coli* peptide transporter YbgH. Structure 22:1152–1160. doi:10.1016/j.str.2014.06.008

Zheng H, Wisedchaisri G, Gonen T (2013) Crystal structure of a nitrate/nitrite exchanger. Nature 497:647–651. doi:10.1038/nature12139

Zhou F, You G (2007) Molecular insights into the structure-function relationship of organic anion transporters OATs. Pharm Res 24:28–36. doi:10.1007/s11095-006-9144-9

How Pathogens Survive Drug Pressure?

Brijendra Kumar Tiwari, Gunjan Kak, Deepika Sharma, and Krishnamurthy Natarajan

Abstract Antibiotic resistance can be a consequence of repeat-induced point (RIP) mutation and even by horizontal gene transfer in the pathogen genome for every chromosomal replication. On the account of a few vital antibiotic agents, point mutation of chromosomally encoded proteins is the essential instrument for resistance. Another procedure that may add to the development of resistance in the course of treatment is adaptive or induced change. Notwithstanding RIP mutation, resistance may likewise be interceded by enzymes that change the antibiotic and the target protein or lessen the intracellular concentration of the antibiotics. These systems of resistance are dispersed between microscopic organisms by horizontal gene transfer. Drug resistance grants bacterial development in the nearness of an antibiotic; in any case, it is by all account not the only variable adding to treatment failure. The resistance is also reflected in cases wherein the antibiotic fails to clear the infection regardless of the absence of resistant microbes. These microbes are tolerant, and clinical reports advocate that the level of tolerance to treatment failure and mortality in a few diseases can be as crucial as the nature of antibiotic resistance. Intelligent methodologies and awareness of potential harmful effects of drugs will expect to promise continuous worldwide access to efficient antibiotics.

1 Introduction

Antibiotic discovery was one of the momentous advances in the field of modern medicine. Antibiotics remain as a mainstay in therapeutic regimes. Antibiotics have saved numerous lives afflicted with bacterial infections and other life-threatening infections. The successful discovery of the first β-lactam, penicillin G, prompted the exploration and subsequent development of additional and effective antibiotics. This quest eventually led to the production of countless antimicrobial compounds,

B.K. Tiwari (✉) • G. Kak • D. Sharma • K. Natarajan
Infectious Disease Immunology Laboratory, Dr. B. R. Ambedkar Centre for Biomedical Research, University of Delhi (North Campus), Delhi 110007, India
e-mail: brijendratiwari@gmail.com

G. Arora et al. (eds.), *Drug Resistance in Bacteria, Fungi, Malaria, and Cancer*,
DOI 10.1007/978-3-319-48683-3_5

which are in widespread usage today. However, the major obstacle towards this end was the rapid emergence of drug resistance amongst various microorganisms, which thwarted the efficacy of therapeutic interventions. The first known resistance was seen in *Staphylococcus aureus* alongside penicillin (Keeney et al. 1979).This ultimately led to reduced usefulness of the drug and consequently limited therapeutic options. Till date, almost all known drugs have counterattack by their target microorganisms (Table 1). According to various reports, steady rise of resistance in *Plasmodium falciparum* against artemisinin-based combination therapy, spread of methicillin-resistant *Staphylococcus aureus* (MRSA) amongst hospital-acquired infections, resistance of *Neisseria gonorrhoeae* to cephalosporins and *Escherichia coli* to fluoroquinolones, etc. are emerging threats to the healthcare of modern society (Espadinha et al. 2013; Johnson et al. 2013; Ashley et al. 2014; Bharara et al. 2015; Pham et al. 2015). Resistance leads to life-threatening disease conditions and prolongs infection and prognosis. It aggravates mortality and cost of treatment, too. Drug resistance refers to the phenomenon when microorganisms such as bacteria, viruses, fungi and parasites alter ways such that the medications used to cure the infections are rendered ineffective. When the microorganisms become resistant to most antimicrobial compounds, they are often referred as "superbugs" (Khan and Khan 2016). This can give rise to major health crisis because such a deadly resistance may culminate into a fatal infection and lead to spread to others imposing huge treatment costs. Thus, it can have severe impact on healthcare and wreak havoc on individuals and society.

Microbes conquer antibiotic drug pressure by various biological processes (Fig. 1), which could be of two main types. First, it occurs when the microorganism has never encountered the drug against which it exhibits resistance. Second, as acquired resistance and it manifests itself following drug exposure. Antimicrobial drugs usually target a vital metabolic pathway or cell wall synthesis. To combat this threat, the microbe may either alter the target site of the drug (seen in fungi by altering cell wall composition rendering resistance to antifungal compounds) or might chemically modify it (e.g. aminoglycoside modification) (Shi et al. 2013), or plasmid-encoded degradative enzymes such as β-lactamases cleave the drug, thus hampering its action (Renneberg and Walder 1989). Genes of resistance are either plasmid-borne (Perlin and Lerner 1979) or present on mobile genetic elements (transposons) (Domingues et al. 2012) which may easily disseminate through conjugation, etc. This leads to spread of resistance in a population.

2 Aminoglycoside Resistance

Aminoglycosides represent important class of drugs. Chemically these contain an amino sugar attached to the aminocyclitol. These antibiotics are so essential attributable to their expansive range of movement against range of microscopic organisms. Amikacin, gentamicin, kanamycin, streptomycin and tobramycin are some examples and most effective in treatment of gram-negative and gram-positive

Table 1 Outline of resistance mechanism(s) of various classes of antibiotics

Antibiotic class	Target	Mechanism(s) of resistance	Example(s)	Reference (s)
Aminoglycosides	Translation	Phosphorylation, acetylation, nucleotidylation, efflux, altered target	Gentamicin, streptomycin, spectinomycin	Bryan and Kwan (1983), Busse et al. (1992), Mahbub et al. (2005)
β-Lactams	Peptidoglycan synthesis	Hydrolysis, efflux, altered target	Penicillins (ampicillins), cephalosporins (cephamycins), penems (meropenems)	Spratt and Cromie (1988), Philippon et al. (1989), Pournaras et al. (2005)
Cationic peptides	Cell membrane	Efflux, altered target	Colistin	Cai et al. (2012)
Glycopeptides	Peptidoglycan synthesis	Reprogramming peptidoglycan synthesis	Vancomycin, teicoplanin	Arthur and Courvalin (1993)
Lincosamides	Translation	Nucleotidylation, efflux, altered target	Clindamycin	Leclercq and Courvalin (1991), Leclercq (2002)
Lipopeptides	Cell membrane	Altered target	Daptomycin	Tenover (2006), Boucher and Sakoulas (2007)
Macrolides	Translation	Hydrolysis, efflux, altered target, glycosylation, phosphorylation	Erythromycin, azithromycin	Ross et al. (1990), Leclercq and Courvalin (1991), Leclercq (2002)
Oxazolidinones	Translation	Efflux, altered target	Linezolid	Prystowsky et al. (2001), Meka and Gold (2004)
Phenicols	Translation	Acetylation, efflux, altered target	Chloramphenicol	Schwarz et al. (2004), Mingoia et al. (2007)

(continued)

Table 1 (continued)

Antibiotic class	Target	Mechanism(s) of resistance	Example(s)	Reference (s)
Pyrimidines	C1 metabolism	Efflux, altered target	Trimethoprim	Huovinen (2001), Holmes et al. (2016)
Quinolones	DNA replication	Acetylation, efflux, altered target	Ciprofloxacin	Ferrero et al. (1995), Webber and Piddock (2001), Jacoby (2005)
Rifamycins	Transcription	ADP-ribosylation, efflux, altered target	Rifampin	Palomino and Martin (2014)
Streptogramins	Translation	C-O lyase (type B streptogramins), acetylation (type A streptogramins)	Synercid	Jensen et al. (2000)
Sulphonamides	C1 metabolism	Efflux, altered target	Sulphamethoxazole	Huovinen (2001), Sanchez and Martinez (2015)
Tetracyclines	Translation	Monooxygenation, efflux, altered target	Minocycline, tigecycline	Fluit et al. (2005), Linkevicius et al. (2015)

bacterial infection. The drug elicits its effect by binding to the bacterial ribosome irreversibly and hindering protein synthesis though drug interactions (Chen and Murchie 2014; Dunkle et al. 2014; Song et al. 2014). Drug-modifying enzymes make the drug inactive by introducing chemical changes. Several aminoglycoside-modifying enzymes are known (Fluit and Schmitz 1999; Schmitz et al. 1999). Resistance to aminoglycoside drugs can be mediated through enzymatic chemical modifications like phosphorylations, adenylations and acetylations. Phosphorylations are catalysed by ATP-dependent O-phosphorylation (APH), nucleotidyltransferases catalysed O-adenylation (ANT) and acetyltransferases mediate N-acetylation which requires acetyl-coA-cofactor (AAC) (Marengo et al. 1974; Araoz et al. 2000; Chesneau et al. 2007). These transformations make the drug incapable of binding to the ribosome, and hence translation process remains uninhibited. In addition to enzymatic aminoglycoside-modifying enzymes, efflux pumps and ribosomal RNA mutations also contribute to reduced drug susceptibility (Kriengkauykiat et al. 2005; Corcoran et al. 2006; Takaya et al. 2013).

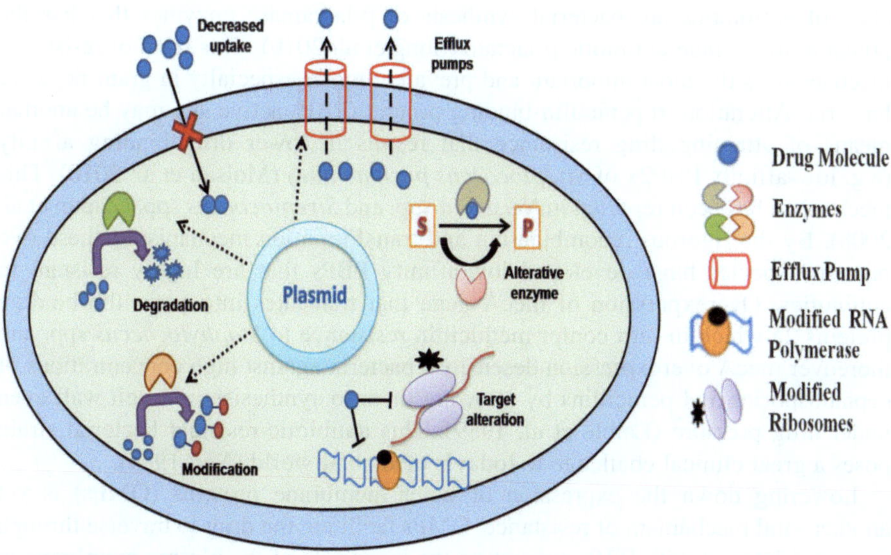

Fig. 1 Diverse biological processes used by microbes against drug pressure

3 Resistance to β-Lactam

The very first β-lactamase was identified from *Escherichia coli* long before penicillin came in clinical use. Kirby (Kirby and Burnell 1954) extracted "penicillin inactivators" from *Staphylococcus aureus*. The surge in the usage of β-lactam antibiotics has put a kind of selection pressure on bacteria, which ultimately resulted in the survival of drug-resistant bacteria that have the capacity to express multiple β-lactamases. Until date, several β-lactamases have been acknowledged. These enzymes have the potential to degrade the antibiotic by hydrolysing the β-lactam ring of antimicrobial drugs like cephalosporins, penicillin, etc. (Kasik and Peacham 1968). Through this mechanism of degradation of β-lactamases, antibiotics lead to lowering down of the efficacy of these molecules, ultimately leading to the survival of the bacterial species in the presence of the drug pressure. Different β-lactamases exhibit different specificities towards the substrate and differ in host range (Hanaki et al. 2007). Various types of β-lactamases are usually secreted by gram-negative bacteria which can degrade some cephalosporins like cephalothin. In few bacterial species, these enzymes are encoded by chromosomes, for example, cephalosporinases of *Pseudomonas*. In other bacteria such as *Enterobacteriaceae*, these enzymes result due to the presence of plasmid that encodes them. As found by Dhara and Tripathi (2014), plasmid-encoded enzymes also degrade a number of penicillins, and this effect can be overcome by the presence of β-lactamase inhibitors like clavulanic acid (Song et al. 2010). β-Lactam antibiotics elicit their response by targeting and inhibiting the action of key enzymes involved in the synthesis of bacterial cell wall. The basic mechanisms linked to resistance of this class of antibiotics are bacterial synthesis of β-lactamase enzymes that has the potential to degrade antibiotic β-lactam (Song et al. 2010). This form of resistance mechanism is the most important and prevalent mode especially in gram-negative bacteria. Alteration in penicillin-binding protein (PBP) active site may be another means of attaining drug resistance that results in lower drug binding affinity (e.g. low-affinity PBP2x of *Streptococcus pneumoniae*) (Moisan et al. 2010). This mechanism has been reported in *Neisseria* spp. and *Streptococcus* spp. (Zapun et al. 2008). By the rigorous recombination and transformation mechanisms, these two bacterial species have developed low-affinity PBPs that are highly resistant to antibiotics. Overexpression of mecA gene that translates into penicillin-binding proteins 2a which in turn confer methicillin resistance to *Staphylococcus* spp. and moreover mecA overexpression desensitise bacteria against high concentrations of cephalosporins and penicillins by allowing them to synthesise new cell wall even under drug pressure (Laible et al. 1989). This antibiotic-resistant bacterial strain poses a great clinical challenge to today's medicinal world (Neu 1984).

Lowering down the expression of outer membrane proteins (OMPs) is yet another vital mechanism of resistance. OMPs facilitate the drug to traverse through them and interact with PBPs present on the inner side of the plasma membrane in gram-negative bacteria. For example, resistance against carbapenems in some *Enterobacteriaceae* (*Klebsiella pneumoniae* and *Enterobacter* spp.) has been developed due to downregulation of these OMPs (Doumith et al. 2009); lowering down the expression of OprD gene is also linked with resistance towards imipenem and decrease in efficacy of meropenem in non-fermenter *Pseudomonas aeruginosa* (Moghoofei et al. 2015; Rodriguez-Beltran et al. 2015). Resistance against meropenem and imipenem has been reported due to downregulation of the CarO outer membrane protein (OMP) in multidrug-resistant clinical isolates of *Acinetobacter baumannii* (Fernandez-Cuenca et al. 2011). Various point mutations or insertion sequences in the genes coding for these porins proteins can produce altered OMPs that have loss in function or retarded function and permeability. Activation of efflux pumps provides intrinsic or acquired resistance phenotype. These efflux pumps are main determinants of multidrug resistance in various gram-negative pathogens, above all in *Acinetobacter* spp. and *Pseudomonas aeruginosa* (Morita et al. 2012). In *P. aeruginosa*, upregulation of the MexA-MexB-OprD framework and organism's low outer membrane permeability have been reported in several reports (Tamber et al. 2006), which have been attributed to formidable drug resistance such as decreased susceptibility against antibiotics like tetracycline (tet), penicillins, cephalosporins, chloramphenicol and quinolones. Moreover, upregulation of efflux pumps (e.g. AdeABC which is resistance-nodulation-division (RND) family sort efflux pump ordinarily found in *A. baumannii*) has been reported to confer carbapenem resistance by synthesising catalytically poor form of β-lactamase (del Mar et al. 2005).

3.1 Penicillin Resistance

The most potent mechanism of evading the action of penicillin by pneumococci is elicited through alteration in the penicillin-binding proteins (PBPs). These proteins are absolutely essential for the cell wall synthesis and serve to enforce the efficacy of β-lactam antibiotics by binding to them; therefore these alterations substantially decrease the affinity of PBPS to the drug and related classes of drugs, hence effectively hampering drug action and effect (Dowson et al. 1990). "Mosaics" comprising of mixed regions of native and acquired foreign DNA segments are responsible for encoding these altered PBPs. More often than not, the DNA from foreign source belongs to the more resistant strains like viridians streptococci. There are evidences about the transfer of such mixed and hybrid genetic elements between pneumococci and gram-positive bacteria like *Streptococcus oralis* (Sibold et al. 1994).

4 Quinolone Resistance

Nalidixic acid was the first discovered quinolone. Many derivatives have been made available since then, fluoroquinolones being the most important ones (Emmerson and Jones 2003). These compounds possess a fluorine substitution at sixth position on the quinolone moiety, making it highly efficient against gram-positive to gram-negative bacteria and anaerobes. Quinolone drug action is brought about by inhibiting an important class of enzymes called as bacterial topoisomerases (DNA gyrase and topoisomerase IV) (Chen et al. 1996). These enzymes play an important role in bacterial DNA replication and are central to the maintenance of bacterial replication fork by modifying the topology of double-stranded DNA. Enzyme structure comprises of two subunits, namely, A and B that are heterotetrameric in nature making it highly efficient against gram-positive to gram-negative bacteria and anaerobes (Chen et al. 1996; Higgins et al. 2003).

Two principal mechanisms can very well explain the resistance seen against quinolones. Firstly by modifying the target enzyme and second by limiting the permeability of the drug (Nikaido 1998; Hernandez et al. 2011). Quite plausibly, most changes are centred at the active domains of the enzyme, which drastically reduce drug binding. DNA gyrase activity is mostly inhibited in gram-negative bacteria, but in gram-positive either DNA gyrase or topoisomerase IV can be inhibited depending on the choice of fluoroquinolones used (Jacoby 2005). In majority of the cases, there is an amino acid substitution in quinolone-resistance-determining region, which introduces a bulky hydrophobic residue instead of a polar hydroxyl group (Mehla and Ramana 2016). Mutations in gyrA gene modify the enzyme-binding site or alter a charge that leads to conformational changes essential in maintaining drug enzyme interaction. Alterations in the outer membrane structure culminate into resistance as displayed by most gram-positive bacteria (Ferrero et al. 1994). Consequently, there is reduced drug influx and uptake. More unexpectedly, resistance mechanism spread by transmission leading to fluoroquinolone inactivation has also surfaced. This mechanism has cropped up because of the ability of aminoglycoside N-acetyltransferases to modify a secondary amine on the fluoroquinolones thus leading to lowered activity. The latter mechanism confers a low-level tolerance favouring the selection of resistance mutants (Robicsek et al. 2006). *Mycobacterium smegmatis* and *Mycobacterium bovis* elicit a basal and low-level resistance to several fluoroquinolones (Montero et al. 2001). The chromosomal gene called MfpA expression is plasmid encoded, the plasmid being present in multiple copies within the bacterium. Conversely, MfpA gene disruptions enhance drug efficacy and making wild type *M. smegmatis* more prone to drug action. Hence, drug susceptibility is directly linked to MfpA expression level.

A very fascinating mechanism of resistance against fluoroquinolones has been investigated in *Mycobacterium tuberculosis*. Studies elucidating *M. tuberculosis* MfpA structure have unfolded a unique three-dimensional structure of MfpA that shows similarity to bacterial DNA double helix. It is speculated that MfpA could serve to sequester the entire drug and free the bacterial DNA from drug effect. Therefore, target mimicry seen in mycobacterium affords protection against fluoroquinolones (Hegde et al. 2005). In addition, point mutations in genes like cytochrome b, or dihydrofolate reductase, are known to cause atovaquone resistance or pyrimethamine resistance, respectively (Meneceur et al. 2008).

5 Tetracycline Resistance

The ease of availability, broad range of activity and cost-effectiveness make tetracyclines as the most favourite and widely used antibiotics. Since their discovery in the 1940s, they have been readily used for therapeutic interventions (Nguyen et al. 2014). The drug elicits its inhibitory effects by impeding bacterial translation through the prevention of aminoacyl-tRNA attachment to the ribosomes (Connell et al. 2013). These antibiotics successfully combat pathogenic challenges from a wide array of microorganisms including gram-positive and gram-negative microbes, atypical life forms, for example, protozoan, *Chlamydiae*, *Rickettsiae* and *Mycoplasma* parasites. Tetracyclines include agents like tetracycline, minocycline, oxytetracycline and doxycycline (Rasmussen et al. 1997). The phenomenon of resistance against this class of drugs can be due to drug efflux, protection of bacterial ribosomes or chemical modification of the drug. Export proteins can contribute to the resistance by mediating drug efflux (Piddock et al. 2000). These gatherings of proteins have a place with the real facilitator superfamily. Tetracycline (tet) efflux pumps encode these fare proteins and subsequently encourage drug efflux (Stavropoulos and Strathdee 2000; Tuckman et al. 2000). The expulsion of drug ultimately lowers the drug concentration, and the inhibitory effects on the ribosomes are diminished. Ribosome protection proteins that are cytoplasmic in nature aid ribosomal protection. This mode of resistance is mostly prevalent in case of doxycycline and minocycline, whereas drug efflux is the major mechanism imparting resistance against most other classes of tetracyclines (Kobayashi et al. 2007). These efflux proteins share a marked homology with other class of efflux proteins that confer multidrug resistance to various other classes of antibiotics (Wang et al. 2004). Large plasmids, encoding for such efflux genes, are transmitted through conjugation and are believed to confer resistance to gram-negative bacteria (Roberts 1997). Another important means of resistance involves enzymatic inactivation of the drug. The role of tet (X) gene has been implicated in altering tetracycline function. The tet (X) gene encodes a 44 kDa product that is capable of modifying tetracycline chemically in the presence of oxygen and NADPH. This resistance gene is present on transposons and found in anaerobic *Bacteroides* species (Speer et al. 1991). More recently tetracycline destructases have been discovered. These are a novel class of inactivating enzymes belonging to flavoenzyme family and catalyse oxidation of the drug. Consequently, there is modification in the structure and function of the drug (Forsberg et al. 2015).

6 Peptide-Based Drug Resistance

The intrinsic and widespread resistance to most common and rampantly used antibiotics has led to the emergence of enterococci garnering the ability to survive in a hospital-borne environment (Canton et al. 1999). Their extensive survival and resilience to the most front-line drugs used in trauma care and hospital can be attributed to wide arrays of genomic changes such as mutations, acquisition of foreign genetic element harbouring resistance genes, plasmid transfer, transposons, etc. (Rossi et al. 2014; Hu et al. 2015a). The commonest resistance ensues against drugs of classes β-lactam and glycopeptides. Synergising antibiotics such as glycopeptides with an aminoglycoside can prove to be extremely fruitful in circumventing the deleterious emergence of hospital-borne antibiotic resistance (Hu et al. 2015b). A highly regulated clustered gene unit termed as operon is believed to mediate the acquisition of glycopeptide resistance in enterococci (James et al. 2012). This operon encodes an alternative pathway responsible for the production of a transformed cell wall component. Consequently, vancomycin binds to this modified precursor peptidoglycan more readily as the normal substrate remains unaffected and available for cell wall synthesis. Thus, the progression of the normal biosynthetic pathway remains unhindered (Fraise et al. 1997). Innate resistance to vancomycin can be attributed to two types of gene cluster designated as vanA and vanB gene clusters (Grissom-Arnold et al. 1997; Baptista et al. 1997). These confer resistance by modifying target from D-alanine-D-alanine to D-alanine-D-lactate (Marshall et al. 1997).

7 Resistance to Echinocandins and Azoles

Reduced susceptibility to echinocandins can be linked to genetic events such as mutations or instigation of an adaptive stress response (Astruey-Izquierdo et al. 2011). Mutations are mostly centred around regions known as "hot spots". These are much conserved gene clusters and are hubs of intrinsic point mutations. Mutation in FKS gene encoding, fungal FKS subunits of β(1,3)D-glucan synthase lead to cross resistance and decreased drug efficacy (Marti-Carrizosa et al. 2015; Dichtl et al. 2015). The drug-induced threat is bypassed by fungal cells through commencement of a stress response. This compensates for the drug-induced loss of a cell wall component by overproducing one or more other constituents.

The cell wall synthesis is highly regulated. In response to the drug, chitin levels are unregulated to balance the inhibition by echinocandins (Prasad et al. 2016). These elaborate metabolic changes are believed to be mediated via high-osmolarity glycerol, protein kinase C responses and Ca^{2+}-calcineurin signalling pathways. This helps in negating the fatal effects of echinocandins (Walker et al. 2010). A pivotal role is also played by genomic plasticity in aggravating resistance. This is achieved by loss of heterozygosity and is acquired by genetic rearrangements and amplifications majorly at genetic regions that are linked with resistance (Niimi et al. 2010). Frequent and rampant use of antifungal compounds, particularly fluconazole, has prompted the rise of resistance amongst different types of *Candida* species. These organisms display varied levels of susceptibility depending on the amount of selection pressure and the prevalence of infections (Mane et al. 2016). The activity of an important enzyme catalyst, i.e. lanosterol 14-α-sterol demethylase involved in cell wall biosynthesis, is inhibited by azoles (Warrilow et al. 2012). Additionally, accrual of a toxic by-product, namely, 14-α-methyl-3,6-diol, further contributes to the inhibitory effects of the antifungal agent (Warrilow et al. 2012). As a result, cell wall structure, with regard to ergosterol content, is altered leading to disruptions in membrane integrity and functioning.

Candida species are adept in manifesting resistance and can successfully evade prophylactic and therapeutic regimes. Three major mechanisms dictate this first mechanism involves upregulation and overexpression of efflux pumps. This leads to significantly lowered drug levels inside the cells (Niimi 2004). Efflux pumps of *Candida* spp. are CDR gene encoded belonging to ATP-binding superfamily or are products of MDR1 locus which encode major facilitator superfamily (MFS) proteins (Shao et al. 2016). Enhanced expression of CDR gene product culminating into expanded no. of efflux pumps presents imperviousness to all known azoles. Nevertheless, MDR-encoded efflux pumps bolster just fluconazole resistance. Examples of *Candida* spp. evoking azole resistance are as per the following: *Candida glabrata* (CgCDR1, CgCDR2) and *C. albicans* (CDR1, CDR2, MDR1) (Shao et al. 2016). One more resistance mechanism comprises the alteration or overproduction of the target molecule. A key player determining this mode of resistance is ERG11 gene that codes for the enzyme lanosterol 14-a-demethylase. Mutations pertaining to this gene lead to subtle modifications in the target enzyme that change the binding affinity of the enzyme to the drug (Martel et al. 2010). Lowered susceptibility of ERG11p to fluconazole as seen in *C. krusei* is due to this kind of altered binding (Martel et al. 2010). Overproduction of ERG11p makes the antifungal agent ineffective at its normal dosage, which is due to the outnumbering of target molecules with reference to drug molecules. Consequently, drug is present in insufficient amounts as compared to its target to successfully exhibit the inhibitory effects (Wang et al. 2009). ERG3 gene mutations circumvent the accumulation of toxic metabolic products and lead to formation of proper and functional cell membrane. Devising an alternate/bypass biosynthetic pathway paves way for another mechanism of resistance in fungi (Wang et al. 2009; Lo et al. 2015).

8 Methicillin-Resistant *Staphylococcus aureus*

The presence of MRSA has been detected in both the community-acquired and hospital settings. These are found to express mecA gene, which confers them resistance against methicillin and other β-lactams (Neu 1984). However, mecA gene's genetic environment is found to be different for the hospital-acquired and community-acquired isolates. Nosocomial MRSA is an example of multidrug resistance (Panda et al. 2016). Imperviousness to methicillin and a few other β-lactams has been connected to the ability of mecA to encode low-affinity penicillin-binding protein PBP2a (Roychoudhury et al. 1994). PBP2a-encoding mecA gene is located on a genetically mobile element that is termed as staphylococcal chromosomal cassette (SCC-mec) (Hososaka et al. 2007). In particular, resistance towards fluoroquinolone is regarded as a hallmark for nosocomial MRSA. Expression of EMRSA-17 lead to development of resistance towards a wide range of antibiotics like methicillin, macrolides (erythromycin), fluoroquinolones (ciprofloxacin), tetracycline, fusidic acid, aminoglycosides (kanamycin, streptomycin, gentamicin and neomycin) and rifampicin (Aucken et al. 2002).

9 Prospective

Augmentation in a number of diabetic patients and burn patients increases the susceptibility to acquired infections and spread of resistance. The resistant strains may evolve naturally when microorganisms replicate themselves in an erroneous fashion or by the exchange of resistant traits between them. Our abuse of antibiotics in people incomprehensibly quickened the procedure of drug resistance; however drug administration in intensive care units of hospitals and treatment of immunocompromised patients further leads to expansion of multidrug resistance and prevalence of nosocomial infections. To contain antibiotic resistance, motivating forces for drug organisation, clinics, specialist and patients have to be devised to act in ways that may restrain the exhaustion of antimicrobial efficacy.

Acknowledgements The authors wish to thank the Director of Dr. B.R. Ambedkar Center for Biomedical Research, University of Delhi (North Campus), Delhi, India, for providing necessary facilities. Brijendra K. Tiwari is thankful to UGC for granting Dr. D.S. Kothari Postdoctoral Fellowship.

References

Araoz R, Anhalt E, Rene L, Badet-Denisot MA, Courvalin P, Badet B (2000) Mechanism-based inactivation of VanX, a D-alanyl-D-alanine dipeptidase necessary for vancomycin resistance. Biochemistry 39:15971–15979. doi:10.1021/bi001408b

Arthur M, Courvalin P (1993) Genetics and mechanisms of glycopeptide resistance in enterococci. Antimicrob Agents Chemother 37:1563–1571. doi:10.1128/AAC.37.8.1563

Ashley EA, Dhorda M, Fairhurst RM, Amaratunga C, Lim P, Suon S, Sreng S, Anderson JM, Mao S, Sam B, Sopha C, Chuor CM, Nguon C, Sovannaroth S, Pukrittayakamee S, Jittamala P, Chotivanich K, Chutasmit K, Suchatsoonthorn C, Runcharoen R, Hien TT, Thuy-Nhien NT, Thanh NV, Phu NH, Htut Y, Han KT, Aye KH, Mokuolu OA, Olaosebikan RR, Folaranmi OO, Mayxay M, Khanthavong M, Hongvanthong B, Newton PN, Onyamboko MA, Fanello CI, Tshefu AK, Mishra N, Valecha N, Phyo AP, Nosten F, Yi P, Tripura R, Borrmann S, Bashraheil M, Peshu J, Faiz MA, Ghose A, Hossain MA, Samad R, Rahman MR, Hasan MM, Islam A, Miotto O, Amato R, MacInnis B, Stalker J, Kwiatkowski DP, Bozdech Z, Jeeyapant A, Cheah PY, Sakulthaew T, Chalk J, Intharabut B, Silamut K, Lee SJ, Vihokhern B, Kunasol C, Imwong M, Tarning J, Taylor WJ, Yeung S, Woodrow CJ, Flegg JA, Das D, Smith J, Venkatesan M, Plowe CV, Stepniewska K, Guerin PJ, Dondorp AM, Day NP, White NJ (2014) Spread of artemisinin resistance in *Plasmodium falciparum* malaria. N Engl J Med 371:411–423. doi:10.1056/NEJMoa1314981

Astruey-Izquierdo A, Cuesta I, Ros L, Mellado E, Rodriguez-Tudela JL (2011) Antifungal susceptibility profile of clinical *Alternaria spp.* identified by molecular methods. J Antimicrob Chemother 66:2585–2587. doi:10.1093/jac/dkr365

Aucken HM, Ganner M, Murchan S, Cookson BD, Johnson AP (2002) A new UK strain of epidemic methicillin-resistant *Staphylococcus aureus* (EMRSA-17) resistant to multiple antibiotics. J Antimicrob Chemother 50:171–175. doi:10.1093/jac/dkf117

Baptista M, Depardieu F, Reynolds P, Courvalin P, Arthur M (1997) Mutations leading to increased levels of resistance to glycopeptide antibiotics in VanB-type enterococci. Mol Microbiol 25:93–105. doi:10.1046/j.1365-2958.1997.4401812.x

Bharara T, Bhalla P, Rawat D, Garg VK, Sardana K, Chakravarti A (2015) Rising trend of antimicrobial resistance among Neisseria gonorrhoeae isolates and the emergence of *N. gonorrhoeae* isolate with decreased susceptibility to ceftriaxone. Indian J Med Microbiol 33:39–42. doi:10.4103/0255-0857.148374

Boucher HW, Sakoulas G (2007) Perspectives on Daptomycin resistance, with emphasis on resistance in *Staphylococcus aureus*. Clin Infect Dis 45:601–608. doi:10.1086/520655

Bryan LE, Kwan S (1983) Roles of ribosomal binding, membrane potential, and electron transport in bacterial uptake of streptomycin and gentamicin. Antimicrob Agents Chemother 23:835–845. doi:10.1128/AAC.23.6.835

Busse HJ, Wostmann C, Bakker EP (1992) The bactericidal action of streptomycin: membrane permeabilization caused by the insertion of mistranslated proteins into the cytoplasmic membrane of Escherichia coli and subsequent caging of the antibiotic inside the cells due to degradation of these proteins. J Gen Microbiol 138:551–561. doi:10.1099/00221287-138-3-551

Cai Y, Chai D, Wang R, Liang B, Bai N (2012) Colistin resistance of *Acinetobacter baumannii*: clinical reports, mechanisms and antimicrobial strategies. J Antimicrob Chemother 67:1607–1615. doi:10.1093/jac/dks084

Canton R, Mir N, Sanchez M, Baquero F (1999) MIC distribution and inoculum effect of LY333328: a study of vancomycin-susceptible and VanA-type and VanC-type enterococci obtained from intensive care unit patient surveillance cultures. Clin Microbiol Infect 5:554–559. doi:10.1111/j.1469-0691.1999.tb00434.x

Chen D, Murchie AI (2014) An aminoglycoside sensing riboswitch controls the expression of aminoglycoside resistance acetyltransferase and adenyltransferases. Biochim Biophys Acta 1839:951–958. doi:10.1016/j.bbagrm.2014.02.019

Chen CR, Malik M, Snyder M, Drlica K (1996) DNA gyrase and topoisomerase IV on the bacterial chromosome: quinolone-induced DNA cleavage. J Mol Biol 258:627–637. doi:10.1006/jmbi.1996.0274

Chesneau O, Tsvetkova K, Courvalin P (2007) Resistance phenotypes conferred by macrolide phosphotransferases. FEMS Microbiol Lett 269:317–322. doi:10.1111/j.1574-6968.2007.00643

Connell SR, Tracz DM, Nierhaus KH, Taylor DE (2013) Ribosomal protection proteins and their mechanism of tetracycline resistance. Antimicrob Agents Chemother 47:3675–3681. doi:10.1128/AAC.47.12.3675-3681.2003

Corcoran D, Quinn T, Cotter L, Fanning S (2006) An investigation of the molecular mechanisms contributing to high-level erythromycin resistance in Campylobacter. Int J Antimicrob Agents 27:40–45. doi:10.1016/j.ijantimicag.2005.08.019

del Mar TM, Beceiro A, Perez A, Velasco D, Moure R, Villanueva R, Martinez-Beltran J, Bou G (2005) Cloning and functional analysis of the gene encoding the 33- to 36-kilodalton outer membrane protein associated with carbapenem resistance in *Acinetobacter baumannii*. Antimicrob Agents Chemother 49:5172–5175. doi:10.1128/AAC.49.12.5172-5175.2005

Dhara L, Tripathi A (2014) Genetic and structural insights into plasmid-mediated extended-spectrum beta-lactamase activity of CTX-M and SHV variants among pathogenic Enterobacteriaceae infecting Indian patients. Int J Antimicrob Agents 43:518–526. doi:10.1016/j.ijantimicag.2014.03.002

Dichtl K, Samantaray S, Aimanianda V, Zhu Z, Prevost MC, Latge JP, Ebel F, Wagener J (2015) *Aspergillus fumigatus* devoid of cell wall beta-1,3-glucan is viable, massively sheds galactomannan and is killed by septum formation inhibitors. Mol Microbiol 95:458–471. doi:10.1111/mmi.12877

Domingues S, da Silva GJ, Nielsen KM (2012) Integrons: vehicles and pathways for horizontal dissemination in bacteria. Mob Genet Elements 2:211–223. doi:10.4161/mge.22967

Doumith M, Ellington MJ, Livermore DM, Woodford N (2009) Molecular mechanisms disrupting porin expression in ertapenem-resistant Klebsiella and *Enterobacter spp*. clinical isolates from the UK. J Antimicrob Chemother 63:659–667. doi:10.1093/jac/dkp029

Dowson CG, Hutchison A, Woodford N, Johnson AP, George RC, Spratt BG (1990) Penicillin-resistant viridans streptococci have obtained altered penicillin-binding protein genes from penicillin-resistant strains of *Streptococcus pneumoniae*. Proc Natl Acad Sci U S A 87:5858–5862. doi:10.1073/pnas.87.15.5858

Dunkle JA, Vinal K, Desai PM, Zelinskaya N, Savic M, West DM, Conn GL, Dunham CM (2014) Molecular recognition and modification of the 30S ribosome by the aminoglycoside-resistance methyltransferase NpmA. Proc Natl Acad Sci U S A 111:6275–6280. doi:10.1073/pnas.1402789111

Emmerson AM, Jones AM (2003) The quinolones: decades of development and use. J Antimicrob Chemother 51:13–20. doi:10.1093/jac/dkg208

Espadinha D, Faria NA, Miragaia M, Lito LM, Melo-Cristino J, de Lencastre H (2013) Extensive dissemination of methicillin-resistant *Staphylococcus aureus* (MRSA) between the hospital and the community in a country with a high prevalence of nosocomial MRSA. PLoS One 8, e59960. doi:10.1371/journal.pone.0059960

Fernandez-Cuenca F, Smani Y, Gomez-Sanchez MC, Docobo-Perez F, Caballero-Moyano FJ, Dominguez-Herrera J, Pascual A, Pachon J (2011) Attenuated virulence of a slow-growing pandrug-resistant *Acinetobacter baumannii* is associated with decreased expression of genes encoding the porins CarO and OprD-like. Int J Antimicrob Agents 38:548–549. doi:10.1016/j.ijantimicag.2011.08.002

Ferrero L, Cameron B, Manse B, Lagneaux D, Crouzet J, Famechon A, Blanche F (1994) Cloning and primary structure of *Staphylococcus aureus* DNA topoisomerase IV: a primary target of fluoroquinolones. Mol Microbiol 13:641–653. doi:10.1111/j.1365-2958.1994.tb00458.x

Ferrero L, Cameron B, Crouzet J (1995) Analysis of gyrA and grlA mutations in stepwise-selected ciprofloxacin-resistant mutants of *Staphylococcus aureus*. Antimicrob Agents Chemother 39:554–1558. doi:10.1128/AAC.39.7.1554

Fluit AC, Schmitz FJ (1999) Class 1 integrons, gene cassettes, mobility, and epidemiology. Eur J Clin Microbiol Infect Dis 18:761–770. doi:10.1007/s100960050398

Fluit AC, Florijn A, Verhoef J, Milatovic D (2005) Presence of tetracycline resistance determinants and susceptibility to tigecycline and minocycline. Antimicrob Agents Chemother 49:1636–1638. doi:10.1128/AAC.49.4.1636-1638.2005

Forsberg KJ, Patel S, Wencewicz TA, Dantas G (2015) The tetracycline destructases: a novel family of tetracycline-inactivating enzymes. Chem Biol 22:888–897. doi:10.1016/j.chembiol.2015.05.017

Fraise AP, Andrews J, Wise R (1997) Activity of a new glycopeptide antibiotic (LY333328) against enterococci and other resistant gram-positive organisms. J Antimicrob Chemother 40:423–425. doi:10.1093/jac/40.3.423

Grissom-Arnold J, Alborn WE Jr, Nicas TI, Jaskunas SR (1997) Induction of VanA vancomycin resistance genes in *Enterococcus faecalis*: use of a promoter fusion to evaluate glycopeptide and nonglycopeptide induction signals. Microb Drug Resist 3:53–64. doi:10.1089/mdr.1997.3.53

Hanaki H, Koide Y, Yamazaki H, Kubo R, Nakano T, Atsuda K, Sunakawa K (2007) Substrate specificity of HMRZ-86 for beta-lactamases, including extended-spectrum beta-lactamases (ESBLs). J Infect Chemother 13:390–395. doi:10.1007/s10156-007-0563-2

Hegde SS, Vetting MW, Roderick SL, Mitchenall LA, Maxwell A, Takiff HE, Blanchard JS (2005) A fluoroquinolone resistance protein from *Mycobacterium tuberculosis* that mimics DNA. Science 308:1480–1483. doi:10.1126/science.1110699

Hernandez A, Sanchez MB, Martinez JL (2011) Quinolone resistance: much more than predicted. Front Microbiol 2:22. doi:10.3389/fmicb.2011.00022

Higgins PG, Fluit AC, Schmitz FJ (2003) Fluoroquinolones: structure and target sites. Curr Drug Targets 4:181–190. doi:10.2174/1389450033346920

Holmes AH, Moore LS, Sundsfjord A, Steinbakk M, Regmi S, Karkey A, Guerin PJ, Piddock LJ (2016) Understanding the mechanisms and drivers of antimicrobial resistance. Lancet 387:176–187. doi:10.1016/S0140-6736(15)00473-0

Hososaka Y, Hanaki H, Endo H, Suzuki Y, Nagasawa Z, Otsuka Y, Nakae T, Sunakawa K (2007) Characterization of oxacillin-susceptible mecA-positive Staphylococcus aureus: a new type of MRSA. J Infect Chemother 13:79–86. doi:10.1007/s10156-006-0502-7

Hu J, Zhang X, Liu X, Chen C, Sun B (2015a) Mechanism of reduced vancomycin susceptibility conferred by walK mutation in community-acquired methicillin-resistant Staphylococcus aureus strain MW2. Antimicrob Agents Chemother 59:1352–1355. doi:10.1128/AAC.04290-14

Hu Y, Liu A, Vaudrey J, Vaiciunaite B, Moigboi C, McTavish SM, Kearns A, Coates A (2015b) Combinations of beta-lactam or aminoglycoside antibiotics with plectasin are synergistic against methicillin-sensitive and methicillin-resistant *Staphylococcus aureus*. PLoS One 10, e0117664. doi:10.1371/journal.pone.0117664

Huovinen P (2001) Resistance to trimethoprim-sulfamethoxazole. Clin Infect Dis 32:1608–1614. doi:10.1086/320532

Jacoby GA (2005) Mechanisms of resistance to quinolones. Clin Infect Dis 41(Suppl 2):S120–S126. doi:10.1086/428052

James RC, Pierce JG, Okano A, Xie J, Boger DL (2012) Redesign of glycopeptide antibiotics: back to the future. ACS Chem Biol 7:797–804. doi:10.1021/cb300007j

Jensen LB, Hammerum AM, Aarestrup FM (2000) Linkage of vat(E) and erm(B) in streptogamin-resistant *Enterococcus faecium* isolates from Europe. Antimicrob Agents Chemother 44:2231–2232. doi:10.1128/AAC.44.8.2231-2232.2000

Johnson R, Tchesnokova V, Johnston B, Clabots C, Roberts PL, Billig M, Riddell K, Rogers P, Qin X, Butler-Wu S, Price LB, Aziz M, Nicolas-Chanoine MH, Debroy C, Robicsek A, Hansen G, Urban C, Platell J, Trott DJ, Zhanel G, Weissman SJ, Cookson BT, Fang FC, Limaye AP, Scholes D, Chattopadhyay S, Hooper DC, Sokurenko EV (2013) Abrupt emergence of a single dominant multidrug-resistant strain of *Escherichia coli*. J Infect Dis 207:919–928. doi:10.1093/infdis/jis933

Kasik JE, Peacham L (1968) Properties of beta-lactamases produced by three species of mycobacteria. Biochem J 107:675–682. doi:10.1042/bj1070675

Keeney RE, Seamans ML, Russo RM, Gururaj VJ, Allen JE (1979) The comparative efficacy of minocycline and penicillin-V in *Staphylococcus aureus* skin and soft tissue infections. Cutis 23:711–718

Khan SN, Khan AU (2016) Breaking the spell: combating multidrug resistant 'superbugs'. Front Microbiol 7:174. doi:10.3389/fmicb.2015.01574

Kirby WM, Burnell JM (1954) Effect of combinations of antibiotics on lysis of *Staphylococcus aureus* by penicillin. J Bacteriol 67:50–52

Kobayashi T, Nonaka L, Maruyama F, Suzuki S (2007) Molecular evidence for the ancient origin of the ribosomal protection protein that mediates tetracycline resistance in bacteria. J Mol Evol 65:228–235. doi:10.1007/s00239-007-9006-z

Kriengkauykiat J, Porter E, Lomovskaya O, Wong-Beringer A (2005) Use of an efflux pump inhibitor to determine the prevalence of efflux pump-mediated fluoroquinolone resistance and multidrug resistance in *Pseudomonas aeruginosa*. Antimicrob Agents Chemother 49:565–570. doi:10.1128/AAC.49.2.565-570.2005

Laible G, Hakenbeck R, Sicard MA, Joris B, Ghuysen JM (1989) Nucleotide sequences of the pbpX genes encoding the penicillin-binding proteins 2x from *Streptococcus pneumoniae* R6 and a cefotaxime-resistant mutant, C506. Mol Microbiol 3:1337–1348. doi:10.1111/j.1365-2958.1989.tb00115.x

Leclercq R (2002) Mechanisms of resistance to macrolides and lincosamides: nature of the resistance elements and their clinical implications. Clin Infect Dis 34:482–492. doi:10.1086/324626

Leclercq R, Courvalin P (1991) Bacterial resistance to macrolide, lincosamide, and streptogramin antibiotics by target modification. Antimicrob Agents Chemother 35:1267–1272. doi:10.1128/AAC.35.7.1267

Linkevicius M, Sandegren L, Andersson DI (2015) Potential of tetracycline resistance proteins to evolve tigecycline resistance. Antimicrob Agents Chemother 60:789–796. doi:10.1128/AAC.02465-15

Lo HJ, Tseng KY, Kao YY, Tsao MY, Lo HL, Yang YL (2015) Cph1p negatively regulates MDR1 involved in drug resistance in *Candida albicans*. Int J Antimicrob Agents 45:617–621. doi:10.1016/j.ijantimicag.2015.01.017

Mahbub AM, Kobayashi N, Ishino M, Sumi A, Kobayashi K, Uehara N, Watanabe N (2005) Detection of a novel aph(2″) allele (aph[2″]-Ie) conferring high-level gentamicin resistance and a spectinomycin resistance gene ant(9)-Ia (aad 9) in clinical isolates of enterococci. Microb Drug Resist 11:239–247. doi:10.1089/mdr.2005.11.239

Mane A, Vidhate P, Kusro C, Waman V, Saxena V, Kulkarni-Kale U, Risbud A (2016) Molecular mechanisms associated with Fluconazole resistance in clinical *Candida albicans* isolates from India. Mycoses 59:93–100. doi:10.1111/myc.12439

Marengo PB, Chenoweth ME, Overturf GD, Wilkins J (1974) Phosphorylation of kanamycin, lividomycin A, and butirosin B by *Providencia stuartii*. Antimicrob Agents Chemother 6:821–824. doi:10.1128/AAC.6.6.821

Marshall CG, Broadhead G, Leskiw BK, Wright GD (1997) D-Ala-D-Ala ligases from glycopeptide antibiotic-producing organisms are highly homologous to the enterococcal vancomycin-resistance ligases VanA and VanB. Proc Natl Acad Sci U S A 94:6480–6483. doi:10.1073/pnas.94.12.6480

Martel CM, Parker JE, Bader O, Weig M, Gross U, Warrilow AG, Kelly DE, Kelly SL (2010) A clinical isolate of *Candida albicans* with mutations in ERG11 (encoding sterol 14alpha-demethylase) and ERG5 (encoding C22 desaturase) is cross resistant to azoles and amphotericin B. Antimicrob Agents Chemother 54:3578–3583. doi:10.1128/AAC.00303-10

Marti-Carrizosa M, Sanchez-Reus F, March F, Canton E, Coll P (2015) Implication of Candida parapsilosis FKS1 and FKS2 mutations in reduced echinocandin susceptibility. Antimicrob Agents Chemother 59:3570–3573. doi:10.1128/AAC.04922-14

Mehla K, Ramana J (2016) Structural signature of Ser83Leu and Asp87Asn mutations in DNA gyrase from enterotoxigenic *Escherichia coli* and impact on quinolone resistance. Gene 576:28–35. doi:10.1016/j.gene.2015.09.063

Meka VG, Gold HS (2004) Antimicrobial resistance to linezolid. Clin Infect Dis 39:1010–1015. doi:10.1086/423841

Meneceur P, Bouldouyre MA, Aubert D, Villena I, Menotti J, Sauvage V, Garin JF, Derouin F (2008) In vitro susceptibility of various genotypic strains of *Toxoplasma gondii* to pyrimethamine, sulfadiazine and atovaquone. Antimicrob Agents Chemother 52:1269–1277. doi:10.1128/AAC.01203-07

Mingoia M, Vecchi M, Cochetti I, Tili E, Vitali LA, Manzin A, Varaldo PE, Montanari MP (2007) Composite structure of *Streptococcus pneumoniae* containing the erythromycin efflux resistance gene mefI and the chloramphenicol resistance gene catQ. Antimicrob Agents Chemother 51:3983–3987. doi:10.1128/AAC.00790-07

Moghoofei M, Fazeli H, Poursina F, Nasr EB, Moghim S, Vaez H, Hadifar S, Ghasemian SH (2015) Morphological and bactericidal effects of amikacin, meropenem and imipenem on *Pseudomonas aeruginosa*. Jundishapur J Microbiol 8, e25250. doi:10.5812/jjm.25250

Moisan H, Pruneau M, Malouin F (2010) Binding of ceftaroline to penicillin-binding proteins of Staphylococcus aureus and *Streptococcus pneumoniae*. J Antimicrob Chemother 65:713–716. doi:10.1093/jac/dkp503

Montero C, Mateu G, Rodriguez R, Takiff H (2001) Intrinsic resistance of *Mycobacterium smegmatis* to fluoroquinolones may be influenced by new pentapeptide protein MfpA. Antimicrob Agents Chemother 45:3387–3392. doi:10.1128/AAC.45.12.3387-3392.2001

Morita Y, Tomida J, Kawamura Y (2012) MexXY multidrug efflux system of *Pseudomonas aeruginosa*. Front Microbiol 3:408. doi:10.3389/fmicb.2012.00408

Neu HC (1984) Changing patterns of hospital infections: implications for therapy. Changing mechanisms of bacterial resistance. Am J Med 77:11–23. doi:10.1016/j.eimc.2008.11.001

Nguyen F, Starosta AL, Arenz S, Sohmen D, Donhofer A, Wilson DN (2014) Tetracycline antibiotics and resistance mechanisms. Biol Chem 395:559–575. doi:10.1515/hsz-2013-0292

Niimi M (2004) An efficient system for functional hyper-expression of multidrug efflux pumps in *Saccharomyces cerevisiae*. Nihon Ishinkin Gakkai Zasshi 45:63–69. doi:10.3314/jjmm.45.63

Niimi K, Monk BC, Hirai A, Hatakenaka K, Umeyama T, Lamping E, Maki K, Tanabe K, Kamimura T, Ikeda F, Uehara Y, Kano R, Hasegawa A, Cannon RD, Niimi M (2010) Clinically significant micafungin resistance in *Candida albicans* involves modification of a glucan synthase catalytic subunit GSC1 (FKS1) allele followed by loss of heterozygosity. J Antimicrob Chemother 65:842–852. doi:10.1093/jac/dkq073

Nikaido H (1998) Antibiotic resistance caused by gram-negative multidrug efflux pumps. Clin Infect Dis 27:S32–S41. doi:10.1086/514920

Palomino JC, Martin A (2014) Drug resistance mechanisms in *Mycobacterium tuberculosis*. Antibiotics 3:317–340. doi:10.3390/antibiotics3030317

Panda RK, Mahapatra A, Mallick B, Chayani N (2016) Evaluation of genotypic and phenotypic methods for detection of methicillin resistant *Staphylococcus aureus* in a tertiary care hospital of Eastern Odisha. J Clin Diagn Res 10:DC19–DC21. doi:10.7860/JCDR/2016/17476.7278

Perlin MH, Lerner SA (1979) Amikacin resistance associated with a plasmid-borne aminoglycoside phosphotransferase in *Escherichia coli*. Antimicrob Agents Chemother 16:598–604. doi:10.1128/AAC.16.5.598

Pham QD, Do NT, Le YN, Nguyen TV, Nguyen DB, Huynh TK, Bui DD, Van KN, Nguyen PD, Luong AQ, Bui HT, Nguyen HH, McConnell M, Nguyen LT, Zhang L, Truong LX (2015) Pretreatment HIV-1 drug resistance to first-line drugs: results from a baseline assessment of a large cohort initiating ART in Vietnam, 2009-10. J Antimicrob Chemother 70:941–947. doi:10.1093/jac/dku473

Philippon A, Labia R, Jacoby G (1989) Extended-spectrum beta-lactamases. Antimicrob Agents Chemother 33:1131–1136. doi:10.1128/AAC.33.8.1131

Piddock LJ, White DG, Gensberg K, Pumbwe L, Griggs DJ (2000) Evidence for an efflux pump mediating multiple antibiotic resistance in *Salmonella enterica serovar Typhimurium*. Antimicrob Agents Chemother 44:3118–3121. doi:10.1128/AAC.44.11.3118-3121.2000

Pournaras S, Maniati M, Spanakis N, Ikonomidis A, Tassios PT, Tsakris A, Legakis NJ, Maniatis AN (2005) Spread of efflux pump-overexpressing, non-metallo-beta-lactamase-producing, meropenem-resistant but ceftazidime-susceptible *Pseudomonas aeruginosa* in a region with blaVIM endemicity. J Antimicrob Chemother 56:761–764. doi:10.1093/jac/dki296

Prasad R, Shah AH, Rawal MK (2016) Antifungals: mechanism of action and drug resistance. Adv Exp Med Biol 892:327–349. doi:10.1007/978-3-319-25304-6_14

Prystowsky J, Siddiqui F, Chosay J, Shinabarger DL, Millichap J, Peterson LR, Noskin GA (2001) Resistance to linezolid: characterization of mutations in rRNA and comparison of their occurrences in vancomycin-resistant enterococci. Antimicrob Agents Chemother 45:2154–2156. doi:10.1128/AAC.45.7.2154-2156.2001

Rasmussen BA, Bush K, Tally FP (1997) Antimicrobial resistance in anaerobes. Clin Infect Dis 24:S110–S120. doi:10.1093/clinids/24.Supplement_1.S110

Renneberg J, Walder M (1989) The role of beta-lactamase in mixed infections in mice in relation to treatment with ampicillin. J Infect Dis 160:337–341. doi:10.1093/infdis/160.2.337

Roberts MC (1997) Genetic mobility and distribution of tetracycline resistance determinants. Ciba Found Symp 207:206–218. doi:10.1002/9780470515358.ch13

Robicsek A, Strahilevitz J, Jacoby GA, Macielag M, Abbanat D, Park CH, Bush K, Hooper DC (2006) Fluoroquinolone-modifying enzyme: a new adaptation of a common *aminoglycoside acetyltransferase*. Nat Med 12:83–88. doi:10.1038/nm1347

Rodriguez-Beltran J, Cabot G, Valencia EY, Costas C, Bou G, Oliver A, Blazquez J (2015) N-acetylcysteine selectively antagonizes the activity of imipenem in Pseudomonas aeruginosa by an OprD-mediated mechanism. Antimicrob Agents Chemother 59:3246–3251. doi:10.1128/AAC.00017-15

Ross JI, Eady EA, Cove JH, Cunliffe WJ, Baumberg S, Wootton JC (1990) Inducible erythromycin resistance in staphylococci is encoded by a member of the ATP-binding transport super-gene family. Mol Microbiol 4:1207–1214. doi:10.1111/j.1365-2958.1990.tb00696.x

Rossi F, Diaz L, Wollam A, Panesso D, Zhou Y, Rincon S, Narechania A, Xing G, Di Gioia TS, Doi A, Tran TT, Reyes J, Munita JM, Carvajal LP, Hernandez-Roldan A, Brandao D, van der Heijden I, Murray BE, Planet PJ, Weinstock GM, Arias CA (2014) Transferable vancomycin resistance in a community-associated MRSA lineage. N Engl J Med 370:1524–1531. doi:10.1056/NEJMoa1303359

Roychoudhury S, Dotzlaf JE, Ghag S, Yeh WK (1994) Purification, properties, and kinetics of enzymatic acylation with beta-lactams of soluble penicillin-binding protein 2a. A major factor in methicillin-resistant Staphylococcus aureus. J Biol Chem 269:12067–12073

Sanchez MB, Martinez JL (2015) The efflux pump SmeDEF contributes to trimethoprim-sulfamethoxazole resistance in *Stenotrophomonas maltophilia*. Antimicrob Agents Chemother 59:4347–4348. doi:10.1128/AAC.00714-15

Schmitz FJ, Fluit AC, Gondolf M, Beyrau R, Lindenlauf E, Verhoef J, Heinz HP, Jones ME (1999) The prevalence of aminoglycoside resistance and corresponding resistance genes in clinical isolates of staphylococci from 19 European hospitals. J Antimicrob Chemother 43:253–259. doi:10.1093/jac/43.2.253

Schwarz S, Kehrenberg C, Doublet B, Cloeckaert A (2004) Molecular basis of bacterial resistance to chloramphenicol and florfenicol. FEMS Microbiol Rev 28:519–542. doi:10.1016/j.femsre.2004.04.001

Shao J, Zhang M, Wang T, Li Y, Wang C (2016) The roles of CDR1, CDR2, and MDR1 in kaempferol-induced suppression with fluconazole-resistant Candida albicans. Pharm Biol 54:984–992. doi:10.3109/13880209.2015.1091483

Shi K, Caldwell SJ, Fong DH, Berghuis AM (2013) Prospects for circumventing aminoglycoside kinase mediated antibiotic resistance. Front Cell Infect Microbiol 3:22. doi:10.3389/fcimb.2013.00022

Sibold C, Henrichsen J, Konig A, Martin C, Chalkley L, Hakenbeck R (1994) Mosaic pbpX genes of major clones of penicillin-resistant *Streptococcus pneumoniae* have evolved from pbpX genes of a penicillin-sensitive Streptococcus oralis. Mol Microbiol 12:1013–1023. doi:10.1111/j.1365-2958.1994.tb01089.x

Song JY, Jensen SE, Lee KJ (2010) Clavulanic acid biosynthesis and genetic manipulation for its overproduction. Appl Microbiol Biotechnol 88:659–669. doi:10.1007/s00253-010-2801-2

Song W, Kim YH, Sim SH, Hwang S, Lee JH, Lee Y, Bae J, Hwang J, Lee K (2014) Antibiotic stress-induced modulation of the endoribonucleolytic activity of RNase III and RNase G confers resistance to aminoglycoside antibiotics in *Escherichia coli*. Nucleic Acids Res 42:4669–4681. doi:10.1093/nar/gku093

Speer BS, Bedzyk L, Salyers AA (1991) Evidence that a novel tetracycline resistance gene found on two Bacteroides transposons encodes an NADP-requiring oxidoreductase. J Bacteriol 173:176–183

Spratt BG, Cromie KD (1988) Penicillin-binding proteins of gram-negative bacteria. Rev Infect Dis 10:699–711. doi:10.1093/clinids/10.4.699

Stavropoulos TA, Strathdee CA (2000) Expression of the tetA(C) tetracycline efflux pump in *Escherichia coli* confers osmotic sensitivity. FEMS Microbiol Lett 190:147–150

Takaya A, Sato Y, Shoji T, Yamamoto T (2013) Methylation of 23S rRNA nucleotide G748 by RlmAII methyltransferase renders *Streptococcus pneumoniae* telithromycin susceptible. Antimicrob Agents Chemother 57:3789–3796. doi:10.1128/AAC.00164-13

Tamber S, Ochs MM, Hancock RE (2006) Role of the novel OprD family of porins in nutrient uptake in *Pseudomonas aeruginosa*. J Bacteriol 188:45–54. doi:10.1128/JB.188.1.45-54.2006

Tenover FC (2006) Mechanisms of antimicrobial resistance in bacteria. Am J Med 119:S3–S10. doi:10.1016/j.amjmed.2006.03.011

Tuckman M, Petersen PJ, Projan SJ (2000) Mutations in the interdomain loop region of the tetA (A) tetracycline resistance gene increase efflux of minocycline and glycylcyclines. Microb Drug Resist 6:277–282. doi:10.1089/mdr.2000.6.277

Walker LA, Gow NA, Munro CA (2010) Fungal echinocandin resistance. Fungal Genet Biol 47:117–126. doi:10.1016/j.fgb.2009.09.003

Wang Y, Shoemaker NB, Salyers AA (2004) Regulation of a Bacteroides operon that controls excision and transfer of the conjugative transposon CTnDOT. J Bacteriol 186:2548–2557. doi:10.1128/JB.186.9.2548-2557.2004

Wang H, Kong F, Sorrell TC, Wang B, McNicholas P, Pantarat N, Ellis D, Xiao M, Widmer F, Chen SC (2009) Rapid detection of ERG11 gene mutations in clinical *Candida albicans* isolates with reduced susceptibility to fluconazole by rolling circle amplification and DNA sequencing. BMC Microbiol 9:167. doi:10.1186/1471-2180-9-167

Warrilow AG, Mullins JG, Hull CM, Parker JE, Lamb DC, Kelly DE, Kelly SL (2012) S279 point mutations in *Candida albicans* Sterol 14-alpha demethylase (CYP51) reduce in vitro inhibition by fluconazole. Antimicrob Agents Chemother 56:2099–2107. doi:10.1128/AAC.05389-11

Webber MA, Piddock LJ (2001) Absence of mutations in marRAB or soxRS in acrB-overexpressing fluoroquinolone-resistant clinical and veterinary isolates of *Escherichia coli*. Antimicrob Agents Chemother 45:1550–1552. doi:10.1128/AAC.45.5.1550-1552.2001

Zapun A, Contreras-Martel C, Vernet T (2008) Penicillin-binding proteins and beta-lactam resistance. FEMS Microbiol Rev 32:361–385. doi:10.1111/j.1574-6976.2007.00095.x

Functional Diversity and Applications of Mobile Group II Introns

Bhagwan Rekadwad, Juan M. Gonzalez, and Chandrahasya N. Khobragade

Abstract *Bacteria*, *Archaea*, and *Eukarya* include numerous thermophiles that are ubiquitous and have been detected in a variety of environments covering a really broad range of temperatures among factors. This suggests a great adaptability to both environmental conditions and nutrient sources which places thermophiles as a major target for environmental and evolutive studies with a great biotechnological potential as source for thermophilic enzymes and the biodegradation of various recalcitrant pollutants. While growth under optimal laboratory conditions is well studied, the potential for thriving under nonoptimal conditions, far from those considered ideal for a microorganism, remains to be studied. This chapter highlights possible development of novel methodology for the analysis of thermophilic microorganisms, which is applicable to other organisms, under natural conditions and for a broad range of extreme environments ranging from cold to hot temperatures, water activity, pH, and salinity as major naturally occurring extreme events. The mobile group II intron (MGI) functional diversity and abundance are assumed to represent a key feature indicator for the use of the full potential of microbial enzymes and a basic physiological process needed to understand microbial capabilities to grow and thrive under extreme conditions.

1 Mobile Group II Introns: A Novel Tool for Biotechnology

Mobile group II introns (MGIs) are "autocatalytic (self-splicing) genetic elements—'ribozymes'—found in bacterial and organellar DNAs of few eukaryotic organisms." These are evolutionary ancestors of spliceosomal introns, retrotransposons, and spliceosomes and in higher organisms, i.e., eukaryotes.

B. Rekadwad (✉) • C.N. Khobragade
School of Life Sciences, Swami Ramanand Teerth Marathwada University, Nanded 431606, Maharashtra, India
e-mail: rekadwad@gmail.com

J.M. Gonzalez
Institute of Natural Resources and Agrobiology, Spanish National Research Council, IRNAS-CSIC, Avda. Reina Mercedes 10, 41012 Sevilla, Spain
e-mail: jmgrau@irnase.csic.es

G. Arora et al. (eds.), *Drug Resistance in Bacteria, Fungi, Malaria, and Cancer*,
DOI 10.1007/978-3-319-48683-3_6

These lines present a novel and promising perspective to better understand the ubiquity, genetic diversity, and abundance of microorganisms under the broadest range of environmental conditions described on Earth. This chapter focuses on understanding the diversity and potential of MGI and their relationship to comprehend the microbial capacity to conquer a wide range of environments, including those representing extreme conditions and the ability of microorganisms to thrive and adapt to suboptimal environmental conditions. A listing of currently known group II introns in microorganisms is presented in later part of this chapter. This clearly indicates the broad distribution of these introns in prokaryotes and induces to believe that a large number of additional MGI can be discovered and described in the next years through extensive investigation based on new-generation sequencing platforms and the huge quantities of high-quality sequence data that is becoming available at the present time. Understanding the potential for the environmental and biotechnological application of microorganisms and specifically of their group II introns is a matter for the development in the next few years (Zimmerly 2014), and it deserves to be highlighted.

2 Current State of Mobile Group II Introns

Some research has been carried out on MGI from extreme environments worldwide, but there is scarce work performed on MGI functional diversity and its consequences for microbial growth, evolution, and their applications. Toro et al. (2007) studied the mobility of bacterial MGI and their splicing mechanisms. They have described reported gene-targeting-based recent development in MGI research. They have also discussed on bacterial MGI, phylogeny, and behavior of MGI in prokaryotes (bacteria). Dai and Zimmerly (2002) suggested that bacterial MGI behaves like retroelements, and their fundamental strategy is different from introns found in eukaryotes. Lambowitz and Zimmerly (2004) described the development of programmable and target-specific MGI into "targetrons." Jones et al. (2004) suggested that MGI represents a novel class of agent which has functions in targeted genetic repair. Chee and Takami (2005) reported the presence of active MGI in the recA gene of *Geobacillus kaustophilus*. Pyle (2010) reported the structure and molecular data of MGI-based location of domain six (D-VI) and additional domain sites of the group IIA and IIB introns. Perutka et al. (2004) developed a computer-based algorithm to allow a swift and proficient disruption/simulation of bacterial gene. The algorithm drew target site recognition in limelight, and *E. coli* DExH/D-box protein and DNA helicase disruptants found in *Escherichia coli* were successfully analyzed in relationship to the function of these proteins. Further research on the role and applicability of MGIs in microorganisms, above all, on those thriving under extreme environments, will significantly contribute to our understanding on microbial functional diversity and its evolutionary significance.

3 Diversity of MGI

Numerous new MGIs are recorded in bacteria followed by archaea, in fungi and higher organisms (Fig. 1). More than 324 MGIs were detected in bacteria with ORF domain up to X (i.e., 10). Most of them were located on the chromosome, followed by the variety of plasmids, transposons, and integrons. As compared to bacteria, in archaea (few more than 16) MGIs were discovered. Most of them were located on the chromosome. The ORF-less MGIs also detected in bacteria and archaea were very few in number. All of them were located on the chromosome, followed by the plasmid. Similarly, remarkable numbers of intron fragments were found in bacteria as well as in archaea. All of them were located on chromosome followed by plasmid and transposons. Unlike prokaryotes, mobile group II introns were discovered on eukaryotes on mitochondrial DNA. Eukaryote such as liverwort, green plants, algae, fungi, yeasts, and *Ichthyospora* possesses mitochondrial MGI. Most of these were found in ORF domain RT 2-10 and RT-X. Liverworts, fungi, and yeasts possess MGI fused with upstream exons (Dai et al. 2003; Simon et al. 2008; Candales et al. 2012; Zimmerly 2014).

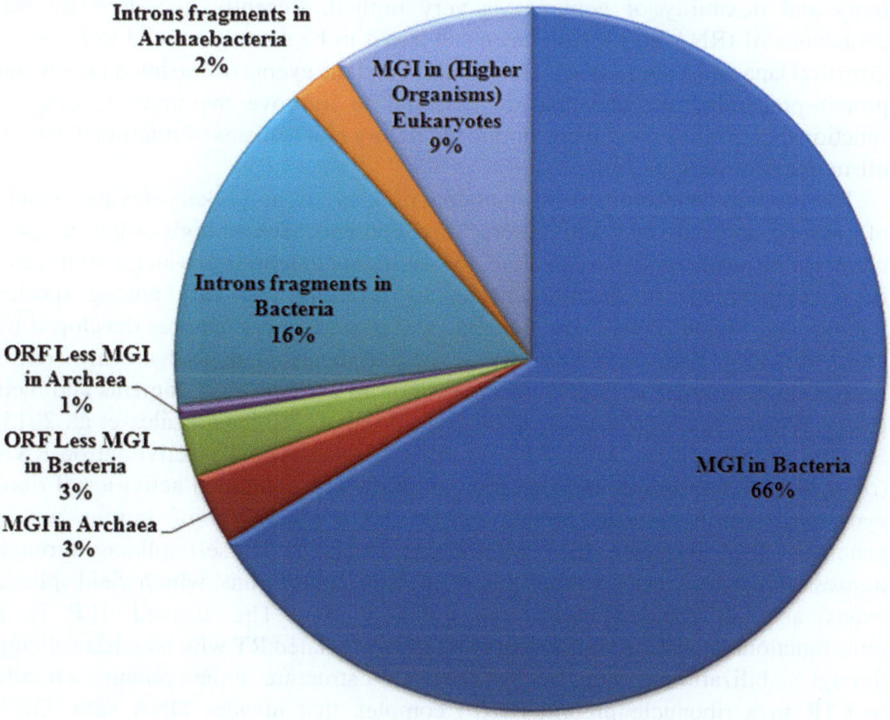

Fig. 1 Distribution and percentage of mobile group II introns among different life forms

4 Applications of MGI

Bacterial species are considered as dispersion-unlimited organisms due to its small dimensions and the broad range of metabolic capabilities. Thus, bacteria are able to reach and grow practically everywhere although the environment selects which species can develop under specific conditions (Prosser 2012). Joined to the huge bacterial diversity (Curtis et al. 2002) in our planet, the potential for developing under highly variable conditions turns into a surprising potential for bacterial growth on Earth and perhaps in other planets. Bacteria under optimum conditions, such as those provided in the laboratory, are able to show fast growth; however, scarce information is available on the functioning of bacteria in the environment and specifically under generally considered extreme conditions such as low or high temperatures (near 0 °C and up to 80 °C), low or high pH values, reduced water content, and high salinity, as the most typical examples occurring in nature. Seasonality is another cause of natural variability reaching extreme conditions above all as a result of current climate change events (Davidson and Janssens 2006; Rekadwad and Khobragade 2015, 2016). Genomes present a pool of all bacterial information available to thrive under easy and hardish conditions. However, our current understanding of the functional diversity of genes and the plasticity and flexibility of genomes is very limited. Recently, the diversity and abundance of tRNA genes have been suggested to be directly related to bacterial growth (Dana and Tuller 2014). Thus, growth and the events occurring on RNA and protein-processing mechanisms are essential to improve the understanding of functional capabilities and the genomic regulatory mechanisms of microbial metabolism and physiology (Table 1).

Thermophilic microorganisms including prokaryotes and eukaryotes are capable of growing under diverse and extreme environments such as high to low temperature, soil to water, acidic to alkaline, low-nutrient content, low water activity, etc. Base compositions in microorganisms are different and vary among species (Gomes and Steiner 2004). The genomic and physiological features developed by these microorganisms represent key adaptative mechanisms that are in need of further study in order to understand adaptation to extreme environments and their potential biotechnological application (Portillo et al. 2012; González et al. 2015; Santana and González, 2015). MGIs consist of a catalytically active intron RNA (ribozyme) and an intron-encoded protein (IEP). The combined activities of ribozyme and intron-encoded protein enable the proliferation of introns within genomes. The ribozyme (i.e., MGI-RNA) catalyzes its self-splicing through transesterification exactly similar that of spliceosomal introns, which yield spliced exons and an excised-intron-lariat RNA (LTR). The formed IEP is a multifunctional non-LTR-retrotransposon RT and related RT which assists splicing through stabilization of catalytically active RNA structure. It then remains a hurdle to LTR in a ribonucleoprotein (RNP) complex that invades DNA sites. DNA invasion caused ribozyme activity of intron (MGI-RNA), which reverses spliced into a host DNA strand. After invading host DNA strand, it's transcribed back into

Table 1 Spotlight on recently identified species having MGI

Species	MGI/enzyme	Ref.
Thermosynechococcus elongatus, Geobacillus stearothermophilus	Ribozyme	Mohr et al. (2013), Collins and Nilsen (2013)
Sinorhizobium meliloti, Sinorhizobium medicae	RmInt1	Toro et al. (2014)
Clostridium thermocellum	Catalytic RNAs, artificial nucleases, nucleic acid analogs, and peptide nucleic acids Xylanase, cellulase	Akinosho et al. (2014), Nakashima and Miyazaki (2014), Thomas et al. (2014)
Lactobacillus lactis	Ll.LtrB	
Podospora anserine	Possesses 15 group I introns and 1 MGI	Aguileta et al. (2014)
Coralline algae, *Gelidium vagum, Gelidium elegans*	MGI in the *chl*B gene	
Gracilaria chilensis, Gracilaria salicornia, Gracilaria tenuistipitata var. *liui, Grateloupia taiwanensis*	MGI in *trn*Me tRNA	Lee et al. (2016)
Liverwort	MGI	Hammani and Giege (2014)

new DNA by the IEP. Repeated cycles/invasion results in RNA splicing, and reverse splicing enables the invading introns to proliferate to new DNA sites and minimal impairing gene expression (Lambowitz and Zimmerly 2011). During the protein synthesis, each tRNA is changed and delivered into the ribosomes. Expression of tRNA genes has implications on the differential expression of different functional and structural proteins. Nowadays, the fast pace development in the field of transcriptomics and genomics has revealed structure and functions of many noncanonical tRNA genes. Disruption, fragmentation, rearrangement, minimization of tRNA and their re-coding, the relevance of non-coding RNAs, and a variety of small RNA sequences are included in these areas (Hartmann et al. 2004; Kanai 2013). At present, tRNA splicing is a boiling topic of hot debate because of their key role in the protein synthesis and influence in biological evolutions (Randau and Soll 2008; Di Giulio 2012).

5 Importance of MGI and Future Perspectives

MGIs are important elements present in microorganisms and organellar DNAs. Nevertheless, there is very limited information on their role and relevance in the adaptation of microorganisms to thrive under broadly variable environments and extreme conditions. Specifically, future perspectives on this topic will be focused on understanding the metabolic and physiological regulatory mechanisms involving mobile group II introns, and this will be carried out in the base to the current large availability of massive sequencing datasets that have daily been added to public genomic (DNA) and transcriptomic (RNA) repositories.

Thermophiles have been reported to represent a highly diverse bacterial group showing extreme adaptability to broad ranges of conditions for different environmental factors. Their ubiquity and genomic diversity make them a major group of interest for biotechnology (including, e.g., the search for highly efficient and stable enzymes) and environmental bioremediation through biodegradation of recalcitrant pollutants. The biotechnological potential of this group is starting to be discovered, and it involves interdisciplinary perspective to understand their roles, their potentials, and their applications (Sakaff et al. 2012; Santana et al. 2013, 2015). Thermophilic MGIs have important roles in a variety of processes such as biorecuperation, bioremediation, and the biotechnological use of their enzymes under the highly diverse set of working conditions. Future research will clarify all these aspects and will contribute decisively to the development of our understanding of microbial evolution and physiology and their biotechnological potential.

The applications of the ongoing research and current knowledge of biotechnology are presumed to expand exponentially in the following years. An example (Fig. 2) is provided on a current application of mobile group II introns. Additional applications are about to be described, and the future opens a wide range of possibilities for the development of MGI-derived technologies. Future studies on MGI must be focused on gene-targeting vector (genetic tools)-based development. The bacterial MGI is independent upon recombination, which makes this

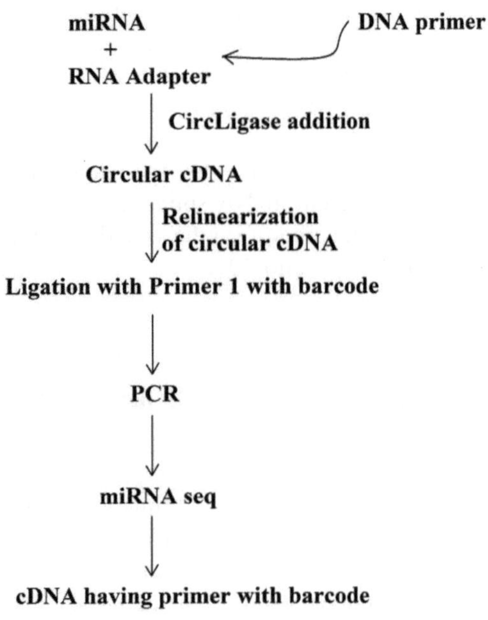

Fig. 2 Thermostable MGI reverse transcriptase (RT) (Enyeart et al. 2014)

technology broadly applicable. This extra feature of retroelements proved themselves as a potential tool for genetic manipulation in higher organisms (plants and animals) with lower chances of recombination. Additionally, certain host factors used by bacterial MGI contribute to intron-RNA folding and their mobility. In the near future, scientists should focus their research on the identification of factors (molecules) involved in increased functionality of bacterial MGI.

Studies on different aspects of MGI involving the architectural and functional organization are continuously helping us to understand the splicing mechanism of actively involved ribozymes and push us for parallel investigations on MGI and MGI-related elements such as retrotransposons and spliceosomes. Recently detected bacterial circular intron supports above statements that more study is needed to be performed in this particular area because MGI may be cosmopolitan in nature than we already thought. The mechanisms lying behind the splicing which occurred in MGI resulting in circle formation—formation of bacterial circular intron—and biological roles of such circles (circular introns) in MGI remain key issues expected to be resolved in the next few years.

References

Aguileta G, de Vienne DM, Ross ON, Hood ME, Giraud T, Petit E, Gabaldón T (2014) High variability of mitochondrial gene order among fungi. Genome Biol Evol 6:451–465. doi:10.1093/gbe/evu028

Akinosho A, Yee K, Close D, Ragauskas A (2014) The emergence of *Clostridium thermocellum* as a high utility candidate for consolidated bioprocessing applications. Front Chem 2:66. doi:10.3389/fchem.2014.00066

Candales MA, Duong A, Hood KS, Li T, Neufeld RAE, Sun R, McNeil BA, Wu L, Jarding AM, Zimmerly S (2012) Database for bacterial group II introns. Nucleic Acids Res 40:D187–D190. doi:10.1093/nar/gkr1043

Chee G, Takami H (2005) Housekeeping recA gene interrupted by group II intron in the thermophilic *Geobacillus kaustophilus*. Gene 363:211–220. doi:10.1016/j.gene.2005.08.017

Collins K, Nilsen TW (2013) Enzyme engineering through evolution: thermostable recombinant group II intron reverse transcriptases provide new tools for RNA research and biotechnology. RNA 19:1017–1018. doi:10.1261/rna.040451.113

Curtis TP, Sloan WT et al (2002) Estimating prokaryotic diversity and its limits. Proc Natl Acad Sci U S A 99:10494–10499. doi:10.1073/pnas.142680199

Dai L, Zimmerly S (2002) Compilation and analysis of group II intron insertions in bacterial genomes: evidence for retroelement behavior. Nucleic Acids Res 30:1091–1102. doi:10.1093/nar/30.5.1091

Dai L, Toor N, Olson R, Keeping A, Zimmerly S (2003) Database for mobile group II introns. Nucleic Acids Res 31:424–426. doi:10.1093/nar/gkg049

Dana A, Tuller T (2014) The effect of tRNA levels on decoding times of mRNA codons. Nucleic Acid Res. doi:10.1093/nar/gku646

Davidson EA, Janssens IA (2006) Temperature sensitivity of soil carbon decomposition and feedbacks to climate change. Nature 440:165–173. doi:10.1038/nature04514

Di Giulio M (2012) The 'recently' split transfer RNA genes may be close to merging the two halves of the tRNA rather than having just separated them. J Theor Biol 310:1–2. doi:10.1016/j.jtbi.2012.06.022

Enyeart PJ, Mohr G, Ellington AD, Lambowitz AM (2014) Biotechnological applications of mobile group II introns and their reverse transcriptases: gene targeting, RNA-seq, and non-coding RNA analysis. Mobile DNA 5:2. doi:10.1186/1759-8753-5-2

Gomes J, Steiner W (2004) The biocatalytic potential of extremophiles and extremozymes. Food Technol Biotechnol 42:223–235

González JM, Portillo MC, Piñeiro-Vidal M (2015) Latitude-dependent underestimation of microbial extracellular enzyme activity in soils. Int J Environ Sci Technol 12:2427–2434. doi:10.1007/s13762-014-0635-7

Hammani K, Giege P (2014) RNA metabolism in plant mitochondria. Trends Plant Sci 19:380–389. doi:10.1016/j.tplants.2013.12.008

Hartmann RK, Morl M, Sprinzl M (2004) The RNA world. RNA 10:344–349. doi:10.1261/rna.5240904

Jones JP III, Kierlin MN, Coon RG, Perutka J, Lambowitz AM, Sullenger BA (2004) Retargeting mobile group II introns to repair mutant genes. Mol Ther 11:687–694. doi:10.1016/j.ymthe.2005.01.014

Kanai A (2013) Molecular evolution of disrupted transfer RNA genes and their introns in archaea. In: Pontarotti P (ed) Evolutionary biology: exobiology and evolutionary mechanisms. Springer, Berlin, pp 181–193

Lambowitz AM, Zimmerly S (2004) Mobile group II introns. Annu Rev Genet 38:135. doi:10.1146/annurev.genet.38.072902.091600

Lambowitz AM, Zimmerly S (2011) Group II introns: mobile ribozymes that invade DNA. Cold Spring Harb Perspect Biol 3:a003616. doi:10.1101/cshperspect.a003616

Lee J, Kim KM, Yang EC, Niller KA, Boo SM, Bhattacharya S, Yoon HS (2016) Reconstructing the complex evolutionary history of mobile plasmids in red algal genomes. Scientific Rep 6:23744. doi:10.1038/srep23744

Mohr G, Hong W, Zhang J, G-z C, Yang Y et al (2013) A targetron system for gene targeting in thermophiles and its application in *Clostridium thermocellum*. PLoS One 8, e69032. doi:10.1371/journal.pone.0069032

Nakashima N, Miyazaki K (2014) Bacterial cellular engineering by genome editing and gene silencing. Int J Mol Sci 2014(15):2773–2793. doi:10.3390/ijms15022773

Perutka J, Wang W, Goerlitz D, Lambowitz AM (2004) Use of computer-designed group ii introns to disrupt *Escherichia coli* DExH/D-box protein and DNA helicase genes. J Mol Biol 336:421–439. doi:10.1016/j.jmb.2003.12.009

Portillo MC, Santana M, González JM (2012) Presence and potential role of thermophilic bacteria in temperate terrestrial environments. Natur wissenschaften 99:43–53. doi:10.1007/s00114-011-0867-z

Prosser J (2012) Ecosystem processes and interactions in a morass of diversity. FEMS Microbiol Ecol 81:507–519. doi:10.1111/j.1574-6941.2012.01435.x

Pyle AM (2010) The tertiary structure of group II introns: implications for biological function and evolution. Crit Rev Biochem Mol Biol 45:215–232. doi:10.3109/10409231003796523

Randau L, Soll D (2008) Transfer RNA genes in pieces. EMBO Rep 9:623–628. doi:10.1038/embor.2008.101

Rekadwad BN, Khobragade CN (2015) A case study on effects of oil spills and tar-ball pollution on beaches of Goa (India). Mar Pollut Bull 100:567–570. doi:10.1016/j.marpolbul.2015.08.019

Rekadwad BN, Khobragade CN (2016) Is the increase in oil pollution a possibility of the presence of diverse microorganisms? A experimental dataset on oil prevalent areas of Goa, India. Data Brief 9:8–12. doi:10.1016/j.dib.2016.07.048

Sakaff MKLM, Rahman AYA, Saito JA, Hou S, Alam M (2012) Complete genome sequence of the thermophilic bacterium *Geobacillus thermoleovorans* CCB_US3_UF5. J Bacteriol 194:1239. doi:10.1128/JB.06580-11

Santana MM, González JM (2015) High temperature microbial activity in upper soil layers. FEMS Microbiol Lett 362. doi:10.1093/femsle/fnv182

Santana MM, Portillo MC, González JM, Clara I (2013) Characterization of new soil thermophilic bacteria potentially involved in soil fertilization. J Plant Nutr Soil Sci 176:47–56. doi:10.1002/jpln.201100382

Santana MM, González JM, Clara MI (2015) Inferring pathways leading to organic-sulfur mineralization in the Bacillales. Crit Rev Microbiol 42:31–45. doi:10.3109/1040841x.2013.877869

Simon DM, Clarke NAC, McNeil BA, Johnson I, Pantuso D, Dai L, Chai D, Zimmerly S (2008) Group II introns in eubacteria and archaea: ORF-less introns and new varieties. RNA 14:1704–1713. doi:10.1261/rna.1056108

Thomas L, Joseph A, Gottumukkala LD (2014) Xylanase and cellulase systems of Clostridium sp.: an insight on molecular approaches for strain improvement. Biores Technol 158:343–350. doi:10.1016/j.biortech.2014.01.140

Toro N, Jimenez-Zurdo JI, Garcıa-Rodrıguez FM (2007) Bacterial group II introns: not just splicing. FEMS Microbiol Rev 31:342–358. doi:10.1111/j.1574-6976.2007.00068.x

Toro N, Rodríguez M, Abarca FM (2014) Insights into the history of a bacterial group II intron remnant from the genomes of the nitrogen-fixing symbionts *Sinorhizobium meliloti* and *Sinorhizobium medicae*. Heredity 113:306–315. doi:10.1038/hdy.2014.32

Zimmerly S (2014) Database for bacterial group II introns. http://webapps2.ucalgary.ca/~groupii/index.html. Accessed 5 June 2016

Bacterial Resistance Against Antibiotics

Anil Kumar and Nikita Chordia

Abstract From the time when discovery of penicillin was done in 1928, antibiotics are considered to be critical for public health that save the lives of millions of people around the world. Antibiotics are considered to be bactericidal which means capable of killing the bacteria. Some people call them as bacteriostatic which cease bacterial multiplication. These act without killing or damaging the body of the person. In recent years, it has been observed that more and more bacteria are becoming resistant to most of the frequently prescribed antibiotics. This situation is getting alarming day by day, and cure for even common diseases is becoming more expensive. Bacteria develop resistance to adapt their environments and ensure their survival. Drug/antibiotic resistance can be innate or acquired; there are many ways to acquire the resistance. It is becoming difficult to effectively treat wide variety of infections due to multidrug resistance. To control the drug resistance, misuse of antibiotics should be stopped, and regulations must be followed. In addition to control of drug resistance, it can be overcome by using additional molecules with antibiotics. Bacteria are finally overrunning our way of defense, so there is an urgent necessity to discover more antibiotics to combat the bacterial infections. To speed up the research, there is a need to advance the microbial informatics, particularly the development of databases and tools. Bioinformatics is the hope to help in easy availability of the information regarding resistance genes, associated proteins, available literature, cluster of orthologs (COG), pathways, and all other information concerning antibiotics.

1 Introduction

We all are exposed to the tiny microbes including bacteria, viruses, fungi, and protozoan. Some of them cause infection and are called pathogens, and others are even harmless inhabitants of our body. Human body has natural defense system against these pathogens, but sometimes it fails to control the infection. This leads to the development of the antibiotics (also called antimicrobials) which interfere with

A. Kumar (✉) • N. Chordia
School of Biotechnology, Devi Ahilya University, Khandwa Rd., Indore 452001, India
e-mail: ak_sbt@yahoo.com

G. Arora et al. (eds.), *Drug Resistance in Bacteria, Fungi, Malaria, and Cancer*,
DOI 10.1007/978-3-319-48683-3_7

the specific life processes of the organism. These are the medicines that kill pathogens without harming humans. These include both synthetic and semisynthetic antibiotics (Davison et al. 2000). Antibiotics are unique therapeutic agents that are directed to invade organism. They block the key reaction(s) in the pathogenic bacteria causing its death and/or inhibiting its multiplication. These are the drugs which can be taken orally, intravenously, or intramuscularly resulting in counter of the infection. This helps the immune system to fight against the infection. However, sometimes, excess use of antibiotics in an individual may affect the microbial ecology of the host (Monroe and Polk 2000). No single antibiotic is effective against all pathogenic bacteria. The antibiotics, viz., gentamicin and amoxicillin, which affect diverse variety of bacteria have been named as broad-spectrum antibiotics, whereas the antibiotics like vancomycin and penicillin which affect selective bacteria have been named as narrow-spectrum antibiotics (Heinemann 1999).

Since the introduction of penicillin in 1940s, more than hundreds of antibiotics have been discovered. Penicillin was hailed as a "miracle drug," and future was predicted as free of infectious diseases (Bentley 2005). The antibiotics are classified based on the spectrum as broad or narrow, similarly based on the route of administration as injectable or oral, and on the type of activity as bactericidal or bacteriostatic. Besides, antibiotics have also been named as β-lactam, macrolides, tetracyclines, fluoroquinolones, sulfonamides, aminoglycosides, imidazoles, peptides, and lincosamides on the basis of structural aspects. Structurally homologous antibiotics exhibit similar antibacterial activity (Kohanski et al. 2010; Wong et al. 2012).

Bacterial resistance can be of two types, viz., intrinsic or innate and acquired resistance. Innate resistance relies on the physiology and biochemistry of the bacteria and is considered to be peculiar property of specific bacteria. However, acquired resistance in a bacteria is developed by different means like transformation of specific gene(s) using bacterial and/or phage vectors, jumping genes, integrons, and site-directed mutagenesis in bacterial gene(s) itself or by a combination of these (Giedraitienė et al. 2011). A list of few resistant bacteria and names of the antibiotics for which bacteria is resistant is given in Table 1.

The frequent prescription of antibiotics by the doctors and habit of taking antibiotics without consulting the doctor especially in the developing countries are considered to be the main causes of resistance in bacteria against a particular antibiotic or a combination of antibiotics. In many cases, it has been observed that initially a particular bacteria remains sensitive to an antibiotic; however, it adapts slowly and ultimately becomes resistant to it. Many times, doctors change the prescription with different antibiotics which may lead in acquiring resistance against one or more drugs, and the condition is called as multidrug resistance (MDR) (Giedraitienė et al. 2011). Nowadays, many multidrug-resistant bacteria are known and to mention a few are specific strains of *Staphylococcus aureus*, *Enterococcus faecalis*, *Klebsiella pneumoniae*, and *Pseudomonas aeruginosa*.

In many bacteria, mode of adaptation for antibiotic resistance has been elucidated which include drug-inactivating enzyme, drug removal from the cell,

Table 1 Bacteria and their resistance to antibiotics

Bacteria	Resistant to antibiotics	References
Acinetobacter baumanii	Imipenem, meropenem, antipseudomonal agents, fluoroquinolones, carbapenems	Lee et al. (2012)
Clostridium difficile	Fluoroquinolone antibiotics, such as ciprofloxacin and levofloxacin	Loo et al. (2005)
Enterococcus faecium/ vancomycin resistant-enterococci (VRE)	Vancomycin, streptomycin, gentamicin, penicillin, ampicillin	Landman and Quale (1997), Arias et al. (2010)
Escherichia coli (ESBL strain)	Oral cephalosporins, TMP/SMX, fluoroquinolones	Prakash et al. (2009)
Klebsiella pneumoniae-extended spectrum beta-lactamases (ESBL)	Second-, third-generation cephalosporins, aztreonam, carbapenem	Paterson et al. (2004), Woodford et al. (2004)
Methicillin-resistant *Staphylococcus aureus* (MRSA)	β-Lactam ringed, viz., ampicillin, amoxicillin, penicillin	Liu et al. (2011)
Multidrug resistant *Mycobacterium tuberculosis* (MDR-TB)	Isoniazid, rifampin, possibly streptomycin	Keshavjee and Farmer (2012)
Mycobacterium tuberculosis	Isoniazid, rifampin	Gillespie (2002)
Neisseria gonorrhoeae	Penicillins, tetracyclines, fluoroquinolones, macrolides, cephalosporins	Farhi et al. (2009)
Pseudomonas aeruginosa	Cephalosporin, cefepime, tobramycin, gentamicin	Lister et al. (2009)
Pseudomonas aeruginosa (multidrug-resistant strains)	Resistance against combination of more than one antibiotics, viz., meropenem, cephalosporins, etc.	Mesaros et al. (2007)
Salmonella enteric	Ampicillin, chloramphenicol, tetracycline, sulfamethoxazole, trimethoprim, fluoroquinolones	Quinn et al. (2006)
Staphylococcus aureus	β-Lactam, fluoroquinolones, gentamicin	Chambers and DeLeo (2009)
Staphylococcus aureus (cMRSA strain)	β-Lactam ringed antibiotics, cephalosporins, erythromycin	Liu et al. (2011)
Staphylococcus aureus (partially resistant to vancomycin, VISA)	Vancomycin, β-lactam ringed antibiotics	Weinstein (2001), Fridkin (2001)
Staphylococcus epidermidis (methicillin resistant)	Penicillin, amoxicillin	Uçkay et al. (2009), Fey and Olson (2010)
Streptococcus pneumoniae (multidrug resistant)	Resistance against combination of more than one antibiotics, viz., penicillin, erythromycin, doxycycline, etc.	Mandell et al. (2007)

acquisition of a target, modification in current target, and reducing cell permeability. Still, critical new aspects of drug-resistance mechanism are continued to be discovered (Livermore 2003).

There is more treatment cost, longer stay as indoor hospital patient, chances of getting more infections, and more chance of death if a person is infected by drug-resistant bacteria (Heymann 2006). Therefore, scientists have realized to discover other ways of treatments such as specific vaccines for common bacterial infection. Unfortunately, in spite of much requirement of new antibiotic therapies, not so many new drugs/antibiotics have been approved by the controlling agencies like FDA. Therefore, there is really a difficult situation due to bacterial resistance against antibiotics.

Now, we have entered in the "post-antibiotic" era as choice for antibiotic is declining (Gandra et al. 2014). The present situation arose due to frequent and not judicious prescription of antibiotics to the patients and manufacturing of the same by the industries. It is considered that antibiotic resistance and multidrug resistance developed in pathogenic bacteria due to underuse and overuse of antibiotics. Besides, the misuse of antibiotics, inadequate diagnostics, and use of antibiotics in farming, aquaculture, and poultry contribute in resistance.

To treat the pathogens which are resistant to available treatment, new antibiotics must be discovered. However, there is a need of new antibiotic, but the number of newly approved drugs is continuously declining since new drugs are being produced either from natural compounds or by chemical modification of existing drugs (Donadio et al. 2010). Many tools and databases are available that contain information about resistant bacteria genome, drug-resistant genes, COG, annotation, and many more. These databases will help in fast discovery of new compounds that can treat the resistant microbes.

2 Mode of Action of Antibiotics

The different antibiotics work by inhibiting different bacterial cellular processes. The main bacterial cellular processes which are inhibited by the different antibiotics are biosynthesis of cell wall, polypeptide (protein), DNA, RNA, metabolic pathways, cell membrane, etc. (Sefton 2002; Kohanski et al. 2010; Wong et al. 2012).

2.1 Cell Wall Biosynthesis Inhibitors

It has been found that many antibiotics work by inhibiting the biosynthesis of cell wall (Tomasz and Waks 1975). The antibiotics commonly target bacterial cell wall formation because animal cells do not have cell walls. Bacterial growth is prevented by inhibiting peptidoglycan synthesis which is an important part of cell wall. This is particularly important when bacteria are dividing, because it is needed for the new cell that is forming. As the bacterium starts to replicate, it first elongates to about twice its normal size. So, more peptidoglycan is made for the extra surface area. But when these antibiotics are present, the peptidoglycan cannot cross-link properly, so

Table 2 Cell wall biosynthesis inhibitory antibiotics

Antibiotic class	Name	Primary target
β-Lactam	Penicillins, carbenicillin, ampicillin, penicillin G, cloxacillin, cephalosporins, monobactams	Penicillin-binding proteins
Glycopeptides	Vancomycin, teicoplanin, telavancin, bleomycin, ramoplanin, and decaplanin	Terminal dipeptide having alanyl moieties
Others	Alafosfalin	Terminal dipeptide having alanyl moieties
	Bacitracid	Prenylation
	Seromycin	Alr and Ddl enzyme
	Monurol/monuril	MurA enzyme
	Tunicamycin	Conversion of the undecaprenyl phosphate to the lipid I intermediate

the cell wall is very weak in places. These bacteria are subjected to osmotic lysis and subsequently die. It means that all of these antibiotics that inhibit peptidoglycan synthesis are bactericidal because they directly kill bacteria (Fisher et al. 2005).

It is found that β-lactam ringed antibiotics generally inhibit biosynthesis of bacterial cell wall. The examples are penicillins, carbenicillin, ampicillin, cloxacillin, cephalosporin, etc. The other classes of cell wall biosynthesis inhibitors are glycopeptides. The examples are bleomycin, vancomycin, decaplanin, etc. These glycopeptides are mono- or polycyclic peptides synthesized without involvement of ribosomes and have bound carbohydrate (glycol) moiety in them. They do this by binding with amino acids within the cell wall that prevents the addition of new units to the peptidoglycan (Silver 2003). Table 2 shows the classes of antibiotics, names, and their primary targets which inhibit cell wall biosynthesis.

2.2 *Nucleic Acid Biosynthesis Inhibitors*

Antibiotics can target nucleic acid (either RNA or DNA) synthesis. Nucleic acids are very important for a cell as these are the instruction manuals of the cells. When a cell divides, it must first replicate its DNA to give the new cell. Therefore, inhibiting nucleic acid synthesis is a good strategy to hinder bacterial growth (Goldberg 1965). The enzymes that carry out DNA and RNA syntheses are different enough between eukaryotic and prokaryotic cells. So there is selective toxicity. Prokaryotic replication and transcription processes include three steps: initiation, elongation, and termination. Antibiotic drugs have been developed to target each of these steps. For example, the antibiotic rifampin inhibits initiation process in RNA biosynthesis by binding with DNA-dependent RNA polymerase which is involved in the biosynthesis of RNA using DNA as a template. The antibiotic molecule is

Table 3 Nucleic acid synthesis inhibitory antibiotics

Antibiotic class	Name	Primary target
Rifamycins	Rifapentine, rifalazil	RNA polymerase (EC 2.7.7.6)
Resistomycins	Resistomycin, resistoflavin	RNA polymerase
Fluoroquinolones	Gemifloxacin, levofloxacin, ofloxacin, moxifloxacin	DNA gyrase
Sulfonamides	Sulfafurazole, sulfacetamide, sulfisomidine	Dihydropteroate synthase
Others	Novobiocin	DNA gyrase

thought to bind to the polymerase in such a way that it creates a wall that prevents the chain of RNA from elongating. In the presence of rifampin, bacteria cannot transcribe any gene that they need to carry out their normal functions, and therefore they die (Boehme et al. 2010).

Another example is quinolones that inhibit DNA synthesis by interfering with the coiling of DNA strands (Khodursky et al. 1995). During DNA replication, DNA gyrase relieves the torsional stress. As the replication fork moves along the bacterial chromosome, the strand of the DNA becomes supercoiled or excessively twisted. DNA gyrase binds to the DNA and cuts one of the strands to untwist before resealing. However, when quinolones are present, DNA gyrase gets inhibited and cannot reseal the strand. This causes the bacterium chromosome to break into smaller fragments and kills bacteria. Table 3 shows the classes of antibiotics, names, and their primary target which inhibit nucleic acid biosynthesis.

2.3 *Protein Biosynthesis Inhibitors*

Many antibiotics inhibit the synthesis of new proteins resulting in inhibition of cell growth/proliferation (Mukhtar and Wright 2005). Antibiotics inhibit bacterial protein synthesis at the ribosomal level and not eukaryotic protein synthesis due to the difference in the prokaryotic and eukaryotic ribosomal structures. The bacterial and eukaryotic ribosomal subunits have differences in RNA to protein ratio, size, sequence, etc. Due to these distinctions, antibiotics destroy microbes by targeting bacterial ribosomal subunits; however, eukaryotic ribosomal subunits are not targeted. Antibiotics work at different levels of translation for inhibiting protein synthesis like initiation, elongation, and termination. For example, tetracyclines bind to the 30S ribosomal subunit at the A site and prevent the attachment of aminoacyl-tRNAs. This hinders the next polypeptide string to be brought onto the ribosome (Brodersen et al. 2000). Another antibiotic, chloramphenicol, interacts with the larger (50S) ribosomal subunit and prevents peptide bond formation. When chloramphenicol is around, the amino acid cannot be linked together into a polypeptide string (Wolfe and Hahn 1965). Table 4 shows the classes of antibiotics, names, and their primary target which inhibit protein biosynthesis.

Table 4 List of antibiotics that inhibit protein biosynthesis

Antibiotic class	Name	Primary target
Tetracyclines	Oxytetracycline, doxycycline, tetracycline, demeclocycline, minocycline	Smaller (30S) ribosomal subunit and subsequently, prevention of binding of aminoacyl-tRNA onto the ribosome
Aminoglycoside	Tobramycin, gentamicin, amikacin, streptomycin, spectinomycin	Smaller (30S) ribosomal subunit and subsequently wrong amino acid incorporation due to misreading
Macrolides	Fidaxomicin, telithromycin, kitasamycin	Larger (50S) ribosomal subunit, interfere in chain elongation by interfering in addition of peptidyl-tRNA to incoming amino acid
Amphenicols	Chloramphenicol, azidamfenicol	Larger (50S) ribosomal subunit, interfere in chain elongation
Lincosamides	Clindamycin, lincomycin	Larger (50S) ribosomal subunit, interfere in chain elongation by interfering in addition of peptidyl-tRNA to incoming amino acid
Pleuromutilins	Valnemulin, azamulin	Larger (50S) ribosomal subunit, interfere in positioning of 3′-end of tRNA resulting in inhibition of peptide bond formation by peptide transferase
Others	Thiostrepton	Inhibit ribosome-dependent EF-Tu and EF-G GTPase

2.4 Metabolic Activity Inhibition

Chemicals that inhibit the essential component of the metabolism are called antimetabolites. These are used as antibiotics and inhibit the use of metabolite. These are analogues of the physiological metabolites. These analogues compete with the physiological metabolites resulting in retardation of cell growth or cell division (Brodie et al. 1958). There are three main types of antimetabolite antibiotics. The first is the antifolates which impair the function of folic acid leading to disruption in the biosynthesis of nucleotides (Kompis et al. 2005). For example, methotrexate is an analogue of folic acid, which inhibits biosynthesis of tetrahydrofolate by binding and inhibiting dihydrofolate reductase enzyme resulting in ultimate inhibition of both DNA and RNA biosynthesis (Hawser et al. 2006).

The second type of antimetabolite antibiotics consists of pyrimidine analogues which mimic the structure of metabolic pyrimidines. Three nucleobases, cytosine (C), thymine (T), and uracil (U), found in nucleic acids are pyrimidine derivatives, and the pyrimidine analogues disrupt their formation and consequently disrupt DNA and RNA synthesis (Kidwai et al. 2003).

The third type of antimetabolite antibiotics is purine analogues. They mimic the structure of metabolic purines. Two of the four bases in nucleic acids, adenine and guanine, are purines. Purine analogues disrupt nucleic acid production. For example, azathioprine is the main immunosuppressive cytotoxic substance that is widely used in transplants to control rejection reactions by inhibiting DNA synthesis in lymphocytes (Plunkett and Saunders 1991).

There are also antimetabolites that are specific for the metabolism of certain bacteria. That makes them suitable to use as an antibiotic against that bacteria. Table 5 shows the classes of antibiotics, names, and their primary target which inhibit metabolic activity.

2.5 *Cell Membrane Alteration*

Antibiotics damage the bacterial plasma membrane resulting in leaking the cell contents and disruption of the cross-membrane potential (ionic gradients) and ultimately leads in cell death (Ernst et al. 2000). Examples of antibiotics that disrupt the cell membrane include gramicidin and polymyxin. Gramicidin is a heterogeneous mixture of six antibiotic compounds. Gramicidin stimulates the movement of monovalent cations like sodium ions through unrestricted regions since bacterial cell membrane becomes more permeable. This leads in destruction of ionic gradient across the cell membrane. Polymyxin interacts with bacterial cell wall phospholipids and damages the structure of the bacterial cell membrane (Mogi and Kita 2009). Table 6 shows the classes of antibiotics, names, and their primary target which cause cell membrane alteration.

Table 5 Metabolic activity inhibitory antibiotics

Antibiotic class	Name	Primary target
Sulfonamides and dapsone	Sulfamethazine, sulfapyridine, sulfamethoxazole, sulfadiazine, sulfamerazine, methotrexate	(1) Dihydropteroate synthase
		(2) Compete with p-amino benzoic acid (PABA) preventing synthesis of folic acid
Pyrimidine analogues	Decitabine, gemcitabine, pentostatin	DNA synthesis
Purine analogues	Azathioprine, mercaptopurine	DNA synthesis

Table 6 Cell membrane altering antibiotics

Antibiotic class	Name	Primary target
Lipopeptides	Polymyxin B	Outer membrane by binding with lipopolysaccharides
Heterogeneous peptide	Gramicidin	Ion channels in the membrane
Others	Valinomycin, nonactin, salinomycin	Membrane ionophore

3 Development of Resistance in Bacteria

Antibiotic acts as a ligand for its target which is a specific molecule of the pathogen. The binding of the antibiotic with the specific target causes killing of the pathogen. As a natural response, antibiotic resistance emerges in the pathogen population either through spontaneous changes or through acquisition of resistant genes from other microbes. Prolonged repeated use of a particular antibiotic leads to a bulk of resistant cells in the pathogen population (Heinemann 1999). The acquired resistance can be attained by any one of the following biochemical mechanisms:

3.1 Production of Drug-Inactivating Enzymes

Bacteria modify the structure of the antibiotic resulting in their protection (Alekshun and Levy 2007). They produce enzymes that destroy or inactivate the antibiotic and thus becoming resistant to that antibiotic. A bacteria becomes resistant to β-lactam ringed antibiotics by producing β-lactamase capable of breaking β-lactam ring in the antibiotic resulting in loss of antibiotic activity. For example, after breaking the β-lactam ring of penicillin, penicilloic acid produced is ineffective in binding to penicillin-binding proteins (PBPs), protecting the process of cell wall synthesis (Livermore 1995).

Some bacteria protect themselves from aminoglycoside antibiotics, viz., neomycin, netilmicin, tobramycin, gentamicin, and amikacin, by secreting a specific enzyme which transfers a specific chemical group like phosphoryl group resulting in loss of antibiotic activity (Mingeot-Leclercq et al. 1999).

3.2 Modification of an Existing Target

Antibiotics work by specifically binding on the target within the bacteria, and even slight alteration in the target may affect binding of the antibiotic on it. Bacteria develop resistance by causing alteration in the target site in the antibiotics. Some bacteria modify their target sites to avoid recognition by the antibiotic. This is the reason that sometimes even without modification in the antibiotic structure, there is no binding of it with the target site in the bacteria resulting in no inhibition (Colas et al. 2000). Change in the bacterial target site generally occurs due to mutagenic change in the gene. Lambert (2005) reported mutation of DNA unfolding enzymes and RNA synthesizing enzyme by quinolones and rifamycins, respectively.

3.3 Alternate Target Production

This mechanism of antibiotic resistance is quite specific. In this mechanism, bacteria produces a substitute target that is not attacked by an antibiotic. Meanwhile, bacteria also produces a native target that is sensitive to antibiotic. Bacteria survives as alternative target adopts the role of the native target. Growth of a specific *Staphylococcus aureus* does not get inhibited by flucloxacillin due to the presence of an additional penicillin-binding protein, PB2′, which does not bind to β-lactams (Otero et al. 2013).

3.4 Reduced Cell Permeability

Antibiotic works only when it enters the bacterial cell and reaches up to its target site where it can interfere with normal functioning of the cell. Antibiotic enters in the bacterial cell through porin channels present in its outer membrane. Therefore, some bacteria lose the porin channels which reduce the uptake of many hydrophilic drugs across the cell wall. This stops the antibiotic from entering across the bacterial cell wall. A variety of microbes which do not retain violet stain modify the cell membrane porin channel frequency, size, and selectivity that reduce the uptake of certain antibiotics, viz., aminoglycosides and β-lactam ringed. The prohibited entry prevents these antibiotics from reaching their intended targets (Nguyen and Gutmann 1994).

3.5 Drug Removal from the Cell

Antibiotics can be effective only when they are present in a certain amount at the target site. Sometimes bacteria throw out the antibiotics almost at the same rate as it can enter with the help of membrane proteins, which act as export or efflux pump. Efflux pumps can be specific to antibiotics. The continuous outward flow of antibiotic from the cell leads to the low concentration of the antibiotic which is insufficient to elicit any response. Most of the efflux pumps are multidrug transporters capable to flow outward many unrelated antibiotics. This causes multidrug resistance (Li et al. 1994; Levy 2002a).

4 Misuse of Antibiotics

Antibiotics are the important drugs. They play important role in counteracting bacterial infection(s), hinder the spread of the disease, and reduce the complication(s) if any due to that particular disease. The overuse and misuse of antibiotics are the main causes for antibiotic resistance including multidrug resistance (Patterson 2001).

Overuse of the drug is caused when an antibiotic is taken for a condition that cannot be treated like viral infection or taking wrong doses. Overuse of drug may happen because of self-prescription and selling of the drug in the open market. This is because of the lack of imposition of legislation laws that leads to the sale of the counterfeit drugs which may contain inappropriate quantities of active ingredients. Overuse may also be because of overprescribing of antibiotics by the doctor that leads to the excessive demand for antibiotics by the population which ultimately results in antibiotic-resistant microbes. It has been observed that especially in the developing countries, antibiotics are mostly prescribed empirically without confirming bacterial confirmation in the pathology/microbiology laboratory. Overuse of antibiotics affects the body's normal flora and disrupts the balance between beneficial bacteria that help digestion (Blaser 2011).

Underuse of the drug can be manifested by not finishing a course of antibiotics as prescribed (stopping the antibiotic before the infection is fully cleared from the body). In developing countries, the unavailability of the drug also leads to the truncated treatment. This leads the infection to persist and proliferate and subsequently may threaten communities with new strains of infectious bacteria. The situation becomes more difficult to cure and more expensive to treat (Gilberg et al. 2003).

Besides exploitation of antibiotics in the treatment of infection in humans, antibiotics are also commonly used in farming, animal husbandry, and aquaculture. The use of antibiotics as pesticides is also done for treating trees and other agricultural products. Besides, antibiotics are added to animal feed for mass prophylaxis against infections or for growth promotion particularly for pigs and poultry farms. The sub-therapeutic doses of antibiotics are also used in water to treat fish diseases (Smith et al. 1994). Excessive application of antibiotics in intensive agricultural and farming units particularly pig and poultry farms is found as a growing threat. This can result in resistant microorganisms, which can spread to humans. Besides, there are indications that microbial resistance may get transmitted among animals including humans through food consumption (Marshall and Levy 2011). Therefore, there is a need to cut unnecessary use of antibiotics in farming. Responsible antibiotic use in industry and good practice for patients and physicians are essential to keep resistant bacterial strains curable and antibiotic treatment affordable to patients (Phillips et al. 2004).

5 Impact of Antibiotic Resistance

It is being observed that the number of antibiotic-resistant bacteria is increasing day by day, and therefore, it has become a matter of great clinical and public health concern (Sacks and Greene 2011). Some of the impacts of the antibiotic resistance bacteria are:

5.1 *Difficult to Treat Infections*

Treatable diseases like pneumonia, tuberculosis, and even minor infections have become incurable because of antibiotic resistance. This leads to the economic and emotional hurdle on families and on our healthcare system. As many strains of bacteria are resistant to several commonly used antibiotics, therefore, physicians will have to think for different antibiotics on trial basis till patient gets relief (Levy 2002b; Klugman 2007).

5.2 *Increased Cost and Length of Treatments*

There is a correlation between spreading of antibiotic resistance, indoor stay in the nursing home/hospital, financial cost of the treatment, and diagnostics (Cosgrove 2006). As bacteria are becoming resistant to antibiotics, drugs are becoming ineffective. So the drugs are replaced with the second line of drugs that are more expensive and may have more side effects. Infection with an organism/pathogen that is resistant to multiple drugs results in expensive treatment due to the use of multiple drugs (Niederman 2001).

5.3 *Increased Morbidity and Mortality*

Antibiotic resistance leads to inadequate or delayed therapy for several diseases that shows adverse outcome of an infection. It also resulted in lesser probability to cure diseases in humans and other eukaryotes including plants and animals. As per Threat Report 2013 of the Center for Disease Control and Prevention, nearly 23,000 persons die annually after infections caused by drug-resistant bacteria in USA. Many more patients die of other conditions complicated by infection with resistant pathogens (Cosgrove 2006).

6 Controlling Antibiotic Resistance

Antibiotic resistance has become a worldwide worry. It is apprehended that after infection of an individual by an antibiotic-resistant microbe, there are chances of its spread in other populations also in addition to tough way of treatment. Therefore, it has become almost necessary to control the spreading of drug-resistant microbes with the judicious use of drugs for prophylaxis and treatment. Antibiotic resistance can be controlled by:

6.1 Prudent Use of Antibiotics

Uncontrolled use of antibiotics must be decreased to retain the potency of existing antibiotics. To decrease the use of these valuable drugs, physicians, pharmacists, and the public must avoid careless use. To control the spread of resistance, antibiotics must be used smartly. Physicians must prescribe antibiotics for microbial loads only and in proper dose for correct amount of time. Besides, doctors should choose narrow-spectrum drugs to avoid killing populations of beneficial bacteria along with the disease-causing bacteria. Antibiotic usage for nontherapeutic purposes in farm animals and agriculture should be discouraged (Phillips 2001; Bergeron 2014).

6.2 Infection Control

Effective infection control that impedes the growth of bacteria can be used. Measures such as cleaning of water supplies, proper sanitation, and reduced overcrowding should be taken to prevent the infection. Other personal preventative measures like frequent hand washing should be taken. It would ensure that people become lesser sick. This will lead to reduce transfer of resistant infections to others (Weinstein 2001).

6.3 Use of Vaccines

Vaccines prevent infections and reduce the need for antibiotics. The need for antibiotics can be reduced by using vaccines that may help in prevention of infections. Vaccine can be used in young children who are more vulnerable to infection. A promising solution of evolving this bacterial resistance is the development of new vaccines. Efforts must be done that vaccines be effective for a longer period if not for lifetime. This would also be a solution to treat other infectious diseases for which there is a lack of efficacious medication (Mishra et al. 2012).

6.4 *Regulations*

Regulations are required for the use of antibiotics, and these may be like a doctor's prescription requirement for a patient to purchase an antibiotic. There must be mandatory requirement of the label on the antibiotic indicating that it must be taken only on the advice of a doctor and for confirmed bacterial infection only. It must not be sold without a photocopy of the prescription by a qualified doctor. Regulations must be such that punitive action must be taken against the pharmacist if found selling it without prescription (Gould 1999; Andersson and Hughes 2010).

7 Overcoming the Drug Resistance

Researchers are engaged to search alternate ways to combat antibiotic resistance which will strengthen the potency of prevailing drugs. It can be done by modifying the antibiotics in such a way that bacterial enzymes responsible to cause modify the antibiotics could not attack them. It may also be done by using additional enhancer like silver that enhances the potency of the antibiotic against gram-negative bacteria (Morones-Ramirez et al. 2013). Reactive oxygen species like superoxide can be used with drugs which can make them more effective as they affect the bacterial physiology (Kohanski et al. 2010). Physicians may be advised to prescribe "decoy" molecules along with the antibiotic. The "decoy" contains the qualities of an antibiotic but is not the actual antibiotic. The antibiotic is hidden behind the decoy molecule, and the secreted bacterial enzyme is unable to attack the antibiotic. Decoy molecules, viz., clavulanic acid, have been tried for making the secreted enzyme ineffective against the β-lactam ringed family. This is the best solution because it does not require the development of a new antibiotic. If more antibiotics are produced and new antibiotics are replacing the older ones, the bacteria will continue to grow resistance against the new antibiotic (Shamnas et al. 2013).

Another approach to solve the problem of antibiotic resistance is to interfere in the mechanism that promotes resistance instead of the attempt to kill bacteria. As an example, if a mechanism that duplicates or moves the bacteria's genetic material is interfered, then it can lead to the elimination of the transfer of resistant genes between bacteria.

8 New Antibiotic: Teixobactin

The fate of modern medicines depends on the potent antibiotics (So et al. 2010). Between 1930 and 1962, there are reports of more than 20 novel classes of antibiotics which have been discovered globally (Coates et al. 2011). These discoveries provided effective cure for many present-day known diseases. However, with increased cases of drug resistance, requirement is being felt for new compounds (Thomson et al. 2004). Almost after 30 years, in 2015 a new antibiotic called teixobactin has been found to treat many common microbial diseases, viz., tuberculosis and septicemia. Ling et al. (2015) showed its mode of action that it binds with the lipid II and lipid III which are precursors of peptidoglycan and cell wall's teichoic acid, respectively. This binding inhibits the biosynthesis of bacterial cell wall.

Teixobactin has been found in soil bacteria and using a high-throughput screening device called the iChip. The iChip has many plates with a large number of wells covered by twofold of semipermeable membranes. It is used for concurrent isolation and growth of uncultured microbes. Teixobactin is produced from the bacteria named *Eleftheria terrae*. Teixobactin has been found effective on mice infected with specific *S. aureus*, antibiotics tolerant strains of *Mycobacterium tuberculosis* and *Streptococcus pneumonia*. The properties of this compound may be exploited in developing new antibiotics likely to avoid development of resistance.

9 Databases and Tools

9.1 CARD

It is a database available at http://arpcard.mcmaster.ca. This database includes a variety of data that describe the resistance genes and associated proteins, antibiotics and their target, and literature concerning antibiotic resistance. CARD is updated continuously and curated using concurrently published data and continuous increasingly Antibiotic Resistance Ontology (ARO). At present, CARD is having a sequence of over 1600 antibiotic-resistant genes. It is integrated with other resources like NCBI and PDB. This may be used for searching all publications related to gene annotations, ontology, and connections with other online databases (McArthur et al. 2013).

9.2 ARDB

Antibiotic Resistance Genes Database (ARDB) contains data about tolerance and nucleic acid sequences annotated with much information regarding linkages to other nucleic acid and protein sequence databases, CDD annotations, COG, ontology, tolerance profile, mode of action, etc. It can be accessed from http://ardb.cbcb.umd.edu/. Till last update, this database has details for 23,137 tolerant genes, 632 genomes, 267 genera, 1737 species, and 2881 vectors and plasmids. Users can access the ARDB either by keyword search and browsing or by BLAST. Any user can identify and annotate new antibiotic-tolerant genes by blasting ARDB sequences. For that, one may use regular BLAST and/or RPS-BLAST tools available within the ARDB database (Liu and Pop 2009).

9.3 ARGO

ARGO stands for Antibiotic Resistance Genes Online that is featured to collect and archive antibiotic-resistance genes in bacteria. The current version of ARGO includes genes that are responsible for resistance to tetracycline, β-lactams, and vancomycin. It is available at http://www.argodb.org/. The ARGO can be searched either by sequence, gene finder, or classification of antibiotic. It is having links of APUA, ROAR, CDC, NRSA, EARSS, and VRSA databases and many more (Scaria et al. 2005).

9.4 MvirDB

It can be accessed at http://mvirdb.llnl.gov/. This contains detailed information regarding microbial virulence factors, antibiotic-tolerant genes, and proteinaceous toxins. The MvirDB collects data from various other online sources like SCORPION, Tox-Prot, the PRINTS virulence factors, VFDB, TVFac, ARGO, Islander, and subset of VIDA. MvirDB provides BLAST tool that can be used to align sequences of proteins and nucleic acids within it. To access the useful information, user can use browser tool. This database has automated system that updates the database weekly. It is having fast annotation system that automatically annotates protein entries (Zhou et al. 2007).

9.5 ARG-ANNOT

This is a tool which identifies the prevailing and suspected new antibiotic-tolerant genes in microbial genomes. It uses local blast program in Bio-Edit software to analyze sequences without user interface. Information required regarding antibiotic-tolerant genetic determinants to run the tool is taken from the literature and databases. To test the software, a database has been built that included 1689 antibiotic-resistant genes. Web interface for ARG-ANNOT is available on http://en.

mediterranee-infection.com/article.php?laref=283%26titre=arg-annot-. This link provides access to Bio-Edit and other tools for BLAST and post-BLAST analyses and tutorials to create a local database (Gupta et al. 2014).

9.6 *ResFinder*

ResFinder identifies acquired antimicrobial-resistant genes in total or partial sequenced isolates of bacteria. It is publically accessible at https://cge.cbs.dtu.dk/services/ResFinder/. ResFinder is updated continuously for newly identified antibiotic-tolerant genes. The tolerant genes can be identified for one or more antimicrobial classes at a same time using BLAST. ResFinder was created using 1862 sequences having 1411 antibiotic-tolerant genes and 23 de novo-sequenced isolates from which it identifies the acquired resistant genes. ResFinder is a web server that easily identifies the acquired antibacterial-tolerant genes in sequenced isolates (Zankari et al. 2012).

10 Conclusion

Development of antibiotic resistance has limited our repertoire of effective drugs, which creates a problematic situation to treat the bacterial infections. This threatens the effective prevention and treatment of resistant microbes.

Hence, there is a requirement of action in society and all government sectors. Although antibiotic resistance is a natural phenomenon, it is rapidly spreading due to human activities. Activities like misuse of antimicrobial drug in animal husbandry and farming support the disclosure and assemblage of antibiotic-tolerant strains. In addition, low-quality preventive measures and ways to curb menaces also aid in development of antibacterial tolerance. As antimicrobial resistance is rapidly growing, it will cause difficulty in treating bacterial infection, and there will be increased cost and length of treatments and more side effects because of the use of multiple and more powerful medications. Resistance can be overcome by strengthening the existing antibiotics either by using an additional molecule or to interfere with the mechanisms that promote resistance. Research is needed to find the best strategies for the optimal use of antibiotics and to find the novel class of antibiotics. To enhance research about antibiotics, various databases and tools are available containing data from bacterial population, genomics, drugs, mechanism of action, ontology, COG, CDD annotations, and many more. Now it is being realized that antibiotics must be taken as special category medicines, and efforts must be done to protect them as these are the wealth for humanity.

Acknowledgments We are grateful to the Department of Biotechnology, Government of India, New Delhi, for providing facilities under the Bioinformatics subcenter which were availed for this work.

References

Alekshun MN, Levy SB (2007) Molecular mechanisms of antibacterial multidrug resistance. Cell 128:1037–1050. doi:10.1016/j.cell.2007.03.004

Andersson DI, Hughes D (2010) Antibiotic resistance and its cost: is it possible to reverse resistance? Nat Rev Microbiol 8:260–271. doi:10.1038/nrmicro2319

Arias CA, Contreras GA, Murray BE (2010) Management of multidrug-resistant enterococcal infections. Clin Microbiol Infect 16:555–562. doi:10.1111/j.1469-0691.2010.03214.x

Bentley R (2005) The development of penicillin: genesis of a famous antibiotic. Perspect Biol Med 48:444–452. doi:10.1353/pbm.2005.0068

Bergeron J (2014) Prudent use of antibiotics. Can Vet J 55:714

Blaser M (2011) Antibiotic overuse: stop the killing of beneficial bacteria. Nature 476:393–394. doi:10.1038/476393a

Boehme CC, Nabeta P, Hillemann D, Nicol MP, Shenai S, Krapp F, Allen J, Tahirli R, Blakemore R, Rustomjee R, Milovic A, Jones M, O'Brien SM, Persing DH, Ruesch-Gerdes S, Gotuzzo E, Rodrigues C, Alland D, Perkins MD (2010) Rapid molecular detection of tuberculosis and rifampin resistance. N Engl J Med 363:1005–1015. doi:10.1056/NEJMoa0907847

Brodersen DE, Clemons WM, Carter AP, Morgan-Warren RJ, Wimberly BT, Ramakrishnan V (2000) The structural basis for the action of the antibiotics tetracycline, pactamycin, and hygromycin B on the 30S ribosomal subunit. Cell 103:1143–1154. doi:10.1016/S0092-8674(00)00216-6

Brodie BB, Gillette JR, La Du BN (1958) Enzymatic metabolism of drugs and other foreign compounds. Annu Rev Biochem 27:427–454. doi:10.1146/annurev.bi.27.070158.002235

Chambers HF, DeLeo FR (2009) Waves of resistance: *Staphylococcus aureus* in the antibiotic era. Nat Rev Microbiol 7:629–641. doi:10.1038/nrmicro2200

Coates AR, Halls G, Hu Y (2011) Novel classes of antibiotics or more of the same? Br J Pharmacol 163:184–194. doi:10.1111/j.1476-5381.2011.01250.x

Colas P, Cohen B, Ferrigno PK, Silver PA, Brent R (2000) Targeted modification and transportation of cellular proteins. Proc Natl Acad Sci U S A 97:13720–13725. doi:10.1073/pnas.97.25.13720

Cosgrove SE (2006) The relationship between antimicrobial resistance and patient outcomes: mortality, length of hospital stay, and health care costs. Clin Infect Dis 42:S82–S89. doi:10.1086/499406

Davison HC, Woolhouse ME, Low JC (2000) What is antibiotic resistance and how can we measure it? Trends Microbiol 8:554–559. doi:10.1016/S0966-842X(00)01873-4

Donadio S, Maffioli S, Monciardini P, Sosio M, Jabes D (2010) Antibiotic discovery in the twenty-first century: current trends and future perspectives. J Antibiot 63:423–430. doi:10.1038/ja.2010.62

Ernst WA, Thoma-Uszynski S, Teitelbaum R, Ko C, Hanson DA, Clayberger C, Krensky AM, Leippe M, Bloom BR, Ganz T, Modlin RL (2000) Granulysin, a T cell product, kills bacteria by altering membrane permeability. J Immunol 165:7102–7108. doi:10.4049/jimmunol.165.12.7102

Farhi D, Hotz C, Poupet H, Gerhardt P, Morand P, Poyart C, Sednaoui P, Avril MF, Dupin N (2009) *Neisseria gonorrhoeae* antibiotic resistance in Paris, 2005 to 2007: implications for treatment guidelines. Acta Derm Venereol 89:484–487. doi:10.2340/00015555-0704

Fey PD, Olson ME (2010) Current concepts in biofilm formation of *Staphylococcus epidermidis*. Future Microbiol 5:917–933. doi:10.2217/fmb.10.56

Fisher JF, Meroueh SO, Mobashery S (2005) Bacterial resistance to β-lactam antibiotics: compelling opportunism, compelling opportunity. Chem Rev 105:395–424. doi:10.1002/chin.200524267

Fridkin SK (2001) Vancomycin-intermediate and-resistant *Staphylococcus aureus*: what the infectious disease specialist needs to know. Clin Infect Dis 32:108–115. doi:10.1086/317542

Gandra S, Barter DM, Laxminarayan R (2014) Economic burden of antibiotic resistance: how much do we really know? Clin Microbiol Infect 20:973–979. doi:10.1111/1469-0691.12798

Giedraitienė A, Vitkauskienė A, Naginienė R, Pavilonis A (2011) Antibiotic resistance mechanisms of clinically important bacteria. Medicina (Kaunas) 47:137–146

Gilberg K, Laouri M, Wade S, Isonaka S (2003) Analysis of medication use patterns: apparent overuse of antibiotics and underuse of prescription drugs for asthma, depression, and CHF. J Manag Care Pharm 9:232–237

Gillespie SH (2002) Evolution of drug resistance in Mycobacterium tuberculosis: clinical and molecular perspective. Antimicrob Agents Chemother 46:267–274. doi:10.1128/AAC.46.2.267-274.2002

Goldberg IH (1965) Mode of action of antibiotics: II. Drugs affecting nucleic acid and protein synthesis. Am J Med 39:722–752. doi:10.1016/0002-9343(65)90094-X

Gould IM (1999) A review of the role of antibiotic policies in the control of antibiotic resistance. J Antimicrob Chemother 43:459–465. doi:10.1093/jac/43.4.459

Gupta SK, Padmanabhan BR, Diene SM, Lopez-Rojas R, Kempf M, Landraud L, Rolain JM (2014) ARG-ANNOT, a new bioinformatic tool to discover antibiotic resistance genes in bacterial genomes. Antimicrob Agents Chemother 58:212–220. doi:10.1128/AAC.01310-13

Hawser S, Lociuro S, Islam K (2006) Dihydrofolate reductase inhibitors as antibacterial agents. Biochem Pharmacol 71:941–948. doi:10.1016/j.bcp.2005.10.052

Heinemann JA (1999) How antibiotics cause antibiotic resistance. Drug Discov Today 4:72–79. doi:10.1016/S1359-6446(98)01294-X

Heymann DL (2006) Resistance to anti-infective drugs and the threat to public health. Cell 124:671–675. doi:10.1016/j.cell.2006.02.009

Keshavjee S, Farmer PE (2012) Tuberculosis, drug resistance, and the history of modern medicine. N Engl J Med 367:931–936. doi:10.1056/NEJMra1205429

Khodursky AB, Zechiedrich EL, Cozzarelli NR (1995) Topoisomerase IV is a target of quinolones in *Escherichia coli*. Proc Natl Acad Sci 92:11801–11805. doi:10.1073/pnas.92.25.11801

Kidwai M, Saxena S, Rastogi S, Venkataramanan R (2003) Pyrimidines as anti-infective agents. Curr Med Chem Anti-Infect Agents 2:269–286. doi:10.2174/1568012033483015

Klugman KP (2007) Clinical impact of antibiotic resistance in respiratory tract infections. Int J Antimicrob Agents 29:S6–S10. doi:10.1016/S0924-8579(07)70004-3

Kohanski MA, Dwyer DJ, Collins JJ (2010) How antibiotics kill bacteria: from targets to networks. Nat Rev Microbiol 8:423–435. doi:10.1038/nrmicro2333

Kompis IM, Islam K, Then RL (2005) DNA and RNA synthesis: antifolates. Chem Rev 105:593–620. doi:10.1021/cr0301144

Lambert PA (2005) Bacterial resistance to antibiotics: modified target sites. Adv Drug Deliv Rev 57:1471–1485. doi:10.1016/j.addr.2005.04.003

Landman D, Quale JM (1997) Management of infections due to resistant enterococci: a review of therapeutic options. J Antimicrob Chemother 40:161–170. doi:10.1093/jac/40.2.161

Lee YT, Kuo SC, Yang SP, Lin YT, Tseng FC, Chen TL, Fung CP (2012) Impact of appropriate antimicrobial therapy on mortality associated with Acinetobacter baumannii bacteremia: relation to severity of infection. Clin Infect Dis 55:209–215. doi:10.1093/cid/cis385

Levy SB (2002a) Active efflux, a common mechanism for biocide and antibiotic resistance. J Appl Microbiol 92:65S–71S. doi:10.1046/j.1365-2672.92.5s1.4.x

Levy SB (2002b) Factors impacting on the problem of antibiotic resistance. J Antimicrob Chemother 49:25–30. doi:10.1093/jac/49.1.25

Li XZ, Livermore DM, Nikaido H (1994) Role of efflux pump(s) in intrinsic resistance of *Pseudomonas aeruginosa*: resistance to tetracycline, chloramphenicol, and norfloxacin. Antimicrob Agents Chemother 38:1732–1741. doi:10.1128/AAC.38.8.1732

Ling LL, Schneider T, Peoples AJ, Spoering AL, Engels I, Conlon BP, Mueller A, Schäberle TF, Hughes DE, Epstein S, Jones M, Lazarides L, Steadman VA, Cohen DR, Felix CR, Fetterman KA, Millett WP, Nitti AG, Zullo AM, Chen C, Lewis K (2015) A new antibiotic kills pathogens without detectable resistance. Nature 517:455–459. doi:10.1038/nature14098

Lister PD, Wolter DJ, Hanson ND (2009) Antibacterial resistant *Pseudomonas aeruginosa*: clinical impact and complex regulation of chromosomally encoded resistance mechanisms. Clin Microbiol Rev 22:582–610. doi:10.1128/CMR.00040-09

Liu B, Pop M (2009) ARDB—antibiotic resistance genes database. Nucleic Acids Res 37:D443–D447. doi:10.1093/nar/gkn656

Liu C, Bayer A, Cosgrove SE, Daum RS, Fridkin SK, Gorwitz RJ, Sheldon LK, Karchmer AW, Levine DP, Murray BE, Rybak MJ, Talan DA, Chambers HF (2011) Clinical practice guidelines by the Infectious Diseases Society of America for the treatment of methicillin-resistant *Staphylococcus aureus* infections in adults and children. Clin Infect Dis 52:e18–e55. doi:10.1093/cid/ciq146

Livermore DM (1995) beta-Lactamases in laboratory and clinical resistance. Clin Microbiol Rev 8:557–584

Livermore DM (2003) Bacterial resistance: origins, epidemiology, and impact. Clin Infect Dis 36: S11–S23. doi:10.1086/344654

Loo VG, Poirier L, Miller MA, Oughton M, Libman MD, Michaud S, Bourgault AM, Nguyen T, Frenette C, Kelly M, Vibien A, Brassard P, Fenn S, Dewar K, Hudson TJ, Horn R, René P, Monczak Y, Dascal A (2005) A predominantly clonal multi-institutional outbreak of Clostridium difficile–associated diarrhea with high morbidity and mortality. N Engl J Med 353:2442–2449. doi:10.1056/NEJMoa051639

Mandell LA, Wunderink RG, Anzueto A, Bartlett JG, Campbell GD, Dean NC, Dowell SF, File TM, Musher DM, Niederman MS, Torres A, Whitney CG (2007) Infectious Diseases Society of America/American Thoracic Society consensus guidelines on the management of community-acquired pneumonia in adults. Clin Infect Dis 44:S27–S72. doi:10.1086/511159

Marshall BM, Levy SB (2011) Food animals and antimicrobials: impacts on human health. Clin Microbiol Rev 24:718–733. doi:10.1128/CMR.00002-11

McArthur AG, Waglechner N, Nizam F, Yan A, Azad MA, Baylay AJ, Bhullar K, Canova MJ, De Pascale G, Ejim L, Kalan L, King AM, Koteva K, Morar M, Mulvey MR, O'Brien JS, Pawlowski AC, Piddock LJ, Spanogiannopoulos P, Sutherland AD, Tang I, Taylor PL, Thaker M, Wang W, Yan M, Yu T, Wright GD (2013) The comprehensive antibiotic resistance database. Antimicrob Agents Chemother 57:3348–3357. doi:10.1128/AAC.00419-13

Mesaros N, Nordmann P, Plésiat P, Roussel-Delvallez M, Van Eldere J, Glupczynski Y, Van Laethem Y, Jacobs F, Lebecque P, Malfroot A, Tulkens PM, Van Bambeke F (2007) Pseudomonas aeruginosa: resistance and therapeutic options at the turn of the new millennium. Clin Microbiol Infect 13:560–578. doi:10.1111/j.1469-0691.2007.01681.x

Mingeot-Leclercq MP, Glupczynski Y, Tulkens PM (1999) Aminoglycosides: activity and resistance. Antimicrob Agents Chemother 43:727–737

Mishra RP, Oviedo-Orta E, Prachi P, Rappuoli R, Bagnoli F (2012) Vaccines and antibiotic resistance. Curr Opin Microbiol 15:596–602. doi:10.1016/j.mib.2012.08.002

Mogi T, Kita K (2009) Gramicidin S and polymyxins: the revival of cationic cyclic peptide antibiotics. Cell Mol Life Sci 66:3821–3826. doi:10.1007/s00018-009-0129-9

Monroe S, Polk R (2000) Antimicrobial use and bacterial resistance. Curr Opin Microbiol 3:496–501. doi:10.1016/S1369-5274(00)00129-6

Morones-Ramirez JR, Winkler JA, Spina CS, Collins JJ (2013) Silver enhances antibiotic activity against gram-negative bacteria. Sci Transl Med 5:190ra81. doi:10.1126/scitranslmed.3006276

Mukhtar TA, Wright GD (2005) Streptogramins, oxazolidinones, and other inhibitors of bacterial protein synthesis. Chem Rev 105:529–542. doi:10.1021/cr030110z

Nguyen VJC, Gutmann L (1994) Resistance to antibiotics caused by decrease of the permeability in gram-negative bacteria. Presse Med 23:522–531

Niederman MS (2001) Impact of antibiotic resistance on clinical outcomes and the cost of care. Crit Care Med 29:N114–N120. doi:10.1097/00003246-200104001-00011

Otero LH, Rojas-Altuve A, Llarrull LI, Carrasco-López C, Kumarasiri M, Lastochkin E, Fishovitz J, Dawley M, Hesek D, Lee M, Johnson JW, Fisher JF, Chang M, Mobashery S, Hermoso JA (2013) How allosteric control of *Staphylococcus aureus* penicillin binding protein 2a enables methicillin resistance and physiological function. Proc Natl Acad Sci U S A 110:16808–16813. doi:10.1073/pnas.1300118110

Paterson DL, Ko WC, Von Gottberg A, Mohapatra S, Casellas JM, Goossens H, Mulazimoglu L, Trenholme G, Klugman KP, Bonomo RA, Rice LB, Wagener MM, McCormack JG, Victor LY (2004) Antibiotic therapy for *Klebsiella pneumoniae* bacteremia: implications of production of extended-spectrum β-lactamases. Clin Infect Dis 39:31–37. doi:10.1086/420816

Patterson JE (2001) Antibiotic utilization: is there an effect on antimicrobial resistance? Chest 119:426S–430S. doi:10.1378/chest.119.2_suppl.426S

Phillips I (2001) Prudent use of antibiotics: are our expectations justified? Clin Infect Dis 33: S130–S132. doi:10.1086/321838

Phillips I, Casewell M, Cox T, De Groot B, Friis C, Jones R, Nightingale C, Preston R, Waddell J (2004) Does the use of antibiotics in food animals pose a risk to human health? A critical review of published data. J Antimicrob Chemother 53:28–52. doi:10.1093/jac/dkg483

Plunkett W, Saunders PP (1991) Metabolism and action of purine nucleoside analogs. Pharmacol Ther 49:239–268. doi:10.1016/0163-7258(91)90057-S

Prakash V, Lewis JS, Herrera ML, Wickes BL, Jorgensen JH (2009) Oral and parenteral therapeutic options for outpatient urinary infections caused by Enterobacteriaceae producing CTX-M extended-spectrum β-lactamases. Antimicrob Agents Chemother 53:1278–1280. doi:10.1128/AAC.01519-08

Quinn T, O'Mahony R, Baird AW, Drudy D, Whyte P, Fanning S (2006) Multi-drug resistance in Salmonella enterica: efflux mechanisms and their relationships with the development of chromosomal resistance gene clusters. Curr Drug Targets 7:849–860. doi:10.2174/138945006777709548

Sacks R, Greene L (2011) Impact of antibiotic resistance. BMJ 342:d3523. doi:10.1136/bmj.d3523

Scaria J, Chandramouli U, Verma SK (2005) Antibiotic Resistance Genes Online (ARGO): a database on vancomycin and β lactam resistance genes. Bioinformation 1:5–7. doi:10.6026/97320630001005

Sefton AM (2002) Mechanisms of antimicrobial resistance: their clinical relevance in the new millennium. Drugs 62:557–566. doi:10.2165/00003495-200262040-00001

Shamnas M, Arya PS, Deepak MG (2013) Optimization of antibiotic chemotherapy: a review. Pharma Innov 2:122–132

Silver LL (2003) Novel inhibitors of bacterial cell wall synthesis. Curr Opin Microbiol 6:431–438. doi:10.1016/j.mib.2003.08.004

Smith P, Hiney MP, Samuelsen OB (1994) Bacterial resistance to antimicrobial agents used in fish farming: a critical evaluation of method and meaning. Annu Rev Fish Dis 4:273–313. doi:10.1016/0959-8030(94)90032-9

So AD, Gupta N, Cars O (2010) Tackling antibiotic resistance. BMJ 340:c2071. doi:10.1136/bmj.c2071

Thomson CJ, Power E, Ruebsamen-Waigmann H, Labischinski H (2004) Antibacterial research and development in the 21st century–an industry perspective of the challenges. Curr Opin Microbiol 7:445–450. doi:10.1016/j.mib.2004.08.009

Tomasz A, Waks S (1975) Mechanism of action of penicillin: triggering of the pneumococcal autolytic enzyme by inhibitors of cell wall synthesis. Proc Natl Acad Sci U S A 72:4162–4166

Uçkay I, Pittet D, Vaudaux P, Sax H, Lew D, Waldvogel F (2009) Foreign body infections due to *Staphylococcus epidermidis*. Ann Med 41:109–119. doi:10.1080/07853890802337045

Weinstein RA (2001) Controlling antimicrobial resistance in hospitals: infection control and use of antibiotics. Emerg Infect Dis 7:188–192. doi:10.3201/eid0702.700188

Wolfe AD, Hahn FE (1965) Mode of action of chloramphenicol. IX. Effects of chloramphenicol upon a ribosomal amino acid polymerization system and its binding to bacterial ribosome. Biochim Biophys Acta 95:146–155. doi:10.1016/0005-2787(65)90219-4

Wong WR, Oliver AG, Linington RG (2012) Development of antibiotic activity profile screening for the classification and discovery of natural product antibiotics. Chem Biol 19:1483–1495. doi:10.1016/j.chembiol.2012.09.014

Woodford N, Tierno PM, Young K, Tysall L, Palepou MFI, Ward E, Painter RE, Suber DF, Shungu D, Silver LL, Inglima K, Kornblum J, Livermore DM (2004) Outbreak of *Klebsiella pneumoniae* producing a new carbapenem-hydrolyzing class A β-lactamase, KPC-3, in a New York medical center. Antimicrob Agents Chemother 48:4793–4799. doi:10.1128/AAC.48.12.4793-4799.2004

Zankari E, Hasman H, Cosentino S, Vestergaard M, Rasmussen S, Lund O, Aarestrup FM, Larsen MV (2012) Identification of acquired antimicrobial resistance genes. J Antimicrob Chemother 67:2640–2644. doi:10.1093/jac/dks261

Zhou CE, Smith J, Lam M, Zemla A, Dyer MD, Slezak T (2007) MvirDB—a microbial database of protein toxins, virulence factors and antibiotic resistance genes for bio-defence applications. Nucleic Acids Res 35:D391–D394. doi:10.1093/nar/gkl791

Drug-Resistant Tuberculosis

Lili Liang, Yun Ma, Xin liu, and Yamin Lv

Abstract The incidence of drug-resistant tuberculosis (TB), particularly multidrug-resistant TB and extensively drug-resistant TB, is increasing and is a major complication in global attempts to control TB. New anti-TB drugs and rapid diagnostics have been developed; however, the pathogenesis of drug resistance remains unclear. Fragmented treatment regimens, efflux pumps, and pharmacokinetic variability may all play a part in the rise of drug-resistant pathogens. Drug-resistant TB continues to be associated with poor treatment outcomes and high mortality rates.

1 Introduction

Tuberculosis (TB) is an ancient disease and one of the world's deadliest communicable diseases. The World Health Organization (WHO) TB estimates for 2013 included approximately 9.0 million cases and 1.5 million deaths. Although global rates of new TB cases have been decreasing since 2005, cases of multidrug-resistant TB (MDR TB) and extensively drug-resistant TB (XDR TB) have been increasing and are out of control in some regions, including Africa. Drug-resistant TB is currently one of the most important threats to global control of the disease (Dye et al. 2002). The proportion of MDR TB is higher among people who have been treated previously (20.5 %) and lower among new cases (<3 %). Eastern European and central Asian countries have the highest levels of MDR TB: 35 % of new cases and 75 % of previously treated cases. Drug-resistance surveillance data from 108 of 144 countries (75 %) indicate that approximately 480,000 individuals developed MDR TB in 2013. Furthermore, approximately 9 % of MDR TB cases were actually

L. Liang (✉) • X. liu • Y. Lv
Henan Provincial Chest Hospital, 1st Weiwu Road, Zhengzhou 450000, China
e-mail: emerald_liang@hotmail.com

Y. Ma
Henan Provincial People's Hospital, Zhengzhou, China

G. Arora et al. (eds.), *Drug Resistance in Bacteria, Fungi, Malaria, and Cancer*,
DOI 10.1007/978-3-319-48683-3_8

XDR TB. More than half of these MDR TB cases were estimated to have occurred in India, China, and the Russian Federation.

Mortality rates for untreated TB are high. Tiemersma et al. (2011) conducted a natural history study of TB and found that around 70 % of HIV-negative patients with sputum smear-positive pulmonary TB and 20 % of patients with culture-positive (but smear-negative) TB died within 10 years.

According to WHO, satisfactory treatment of XDR TB requires an 8-month intensive phase and a 20-month minimum overall treatment duration. Treatment regimens including second-line drugs are more toxic, more expensive, and less convenient than the standard anti-TB regimen. Drug-resistant TB, particularly MDR TB and XDR TB, requires a longer duration of treatment and has worse outcomes than drug-susceptible TB (Dheda et al. 2010a; Jacobson et al. 2010; Kvasnovsky et al. 2011). For instance, only 48 % of patients with MDR TB are treated successfully, with the disease incurring mortality rates of 15 % and a rate of lost of 28 %. Treatment outcomes for patients with XDR TB are worse, with a success rate of 33 % and a mortality rate of 26 % (WHO Global Tuberculosis Report 2013).

Analysis suggests the proportion of new MDR TB cases remained at approximately 3.5 % over the period 2008–2013. However, although drug-resistant TB, especially MDR TB and XDR TB, represents a small proportion of the patients with TB in countries with a high TB burden, these patients consume a large proportion of TB-control resources. In South Africa, the per-patient cost of treating XDR TB was US $26,392 (in 2011), four times greater than that for MDR TB and 103 times greater than that for drug-sensitive TB. Although drug-resistant TB represents only a fraction (2.2 %) of the total case burden, it consumes 32 % of the total estimated national TB budget in South Africa (Pooran et al. 2013). This disproportionate amount of total TB costs is due to the high cost of managing drug-resistant TB, high drug prices (Kang et al. 2006), implementation of the new Xpert MTB/RIF assay as the primary TB diagnostic test (Theron et al. 2011), and the need for extensive supervised patient care, all of which are likely to increase substantially.

2 Pathogenesis and Mechanisms of Drug-Resistant Tuberculosis (TB)

Drug-resistant TB develops in two ways. The first is primary or initial drug resistance: an individual is infected with a strain of *M. tuberculosis* that is already drug resistant. This kind of infection usually occurs in regions with a high prevalence of drug-resistant TB. The second is acquired or secondary drug resistance: resistance to the TB treatment develops as a result of inadequate or incorrect treatment regimens, efflux pumps, and genotype. Treating drug-susceptible TB with monotherapy increases the risk of drug-resistant mutations being selected and eventually becoming the dominant strain.

2.1 Selection of Drug Resistance

Spontaneous gene mutations of chromosomes that encode the target of anti-TB drugs and related *M. tuberculosis* metabolic enzymes are an important cause of single-drug resistance, and MDR is due to a variety of these drug-target gene mutations occurring. Luria and Delbruck (1943) famously showed that resistance-related *M. tuberculosis* genetic mutations were independent of selection pressure. Although the rate of spontaneous mutation is low in individual patients, there is concern that, given the large bacterial burden of up to 10^9 units and the level of bacterial replication, pre-existing *M. tuberculosis* resistant to one anti-TB drug may be possible in some patients.

The probability of pre-existing drug resistance to two or three anti-TB drugs is very small. However, resistance can be acquired, for example, if a patient receives long-term monotherapy, does not comply with treatment regimens, or receives a drug combination that is a pharmacokinetic mismatch (e.g., drugs in the combination have markedly different pharmacokinetic half-lives). The pharmacokinetic mismatch between rifapentine and isoniazid is thought to be a reason for the high rate of acquired rifamycin resistance among patients co-infected with TB and HIV who were treated with once-weekly rifapentine and isoniazid (Vernon et al. 1999). In this situation, patients actually received monotherapy. The initial drugs killed most of the susceptible *M. tuberculosis* subpopulation, and the pre-existing drug-resistant subpopulation was then able to replicate, eventually replacing the drug-resistant population. This form of resistance is known as acquired resistance, and accumulation of acquired resistance may cause MDR and XDR TB.

2.2 Acquired Drug Resistance Based on Drug Concentrations and Efflux Pumps

In some cases, drug-sensitive TB will progress to drug-resistant TB even if patients adhere to treatment (Calver et al. 2010). Some studies have shown that patients with low serum levels of isoniazid and rifampin may have a longer time to culture conversion and a worse overall treatment outcome, with the low concentration of anti-TB drugs possibly playing a role in acquired drug resistance (Blumberg et al. 2003; Jayaram et al. 2004; Weiner et al. 2005; Park et al. 2015). Studies using the hollow fiber system model of TB have shown that when one fluoroquinolone drug is used to kill *M. tuberculosis*, the bacteria easily develop resistance to that drug despite the drug concentration being much higher than the minimum inhibitory concentration (Gumbo et al. 2004, 2005). Other studies using the hollow fiber system model of TB also showed rapid development of resistance to isoniazid, rifampin, pyrazinamide, and ethambutol (Gumbo et al. 2007a, b, 2009). Acquired drug resistance is associated not only with the area under the curve but also with peak drug concentrations. Pasipanodya et al. (2012) conducted a meta-analysis and found that the faster the isoniazid underwent acetylation, the higher the rate of acquired drug resistance. Another study found that drug concentrations and pharmacokinetics varied widely between patients. Among a sample of 142 patients, the ratio of the highest to lowest dose for isoniazid, rifampin, and pyrazinamide was 2.7; the ratios of the highest peak concentration to lowest concentration was 102 for rifampin, 31 for isoniazid, and 63 for pyrazinamide, and peak drug concentration and the area under the curve predicted more than 91 % of treatment failures. Patients with low rifampin and isoniazid peaks and area under the curve concentrations developed acquired drug resistance (Pasipanodya et al. 2013).

Acquired drug resistance is also associated with many resistance efflux pumps, which can protect *M. tuberculosis* replication and enable generation of chromosomal mutations (Srivastava et al. 2010; Pasipanodya and Gumbo 2011). Dose-scheduling studies found that once-weekly therapy regimens, which are associated with more abrupt changes in drug concentrations than are regular daily therapy regimens, were associated with efflux pump-related resistance. Efflux pumps are also a cause of clinically relevant *M. tuberculosis* drug resistance (Jiang et al. 2008; Spies et al. 2008).

2.3 Drug-Resistant TB Genotypes

Given *M. tuberculosis* has a low mutation rate and a slow replication rate, it is unclear how *M. tuberculosis* acquires resistance to multiple anti-TB drugs, especially under treatment with multiple drugs. The target encoding gene mutation of clinical drug-resistant *M. tuberculosis* isolates is closely related to drug resistance. The biological variability of *M. tuberculosis* is the main molecular cause of drug resistance; Table 1 shows the *M. tuberculosis* genes that are resistant to common anti-TB drugs. Recently, whole genome sequencing of clinical *M. tuberculosis* isolates has revealed the importance of mutation in the emergence of drug resistance (Ioerger et al. 2010; Casali et al. 2012). Sequencing of *M. tuberculosis* from patients for whom drug treatment failed revealed that multiple new drug-resistance mutations can occur (Sun et al. 2012). Multidrug resistance may pre-exist in some patients who were initially infected with a drug-susceptible strain of *M. tuberculosis* (Ford et al. 2013). Moreover, several studies have suggested that certain strains of *M. tuberculosis* may be associated with multi-drug resistance (Borrell and Gagneux 2009). Some drug-susceptible TB treated with 'DOTS' (directly observed treatment, short-course) progressed to MDR TB, which might be due to hypermutable *M. tuberculosis* strains in patients who also rapidly metabolize first-line drugs (Gumbo 2013). Mutation rates can differ both between and within genotypes, and the reasons for this are unclear. Some whole genome sequencing studies have shown that target-encoding mutations are relative to compensatory mutations in the *M. tuberculosis* genome (Comas et al. 2012; Sun et al. 2012). It is possible that the drug-resistance encoding mutation could affect both the strain structure and the antigen of *M. tuberculosis*.

Table 1 Genetic mutations related to drug-resistant *Mycobacterium tuberculosis*

Anti-tuberculosis drug	Mutated gene	Minimum inhibitory concentration	Percentage of mutation	Gene product
Isoniazid	*katG* *inhA*	0.02–0.2	50–95	Catalase peroxidase reductase analog
Rifampin	*rpoB*	0.05–1	95	Subunit of RNA polymerase
Pyrazinamide	*pncA*	16–50(PH 5.5)	72–97	Pyrazinamidase
Ethambutol	*embB*	1–5	47–65	Arabinosyltransferase
Streptomycin	*RpsL rrs* *gidB*	2–8	52–59; 8–21	Ribosomal protein S12 16S rRNA
Amikacin	*rrs*	2–4	76	16S rRNA
Capreomycin	*tlyA*			Methyl transferase
Fluoroquinolones	*gyrA* *gyrB*	0.5–2.5	75–94	DNA gyrase A subunit
Ethionamide	*etsA* *etsB*	2.5–10	37	Nitric oxide
Paminosalicylic acid	*inhA* *thyA*	1–8	36–56	Synthesis of thymidine

3 Diagnosis of Drug-Resistant TB

Laboratory testing is important for the confirmation of TB, especially drug-resistant TB. The identification of drug-resistant TB needs to detect *M. tuberculosis*, culture it, and then identify the bacterial species and strains. A drug-sensitivity test (DST) is then conducted using either liquid or solid methods or a WHO-approved molecular method.

3.1 Definitions of Drug-Resistant TB (WHO, 2013)

Mono-resistance: resistance to one first-line anti-TB drug only.

Poly-resistance: resistance to more than one first-line anti-TB drug, other than both isoniazid and rifampin.

MDR: resistance to at least isoniazid and rifampin.

XDR: resistance to any fluoroquinolone and at least one of three injectable second-line drugs (capreomycin, kanamycin, and amikacin), in addition to multidrug-resistance.

Rifampin resistance: resistance to rifampin detected using phenotypic or genotypic methods, with or without resistance to other anti-TB drugs. It includes any resistance to rifampin, whether mono-resistance, poly-resistance, MDR, or XDR.

3.2 *Phenotypic Drug Sensitivity Test (DST)*

3.2.1 Liquid and Solid Methods

Solid DST methods can be used for sputum, other body fluids, and other samples to detect different concentrations of first-line and second-line anti-TB drugs. However, the specificity of the solid method is low, the process is complex, and biological security is difficult to maintain. These problems limit the clinical application of the conventional solid methods (Martin et al. 2008; Visalakshi et al. 2010).

The liquid method can shorten the detection time but still needs 4–6 weeks to obtain a DST result, and the instrument and reagents are expensive (van Kampen et al. 2010).

3.2.2 Drug-Resistance Test by Phage

Subramanyam et al. 2013 reported that the sensitivity and specificity of phage lysin in detecting *M. tuberculosis* from sputum specimens was 90 % and 81 %, respectively, compared with conventional Lowenstein–Jensen (LJ) medium. The agreement between the methods was 87 %, and the rate of contamination was 9.3 %

3.2.3 Microscopic Observation Drug Susceptibility

Microscopic observation drug susceptibility (MODS) entails using an inverted microscope to identify bacteria by observing the structure of the strain in the liquid medium. Anti-TB drugs can be added to the liquid and the DST completed directly. Agarwal et al. 2014 reported that the identification rate between the MODS assay and the reference solid LJ/liquid mycobacteria growth indicator tube (MGIT) culture was 94.8 % (95 % confidence interval 92.3–96.5). Huang et al. (2013) reported that the sensitivity and specificity of the MODS assay to detect resistance to pyrazinamide were 97.8 % and 96.5 %, respectively. MODS is the best method with which to detect pyrazinamide-resistant TB in resource-limited regions.

3.3 *Genotype DST*

The molecular DST (MDST) provides a rapid TB diagnosis and detection of drug resistance with satisfactory sensitivity and specificity. These new molecular tests can detect TB drug resistance within 2 h. The Xpert MTB/RIF assay is a new test that can detect whether the TB is active and whether the *M. tuberculosis* is resistant to rifampin.

3.3.1 Line Probe Assay

The GenoType MTBDRplus assay can rapidly detect *M. tuberculosis* genes that confer resistance to rifampicin and isoniazid; both sensitivity and specificity are satisfied (Crudu et al. 2012; Raveendran et al. 2012; Aubry et al. 2014). WHO recommends this testing for the detection of MDR TB.

3.3.2 Xpert MTB/RIF Assay

The Xpert MTB/RIF assay can both detect *M. tuberculosis* and complete DST for rifampin within 2 h. Its advantages include that it deals directly with sputum specimens, it avoids contamination and biological hazards, and the operation process is simple (Menzies et al. 2012). The Xpert MTB/RIF assay is validated for sputum, and research indicates it can be used to diagnosis extra-pulmonary TB (Causse et al. 2011; Hillemann et al. 2011; Vadwai et al. 2011; Biadglegne et al. 2014).

4 Therapy of Drug-Resistant TB

To gain worldwide control of TB, treatments for drug-resistant TB, especially MDR TB and XDR TB, are urgently needed. Treatment strategies for drug-resistant TB should be based on the specific drug resistance and treatment history, among others. Treatment for drug-resistant TB currently involves an integrated strategy that includes chemotherapy, immunotherapy, interventional therapy, surgery, traditional Chinese medicine, and nutritional support.

4.1 Chemotherapy for Drug-Resistant TB

Chemotherapy remains the primary treatment for drug-resistant TB. The chemotherapy regimen should be based on anti-TB medication history, drug resistance, and the prevalence of *M. tuberculosis* strains in the region.

Mono-resistant TB often involves initial drug resistance or primary drug-resistant TB, and the standard chemotherapy for the particular category of TB will be effective. However, the lack of four effective core drug combinations within the standard chemotherapy regimens means the potential does exist for the cure rate to decrease or the relapse rate to increase. As a result, especially for mono-resistance to rifampin, the chemotherapy regimen should be adjusted appropriately to avoid the possibility of treatment failure and the risk of acquired drug resistance.

MDR or poly-resistant TB are both more complex than mono-resistant TB. Drug resistance takes many forms but usually falls into one of three combinations: resistance to two drugs, resistance to three drugs, or resistance to four drugs. Patients with TB treated with the standard chemotherapy regimen are at greater risk for MDR TB, and treatment regimens should be adjusted to ensure patients receive four drugs that are effective or to which the TB is likely susceptible.

WHO (2011) recommends three basic treatment strategies for MDR TB—standardized, individualized, or empirical—as outlined in the following sections.

4.1.1 Standardized Treatment

Standardized treatment is a group of treatment regimens designed according to DST information and categories of patients within a country or region; patients with the same type of disease should be treated with the same treatment regimen within a country or region.

4.1.2 Individualized Treatment

Individualized treatment is based on the history of anti-TB treatment received and DST results for each patient (often DST is conducted for both first- and second-line drugs). Different patients should receive different individualized treatment regimens.

4.1.3 Empirical Treatment

Each patient's treatment regimen should be determined according to their anti-TB medication history and the DST of a country or region. The treatment regimen should be adjusted according to DST results (often DST is conducted only for a limited number of drugs). This type of treatment is mainly suitable for regions in which individual DST is not available. The basic strategy also applies to other types of drug-resistant TB.

The principles for the treatment of MDR TB with chemotherapy are as follows: (1) Regimens include at least four drugs to which the isolate is (or probably is) susceptible. (2) Regimens include a later-generation fluoroquinolone (e.g., moxifloxacin or levofloxacin) plus an injectable drug (e.g., amikacin or kanamycin), any first-line drug to which the isolate is susceptible, and a fourth drug (e.g., cycloserine, terizidone, ethionamide). (3) Injectable drugs are used for at least 6 months, and the total duration of treatment is 18–24 months.

Effective chemotherapy for the treatment of XDR TB is still lacking. Treatment is often based on nutritional support, symptom relief, improving respiratory function, and other measures to control infection with other pathogens. For disease with low-level resistance to a fluoroquinolone but sensitive to a later-generation

Table 2 Alternative grouping of anti-tuberculosis agents (2011)

Grouping	Drugs
Group 1: first-line oral agents	Isoniazid (H); rifampin (R); ethambutol (E); pyrazinamide (Z); rifabutin (Rfb)
Group 2: injectable agents	Kanamycin (Km); amikacin (Am); capreomycin (Cm); streptomycin (S)
Group 3: fluoroquinolones	Moxifloxacin (Mfx); levofloxacin (Lfx); ofloxacin (Ofx)
Group 4: oral bacteriostatic second-line agents	Ethionamide (Eto); protionamide (Pto); cycloserine (Cs); terizidone (Trd); P-aminosalicylic acid (PAS)
Group 5: agents with unclear efficacy (not recommended by WHO for routine use in patients with MDR TB)	Clofazimine (Cfz); linezolid (Lzd); amoxicillin/clavulanate (Amx/Clv); thioacetazone (Thz); imipenem/cilastatin (Ipm/Cln); high-dose isoniazid (high-dose H); clarithromycin (Clr)

fluoroquinolone (moxifloxacin is often used even when a DST indicates resistance to fluoroquinolones) (Jacobson et al. 2010) and possibly sensitive to a drug such as amikacin or capreomycin (administered via injection), the anti-TB treatment regimen could consist of the above-mentioned later-generation Fluoroquinolone, Amikacin, or Capreomycin and two drugs from the fifth group of anti-TB drugs (see Table 2). Linezolid and bedaquiline may shorten the time of sputum negative conversion for patients with XDR TB, but the cost and toxic effects are significant issues (Lee et al. 2012; Worley and Estrada 2014; Guglielmetti et al. 2015).

4.2 *Immune Therapy*

About 20 % of TB cases self-cured before the age of anti-TB chemotherapy, which supports the theory of immune-mediated clearance of *M. tuberculosis*. Some studies have reported immune-mediated clearance (Eum et al. 2010; Basile et al. 2011; Lindau et al. 2013). The most two active and acceptable immune agents are cytokine and Mycobacterium vaccine.

Many studies have shown the ability of immunomodulatory drugs to improve TB treatment outcomes (Dlugovitzky et al. 2006; Dheda et al. 2010b; Faujdar et al. 2011; Gao et al. 2011; Yang et al. 2011; Butov et al. 2012; Gupta et al. 2012a, b; Skrahin et al. 2014). Immunomodulatory drugs currently in clinical use include mycobacterium vaccae, interferon-γ, recombinant human interleukin 2, steroids, and tumor necrosis factor antagonists, among others. Immune agents are not recommended for patients with mono-resistant TB who are in good physical condition. Patients in poor physical condition can be treated with one kind of immune agent. Patients with MDR TB or XDR TB can be treated with one or two select immune agents depending on their physical and financial status.

4.3 Interventional Therapy

The widespread clinical use of bronchoscopy in recent years means anti-TB drugs administered via percutaneous lung puncture or bronchoscopy have become an effective treatment method for drug-resistant TB, particularly MDR TB.

Interventional therapy is gradually being used as a supplementary treatment method to cure drug-resistant TB (Yang et al. 2012), and we suggest that, as long as conditions permit, interventional therapy should be used as early as possible for drug-resistant TB, particularly MDR TB.

4.4 Surgical Therapy of Drug-Resistant TB

In the past 10 years, the increase in drug-resistant TB has seen a corresponding increase in the number of patients requiring surgical treatment. Surgical therapy has become more important in the treatment of drug-resistant TB, especially MDR TB.

The current theory for surgical therapy is that, for MDR TB, as long as lesions or cavities are confined to one lung or a lung lobe, surgery should be undertaken early to ensure a high cure rate and the lowest possible spread rate (Branscheid et al. 2003; Cummings et al. 2012; Suarez-Garcia and Noguerado 2012; Weyant and Mitchell 2012; Marrone et al. 2013; Calligaro et al. 2014; Mordant et al. 2014). However, surgery is not the end therapy for MDR TB. Generally, patients with MDR TB should receive more than 2 months of anti-TB chemotherapy before surgery as the chemotherapy could reduce spread to the surrounding lung tissue. Patients still require 12–24 months of chemotherapy after surgery.

4.5 Traditional Chinese Medicine Treatment and Nutrition Support

Traditional Chinese medicine can improve the immune function, physical condition, and clinical symptoms of patients with drug-resistant TB (Wang et al. 2015). Treatment of drug-resistant TB can lead to malnutrition, which can lead to worsening of drug-resistant TB. Therefore, patients with drug-resistant TB require nutritional support (Chisti et al. 2013; Hood 2013).

5 Perspectives

The early diagnosis of TB and drug-resistant TB is necessary for the global control of this disease. The slow growth of *M. tuberculosis* is the greatest obstacle to rapid diagnosis and DST. The further development of diagnostic tools, especially molecular methods, mean rapid detection of *M. tuberculosis* and specific chromosome mutations associated with phenotypic resistance to treatment will be possible. However, the cost efficiency of and appropriate settings for these new molecular methods will limit their clinical use. We hope more new laboratory tests and anti-TB drugs will be used clinically to improve drug-resistant TB-related mortality rates and treatment outcomes.

References

Agarwal A, Dhole TN, Sharma YK (2014) Evaluation of p-nitro benzoic acid (pnb) inhibition test to differentiate *Mycobacterium tuberculosis* complex from non-tuberculous mycobacteria using microscopic observation of drug susceptibility (MODS) methodology. Indian J Tuberc 61:232–235

Aubry A, Sougakoff W, Bodzongo P, Delcroix G, Armand S, Millot G, Jarlier V, Courcol R, Lemaitre N (2014) First evaluation of drug-resistant *Mycobacterium tuberculosis* clinical isolates from Congo revealed misdetection of fluoroquinolone resistance by line probe assay due to a double substitution T80A-A90G in GyrA. PLoS One 9, e95083. doi:10.1371/journal.pone.0095083

Basile JI, Geffner LJ, Romero MM, Balboa L, Sabio YGC, Ritacco V, Garcia A, Cuffre M, Abbate E, Lopez B, Barrera L, Ambroggi M, Aleman M, Sasiain MC, de la Barrera SS (2011) Outbreaks of mycobacterium tuberculosis MDR strains induce high IL-17 T-cell response in patients with MDR tuberculosis that is closely associated with high antigen load. J Infect Dis 204:1054–1064. doi:10.1093/infdis/jir460

Biadglegne F, Mulu A, Rodloff AC, Sack U (2014) Diagnostic performance of the Xpert MTB/RIF assay for tuberculous lymphadenitis on fine needle aspirates from Ethiopia. Tuberculosis (Edinb) 94:502–505. doi:10.1016/j.tube.2014.05.002

Blumberg HM, Burman WJ, Chaisson RE, Daley CL, Etkind SC, Friedman LN, Fujiwara P, Grzemska M, Hopewell PC, Iseman MD, Jasmer RM, Koppaka V, Menzies RI, O'Brien RJ, Reves RR, Reichman LB, Simone PM, Starke JR, Vernon AA (2003) American Thoracic Society/Centers for Disease Control and Prevention/Infectious Diseases Society of America: treatment of tuberculosis. Am J Respir Crit Care Med 167:603–662. doi:10.1164/rccm.167.4.603

Borrell S, Gagneux S (2009) Infectiousness, reproductive fitness and evolution of drug-resistant Mycobacterium tuberculosis. Int J Tuberc Lung Dis 13:1456–1466

Branscheid D, Albrecht CM, Diemel KD (2003) Surgical therapy of pulmonary tuberculosis. Internist (Berl) 44:1406–1412. doi:10.1007/s00108-003-1074-6

Butov DA, Efremenko YV, Prihoda ND, Yurchenko LI, Sokolenko NI, Arjanova OV, Stepanenko AL, Butova TS, Zaitzeva SS, Jirathitikal V, Bourinbaiar AS, Kutsyna GA (2012) Adjunct immune therapy of first-diagnosed TB, relapsed TB, treatment-failed TB, multidrug-resistant TB and TB/HIV. Immunotherapy 4:687–695. doi:10.2217/imt.12.59

Calligaro GL, Moodley L, Symons G, Dheda K (2014) The medical and surgical treatment of drug-resistant tuberculosis. J Thorac Dis 6:186–195. doi:10.3978/j.issn.2072-1439.2013.11.11

Calver AD, Falmer AA, Murray M, Strauss OJ, Streicher EM, Hanekom M, Liversage T, Masibi M, van Helden PD, Warren RM, Victor TC (2010) Emergence of increased resistance and extensively drug-resistant tuberculosis despite treatment adherence, South Africa. Emerg Infect Dis 16:264–271. doi:10.3201/eid0210.090968

Casali N, Nikolayevskyy V, Balabanova Y, Ignatyeva O, Kontsevaya I, Harris SR, Bentley SD, Parkhill J, Nejentsev S, Hoffner SE, Horstmann RD, Brown T, Drobniewski F (2012) Microevolution of extensively drug-resistant tuberculosis in Russia. Genome Res 22:735–745. doi:10.1101/gr.128678.111

Causse M, Ruiz P, Gutierrez-Aroca JB, Casal M (2011) Comparison of two molecular methods for rapid diagnosis of extrapulmonary tuberculosis. J Clin Microbiol 49:3065–3067. doi:10.1128/JCM.00491-11

Chisti MJ, Ahmed T, Pietroni MA, Faruque AS, Ashraf H, Bardhan PK, Hossain I, Das SK, Salam MA (2013) Pulmonary tuberculosis in severely-malnourished or HIV-infected children with pneumonia: a review. J Health Popul Nutr 31:308–313

Comas I, Borrell S, Roetzer A, Rose G, Malla B, Kato-Maeda M, Galagan J, Niemann S, Gagneux S (2012) Whole-genome sequencing of rifampicin-resistant *Mycobacterium tuberculosis* strains identifies compensatory mutations in RNA polymerase genes. Nat Genet 44:106–110. doi:10.1038/ng.1038

Crudu V, Stratan E, Romancenco E, Allerheiligen V, Hillemann A, Moraru N (2012) First evaluation of an improved assay for molecular genetic detection of tuberculosis as well as rifampin and isoniazid resistances. J Clin Microbiol 50:1264–1269. doi:10.1128/JCM.05903-11

Cummings I, O'Grady J, Pai V, Kolvekar S, Zumla A (2012) Surgery and tuberculosis. Curr Opin Pulm Med 18:241–245. doi:10.1097/MCP.0b013e3283521391

Dheda K, Shean K, Zumla A, Badri M, Streicher EM, Page-Shipp L, Willcox P, John MA, Reubenson G, Govindasamy D, Wong M, Padanilam X, Dziwiecki A, van Helden PD, Siwendu S, Jarand J, Menezes CN, Burns A, Victor T, Warren R, Grobusch MP, van der Walt M, Kvasnovsky C (2010a) Early treatment outcomes and HIV status of patients with extensively drug-resistant tuberculosis in South Africa: a retrospective cohort study. Lancet 375:1798–1807. doi:10.1016/S0140-6736(10)60492-8

Dheda K, Schwander SK, Zhu B, van Zyl-Smit RN, Zhang Y (2010b) The immunology of tuberculosis: from bench to bedside. Respirology 15:433–450. doi:10.1111/j.1440-1843.2010.01739.x

Dlugovitzky D, Fiorenza G, Farroni M, Bogue C, Stanford C, Stanford J (2006) Immunological consequences of three doses of heat-killed Mycobacterium vaccae in the immunotherapy of tuberculosis. Respir Med 100:1079–1087. doi:10.1016/j.rmed.2005.09.026

Dye C, Williams BG, Espinal MA, Raviglione MC (2002) Erasing the world's slow stain: strategies to beat multidrug-resistant tuberculosis. Science 295:2042–2046. doi:10.1126/science.1063814

Eum SY, Lee YJ, Min JH, Kwak HK, Hong MS, Kong JH, Hwang SH, Park SK, Leblanc JJ, Via LE, Barry CE, Cho SN (2010) Association of antigen-stimulated release of tumor necrosis factor-alpha in whole blood with response to chemotherapy in patients with pulmonary multidrug-resistant tuberculosis. Respiration 80:275–284. doi:10.1159/000283687

Faujdar J, Gupta P, Natrajan M, Das R, Chauhan DS, Katoch VM, Gupta UD (2011) *Mycobacterium indicus pranii* as stand-alone or adjunct immunotherapeutic in treatment of experimental animal tuberculosis. Indian J Med Res 134:696–703. doi:10.4103/0971-5916.90999

Ford CB, Shah RR, Maeda MK, Gagneux S, Murray MB, Cohen T, Johnston JC, Gardy J, Lipsitch M, Fortune SM (2013) *Mycobacterium tuberculosis* mutation rate estimates from different lineages predict substantial differences in the emergence of drug-resistant tuberculosis. Nat Genet 45:784–790. doi:10.1038/ng.2656

Gao XF, Yang ZW, Li J (2011) Adjunctive therapy with interferon-gamma for the treatment of pulmonary tuberculosis: a systematic review. Int J Infect Dis 15:e594–e600. doi:10.1016/j.ijid.2011.05.00

Guglielmetti L, Du DL, Jachym M, Henry B, Martin D, Caumes E, Veziris N, Metivier N, Robert J (2015) Compassionate use of bedaquiline for the treatment of multidrug-resistant and extensively drug-resistant tuberculosis: interim analysis of a French cohort. Clin Infect Dis 60:188–194. doi:10.1093/cid/ciu786

Gumbo T (2013) Biological variability and the emergence of multidrug-resistant tuberculosis. Nat Genet 45:720–721. doi:10.1038/ng.2675

Gumbo T, Louie A, Deziel MR, Parsons LM, Salfinger M, Drusano GL (2004) Selection of a moxifloxacin dose that suppresses drug resistance in *Mycobacterium tuberculosis*, by use of an in vitro pharmacodynamic infection model and mathematical modeling. J Infect Dis 190:1642–1651. doi:10.1086/424849

Gumbo T, Louie A, Deziel MR, Drusano GL (2005) Pharmacodynamic evidence that ciprofloxacin failure against tuberculosis is not due to poor microbial kill but to rapid emergence of resistance. Antimicrob Agents Chemother 49:3178–3181. doi:10.1128/AAC.49.8.3178-3181.2005

Gumbo T, Louie A, Deziel MR, Liu W, Parsons LM, Salfinger M, Drusano GL (2007a) Concentration-dependent *Mycobacterium tuberculosis* killing and prevention of resistance by rifampin. Antimicrob Agents Chemother 51:3781–3788. doi:10.1128/AAC.01533-06

Gumbo T, Louie A, Liu W, Ambrose PG, Bhavnani SM, Brown D, Drusano GL (2007b) Isoniazid's bactericidal activity ceases because of the emergence of resistance, not depletion of *Mycobacterium tuberculosis* in the log phase of growth. J Infect Dis 195:194–201. doi:10.1086/510247

Gumbo T, Dona CS, Meek C, Leff R (2009) Pharmacokinetics-pharmacodynamics of pyrazinamide in a novel in vitro model of tuberculosis for sterilizing effect: a paradigm for faster assessment of new antituberculosis drugs. Antimicrob Agents Chemother 53:3197–3204. doi:10.1128/AAC.01681-08

Gupta A, Ahmad FJ, Ahmad F, Gupta UD, Natarajan M, Katoch VM, Bhaskar S (2012a) Efficacy of *Mycobacterium indicus pranii* immunotherapy as an adjunct to chemotherapy for tuberculosis and underlying immune responses in the lung. PLoS One 7, e39215. doi:10.1371/journal.pone.0039215

Gupta A, Ahmad FJ, Ahmad F, Gupta UD, Natarajan M, Katoch VM, Bhaskar S (2012b) Protective efficacy of *Mycobacterium indicus pranii* against tuberculosis and underlying local lung immune responses in guinea pig model. Vaccine 30:6198–6209. doi:10.1016/j.vaccine.2012.07.061

Hillemann D, Rusch-Gerdes S, Boehme C, Richter E (2011) Rapid molecular detection of extrapulmonary tuberculosis by the automated GeneXpert MTB/RIF system. J Clin Microbiol 49:1202–1205. doi:10.1128/JCM.02268-10

Hood ML (2013) A narrative review of recent progress in understanding the relationship between tuberculosis and protein energy malnutrition. Eur J Clin Nutr 67:1122–1128. doi:10.1038/ejcn.2013.143

Huang ZK, Luo Q, Jiang BX, Li WT, Xu XM, Xiong GL, Li JM (2013) Performance of the microscopic observation drug susceptibility assay in pyrazinamide susceptibility testing for *Mycobacterium tuberculosis*. Chin Med J (Engl) 126:4334–4339

Ioerger TR, Feng Y, Chen X, Dobos KM, Victor TC, Streicher EM, Warren RM, Gey van Pittius NC, Van Helden PD, Sacchettini JC (2010) The non-clonality of drug resistance in Beijing-genotype isolates of Mycobacterium tuberculosis from the Western Cape of South Africa. BMC Genomics 11:670. doi:10.1186/1471-2164-11-670

Jacobson KR, Tierney DB, Jeon CY, Mitnick CD, Murray MB (2010) Treatment outcomes among patients with extensively drug-resistant tuberculosis: systematic review and meta-analysis. Clin Infect Dis 51:6–14. doi:10.1086/653115

Jayaram R, Shandil RK, Gaonkar S, Kaur P, Suresh BL, Mahesh BN, Jayashree R, Nandi V, Bharath S, Kantharaj E, Balasubramanian V (2004) Isoniazid pharmacokinetics-pharmacodynamics in an aerosol infection model of tuberculosis. Antimicrob Agents Chemother 48:2951–2957. doi:10.1128/AAC.48.8.2951-2957.2004

Jiang X, Zhang W, Zhang Y, Gao F, Lu C, Zhang X, Wang H (2008) Assessment of efflux pump gene expression in a clinical isolate *Mycobacterium tuberculosis* by real-time reverse transcription PCR. Microb Drug Resist 14:7–11. doi:10.1089/mdr.2008.0772

Kang YA, Choi YJ, Cho YJ, Lee SM, Yoo CG, Kim YW, Han SK, Shim YS, Yim JJ (2006) Cost of treatment for multidrug-resistant tuberculosis in South Korea. Respirology 11:793–798. doi:10.1111/j.1440-1843.2006.00948.x

Kvasnovsky CL, Cegielski JP, Erasmus R, Siwisa NO, Thomas K, Der Walt ML (2011) Extensively drug-resistant TB in Eastern Cape, South Africa: high mortality in HIV-negative and HIV-positive patients. J Acquir Immune Defic Syndr 57:146–152. doi:10.1097/QAI.0b013e31821190a3

Lee M, Lee J, Carroll MW, Choi H, Min S, Song T, Via LE, Goldfeder LC, Kang E, Jin B, Park H, Kwak H, Kim H, Jeon HS, Jeong I, Joh JS, Chen RY, Olivier KN, Shaw PA, Follmann D, Song SD, Lee JK, Lee D, Kim CT, Dartois V, Park SK, Cho SN, Barry CE (2012) Linezolid for treatment of chronic extensively drug-resistant tuberculosis. N Engl J Med 367:1508–1518. doi:10.1056/NEJMoa1201964

Lindau D, Gielen P, Kroesen M, Wesseling P, Adema GJ (2013) The immunosuppressive tumour network: myeloid-derived suppressor cells, regulatory T cells and natural killer T cells. Immunology 138(2):105–115. doi:10.1111/imm.12036

Luria SE, Delbruck M (1943) Mutations of bacteria from virus sensitivity to virus resistance. Genetics 28:491–511

Marrone MT, Venkataramanan V, Goodman M, Hill AC, Jereb JA, Mase SR (2013) Surgical interventions for drug-resistant tuberculosis: a systematic review and meta-analysis. Int J Tuberc Lung Dis 17:6–16. doi:10.5588/ijtld.12.0198

Martin A, Cubillos-Ruiz A, Von Groll A, Portillo PD, Portaels F, Palomino JC (2008) Nitrate reductase assay for the rapid detection of pyrazinamide resistance in Mycobacterium tuberculosis using nicotinamide. J Antimicrob Chemother 61:123–127. doi:10.1093/jac/dkm418

Menzies NA, Cohen T, Lin HH, Murray M, Salomon JA (2012) Population health impact and cost-effectiveness of tuberculosis diagnosis with Xpert MTB/RIF: a dynamic simulation and economic evaluation. PLoS Med 9, e1001347. doi:10.1371/journal.pmed.1001347

Mordant P, Henry B, Morel S, Robert J, Veziris N, Du DL, Frechet-Jachym M, Similowski T, Caumes E, Riquet M, Pimpec-Barthes FL (2014) Adjuvant surgical resection for multidrug-resistant tuberculosis: a review. Rev Mal Respir 31:511–524. doi:10.1016/j.rmr.2014.01.014

Park JS, Lee JY, Lee YJ, Kim SJ, Cho YJ, Yoon HI, Lee CT, Song J, Lee JH (2015) Serum levels of anti-tuberculosis drugs and their effect on tuberculosis treatment outcome. Antimicrob Agents Chemother. doi:10.1128/AAC.00693-15

Pasipanodya JG, Gumbo T (2011) A new evolutionary and pharmacokinetic-pharmacodynamic scenario for rapid emergence of resistance to single and multiple anti-tuberculosis drugs. Curr Opin Pharmacol 11:457–463. doi:10.1016/j.coph.2011.07.001

Pasipanodya JG, Srivastava S, Gumbo T (2012) Meta-analysis of clinical studies supports the pharmacokinetic variability hypothesis for acquired drug resistance and failure of antituberculosis therapy. Clin Infect Dis 55:169–177. doi:10.1093/cid/cis353

Pasipanodya JG, McIlleron H, Burger A, Wash PA, Smith P, Gumbo T (2013) Serum drug concentrations predictive of pulmonary tuberculosis outcomes. J Infect Dis 208:1464–1473. doi:10.1093/infdis/jit352

Pooran A, Pieterson E, Davids M, Theron G, Dheda K (2013) What is the cost of diagnosis and management of drug resistant tuberculosis in South Africa? PLoS One 8, e54587. doi:10.1371/journal.pone.0054587

Raveendran R, Wattal C, Oberoi JK, Goel N, Datta S, Prasad KJ (2012) Utility of GenoType MTBDRplus assay in rapid diagnosis of multidrug resistant tuberculosis at a tertiary care centre in India. Indian J Med Microbiol 30:58–63. doi:10.4103/0255-0857.93034

Skrahin A, Ahmed RK, Ferrara G, Rane L, Poiret T, Isaikina Y, Skrahina A, Zumla A, Maeurer MJ (2014) Autologous mesenchymal stromal cell infusion as adjunct treatment in patients with multidrug and extensively drug-resistant tuberculosis: an open-label phase 1 safety trial. Lancet Respir Med 2:108–122. doi:10.1016/S2213-2600(13)70234-0

Spies FS, Da Silva PE, Ribeiro MO, Rossetti ML, Zaha A (2008) Identification of mutations related to streptomycin resistance in clinical isolates of *Mycobacterium tuberculosis* and possible involvement of efflux mechanism. Antimicrob Agents Chemother 52:2947–2949. doi:10.1128/AAC.01570-07

Srivastava S, Musuka S, Sherman C, Meek C, Leff R, Gumbo T (2010) Efflux-pump-derived multiple drug resistance to ethambutol monotherapy in Mycobacterium tuberculosis and the pharmacokinetics and pharmacodynamics of ethambutol. J Infect Dis 201:1225–1231. doi:10.1086/651377

Suarez-Garcia I, Noguerado A (2012) Drug treatment of multidrug-resistant osteoarticular tuberculosis: a systematic literature review. Int J Infect Dis 16:e774–e778. doi:10.1016/j.ijid.2012.07.011

Subramanyam B, Sivaramakrishnan G, Dusthackeer A, Kumar V (2013) Phage lysin to control the overgrowth of normal flora in processed sputum samples for the rapid and sensitive detection of Mycobacterium tuberculosis by luciferase reporter phage assay. BMC Infect Dis 13:44. doi:10.1186/1471-2334-13-44

Sun G, Luo T, Yang C, Dong X, Li J, Zhu Y, Zheng H, Tian W, Wang S, Barry CE, Mei J, Gao Q (2012) Dynamic population changes in Mycobacterium tuberculosis during acquisition and fixation of drug resistance in patients. J Infect Dis 206:1724–1733. doi:10.1093/infdis/jis601

Theron G, Peter J, van Zyl-Smit R, Mishra H, Streicher E, Murray S, Dawson R, Whitelaw A, Hoelscher M, Sharma S, Pai M, Warren R, Dheda K (2011) Evaluation of the Xpert MTB/RIF assay for the diagnosis of pulmonary tuberculosis in a high HIV prevalence setting. Am J Respir Crit Care Med 184:132–140. doi:10.1164/rccm.201101-0056OC

Tiemersma EW, Van der Werf MJ, Borgdorff MW, Williams BG, Nagelkerke NJ (2011) Natural history of tuberculosis: duration and fatality of untreated pulmonary tuberculosis in HIV negative patients: a systematic review. PLoS One 6, e17601. doi:10.1371/journal.pone.0017601

Vadwai V, Boehme C, Nabeta P, Shetty A, Alland D, Rodrigues C (2011) Xpert MTB/RIF: a new pillar in diagnosis of extrapulmonary tuberculosis? J Clin Microbiol 49:2540–2545. doi:10.1128/JCM.02319-10

van Kampen SC, Anthony RM, Klatser PR (2010) The realistic performance achievable with mycobacterial automated culture systems in high and low prevalence settings. BMC Infect Dis 10:93. doi:10.1186/1471-2334-10-93

Vernon A, Burman W, Benator D, Khan A, Bozeman L (1999) Acquired rifamycin monoresistance in patients with HIV-related tuberculosis treated with once-weekly rifapentine and isoniazid. Tuberculosis Trials Consortium. Lancet 353:1843–1847. doi:10.1016/S0140-6736(98)11467-8

Visalakshi P, Meharwal SK, Myneedu VP, Behera D (2010) Evaluation of direct method of drug susceptibility testing of Mycobacterium tuberculosis to rifampicin and isoniazid by nitrate reductase assay in a national reference laboratory. Diagn Microbiol Infect Dis 66:148–152. doi:10.1016/j.diagmicrobio.2009.09.008

Wang M, Guan X, Chi Y, Robinson N, Liu JP (2015) Chinese herbal medicine as adjuvant treatment to chemotherapy for multidrug-resistant tuberculosis (MDR-TB): a systematic review of randomised clinical trials. Tuberculosis (Edinb) 95:364–372. doi:10.1016/j.tube.2015.03.003

Weiner M, Benator D, Burman W, Peloquin CA, Khan A, Vernon A, Jones B, Silva-Trigo C, Zhao Z, Hodge T (2005) Association between acquired rifamycin resistance and the pharmacokinetics of rifabutin and isoniazid among patients with HIV and tuberculosis. Clin Infect Dis 40:1481–1491. doi:10.1086/429321

Weyant MJ, Mitchell JD (2012) Multidrug-resistant pulmonary tuberculosis: surgical challenges. Thorac Surg Clin 22:271–276. doi:10.1016/j.thorsurg.2012.04.003

WHO (2011) Guidelines approved by the guidelines review committee. Guidelines for the programmatic management of drug-resistant tuberculosis: 2011 Update. World Health Organization, Geneva

World Health Organization (2013) Definitions and reporting framework for tuberculosis. http://www.who.int/tb/publications/definitions/en/

WHO (2014) Guidelines approved by the guidelines review committee. Companion handbook to the WHO guidelines for the programmatic management of drug-resistant tuberculosis. World Health Organization, Geneva

Worley MV, Estrada SJ (2014) Bedaquiline: a novel antitubercular agent for the treatment of multidrug-resistant tuberculosis. Pharmacotherapy 34(11):1187–1197. doi:10.1002/phar.1482

Yang XY, Chen QF, Li YP, Wu SM (2011) Mycobacterium vaccae as adjuvant therapy to anti-tuberculosis chemotherapy in never-treated tuberculosis patients: a meta-analysis. PLoS One 6, e23826. doi:10.1371/journal.pone.0023826

Yang SH, Zhan P, Sun M, Zhang YP, Ma NL (2012) Perfusing chemotherapy by percutaneous lung puncture in the treatment of extensive drug resistant pulmonary tuberculosis. J Thorac Dis 4:624–628. doi:10.3978/j.issn.2072-1439.2012.06.09

Emergence of Drug Resistance in *Mycobacterium* and Other Bacterial Pathogens: The Posttranslational Modification Perspective

Manu Kandpal, Suruchi Aggarwal, Shilpa Jamwal, and Amit Kumar Yadav

Abstract Microbes portray an immense capacity to colonize their niche by sequestering all resources for themselves and producing antibiotics to thwart the growth of other microbes. Antibiotic resistance by microbes evolved to overcome this challenge. Humans have long exploited the antibiotics to control infectious diseases. This development greatly improved the global health and decreased the mortality rate, thereby increasing average life expectancy. Drug resistance also arose like natural antibiotic resistance and is accelerating at an alarming rate which poses a foreboding challenge for global health. Posttranslational modifications (PTMs) have been recognized for their role in regulating cellular dynamics. Their role in development of drug resistance seems to be hidden but fundamental. For effective drugs against infectious pathogens, it is imperative to understand the basis of drug resistance and persistence. Here, we initially discuss the mechanisms of drug resistance in *Mycobacterium* and other bacterial species and eventually consider how PTMs are involved in emergence of intrinsic or adaptive drug resistance. This chapter aims to motivate the researchers in the field to dig deeper into the proteomes of pathogens to map the role of PTMs in drug resistance. A deep understanding of the roles might be the boost required for designing better antibiotics for tackling disease burden in the future.

M. Kandpal • S. Aggarwal • S. Jamwal • A.K. Yadav, PhD (✉)
Drug Discovery Research Center (DDRC), Translational Health Science and Technology Institute (THSTI), NCR Biotech Science Cluster, 3rd Milestone, Faridabad-Gurgaon Expressway, Faridabad 121001, Haryana, India
e-mail: amit.yadav@thsti.res.in

G. Arora et al. (eds.), *Drug Resistance in Bacteria, Fungi, Malaria, and Cancer*,
DOI 10.1007/978-3-319-48683-3_9

1 Introduction

Antibiotics have been in existence since the antiquity. *Salvarsan*, a synthetically developed antibiotic (Hüntelmann 2012), dates back more than a century (1907) and was subsequently followed by development of *penicillin* and *sulphonamides* (italics is used for drug names). *Penicillin*, discovered serendipitously in 1928 by Alexander Fleming, paved the way for naturally occurring antibiotics. It also inspired several synthetic antibiotics (Cohen 2000), which were developed and used extensively to treat infectious diseases for several decades in the mid-1900s.

The antibiotic boom lasted for a short while due to emergence of drug resistance in pathogens. The emergence of multidrug-resistant pathogens such as *Mycobacterium tuberculosis* (Mtb), *Clostridium difficile*, *Klebsiella pneumoniae*, and many more was a huge blow to the promise of antibiotics (Oldfield and Feng 2014). Parasitic bacteria like Mtb invade the host cells and subvert their defense mechanism and exploit the host metabolism for their own benefit. The first antibiotic against tuberculosis was designed in 1933, but as soon as the cure got popular, a major hurdle arose in the form of antibiotic resistance. Drug resistance in several other bacteria was also observed soon after the introduction of the corresponding antibacterial irrespective of the pathogen. Since most bacterial drugs are antibiotics, we will use the terms "antibiotic" and "drug" interchangeably from here on.

Antibiotic resistance has emerged as a major evolutionary fight back mechanism of pathogens which helps them survive in the face of competitive challenge from other organisms. This weakens an extremely important tool for treating various infectious diseases. There are several modes of combating drug resistance in bacteria but majorly these can be intrinsic or acquired. In intrinsic resistance, resistant bacterial mutants exist before drug/antibiotic treatment and are positively selected if they can survive the drug insult. Another mechanism is adaptive resistance wherein the bacterial pathogens develop resistance by altering its molecular physiology in response to the antibiotic encounter.

Mass spectrometry (MS)-based proteomics has shown that not only eukaryotes but even prokaryotes decorate their proteins with a large number of posttranslational modifications. Recently, phosphoproteomes of two strains of *Mycobacterium* have been thoroughly characterized to identify the differences in their phosphoproteomes of the two strains to characterize their virulence. With the help of such decisive works, it has now been established that PTMs are very important in bacterial physiology and virulence (Szymanski et al. 2002). Methods of PTM enrichment and identification have improved a lot in the last few decades. Bacteria exploit several different mechanisms to modulate its metabolism to escape the effects of drugs. A protein in a single conformation cannot do many functions. Bacteria, being simple unicellular organisms, have small genome and therefore small proteome. To bring about quick changes at molecular level and to increase the repertoire of available proteins, bacteria use posttranslational modifications (PTMs) that are simple chemical additions to already existing proteins that provide an additional layer of functionality. PTMs like phosphorylation and acetylation are involved in various signaling cascades; thus addition and removal of PTMs brings about crucial changes in protein structure and function. Most signaling changes are perceived and transduced by receptors aided by their PTM level changes. Bacteria are known to regulate their metabolism and virulence through posttranslational modifications (Broberg and Orth 2010). Resistant strains of bacteria have also devised alternate ways to regulate the metabolism through manipulating the PTMs, which help in developing drug resistance by diverse mechanisms. PTMs are thus important not only in developing resistance but also in deciding the bacterial sensitivity against the drugs.

There are several known mechanisms of drug resistance in bacteria. The major ones include: thickening the cell wall to prevent drug entry, active efflux pumps, modification of the drugs, horizontal gene transfer of resistance related genes, rewiring of metabolism, DNA damage response, etc. In this chapter, we enumerate examples to explore how PTM modulation can induce drug resistance in bacteria. In all the mechanisms discussed for bacterial drug resistance, we broadly cover simple chemical modifications on either the proteins or another bacterial cell component (DNA, lipopolysaccharide (LPS), etc.) or sometimes even on the drug itself that can help bacteria develop resistance. Although all these chemical modifications are not specifically PTMs in the classical sense, we focus majorly on PTM level modulations but also discuss few non-PTM changes for ensuring broad coverage of the resistance topic. Knowledge of the role PTMs play in drug resistance will help us in designing new therapeutic cures for bacterial diseases and counter the emerging resistance problem in efficient ways.

2 Mechanisms of Bacterial Drug Resistance

The two basic mechanisms of drug resistance are intrinsic or acquired. Drug resistance can be imparted either directly or indirectly by the gain or loss of PTMs. The role of PTMs has been observed in almost all known drug resistance mechanisms like resistance through reduced permeability to antibiotics, alteration in target sites, acquisition of alternative metabolic pathways, enzymatic degradation/modification of the antibiotics, and active efflux of antibiotics from the cell. Apart from these, sometimes even the DNA damage repair mechanisms induced by the drug may lead to mutations in pathogens which may lead to drug resistance. The PTMs involved in this process are critical for the pathogens' survival. Also, the bacterial SOS mechanism makes all the effort to keep it alive, and if the proteins inducing SOS are modified, apoptosis may be triggered in the cell. This shows that PTMs are not only involved deeply in drug resistance mechanisms known traditionally but also the pro-survival mechanisms that are recently linked to resistance. Since the role of PTMs in bacteria is beginning to be appreciated, not much attention was given to the role of PTMs apart from few sporadic studies (Cain et al. 2014; Grangeasse et al. 2015). Although we make an attempt to discuss PTM-specific cases, some examples may not be direct. These speculative examples will be based on indirect

Table 1 Mechanisms of drug resistance in bacteria and role of PTMs

Mechanism	PTM	Drugs	Modified protein/molecule	Bacterial species	Effect
Alterations in cell wall permeability	Mycoloylation	*Beta-lactam, fluoroquinolone*	Porins	*C. glutamicum, C. diphtheriae, C. efficiens*	Decreases the permeability of the cell wall by decreasing the pore size
	Phosphorylation	*Rifampicin* and *novobiocin*	Wag31	*M. smegmatis, M. tuberculosis*	Increases the thickness of cell wall
	Palmitate	*Polymyxin*	Lipid-A	*S. typhimurium, E. coli*	Makes the cell wall compact
Active efflux of antibiotics	Phosphorylation	*Fluoroquinolone, norfloxacin, ciprofloxacin*	mgrA	*S. aureus*	Increases the drug efflux
Direct modification of antibiotics	Acetylation	*Aminoglycosides*	Aminoglycosides	*E. coli*	Blocks interaction of drug with 16s rRNA
	Phosphorylation	*Aminoglycoside, vancomycin, erythromycin*	Aminoglycosides, vancomycin, erythromycin	*E. coli*	Blocks the interaction of drug with the target site
	Glutathione	*Fosfomycin*	Fosfomycin	*M. tuberculosis, E. coli*	Opens the cyclic structure of the drug making it inactive
	Nucleotide monophosphate	*Aminoglycosides, lincomycin*	Aminoglycosides, lincomycin	*S. aureus, S. haemolyticus*	Inactivates the drug
	Ribosylation, glycosylation	*Rifampin*	Rifampin	*M. tuberculosis, Nocardia* spp.	Inactivates the drug

Alterations in target site	Succinylation	*Isoniazid*	KatG	*M. tuberculosis*	Blocks the drug activation
	Phosphorylation	*Vancomycin, ceftriaxone*	SarA	*M. tuberculosis*	Blocks the binding of drug to the target site
	Methylation	*Phenicols, pleuromutilins, streptogramins, lincosamides, oxazolidonones, lincosamides, macrolides, type B streptogramins*	23S rRNA	*S. aureus, E. coli*	Prevents drug binding to 23S rRNA, thus unblocking translation
	Phosphoethanolamine	*Colistin*	Lipid-A	*A. baumannii*	Decrease affinity of drug binding
	Rhamnosylation	*Polymyxin B, cefotaxime, piperacillin*	EF-P	*E. coli, S. enterica, P. aeruginosa*	Blocks the binding of drug activation of isoniazid
Acquisition of alternative metabolic pathways	Phosphorylation	*Vancomycin*	D-Ala-D-Ala (cell wall synthesis)	*Streptococcus, Staphylococcus*	Reduces the binding affinity of drug with the cell wall depsipeptide substrates
DNA damage response	Phosphorylation and unknown modifications	*Fluoroquinolone, metronidazole*	RecA, LexA, mazE, mazF, PezT	All bacteria	Initiates DNA repair mechanism in bacteria

evidences and comparative studies across different genera and have been carefully crafted from critical literature appraisal and, therefore, worth pursuing future research directions for interested researchers in the field. The following subsections deal with each one of the mechanisms in detail and the possible role of PTMs that were hidden from the focus until now. A summary of these mechanisms and role of PTMs is also tabulated in Table 1.

2.1 Alterations in Cell Wall Permeability

Outer membranes or the cell walls of bacteria are the primary defense mechanisms against any antibacterial attack. These structures are exceptionally elaborate macromolecular organizations that provide the necessary defense to the bacteria and still allow the exchange of selective nutrients. Reducing the permeability of cells to drug is the first line of defense mechanism of bacteria. This is triggered by harsh or stressful environment, leading to expression of multitude of virulence expression factors. In *Mycobacterium*, persistence of the bug in a metabolically inactive state is a major cause of resistance since the drugs are designed to target a specific biological/metabolic function in which the bug shuts down and evades it (Koul et al. 2008). Large amounts of concerted research efforts have been traditionally directed at resolving this outer wall and membrane structure to comprehend the persistence mechanism (Jarlier and Nikaido 1994). It is clear that this permeability check has a profound impact on antibiotic susceptibility of bacteria (Jarlier and Nikaido 1994).

In case of Mtb, the enzyme Rel (GTP pyrophosphokinase) is known to increase resilience to antimicrobial mechanisms and brings about alteration in metabolism to modulate the host immune response through phosphorylation. The enzyme exhibits both synthetase and hydrolase activity. Mtb Rel synthetase is known to be significantly expressed during stress and latency. Rel downregulates translational apparatus modulated by phosphorylations from the polyphosphate [poly (p)] molecules and influences the virulence-associated factors, cell wall modification factors, polyketide synthesis, antigenic variation, potent mycobacterial antigens, and so on (Kornberg et al. 1999; Thayil et al. 2011; Morrissey et al. 2012).

The accumulating levels of poly (P) are controlled by phosphate kinases and exophosphatases, mediated by controlled expression of Rel. Many mycobacterial exophosphatases such as Rv1026/PPX2 and MT0516) can hydrolyze long-chain poly (P), which are otherwise inhibited by (p) ppGpp as in many bacteria. Accumulation of poly (P) has been seen in Mtb during starvation and osmotic stress and also in antibiotic presence. The expression of exophosphatases is also regulated by stress conditions that further increase the accumulation of poly (P). The increased poly (P) levels increase cell wall thickness and reduce drug permeability. The dynamic interplay of these kinases and exophosphatases control bacterial resistance and persistence of microbes (Chuang et al. 2015). Resistance toward multiple drugs may also result due to biofilm formation by decreasing permeability although the mechanisms remain unclear. Biofilms limit the access to nutrients, and this is probably recognized as a stress situation leading to accumulation of (p) ppGpp in the cytoplasm. Tolerance to otherwise toxic drugs in some bacterial species is dependent on the biofilm induced levels of (p) ppGpp (Sahal and Bilkay 2014; Sahal et al. 2015).

Phosphorylation of Wag31, which is a DivIVA homolog, has also been associated with drug resistance in *M. smegmatis*. Wag 31 is mainly responsible for maintaining cell shape and cell wall synthesis in mycobacteria, and the activity of Wag31 can be modulated with change in the environmental stress by phosphorylation. It also interacts with ACCase enzyme subunit AccA3 and regulates lipid and mycolic acid synthesis. AccA3 plays an essential role in elongation of acyl chains by converting acetyl-CoA to malonyl-CoA. AccA3 overexpression in *M. smegmatis* is seen to increase the resistance to *rifampicin* and *novobiocin* leading to decreased permeability. Wag31-AccA3 interaction has been implicated in facilitating the complex stability of cell wall and also being considered as a potential drug target (Xu et al. 2014). The mycobacterial kinase, PknA, is associated with phosphorylating Wag31 regulating morphological changes (Chaba et al. 2002). Wag31 phosphorylation helps in its protein-protein interaction. It also regulates and contributes to peptidoglycan (PG) biosynthesis (Lee et al. 2014). While the role of Wag31 in drug resistance has been shown in *M. smegmatis* but not yet in *M. tuberculosis*, it will be an interesting lead to follow for researchers in the field. The role may be similar owing to the protein homology in the two species, or the differences, if any, might reflect important species-specific patterns of resistance development.

Phosphorylation regulates PG biosynthetic pathway either directly or indirectly by interacting with enzymes in the pathway or by affecting the stability of cross-linking PG moieties (Jani et al. 2010). Wag31 is also reported to interact with cwsA to regulate polar PG synthesis (Plocinski et al. 2012). The key mycobacterial cell wall components are long-chain mycolic acids ranging from C60 to C90 alpha-alkyl beta hydroxyl fatty acids produced by all mycobacteria. The mermycolate chain up to C56 and a long saturated alpha branch of carbon 24–26 are the chemical modifications decorating the mermycolate. There are several enzymes in mycolic acid synthesis pathway that includes KasA and KasB (condensing enzymes), MabA (keto reductase), dehydratase (uncharacterized), and InhA (enoyl reductase). The mycolic acid pathway requires the phosphorylation of these enzymes by mycobacterial kinases to trigger and control the whole process (Bhatt et al. 2007).

The complex integration of proteins with sugar and lipid decorations plays a key role in their sensitivity toward antibiotics and development of resistance. Antibiotics gain entry across the outer membrane via diffusion through porin proteins or translocation mediated by lipid chains. Another potential mode of entry is by permeabilizing outer membranes. To counter these, bacteria have evolved mechanisms to disrupt their entry. The pathogens mount a formidable resistance to antibiotic entry as reflected by the modifications in the outer membrane structures by either thickening it or by adding components that increase the cross-linking of outer membrane components to form a compact packing. Contrarily, porins aid the movement of hydrophilic compounds across the membrane. The central hydrophilic pore is restrictive in allowing the ions to pass through based on their size. As revealed by several studies, these structures are not fixed and the functions may be altered in response to specific modulators. In drug-resistant microbes, the pore size has been observed to be narrower, restricting the entry of antibiotics (Fernandez and Hancock 2012). Alterations of porin-mediated diffusion can arise by either the absence of porin or constriction of the hydrophilic pore structures (Bornet et al. 2000).

Recently, posttranslational modification of porins in cell envelops of *Corynebacterium glutamicum* was shown, wherein mycoloylation of PorA was observed at serine 15. The presence of mycoloylated PorA was also confirmed in other species of the order *Corynebacteriales* such as *C. efficiens* and multidrug-resistant *C. diphtheriae*. The modification is implicated to be important for the pore-forming activity of porin and may play a role in multidrug-resistant nature of the microbe (Huc et al. 2010). Mycoloylations are carried by mycoloyltransferases that carry out the fatty acid transfer to generate O-mycoloylated polypeptides (Huc et al. 2010). Mycolic acids are known to play roles in rendering low permeability and resistance to antibiotics in mycobacteria as well (Collins et al. 1982; Brennan and Nikaido 1995; De Smet et al. 1999) (Fig. 1A).

The resistance to *polymyxin* has been observed in case of *Salmonella typhimurium* and even *Escherichia coli* (Dame and Shapiro 1976; Zhou et al. 2001). *Polymyxin*-resistant *S. typhimurium* strains bind only 25 % of *polymyxin* in comparison to wild type (Dame and Shapiro 1976; Vaara et al. 1979). LPS of the *polymyxin B*-resistant strain has decreased negative charge due to increase in 4-arabinose and phosphoethanolamine. Decreased negativity leads to decrease in the repulsion between adjacent LPS molecules making the layer compact causing resistance to *polymyxin B*, PMBN, EDTA, and cationic peptides. This modification of lipid-A under antibiotic stress is controlled by PmrA/PmrB two-component system which is regulated by PhoP/PhoQ (Zhou et al. 2001; Nikaido 2009). It also plays a role in providing resistance against cationic antimicrobial peptides and *polymyxin* in *Salmonella typhimurium* (Moskowitz et al. 2004).

Cationic antimicrobial peptides are about 20–40 amino acids in length. Their mechanism of action is not yet clear, though it is speculated that they kill the bacteria by disrupting the cell membrane (Moskowitz et al. 2004). Activation of PhoQ by a cationic peptide that binds to its periplasmic sensor domain results in its autophosphorylation which phosphorylates PhoP to activate it. Phosphorylated PhoP regulates the transcription of *pag* and *prg* (pho-P activated and repressed genes respectively), which control pagB-pmrAB operon. Activation of pagP encodes palmitate acyl transferase that leads to addition of palmitate to lipid. Palmitate together with aminoarabinose and phosphoethanolamine increases the resistance of bacteria (Guo et al. 1998; Kato and Groisman 2004).

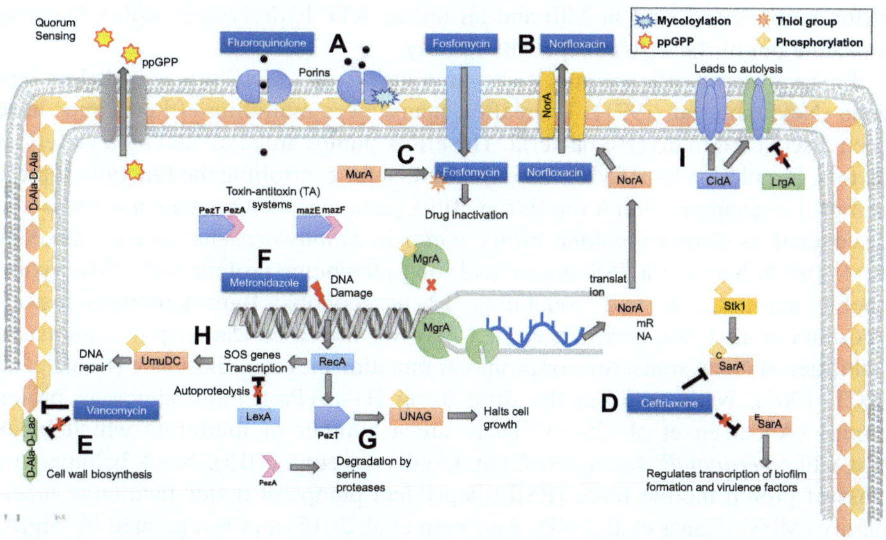

Fig. 1 Mechanisms of drug resistance and their regulation by posttranslational modifications (PTMs) in bacteria. (A) Mycoloylations of porin proteins reduces permeability to entry of drug molecules. (B) Non-phosphorylated mgrA (transcription factor) controls NorA expression. NorA is an efflux pump that expels the drug out of the cell. (C) MurA catalyzes thiol group transfer to *fosfomycin* that inactivates the drug. (D) SarA is a global transcription factor controlling biofilm formation and virulence factors which is inhibited by *ceftriaxone* binding. Loss of phosphorylation site in SarA due to C to E mutation decreases the drug binding affinity while keeping normal function of the protein intact. (E) *Vancomycin* attaches to D-Ala-D-Ala depsipeptide which blocks cell wall biosynthesis. Bacteria replace D-Ala-D-Ala with D-Ala-D-Lac in the peptidoglycan causing resistance. (F) DNA damage triggers RecA-mediated SOS mechanism. (G) RecA induces by UNAG phosphorylation mediated by PezT, which halts the cell growth. (H) Increase in transcription of SOS genes results in phosphorylation of UmuDC, inducing DNA repair. (I) Failure of all DNA repair mechanism leads to CidA activation ultimately leading to autolysis of the cell. (The open source DNA clipart has been used from https://openclipart.org/detail/1941/dna-helix)

2.2 *Active Efflux of Antibiotics*

Efflux pumps in bacteria are used to expel the toxic substances and antibiotics. Of the entire bacterial genome, ~5–10 % is estimated to be involved in transport and a large ratio encodes for efflux pumps (Lomovskaya et al. 2001; Saier and Paulsen 2001). These efflux pumps are specific to one or more antibiotics leading to resistance.

There are five major classes of drug efflux pumps in bacteria: (1) the MFS pumps (major facilitator superfamily), (2) the SMR pumps (small multidrug-resistant superfamily), (3) the MATE pumps (multi-antimicrobial extrusion protein superfamily), (4) the ABC transporter pumps (ATP-binding cassette superfamily), and (5) the RND pumps (resistance-nodulation-cell division superfamily). All of these efflux pumps are found in Mtb and all utilize ATP hydrolysis to efflux the drug molecule out of the cell, except ABC family.

First multidrug efflux pump, discovered in *Bacillus subtilis*, is encoded by *bmr* gene (Neyfakh et al. 1991). Thereafter, many more multidrug efflux pumps have been discovered in diverse bacteria. The efflux pumps are regulated by a variety of factors. Regulation by PTMs is an important factor controlling the function of these pumps. For example, NorA multidrug efflux pump encoded by gene *norA* was first discovered as *fluoroquinolone* efflux pump in *Staphylococcus aureus* (Ubukata et al. 1989). NorA is a 388-amino acid-long membrane protein with 45 % hydrophobic amino acids and containing 12 hydrophobic transmembrane regions (Yoshida et al. 1990; Neyfakh et al. 1993). By increased transcription, increased stabilities of NorA transcripts or promoter mutation elevates resistance (Kaatz et al. 2003, 2005). NorA effluxes the drug using H+-ATPase transmembrane proton pumps (Andersen et al. 2015). There are a number of inhibitors which block NorA like *verapamil*, *kaempferol*, etc. (Andersen et al. 2015). NorA belongs to a class of proton motive force (PMF)-dependent pumps of major facilitator superfamily (MFS) (Kaatz et al. 2005; Andersen et al. 2015) and is regulated by MgrA. Truong-Bolduc et al. showed that the phosphorylation of MgrA at Ser110 and Ser 113 by serine/threonine kinase PKnB decreases the expression of NorA and the dephosphorylation of phospho-MgrA by RsbU increases the expression. MgrA in phosphorylated form functions as the repressor of NorA gene transcription, as phosphorylated MgrA loses its NorA promoter binding, causing deactivation of transcription of *norA* (Truong-Bolduc and Hooper 2010) (Fig. 1B).

Another efflux pump of *norfloxacin* and *ciprofloxacin* called NorB, a transmembrane protein, is also regulated by MgrA but in an antagonistic manner. Here MgrA acts as a deactivator of transcription while phosphorylated MrgA marks active transcription of NorB (Truong-Bolduc and Hooper 2010).

2.3 Direct Modification of Antibiotics

This type of resistance is commonly observed against the self-made antibiotic. The enzyme involved modifies the drug by transferring some chemical group to the drug. This modification deactivates the drug, by decreasing its binding affinity to the target. The enzyme family that transfers these chemical groups is the largest family of resistance causing enzymes that includes acetyltransferases, phosphotransferases, thioltransferases, nucleotidytransferases, ADP-ribosyltransferases, and glycosyltransferases (Wright 2005). We enumerate the major modifying enzymes although these are not PTMs in a classical sense.

2.3.1 Acetyltransferases

Acetyltransferases transfer acetyl group to the hydroxyl or the amine group of the antibiotics (Wright 2005). For example, in the case of *aminoglycosides* which block translation, the aminoglycoside acetyltransferase (AAC) modifies the amine and hydroxyl group changing the net charge of antibiotic (Nikaido 2009). This blocks the interaction of *aminoglycoside* with the 16s rRNA, at the A-site of the ribosome. As a result, the ribosome can perform normal translation leading to resistance to *aminoglycosides*.

2.3.2 Phosphotransferases

Aminoglycosides, *vancomycin*, *erythromycin*, etc. get deactivated by the action of phosphotransferases. Phosphotransferases are kinases that catalyze transfer of phosphate with nucleoside trinucleotide acting as a cofactor (Wright 2005). MPH (2′) 302-amino acid-long protein, coded by *mphB* gene in *E. coli*, inactivates *erythromycin* (Trieu-Cuot and Courvalin 1983). This protein is highly specific for inactivating the 14-member-ring macrolides using ATP, ITP, and GTP as cofactors (O'Hara et al. 1989).

In case of *aminoglycosides*, the kinases known as aminoglycoside phosphotransferases (APH) are generally encoded by plasmid genes, probably shared between various bacterial species by horizontal gene transfer. The phosphorylation of antibiotic blocks the binding to the A-site of the ribosome (Wright 2005) rendering it ineffective. APH(3′) has been characterized in seven classes based on resistant mechanism. This diversity of these enzymes has been the major cause of resistance to *aminoglycosides* and a great example of accumulation of resistance genes through horizontal gene transfer (Shaw et al. 1993; Hastings et al. 2004).

2.3.3 Thioltransferases

UDP-N-acetylglucosamineenolpyruvyl transferase (MurA) catalyzes the first step in cell wall biosynthesis in a number of bacteria, e.g., Mtb (Quan et al. 1997) and *E. coli* (Skarzynski et al. 1996). This step is inhibited by antibiotic *fosfomycin*, which enters the bacterial cell using the glycerol phosphate and glucose-6-phosphate transport system. The enzyme-encoding *fosfomycin* resistance gene is also found on plasmid (Trieu-Cuot and Courvalin 1983). Two well-characterized genes responsible for *fosfomycin* resistance are *fosA* and *fosB*. *FosB* gene encodes for a metalloenzyme which confers resistance by acting in the presence of thiol substrate and divalent metal ion Mg2+. In this reaction, the nucleophilic ring of *fosfomycin* opens up making the antibiotic inactive (Wright 2005) (Fig. 1C).

2.3.4 Nucleotidytransferases

These enzymes transfer nucleotide monophosphate from the nucleotide triphosphate to the hydroxy group of the antibiotic. They modify *aminoglycosides* and *lincomycin* by O-adenosyl transferases (ANTs) in *S. aureus* and Lin protein in *S. haemolyticus*, respectively (Wright 2005).

2.3.5 ADP-Ribosyltransferases

These enzymes donate ADP-ribosyl group to antibiotics, thus blocking their attachment to the target. For example, in *M. smegmatis*, ADP-ribosyltransferases donate ADP-ribosyl using NAD as cofactor to *rifampin. Rifampin* is a semisynthetic antibiotic that blocks beta-subunit of RNA polymerase (Wright 2005). Ribosylation of *rifampin* makes the antibiotic inactive decreasing the susceptibility of *M. smegmatis* to this antibiotic (Brown-Elliott et al. 2012). This process of ribosylation is catalyzed by Arr enzyme. Old form of *rifampin* as well as natural product of *rifamycin SV* and new modified agents like *rifaximin* are all the substrates of Arr enzyme (Stallings et al. 2011). Most modified forms of *rifampin* are generated by modifying naphthyl ring in the drug and ansa-bridge at position 23 of the drug. Arr enzyme is tolerant to modification in this ring. As every available form of *rifampin* is deactivated by this enzyme, potential Arr inhibitors should be used to overcome the risk of next-generation *rifampin* drug inactivation (Brown-Elliott et al. 2012).

2.3.6 Glycosyltransferases

Glycosyltransferases can modify the antibiotic by transferring a glycosyl group to the antibiotic that attenuates its binding capacity, thus causing resistance. Glycosylation of *rifampin* is reported in *Nocardia* which was first described in 1994 (Yazawa et al. 1994). This inactivation mechanism for *rifampin* is not reported in any bacteria outside *Nocardia* genus.

2.4 Alterations in Target Site

This is the most common mechanism used by bacteria for acquiring resistance against drugs. Several bacterial species alter the drug targets for developing resistance, e.g., *Haemophilus influenza*, *Helicobacter pylori*, *Proteus mirabilis*, *Acinetobacter baumannii*, *Pseudomonas aeruginosa*, *Streptococcus pyogenes*, and *Listeria monocytogenes*. Modification of the target site decreases or blocks the affinity for drug binding but the normal activity of the target is not disturbed.

Isoniazid (INH), a prodrug used most widely against TB is activated by KatG, an enzyme having both catalase and peroxidase property. It has a critical role in oxidative stress management of bacteria. The KatG protein oxidizes *INH* to an acylated species, i.e., to acyl-nicotinamide adenine dinucleotide adduct, an activated form of *INH* which is bactericidal in nature. Natural mutants of KatG contain a S315T mutation which may result in a loss of function of the protein, but succinylation at K310 near this mutation assists the enzyme to retain its native activity, i.e., catalase-peroxidase, while the *INH* activating property of KatG is reduced by almost 30 %. By decreasing *INH* activation KatG is able to increase the minimum inhibitory concentration (MIC) of bacteria up to 200-folds. This protein has been reported to have nine succinylation sites, and succinylation at K310 of KatG is responsible for drug resistance in Mtb (Xie et al. 2015).

SarA is a global transcription regulator of Mtb. Antibiotics targeting the cell wall, like *vancomycin* and *ceftriaxone*, decrease cysteine phosphorylation of SarA which inhibits its interactions with the promoters of transcriptional targets. SarA is phosphorylated by Stk1. The kinase activity of Stk1 is regulated by stp1. An in vivo mouse model study reports elevated phosphorylation of SarA reducing the virulence. To resist this process, bacteria convert Cys to Glu in SarA, mimicking the phosphorylation of SarA and making bacteria resistant (Cain et al. 2014). Another way Mtb protects itself is by the modification of target site making it inaccessible for drug action. *Chloramphenicol-florfenicol* resistance (*cfr*) methyltransferase adds methyl group to the A2503 in the 23S rRNA of *S. aureus* and provides resistance against a wide range of drugs like *phenicols*, *lincosamides*, *oxazolidonones*, *pleuromutilins*, and *streptogramins*. The phenotype for resistance to above classes of drugs is named PhLOPSA (Long et al. 2006) (Fig. 1D).

The phosphorylated elongation factor (EF-P) is a protein that modulates antibiotic resistance in several bacteria (*Escherichia coli*, *Salmonella enterica*, etc.), thus contributing to their virulence, causing many infectious diseases. The post-translational modification of beta-lysylation at Lys-34 activates EF-P and also provides resistance. EF-P is required for proficient protein synthesis, specifically for poly-proline motif synthesis during translation. *P. aeruginosa* lack any PoxA protein which is needed for beta-lys transfer to EF-P. *P. aeruginosa* have evolved a special mechanism, which posttranslationally modify Arg-32 containing EF-P by attaching rhamnose moiety. This PTM has similar effect as beta-lys in *E. coli* and *S. enterica*, thus providing drug resistance to the organism as the drug is designed to target phosphorylation and not rhamnose. The glycosyltransferase EarP attaches rhamnose to R32. Lack of glycosylation at phosphorylated EF increases its antibiotic susceptibility, as a result of decreased expression of poly-proline. Since the glycosylated EF-P might be required for MexA synthesis, an essential component for *beta-lactam* specificity, this PTM is responsible for drug resistance in its absence (Rajkovic et al. 2015).

E. coli has developed resistance to three classes of drugs, i.e., (1) *macrolides*, (2) *lincosamides*, and (3) type B *streptogramins*, called the MLS phenotype. N-methyltransferase of erm (erythromycin ribosome methylation) gene class usually coded by plasmid transfers methyl group to the N6-purine ring of 23sRNA at

A2058 blocking antibiotic binding, because of loss of its H-bonding property (Brodersen et al. 2000).

Colistin attaches to the negative LPS layer at lipid-A in *A. baumannii* killing the bacteria. However, the resistant strains contain pmrB gene mutation which induces the upregulation of pmrAB. This protein helps in increased expression of pmrC, which modifies lipid-A by phosphoethanolamine addition. This addition decreases the negativity of LPS layer, thereby making the attachment of positively charged *colistin* unfavorable, causing resistance (Beceiro et al. 2011; Cai et al. 2012).

2.5 Metabolic Reprogramming (Acquisition of Alternative Metabolic Pathways)

In addition to other mechanisms of antibiotic resistance, microbes evolve alternate metabolic pathways to bypass the drug effect.

In *Enterococcus faecium* and *E. faecalis*, the changes in their peptidoglycan layer help them escape the effects of *vancomycin*. In the growing peptidoglycan cell wall, the drug binds to D-Ala-D-Ala depsipeptide forming a complex through five hydrogen bonds. Production of acyl-D-alanyl-D-lactate, an isosteric depsipeptide, provides resistance against the drug. It replaces the amide bond of D-alanyl-D-alanine with an ester linkage hindering the hydrogen bond formation which inhibits the attachment of drug (Wright 2011).

This resistance mechanism involves five proteins, of which two (vanS and vanR) are responsible for regulation. VanS catalyzes ATP-dependent autophosphorylation of His-164 and His-233 mediated through its cytoplasmic domain. The transcriptional control element (VanR) is thus phosphorylated at Asp-53. VanS has both kinase and phosphatase activity for VanR. In the presence of antibiotics, VanR induces the expression of VanH, VanA, and VanX, the other three proteins. VanX specifically cleaves D-alanyl-D-alanine, VanH is a D-lactate dehydrogenase that generates D-lactate, and VanA is a ligase that catalyzes the synthesis of acyl-D-alanyl-D-lactate. The components generated by this reprogramming now get incorporated in new cell wall and provide resistance (Depardieu et al. 2007; Wright 2011) (Fig. 1E).

2.6 DNA Damage Response

Bacterial cells have evolved around a toxin-antitoxin (TA) system for their survival. This system usually corresponds to the "addiction modules" presented by the plasmids. In this system, a toxin which is generally a long-lasting protein is continuously being neutralized by the antitoxins which have small life due to early degradation by serine proteases. Both the toxin and the corresponding antitoxin are encoded by the plasmids. Bacterial cells have used this TA system to respond to DNA-damaging agents like *fluoroquinolone* (Neyfakh et al. 1993). When DNA damage is observed in the bacterial cell, the cellular processes are halted with the help of "mazEF" TA system to initiate the SOS response in the cell (Yarmolinsky 1995; Moritz and Hergenrother 2007; Mutschler et al. 2011; Sadeghifard et al. 2014; Wang et al. 2014). mazEF system was extensively observed and studied in *E. coli* but the functionality is also observed in Mtb and other bacteria (Kolodkin-Gal and Engelberg-Kulka 2009; Fortuin et al. 2015; Ramisetty et al. 2015). In this system, mazF protein acts as a toxin and halts the growth process. mazF is a endonuclease that cuts the mRNA at a specific site between A and C in the ACA sequence, thus controlling the translation process by chopping off mRNA. In normal conditions, the bacteria keep growing by blocking mazF with mazE, its antitoxin. Interestingly, when DNA damage is observed in bacteria, mazEF interact in a unique way to stop mazE synthesis, thus letting mazF chop off mRNAs and stop cell growth. Once growth is halted, all energy and proteins are directed toward the repair of the DNA (Kreuzer 2013; Bayles 2014; Ramisetty et al. 2015).

In another TA system PezT-PezA, the degradation of antitoxin PezA releases PezT kinase in stress conditions. PezT phosphorylates UNAG and converts it to UNAG-3P, which inhibits peptidoglycan synthesis (Sedwick 2011). Slow dividing bacteria can survive this condition in contrast to rapidly growing bacteria which undergo cell lysis. Autolysis of the cell releases a toxin, pneumolysin, which is a major factor for *S. pneumoniae* virulence. This fractional autolysis and inhibition of polysaccharides of murine layer by UNAG-3P favors biofilm formation (Mutschler et al. 2011) (Fig. 1G). The biofilm formation of the bacteria reduces bacterial susceptibility to number of drugs. This reduced susceptibility to drugs is achieved by several processes (Hoiby et al. 2010) as mentioned below:

(a) Reducing the rate of penetration, the antibiotic inactivating enzyme plays a critical role in providing resistance to the pathogen.
(b) Most of the bacterial cell in a biofilm is in a stationary or the slow growing phase which helps the bacteria to escape the effect of the antibiotics like *penicillin*, and the factors which are involved in slow growth of the cell also alter antibiotics. For example, availability of oxygen alters the action of *aminoglycosides*.
(c) Adaptive response of bacteria helps the cell to live in stress conditions, like increasing the expression of drug efflux pumps.
(d) Some of the bacterial cells survive even after long period of antibiotic treatment as they enter in a highly safe or spore-like state.

There are 79 TA systems known in *Mycobacterium* till now that aid in regulation of various cellular processes to aid in protection and survival. The regulation of DDR in higher eukaryotes has depicted plentiful examples of regulation by PTMs to refactor the metabolism toward DNA repair. With the ever-increasing examples of PTM regulation in bacteria (Broberg and Orth 2010; Cain et al. 2014) and their roles in virulence and resistance (Sun et al. 2012; Rajkovic et al. 2015), it is not unimaginable to draw a simile here to speculate that many PTMs might be involved in bacterial DDR as well, waiting to be discovered.

2.6.1 SOS Response in Bacteria

When a bacterial cell is exposed to DNA-damaging elements like UV radiation or chemical agents like *metronidazole*, a SOS response is triggered to evade the effects of the stress introduced. The SOS response is a pathway to repair DNA in most bacterial species including *Mycobacterium*. It consists of two major proteins that start the DNA repair pathway—the inducer RecA and the repressor LexA. In normal stress-free condition, LexA forms a dimer and binds to the SOS boxes on the bacterial DNA to stop the transcription of SOS genes. When stress is introduced in the bacterial cell, RecA is activated and it binds to LexA to induce autoproteolysis of LexA leading to the removal of LexA from SOS boxes and causing the expression of SOS genes. Once the SOS response is activated, DNA polymerases start the error-prone repair of bacterial DNA (Kreuzer 2013; Zgur-Bertok 2013) (Fig. 1F).

In *H. pylori*, it has been shown that RecA is posttranslationally modified for the protein to function as an inducer. On initial sequencing, a glycosylation site similar to *C. jejuni* was found in its sequence, suggesting that it may be a potential site for the modification in RecA protein, which was confirmed using Western blotting. It was also experimentally observed that the modification is required for the fully functional protein in DNA repair mechanism (Fischer and Haas 2004).

To properly understand the autoproteolysis activity of LexA protein, Oliviera and Lindblad performed a series of experiments to understand the mechanism of autoproteolysis in LexA. They observed that LexA is both posttranscriptionally and posttranslationally modified during the SOS response. The authors were not able to identify the modification due to low mass spectrometric resolution and low coverage, but could observe the mass difference was small. They also found two more forms of LexA apart from the normal one, confirming the presence of modified LexA during proteolysis (Oliveira and Lindblad 2011).

The autoproteolysis of LexA is due to posttranslational modifications, and similar observation was also made for DNA polymerase V activity in SOS response. During SOS response, DNA polymerase V, also known as umoDC, is present in two forms—umoD that participates in a checkpoint for DNA damage and umoD′ which forms a complex with umoC to initiate DNA damage repair. The umoD form is 24 bp larger than the umoD′. Initially the polymerase is present in its umoD form, but when a lesion is identified in the DNA, it is cleaved by the RecA/ss-DNA-mediated autodigestion to form umoD′. This posttranslational modification acts as a molecular switch between the two forms of DNA polymerase V to regulate its functions in DNA damage repair (Sutton et al. 2002) (Fig. 1H).

Apart from the above examples for the posttranslational modifications driving the DNA damage response in bacterial cells, the differences in the phosphoproteomes of two *Mycobacterium* species *M. smegmatis* and *M. bovis* BCG highlighting a number of phosphorylation sites present in DNA polymerases of stress response proteins have been recently enumerated (Nakedi et al. 2015). These phosphosites might be of interest for studying their role in the DNA damage response and can be explored by interested researchers.

2.6.2 Programmed Cell Death

When the damage in the cell is beyond the repair, it undergoes a cellular suicide mechanism widely known as programmed cell death (PCD). Although it is debatable whether PCD exists for a unicellular organism, compounding evidences suggest that bacterial cells have mechanisms to undergo programmed cell death, indicating the evolutionary selection of apoptosis from a unicellular organism to complex mammals (Bayles 2014). It is an altruistic mechanism rather than being beneficial to the individual cell directly. A number of bactericidal antibiotics induce DNA fragmentation, chromosome condensation, and phosphatidylserine exposure at outer membrane (biochemical marker for apoptosis).

Apparently, some myxobacteria undergo PCD to provide raw material to the fruiting body (Yarmolinsky 1995); others use it as a mechanism to produce raw material for the formation of biofilms to protect the colony. A mechanism of quorum sensing is also used in the conditions where a stress signal (ppGpp) has to be generated by the cell to warn the colony members of the introduced cell signal. But for PCD to get activated, the cell should reach a damaging condition in which any amount of repair cannot restart the cellular machinery. In such a case of point of no return, BapE (bacterial apoptosis endonuclease) is used to induce apoptosis like cell death or autolysis, based on the signal induced. BapE is an endonuclease that will fragment the bacterial DNA to promote apoptosis like cell death in bacteria.

Another mechanism by which cell death can occur in the bacterium is through the autolysis of the bacterial cell. In this context, holin-antiholin systems have been extensively studied for all bacterial forms. Here, a holin protein Cida dissipates membrane potential to activate murein hydrolase activity that degrades the cell wall and thus causes lysis. A counter mechanism also exists in the form of antiholin (lrgA), which oligomerizes with the CidA and inhibits its activity until the signal is received (Rice et al. 2003; Yang et al. 2005; Ranjit et al. 2011) (Fig. 1I).

3 Conclusion

Apart from phosphorylation, the discovery of PTMs in bacterial regulation has been recent. The hidden layer of metabolic regulation has sprung up many surprises as their roles have discovered in a wide array of functions in a short time span. PTMs are now known to regulate bacterial growth, survival, and virulence (Broberg and Orth 2010; Cain et al. 2014; Grangeasse et al. 2015). We have just started to understand the role of PTMs and the technological advances in mass spectrometry-based quantitative proteomics can help us elucidate many more PTMs and their functions. It is time the research community focuses the efforts toward understanding their roles in developing drug resistance, a major economic and health burden in modern society. We have attempted to depict their importance by discussing classical and other known examples in a new light. Combinations of drugs developed from a deep understanding of metabolic connections that can block alternate routes for drugs helped by the modifications can be the next shot in the arm for developing next-generation antibiotics.

Acknowledgment AKY is supported by the Innovative Young Biotechnologist Award (IYBA), and MK is supported by the IYBA Junior Research Fellowship (IYBAJRF) from the Department of Biotechnology, India. SA is supported by the Senior Research Fellowship (SRF), and from the Indian Council of Medical Research (ICMR), India. AKY and SJ also acknowledge DDRC-SFC grant from the Department of Biotechnology, India.

References

Andersen JL, He GX, Kakarla P, KC R, Kumar S, Lakra WS, Mukherjee MM, Ranaweera I, Shrestha U, Tran T (2015) Multidrug efflux pumps from enterobacteriaceae, *Vibrio cholerae* and *Staphylococcus aureus* bacterial food pathogens. Int J Environ Res Pub Health 12: 1487–1547. doi:10.3390/ijerph120201487

Bayles KW (2014) Bacterial programmed cell death: making sense of a paradox. Nat Rev Microbiol 12:63–69. doi:10.1038/nrmicro3136

Beceiro A, Llobet E, Aranda J, Bengoechea JA, Doumith M, Hornsey M, Dhanji H, Chart H, Bou G, Livermore DM, Woodford N (2011) Phosphoethanolamine modification of lipid A in colistin-resistant variants of *Acinetobacter baumannii* mediated by the pmrAB two-component regulatory system. Antimicrob Agents Chemother 55:3370–3379. doi:10.1128/AAC.00079-11

Bhatt A, Molle V, Besra GS, Jacobs WR Jr, Kremer L (2007) The *Mycobacterium tuberculosis* FAS-II condensing enzymes: their role in mycolic acid biosynthesis, acid-fastness, pathogenesis and in future drug development. Mol Microbiol 64:1442–1454. doi:10.1111/j.1365-2958.2007.05761.x

Bornet C, Davin-Regli A, Bosi C, Pages JM, Bollet C (2000) Imipenem resistance of *Enterobacter aerogenes* mediated by outer membrane permeability. J Clin Microbiol 38:1048–1052

Brennan PJ, Nikaido H (1995) The envelope of mycobacteria. Annu Rev Biochem 64:29–63. doi:10.1146/annurev.bi.64.070195.000333

Broberg CA, Orth K (2010) Tipping the balance by manipulating post-translational modifications. Curr Opin Microbiol 13:34–40. doi:10.1016/j.mib.2009.12.004

Brodersen DE, Clemons WM Jr, Carter AP, Morgan-Warren RJ, Wimberly BT, Ramakrishnan V (2000) The structural basis for the action of the antibiotics tetracycline, pactamycin, and hygromycin B on the 30S ribosomal subunit. Cell 103:1143–1154. doi:10.1016/S0092-8674(00)00216-6

Brown-Elliott BA, Nash KA, Wallace RJ Jr (2012) Antimicrobial susceptibility testing, drug resistance mechanisms, and therapy of infections with nontuberculous mycobacteria. Clin Microbiol Rev 25:545–582. doi:10.1128/CMR.05030-11

Cai Y, Chai D, Wang R, Liang B, Bai N (2012) Colistin resistance of *Acinetobacter baumannii*: clinical reports, mechanisms and antimicrobial strategies. J Antimicrob Chemother 67: 1607–1615. doi:10.1093/jac/dks084

Cain JA, Solis N, Cordwell SJ (2014) Beyond gene expression: the impact of protein post-translational modifications in bacteria. J Proteomics 97:265–286. doi:10.1016/j.jprot.2013.08.012

Chaba R, Raje M, Chakraborti PK (2002) Evidence that a eukaryotic-type serine/threonine protein kinase from *Mycobacterium tuberculosis* regulates morphological changes associated with cell division. Eur J Biochem 269:1078–1085. doi:10.1046/j.1432-1033.2002.02778.x

Chuang YM, Bandyopadhyay N, Rifat D, Rubin H, Bader JS, Karakousis PC (2015) Deficiency of the novel exopolyphosphatase Rv1026/PPX2 leads to metabolic downshift and altered cell wall permeability in *Mycobacterium tuberculosis*. MBio 6, e02428. doi:10.1128/mBio.02428-14

Cohen ML (2000) Changing patterns of infectious disease. Nature 406:762–767. doi:10.1038/35021206

Collins MD, Goodfellow M, Minnikin DE (1982) Fatty acid composition of some mycolic acid-containing coryneform bacteria. J Gen Microbiol 128:2503–2509. doi:10.1099/00221287-128-11-2503

Dame JB, Shapiro BM (1976) Use of polymyxin B, levallorphan, and tetracaine to isolate novel envelope mutants of *Escherichia coli*. J Bacteriol 127:961–972

De Smet KA, Kempsell KE, Gallagher A, Duncan K, Young DB (1999) Alteration of a single amino acid residue reverses fosfomycin resistance of recombinant MurA from *Mycobacterium tuberculosis*. Microbiology 145:3177–3184. doi:10.1099/00221287-145-11-3177

Depardieu F, Podglajen I, Leclercq R, Collatz E, Courvalin P (2007) Modes and modulations of antibiotic resistance gene expression. Clin Microbiol Rev 20:79–114. doi:10.1128/CMR.00015-06

Fernandez L, Hancock RE (2012) Adaptive and mutational resistance: role of porins and efflux pumps in drug resistance. Clin Microbiol Rev 25:661–681. doi:10.1128/CMR.00043-12

Fischer W, Haas R (2004) The RecA protein of *Helicobacter pylori* requires a posttranslational modification for full activity. J Bacteriol 186:777–784. doi:10.1128/JB.186.3.777-784.2004

Fortuin S, Tomazella GG, Nagaraj N, Sampson SL, Gey van Pittius NC, Soares NC, Wiker HG, de Souza GA, Warren RM (2015) Phosphoproteomics analysis of a clinical *Mycobacterium tuberculosis* Beijing isolate: expanding the mycobacterial phosphoproteome catalog. Front Microbiol 6:6. doi:10.3389/fmicb.2015.00006

Grangeasse C, Stulke J, Mijakovic I (2015) Regulatory potential of post-translational modifications in bacteria. Front Microbiol 6:500. doi:10.3389/fmicb.2015.00500

Guo L, Lim KB, Poduje CM, Daniel M, Gunn JS, Hackett M, Miller SI (1998) Lipid A acylation and bacterial resistance against vertebrate antimicrobial peptides. Cell 95:189–198. doi:10.1016/S0092-8674(00)81750-X

Hastings PJ, Rosenberg SM, Slack A (2004) Antibiotic-induced lateral transfer of antibiotic resistance. Trends Microbiol 12:401–404. doi:10.1016/j.tim.2004.07.003

Hoiby N, Bjarnsholt T, Givskov M, Molin S, Ciofu O (2010) Antibiotic resistance of bacterial biofilms. Int J Antimicrob Agents 35:322–332. doi:10.1016/j.ijantimicag.2009.12.011

Huc E, Meniche X, Benz R, Bayan N, Ghazi A, Tropis M, Daffe M (2010) O-mycoloylated proteins from *Corynebacterium*: an unprecedented post-translational modification in bacteria. J Biol Chem 285:21908–21912. doi:10.1074/jbc.C110.133033

Hüntelmann A (2012) Making salvarsan. Experimental therapy and the development and marketing of salvarsan at the interface between science, clinic, industry and public health. In: Gaudillière J-P, Hess V (eds) Ways of regulating drugs in the 19th and 20th centuries. Palgrave Macmillan, New York, pp 43–65

Jani C, Eoh H, Lee JJ, Hamasha K, Sahana MB, Han JS, Nyayapathy S, Lee JY, Suh JW, Lee SH, Rehse SJ, Crick DC, Kang CM (2010) Regulation of polar peptidoglycan biosynthesis by Wag31 phosphorylation in mycobacteria. BMC Microbiol 10:327. doi:10.1186/1471-2180-10-327

Jarlier V, Nikaido H (1994) Mycobacterial cell wall: structure and role in natural resistance to antibiotics. FEMS Microbiol Lett 123:11–18. doi:10.1111/j.1574-6968.1994.tb07194.x

Kaatz GW, Moudgal VV, Seo SM, Kristiansen JE (2003) Phenothiazines and thioxanthenes inhibit multidrug efflux pump activity in *Staphylococcus aureus*. Antimicrob Agents Chemother 47: 719–726. doi:10.1128/AAC.47.2.719-726.2003

Kaatz GW, Thyagarajan RV, Seo SM (2005) Effect of promoter region mutations and mgrA overexpression on transcription of norA, which encodes a *Staphylococcus aureus* multidrug efflux transporter. Antimicrob Agents Chemother 49:161–169. doi:10.1128/AAC.49.1.161-169.2005

Kato A, Groisman EA (2004) Connecting two-component regulatory systems by a protein that protects a response regulator from dephosphorylation by its cognate sensor. Genes Dev 18: 2302–2313. doi:10.1101/gad.1230804

Kolodkin-Gal I, Engelberg-Kulka H (2009) The stationary-phase sigma factor sigma(S) is responsible for the resistance of *Escherichia* coli stationary-phase cells to mazEF-mediated cell death. J Bacteriol 191:3177–3182. doi:10.1128/JB.00011-09

Kornberg A, Rao NN, Ault-Riche D (1999) Inorganic polyphosphate: a molecule of many functions. Annu Rev Biochem 68:89–125. doi:10.1146/annurev.biochem.68.1.89

Koul A, Vranckx L, Dendouga N, Balemans W, Van den Wyngaert I, Vergauwen K, Gohlmann HW, Willebrords R, Poncelet A, Guillemont J, Bald D, Andries K (2008) Diarylquinolines are bactericidal for dormant mycobacteria as a result of disturbed ATP homeostasis. J Biol Chem 283:25273–25280. doi:10.1074/jbc.M803899200

Kreuzer KN (2013) DNA damage responses in prokaryotes: regulating gene expression, modulating growth patterns, and manipulating replication forks. Cold Spring Harb Perspect Biol 5: a012674. doi:10.1101/cshperspect.a012674

Lee JJ, Kan CM, Lee JH, Park KS, Jeon JH, Lee SH (2014) Phosphorylation-dependent interaction between a serine/threonine kinase PknA and a putative cell division protein Wag31 in *Mycobacterium tuberculosis*. New Microbiol 37:525–533

Lomovskaya O, Warren MS, Lee A, Galazzo J, Fronko R, Lee M, Blais J, Cho D, Chamberland S, Renau T (2001) Identification and characterization of inhibitors of multidrug resistance efflux pumps in *Pseudomonas aeruginosa*: novel agents for combination therapy. Antimicrob Agents Chemother 45:105–116. doi:10.1128/AAC.45.1.105-116.2001

Long KS, Poehlsgaard J, Kehrenberg C, Schwarz S, Vester B (2006) The Cfr rRNA methyltransferase confers resistance to Phenicols, Lincosamides, Oxazolidinones, Pleuromutilins, and Streptogramin A antibiotics. Antimicrob Agents Chemother 50:2500–2505. doi:10.1128/AAC.00131-06

Moritz EM, Hergenrother PJ (2007) Toxin-antitoxin systems are ubiquitous and plasmid-encoded in vancomycin-resistant enterococci. Proc Natl Acad Sci U S A 104:311–316. doi:10.1073/pnas.0601168104

Morrissey JH, Choi SH, Smith SA (2012) Polyphosphate: an ancient molecule that links platelets, coagulation, and inflammation. Blood 119:5972–5979. doi:10.1182/blood-2012-03-306605

Moskowitz SM, Ernst RK, Miller SI (2004) PmrAB, a two-component regulatory system of *Pseudomonas aeruginosa* that modulates resistance to cationic antimicrobial peptides and addition of aminoarabinose to lipid A. J Bacteriol 186:575–579. doi:10.1128/JB.186.2.575-579.2004

Mutschler H, Gebhardt M, Shoeman RL, Meinhart A (2011) A novel mechanism of programmed cell death in bacteria by toxin-antitoxin systems corrupts peptidoglycan synthesis. PLoS Biol 9, e1001033. doi:10.1371/journal.pbio.1001033

Nakedi KC, Nel AJ, Garnett S, Blackburn JM, Soares NC (2015) Comparative Ser/Thr/Tyr phosphoproteomics between two mycobacterial species: the fast growing *Mycobacterium smegmatis* and the slow growing *Mycobacterium bovis* BCG. Front Microbiol 6:237. doi:10.3389/fmicb.2015.00237

Neyfakh AA, Bidnenko VE, Chen LB (1991) Efflux-mediated multidrug resistance in *Bacillus subtilis*: similarities and dissimilarities with the mammalian system. Proc Natl Acad Sci U S A 88:4781–4785

Neyfakh AA, Borsch CM, Kaatz GW (1993) Fluoroquinolone resistance protein NorA of *Staphylococcus aureus* is a multidrug efflux transporter. Antimicrob Agents Chemother 37:128–129. doi:10.1128/AAC.37.1.128

Nikaido H (2009) Multidrug resistance in bacteria. Annu Rev Biochem 78:119–146. doi:10.1146/annurev.biochem.78.082907.145923

O'Hara K, Kanda T, Ohmiya K, Ebisu T, Kono M (1989) Purification and characterization of macrolide 2′-phosphotransferase from a strain of *Escherichia coli* that is highly resistant to erythromycin. Antimicrob Agents Chemother 33:1354–1357

Oldfield E, Feng X (2014) Resistance-resistant antibiotics. Trends Pharmacol Sci 35:664–674. doi:10.1016/j.tips.2014.10.007

Oliveira P, Lindblad P (2011) Novel insights into the regulation of LexA in the cyanobacterium *Synechocystis sp*. Strain PCC 6803. J Bacteriol 193:3804–3814. doi:10.1128/JB.00289-11

Plocinski P, Arora N, Sarva K, Blaszczyk E, Qin H, Das N, Plocinska R, Ziolkiewicz M, Dziadek J, Kiran M, Gorla P, Cross TA, Madiraju M, Rajagopalan M (2012) *Mycobacterium tuberculosis* CwsA interacts with CrgA and Wag31, and the CrgA-CwsA complex is involved in peptidoglycan synthesis and cell shape determination. J Bacteriol 194:6398–6409. doi:10.1128/JB.01005-12

Quan S, Venter H, Dabbs ER (1997) Ribosylative inactivation of rifampin by *Mycobacterium smegmatis* is a principal contributor to its low susceptibility to this antibiotic. Antimicrob Agents Chemother 41:2456–2460

Rajkovic A, Erickson S, Witzky A, Branson OE, Seo J, Gafken PR, Frietas MA, Whitelegge JP, Faull KF, Navarre W, Darwin AJ, Ibba M (2015) Cyclic rhamnosylated elongation factor P establishes antibiotic resistance in *Pseudomonas aeruginosa*. MBio 6, e00823. doi:10.1128/mBio.00823-15

Ramisetty BC, Natarajan B, Santhosh RS (2015) mazEF-mediated programmed cell death in bacteria: "what is this?". Crit Rev Microbiol 41:89–100. doi:10.3109/1040841X.2013.804030

Ranjit DK, Endres JL, Bayles KW (2011) *Staphylococcus aureus* CidA and LrgA proteins exhibit holin-like properties. J Bacteriol 193:2468–2476. doi:10.1128/JB.01545-10

Rice KC, Firek BA, Nelson JB, Yang SJ, Patton TG, Bayles KW (2003) The *Staphylococcus aureus* cidAB operon: evaluation of its role in regulation of murein hydrolase activity and penicillin tolerance. J Bacteriol 185:2635–2643. doi:10.1128/JB.185.8.2635-2643.2003

Sadeghifard N, Soheili S, Sekawi Z, Ghafourian S (2014) Is the mazEF toxin-antitoxin system responsible for vancomycin resistance in clinical isolates of *Enterococcus faecalis*? GMS Hyg Infect Control 9:Doc05. doi:10.3205/dgkh000225

Sahal G, Bilkay IS (2014) Multi drug resistance in strong biofilm forming clinical isolates of *Staphylococcus epidermidis*. Braz J Microbiol 45:539–544. doi:10.1590/S1517-83822014005000042

Sahal G, Nasseri B, Bilkay IS, Piskin E (2015) Anti-biofilm effect of nanometer scale silver (NmSAg) coatings on glass and polystyrene surfaces against *P. mirabilis, C. glabrata* and *C. tropicalis* strains. J Appl Biomater Funct Mater. doi:10.5301/jabfm.5000248

Saier MH Jr, Paulsen IT (2001) Phylogeny of multidrug transporters. Semin Cell Dev Biol 12:205–213. doi:10.1006/scdb.2000.0246

Sedwick C (2011) PezT: a bacterial suicide gene. PLoS Biol 9, e1001036. doi:10.1371/journal.pbio.1001036

Shaw KJ, Rather PN, Hare RS, Miller GH (1993) Molecular genetics of aminoglycoside resistance genes and familial relationships of the aminoglycoside-modifying enzymes. Microbiol Rev 57:138–163

Skarzynski T, Mistry A, Wonacott A, Hutchinson SE, Kelly VA, Duncan K (1996) Structure of UDP-N-acetylglucosamine enolpyruvyl transferase, an enzyme essential for the synthesis of bacterial peptidoglycan, complexed with substrate UDP-N-acetylglucosamine and the drug fosfomycin. Structure 4:1465–1474. doi:10.1016/S0969-2126(96)00153-0

Stallings CL, Chu L, Li LX, Glickman MS (2011) Catalytic and non-catalytic roles for the mono-ADP-ribosyltransferase Arr in the mycobacterial DNA damage response. PLoS One 6, e21807. doi:10.1371/journal.pone.0021807

Sun F, Ding Y, Ji Q, Liang Z, Deng X, Wong CCL, Yi C, Zhang L, Xie S, Alvarez S, Hicks LM, Luo C, Jiang H, Lan L, He C (2012) Protein cysteine phosphorylation of SarA/MgrA family transcriptional regulators mediates bacterial virulence and antibiotic resistance. Proc Natl Acad Sci U S A 109:15461–15466. doi:10.1073/pnas.1205952109

Sutton MD, Narumi I, Walker GC (2002) Posttranslational modification of the umuD-encoded subunit of *Escherichia coli* DNA polymerase V regulates its interactions with the beta processivity clamp. Proc Natl Acad Sci U S A 99:5307–5312. doi:10.1073/pnas.082322099

Szymanski CM, Burr DH, Guerry P (2002) *Campylobacter* protein glycosylation affects host cell interactions. Infect Immun 70:2242–2244. doi:10.1128/IAI.70.4.2242-2244.2002

Thayil SM, Morrison N, Schechter N, Rubin H, Karakousis PC (2011) The role of the novel exopolyphosphatase MT0516 in *Mycobacterium tuberculosis* drug tolerance and persistence. PLoS One 6, e28076. doi:10.1371/journal.pone.0028076

Trieu-Cuot P, Courvalin P (1983) Nucleotide sequence of the *Streptococcus faecalis* plasmid gene encoding the 3′5″-aminoglycoside phosphotransferase type III. Gene 23:331–341. doi:10.1016/0378-1119(83)90022-7

Truong-Bolduc QC, Hooper DC (2010) Phosphorylation of MgrA and its effect on expression of the NorA and NorB efflux pumps of *Staphylococcus aureus*. J Bacteriol 192:2525–2534. doi:10.1128/JB.00018-10

Ubukata K, Itoh-Yamashita N, Konno M (1989) Cloning and expression of the norA gene for fluoroquinolone resistance in *Staphylococcus aureus*. Antimicrob Agents Chemother 33: 1535–1539. doi:10.1128/AAC.33.9.1535

Vaara M, Vaara T, Sarvas M (1979) Decreased binding of polymyxin by polymyxin-resistant mutants of *Salmonella typhimurium*. J Bacteriol 139:664–667

Wang J, Stephan R, Zurfluh K, Hachler H, Fanning S (2014) Characterization of the genetic environment of bla ESBL genes, integrons and toxin-antitoxin systems identified on large transferrable plasmids in multi-drug resistant *Escherichia coli*. Front Microbiol 5:716. doi:10.3389/fmicb.2014.00716

Wright GD (2005) Bacterial resistance to antibiotics: enzymatic degradation and modification. Adv Drug Deliv Rev 57:1451–1470. doi:10.1016/j.addr.2005.04.002

Wright GD (2011) Molecular mechanisms of antibiotic resistance. Chem Commun (Camb) 47: 4055–4061. doi:10.1039/c0cc05111j

Xie L, Liu W, Li Q, Chen S, Xu M, Huang Q, Zeng J, Zhou M, Xie J (2015) First succinyl-proteome profiling of extensively drug-resistant *Mycobacterium tuberculosis* revealed involvement of succinylation in cellular physiology. J Proteome Res 14:107–119. doi:10.1021/pr500859a

Xu WX, Zhang L, Mai JT, Peng RC, Yang EZ, Peng C, Wang HH (2014) The Wag31 protein interacts with AccA3 and coordinates cell wall lipid permeability and lipophilic drug resistance in *Mycobacterium smegmatis*. Biochem Biophys Res Commun 448:255–260. doi:10.1016/j.bbrc.2014.04.116

Yang SJ, Rice KC, Brown RJ, Patton TG, Liou LE, Park YH, Bayles KW (2005) A LysR-type regulator, CidR, is required for induction of the *Staphylococcus aureus* cidABC operon. J Bacteriol 187:5893–5900. doi:10.1128/JB.187.17.5893-5900.2005

Yarmolinsky MB (1995) Programmed cell death in bacterial populations. Science 267:836–837. doi:10.1126/science.7846528

Yazawa K, Mikami Y, Sakamoto T, Ueno Y, Morisaki N, Iwasaki S, Furihata K (1994) Inactivation of the macrolide antibiotics erythromycin, midecamycin, and rokitamycin by pathogenic Nocardia species. Antimicrob Agents Chemother 38:2197–2199. doi:10.1128/AAC.38.9.2197

Yoshida H, Bogaki M, Nakamura S, Ubukata K, Konno M (1990) Nucleotide sequence and characterization of the *Staphylococcus aureus* norA gene, which confers resistance to quinolones. J Bacteriol 172:6942–6949

Zgur-Bertok D (2013) DNA damage repair and bacterial pathogens. PLoS Pathog 9, e1003711. doi:10.1371/journal.ppat.1003711

Zhou Z, Ribeiro AA, Lin S, Cotter RJ, Miller SI, Raetz CR (2001) Lipid A modifications in polymyxin-resistant *Salmonella typhimurium*: PMRA-dependent 4-amino-4-deoxy-L-arabinose, and phosphoethanolamine incorporation. J Biol Chem 276:43111–43121. doi:10.1074/jbc.M106960200

Implications of Chromosomal Mutations for Mycobacterial Drug Resistance

Gail E. Louw and Samantha L. Sampson

Abstract Tuberculosis (TB) remains a global health concern, despite availability of antituberculosis drugs. Drug-resistant *Mycobacterium tuberculosis* strains were identified shortly after the discovery and introduction of streptomycin for the treatment of this disease. Subsequently, multidrug therapy was implemented for TB treatment; however, this was soon followed by reports of multi-, extensively, and totally drug-resistant tuberculosis cases globally. The amplification of this drug resistance is due to the sequential accumulation of chromosomal alterations in target genes in the *Mycobacterium tuberculosis* genome. It is also evident that the presence of mutations that confer drug resistance results in the emergence of compensatory mechanisms which restore bacterial fitness. The recent approval by the Food and Drug Administration for bedaquiline as an antituberculosis drug provided some hope. However, clinical resistance to this new drug has already been reported. This underscores that it is imperative to understand drug resistance and its associated mechanisms in order to direct research efforts to the development of antituberculosis regimens with novel mechanisms of actions.

1 Introduction

In 2015 the World Health Organization (WHO) reported 9.6 million new cases of tuberculosis (TB), with 3.3 % of these and 20 % of previously treated cases infected with a multidrug-resistant (MDR) strain of *Mycobacterium tuberculosis* (WHO

G.E. Louw (✉)
Faculty of Health Sciences, Division of Medical Microbiology, Department of Pathology, Institute of Infectious Disease and Molecular Medicine, University of Cape Town, Observatory, Cape Town, South Africa
e-mail: gail.louw@uct.ac.za

S.L. Sampson
DST/NRF Centre of Excellence for Biomedical Tuberculosis Research/SA MRC Centre for Molecular and Cellular Biology, Division of Molecular Biology and Human Genetics, Department of Biomedical Sciences, Faculty of Health Sciences, Stellenbosch University, Stellenbosch, South Africa

G. Arora et al. (eds.), *Drug Resistance in Bacteria, Fungi, Malaria, and Cancer*,
DOI 10.1007/978-3-319-48683-3_10

2015). More alarmingly, the average proportion of MDR-TB cases with extensively drug-resistant TB (XDR-TB) is 9.7 % (WHO 2015). Resistance to the first effective anti-TB drug, streptomycin (STR), was observed shortly after its introduction in 1944 (Nachega and Chaisson 2003; Keshavjee and Farmer 2012), and this trend had continued for many other TB drugs (Fig. 1). Numerous MDR-TB outbreaks were identified in the early 1990s, emphasizing TB as a global health problem (Nachega and Chaisson 2003) (Fig. 1). MDR-TB is characterized by a mycobacterial infection with *M. tuberculosis* strains that are resistant to rifampicin (RIF) and isoniazid (INH) (Gupta et al. 2003). Outbreaks of XDR-TB have been reported globally (Gandhi et al. 2006; Migliori et al. 2007b; Masjedi et al. 2010; Klopper et al. 2013; Cohen et al. 2015), with XDR-TB defined as an infection with an MDR-TB strain with further resistance to a fluoroquinolone (FQ) and one injectable drug, amikacin (AMI), kanamycin (KANA), and capreomycin (CAP) (Holtz 2007; Holtz and Cegielski 2007; Louw et al. 2009). Recently, *M. tuberculosis* strains resistant to all available anti-TB drugs have been identified globally and have been named totally drug-resistant TB (TDR-TB) (Migliori et al. 2007a; Velayati et al. 2009; Udwadia 2012; Udwadia et al. 2012; Klopper et al. 2013; Udwadia and Vendoti 2013) (Fig. 1). Although this term is somewhat controversial, TDR-TB has been defined as *M. tuberculosis* strains with in vitro resistance to all available first- and second-line drugs tested (INH, RIF, STR, EMB, PZA, ETH, PAS, DCS, OFL, AMI, CIP, CAP, KANA) (Parida et al. 2015). Factors fueling the drug-resistant TB epidemic include the inadequacies of TB control in combination with HIV coinfection.

The WHO recommends that new patients with pulmonary TB receive intensive phase treatment (2 months duration) which consists of INH, RIF, PZA, and EMB. Subsequently, a patient infected with a drug-sensitive *M. tuberculosis* strain is treated with INH and RIF during the 4-month continuation phase treatment (WHO 1997). Patient noncompliance is a consequence of the long treatment duration, and these factors fuel the development of drug resistance. An 8-month

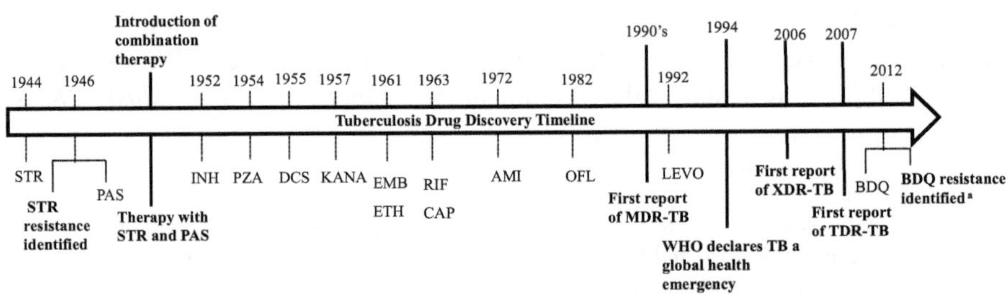

Fig. 1 Illustration of the tuberculosis drug discovery timeline and drug resistance development reports: *STR* streptomycin, *PAS* para-aminosalicylic acid, *INH* isoniazid, *PZA* pyrazinamide, *DCS* D-cycloserine, *KANA* kanamycin, *EMB* ethambutol, *ETH* ethionamide, *RIF* rifampicin, *CAP* capreomycin, *AMI* amikacin, *OFL* ofloxacin, *LEVO* levofloxacin, *BDQ* bedaquiline, *MDR-TB* multidrug-resistant tuberculosis, *WHO* World Health Organization, *XDR-TB* extensively drug-resistant tuberculosis, *TDR-TB* totally drug-resistant tuberculosis. [a]First report of BDQ resistance identified in a TB patient (Bloemberg et al. 2015)

Table 1 Treatment for the various forms of TB (WHO 2008)

Resistance pattern	Treatment
Sensitive	INH-RIF-PZA-EMB
INH	RIF-PZA-EMB
INH-RIF	PZA-STR-LEVO-ETH-DCS-PAS
INH-RIF-EMB	
INH-RIF-PZA-EMB	STR-LEVO-ETH-DCS-PAS
INH-RIF-STR	KANA-LEVO-ETH-DCS-PAS
INH-RIF-EMB-STR	
INH-RIF-EMB-PZA-STR	

INH isoniazid, *RIF* rifampicin, *EMB* ethambutol, *PZA* pyrazinamide, *STR* streptomycin, *LEVO* levofloxacin, *ETH* ethionamide, *DCS* cycloserine, *PAS* para-aminosalicylic acid, *KANA* anamycin

retreatment regimen with first-line anti-TB drugs for previously treated patients awaiting DST results consists of 2 months with INH, RIF, PZA, EMB, and STR; 1 month with INH, RIF, PZA, and EMB; and 5 months with INH, RIF, and EMB. Treatment of MDR-TB requires a regimen with second-line drugs administered over 18–24 months (Mukherjee et al. 2004). Recommendations for the treatment of various forms of drug-resistant TB are tabulated in Table 1. Current drugs used for TB treatment have limited efficacy against drug-resistant *M. tuberculosis* strains. However, new anti-TB drugs in development, specifically drugs with different modes of actions than the current drugs, could be effective against both drug-sensitive and drug-resistant TB.

2 Mode of Action and Mycobacterial Drug Resistance Mechanism

2.1 Cell Wall Synthesis Inhibitors

2.1.1 Isoniazid

INH is a prodrug that inhibits mycolic acid biosynthesis (Vilcheze and Jacobs 2007). This inhibition occurs via multiple mechanisms and results in the loss of trehalose monomycolate, trehalose dimycolate, and mycolates (Vilcheze and Jacobs 2007). INH is activated by KatG, which is a catalase-peroxidase, encoded by the *katG* gene. Upon activation, INH forms an adduct with NAD (Rozwarski et al. 1998) and binds and inhibits *inhA*, encoded by the enoyl-acyl carrier protein reductase InhA (NADH dependent), which is part of the fatty acid synthase type II system (Marrakchi et al. 2000). The INH-NAD adducts inhibit the activity of InhA, thereby resulting in intracellular accumulation of long-chain fatty acids, decreased mycolic acid biosynthesis, and subsequent cell death.

The loss of activation of INH by KatG is one of the mechanisms of INH resistance in mycobacteria. Mutations in the *katG* gene lead to a reduction in catalase activity. This results in a decrease in activated INH and a decreased capacity to form the INH-NAD adduct to inhibit InhA and subsequent high-level INH resistance (Heym et al. 1999; Ramaswamy et al. 2003). The Ser315Thr mutation in the *katG* gene is reported to be the most frequent mutation found in clinical *M. tuberculosis* strains resistant to INH (Seifert et al. 2015). Mutations within the *inhA* promoter (−15T and −8A loci) result in overexpression or modification of *inhA* and subsequently confer low-level INH resistance and ETH cross-resistance (Banerjee et al. 1994). Mutations in the structural gene are less frequent, but the Ser94Ala *inhA* mutation has been reported to be associated with low-level INH resistance (Quemard et al. 1995). Approximately 10 % of INH resistance is not attributed to mutations in *katG* and *inhA*, suggesting that additional resistance mechanisms contribute to INH resistance in mycobacteria. Additional genes (*kasA*, *ahpC*, *ndh*, and the *ahpC-oxyR* intergenic region) have been implicated in INH resistance; however, their direct impact on clinical INH resistance is not fully understood (Vilcheze et al. 2005; Vilcheze and Jacobs 2007; Campbell et al. 2011).

2.1.2 Ethionamide

The second-line drug, ETH, has a common molecular target to INH, namely, InhA of the FAS II system (Banerjee et al. 1994; Marrakchi et al. 2000). ETH is a prodrug and INH structural analog, which also inhibits mycolic acid biosynthesis. It was shown that *M. tuberculosis* strains with low-level INH resistance also exhibit resistance to ETH (Banerjee et al. 1994). ETH is activated by the monooxygenase, *ethA*, with subsequent formation of an ETH-NAD adduct. Even though the ETH-NAD adduct inhibits InhA, in the same manner as the INH-NAD adduct, the activating enzymes of the different compounds are distinct.

Numerous mutations in the *ethA* gene, resulting in a failure to activate ETH, have been reported to contribute to ETH resistance (Morlock et al. 2003; Brossier et al. 2011). The TetR-like repressor, EthR, negatively regulates the expression of *ethA* and interacts directly with the *ethA* promoter region, and EthR overexpression leads to ETH resistance (Baulard et al. 2000; DeBarber et al. 2000). Intragenic *inhA* mutations (Ser94Ala, Ser94Trp, Leu11Val) in addition to *inhA* promoter mutations (−102A and −47C) have also been identified in ETH-resistant *M. tuberculosis* isolates (Morlock et al. 2003; Brossier et al. 2011).

Approximately 50 % of ETH-resistant *M. tuberculosis* strains exhibit an absence of mutations in *inhA* or *ethA*, suggesting an alternative resistance mechanism (Boonaiam et al. 2010). Recently, mutations in the *mshA* gene (including a Val171Gly-Ala187Val double mutation) were identified in ETH-resistant isolates (Vilcheze et al. 2008; Brossier et al. 2011). MshA is a glycosyltransferase that is involved in mycothiol biosynthesis, and mutations in *mshA* have been proposed to result in the failure to activate ETH (Vilcheze et al. 2008). Interestingly, it was also observed that mutations in *ndh* resulted in defects in NdhII activity, subsequently leading to increased intracellular NADH/NAD+ ratio (Vilcheze et al. 2005). The increase in the NADH levels protects against InhA inhibition by either the INH-NAD or ETH-NAD formed when INH and ETH is activated, subsequently leading to ETH and INH co-resistance (Vilcheze et al. 2005). Even with the identification of the additional gene mutations, it is evident that additional resistance mechanisms exist that could contribute to ETH resistance.

2.1.3 Ethambutol

EMB is a bacteriostatic agent that targets the integral membrane arabinosyltransferases involved in polymerizing arabinose into arabinan components of arabinogalactan (Takayama and Kilburn 1989; Zhu et al. 2004; Wolucka 2008; Xu et al. 2015). Resistance to EMB is primarily attributed to mutations in the arabinosyltransferases encoded by *embB*, with 60 % of EMB-resistant isolates carrying a mutation at *embB*306 (Ramaswamy et al. 2000; Zhang and Yew 2009; Safi et al. 2013; Xu et al. 2015). However, several studies report discordance between genotypic and phenotypic resistance testing; this could be due to inaccurate diagnostic tests that are dependent on the medium used (Sreevatsan et al. 1997; Johnson et al. 2006a; Plinke et al. 2010; Xu et al. 2015).

Mutations in the *embC*, *embA*, and *embR* genes have also been implicated in EMB resistance, with alterations located in the embC-*embA* intergenic region conferring high-level EMB resistance (Cui et al. 2014; Xu et al. 2015). *embR* has been reported to modulate the level of arabinosyltransferase activity in vitro in a phosphorylation-dependent manner, acting downstream of the Ser/Thr-kinase PknH (Belanger et al. 1996). Interestingly, mutations were identified in the *ubiA* gene in EMB-resistant XDR-TB isolates lacking shared *embB* mutations (Motiwala et al. 2010; He et al. 2015), and these mutations were associated with high-level EMB resistance (Safi et al. 2013). The *ubiA* gene is essential for growth of *M. tuberculosis* and is involved in the synthesis of decaprenylphosphoryl-D-arabinose (Huang et al. 2005). It was recently reported that overexpression of wild-type *ubiA* gene resulted in an increase in EMB resistance in *M. tuberculosis* (He et al. 2015). This indicates that multiple mechanisms could result in the EMB resistance phenotype in mycobacteria.

2.1.4 SQ109

One of the newer anti-TB drugs, SQ109, was identified by screening a library of EMB derivatives based on the upregulation of the *iniBAC* operon promoter (Lee et al. 2003; Protopopova et al. 2005). Exposure of mycobacteria to SQ109 leads to the inhibition of trehalose dimycolate production and concomitant upregulation of trehalose monomycolate levels (Li et al. 2014b). This results in failure to attach mycolic acids to the cell wall arabinogalactan (Grzegorzewicz et al. 2012; Tahlan et al. 2012). The MIC for SQ109 ranges from 0.16 to 0.78 μg/ml for all

M. tuberculosis strains tested (Jia et al. 2005), and synergy was observed between INH/RIF and SQ109 in in vitro and in vivo analysis (Nikonenko et al. 2007). *M. tuberculosis* has a low spontaneous mutation rate of 2.55×10^{-11} for SQ109 resistance (Sacksteder et al. 2012).

The mycobacterial transport protein responsible for trehalose dimycolate transport, MmpL3, has been identified as the target of SQ109 (Sacksteder et al. 2012; Tahlan et al. 2012). Attempts to generate mutants against SQ109 have been unsuccessful. However, whole genome sequencing of in vitro mutants generated against analogs of SQ109 revealed that mutations in the *mmpL3* gene led to SQ109 and SQ109 analog resistance without cross-resistance to EMB (Tahlan et al. 2012). Mmpl3 mutations (Ala700Thr, Gln40Arg, and Leu567Pro) were reported to result in a greater than fourfold increase in SQ109 resistance level (Tahlan et al. 2012), with cross-resistance being observed between other MmpL3 inhibitors (Li et al. 2014b). Recently, it was observed that SQ109 inhibits enzymes involved in menaquinone synthesis, respiration, and therefore ATP synthesis (Li et al. 2014a). Additionally, SQ109 disrupts the proton motive force, thereby acting as an uncoupler (Li et al. 2014b). This effect on the proton motive force may also impact MmpL proteins, since it is suggested that the resistance-nodulation-division transporters catalyze the export of substrates via a proton anti-port mechanism (Li et al. 2014b).

2.1.5 D-Cycloserine

DCS is recommended by the WHO for the treatment of drug-resistant TB, despite severe side effects (WHO 2000). Resistance to DCS is attributed to overexpression of *alrA* in *M. smegmatis* (Caceres et al. 1997). AlrA encodes for D-alanine racemase that is involved in D-alanine synthesis. D-Alanine is an integral component of peptidoglycan which is an essential component of the cell wall. L-Alanine is converted to D-alanine by the catalytic activity of AlrA (Chacon et al. 2002). Subsequently, the D-alanine/D-alanine ligase (Ddl) catalyzes the dimerization of D-alanine into D-alanyl-D-alanine (Chacon et al. 2002). Studies indicate that *alrA* overexpression is a result of a G→T transversion in the *alrA* promoter (Caceres et al. 1997). These reports also show that *M. smegmatis alrA* null mutants have the ability to grow in the absence of D-alanine, suggesting the presence of another pathway of D-alanine biosynthesis (Chacon et al. 2002). Moreover, these *alrA* null mutants were more susceptible to DCS. It was also observed that a mutation (Gly122Ala) in the *cycA* gene, which encodes a D-serine/alanine/glycine transporter, partially contributes to the DCS resistance phenotype in *M. bovis* BCG vaccine strains (Chen et al. 2012). From these reports it is evident that more research needs to be done on DCS in order to elucidate and understand its resistance mechanisms fully.

2.2 *Inhibitors of DNA Replication*

2.2.1 Fluoroquinolones

Quinolones are synthetic compounds active on the enzymes essential for DNA replication, the DNA gyrases (Ginsburg et al. 2003). By interfering with DNA gyrase activity, the FQs disrupt DNA supercoiling, thereby inhibiting cell division and gene expression. DNA gyrase is comprised of two alpha and two beta subunits, encoded by the *gyrA* and *gyrB* genes, respectively (Takiff et al. 1994). Scientific reports indicate that spontaneous mutations develop at a frequency of 2×10^{-6} to 10^{-8} (Alangaden et al. 1995).

Approximately 90 % of FQ resistance in *M. tuberculosis* is attributed to mutations in a region named the quinolone-resistance-determining region (QRDR) in the *gyrA* and the *gyrB* gene (Takiff et al. 1994; Aubry et al. 2006). Mutations at codons 90 and 94 in the *gyrA* gene are most commonly observed among clinical isolates (Aubry et al. 2006), along with a Ser95Thr polymorphism in *gyrA* that is also present in FQ-sensitive clinical isolates (Maruri et al. 2012). Double mutations in *gyrA* and *gyrB* have been reported to exhibit high-level OFL resistance (Isaeva et al. 2013; Nosova et al. 2013). Mutations in *gyrA* (e.g., Ser91Pro, Asp94Ala, Ala90Val) also result in OFL, MOXI, and LFX cross-resistance with MIC90 > 4 μg/ml (Kambli et al. 2015; Willby et al. 2015). Although the majority of clinical FQ resistance is attributed to mutations in the *gyrA* and *gyrB* genes, additional mechanisms that can contribute to FQ resistance include efflux and DNA mimicry (Pasca et al. 2004). The clinical significance of these mechanisms has not been extensively investigated yet.

2.3 *Inhibitors of Transcription*

2.3.1 Rifampicin

RIF is a highly effective rifamycin that interferes with transcription by inhibiting the DNA-dependent RNA polymerase (RNAP) enzyme (McClure and Cech 1978). The majority of RIF-resistant *M. tuberculosis* strains harbor mutations in an 81 bp RIF resistance-determining region (RRDR) of the *rpoβ* gene, which encodes the β-subunit of RNAP (Telenti et al. 1993). Mutations at different loci in the RRDR of the *rpoβ* gene result in different RIF resistance levels (Louw et al. 2011), with His526Arg, His526Asp, His526Pro, His526Tyr, and Ser531Leu mutations being among the most common among RIF-resistant *M. tuberculosis* isolates (Telenti et al. 1993; Bodmer et al. 1995). Mutations in the RRDR are not the sole contributors to RIF resistance; mutations outside of the RRDR (Heep et al. 2001; Siu et al. 2011), along with the significant upregulation of efflux pumps upon RIF exposure (Louw et al. 2011), have been associated with RIF resistance. In 2011, the WHO endorsed the implementation of an automated test, Xpert® MTB/RIF assay, to rapidly detect TB and RIF-resistant TB (Friedrich et al. 2013). Assessments of the assay indicates that despite the cost limitations, it does provide rapid results, and it significantly increases detection of TB and RIF resistance in culture-confirmed cases, compared to smear microscopy (Steingart et al. 2014).

2.4 Inhibitors of Translation

2.4.1 Aminoglycosides

2.4.1.1 Streptomycin, Amikacin, and Kanamycin

The aminoglycosides inhibit protein synthesis by binding to the 30S subunit of the mycobacterial ribosome (Ramaswamy and Musser 1998), with mutations in the *rpsL*, *rrs*, *gidB*, and *eis* genes implicated in aminoglycoside resistance (Maus et al. 2005a; Zaunbrecher et al. 2009; Georghiou et al. 2012; Reeves et al. 2013). Mutations in the essential *rpsL* gene, which encodes the 12S protein, result in resistance to STR, with the most common *rpsL* mutations being K43R and K88R (Ali et al. 2015). Mutations in the *rrs* gene, encoding for 16S rRNA, result in high-level resistance to STR, AMI, and KANA, with the A1401G mutation being the most frequently observed in AMI and KANA co-resistance (Campbell et al. 2011). Various different mutations in the *gidB* gene, which encodes a 7-methylguanosine methyltransferase that specifically modifies residues on 16S rRNA, have been identified in STR-resistant *M. tuberculosis* strains. These mutations result in the failure to methylate specific residues on the 16S rRNA molecule, thereby leading to resistance conferred by loss-of-function mutations (Ali et al. 2015). It was reported that promoter mutations in the 5′ untranslated region of the *eis* gene, encoding an aminoglycoside acetyltransferase, confer clinical low-level resistance to KANA. This acetyltransferase acetylates KANA, thereby leading to its inactivation, which subsequently prevents the drug from binding to the 30S ribosome (Zaunbrecher et al. 2009). To date, these mutations have been relatively selective for KANA resistance; therefore many strains with *eis* mutations would be classified as AMI susceptible. Interestingly, it has recently been reported that mutations in the 5′ untranslated region of the *eis* transcriptional activator, *whiB7*, also results in KANA resistance. These mutations in *whiB7* lead to an upregulation of *eis*, thereby resulting in KANA degradation and subsequent resistance (Reeves et al. 2013).

2.4.2 Cyclic Peptides

2.4.2.1 Capreomycin and Viomycin

CAP and VIO are cyclic peptides that inhibit protein synthesis. VIO has been shown to bind both the 30S and 50S ribosome subunits and to inhibit ribosomal translocation by interference with the peptidyl tRNA acceptor site (Yamada et al. 1978). VIO and CAP cross-resistance occurs in *M. tuberculosis*. Cross-resistance between CAP and AMI/KANA has been reported, but cross-resistance between CAP and STR is rare (Maus et al. 2005a). Mutations at A1401G, C1402T, and G1484T are associated with CAP resistance, with additional mutations at various positions in the *tlyA* gene, an rRNA methyltransferase reported to exhibit VIO and CAP resistance (Maus et al. 2005a, b).

2.4.3 Oxazolidinones

2.4.3.1 Linezolid

Linezolid (LIN) was first introduced to treat gram-positive infections, including staphylococcal and streptococcal infections (Perry and Jarvis 2001). In vitro linezolid MICs for susceptible *M. tuberculosis* strains ranged from 0.25 to 1 μg/ml with an MIC90 of 0.5 μg/ml. Development of resistance against linezolid was considered to be rare (Richter et al. 2007). Reported in vitro frequencies for linezolid resistant mutants were 2×10^{-8} to 5×10^{-9} (Hillemann et al. 2008). Sequencing of the 23S rRNA gene in linezolid resistant mutants revealed the presence of a G to T nucleotide substitution at either position 2061 or position 2576 (Richter et al. 2007). The level of resistance for LIN mutants with the nucleotide substitution at position 2061 was 32 μg/ml, whereas those with a nucleotide substitution at position 2576 had a resistance level of 16 μg/ml (Richter et al. 2007). Interestingly, the predominant mutation identified in clinical and in vitro selected LIN mutants was in the *rplC* gene, encoding the L3 ribosomal protein, at T460C (Beckert et al. 2012).

2.5 *Anti-TB Drugs That Target Energy Metabolism*

2.5.1 Pyrazinamide

Pyrazinamide (PZA) susceptibility testing is technically difficult due to the acidic medium required for DST tests (Hoffner et al. 2013). PZA-resistant *M. tuberculosis* strains emerge due to a lack of pyrazinamidase (PZase) activity. PZase is required to convert PZA to its active form pyrazinoic acid (POA) (Konno et al. 1967). The protonated form, HPOA, enters the cell, accumulates, and eventually kills the cell (Zhang and Mitchison 2003). The PZA MIC of M. *tuberculosis* ranges from 6.25 to 50 μg/ml at pH 5.5 (Stottmeier et al. 1967). However, a PZA MIC > 2000 μg/ml has been reported for *M. avium* and *M. smegmatis* due to intrinsic PZA resistance as a result of efflux. *M. bovis* is also naturally resistant to PZA due to C→Gnt169 in *pncA*, whereas *M. kansasii* has weak PZase activity and exhibits an MIC of 250 μg/ml (Ramirez-Busby and Valafar 2015). PZA resistance in *M. tuberculosis* is mostly due to mutations in the *pncA* gene (Whitfield et al. 2015a); however, *pncA* polymorphisms that do not confer the PZA-resistant phenotype have also been identified (Whitfield et al. 2015b). Mutations in *rspA*, involved in trans-translation, have also been identified in PZA-resistant strains (Louw et al. 2006; Shi et al. 2011; Feuerriegel et al. 2013; Simons et al. 2013b; Tan et al. 2014). Interestingly, *M. canetti* is naturally resistant to PZA due to a mutation (Met117Thr) in *panD* (Zhang et al. 2013). Subsequently, *panD* mutations in PZA-resistant *M. tuberculosis* strains lacking *rpsA* or *pncA* mutations have also been identified (Shi et al. 2014). POA inhibits enzymatic activity of *panD*, and it was observed that anti-TB activity of POA could be antagonized by B-alanine or pantothenate (Dillon et al. 2014).

2.5.2 Bedaquiline

Bedaquiline (BDQ) (Sirturo or TMC207) is the first anti-TB drug in 40 years to be FDA approved for treatment of sensitive and MDR-TB. The use of BDQ in addition to the standard TB therapy in the murine model accelerated the bactericidal effect (Andries et al. 2005; Lounis et al. 2006; Ibrahim et al. 2007). The minimum inhibitory concentrations of BDQ for *M. tuberculosis* H37Rv and drug-susceptible strains ranged from 0.03 to 0.12 μg/ml (Table 1) (Andries et al. 2005). Computational models suggest that BDQ restricts the rotational activity of ATP synthase, thereby inhibiting ATP production (deJonge et al. 2007). Spontaneous mutant selection and subsequent whole genome sequence analysis of the resistant *M. tuberculosis* and *M. smegmatis* mutants identified mutations (Ala63Pro and Asp32Val) in the c-subunit of ATP synthase encoded by the *atpE* gene (Andries et al. 2005; Koul et al. 2007). Mutations in *atpE* partially account for the BDQ resistance phenotype, with the report of spontaneous mutants without *atpE* gene mutations (Andries et al. 2005; Huitric et al. 2007, 2010). Recently, clofazimine (CFZ)-BDQ cross-resistance was observed in CFZ-resistant in vitro mutants. In the absence of *atpE* mutations, these mutants harbored mutations in the transcriptional repressor, Rv0678, which subsequently resulted in the upregulation of the *Rv0678* and the *mmpL5-mmpS5* efflux system (Milano et al. 2009; Hartkoorn et al. 2014). This upregulation led to a four- to eightfold increase in the level of resistance for CFZ and BDQ, which could be reversed with the addition of verapamil and reserpine (Andries et al. 2014; Hartkoorn et al. 2014).

2.6 *Multi-target Drugs*

2.6.1 PA-824/Pretomanid

PA-824 is a member of the nitroimidazole family containing a nitroimidazopyran nucleus. The MIC for PA-824 ranges from 0.039 to 0.25 μg/ml for sensitive strains compared to 0.015–0.513 μg/ml for drug-resistant strains, with a mutation frequency of 1.9×10^{-5} to 6.38×10^{-7} (Stover et al. 2000). PA-824 is a prodrug that is activated to its toxic form, by the mycobacterial membrane-bound nitroreductase Ddn, a deazaflavin F420-dependent enzyme. This activation leads to the inhibition of mycolic acid synthesis, resulting in cell death (Singh et al. 2008). Investigation on the modes of action of PA-824 has shown that intermediate metabolites of PA-824 act as intracellular nitric oxide donors, therefore encouraging intracellular killing of *M. tuberculosis* in anaerobic conditions (Singh et al. 2008; Manjunatha et al. 2009). When bacteria are in a hypoxic nonreplicating state, PA-824 kills as a nitrous donor (Manjunatha et al. 2009). Interestingly, *M. leprae* is intrinsically resistant to PA-824 due to the lack of the *ddn* gene (Manjunatha et al. 2006).

Another mode of action for PA-824 is suggested by the observation that an *fbiC* knockout mutant in H37Rv, which is deficient for F420 production, is hypersensitive to oxidative stress and INH, moxifloxacin, and CFZ (Gurumurthy et al. 2013). By isolating PA-824-resistant mutants from the H37Rv *M. tuberculosis* background, it was observed that 29 % of isolates harbored mutations in the *ddn* gene and 26 % (*fbiC*), 19 % (*fbiA*), 7 % (*fgd1*), and 2 % in the *fbiA* gene. The mutation Ser11STOP in *ddn* gene conferred high-level PA-824 resistance; however, approximately 17 % of mutants lacked mutations in target genes screened, suggesting a different resistance mechanism (Haver et al. 2015).

2.6.2 OPC67683/Delamanid

Delamanid belongs to the nitro-dihydro-imidazooxazole class of antibiotics that inhibit mycolic acid synthesis (Barry and O'Connor 2007). Delamanid has an MIC90 of 0.006–0.05 μg/ml (Diacon et al. 2011), with an in vitro mutation frequency of 6.44×10^{-6} to 4.19×10^{-5} (Szumowski and Lynch 2015). Mutations in F420 biosynthetic genes also result in PA-824-delamanid cross-resistance.

2.6.3 Clofazimine

CFZ is lipophilic riminophenazine developed in 1957 for the treatment of MR-TB (Van Deun et al. 2010). It is a prodrug that is reduced by NADH dehydrogenase (Ndh2), and subsequently re-oxidized by O_2, to release reactive oxygen species (ROS). The production of ROS and subsequent cell death have been reported in *M. smegmatis* treated with CFZ and CFZ analogs (Yano et al. 2011). In vitro isolation of CFZ mutants reported cross-resistance to BDQ due to the presence of mutations in the transcriptional repressor, *Rv0678*, and subsequent upregulation of efflux pumps *mmpL5-mmpS5* (Hartkoorn et al. 2014). Recently, whole genome sequence analysis of spontaneous CFZ mutants revealed mutations in two additional genes that conferred the CFZ-resistant phenotype. These mutations were Glu89STOP in the putative peptidase, PepQ, resulting in the inactivation of this protein (Zhang et al. 2015a). The authors suggest that PepQ could be involved in CFZ activation. The additional mutation, Val351Ala, was identified in a possible permease, Rv1979c, which is involved in amino acid transport (Zhang et al. 2015a). Although it is suggested that this protein could be involved in CFZ uptake and transport, it is evident that the direct effect of these two additional genes on CFZ resistance should be investigated further.

2.7 *Anti-TB Drugs That Target Pathways*

2.7.1 Para-aminosalicylic Acid

PAS is used as a second-line drug that targets the mycobacterial folate pathway (WHO 2000; Chakraborty et al. 2013). This prodrug is a structural analog of para-aminobenzoic acid (PABA) that is the substrate of the dihydropteroate synthase, encoded by *folP1/folP2*. The condensation of PABA and 6-hydroxymethyl-7,8-dihydropterin pyrophosphate to 7,8-dihydropteroate is catalyzed by dihydropteroate synthase. This is subsequently converted to dihydrofolate and reduced by dihydrofolate reductase, encoded by *dfrA*, to produce tetrahydrofolate (Table 2).

Rengarajan and colleagues showed that PAS resistance is attributed to mutations in the *thyA* gene, encoded for by thymidylate synthase A, which is essential for thymine synthesis. In addition, *thyA* gene mutations were also present in clinical *M. tuberculosis* isolates resistant to PAS, indicating that PAS functions as a folate antagonist (Rengarajan et al. 2004; Fivian-Hughes et al. 2012). The dihydrofolate synthase, FolC, is essential for the activation of PAS, and mutations in *folC* have been reported to result in the PAS-resistant phenotype (Zhao et al. 2014). In addition, mutations in *ribD*, encoded for by the alternate dihydrofolate reductase, have been reported to result in its overexpression, thereby leading to PAS resistance (Zheng et al. 2013; Zhao et al. 2014; Zhang et al. 2015b). It was suggested that overexpression of *ribD* confers resistance by compensating for the inhibition of DfrA function.

3 Drug Resistance Mechanisms Other Than Chromosomal Mutations

Drug resistance in *M. tuberculosis* is not attributed to horizontal gene transfer, due to the lack of plasmids in this bacillus (Zainuddin and Dale 1990). Alternative mechanisms that contribute to mycobacterial drug resistance include (a) the production of drug-modifying enzymes, (b) the production of enzymes that inactivated the drug, (c) low cell wall permeability resulting in a decrease in drug influx, and (d) efflux-related mechanisms leading to a reduction in intracellular drug concentration (Davies and Courvalin 1977; Dabbs et al. 1995; Liu et al. 1996; Takiff et al.

Table 2 Characteristics of the anti-TB drugs, its associated MICs, and drug targets

Mode of action	Drug	MIC range for *M. tuberculosis* (μg/ml) Sensitive	Resistant	In vitro mutation frequency	Target	References
Cell wall synthesis inhibitors	INH	0.5	>100	3.5×10^{-6}	*katG*, *inhA*, *furA*, *ahpC*, *kasA*, *ndh*	Vilcheze et al. (2005), Siddiqi et al. (2007), Vilcheze and Jacobs (2007), Zhang and Yew (2009)
	ETH	0.25	>100	$1–4 \times 10^{-8}$	*ethA*, *ethR*, *inhA*, *ndh*	Rastogi et al. (1996), Baulard et al. (2000), DeBarber et al. (2000), Morlock et al. (2003), Vilcheze et al. (2005), Vilcheze et al. (2008), Boonaiam et al. (2010), Brossier et al. (2011)
	EMB	5–15	4–40	3.1×10^{-5}	*embR*, *embC*, *embA*, *embB*, *ubiA*	Sreevatsan et al. (1997), Zhu et al. (2004), Safi et al. (2013), Cui et al. (2014)
	SQ109	0.16–0.78	2–4	10^{-7} (for analogs)	*mmpL3*	Jia et al. (2005), Sacksteder et al. (2012), Tahlan et al. (2012), Li et al. (2014b)
	DCS	8–25	>100	10^{-6} to 10^{-7}	*cycA*, *ddlA*, *alr*	Rastogi et al. (1996), Caceres et al. (1997), Chacon et al. (2002), Chen et al. (2012)
Transcriptional inhibitors	RIF	>0.5	>256	3.1×10^{-8}	*rpoβ*, *rpoA*, *rpoC*	Johnson et al. (2006b), Louw et al. (2011)
Translational inhibitors	STR, CAP, AMI, KANA, VIO	1	>16	3.80×10^{-6}	*rpsL*, *rrs*, *rrl*, *tlyA*, *eis*, *whiB7*, *gidB*	Rastogi et al. (1996), Campbell et al. (2011), Reeves et al. (2013)
	LIN, STZ	0.25–1	>64	2×10^{-8} to 5×10^{-9}	*rplC*, *rrl*	Richter et al. (2007), Hillemann et al. (2008), Beckert et al. (2012), Adams et al. (2014)
DNA gyrase inhibitors	CIP, MOXI, OFL, GATI	0.5	>1	2×10^{-6} to 10^{-8}	*gyrA*, *gyrB*	Avalos et al. (2015), Singh et al. (2015), Willby et al. (2015)
Energy metabolism	BDQ	0.030–0.12	0.12–3.84	2×10^{-8}	*atpE*	Andries et al. (2005), Petrella et al. (2006), Huitric et al. (2007)
	PZA	<100	>400	1×10^{-5}	*pncA*, *panD*, *rspA*	Louw et al. (2006)

(continued)

Table 2 (continued)

Mode of action	Drug	MIC range for *M. tuberculosis* (μg/ml)		In vitro mutation frequency	Target	References
		Sensitive	Resistant			
Multi-targeted	PA-824, delaminid	0.039–0.25	0.36–1.79	1.9×10^{-5} to 6.38×10^{-7}	*ddn*, *fbiA*, *fbiB*, *fbiC*, *fgd1*	Stover et al. (2000), Haver et al. (2015)
	CFZ	0.44–0.25	0.108–8	5×10^{-6}	*ndh2*, *Rv0678*, *mmpL5-mmpS5*, *Rv1979c*, *Rv2535c*	Reddy et al. (1999), Lu et al. (2011), Gupta et al. (2014), Zhang et al. (2015a)
Folate pathway inhibition	PAS	<0.125	>128	2×10^{-7} to 2×10^{-8}	*folC*, *folP1*, *thyA*, *ribD*, *folP2*	Rengarajan et al. (2004), Zheng et al. (2013), Minato et al. (2015)

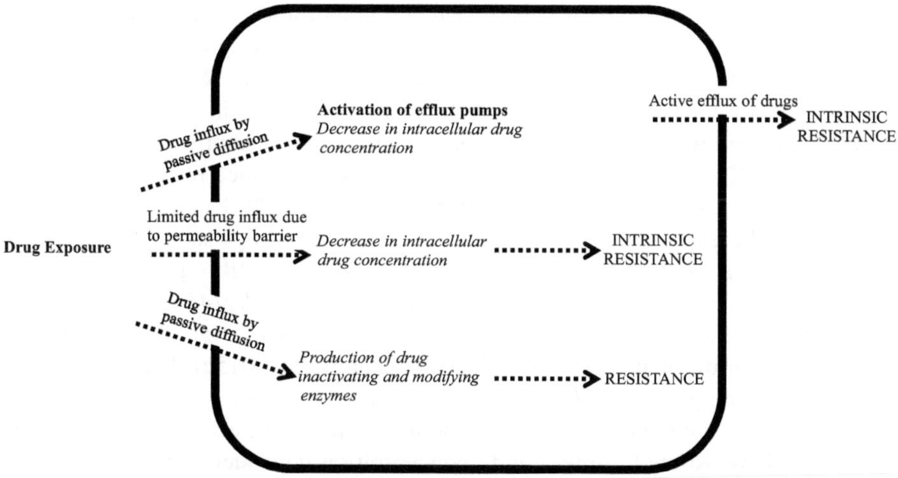

Fig. 2 Mycobacterial drug resistance mechanisms other than chromosomal alterations. Mechanisms such as the activation of efflux pumps and limited drug influx due to the decreased drug permeability lead to a reduction in the intracellular drug concentration and subsequent intrinsic resistance. The production of drug-inactivating and drug-modifying enzymes also results in drug resistance

1996; Davies and Wright 1997; Quan et al. 1997; Imai et al. 1999; Nikaido 2001; Brennan 2003; Draker et al. 2003; Li et al. 2004; Ashenafi et al. 2014) (Fig. 2).

3.1 Permeability Barrier and Activation of Efflux Pumps

Certain mycobacterial species exhibit an intrinsic drug-resistant phenotype that is not the result of antibiotic exposure (Fajardo et al. 2008). Intrinsic drug resistance is attributed to the activation of efflux pumps and an inherently low permeability of the mycobacterial cell wall (Nikaido 2001; Borges-Walmsley et al. 2003; Louw et al. 2009). Recently, knockdown of Rv1026 (*ppx2*), an exopolyphosphatase, was shown to result in increased bacterial cell wall thickness and decreased INH permeability (Nikaido 2001; Brennan 2003; Chuang et al. 2015). This indicated a molecular basis contributing to decreased permeability and intrinsic drug resistance.

Whole genome sequencing of *M. tuberculosis* revealed the presence of various efflux pumps that may enable the bacilli to evade the antimycobacterial killing action. Efflux pumps export various toxic compounds including antibiotics and metabolites, resulting in a decrease in intracellular concentration (Pages et al. 2005; Gupta et al. 2006). This phenomenon has been extensively studied in mycobacteria recently (Li et al. 2004; Morris et al. 2005; Buroni et al. 2006; Zechini and Versace 2009; Adams et al. 2011; Louw et al. 2011; Rodrigues et al. 2011, 2012; Balganesh et al. 2012; Hartkoorn et al. 2014).

In vivo and in vitro *studies* have revealed that antibiotic exposure of mycobacterial cells resulted in the significant upregulation of efflux pumps. It was shown that exposure to RIF resulted in an increase in expression of *Rv1258c*, which is a tap-like efflux pump (Adams et al. 2011, 2014), and arise in the RIF resistance level. Treatment with efflux pump inhibitors, verapamil, reserpine, and tetrandrine, along with RIF, INH, and EMB, could reverse the resistance phenotype of these anti-TB drugs (Adams et al. 2011, 2014; Louw et al. 2011). Studies have also shown that the exposure of *M. tuberculosis* to anti-TB drugs such as EMB, INH, RIF, OFL, STR, and PAS results in the upregulation of efflux pumps like *drrA*, *drrB*, *efpA*, *mmr*, *jefA*, *Rv1634*, *whiB7*, *Rv1456c-Rv1457c-Rv1458c*, *Rv1258c*, and *pstB* (Morris et al. 2005; Ramon-Garcia et al. 2012; Gupta et al. 2014; Hartkoorn et al. 2014; Garima et al. 2015; Li et al. 2015; Zhang et al. 2015c). The upregulation of the efflux pumps results in an MDR phenotype. Interestingly, the organosilicon compound, SILA-421 and thioradazine, both shown to have efflux pump inhibitory activity, demonstrated time- and concentration-dependent activity against *M. tuberculosis* as well as the enhanced killing of intracellular XDR-TB (Martins et al. 2009; Simons et al. 2013a; de Knegt et al. 2014, 2015). These compounds also enhanced the activity of INH and RIF in vitro and prevented the emergence of INH- and RIF-resistant mutants. However, they did not show in vivo activity enhancement of INH and RIF in *M. tuberculosis*-infected mice treated with INH-RIF-PZA for 13 weeks (de Knegt et al. 2014, 2015).

3.2 Production of Drug-Modifying and Inactivating Enzymes

M. smegmatis has been confirmed to be naturally resistant to RIF due the rifampin ADP-ribosyltransferase (Arr-ms), encoded by the chromosome, which assists in covalently adding a ribose group to RIF. This addition modifies and inactivates RIF, thus resulting in intrinsic resistance in *M. smegmatis* to RIF (Dabbs et al. 1995; Imai et al. 1999; Quan et al. 1997; Baysarowich et al. 2008).

The production of inactivating enzymes, e.g., the acetyltransferase AAC (2′)—Ic and the phosphotransferase encoded by the *Rv3225c* gene, APH (6)-la and APH (6)-ld from producer strain *Streptomyces griseus*, has been associated with STR resistance (Davies and Courvalin 1977; Davies and Wright 1997; Draker et al. 2003; Ashenafi et al. 2014). Similarly, the lack of antimicrobial activity in *M. abscessus* of aminoglycosides could be reversed by disruption of the chromosomally encoded *aac(2′)* gene (Maurer et al. 2014, 2015) (Fig. 2). By using *M. smegmatis*, it was shown that the activity of acetyltransferase was significantly induced in response to aminoglycoside, thereby resulting in the inhibition of protein synthesis.

4 Compensatory Mechanisms, Fitness, and Drug Resistance

Some resistance-causing mutations have been found to incur a fitness cost (Gagneux et al. 2006). The fitness cost may be compensated for by the acquisition of secondary mutations at a different site during the evolution of resistant bacteria (Bjorkman et al. 2000). The mutant carrying the chromosomal alteration can become extinct, or the mutations might be fixed in the population by means of compensatory evolution (Bottger and Springer 2008). These compensatory mechanisms can reduce the cost by restoring physiological functions impaired by the resistance mutations without altering the level of bacterial resistance (Schrag and Perrot 1996).

Recently, whole genome sequencing of RIF-resistant *M. tuberculosis* strains with *rpoB* mutations revealed novel mutations in *rpoA* and *rpoC* that emerged over time. Strains with these mutations exhibit high competitive fitness in vitro and in vivo and lead to MDR strains with high fitness (Comas et al. 2012). Previously, it was shown by in vitro pair-wise competition experiments that the wild-type *rpoB* *M. tuberculosis* strains outcompeted strains harboring the Ser522Leu, His526Tyr, and Ser531Trp mutations (Billington et al. 1999; Mariam et al. 2004). The extent of fitness loss was dependent on the specific *rpoB* mutation, with the Ser531Leu *rpoB* mutation only exhibiting a minor fitness defect compared to other mutations (Billington et al. 1999; Gagneux et al. 2006; Mariam et al. 2004). Additionally, mutations in *rpoC* illustrated that epistatic interactions between mutations that confer drug resistance, compensatory mutations, and diverse strain genetic background might influence compensatory evolution (de Vos et al. 2012).

In INH resistance, mutations in *katG* eliminate catalase-peroxidase activity, thereby preventing the activation of INH (Heym et al. 1999). It was shown that the expression of KatG or the alkyl hydroperoxidase, AhpC, exhibited a protective effect against organic peroxides in bacilli. The overexpression of AhpC, due to the presence of a mutation in *ahpC*, enabled INH-resistant *katG* mutants to survive during infection (Sherman et al. 1996).

These alternative mechanisms compensating for the loss of fitness caused by genetic mutations are difficult to detect using PCR-based methods as these methods only target mutation hotspots associated with drug resistance. Thus, it is imperative to also consider these compensatory mechanisms upon designing and developing new drugs and treatment regimens.

5 Perspectives

The history of TB drug development and use provides numerous examples of chromosomally encoded resistance, which often emerges very rapidly after the introduction of new drugs. This highlights the need for a diverse product portfolio entering the TB drug development pipeline. Fortunately, there are several promising new drugs at various stages within the TB drug development pipeline. These include bactericidal compounds in the benzothiazinone class, targeting the enzyme decaprenylphosphoryl-β-D-ribose 2′-oxidase (DprE1), which is essential for cell wall synthesis (Makarov et al. 2015). However, as with all TB drugs, there is a need for a better understanding of mechanisms of drug resistance and consequences of mutations that confer drug resistance. The emergence of compensatory mechanisms following the evolution of drug resistance-conferring mutations, after selective pressure, is an additional factor to consider upon rational drug design. Recently, bacterial collateral resistance and sensitivity to various combinations of anti-TB drugs have been reported. However, it is evident that the collateral sensitivity and resistance networks are complex, thereby complicating tailoring specific treatment regimens based on existing drug treatments. It would be desirable to explore alternative approaches to treatment, including the inclusion of efflux pump inhibitors or immunomodulators. Ideal treatment regimens would eliminate the formation of bacterial persisters, reduce the selection of resistant mutants, and ultimately offer a much-reduced treatment regime, to increase compliance.

Acknowledgements This work was funded, in part, by the Intramural Research Program of the NIAID and by grants from the Foundation for the National Institutes of Health. SLS is funded by the South African Research Chairs Initiative of the Department of Science and Technology and National Research Foundation (NRF) of South Africa, award number UID 86539. The content is solely the responsibility of the authors and does not necessarily represent the official views of the NRF.

References

Adams KN, Takaki K, Connolly LE, Wiedenhoft H, Winglee K, Humbert O, Edelstein PH, Cosma CL, Ramakrishnan L (2011) Drug tolerance in replicating mycobacteria mediated by a macrophage-induced efflux mechanism. Cell 145:39–53. doi:10.1016/j.cell.2011.02.022

Adams KN, Szumowski JD, Ramakrishnan L (2014) Verapamil, and its metabolite norverapamil, inhibit macrophage-induced, bacterial efflux pump-mediated tolerance to multiple anti-tubercular drugs. J Infect Dis 210:456–466. doi:10.1093/infdis/jiu095

Alangaden GJ, Manavathu EK, Vakulenko SB, Zvonok NM, Lerner SA (1995) Characterization of fluoroquinolone-resistant mutant strains of *Mycobacterium tuberculosis* selected in the laboratory and isolated from patients. Antimicrob Agents Chemother 39:1700–1703. doi:10.1128/AAC.39.8.1700

Ali A, Hasan Z, McNerney R, Mallard K, Hill-Cawthorne G, Coll F, Nair M, Pain A, Clark TG, Hasan R (2015) Whole genome sequencing based characterization of extensively drug-resistant *Mycobacterium tuberculosis* isolates from Pakistan. PLoS One 10, e0117771. doi:10.1371/journal.pone.0117771

Andries K, Verhasselt P, Guillemont J, Gohlmann HW, Neefs JM, Winkler H, Van Gestel J, Timmerman P, Zhu M, Lee E, Williams P, de Chaffoy D, Huitric E, Hoffner S, Cambau E, Truffot-Pernot C, Lounis N, Jarlier V (2005) A diarylquinoline drug active on the ATP synthase of *Mycobacterium tuberculosis*. Science 307:223–227. doi:10.1126/science.1106753

Andries K, Villellas C, Coeck N, Thys K, Gevers T, Vranckx L, Lounis N, de Jong BC, Koul A (2014) Acquired resistance of *Mycobacterium tuberculosis* to bedaquiline. PLoS One 9, e102135. doi:10.1371/journal.pone.0102135

Ashenafi M, Ammosova T, Nekhai S, Byrnes WM (2014) Purification and characterization of aminoglycoside phosphotransferase APH(6)-Id, a streptomycin-inactivating enzyme. Mol Cell Biochem 387:207–216. doi:10.1007/s11010-013-1886-1

Aubry A, Veziris N, Cambau E, Truffot-Pernot C, Jarlier V, Fisher LM (2006) Novel gyrase mutations in quinolone-resistant and -hypersusceptible clinical isolates of *Mycobacterium tuberculosis*: functional analysis of mutant enzymes. Antimicrob Agents Chemother 50:104–112. doi:10.1128/AAC.50.1.104-112.2006

Avalos E, Catanzaro D, Catanzaro A, Ganiats T, Brodine S, Alcaraz J, Rodwell T (2015) Frequency and geographic distribution of *gyrA* and *gyrB* mutations associated with fluoroquinolone resistance in clinical *Mycobacterium tuberculosis* isolates: a systematic review. PLoS One 10, e0120470. doi:10.1371/journal.pone.0120470

Balganesh M, Dinesh N, Sharma S, Kuruppath S, Nair AV, Sharma U (2012) Efflux pumps of *Mycobacterium tuberculosis* play a significant role in antituberculosis activity of potential drug candidates. Antimicrob Agents Chemother 56:2643–2651. doi:10.1128/AAC.06003-11

Banerjee A, Dubnau E, Quemard A, Balasubramanian V, Um KS, Wilson T, Collins D, de Lisle G, Jacobs WR Jr (1994) *inhA*, a gene encoding a target for isoniazid and ethionamide in *Mycobacterium tuberculosis*. Science 263:227–230. doi:10.1126/science.8284673

Barry PJ, O'Connor TM (2007) Novel agents in the management of *Mycobacterium tuberculosis* disease. Curr Med Chem 14:2000–2008. doi:10.2174/092986707781368496

Baulard AR, Betts JC, Engohang-Ndong J, Quan S, McAdam RA, Brennan PJ, Locht C, Besra GS (2000) Activation of the pro-drug ethionamide is regulated in mycobacteria. J Biol Chem 275:28326–28331

Baysarowich J, Koteva K, Hughes DW, Ejim L, Griffiths E, Zhang K, Junop M, Wright GD (2008) Rifamycin antibiotic resistance by ADP-ribosylation: structure and diversity of Arr. Proc Natl Acad Sci U S A 105:4886–4891. doi:10.1073/pnas.0711939105

Beckert P, Hillemann D, Kohl TA, Kalinowski J, Richter E, Niemann S, Feuerriegel S (2012) rplC T460C identified as a dominant mutation in linezolid-resistant *Mycobacterium tuberculosis* strains. Antimicrob Agents Chemother 56:2743–2745. doi:10.1128/AAC.06227-11

Belanger AE, Besra GS, Ford ME, Mikusova K, Belisle JT, Brennan PJ, Inamine JM (1996) The *embAB* genes of *Mycobacterium avium* encode an arabinosyl transferase involved in cell wall arabinan biosynthesis that is the target for the antimycobacterial drug ethambutol. Proc Natl Acad Sci U S A 93:11919–11924

Billington OJ, McHugh TD, Gillespie SH (1999) Physiological cost of rifampin resistance induced *in vitro* in *Mycobacterium tuberculosis*. Antimicrob Agents Chemother 43:1866–1869

Bjorkman J, Nagaev I, Berg OG, Hughes D, Andersson DI (2000) Effects of environment on compensatory mutations to ameliorate costs of antibiotic resistance. Science 287:1479–1482. doi:10.1126/science.287.5457.1479

Bloemberg GV, Keller PM, Stuckia D, Trauner A, Borrell S, Latshang T, Coscolla M, Rothe T, Homke R, Ritter C, Feldmann J, Schulthess B, Gagneux S, Bottger EC (2015) Acquired resistance to bedaquiline and delamanid in therapy for tuberculosis. N Engl J Med 373:1986–1988. doi:10.1056/NEJMc1505196

Bodmer T, Zurcher G, Imboden P, Telenti A (1995) Mutation position and type of substitution in the beta-subunit of the RNA polymerase influence *in-vitro* activity of rifamycins in rifampicin-resistant *Mycobacterium tuberculosis*. J Antimicrob Chemother 35:345–348. doi:10.1093/jac/35.2.345

Boonaiam S, Chaiprasert A, Prammananan T, Leechawengwongs M (2010) Genotypic analysis of genes associated with isoniazid and ethionamide resistance in MDR-TB isolates from Thailand. Clin Microbiol Infect 16:396–399. doi:10.1111/j.1469-0691.2009.02838.x

Borges-Walmsley MI, McKeegan KS, Walmsley AR (2003) Structure and function of efflux pumps that confer resistance to drugs. Biochem J 376:313–338. doi:10.1042/bj20020957

Bottger EC, Springer B (2008) Tuberculosis: drug resistance, fitness, and strategies for global control. Eur J Pediatr 167:141–148

Brennan PJ (2003) Structure, function, and biogenesis of the cell wall of *Mycobacterium tuberculosis*. Tuberculosis (Edinb) 83:91–97

Brossier F, Veziris N, Truffot-Pernot C, Jarlier V, Sougakoff W (2011) Molecular investigation of resistance to the antituberculous drug ethionamide in multidrug-resistant clinical isolates of *Mycobacterium tuberculosis*. Antimicrob Agents Chemother 55:355–360. doi:10.1128/AAC.01030-10

Buroni S, Manina G, Guglierame P, Pasca MR, Riccardi G, De Rossi E (2006) LfrR is a repressor that regulates expression of the efflux pump LfrA in *Mycobacterium smegmatis*. Antimicrob Agents Chemother 50:4044–4052. doi:10.1128/AAC.00656-06

Caceres NE, Harris NB, Wellehan JF, Feng Z, Kapur V, Barletta RG (1997) Overexpression of the D-alanine racemase gene confers resistance to D-cycloserine in *Mycobacterium smegmatis*. J Bacteriol 179:5046–5055

Campbell PJ, Morlock GP, Sikes RD, Dalton TL, Metchock B, Starks AM, Hooks DP, Cowan LS, Plikaytis BB, Posey JE (2011) Molecular detection of mutations associated with first- and second-line drug resistance compared with conventional drug susceptibility testing of *Mycobacterium tuberculosis*. Antimicrob Agents Chemother 55:2032–2041. doi:10.1128/AAC.01550-10

Chacon O, Feng Z, Harris NB, Caceres NE, Adams LG, Barletta RG (2002) *Mycobacterium smegmatis* D-Alanine Racemase mutants are not dependent on D-Alanine for growth. Antimicrob Agents Chemother 46:47–54. doi:10.1128/AAC.46.2.47-54.2002

Chakraborty S, Gruber T, Barry CE III, Boshoff HI, Rhee KY (2013) Para-aminosalicylic acid acts as an alternative substrate of folate metabolism in *Mycobacterium tuberculosis*. Science 339:88–91. doi:10.1126/science.1228980

Chen JM, Uplekar S, Gordon SV, Cole ST (2012) A point mutation in *cycA* partially contributes to the D-cycloserine resistance trait of *Mycobacterium bovis* BCG vaccine strains. PLoS One 7, e43467. doi:10.1371/journal.pone.0043467

Chuang YM, Bandyopadhyay N, Rifat D, Rubin H, Bader JS, Karakousis PC (2015) Deficiency of the novel exopolyphosphatase Rv1026/PPX2 leads to metabolic downshift and altered cell wall permeability in *Mycobacterium tuberculosis*. mBio 6, e02428. doi:10.1128/mBio.02428-14

Cohen KA, Abeel T, Manson McGuire A, Desjardins CA, Munsamy V, Shea TP, Walker BJ, Bantubani N, Almeida DV, Alvarado L, Chapman SB, Mvelase NR, Duffy EY, Fitzgerald MG, Govender P, Gujja S, Hamilton S, Howarth C, Larimer JD, Maharaj K, Pearson MD, Priest ME, Zeng Q, Padayatchi N, Grosset J, Young SK, Wortman J, Mlisana KP, O'Donnell MR, Birren BW, Bishai WR, Pym AS, Earl AM (2015) Evolution of extensively drug-resistant tuberculosis over four decades: whole genome sequencing and dating analysis of *Mycobacterium tuberculosis* isolates from KwaZulu-Natal. PLoS Med 12, e1001880. doi:10.1371/journal.pmed.1001880

Comas I, Borrell S, Roetzer A, Rose G, Malla B, Kato-Maeda M, Galagan J, Niemann S, Gagneux S (2012) Whole-genome sequencing of rifampicin-resistant *Mycobacterium tuberculosis* strains identifies compensatory mutations in RNA polymerase genes. Nat Genet 44:106–110. doi:10.1038/ng.1038

Cui Z, Li Y, Cheng S, Yang H, Lu J, Hu Z, Ge B (2014) Mutations in the embC-embA intergenic region contribute to *Mycobacterium tuberculosis* resistance to ethambutol. Antimicrob Agents Chemother 58:6837–6843. doi:10.1128/AAC.03285-14

Dabbs ER, Yazawa K, Mikami Y, Miyaji M, Morisaki N, Iwasaki S, Furihata K (1995) Ribosylation by mycobacterial strains as a new mechanism of rifampin inactivation. Antimicrob Agents Chemother 39:1007–1009

Davies J, Courvalin P (1977) Mechanisms of resistance to aminoglycosides. Am J Med 62:868–872

Davies J, Wright GD (1997) Bacterial resistance to aminoglycoside antibiotics. Trends Microbiol 5:234–240. doi:10.1016/S0966-842X(97)01033-0 de Knegt GJ, Ten Kate MT, van Soolingen D, Aarnoutse R, Boeree MJ, Bakker-Woudenberg IA, de Steenwinkel JE (2014) Enhancement of *in vitro* activity of tuberculosis drugs by addition of thioridazine is not reflected by improved *in vivo* therapeutic efficacy. Tuberculosis (Edinb). doi:10.1016/j.tube.2014.09.002

de Knegt GJ, Bakker-Woudenberg IA, van Soolingen D, Aarnoutse R, Boeree MJ, de Steenwinkel JE (2015) SILA-421 activity *in vitro* against rifampicin-susceptible and rifampicin-resistant *Mycobacterium tuberculosis*, and *in vivo* in a murine tuberculosis model. Int J Antimicrob Agents 46:66–72. doi:10.1016/j.ijantimicag.2015.02.025

de Vos M, Muller B, Borrell S, Black P, van Helden P, Warren R, Gagneux S, Victor T (2012) Putative compensatory mutations in the rpoC gene of rifampicin-resistant *Mycobacterium tuberculosis* are associated with ongoing transmission. Antimicrob Agents Chemother 57:827–832. doi:10.1128/AAC.01541-12

DeBarber AE, Mdluli K, Bosman M, Bekker LG, Barry CE III (2000) Ethionamide activation and sensitivity in multidrug-resistant *Mycobacterium tuberculosis*. Proc Natl Acad Sci U S A 97:9677–9682. doi:10.1073/pnas.97.17.9677

deJonge MR, Koymans LH, Guillemont JE, Koul A, Andries K (2007) A computational model of the inhibition of *Mycobacterium tuberculosis* ATPase by a new drug candidate R207910. Proteins 67:971–980. doi:10.1002/prot.21376

Diacon AH, Dawson R, Hanekom M, Narunsky K, Venter A, Hittel N, Geiter LJ, Wells CD, Paccaly AJ, Donald PR (2011) Early bactericidal activity of delamanid (OPC-67683) in smear-positive pulmonary tuberculosis patients. Int J Tuberc Lung Dis 15:949–954. doi:10.5588/ijtld.10.0616

Dillon NA, Peterson ND, Rosen BC, Baughn AD (2014) Pantothenate and pantetheine antagonize the antitubercular activity of pyrazinamide. Antimicrob Agents Chemother 58:7258–7263. doi:10.1128/AAC.04028-14

Draker KA, Boehr DD, Elowe NH, Noga TJ, Wright GD (2003) Functional annotation of putative aminoglycoside antibiotic modifying proteins in *Mycobacterium tuberculosis* H37Rv. J Antibiot (Tokyo) 56:135–142

Fajardo A, Martinez-Martin N, Mercadillo M, Galan JC, Ghysels B, Matthijs S, Cornelis P, Wiehlmann L, Tummler B, Baquero F, Martinez JL (2008) The neglected intrinsic resistome of bacterial pathogens. PLoS One 3, e1619. doi:10.1371/journal.pone.0001619

Feuerriegel S, Koser CU, Richter E, Niemann S (2013) *Mycobacterium canettii* is intrinsically resistant to both pyrazinamide and pyrazinoic acid. J Antimicrob Chemother 68:1439–1440. doi:10.1093/jac/dkt042

Fivian-Hughes AS, Houghton J, Davis EO (2012) *Mycobacterium tuberculosis* thymidylate synthase gene *thyX* is essential and potentially bifunctional, while *thyA* deletion confers resistance to p-aminosalicylic acid. Microbiology 158:308–318. doi:10.1099/mic.0.053983-0

Friedrich SO, Rachow A, Saathoff E, Singh K, Mangu CD, Dawson R, Phillips PP, Venter A, Bateson A, Boehme CC, Heinrich N, Hunt RD, Boeree MJ, Zumla A, McHugh TD, Gillespie SH, Diacon AH, Hoelscher M, Pan African Consortium for the Evaluation of Anti-tuberculosis A (2013) Assessment of the sensitivity and specificity of Xpert MTB/RIF assay as an early sputum biomarker of response to tuberculosis treatment. Lancet Respir Med 1:462–470. doi:10.1016/S2213-2600(13)70119-X

Gagneux S, Long CD, Small PM, Van T, Schoolnik GK, Bohannan BJ (2006) The competitive cost of antibiotic resistance in *Mycobacterium tuberculosis*. Science 312:1944–1946. doi:10.1126/science.1124410

Gandhi NR, Moll A, Sturm AW, Pawinski R, Govender T, Lalloo U, Zeller K, Andrews J, Friedland G (2006) Extensively drug-resistant tuberculosis as a cause of death in patients co-infected with tuberculosis and HIV in a rural area of South Africa. Lancet 368:1575–1580. doi:10.1016/S0140-6736(06)69573-1

Garima K, Pathak R, Tandon R, Rathor N, Sinha R, Bose M, Varma-Basil M (2015) Differential expression of efflux pump genes of *Mycobacterium tuberculosis* in response to varied subinhibitory concentrations of antituberculosis agents. Tuberculosis (Edinb) 95:155–161. doi:10.1016/j.tube.2015.01.005

Georghiou SB, Magana M, Garfein RS, Catanzaro DG, Catanzaro A, Rodwell TC (2012) Evaluation of genetic mutations associated with *Mycobacterium tuberculosis* resistance to amikacin, kanamycin and capreomycin: a systematic review. PLoS One 7, e33275. doi:10.1371/journal.pone.0033275

Ginsburg AS, Grosset JH, Bishai WR (2003) Fluoroquinolones, tuberculosis, and resistance. Lancet Infect Dis 3:432–442. doi:10.1016/S1473-3099(03)00671-6

Grzegorzewicz AE, Pham H, Gundi VA, Scherman MS, North EJ, Hess T, Jones V, Gruppo V, Born SE, Kordulakova J, Chavadi SS, Morisseau C, Lenaerts AJ, Lee RE, McNeil MR, Jackson M (2012) Inhibition of mycolic acid transport across the *Mycobacterium tuberculosis* plasma membrane. Nat Chem Biol 8:334–341. doi:10.1038/nchembio.794

Gupta R, Espinal M, Stop TBWGoD-PfMDRTB (2003) A prioritised research agenda for DOTS-Plus for multidrug-resistant tuberculosis (MDR-TB). Int J Tuberc Lung Dis 7:410–414

Gupta AK, Chauhan DS, Srivastava K, Das R, Batra S, Mittal M, Goswami P, Singhal N, Sharma VD, Venkatesan K, Hasnain SE, Katoch VM (2006) Estimation of efflux mediated multi-drug resistance and its correlation with expression levels of two major efflux pumps in mycobacteria. J Commun Dis 38:246–254

Gupta S, Cohen KA, Winglee K, Maiga M, Diarra B, Bishai WR (2014) Efflux inhibition with verapamil potentiates bedaquiline in *Mycobacterium tuberculosis*. Antimicrob Agents Chemother 58:574–576. doi:10.1128/AAC.01462-13

Gurumurthy M, Rao M, Mukherjee T, Rao SP, Boshoff HI, Dick T, Barry CE III, Manjunatha UH (2013) A novel F(420)-dependent anti-oxidant mechanism protects *Mycobacterium tuberculosis* against oxidative stress and bactericidal agents. Mol Microbiol 87:744–755. doi:10.1111/mmi.12127

Hartkoorn RC, Uplekar S, Cole ST (2014) Cross-resistance between clofazimine and bedaquiline through upregulation of MmpL5 in *Mycobacterium tuberculosis*. Antimicrob Agents Chemother 58:2979–2981. doi:10.1128/AAC.00037-14

Haver HL, Chua A, Ghode P, Lakshminarayana SB, Singhal A, Mathema B, Wintjens R, Bifani P (2015) Mutations in genes for the F420 biosynthetic pathway and a nitroreductase enzyme are the primary resistance determinants in spontaneous in vitro-selected PA-824-resistant mutants of *Mycobacterium tuberculosis*. Antimicrob Agents Chemother 59:5316–5323. doi:10.1128/AAC.00308-15

He L, Wang X, Cui P, Jin J, Chen J, Zhang W, Zhang Y (2015) *ubiA* (*Rv3806c*) encoding DPPR synthase involved in cell wall synthesis is associated with ethambutol resistance in *Mycobacterium tuberculosis*. Tuberculosis (Edinb) 95:149–154. doi:10.1016/j.tube.2014.12.002

Heep M, Brandstatter B, Rieger U, Lehn N, Richter E, Rusch-Gerdes S, Niemann S (2001) Frequency of *rpoB* mutations inside and outside the cluster I region in rifampin-resistant clinical *Mycobacterium tuberculosis* isolates. J Clin Microbiol 39:107–110. doi:10.1128/JCM.39.1.107-110.2001

Heym B, Saint-Joanis B, Cole ST (1999) The molecular basis of isoniazid resistance in *Mycobacterium tuberculosis*. Int J Tuberc Lung Dis 79:267–271

Hillemann D, Rusch-Gerdes S, Richter E (2008) In vitro-selected linezolid-resistant *Mycobacterium tuberculosis* mutants. Antimicrob Agents Chemother 52:800–801. doi:10.1128/AAC.01189-07

Hoffner S, Angeby K, Sturegard E, Jonsson B, Johansson A, Sellin M, Werngren J (2013) Proficiency of drug susceptibility testing of *Mycobacterium tuberculosis* against pyrazinamide: the Swedish experience. Int J Tuberc Lung Dis 17:1486–1490. doi:10.5588/ijtld.13.0195

Holtz TH (2007) XDR-TB in South Africa: revised definition. PLoS Med 4, e161. doi:10.1371/journal.pmed.0040161

Holtz TH, Cegielski JP (2007) Origin of the term XDR-TB. Eur Respir J 30:396. doi:10.1183/09031936.00042607

Huang H, Scherman MS, D'Haeze W, Vereecke D, Holsters M, Crick DC, McNeil MR (2005) Identification and active expression of the *Mycobacterium tuberculosis* gene encoding 5-phospho-{alpha}-d-ribose-1-diphosphate:decaprenyl-phosphate5-phosphoribosyltransferase, the first enzyme committed to decaprenylphosphoryl-d-arabinose synthesis. J Biol Chem 280:24539–24543. doi:10.1074/jbc.M504068200

Huitric E, Verhasselt P, Andries K, Hoffner SE (2007) In vitro antimycobacterial spectrum of a diarylquinoline ATP synthase inhibitor. Antimicrob Agents Chemother 51:4202–4204. doi:10.1128/AAC.00181-07

Huitric E, Verhasselt P, Koul A, Andries K, Hoffner S, Andersson DI (2010) Rates and mechanisms of resistance development in *Mycobacterium tuberculosis* to a novel diarylquinoline ATP synthase inhibitor. Antimicrob Agents Chemother 54:1022–1028. doi:10.1128/AAC.01611-09

Ibrahim M, Andries K, Lounis N, Chauffour A, Truffot-Pernot C, Jarlier V, Veziris N (2007) Synergistic activity of R207910 combined with pyrazinamide against murine tuberculosis. Antimicrob Agents Chemother 51:1011–1015. doi:10.1128/AAC.00898-06

Imai T, Watanabe K, Mikami Y, Yazawa K, Ando A, Nagata Y, Morisaki N, Hashimoto Y, Furihata K, Dabbs ER (1999) Identification and characterization of a new intermediate in the ribosylative inactivation pathway of rifampin by *Mycobacterium smegmatis*. Microb Drug Resist 5:259–264. doi:10.1089/mdr.1999.5.259

Isaeva Y, Bukatina A, Krylova L, Nosova E, Makarova M, Moroz A (2013) Determination of critical concentrations of moxifloxacin and gatifloxacin for drug susceptibility testing of Mycobacterium tuberculosis in the BACTEC MGIT 960 system. J Antimicrob Chemother 68:2274–2281. doi:10.1093/jac/dkt202

Jia L, Tomaszewski JE, Hanrahan C, Coward L, Noker P, Gorman G, Nikonenko B, Protopopova M (2005) Pharmacodynamics and pharmacokinetics of SQ109, a new diamine-based antitubercular drug. Br J Pharmacol 144:80–87. doi:10.1038/sj.bjp.0705984

Johnson R, Jordaan AM, Pretorius L, Engelke E, van der Spuy G, Kewley C, Bosman M, van Helden PD, Warren R, Victor TC (2006a) Ethambutol resistance testing by mutation detection. Int J Tuberc Lung Dis 10:68–73

Johnson R, Streicher EM, Louw GE, Warren RM, van Helden PD, Victor TC (2006b) Drug resistance in *Mycobacterium tuberculosis*. Curr Issues Mol Biol 8:97–111

Kambli P, Ajbani K, Sadani M, Nikam C, Shetty A, Udwadia Z, Rodwell TC, Catanzaro A, Rodrigues C (2015) Correlating Minimum Inhibitory Concentrations of ofloxacin and moxifloxacin with *gyrA* mutations using the genotype MTBDRsl assay. Tuberculosis (Edinb) 95:137–141. doi:10.1016/j.tube.2014

Keshavjee S, Farmer PE (2012) Tuberculosis, drug resistance, and the history of modern medicine. N Engl J Med 367:931–936. doi:10.1056/NEJMra1205429

Klopper M, Warren RM, Hayes C, Gey van Pittius NC, Streicher EM, Muller B, Sirgel FA, Chabula-Nxiweni M, Hoosain E, Coetzee G, David van Helden P, Victor TC, Trollip AP (2013) Emergence and spread of extensively and totally drug-resistant tuberculosis, South Africa. Emerg Infect Dis 19:449–455. doi:10.3201/EID1903.120246

Konno K, Feldmann FM, McDermott W (1967) Pyrazinamide susceptibility and amidase activity of tubercle bacilli. Am Rev Respir Dis 95:461–469

Koul A, Dendouga N, Vergauwen K, Molenberghs B, Vranckx L, Willebrords R, Ristic Z, Lill H, Dorange I, Guillemont J, Bald D, Andries K (2007) Diarylquinolines target subunit c of mycobacterial ATP synthase. Nat Chem Biol 3:323–324. doi:10.1038/nchembio884

Lee RE, Protopopova M, Crooks E, Slayden RA, Terrot M, Barry CE III (2003) Combinatorial lead optimization of [1,2]-diamines based on ethambutol as potential antituberculosis preclinical candidates. J Comb Chem 5:172–187. doi:10.1021/cc020071p

Li XZ, Zhang L, Nikaido H (2004) Efflux pump-mediated intrinsic drug resistance in *Mycobacterium smegmatis*. Antimicrob Agents Chemother 48:2415–2423. doi:10.1128/AAC.48.7.2415-2423.2004

Li K, Schurig-Briccio LA, Feng X, Upadhyay A, Pujari V, Lechartier B, Fontes FL, Yang H, Rao G, Zhu W, Gulati A, No JH, Cintra G, Bogue S, Liu YL, Molohon K, Orlean P, Mitchell DA, Freitas-Junior L, Ren F, Sun H, Jiang T, Li Y, Guo RT, Cole ST, Gennis RB, Crick DC, Oldfield E (2014a) Multitarget drug discovery for tuberculosis and other infectious diseases. J Med Chem 57:3126–3139. doi:10.1021/jm500131s

Li W, Upadhyay A, Fontes FL, North EJ, Wang Y, Crans DC, Grzegorzewicz AE, Jones V, Franzblau SG, Lee RE, Crick DC, Jackson M (2014b) Novel insights into the mechanism of inhibition of MmpL3, a target of multiple pharmacophores in *Mycobacterium tuberculosis*. Antimicrob Agents Chemother 58:6413–6423. doi:10.1128/AAC.03229-14

Li G, Zhang J, Guo Q, Jiang Y, Wei J, Zhao LL, Zhao X, Lu J, Wan K (2015) Efflux pump gene expression in multidrug-resistant *Mycobacterium tuberculosis* clinical isolates. PLoS One 10, e0119013. doi:10.1371/journal.pone.0119013

Liu J, Takiff HE, Nikaido H (1996) Active efflux of fluoroquinolones in *Mycobacterium smegmatis* mediated by LfrA, a multidrug efflux pump. J Bacteriol 178:3791–3795

Lounis N, Veziris N, Chauffour A, Truffot-Pernot C, Andries K, Jarlier V (2006) Combinations of R207910 with drugs used to treat multidrug-resistant tuberculosis have the potential to shorten treatment duration. Antimicrob Agents Chemother 50:3543–3547. doi:10.1128/AAC.00766-06

Louw GE, Warren RM, Donald PR, Murray MB, Bosman M, Van Helden PD, Young DB, Victor TC (2006) Frequency and implications of pyrazinamide resistance in managing previously treated tuberculosis patients. Int J Tuberc Lung Dis 10:802–807

Louw GE, Warren RM, Gey van Pittius NC, McEvoy CR, Van Helden PD, Victor TC (2009) A balancing act: efflux/influx in mycobacterial drug resistance. Antimicrob Agents Chemother 53:3181–3189. doi:10.1128/AAC.01577-08

Louw GE, Warren RM, Gey van Pittius NC, Leon R, Jimenez A, Hernandez-Pando R, McEvoy CR, Grobbelaar M, Murray M, van Helden PD, Victor TC (2011) Rifampicin reduces susceptibility to ofloxacin in rifampicin-resistant *Mycobacterium tuberculosis* through efflux. Am J Respir Crit Care Med 184:269–276. doi:10.1164/rccm.201011-1924OC

Lu Y, Zheng M, Wang B, Fu L, Zhao W, Li P, Xu J, Zhu H, Jin H, Yin D, Huang H, Upton AM, Ma Z (2011) Clofazimine analogs with efficacy against experimental tuberculosis and reduced potential for accumulation. Antimicrob Agents Chemother 55:5185–5193. doi:10.1128/AAC.00699-11

Makarov V, Neres J, Hartkoorn RC, Ryabova OB, Kazakova E, Sarkan M, Huszar S, Piton J, Kolly GS, Vocat A, Conroy TM, Mikusova K, Cole ST (2015) The 8-pyrrole-benzothiazinones are noncovalent inhibitors of DprE1 from *Mycobacterium tuberculosis*. Antimicrob Agents Chemother 59:4446–4452. doi:10.1128/AAC.00778-15

Manjunatha UH, Boshoff H, Dowd CS, Zhang L, Albert TJ, Norton JE, Daniels L, Dick T, Pang SS, Barry CE III (2006) Identification of a nitroimidazo-oxazine-specific protein involved in PA-824 resistance in *Mycobacterium tuberculosis*. Proc Natl Acad Sci U S A 103:431–436. doi:10.1073/pnas.0508392103

Manjunatha U, Boshoff HI, Barry CE (2009) The mechanism of action of PA-824: novel insights from transcriptional profiling. Commun Integr Biol 2:215–218

Mariam DH, Mengistu Y, Hoffner SE, Andersson DI (2004) Effect of *rpoB* mutations conferring rifampin resistance on fitness of *Mycobacterium tuberculosis*. Antimicrob Agents Chemother 48:1289–1294. doi:10.1128/AAC.48.4.1289-1294.2004

Marrakchi H, Laneelle G, Quemard A (2000) InhA, a target of the antituberculous drug isoniazid, is involved in a mycobacterial fatty acid elongation system, FAS-II. Microbiology 146:289–296. doi:10.1099/00221287-146-2-289

Martins M, Viveiros M, Ramos J, Couto I, Molnar J, Boeree M, Amaral L (2009) SILA 421, an inhibitor of efflux pumps of cancer cells, enhances the killing of intracellular extensively drug-resistant tuberculosis (XDR-TB). Int J Antimicrob Agents 33:479–482. doi:10.1016/j.ijantimicag.2008.10.028

Maruri F, Sterling TR, Kaiga AW, Blackman A, van der Heijden YF, Mayer C, Cambau E, Aubry A (2012) A systematic review of gyrase mutations associated with fluoroquinolone-resistant *Mycobacterium tuberculosis* and a proposed gyrase numbering system. J Antimicrob Chemother 67:819–831. doi:10.1093/jac/dkr566

Masjedi MR, Tabarsi P, Baghaei P, Jalali S, Farnia P, Chitsaz E, Amiri M, Mansouri D, Velayati AA (2010) Extensively drug-resistant tuberculosis treatment outcome in Iran: a case series of seven patients. Int J Infect Dis 14:e399–e402. doi:10.1371/journal.pmed.0030466

Maurer FP, Bruderer VL, Ritter C, Castelberg C, Bloemberg GV, Bottger EC (2014) Lack of antimicrobial bactericidal activity in *Mycobacterium abscessus*. Antimicrob Agents Chemother 58:3828–3836. doi:10.1128/AAC.02448-14

Maurer FP, Bruderer VL, Castelberg C, Ritter C, Scherbakov D, Bloemberg GV, Bottger EC (2015) Aminoglycoside-modifying enzymes determine the innate susceptibility to aminoglycoside antibiotics in rapidly growing mycobacteria. J Antimicrob Chemother 70:1412–1419. doi:10.1093/jac/dku550

Maus CE, Plikaytis BB, Shinnick TM (2005a) Molecular analysis of cross-resistance to capreomycin, kanamycin, amikacin, and viomycin in *Mycobacterium tuberculosis*. Antimicrob Agents Chemother 49:3192–3197. doi:10.1128/AAC.49.8.3192-3197.2005

Maus CE, Plikaytis BB, Shinnick TM (2005b) Mutation of *tlyA* confers capreomycin resistance in *Mycobacterium tuberculosis*. Antimicrob Agents Chemother 49:571–577. doi:10.1128/AAC.49.2.571-577.2005

McClure WR, Cech CL (1978) On the mechanism of rifampicin inhibition of RNA synthesis. J Biol Chem 253:8949–8956

Migliori GB, De Iaco G, Besozzi G, Centis R, Cirillo DM (2007a) First tuberculosis cases in Italy resistant to all tested drugs. Euro Surveill 12:E070517 070511

Migliori GB, Ortmann J, Girardi E, Besozzi G, Lange C, Cirillo DM, Ferrarese M, De Iaco G, Gori A, Raviglione MC, Group STS (2007b) Extensively drug-resistant tuberculosis, Italy and Germany. Emerg Infect Dis 13:780–782. doi:10.3201/eid1305.060200

Milano A, Pasca MR, Provvedi R, Lucarelli AP, Manina G, Ribeiro AL, Manganelli R, Riccardi G (2009) Azole resistance in *Mycobacterium tuberculosis* is mediated by the MmpS5-MmpL5 efflux system. Tuberculosis (Edinb) 89:84–90. doi:10.1016/j.tube.2008.08.003

Minato Y, Thiede JM, Kordus SL, McKlveen EJ, Turman BJ, Baughn AD (2015) *Mycobacterium tuberculosis* folate metabolism and the mechanistic basis for para-aminosalicylic acid susceptibility and resistance. Antimicrob Agents Chemother 59:5097–5106. doi:10.1128/AAC.00647-15

Morlock GP, Metchock B, Sikes D, Crawford JT, Cooksey RC (2003) *ethA*, *inhA*, and *katG* loci of ethionamide-resistant clinical *Mycobacterium tuberculosis* isolates. Antimicrob Agents Chemother 47:3799–3805. doi:10.1128/AAC.47.12.3799-3805.2003

Morris RP, Nguyen L, Gatfield J, Visconti K, Nguyen K, Schnappinger D, Ehrt S, Liu Y, Heifets L, Pieters J, Schoolnik G, Thompson CJ (2005) Ancestral antibiotic resistance in *Mycobacterium tuberculosis*. Proc Natl Acad Sci U S A 102:12200–12205. doi:10.1073/pnas.0505446102

Motiwala AS, Dai Y, Jones-Lopez EC, Hwang SH, Lee JS, Cho SN, Via LE, Barry CE, Alland D (2010) Mutations in extensively drug-resistant *Mycobacterium tuberculosis* that do not code for known drug-resistance mechanisms. J Infect Dis 201:881–888. doi:10.1086/650999

Mukherjee JS, Rich ML, Socci AR, Joseph JK, Viru FA, Shin SS, Furin JJ, Becerra MC, Barry DJ, Kim JY, Bayona J, Farmer P, Smith Fawzi MC, Seung KJ (2004) Programmes and principles in treatment of multidrug-resistant tuberculosis. Lancet 363:474–481. doi:10.1016/S0140-6736(04)15496-2

Nachega JB, Chaisson RE (2003) Tuberculosis drug resistance: a global threat. Clin Infect Dis 36 (Suppl 1):S24–S30

Nikaido H (2001) Preventing drug access to targets: cell surface permeability barriers and active efflux in bacteria. Semin Cell Dev Biol 12:215–223

Nikonenko BV, Protopopova M, Samala R, Einck L, Nacy CA (2007) Drug therapy of experimental tuberculosis (TB): improved outcome by combining SQ109, a new diamine antibiotic,

with existing TB drugs. Antimicrob Agents Chemother 51:1563–1565. doi:10.1128/AAC.01326-06
Nosova EY, Bukatina AA, Isaeva YD, Makarova MV, Galkina KY, Moroz AM (2013) Analysis of mutations in the *gyrA* and *gyrB* genes and their association with the resistance of Mycobacterium tuberculosis to levofloxacin, moxifloxacin and gatifloxacin. J Med Microbiol 62:108–113. doi:10.1099/jmm.0.046821-0
Pages JM, Masi M, Barbe J (2005) Inhibitors of efflux pumps in Gram-negative bacteria. Trends Mol Med 11:382–389. doi:10.1016/j.molmed.2005.06.006
Parida SK, Axelsson-Robertson R, Rao MV, Singh N, Master I, Lutckii A, Keshavjee S, Andersson J, Zumla A, Maeurer M (2015) Totally drug-resistant tuberculosis and adjunct therapies. J Intern Med 277:388–405. doi:10.1111/joim.12264
Pasca MR, Guglierame P, Arcesi F, Bellinzoni M, De Rossi E, Riccardi G (2004) Rv2686c-Rv2687c-Rv2688c, an ABC fluoroquinolone efflux pump in *Mycobacterium tuberculosis*. Antimicrob Agents Chemother 48:3175–3178. doi:10.1128/AAC.48.8.3175-3178.2004
Perry CM, Jarvis B (2001) Linezolid: a review of its use in the management of serious gram-positive infections. Drugs 61:525–551
Petrella S, Cambau E, Chauffour A, Andries K, Jarlier V, Sougakoff W (2006) Genetic basis for natural and acquired resistance to the diarylquinoline R207910 in mycobacteria. Antimicrob Agents Chemother 50:2853–2856. doi:10.1128/AAC.00244-06
Plinke C, Cox HS, Zarkua N, Karimovich HA, Braker K, Diel R, Rusch-Gerdes S, Feuerriegel S, Niemann S (2010) embCAB sequence variation among ethambutol-resistant *Mycobacterium tuberculosis* isolates without *embB*306 mutation. J Antimicrob Chemother 65:1359–1367. doi:10.1093/jac/dkq120
Protopopova M, Hanrahan C, Nikonenko B, Samala R, Chen P, Gearhart J, Einck L, Nacy CA (2005) Identification of a new antitubercular drug candidate, SQ109, from a combinatorial library of 1,2-ethylenediamines. J Antimicrob Chemother 56:968–974. doi:10.1093/jac/dki319
Quan S, Venter H, Dabbs ER (1997) Ribosylative inactivation of rifampin by *Mycobacterium smegmatis* is a principal contributor to its low susceptibility to this antibiotic. Antimicrob Agents Chemother 41:2456–2460
Quemard A, Sacchettini JC, Dessen A, Vilcheze C, Bittman R, Jacobs WR Jr, Blanchard JS (1995) Enzymatic characterization of the target for isoniazid in *Mycobacterium tuberculosis*. Biochemistry 34:8235–8241
Ramaswamy S, Musser JM (1998) Molecular genetic basis of antimicrobial agent resistance in *Mycobacterium tuberculosis*: 1998 update. Int J Tuberc Lung Dis 79:3–29
Ramaswamy SV, Amin AG, Goksel S, Stager CE, Dou SJ, El Sahly H, Moghazeh SL, Kreiswirth BN, Musser JM (2000) Molecular genetic analysis of nucleotide polymorphisms associated with ethambutol resistance in human isolates of *Mycobacterium tuberculosis*. Antimicrob Agents Chemother 44:326–336. doi:10.1128/AAC.44.2.326-336.2000
Ramaswamy SV, Reich R, Dou SJ, Jasperse L, Pan X, Wanger A, Quitugua T, Graviss EA (2003) Single nucleotide polymorphisms in genes associated with isoniazid resistance in *Mycobacterium tuberculosis*. Antimicrob Agents Chemother 47:1241–1250. doi:10.1128/AAC.47.4.1241-1250.2003
Ramirez-Busby SM, Valafar F (2015) Systematic review of mutations in pyrazinamidase associated with pyrazinamide resistance *in Mycobacterium tuberculosis* clinical isolates. Antimicrob Agents Chemother 59:5267–5277. doi:10.1128/AAC.00204-15
Ramon-Garcia S, Mick V, Dainese E, Martin C, Thompson CJ, De Rossi E, Manganelli R, Ainsa JA (2012) Functional and genetic characterization of the tap efflux pump in *Mycobacterium bovis* BCG. Antimicrob Agents Chemother 56:2074–2083. doi:10.1128/AAC.05946-11
Rastogi N, Labrousse V, Goh KS (1996) *In vitro* activities of fourteen antimicrobial agents against drug susceptible and resistant clinical isolates of *Mycobacterium tuberculosis* and comparative intracellular activities against the virulent H37Rv strain in human macrophages. Curr Microbiol 33:167–175. doi:10.1007/s002849900095

Reddy VM, O'Sullivan JF, Gangadharam PR (1999) Antimycobacterial activities of riminophenazines. J Antimicrob Chemother 43:615–623. doi:10.1093/jac/43.5.615

Reeves AZ, Campbell PJ, Sultana R, Malik S, Murray M, Plikaytis BB, Shinnick TM, Posey JE (2013) Aminoglycoside cross-resistance in *Mycobacterium tuberculosis* due to mutations in the 5′ untranslated region of *whiB7*. Antimicrob Agents Chemother 57:1857–1865. doi:10.1128/AAC.02191-12

Rengarajan J, Sassetti CM, Naroditskaya V, Sloutsky A, Bloom BR, Rubin EJ (2004) The folate pathway is a target for resistance to the drug para-aminosalicylic acid (PAS) in mycobacteria. Mol Microbiol 53:275–282. doi:10.1111/j.1365-2958.2004.04120.x

Richter E, Rusch-Gerdes S, Hillemann D (2007) First linezolid-resistant clinical isolates of *Mycobacterium tuberculosis*. Antimicrob Agents Chemother 51:1534–1536. doi:10.1128/AAC.01113-06

Rodrigues L, Ainsa JA, Amaral L, Viveiros M (2011) Inhibition of drug efflux in mycobacteria with phenothiazines and other putative efflux inhibitors. Recent Pat Antiinfect Drug Discov 6:118–127. doi:10.2174/157489111796064579

Rodrigues L, Machado D, Couto I, Amaral L, Viveiros M (2012) Contribution of efflux activity to isoniazid resistance in the *Mycobacterium tuberculosis* complex. Infect Genet Evol 12:695–700. doi:10.1016/j.meegid.2011.08.009

Rozwarski DA, Grant GA, Barton DH, Jacobs WR Jr, Sacchettini JC (1998) Modification of the NADH of the isoniazid target (InhA) from *Mycobacterium tuberculosis*. Science 279:98–102. doi:10.1126/science.279.5347.98

Sacksteder KA, Protopopova M, Barry CE III, Andries K, Nacy CA (2012) Discovery and development of SQ109: a new antitubercular drug with a novel mechanism of action. Future Microbiol 7:823–837. doi:10.2217/fmb.12.56

Safi H, Lingaraju S, Amin A, Kim S, Jones M, Holmes M, McNeil M, Peterson SN, Chatterjee D, Fleischmann R, Alland D (2013) Evolution of high-level ethambutol-resistant tuberculosis through interacting mutations in decaprenylphosphoryl-beta-D-arabinose biosynthetic and utilization pathway genes. Nat Genet 45:1190–1197. doi:10.1038/ng.2743

Schrag SJ, Perrot V (1996) Reducing antibiotic resistance. Nature 381:120–121. doi:10.1038/381120b0

Seifert M, Catanzaro D, Catanzaro A, Rodwell TC (2015) Genetic mutations associated with isoniazid resistance in *Mycobacterium tuberculosis*: a systematic review. PLoS One 10, e0119628. doi:10.1371/journal.pone.0119628

Sherman DR, Mdluli K, Hickey MJ, Arain TM, Morris SL, Barry CE III, Stover CK (1996) Compensatory *ahpC* gene expression in isoniazid-resistant *Mycobacterium tuberculosis*. Science 272:1641–1643. doi:10.1126/science.272.5268.1641

Shi W, Zhang X, Jiang X, Yuan H, Lee JS, Barry CE III, Wang H, Zhang W, Zhang Y (2011) Pyrazinamide inhibits trans-translation in *Mycobacterium tuberculosis*. Science 333:1630–1632. doi:10.1126/science.1208813

Shi W, Chen J, Feng J, Cui P, Zhang S, Weng X, Zhang W, Zhang Y (2014) Aspartate decarboxylase (PanD) as a new target of pyrazinamide in *Mycobacterium tuberculosis*. Emerg Microbes Infect 3, e58. doi:10.1038/emi.2014.61

Siddiqi S, Takhar P, Baldeviano C, Glover W, Zhang Y (2007) Isoniazid induces its own resistance in nonreplicating *Mycobacterium tuberculosis*. Antimicrob Agents Chemother 51:2100–2104. doi:10.1128/AAC.00086-07

Simons SO, Kristiansen JE, Hajos G, van der Laan T, Molnar J, Boeree MJ, van Ingen J, Christensen JB, Viveiros M, Riedl Z, Amaral L, van Soolingen D (2013a) Activity of the efflux pump inhibitor SILA 421 against drug-resistant tuberculosis. Int J Antimicrob Agents 41:488–489. doi:10.1016/j.ijantimicag.2013.01.001

Simons SO, Mulder A, van Ingen J, Boeree MJ, van Soolingen D (2013b) Role of *rpsA* gene sequencing in diagnosis of pyrazinamide resistance. J Clin Microbiol 51:382. doi:10.1128/JCM.02739-12

Singh R, Manjunatha U, Boshoff HI, Ha YH, Niyomrattanakit P, Ledwidge R, Dowd CS, Lee IY, Kim P, Zhang L, Kang S, Keller TH, Jiricek J, Barry CE III (2008) PA-824 kills nonreplicating *Mycobacterium tuberculosis* by intracellular NO release. Science 322:1392–1395. doi:10.1126/science.1164571

Singh P, Jain A, Dixit P, Prakash S, Jaiswal I, Venkatesh V, Singh M (2015) Prevalence of gyrA and B gene mutations in fluoroquinolone-resistant and -sensitive clinical isolates of *Mycobacterium tuberculosis* and their relationship with MIC of ofloxacin. J Antibiot (Tokyo) 68:63–66. doi:10.1038/ja.2014.95

Siu GK, Zhang Y, Lau TC, Lau RW, Ho PL, Yew WW, Tsui SK, Cheng VC, Yuen KY, Yam WC (2011) Mutations outside the rifampicin resistance-determining region associated with rifampicin resistance in *Mycobacterium tuberculosis*. J Antimicrob Chemother 66:730–733. doi:10.1093/jac/dkq519

Sreevatsan S, Stockbauer KE, Pan X, Kreiswirth BN, Moghazeh SL, Jacobs WR Jr, Telenti A, Musser JM (1997) Ethambutol resistance in *Mycobacterium tuberculosis*: critical role of embB mutations. Antimicrob Agents Chemother 41:1677–1681

Steingart KR, Schiller I, Horne DJ, Pai M, Boehme CC, Dendukuri N (2014) Xpert(R) MTB/RIF assay for pulmonary tuberculosis and rifampicin resistance in adults. Cochrane Database Syst Rev 1, CD009593. doi:10.1002/14651858.CD009593.pub3

Stottmeier KD, Beam RE, Kubica GP (1967) Determination of drug susceptibility of mycobacteria to pyrazinamide in 7H10 agar. Am Rev Respir Dis 96:1072–1075

Stover CK, Warrener P, Van Devanter DR, Sherman DR, Arain TM, Langhorne MH, Anderson SW, Towell JA, Yuan Y, McMurray DN, Kreiswirth BN, Barry CE, Baker WR (2000) A small-molecule nitroimidazopyran drug candidate for the treatment of tuberculosis. Nature 405:962–966. doi:10.1038/35016103

Szumowski JD, Lynch JB (2015) Profile of delamanid for the treatment of multidrug-resistant tuberculosis. Drug Des Devel Ther 9:677–682. doi:10.2147/DDDT.S60923

Tahlan K, Wilson R, Kastrinsky DB, Arora K, Nair V, Fischer E, Barnes SW, Walker JR, Alland D, Barry CE III, Boshoff HI (2012) SQ109 targets MmpL3, a membrane transporter of trehalose monomycolate involved in mycolic acid donation to the cell wall core of *Mycobacterium tuberculosis*. Antimicrob Agents Chemother 56:1797–1809. doi:10.1128/AAC.05708-11

Takayama K, Kilburn JO (1989) Inhibition of synthesis of arabinogalactan by ethambutol in Mycobacterium smegmatis. Antimicrob Agents Chemother 33:1493–1499. doi:10.1128/AAC.33.9.1493

Takiff HE, Salazar L, Guerrero C, Philipp W, Huang WM, Kreiswirth B, Cole ST, Jacobs WR Jr, Telenti A (1994) Cloning and nucleotide sequence of *Mycobacterium tuberculosisgyrA* and *gyrB* genes and detection of quinolone resistance mutations. Antimicrob Agents Chemother 38:773–780. doi:10.1128/AAC.38.4.773

Takiff HE, Cimino M, Musso MC, Weisbrod T, Martinez R, Delgado MB, Salazar L, Bloom BR, Jacobs WR Jr (1996) Efflux pump of the proton antiporter family confers low-level fluoroquinolone resistance in *Mycobacterium smegmatis*. Proc Natl Acad Sci U S A 93:362–366

Tan Y, Hu Z, Zhang T, Cai X, Kuang H, Liu Y, Chen J, Yang F, Zhang K, Tan S, Zhao Y (2014) Role of *pncA* and *rpsA* gene sequencing in detection of pyrazinamide resistance in *Mycobacterium tuberculosis* isolates from southern China. J Clin Microbiol 52:291–297. doi:10.1128/JCM.01903-13

Telenti A, Imboden P, Marchesi F, Lowrie D, Cole S, Colston MJ, Matter L, Schopfer K, Bodmer T (1993) Detection of rifampicin-resistance mutations in *Mycobacterium tuberculosis*. Lancet 341:647–650. doi:10.1016/0140-6736(93)90417-F

Udwadia ZF (2012) Totally drug-resistant tuberculosis in India: who let the djinn out? Respirology 17:741–742. doi:10.1111/j.1440-1843.2012.02192.x

Udwadia Z, Vendoti D (2013) Totally drug-resistant tuberculosis (TDR-TB) in India: every dark cloud has a silver lining. J Epidemiol Community Health 67:471–472. doi:10.1136/jech-2012-201640

Udwadia ZF, Amale RA, Ajbani KK, Rodrigues C (2012) Totally drug-resistant tuberculosis in India. Clin Infect Dis 54:579–581. doi:10.1093/cid/cir889

Van Deun A, Maug AK, Salim MA, Das PK, Sarker MR, Daru P, Rieder HL (2010) Short, highly effective, and inexpensive standardized treatment of multidrug-resistant tuberculosis. Am J Respir Crit Care Med 182:684–692. doi:10.1164/rccm.201001-0077OC

Velayati AA, Masjedi MR, Farnia P, Tabarsi P, Ghanavi J, Ziazarifi AH, Hoffner SE (2009) Emergence of new forms of totally drug-resistant tuberculosis bacilli: super extensively drug-resistant tuberculosis or totally drug-resistant strains in Iran. Chest 136:420–425. doi:10.1378/chest.08-2427

Vilcheze C, Jacobs WR Jr (2007) The mechanism of isoniazid killing: clarity through the scope of genetics. Annu Rev Microbiol 61:35–50. doi:10.1146/annurev.micro.61.111606.122346

Vilcheze C, Weisbrod TR, Chen B, Kremer L, Hazbon MH, Wang F, Alland D, Sacchettini JC, Jacobs WR Jr (2005) Altered NADH/NAD+ ratio mediates coresistance to isoniazid and ethionamide in mycobacteria. Antimicrob Agents Chemother 49:708–720. doi:10.1128/AAC.49.2.708-720.2005

Vilcheze C, Av-Gay Y, Attarian R, Liu Z, Hazbon MH, Colangeli R, Chen B, Liu W, Alland D, Sacchettini JC, Jacobs WR Jr (2008) Mycothiol biosynthesis is essential for ethionamide susceptibility in *Mycobacterium tuberculosis*. Mol Microbiol 69:1316–1329. doi:10.1111/j.1365-2958.2008.06365.x

Whitfield MG, Soeters HM, Warren RM, York T, Sampson SL, Streicher EM, van Helden PD, van Rie A (2015a) A global perspective on pyrazinamide resistance: systematic review and meta-analysis. PLoS One 10, e0133869. doi:10.1371/journal.pone.0133869

Whitfield MG, Warren RM, Streicher EM, Sampson SL, Sirgel FA, van Helden PD, Mercante A, Willby M, Hughes K, Birkness K, Morlock G, van Rie A, Posey JE (2015b) *Mycobacterium tuberculosis pncA* polymorphisms that do not confer pyrazinamide resistance at a breakpoint concentration of 100 Micrograms per Milliliter in MGIT. J Clin Microbiol 53:3633–3635. doi:10.1128/JCM.01001-15

WHO (1997) Treatment of tuberculosis: guidelines for national programmes. WHO/TB/97.220. Accessed 26 Oct 2015

WHO (2000) Guidelines for establishing DOTS-PLUS pilot projects for the management of multidrug-resitant tuberculosis (MDR-TB). WHO/CDS/TB/2000.279. Accessed 3 Nov 2015

WHO (2008) Management of MDR-TB: a field guide: a companion document to guidelines for programmatic management of drug-resistant TB: integrated management of adolescent and adult illness (IMAI). WHO/HTM/TB/2008.402a. Accessed 3 Nov 2015

WHO (2015) Global tuberculosis report. WHO/HTM/TB/2015.22. Accessed 5 Nov 2015

Willby M, Sikes RD, Malik S, Metchock B, Posey JE (2015) Correlation between GyrA substitutions and ofloxacin, levofloxacin, and moxifloxacin cross-resistance in *Mycobacterium tuberculosis*. Antimicrob Agents Chemother 59:5427–5434. doi:10.1128/AAC.00662-15

Wolucka BA (2008) Biosynthesis of D-arabinose in mycobacteria – a novel bacterial pathway with implications for antimycobacterial therapy. FEBS J 275:2691–2711. doi:10.1111/j.1742-4658.2008.06395.x

Xu Y, Jia H, Huang H, Sun Z, Zhang Z (2015) Mutations found in *embCAB*, *embR*, and *ubiA* genes of ethambutol-sensitive and -resistant *Mycobacterium tuberculosis* clinical isolates from China. Biomed Res Int 2015:951706. doi:10.1155/2015/951706

Yamada T, Mizugichi Y, Nierhaus KH, Wittmann HG (1978) Resistance to viomycin conferred by RNA of either ribosomal subunit. Nature 275:460–461. doi:10.1038/275460a0

Yano T, Kassovska-Bratinova S, Teh JS, Winkler J, Sullivan K, Isaacs A, Schechter NM, Rubin H (2011) Reduction of clofazimine by mycobacterial type 2 NADH:quinone oxidoreductase: a pathway for the generation of bactericidal levels of reactive oxygen species. J Biol Chem 286:10276–10287. doi:10.1074/jbc.M110.200501

Zainuddin ZF, Dale JW (1990) Does *Mycobacterium tuberculosis* have plasmids? Tubercle 71:43–49

Zaunbrecher MA, Sikes RD Jr, Metchock B, Shinnick TM, Posey JE (2009) Overexpression of the chromosomally encoded aminoglycoside acetyltransferase *eis* confers kanamycin resistance in *Mycobacterium tuberculosis*. Proc Natl Acad Sci U S A 106:20004–20009. doi:10.1073/pnas.0907925106

Zechini B, Versace I (2009) Inhibitors of multidrug resistant efflux systems in bacteria. Recent Pat Antiinfect Drug Discov 4:37–50. doi:10.2174/157489109787236256

Zhang Y, Mitchison D (2003) The curious characteristics of pyrazinamide: a review. Int J Tuberc Lung Dis 7:6–21

Zhang Y, Yew WW (2009) Mechanisms of drug resistance in *Mycobacterium tuberculosis*. Int J Tuberc Lung Dis 13:1320–1330

Zhang S, Chen J, Shi W, Liu W, Zhang W, Zhang Y (2013) Mutations in *panD* encoding aspartate decarboxylase are associated with pyrazinamide resistance in *Mycobacterium tuberculosis*. Emerg Microbes Infect 2, e34. doi:10.1038/emi.2013.38

Zhang S, Chen J, Cui P, Shi W, Zhang W, Zhang Y (2015a) Identification of novel mutations associated with clofazimine resistance in *Mycobacterium tuberculosis*. J Antimicrob Chemother 70:2507–2510. doi:10.1093/jac/dkv150

Zhang X, Liu L, Zhang Y, Dai G, Huang H, Jin Q (2015b) Genetic determinants involved in p-aminosalicylic acid resistance in clinical isolates from tuberculosis patients in northern China from 2006 to 2012. Antimicrob Agents Chemother 59:1320–1324. doi:10.1128/AAC.03695-14

Zhang Z, Yan J, Xu K, Ji Z, Li L (2015c) Tetrandrine reverses drug resistance in isoniazid and ethambutol dual drug-resistant *Mycobacterium tuberculosis* clinical isolates. BMC Infect Dis 15:153. doi:10.1186/s12879-015-0905-0

Zhao F, Wang XD, Erber LN, Luo M, Guo AZ, Yang SS, Gu J, Turman BJ, Gao YR, Li DF, Cui ZQ, Zhang ZP, Bi LJ, Baughn AD, Zhang XE, Deng JY (2014) Binding pocket alterations in dihydrofolate synthase confer resistance to para-aminosalicylic acid in clinical isolates of *Mycobacterium tuberculosis*. Antimicrob Agents Chemother 58:1479–1487. doi:10.1128/AAC.01775-13

Zheng J, Rubin EJ, Bifani P, Mathys V, Lim V, Au M, Jang J, Nam J, Dick T, Walker JR, Pethe K, Camacho LR (2013) para-Aminosalicylic acid is a prodrug targeting dihydrofolatereductase in *Mycobacterium tuberculosis*. J Biol Chem 288:23447–23456. doi:10.1074/jbc.M113.475798

Zhu M, Burman WJ, Starke JR, Stambaugh JJ, Steiner P, Bulpitt AE, Ashkin D, Auclair B, Berning SE, Jelliffe RW, Jaresko GS, Peloquin CA (2004) Pharmacokinetics of ethambutol in children and adults with tuberculosis. Int J Tuberc Lung Dis 8:1360–1367

Glycogen as Key Energy Storehouse and Possibly Responsible for Multidrug Resistance in *Mycobacterium tuberculosis*

Anil Kumar Gupta, Amit Singh, and Sarman Singh

Abstract Tuberculosis (TB) is a major public health problem with a high mortality rate worldwide due to *Mycobacterium tuberculosis* (*M. tuberculosis*) pathogen, claiming 9.6 million total cases were estimated in 2014 and more than 1.5 million people dead. *M. tuberculosis* and other pathogenic mycobacterial species produce a variety of glycogen or glycogen-associated molecules like lipoarabinomannan (LAM), trehalose monomycolate (TMM), phenolic glycolipids (PGLs), trehalose dimycolate (TDM), phosphatidylinositol-containing mannosides (PIMs), etc., that represent as major glycans present in the outermost layer of *M. tuberculosis*. The *M. tuberculosis* accumulate glycogen during harsh environmental condition, *i.e.* presence of reactive oxygen and nitrogen intermediates, limited nutrients availability and depletion of other essential elements required for their survival within the host. The glycosyltransferases (GTs) enzyme involves two families, glycogen transferase-3 (eukaryotes) and *GTs*-5 (eubacterial and archaeal), that play a major role in the regulation of glycogen metabolism. In bacteria, regulation of glycogen anabolism involves several glycogen synthase enzymes, *i.e.* α-D-glycogen synthase A (glgA), 1,4-α-D-glucan 6-glucosyltransferase (glgB) and glucose-1-phosphate adenylyltransferase (glgC), while catabolism involves glycogen phosphorylase (glgP) enzyme. In recent years, role of glycogen was investigated enormously in the pathogenesis of *M. tuberculosis*. Two major glycogen conjugates present in the cell wall of *M. tuberculosis* are TDM and TMM. These conjugates serve as precursors for the synthesis of mycolic acid that plays a key role in the invasion and pathogenesis of *M. tuberculosis*. This chapter summarizes the current updates of the presence of glycogen/glycoconjugates and their physiological role in the survival and pathogenesis mechanisms of *M. tuberculosis* during antagonistic conditions. Also, the chapter summarizes evidence of the putative GTs in the *Mycobacterium* spp.

A.K. Gupta • A. Singh • S. Singh, MD, FRSC, FRSTMH, FAMS (✉)
Division of Clinical Microbiology and Molecular Medicine, All India Institute of Medical Sciences, New Delhi 110029, India
e-mail: sarman_singh@yahoo.com; sarman.singh@gmail.com

G. Arora et al. (eds.), *Drug Resistance in Bacteria, Fungi, Malaria, and Cancer*,
DOI 10.1007/978-3-319-48683-3_11

Abbreviations

ACP reductase	Enoyl-acyl carrier protein reductase
ADP	Adenosine diphosphate
AM	Arabinomannan
CR-3	Complement receptor 3
G1P	Glucose-1-phosphate
G6P	Glucose-6-phosphate
glgA	*α*-D-Glycogen synthase A
glgB	1,4-α-D-Glucan 6-glucosyltransferase
glgC	Glucose-1-phosphate adenylyltransferase
glgP	Glycogen phosphorylase
glgX	Glucan hydrolase
GS	Glycogen synthase
GTB	Glycosyltransferase B
GTs	Glycosyltransferases
KGD	α-Ketoglutarate decarboxylase
LAM	Lipoarabinomannan
LPS	Lipopolysaccharide
M. tuberculosis	*Mycobacterium tuberculosis*
MGLP	6-O-Methylglucosyl-containing lipopolysaccharides
NPP	Nucleotide pyrophosphate
ODHc	2-Oxoglutarate dehydrogenase complex
PAMP	Pathogen-associated molecular patterns
PGLs	Phenolic glycolipids
PGM	Phosphoglucomutase
PI	Phosphatidyl-myo-inositol
PIMs	Phosphatidylinositol-containing mannosides
STRE	*Cis*-element stress response element
TB	Tuberculosis
TLR2	Toll-like receptor
TMM	Trehalose monomycolate
UDP	Uridine diphosphate

1 Introduction

All living things store glucose (energy source) as glycogen (kingdom Monera, Protista, Fungi and Animalia) or starch (kingdom Plantae, some fungi and protists) (Roach et al. 1998; Gupta et al. 2014; Asención-Diez et al. 2015). Glycogen, a polysaccharide present in the cytosol of the cell, is principally used by bacteria, fungi or animals, while starch is synthesized and stored in plastids by plants, some protists or planktons (Calder 1991; François and Parrou 2001; Ball and Morell 2003; Preiss 2006). Glycogen has similar structure to amylopectin and is mostly branched but more compact than starch. Glycogen is present in the cytosol/cytoplasm of cells in granular form and plays a significant role in the glucose metabolic cycle. Glycogen acts as energy reserve that can be quickly mobilized to overcome the quick need of glucose, because glycogen is less compact than other energy reserves such as triglycerides (lipids).

In humans, glycogen is produced and stored primarily in the liver and muscle cells in the hydrated form with three or four parts of water (Preiss 2006). The glycogen is stored as the chief energy store in the liver adipose tissues. In the liver cells (known as hepatocytes), glycogen can comprise up to 8 % of its dry weight (100–120 g in an adult) soon after a meal (Campbell et al. 2006), whereas only small amount (1–2 %) is present in the muscle cells. The common glycogen storage organs/tissues other than the liver and muscles are red blood cell (Moses et al. 1972; Ingermann and Virgin 1987). Nevertheless, in mammals, glucose uptake and its utilization is well regulated. Any error in the glucose regulation pathways results in the induction of different glycogen storage diseases (Buschiazzo et al. 2004).

The presence of glycogen molecules/granules in the cytoplasm has been described in more than 40 species of bacteria (Preiss and Romeo 1994). Both sugar molecules (glycogen and starch) are compact glucose polymers and act as reservoir of spontaneously accessible carbon and energy source in diverse organisms representing archaea, eubacteria, yeasts, plants and animals (Henrissat et al. 2002). In bacteria, it is usually synthesized when growth is limited by the nitrogen source, and the presence of excess amount of carbon source in in-vitro conditions. Hence, accumulation of glycogen supports the survival of bacteria under nutritional stresses. The inverse relation is observed between growth rate and the amount of the glycogen accumulation during the nitrogen restrictions. Its accumulation is quite rapid just prior to sporulation initiation in *Bacillus cereus* and exopolysaccharide production in all Corynebacteriaceae family members such as *Corynebacterium diphtheriae* and *M. tuberculosis* (Schwebach et al. 2002).

M. tuberculosis is an aerobic, gram positive and acid-fast bacillus that causes TB in humans. It accumulates glycogen during the unreceptive conditions—accumulation of reactive oxygen and intermediates of nitrogen, depletion of nutrients, low pH and during starvation for their survival within the host (Antoine and Tepper 1969). However, the accumulation of glycogen does not happen during the exponential growth of *M. tuberculosis* under laboratory conditions, but its presence may enhance the sustainability of *M. tuberculosis* during these hostile conditions. The role of glycogen in the pathogenic behaviour of *M. tuberculosis* has been reported by Pal et al. (2010). It was validated that if the bacteria are physically inactivated for long period of time, their stored sugar fulfils all energy needs of bacterium and becomes very important molecules for its survival. Various research groups have reported that glycogen or its intermediates may regulate biological pathways by binding glycogen molecules to their respective proteins/enzymes and that of lipids during the process of post-translational modifications. Quite a few uncharacterized glycosyltransferase (GTs) of *M. tuberculosis* are of major interest and will be discussed in detail in following prokaryotes.

The chapter presents the current status of the information about the various enzymatic processes leading to bacterial glycogen synthesis or degradation pathways that play critical role in the survival and possible drug resistance mechanisms in *M. tuberculosis*. Also, the chapter summarizes evidences of the putative GTs in the *Mycobacterium* spp.

2 Existence and Structure of Glycogen

The storage materials are usually accumulated as granular form in the cytosol of the bacteria. This accumulation conventionally takes place in response to depletion of nutrients or in the presence of excess of the relevant harmful substrates. These granules are superficially dispersed within the cytoplasm of bacteria. The size of granules is 20–100 nm in diameter and exists in the form of non-membrane-enclosed inclusions. However variations in the shape and size of the granules have been reported (Preiss 2006). In Cyanobacteria, glycogen stores may occur in the form of crystals, spherical or in rod forms. Glycogen stores can also be found in membrane-enclosed polyglucose inclusion bodies in several strains of Clostridia (Preiss 2006) with granule sizes ranging from 160 to 300 nm in diameter.

Glycogen accumulation is generally initiated in the stationary phase (Wilson et al. 2010). In a large number of bacteria, glycogen accumulates, while bacterial growth reaches to stationary phase due to the limitations of essential nutrients such as phosphorus, nitrogen or sulphur. Glycogen structure resembles that of amylopectin (a component of starch) with alpha glycosidic linkages. However, it is even more branched and has extra glucose units than amylopectin (Ball and Morell 2003). The complex molecules—glycogen and amylopectin—comprise of α-1-4-linked D-glucose unit with α-1-6-branching point. Branches are covalently linked to the main chain from which they are branching off by the glycosidic bonds. The glycosidic bond is formed between the first glucose (C-1) of the new branch and sixth glucose (C-6) on the stem chain. The branching takes place at every interval of 8–10 units of glucose, whereas in amylopectin 12–20 glucose units segregate the branches.

The variation in the size of branches (length and number) depends on the organism and size of the glycogen granules (Belanger and Hatfull 1999; Cid et al. 2002). In *Mycobacterium smegmatis* (*M. smegmatis*), the degree of branching is less and the rate of sedimentation coefficient is more due to bigger size of glycogen granules. *Mycobacterium phlei* (*M. phlei*) possesses a glycogen molecular weight 1.2×10^8 (Antoine and Tepper 1969) in comparison with 8.2×10^7 of *Escherichia coli* (*E. coli*) (Preiss and Romeo 1994) (Fig. 1). The branches of glycogen are extremely important to the quick response to metabolic needs, because the biosynthesis and degradation of glycogen molecules only happen from the non-reducing ends of amylose chain. Therefore, extremely branched glycogen has higher number of reducing ends per molecules of glycogen, which generate more glucose

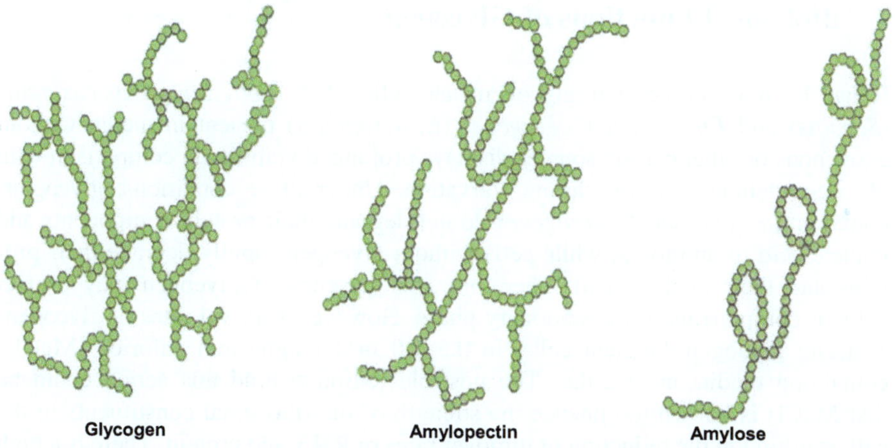

Fig. 1 Branched structure of glycogen and its component showing the helical structure known as amylose (α-1,4 linkage) and branched amylopectin (α-1,6-glucosyl linkages) structure

molecules at single time. Branches also increase solubility of glycogen in the water (Zmasek and Godzik 2014).

3 Glycogen: As Energy Storehouse

Many bacteria accumulate glucose in the form of glycogen, a polysaccharide comprising the glucose connected with α-1,4 glycosidic linkage and α-1,6 linked branched oligosaccharide chains with molecular weight of about 107–108 kDa. It acts as chief carbon source and provides energy to the organisms. Wilson et al. (2010) expressed that the compounds with energy-storage functions should fulfil three criteria:

1. Be accumulated intracellularly, in the presence of excess energy supplements.
2. Be consumed when the limited external carbon supplies are present for the growth maintenance or related processes necessary for the cell growth. The storage compound should be degradable to the energy for consumption by the cell for their survival or respond well to the environment.

Since glycogen meets all these criteria, it is one of the best storage materials in living things.

4 Biological Functions of Glycogen

It has been observed that glycogen-rich cells of *E. coli*, *Streptococcus mitis* (*S. mitis*) and *Enterobacter aerogenes* (*E. aerogenes*) present in media without exogenous or other carbon sources displays prolonged viability in comparison with glycogen-deficient strains during starvation. Under these conditions, glycogens containing *E. coli* and *E. aerogenes* do not degrade their protein components and nucleic acid to ammonia, while cells without glycogen rapidly degrade their proteins and their nucleic acid. Therefore, the presence of glycogen may protect cellular components in the stationary phase. However, survival rates of glycogen-rich and glycogen-deficient cells, in 0.5–1.0 mM magnesium chloride ($MgCl_2$) containing media, are similar. The possible reason behind this action could be that $MgCl_2$ is known to enhance the strength of the ribosomal constituents in the cell, resulting in the reduction of turnover rates of RNA and protein. There is a high probability that glycogen may prove effective in preserving and conserving the intracellular Mg^{2+} concentration. In Clostridia, up to 60 % of the cell dry mass may be accumulated as glycogen, prior to the beginning of sporulation (Shively 1974). This polysaccharide degrades rapidly during spore formation which suggests that glycogen serves as a solitary carbon and energy source for the development of spores and their maturation. Similarly, in sporulating hyphae of *Streptomyces viridochromeogenes* (*S. viridochromeogenes*), glycogen granules reach a maximal number in the beginning of maturation. In later stages of maturation, granules decrease in the fungal hyphae and accumulate in mature spores. Glycogen accumulation also occurs in *Bacillus cereus* (*B. cereus*) during early stage of sporulation and degrades during its maturation (Slock and Stahly 1974). Hence, it has been presumed in the sporulating microorganisms that glycogen acts as carbon and energy source. Several investigators suggested that glycogen helps bacteria to survive during starvation. However, some other observations are contrary to the concept of glycogen being an energy-storage house. As revealed earlier, the presence of higher amount of magnesium enhances survival rates of *E. aerogenes* and *E. coli* cells and is independent of the presence of glycogen (Shively 1974), while *Sarcina lutea* (*S. lutea*) dies at a faster rate in glycogen-rich conditions during starvation in the phosphate media (Rose and Tempest 1989). More efforts are needed to elucidate the role of glycogen in bacteria.

5 Glycosyltransferases: An Enzyme Vital for Glycogen Synthesis

Glycogen biosynthesis is facilitated by the action of GT enzymes (Rini et al. 2009). A large number of enzymes grouped in GTs are involved in the biosynthesis of oligosaccharides, polysaccharides and other glycogen conjugates (Pederson et al. 2000; Pedersen et al. 2003). The GTs have immense diversity but mediate a comprehensive variety of functions—structural and storage— important for molecular signalling. These enzymes are widely present in prokaryotic organisms and in eukaryotic organisms also, but illustrate enormous specificity for the glycosyl acceptor and donor molecules. In eukaryotes, many structural oligosaccharides are produced during the glycosylation process in Golgi apparatus (Breton et al. 2006; Possner et al. 2015). Prokaryotes produce different glycol conjugates and other polysaccharides, which vary in structures and complexity. In *E. coli*, glucose-1-phosphate adenylyltransferase *(glgC)*, *alpha*-D-glycogen synthase *(glgA)* and 1,4-α-D-glucan 6-glucosyltransferase *(glgB)* genes encode for enzymes responsible for its synthesis, while glycogen phosphorylase *(glgP)* and glucan hydrolase *(glgX)* genes encode for its degradation. *glgC* gene is responsible for creation of charged glucose nucleotide pyrophosphate (NPP) and *glgA* gene for linear glucan chain. Glycogen branching enzyme (glgB), transfers 6–7 glucose units from the hydroxyl group of carbon number 6 of the non-reducing end, either on the similar or adjacent chains. GlgB in bacteria (Seibold et al. 2011) and fungi is responsible for glycogen branches. Additionally, bacterial glycans that include numerous unusual sugars such as Kdo, heptoses and modified hexoses (absent in vertebrates) take part in the pathogenesis of bacteria. Also, some other molecules, *i.e.* lipids coupled with glucose/mannose or a precursor of dolichol oligosaccharide, are responsible for the peptidoglycan, lipopolysaccharide (LPS) and capsules assembly (Rini et al. 2009).

6 Functional Classification of Glycosyltransferase Enzymes

According to nucleotide sequence and structural comparisons, glycogen synthase (GS) enzymes have been categorized as glycosyltransferase B (GTB). The structural characteristics of GS include two Rossmann fold domains among the catalytic and substrate-binding sites. Further, enzymes of GTB have been subclassified into GT3 and GT5 families (Fig. 2). The GT5 family consists of both eubacterial and archaeal GS enzymes, and GT3 family consists of eukaryotic enzymes and is regulated by the inhibitory phosphorylation and allosteric activator (G6P) (Unligil and Rini 2000).

The major difference between prokaryotic and eukaryotic enzymes is the use of adenosine diphosphate (ADP) glucose in bacteria and uridine diphosphate (UDP)-glucose in the eukaryotes. Apart from that, archaeal enzymes are proficient in utilizing ADP and UDP-glucose as substrates. Three-dimensional structures of the GT5 family members (three) have been determined in *E. coli* as monomeric, *Agrobacterium tumefaciens* (*A. tumefaciens*) as dimeric and in *Pyrococcus abyssi (P. abyssi)* as trimeric enzymes. This information is not proving very effective on elucidating the regulatory mechanisms in eukaryotes.

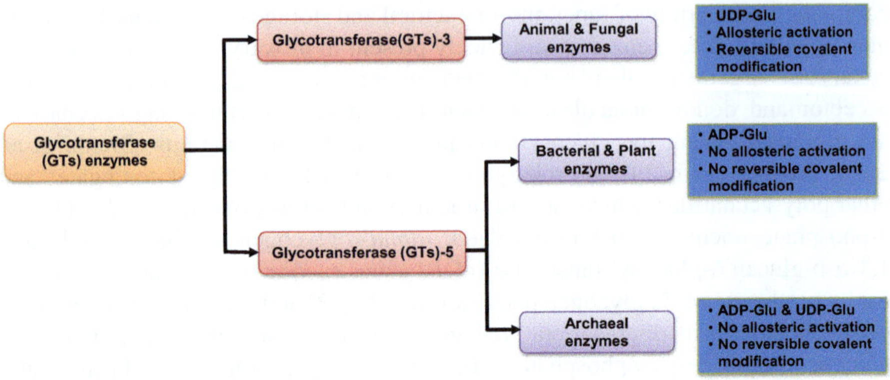

Fig. 2 Classification of prokaryotic and eukaryotic glycotransferase (GTs) enzymes, based on sugar donor molecule

7 Mechanisms of Action of Glycotransferase Enzyme

The mode of action of GT enzymes depends upon the activated donor, like NPP sugar, nucleoside monophosphosugar or lipid phosphor sugar and hydroxyl group of amino acid that acts as acceptor molecules. Monosaccharide component of activated nucleotide sugar is transferred to the glycosyl and forms glycosidic bonds. Inversion of GTs is supposed to resemble the inverting glycosyl hydrolase enzymes with an acidic amino acid that is responsible for the activation of the acceptor OH group by the deprotonation (Breton et al. 2006).

8 Synthesis and Regulation of Glycogen in Eukaryotes

In higher eukaryotic organisms, glycogen is synthesized when the nutrients are non-unlimiting. In eukaryotes, skeletal muscles, the liver, and red blood cells (RBCs) to some extent store glycogen. Several other organs like brain, adipose tissue, pancreas and kidney also synthesize minute amount of glycogen. In the skeletal muscles, glycogen is converted into glucose-6-phosphate (G6P) and enters into glycolysis cycle for the generation of ATP molecules, which are chiefly consumed as energy source for muscular contraction. Liver glycogen plays significant role in glucose homeostasis during fasting. Genetic or functional changes in the enzymes responsible for metabolizing glycogen result in the development of various glycogen storage diseases, affecting the liver, muscle, etc., and may be life-threatening (Gupta et al. 2014).

Yeast (*Saccharomyces cerevisiae*) is an unicellular eukaryotic microorganism, which is commonly present in the sugar-rich ingredients such as fruits, berries, plant exudates, etc. Yeast glycogen has structural similarity to other eukaryotic organisms (Wilson et al. 2010). Yeast glycogen is one of the chief reservoirs of carbohydrate like several bacteria and it covers almost 20 % of the yeast cell mass. The quantity of glycogen accumulation increases exponentially in the stationary phase or in enervation of vital nutrients like nitrogen and phosphorus. Moreover, several studies supported that glycogen accumulation occurs in exponentially growing yeast when exposed to high temperature, salt, ethanol or oxidizing agents. Here, the yeast uses stored glycogen for survival (Silljé et al. 1999; Baskaran et al. 2010). The regulation of glycogen is done by various enzymes such as glycogen synthase (Farkas et al. 1991; Baskaran et al. 2010), glycogenin (Cheng et al. 1995), branching and de-branching enzymes (Wilson et al. 2010) and numerous other mechanisms including covalent modification (Lin et al. 1995), functioning of allosteric activators and translocation inside the cells. The regulation of GTs occurs via phosphorylation (Ramaswamy et al. 1998) and allosteric activation by G6P. However, the biological processes that inhibit these regulatory controls vary from tissue to tissue of the same or different organisms. Yeast has two isoforms of GS, which are designated as GSY-1 and GSY-2, from which the nutritionally regulated isoform-2 (GSY-2) is an utmost essential enzyme for glycogen accumulation in the cells. Predominantly, transcriptional and enzymatic mechanisms are involved in regulating glycogen metabolism. The transcription-mediated regulation is typically dependent on the promoter region of the genes, *i.e. cis*-element stress response element (STRE), while enzyme-mediated regulation of glycogen accumulation is completed by activation of the GS via G6P and inactivation of GPH via phosphorylation. The introduction of nutrients to the starved cells stimulates GPH, resulting in inhibition of GS and vice-versa.

9 Synthesis and Regulation of Glycogen in Prokaryotes

Biosynthesis and degradation processes of glycogen is highly conserved in prokaryotes (Ballicora et al. 2003; Preiss 2006). The enzymatic action of carbohydrate phosphotransferase system (PTS) takes up the extracellular glucose and converts it into G6P. Further G6P is then transformed into G1P via phosphoglucomutase (PGM) enzyme and in the end into ADP glucose (ADPG) in the presence of ATP molecules and Mg^{2+} (Ballicora et al. 2003). glgA utilizes ADPG as sugar donor nucleotide and produces linear glucose chain (amylose). Thereafter, development of branched oligosaccharide chain is initiated by the action of glgB by the formation of α-1,6-glucosidic linkages (Preiss 2006). Based on the genetic evidences of glycogen synthesis, it has been proposed that glgC is the solitary enzyme for generating ADPG (Leung et al. 1986; Ballicora et al. 2003).

The glycogen metabolism regulation has been extensively studied in *E. coli*. The regulation of glycogen synthesis and degradation contain a complex set of factors, which adjust the biological and energy level of the cell (Alonso-Casajús et al. 2006; Montero et al. 2009), expression of analogous genes and communication between cells (Morán-Zorzano et al. 2008). At genomic level, numerous factors control glycogen accumulation in the bacteria such as allosteric regulation, etc. (Deutscher et al. 2006; Preiss 2006). The higher level of glgC protein indicate the presence of higher amount of carbon and energy contents, while the existence of higher amount of inhibitors in growth medium may represent low metabolic energy levels inside the cells. The allosteric control of glgC has been broadly reviewed recently, which included structural and functional connections between glgA, glgB and glgC (Ballicora et al. 2003; Preiss 2006).

The glgP takes part in glycogen degradation pathway. The enzyme eliminates glucose monomers from the non-reducing ends (Dauvillée et al. 2005; Alonso-Casajús et al. 2006). These glucose-6-phosphate molecules enters in the glycolysis cycle. Through surface plasmon resonance (SPR) ligand fishing analysis (technique used for the detection and characterization of molecular interactions between interactive partners), it has been observed that glgP shows the specific interaction between glgP and HPr (Deutscher et al. 2006). Once HPr is wholly phosphorylated, it reduces the activity of glgP enzyme in log phase of bacteria and vice-versa. At this stage, binding of glgP to HPr is maximal. It has been projected that glgP activity is controlled by the phosphorylation level of HPr, which allows the glycogen to accumulate at the beginning of the stationary phase especially under situations where glucose is in excess (Seok et al. 2001; Deutscher et al. 2006).

10 Glycogen Metabolism in *Mycobacterium tuberculosis*

The biosynthesis of glycogen is an endergonic reaction involving monomers of uridine diphosphate (UDP)-glucose. The biosynthesis and degradation of glycogen pathways in *M. tuberculosis* are similar to *E. coli* that is widely studied. Three genes reported as *glgA*, *glgB* and *glgC* encode glycogen biosynthesis enzymes, and other two genes *glgX* and *glgP* encode enzymes for glycogen degradation (Dauvillée et al. 2005; Bourassa and Camilli 2009). It is assumed that *M. tuberculosis* synthesizes glycogen through glgC-glgA pathway as shown in Fig. 3. The nucleotide diphosphoglucose pyrophosphorylase (glgC) utilizes G1P phosphate and generates activated glucose nucleotide diphosphate which is followed by generation of linear glucans by the action of glgA enzyme (Ball and Morell 2003; Chandra et al. 2011). After that, glgB enzyme transforms linear glucan into glycogen via addition of oligoglucan to the non-reducing end of the residual chain (at position 6) for the elongation of side chains (Palomo et al. 2009; Chandra et al. 2011). The regulation of gene expressions of *glgA* and *glgC* happens in bacteria via intracellular signals, which indicates the energy status of the cell (Fig. 4) (McMeechan et al. 2005). Any defect, or mutations in the *glgC* gene, prevents the synthesis of glycogen (Preiss and Romeo 1994). Few scientific reports are available in the database, which suggest that *glgC*-deleted mutant strains could be able to synthesize small amount of glycogen during the growing stage of bacteria under specific conditions (Leung et al. 1986; Bourassa and Camilli 2009). Another enzyme glgS is also involved in the glycogen synthesis pathway; however its role is still not clear. Recent studies suggest that it could play significant role in glycogen accumulation in *E. coli*

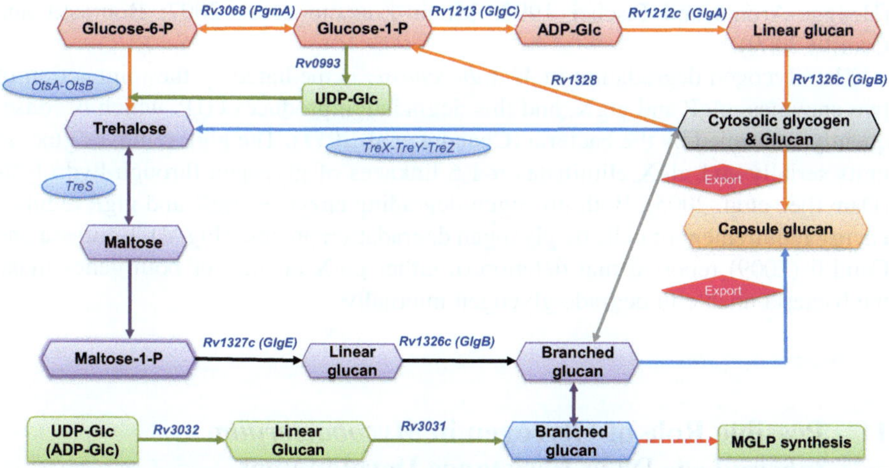

Fig. 3 Mycobacterial glucan pathways. In classical glycogen pathways *Rv1213* (*GlgC*) and *Rv1212c* (*GlgA*) are in central. The new Rv3032 pathway is connected to the MGLP. The recently identified Rv1327c (GlgE) pathway (*purple*) may contribute to capsular glucan, cytosolic glycogen and/or MGLP (Gupta et al. 2014)

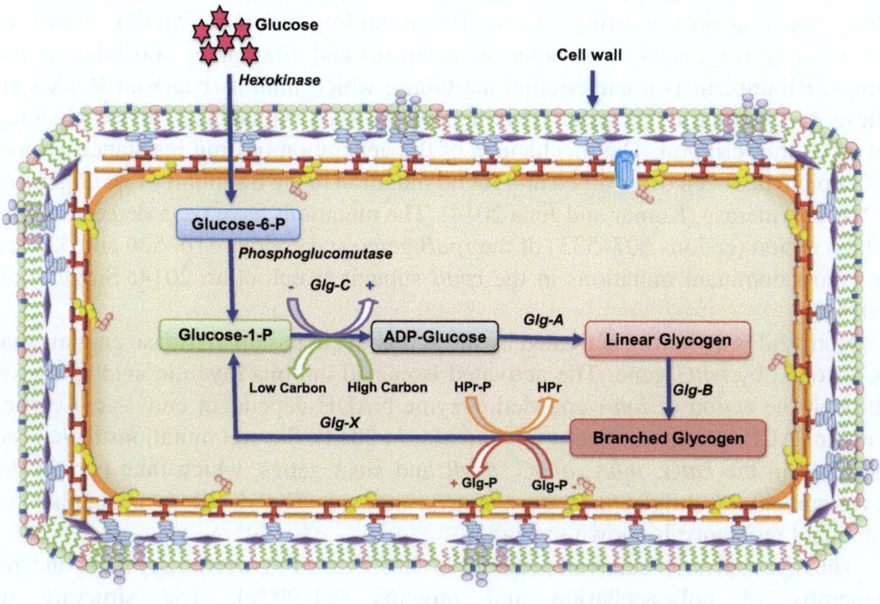

Fig. 4 Regulatory pathway of glycogen synthesis in *Mycobacterium tuberculosis*

(Hengge-Aronis and Fischer 1992; Morán-Zorzano et al. 2007; Bourassa and Camilli 2009).

The glycogen degradation in *M. tuberculosis* is mediated by the joint action of two enzymes, glgP and glgX, and this degradation produces G1P, which is consequently consumed by the bacteria (Chandra et al. 2011). The glgP removes glucose units serially and glgX eliminates α-1,6 linkages of glycogen through hydrolysis (Dauvillée et al. 2005). Both glycogen-degrading enzymes glgX and glgP regulate energy requirement of cells by glycogen degradation process (Fig. 4). Bourassa and Camilli (2009) reported that deletion of either glgX or glgP or both genes make the bacteria unable to degrade glycogen internally.

11 Possible Role of Glycogen in *Mycobacterium tuberculosis* Drug Resistance Development

The emergence of drug-resistant tuberculosis has increased significantly during the past decade in several countries. Generation of resistance to individual antitubercular drug occures after mutations in the concerned genes or chromosomal genes, resulting overproduction of particular gene product (protein), which alter the drug target against existing drugs. The multidrug-resistant (MDR) strains of *M. tuberculosis* display resistance to isoniazid and rifampicin anti-tuberculosis drugs. Rifampicin is a bactericidal antibiotic, which inhibits bacterial RNA synthesis by obstructing bacterial DNA-dependent RNA polymerase, hence blocking of RNA transcription. The mechanism of the generation of drug resistance against rifampicin has been determined and found mutation in the ß subunit of *rpoB* gene of RNA polymerase (Kumar and Jena 2014). The mutations have been detected in the 81-bp region (codons 507–533) of the *rpoB* gene, and codons 516, 526 and 531 are the most dominant mutations in the *rpoB* subunit (Koch et al. 2014; Singh et al. 2014).

Isoniazid is a prodrug activated by the action of catalase/peroxidase enzyme that is encoded by *katG* gene. The activated isoniazid inhibits mycolic acid synthesis through the action of *inhA*-encoded enzyme NADH-dependent enoyl-acyl carrier protein (ACP) reductase (Palomino and Martin 2014). Several mutations have been detected in the *katG*, *inhA*, *ahpC*, *oxyR* and *kasA* genes, which take part in the development of drug resistance against isoniazid. The detailed mechanism of isoniazid resistance is shown in Fig. 5.

The mycobacterial cell wall accounts around 2–3 % of bacterial dry mass and are generally of polysaccharide and proteins (94–99 %). The structure of *M. tuberculosis* cell wall has been represented as pathogen-associated molecular patterns (PAMP), which includes glycolipids, LAM, lipopeptides, etc. The arrangement of triglycerides and glycolipids on the outer surface of the mycobacterial cell wall protects the bacilli against degradation by host enzymes, impenetrability to

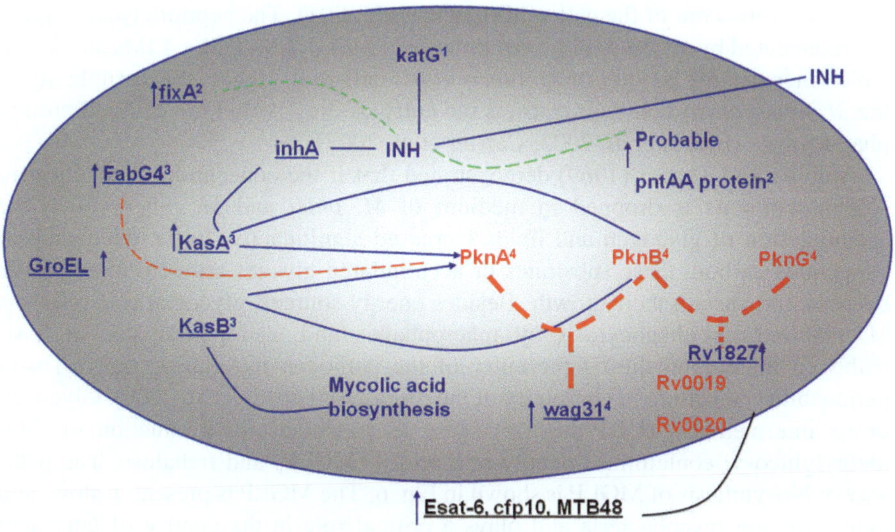

Fig. 5 Schematic illustration of proposed isoniazid drug resistance mechanism of *Mycobacterium tuberculosis*. Alternate mechanism of activation of INH (*dashed green lines*), normal mechanism of activation of INH (*blue lines*) and possible mechanism involved in drug resistance (*dashed red lines*). [1]*KatG* (encoding catalase and peroxidase) gene harbouring mutation resulting in compromised activation of prodrug molecule of INH into active form. The expression of superoxide dismutase may activate INH to overcome the oxidative stress caused by the INH action. [2]*fixA* gene is reported to be involved in INH activation in a cell-harbouring *KatG* gene mutation. Probable pntAA protein and PEP kinases are stress-related protein, and overexpression of pntAA protein may favour the INH resistance. [3]The kasA, kasB and fabG4 are primarily involved in fatty acid metabolism, which are known primary target of INH. The overexpression of these proteins supports that INH mechanism of action. [4]wag31 (cell division protein) and Rv1827 (FHA domain-containing protein associated with exponential phase growth and glycogen accumulation) were overexpressed while acquiring resistance to INH and RIF. Both these proteins regulate cell growth, size and morphology through signal transduction pathway primarily regulated by serine/threonine protein kinases. Only three (PknA, PknB and PknG) out of 11 STPKs in *M. tuberculosis* are essential for sustained growth. Rv1827 protein having a unique phosphorylation site for PknB and PknG and playing a regulatory role in glycogen metabolism (Unpublished data from Amit Singh, Krishnamoorthy Gopinath and Sarman Singh)

poisonous macromolecules (antibiotics), inactivation of reactive oxygen and nitrogen derivatives (Korf et al. 2005).

Trehalose, a disaccharide typically found in mycobacteria, acts as storage compound and is used as energy reservoir, and a stress protectant such as survival under desiccation, cold, osmotic and other traumatic situations (Argüelles 2000; Bolat 2008) consists of α-1-1 linkage of di-glucose. In *M. tuberculosis*, trehalose is synthesized from all three pathways; these are using TreY-TreZ, TreS, (De Smet et al. 2000) and GalU-OtsA-OtsB pathway (Pan et al. 2004). It serves as the precursor for the synthesis of mycolyl acetyl trehalose (also known as cord factor or mycolic acid) (Gibson et al. 2002; Chen and Haddad 2004; Takayama et al. 2005). The combined association of mycolic acid and peptidoglycan has established the core of the cell wall (Crick et al. 2001). The peptidoglycan layer is interconnected by a variety of glycolipids, *i.e.* TMM, LAM, PGLs, PIMs and TDM, which increase the rigidity of *M. tuberculosis* cell wall. Hence, this complexity of the *M. tuberculosis* cell wall restricts the entry of drug within the cell and prompt phagocytosis (Elbein et al. 2003; Carroll et al. 2007).

Antoine and Tepper (1969) demonstrated that if the concentration of nitrogen/ sulphur contents is dropped in medium of *M. phlei* and *M. tuberculosis*, the accumulation of glycogen and lipids increased significantly. In the deficiency of exogenous carbon, these substrates or accumulated glycans are consumed by the bacteria and sustain their growth. Besides energy sources, glycogen also prevents *M. tuberculosis* phagocytosis by macrophages and also participates in host-pathogen interaction during the entry of the pathogen by altering the cell wall permeability, resulting in chronicity of the disease (Geurtsen et al. 2009). Glycogen or its intermediates (UDP-glucose) serve as precursor for production of 6-O-methylglucosyl-containing lipopolysaccharides (MGLP) and trehalose. The pathway of biosynthesis of MGLP is shown in Fig. 6. The MGLP is present in slow- and rapid-growing mycobacteria and plays a critical role in the control of fatty acid synthesis. A cluster of genes has been identified in *M. smegmatis* and *M. tuberculosis* and plays the main role in the MGLP biosynthesis. It has been observed that overexpression of the putative GT enzyme (encoded by *Rv3032* and *Rv3030*) in *M. tuberculosis* significantly increased MGLP production, and disruption of *Rv3032* gene dramatically reduces the amounts of MGLP vice-versa, which results in accumulation of UDP-glucose in the *M. tuberculosis* and ultimately reduction of glycogen contents in *M. tuberculosis* (Stadthagen et al. 2007). Significant changes were found in the glycogen contents between MDR-TB strains and drug-sensitive TB strains grown under drug pressure in initial stage (Singh et al. 2015). The mRNA level of corresponding gene (*GarA*) was confirmed by real-time PCR. The glycogen accumulation was relatively higher from the 7th day to 15th day only (Figs. 7 and 8), but no significant changes were observed after 15th day of growth (Singh et al. 2015). This observation confirmed the role of *GarA* gene in the multidrug-resistant tuberculosis at initial stage.

Apart from that, we have found that protein kinases and GarA of *M. tuberculosis* $H_{37}Rv$ play an important role in acquisition of drug resistance (Singh et al. 2015). Significant amount of glutamate production was also observed when *pknG* deletion mutants were retreated with ethambutol. The positive impact of deletion of *PknG* on glutamate formation may be a consequence of an increased level of unphosphorylated odhI and results in inhibiting 2-oxoglutarate dehydrogenase complex (ODHc) action and an elevated efflux of 2-oxoglutrate towards glutamate. pknG may also be involved probably with other kinases in the switch-on or switch-off mechanisms among active (unphosphorylated) and inactive (phosphorylated) forms of *GarA* gene. GarA also modulates the activities of α-ketoglutarate decarboxylase (KGD) and glutamate dehydrogenase, and the OdhI phosphorylation is grossly determined by PknG (Fig. 9). The proteomic analysis of *M. smegmatis* has revealed glutamate-1-semialdehyde-2 and 1-aminomutase (*hem*L) overexpressed in exposure to EMB which suggested its role in glutamate efflux. Interestingly, our

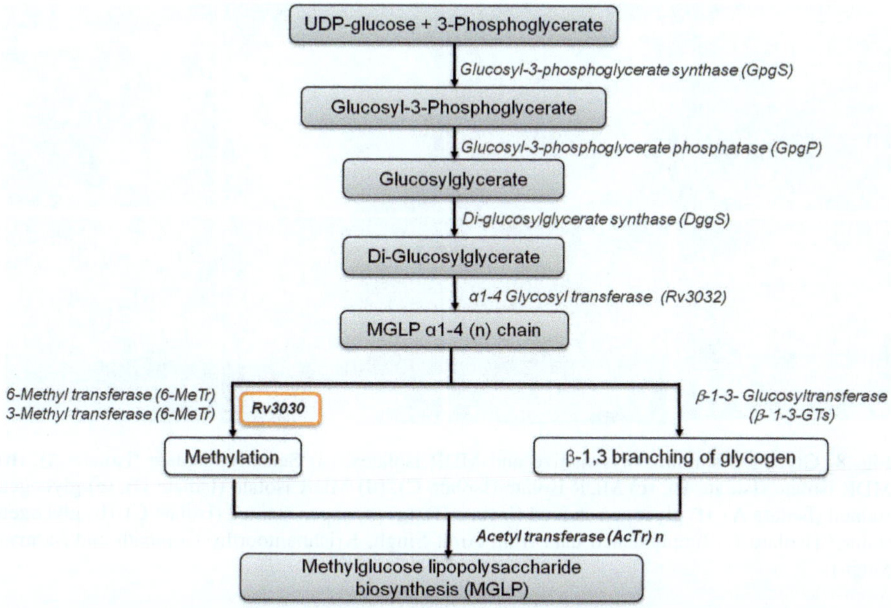

Fig. 6 MGLP synthesis pathway in *M. tuberculosis*. *Grey shades* represent confirmed activities, *white boxes* indicate putative/deduced enzyme activities. *Orange box* carries information on genes linked to the MGLP pathway: *GpgS* (*Rv1208*), *GpgP* (Rv2419), *DggS* and glucosyltransferases (Gupta et al. 2014)

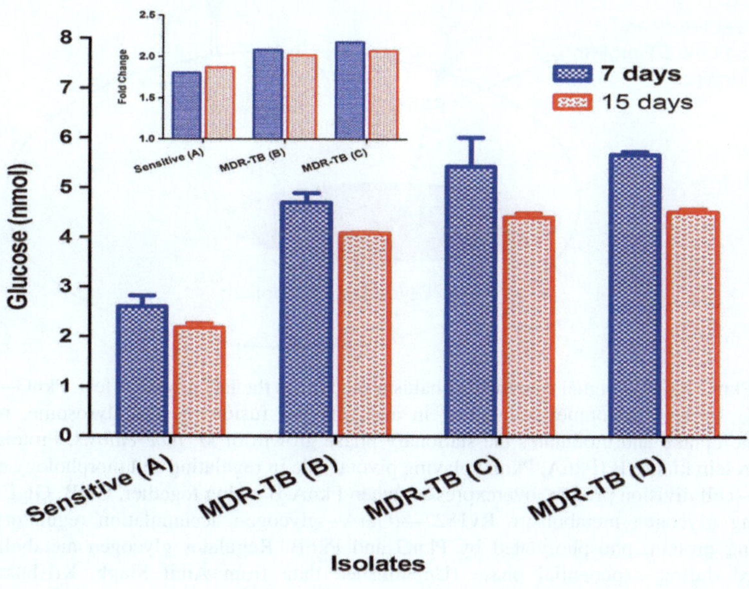

Fig. 7 Comparison of glycogen content in sensitive vs. MDR strain of *M. tuberculosis* (Singh et al. 2015)

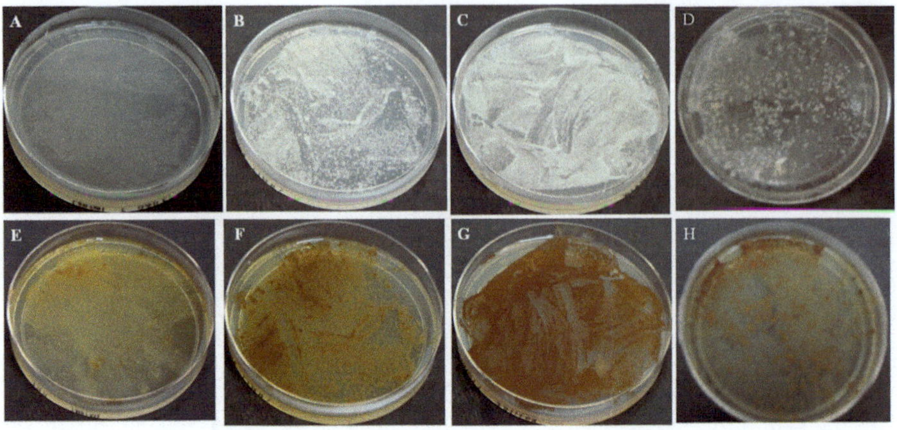

Fig. 8 Glycogen staining of sensitive and MDR isolates. (**a**) Sensitive isolate (Isolate A), (**b**) MDR isolate (Isolate B), (**c**) MDR isolate (Isolate C), (**d**) MDR isolate (Isolate D), (**e**) glycogen stained (Isolate A), (**f**) glycogen stained (Isolate B), (**g**) glycogen stained (Isolate C), (**h**) glycogen stained (Isolate D) (unpublished data from Amit Singh, Krishnamoorthy Gopinath and Sarman Singh)

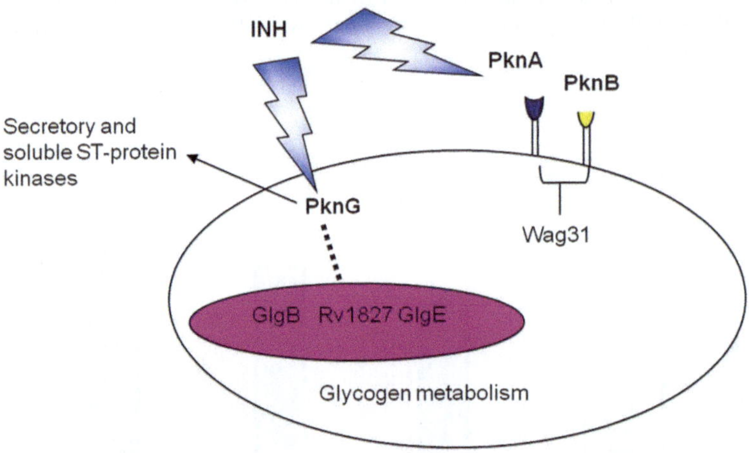

Fig. 9 PknG-Rv1827 signal cascade mechanism modulates the late growth effect. PknG—protein kinase G senses environment, involved in arresting the fusion of phagolysosome, regulates glutamine uptake and modulates the stationary phase growth of *M. tuberculosis*. Protein kinase A and protein kinase B (PknA, PknB) playing pivotal role in regulating cell morphology and size. Wag31—cell division protein, overexpressed when PknA-B acting together. GlgB, GlgE—genes regulating glycogen metabolism. Rv1827—(GarA—glycogen accumulation regulator)—FHA containing protein, phosphorylated by PknG and PknB. Regulates glycogen metabolism and expressed during exponential phase (Unpublished data from Amit Singh, Krishnamoorthy Gopinath and Sarman Singh)

study identified glutamyl-tRNA (gatA, Rv3011c, spot 11) overexpressed in MDR strains (Singh et al. 2015).

12 Glycogen in the Pathogenesis of *Tuberculosis*

Glycogen not only offers major nutrition to organisms but also plays significant role during host-pathogen interaction (Pal et al. 2010). Glycogen plays a minor role in colonization *of M. tuberculosis* within the host, but plays significant role in the intracellular survival of *M. tuberculosis* (McMeechan et al. 2005; Pal et al. 2010; Chauhan et al. 2012). The arabinomannan (AM) and α-glucans, which are the major composition of mycobacterial capsule and are located on the outer side of the mycolic acid layers in the cell wall and serve as receptor during host-pathogen interactions (Berg et al. 2007). The capsule mediates the adhesion to the host cell receptors and penetration of bacilli into the host cells (Daffé and Etienne 1999). Thereafter, mycobacterium modulates the environment conditions inside the host cell for their survival (Korf et al. 2005).

13 Immunology of *Mycobacterium tuberculosis* Glycans in the Host-Pathogen Interaction

Capsule of *M. tuberculosis* contains glycans, which can be up to 80 % of the extracellular polysaccharides, and is composed of α-4-D-Glc-1 core, which is branched at position 6 of each 5/6 residues by 4-α-D-Glc-1 oligoglucosides (Lemassu and Daffé 1994; Ortalo-Magné et al. 1995; Dinadayala et al. 2008). Since the discovery of capsular carbohydrates, with their role in bacterial pathogenesis, the focus has been on the macrophage receptors, which take part in the binding and phagocytosis of this microbe. The utility of carbohydrates in pathogenic mycobacterial species has followed the findings of the mycobacterial capsule (Dinadayala et al. 2008; Mendes et al. 2011). The AG, a disaccharide, contains 2–3 branched sugar chains and helps in the pathogenic role played by mycolic acid. These branched chains are connected at position 5 to Galf residue of the galactan's chain towards the reducing ends, consisting 22 Araf residues in D-arabinan chain (Besra et al. 1995). The D-arabinan contains backbone of α-1,5-linked Araf with numerous α-1,3-linkages and non-reducing ends are always terminated by β-1,2-Araf, resulting in hexa-arabinoside (Ara6) motifs shifted to the AG, and the dimers [β-D-Araf-1,2-α-D-Araf] form connecting site of mycolic acid. Both peptidoglycan and arabinoglycans form together covalently linked between mycolic acid layer and plasma membrane. Due to this, the cell wall of mycobacteria is enormously robust and difficult to penetrate by the drugs (Berg et al. 2007).

Unlike AG, cell envelope components are non-covalently bound to the LAM and might be linked to the mycolic acid layer or plasma membrane or both via phosphatidyl-myo-inositol (PI) unit. The reducing ends of the LAM show structural resemblances to the PIMs. The inositol residues of the phosphatidylinositol are mannosylated at the positions 2 and 6 of the LAM (Berg et al. 2007). The mycobacterial cell wall moieties (LAM) bind to the macrophage and prevent the attachment of mycobacteria to the complement receptors (Stokes et al. 2004). The capsular polysaccharides of *M. tuberculosis* facilitate non-opsonic binding of the bacteria to the complement receptor (CR-3) (Cywes et al. 1997; Dinadayala et al. 2008). Additionally, these glycans are able to generate innate immune responses through the attachment to the toll-like receptor (TLR2), CD14 and myeloid differentiation primary response gene-88 (MyD88-TLR) receptors (Bittencourt et al. 2006). Similarly, capsular components of the *M. tuberculosis* showed anti-phagocytic characteristics like that of macrophages (Stokes et al. 2004) and induced monocytes for proliferation and transformed into altered dendritic cells. These modified dendritic cells could not present lipid antigens to the CD1-restricted T cells (Gagliardi et al. 2007).

14 Glycans: Novel Drug Targets for MDR-TB

The emergence of MDR strains of *M. tuberculosis* (Koch et al. 2014) emphasizes the urgent need to find out the novel drug targets for *M. tuberculosis* (De Smet et al. 2000; Shriver et al. 2004). To find out the effective drugs, it is essential that appropriate drug targets are discussed before. The enzymes that take part in glycogen metabolism or those that mediate in the biosynthesis of vital components of *M. tuberculosis* cell envelope, cell wall or its survival machinery could be promising drug targets against mycobacteria. It has been validated that glgB auxotrophic strain of *M. tuberculosis* accumulates toxic polymers within the cells and induces cell death in *M. tuberculosis*. The deficiency of glycogen did not disturb macrophage infection by mycobacterium mutants; however, its presence showed the defending role during hostile stage of mycobacteria infections (Kalscheuer et al. 2010). Moreover, glgE-dependent pathway of glycogen synthesis was identified in mycobacteria. The product of glgE gene transfers activated glucose molecule to the maltose-1-phosphate through α-1-4 glycosidic linkage. Another enzyme Pep2 (encoded by Rv0127) is known to phosphorylate maltose and activate glycogen polymerization. Hence, the glgE enzyme induces cell death by two processes, the one known as glgE-dependent (induce self-poisoning via accumulation of maltose-1-phosphate followed by feedback inhibition of glgE) and the other glgE-independent mechanisms. Therefore, obstructing glgE activation could be an exciting drug target (Kalscheuer et al. 2010; Leiba et al. 2013).

Apart from that, biosynthesis of trehalose from glycogen is extensively studied in mycobacteria, and involved enzymes could also be possible drug targets, because of its significant role in bacterial cytosol and appearance in toxic glycolipids

(De Smet et al. 2000). It has been observed that any disruption in the enzyme—trehalose mycolyltransferase through 6-azido-6-deoxy-*a*,*a*-trehalose—inhibits growth of *M. tuberculosis* in-vitro (Belisle et al. 1997). In summary, the glycotransferase enzymes that are responsible for the synthesis of essential elements of the cell envelope in *M. tuberculosis* need to be explored as potential novel drug targets against mycobacterial pathogen.

References

Alonso-Casajús N, Dauvillée D, Viale AM, Muñoz FJ, Baroja-Fernández E, Morán-Zorzano MT, Eydallin G, Ball S, Pozueta-Romero J (2006) Glycogen phosphorylase, the product of the *glgP* Gene, catalyzes glycogen breakdown by removing glucose units from the nonreducing ends in *Escherichia coli*. J Bacteriol 188:5266–5272. doi:10.1128/JB.01566-05

Antoine AD, Tepper BS (1969) Environmental control of glycogen and lipid content of *Mycobacterium phlei*. J Gen Microbiol 55:217–226. doi:10.1099/00221287-55-2-217

Argüelles JC (2000) Physiological roles of trehalose in bacteria and yeasts: a comparative analysis. Arch Microbiol 174:217–224

Asención Diez MD, Demonte AM, Syson K, Arias DG, Gorelik A, Guerrero SA, Bornemann S, Iglesias AA (2015) Allosteric regulation of the partitioning of glucose-1-phosphate between glycogen and trehalose biosynthesis in *Mycobacterium tuberculosis*. Biochim Biophys Acta 1850:13–21. doi:10.1016/j.bbagen.2014.09.023

Ball SG, Morell MK (2003) From bacterial glycogen to starch: understanding the biogenesis of the plant starch granule. Annu Rev Plant Biol 54:207–233. doi:10.1146/annurev.arplant.54.031902.134927

Ballicora MA, Iglesias AA, Preiss J (2003) ADP-glucose pyrophosphorylase, a regulatory enzyme for bacterial glycogen synthesis. Microbiol Mol Biol Rev 67:213–225. doi:10.1128/MMBR.67.2.213-225.2003

Baskaran S, Roach PJ, DePaoli-Roach AA, Hurley TD (2010) Structural basis for glucose-6-phosphate activation of glycogen synthase. Proc Natl Acad Sci U S A 107:17563–17568. doi:10.1073/pnas.1006340107

Belanger AE, Hatfull GF (1999) Exponential-phase glycogen recycling is essential for growth of *Mycobacterium smegmatis*. J Bacteriol 181:6670–6678

Belisle JT, Vissa VD, Sievert T, Takayama K, Brennan PJ, Besra GS (1997) Role of the major antigen of *Mycobacterium tuberculosis* in cell wall biogenesis. Science 276:1420–1422

Berg S, Kaur D, Jackson M, Brennan PJ (2007) The glycosyltransferases of *Mycobacterium tuberculosis* – roles in the synthesis of arabinogalactan, lipoarabinomannan, and other glycoconjugates. Glycobiology 17:35R–56R. doi:10.1093/glycob/cwm010

Besra GS, Khoo KH, McNeil MR, Dell A, Morris HR, Brennan PJ (1995) A new interpretation of the structure of the mycolyl-arabinogalactan complex of *Mycobacterium tuberculosis* as revealed through characterization of oligoglycosylalditol fragments by fast-atom bombardment mass spectrometry and 1H nuclear magnetic resonance spectroscopy. Biochemistry (Mosc) 34:4257–4266

Bittencourt VCB, Figueiredo RT, da Silva RB, Mourão-Sá DS, Fernandez PL, Sassaki GL, Mulloy B, Bozza MT, Barreto-Bergter E (2006) An alpha-glucan of Pseudallescheria boydii is involved in fungal phagocytosis and Toll-like receptor activation. J Biol Chem 281:22614–22623. doi:10.1074/jbc.M511417200

Bolat I (2008) The importance of trehalose in brewing yeast survival. Innov Romanian Food Biotechnol 2:1–10

Bourassa L, Camilli A (2009) Glycogen contributes to the environmental persistence and transmission of *Vibrio cholerae*. Mol Microbiol 72:124–138. doi:10.1111/j.1365-2958.2009.06629

Breton C, Snajdrová L, Jeanneau C, Koca J, Imberty A (2006) Structures and mechanisms of glycosyltransferases. Glycobiology 16:29R–37R. doi:10.1093/glycob/cwj016

Buschiazzo A, Ugalde JE, Guerin ME, Shepard W, Ugalde RA, Alzari PM (2004) Crystal structure of glycogen synthase: homologous enzymes catalyze glycogen synthesis and degradation. EMBO J 23:3196–3205. doi:10.1038/sj.emboj.7600324

Calder PC (1991) Glycogen structure and biogenesis. Int J Biochem 23:1335–1352. doi:10.1016/0020-711X(91)90274-Q

Campbell NA, Williamson B, Heyden RJ (2006) Biology: Exploring Life. Pearson Prentice Hall, Upper Saddle River, NJ, 492 p. ISBN 0-13-250882-6

Carroll JD, Pastuszak I, Edavana VK, Pan YT, Elbein AD (2007) A novel trehalase from *Mycobacterium smegmatis* – purification, properties, requirements. FEBS J 274:1701–1714. doi:10.1111/j.1742-4658.2007.05715

Chandra G, Chater KF, Bornemann S (2011) Unexpected and widespread connections between bacterial glycogen and trehalose metabolism. Microbiology 157:1565–1572. doi:10.1099/mic.0.044263-0

Chauhan DS, Chandra S, Gupta A, Singh TR (2012) Molecular modelling, docking and interaction studies of human-plasmogen and salmonella enolase with enolase inhibitors. Bioinformation 8:185–188. doi:10.6026/97320630008185

Chen Q, Haddad GG (2004) Role of trehalose phosphate synthase and trehalose during hypoxia: from flies to mammals. J Exp Biol 207:3125–3129. doi:10.1242/jeb.01133

Cheng C, Mu J, Farkas I, Huang D, Goebl MG, Roach PJ (1995) Requirement of the self-glucosylating initiator proteins Glg1p and Glg2p for glycogen accumulation in *Saccharomyces cerevisiae*. Mol Cell Biol 15:6632–6640. doi:10.1128/MCB.15.12.6632

Cid E, Geremia RA, Guinovart JJ, Ferrer JC (2002) Glycogen synthase: towards a minimum catalytic unit? FEBS Lett 528:5–11. doi:10.1016/S0014-5793(02)03313-6

Crick DC, Mahapatra S, Brennan PJ (2001) Biosynthesis of the arabinogalactan-peptidoglycan complex of *Mycobacterium tuberculosis*. Glycobiology 11:107R–118R. doi:10.1093/glycob/11.9.107R

Cywes C, Hoppe HC, Daffé M, Ehlers MR (1997) Nonopsonic binding of *Mycobacterium tuberculosis* to complement receptor type 3 is mediated by capsular polysaccharides and is strain dependent. Infect Immun 65:4258–4266

Daffé M, Etienne G (1999) The capsule of *Mycobacterium tuberculosis* and its implications for pathogenicity. Tuberc Lung Dis 79:153–169

Dauvillée D, Kinderf IS, Li Z, Kosar-Hashemi B, Samuel MS, Rampling L, Ball S, Morell MK (2005) Role of the *Escherichia coli* glgX gene in glycogen metabolism. J Bacteriol 187:1465–1473. doi:10.1128/JB.187.4.1465-1473.2005

De Smet KA, Weston A, Brown IN, Young DB, Robertson BD (2000) Three pathways for trehalose biosynthesis in mycobacteria. Microbiology 146(Pt 1):199–208. doi:10.1099/00221287-146-1-199

Deutscher J, Francke C, Postma PW (2006) How phosphotransferase system-related protein phosphorylation regulates carbohydrate metabolism in bacteria. Microbiol Mol Biol Rev 70:939–1031. doi:10.1128/MMBR.00024-06

Dinadayala P, Sambou T, Daffé M, Lemassu A (2008) Comparative structural analyses of the alpha-glucan and glycogen from *Mycobacterium bovis*. Glycobiology 18:502–508. doi:10.1093/glycob/cwn031

Elbein AD, Pan YT, Pastuszak I, Carroll D (2003) New insights on trehalose: a multifunctional molecule. Glycobiology 13:17R–27R. doi:10.1093/glycob/cwg047

Farkas I, Hardy TA, Goebl MG, Roach PJ (1991) Two glycogen synthase isoforms in *Saccharomyces cerevisiae* are coded by distinct genes that are differentially controlled. J Biol Chem 266:15602–15607

François J, Parrou JL (2001) Reserve carbohydrates metabolism in the yeast *Saccharomyces cerevisiae*. FEMS Microbiol Rev 25:125–145. doi:10.1111/j.1574-6976.2001.tb00574.x

Gagliardi MC, Lemassu A, Teloni R, Mariotti S, Sargentini V, Pardini M, Daffé M, Nisini R (2007) Cell wall-associated alpha-glucan is instrumental for *Mycobacterium tuberculosis* to block CD1 molecule expression and disable the function of dendritic cell derived from infected monocyte. Cell Microbiol 9:2081–2092. doi:10.1111/j.1462-5822.2007.00940.x

Geurtsen J, Chedammi S, Mesters J, Cot M, Driessen NN, Sambou T, Kakutani R, Ummels R, Maaskant J, Takata H, Baba O, Terashima T, Bovin N, Vandenbroucke-Grauls CMJE, Nigou J, Puzo G, Lemassu A, Daffé M, Appelmelk BJ (2009) Identification of mycobacterial alpha-glucan as a novel ligand for DC-SIGN: involvement of mycobacterial capsular polysaccharides in host immune modulation. J Immunol 183:5221–5231. doi:10.4049/jimmunol.0900768

Gibson RP, Turkenburg JP, Charnock SJ, Lloyd R, Davies GJ (2002) Insights into trehalose synthesis provided by the structure of the retaining glucosyltransferase OtsA. Chem Biol 9:1337–1346. doi:10.1016/S1074-5521(02)00292-2

Gupta AK, Singh A, Singh S (2014) Glycogenomics of *Mycobacterium tuberculosis*. Mycobact Dis 4:175. doi:10.4172/2161-1068.1000175

Hengge-Aronis R, Fischer D (1992) Identification and molecular analysis of glgS, a novel growth-phase-regulated and rpoS-dependent gene involved in glycogen synthesis in *Escherichia coli*. Mol Microbiol 6:1877–1886. doi:10.1111/j.1365-2958.1992.tb01360.x

Henrissat B, Deleury E, Coutinho PM (2002) Glycogen metabolism loss: a common marker of parasitic behaviour in bacteria? Trends Genet TIG 18:437–440. doi:10.1016/S0168-9525(02)02734-8

Ingermann RL, Virgin GL (1987) Glycogen content and release of glucose from red blood cells of the sipunculan worm themiste dyscrita. J Exp Biol 129:141–149

Kalscheuer R, Syson K, Veeraraghavan U, Weinrick B, Biermann KE, Liu Z, Sacchettini JC, Besra G, Bornemann S, Jacobs WR (2010) Self-poisoning of *Mycobacterium tuberculosis* by targeting GlgE in an alpha-glucan pathway. Nat Chem Biol 6:376–384. doi:10.1038/nchembio.340

Koch A, Mizrahi V, Warner DF (2014) The impact of drug resistance on *Mycobacterium tuberculosis* physiology: what can we learn from rifampicin? Emerg Microbes Infect 3, e17. doi:10.1038/emi.2014.17

Korf J, Stoltz A, Verschoor J, De Baetselier P, Grooten J (2005) The *Mycobacterium tuberculosis* cell wall component mycolic acid elicits pathogen-associated host innate immune responses. Eur J Immunol 35:890–900. doi:10.1002/eji.200425332

Kumar S, Jena L (2014) Understanding rifampicin resistance in tuberculosis through a computational approach. Genom Inform 12:276–282. doi:10.5808/GI.2014.12.4.276

Leiba J, Syson K, Baronian G, Zanella-Cléon I, Kalscheuer R, Kremer L, Bornemann S, Molle V (2013) *Mycobacterium tuberculosis* maltosyltransferase GlgE, a genetically validated antituberculosis target, is negatively regulated by Ser/Thr phosphorylation. J Biol Chem 288:16546–16556. doi:10.1074/jbc.M112.398503

Lemassu A, Daffé M (1994) Structural features of the exocellular polysaccharides of *Mycobacterium tuberculosis*. Biochem J 297(Pt 2):351–357. doi:10.1042/bj2970351

Leung P, Lee YM, Greenberg E, Esch K, Boylan S, Preiss J (1986) Cloning and expression of the *Escherichia coli* glgC gene from a mutant containing an ADPglucose pyrophosphorylase with altered allosteric properties. J Bacteriol 167:82–88

Lin K, Hwang PK, Fletterick RJ (1995) Mechanism of regulation in yeast glycogen phosphorylase. J Biol Chem 270:26833–26839. doi:10.1074/jbc.270.45.26833

McMeechan A, Lovell MA, Cogan TA, Marston KL, Humphrey TJ, Barrow PA (2005) Glycogen production by different *Salmonella enterica* serotypes: contribution of functional glgC to virulence, intestinal colonization and environmental survival. Microbiology 151:3969–3977. doi:10.1099/mic.0.28292-0

Mendes V, Maranha A, Alarico S, da Costa MS, Empadinhas N (2011) *Mycobacterium tuberculosis* Rv2419c, the missing glucosyl-3-phosphoglycerate phosphatase for the second step in methylglucose lipopolysaccharide biosynthesis. Sci Rep 1:177. doi:10.1038/srep00177

Montero M, Eydallin G, Viale AM, Almagro G, Muñoz FJ, Rahimpour M, Sesma MT, Baroja-Fernández E, Pozueta-Romero J (2009) Escherichia coli glycogen metabolism is controlled by the PhoP-PhoQ regulatory system at submillimolar environmental Mg2+ concentrations, and is highly interconnected with a wide variety of cellular processes. Biochem J 424:129–141. doi:10.1042/BJ20090980

Morán-Zorzano MT, Alonso-Casajús N, Muñoz FJ, Viale AM, Baroja-Fernández E, Eydallin G, Pozueta-Romero J (2007) Occurrence of more than one important source of ADPglucose linked to glycogen biosynthesis in *Escherichia coli* and *Salmonella*. FEBS Lett 581:4423–4429. doi:10.1016/j.febslet.2007.08.017

Morán-Zorzano MT, Montero M, Muñoz FJ, Alonso-Casajús N, Viale AM, Eydallin G, Sesma MT, Baroja-Fernández E, Pozueta-Romero J (2008) Cytoplasmic *Escherichia coli* ADP sugar pyrophosphatase binds to cell membranes in response to extracellular signals as the cell population density increases. FEMS Microbiol Lett 288:25–32. doi:10.1016/j.febslet.2007.08.017

Moses SW, Bashan N, Gutman A (1972) Glycogen metabolism in the normal red blood cell. Blood 40:836–843

Ortalo-Magné A, Dupont MA, Lemassu A, Andersen AB, Gounon P, Daffé M (1995) Molecular composition of the outermost capsular material of the tubercle bacillus. Microbiology 141 (Pt 7):1609–1620. doi:10.1099/13500872-141-7-1609

Pal K, Kumar S, Sharma S, Garg SK, Alam MS, Xu HE, Agrawal P, Swaminathan K (2010) Crystal structure of full-length *Mycobacterium tuberculosis* H37Rv glycogen branching enzyme: insights of N-terminal beta-sandwich in substrate specificity and enzymatic activity. J Biol Chem 285:20897–20903. doi:10.1074/jbc.M110.121707

Palomino JC, Martin A (2014) Drug resistance mechanisms in *Mycobacterium tuberculosis*. Antibiot Basel Switz 3:317–340. doi:10.3390/antibiotics3030317

Palomo M, Kralj S, van der Maarel MJEC, Dijkhuizen L (2009) The unique branching patterns of Deinococcus glycogen branching enzymes are determined by their N-terminal domains. Appl Environ Microbiol 75:1355–1362. doi:10.1128/AEM.02141-08

Pan YT, Koroth Edavana V, Jourdian WJ, Edmondson R, Carroll JD, Pastuszak I, Elbein AD (2004) Trehalose synthase of *Mycobacterium smegmatis*: purification, cloning, expression, and properties of the enzyme. Eur J Biochem FEBS 271:4259–4269. doi:10.1111/j.1432-1033.2004.04365.x

Pedersen LC, Dong J, Taniguchi F, Kitagawa H, Krahn JM, Pedersen LG, Sugahara K, Negishi M (2003) Crystal structure of an alpha 1,4-N-acetylhexosaminyltransferase (EXTL2), a member of the exostosin gene family involved in heparan sulfate biosynthesis. J Biol Chem 278:14420–14428. doi:10.1074/jbc.M210532200

Pederson BA, Cheng C, Wilson WA, Roach PJ (2000) Regulation of glycogen synthase. Identification of residues involved in regulation by the allosteric ligand glucose-6-P and by phosphorylation. J Biol Chem 275:27753–27761. doi:10.1074/jbc.M003342200

Possner DDD, Claesson M, Guy JE (2015) Structure of the glycosyl transferase Ktr4p from *Saccharomyces cerevisiae*. PLoS One 10, e0136239. doi:10.1371/journal.pone.0136239

Preiss J (2006) Bacterial glycogen inclusions: enzymology and regulation of synthesis. In: Shively DJM (ed) Inclusions in prokaryotes. Springer, Berlin, pp 71–108. doi:10.1007/7171_004

Preiss J, Romeo T (1994) Molecular biology and regulatory aspects of glycogen biosynthesis in bacteria. Prog Nucleic Acid Res Mol Biol 47:299–329. doi:10.1016/S0079-6603(08)60255-X

Ramaswamy NT, Li L, Khalil M, Cannon JF (1998) Regulation of yeast glycogen metabolism and sporulation by Glc7p protein phosphatase. Genetics 149:57–72

Rini J, Esko J, Varki A (2009) Glycosyltransferases and glycan-processing enzymes. In: Varki A (ed) Essentials of glycobiology, 2nd edn. Cold Spring Harbor Laboratory Press, Cold Spring Harbor, NY. ISBN 978-1-621821-32-8

Roach PJ, Cheng C, Huang D, Lin A, Mu J, Skurat AV, Wilson W, Zhai L (1998) Novel aspects of the regulation of glycogen storage. J Basic Clin Physiol Pharmacol 9:139–151. doi:10.1515/JBCPP.1998.9.2-4.139

Rose AH, Tempest DW (1989) Advances in microbial physiology. Academic Press, London. ISBN 0-12-0277730-1

Schwebach JR, Glatman-Freedman A, Gunther-Cummins L, Dai Z, Robbins JB, Schneerson R, Casadevall A (2002) Glucan is a component of the *Mycobacterium tuberculosis* surface that is expressed in-vitro and in-vivo. Infect Immun 70:2566–2575. doi:10.1128/IAI.70.5.2566-2575.2002

Seibold GM, Breitinger KJ, Kempkes R, Both L, Krämer M, Dempf S, Eikmanns BJ (2011) The glgB-encoded glycogen branching enzyme is essential for glycogen accumulation in *Corynebacterium glutamicum*. Microbiology 157:3243–3251. doi:10.1099/mic.0.051565-0

Seok YJ, Koo BM, Sondej M, Peterkofsky A (2001) Regulation of *E. coli* glycogen phosphorylase activity by HPr. J Mol Microbiol Biotechnol 3:385–393

Shively JM (1974) Inclusion bodies of prokaryotes. Annu Rev Microbiol 28:167–187. doi:10.1146/annurev.mi.28.100174.001123

Shriver Z, Raguram S, Sasisekharan R (2004) Glycomics: a pathway to a class of new and improved therapeutics. Nat Rev Drug Discov 3:863–873. doi:10.1038/nrd1521

Silljé HH, Paalman JW, ter Schure EG, Olsthoorn SQ, Verkleij AJ, Boonstra J, Verrips CT (1999) Function of trehalose and glycogen in cell cycle progression and cell viability in *Saccharomyces cerevisiae*. J Bacteriol 181:396–400

Singh A, Gopinath K, Singh N, Singh S (2014) Deciphering the sequential events during in-vivo acquisition of drug resistance in *Mycobacterium tuberculosis*. Int J Mycobacteriol 3:36–40. doi:10.1016/j.ijmyco.2013.10.006

Singh A, Gopinath K, Sharma P, Bisht D, Sharma P, Singh N, Singh S (2015) Comparative proteomic analysis of sequential isolates of *Mycobacterium tuberculosis* from a patient pulmonary tuberculosis turning from drug sensitive to multidrug resistant. Indian J Med Res 141:27–45. doi:10.4103/0971-5916.154492

Slock JA, Stahly DP (1974) Polysaccharide that may serve as a carbon and energy storage compound for sporulation in Bacillus cereus. J Bacteriol 120:399–406

Stadthagen G, Sambou T, Guerin M et al (2007) Genetic basis for the biosynthesis of methylglucose lipopolysaccharides in *Mycobacterium tuberculosis*. J Biol Chem 282:27270–27276. doi:10.1074/jbc.M702676200

Stokes RW, Norris-Jones R, Brooks DE, Beveridge TJ, Doxsee D, Thorson LM (2004) The glycan-rich outer layer of the cell wall of *Mycobacterium tuberculosis* acts as an antiphagocytic capsule limiting the association of the bacterium with macrophages. Infect Immun 72:5676–5686. doi:10.1128/IAI.72.10.5676-5686.2004

Takayama K, Wang C, Besra GS (2005) Pathway to synthesis and processing of mycolic acids in *Mycobacterium tuberculosis*. Clin Microbiol Rev 18:81–101. doi:10.1128/CMR.18.1.81-101.2005

Unligil UM, Rini JM (2000) Glycosyltransferase structure and mechanism. Curr Opin Struct Biol 10:510–517. doi:10.1016/S0959-440X(00)00124-X

Wilson WA, Roach PJ, Montero M, Baroja-Fernández E, Muñoz FJ, Eydallin G, Viale AM, Pozueta-Romero J (2010) Regulation of glycogen metabolism in yeast and bacteria. FEMS Microbiol Rev 34:952–985. doi:10.1111/j.1574-6976.2010.00220

Zmasek CM, Godzik A (2014) Phylogenomic analysis of glycogen branching and debranching enzymatic duo. BMC Evol Biol 14:183. doi:10.1186/s12862-014-0183-2

Role of External and Environmental Factors in Drug Resistance Emergence: Gut Microbiota

Daniel Ryan, Sangeeta Jaiswal, and Mrutyunjay Suar

Abstract The gut possesses a diverse microbial community comprising bacteria, eukaryotes, archaea, and viruses, all of which make up the gut microbiome. The interactions between these various organisms are complex and difficult to model; however, they greatly influence our human health in a variety of ways. Commensals form the majority of this community and have a great impact on our immunity and resistance to disease. Consequently, the genetic pool or "metagenome" of the gut microbiome is a valuable resource into studies on human health. Metagenomic studies have revealed the presence of several genes contributing to drug resistance in the microbiome. These may have arisen either as a by-product of an essential survival pathway for the microbe or through spontaneous mutations. Another possible mode of entry is through pathogens carrying drug-resistant genes that may be introduced into the gut environment in a variety of ways, food being a significant point of entry. Consequently, all of the above factors contribute to an increasing number of drug-resistant genes in the gut microbiome. To add to this phenomenon, transmission of these genes through members of the microbiome may occur by horizontal gene transfer mechanisms adding to the diversity of organisms exhibiting resistance. Moreover, the administration of antibiotics for routine treatments has been found to further exacerbate this by deleting the beneficial commensal pool. Thus, it is of utmost importance to investigate and impede the emergence of resistance in the gut microbiome to benefit long-term human health.

1 Introduction

The human gut possesses a vast microbial community consisting of several trillion cells far outnumbering the number of human cells by a considerable margin. This community of microbes resident in the gut is termed the gut microbiome and

D. Ryan • S. Jaiswal • M. Suar (✉)
School of Biotechnology, KIIT University, Campus 11, Patia, Bhubaneswar 751024, Odisha, India
e-mail: msuar@kiitbiotech.ac.in

G. Arora et al. (eds.), *Drug Resistance in Bacteria, Fungi, Malaria, and Cancer*, DOI 10.1007/978-3-319-48683-3_12

includes bacteria, archaea, eukarya, and viruses. The human microbiome refers to the entire set of genetic elements of all the organisms comprising the microbiota. The term metagenomics refers to a study of the functions and interactions of all of these organisms inferred from their genomic data. The gut microbiome plays several essential functions, namely, the development of both adaptive and innate immunity, maintaining the integrity of the intestinal lining, energy production, synthesis of several vitamins, and factors and protection from colonization of invasive pathogens. To this effect, the Human Microbiome Project was constituted by the National Institute of Health to gain a deeper understanding of this diverse, complex microbial community with the following points of focus:

1. Complete characterization of the resident communities.
2. Is there a core community shared by all?
3. To study the effects of changes to this community on human health (The NIHHMPWG et al. 2009).

The use of antibiotics affects not only a specific target but also a broader population, some of which are beneficial to the host. Such is the effect exerted by antibiotics on the gut microbiome. The use of antibiotics or consumption of plant, animal, and dairy products containing antibiotics exerts an impact on these microbiota resulting in their destabilization or emergence of resistance. Destabilization has consequences on the host resulting in diarrhea and other opportunistic infections, while emergence of resistance leads to further complication for much longer periods of time. Thus, a study of the gut microbiome and the effects of drugs on its constituent communities are essential in determining the benefits and consequences to human health.

2 The Normal Gut Microbiota

The terms "normal" and "healthy" in context of the human gut microbiome are somewhat difficult to define primarily due to the significant interindividual differences that are prevalent. Thus, defining a "core microbiota" poses a greater challenge as opposed to defining "core functions" as a more unifying term. Several factors have been attributed to differences in the human gut microbiome, namely, age, gender, ethnic background, and environmental factors (diet, medication, stress, smoking, and infections) (Mueller et al. 2006).

A large proportion of gut bacteria ($\approx 80\,\%$) cannot be cultured in vitro by standard microbiological techniques, primarily due to their stringent nutrient requirements, anaerobic nature of their niche, and interdependence on one other (Dethlefsen et al. 2007). Consequently, several molecular techniques such as 16S rRNA sequencing, terminal restriction fragment length polymorphism (TRFLP), denaturing gradient gel electrophoresis (DGGE), and fluorescent in situ hybridization have proved useful in this regard for the identification and characterization of constituent members of the human microbiome. Studies have revealed large, diverse microbial communities to reside in the oral cavity and distal gastrointestinal (GI) tract with simpler groups residing in the esophagus, stomach, and small intestine. There have been very few studies on the latter group, and thus identifying them as resident bacteria or just transient travelers presents a significant challenge. Predominant members of the oral cavity include *Firmicutes*, *Bacteroidetes*, *Proteobacteria*, *Fusobacteria*, and *Actinobacteria* constituting about 99 % of the total members present. The remainder is composed of *Cyanobacteria*, *Spirochaetes*, TM7, and several others. The distal esophagus harbors a microbial community that is similar to the oral cavity with exception that most of the members could be cultured microbiologically (Ahn et al. 2011). Earlier, it was believed that the stomach with its extremely low pH possessed a transient microbial community with a simpler communal structure; however, gastric biopsies of 23 human patients presented a 16S rRNA library comprising diverse microbes, including members of *Actinobacteria*, *Proteobacteria*, *Fusobacteria*, and more (Bik et al. 2006). This community was found to differ considerably from communities of the mouth and esophagus. The microbial constitution of the small intestine was found to be rich in facultative anaerobic species; however, there are relatively few studies, and thus much more work needs to be done in this area. The colon on the other hand presented the greatest concentration of microbes with approximately 10^{11} to 10^{12} cells per gram of feces and a twofold greater number of microbial genes as compared to the human genome itself (Ley et al. 2006). The dominant phyla in the colon include *Firmicutes* and *Bacteroides* with seven other groups (*Fusobacteria*, *Actinobacteria*, *Verrucomicrobia*, *Proteobacteria*, *Spirochaetes*, *Cyanobacteria*, and VadinBE97) also serving as residents (Backhed et al. 2005).

The Metagenome of the Human Intestinal Tract consortium studied the metagenomes of 124 fecal specimens utilizing the Illumina Genome Analyzer. Specimens were isolated from healthy, overweight, and obese individuals and those suffering from inflammatory bowel disease (IBD) residing in Spain or Denmark (Arumugam et al. 2011). Data from the study revealed the human gut microbiota to comprise of about 1150 different species (Arumugam et al. 2011). Additionally, they identified 536,112 unique genes of which 99 % were found to be bacterial. At least 40 % of the bacterial genes from each specimen were found to be present within about half of the remaining specimens. These conserved genes were found to cluster to pathways involved in digestion and degradation of complex sugars, short-chain fatty acid production, and vitamin biosynthesis. The study also revealed the presence of a set of core functions that the microbiome performs, rather than a core group of organisms.

The establishment of the gut microbiome occurs at birth and continues to evolve and develop throughout the lifetime of the host. Studies have found variations in the gut microbiome of infants depending on the mode of delivery. 16S rRNA pyrosequencing revealed that infants delivered naturally possessed bacterial communities that resembled the vaginal microbiota of their mothers (*Lactobacillus*, *Prevotella*, etc.), while those delivered by Caesarian section possessed communities resembling those on the surface of the skin (*Staphylococcus*, *Propionibacterium*, etc.). These initial species termed as "founder species" undergo evolution within each host under both extrinsic and intrinsic factors that may be genetic, physiological, and environmental and ultimately constitute the gut microbiome.

Gut microbiota dysbiosis could lead to several pathological conditions. Diarrheal infections pose a significant challenge worldwide with a large proportion of them being food-borne. The most common bacterial infections are caused by *Salmonella*, *Shigella*, *Campylobacter*, *Vibrio cholerae*, etc., while viral pathogens include *Norovirus*, *Rotavirus*, etc. and *Giardia*, *Cryptosporidia*, etc. constitute the parasites. Antibiotic-associated diarrheal infections such as those caused by *Clostridium difficile* are due to a dysbiosis induced by the use of antibiotics. This leads to a loss of colonization resistance normally provided by the microbiota that subsequently allows overgrowth of *Clostridium difficile* and production of toxins A and B that bring about colonic damage, diarrhea, and, in severe cases, even death. Short bowel syndrome is a disorder that causes malabsorption due to either dysfunction of a large portion of the intestine or surgical removal of part of the intestine, or in some cases, it may be congenital. In such patients, the colon serves as an extremely important organ for energy salvage; however, in some cases, excess bacteria from the colon may lead to overgrowth, causing small intestinal bacterial overgrowth (SIBO). Studies have revealed a link between interaction of the host and microbiota in inflammatory bowel syndrome (IBS) pathology. Thus, the focus is now underway as to what antigens are present in the normal microbiota that may drive chronic inflammation in predisposed individuals.

3 Factors Affecting Drug Resistance

The spread of bacteria and genes responsible for the emergence of antibiotic resistance in the microbiota of the gut is dependent on a variety of factors, the most important of which is antibiotic use. Additional factors that play a role in the emergence and persistence of drug-resistant strains include their rate of adaptability, frequency of mutations that confer resistant traits, and ability to undergo various mechanisms of horizontal gene transfer (HGT) such as transformation, transduction, and conjugation. When exposed to a particular antibiotic, the acquisition of resistance by the bacterium endows it with an advantage in that particular environment. On the flipside, if the particular selective pressure, in this case an antibiotic, is removed, the resistant bacterium may not survive as well as its susceptible other.

A number of studies investigating the effects of antibiotic use on gut flora have shown resistant genes to be detected in members of the microbiota several years after treatment. Thus, bacteria are able to persist and transfer resistance traits for a considerable amount of time even after antibiotic use has been discontinued. For example, a study by Sjolund et al. revealed the *ermB* gene responsible for conferring clarithromycin resistance to be present in enterococci up to 1 year posttreatment (Sjolund et al. 2003). Interestingly, one particular patient was detected to have a resistant clone 3 years posttreatment. Such results reveal crucial clinical consequences since there remains an extended window during which resistance genes may be transferred to other species as well.

The intestine presents a highly suitable environment for the transfer of resistant genes. It provides a warm moisture-laden environment having an abundance of nutrients and a diverse bacterial population that could serve as a source and targets for acquiring resistance. Once particular resistance genes have been selected in the commensals, they may then be transferred to more pathogenic bacteria, thus posing a severe risk to health. The selection pressure provides a double blow with an increase in the number of bacteria possessing the resistant gene that could in turn further transfer it through the microbiota. Several bacteria are not residents rather passing through the intestine along with food. These may well serve as additional sources of resistant genes that could be transferred to commensal bacteria. Another potential source of selective pressure could be from agriculture. The use of certain compounds may impose selective pressure leading to the emergence of resistant genes in the microbiota when foods containing them are consumed.

4 External and Environmental Factors

4.1 Antibiotics

There is a delicate ecological balance that exists between the gut-associated microorganisms and their human host. This balance is however quite easily disturbed by a number of factors, the most important of which is antibiotic use. Major consequences of this are a decreased ability of commensal bacteria to resist colonization of an invasive species and the emergence of drug-resistant strains. While there are several studies investigating the short-term effects of the use of such antibiotics, their long-term effects are still relatively less studied. The effects induced on the microbiota by an antibiotic depend on a number of factors such as the dosage, administration route, time course, spectrum (broad or narrow), and mode of action. Antibiotic use that results in discharge into the gut may gravely affect the resident microbiota and consequently have an adverse impact on health by creating unforeseen complication such as *Clostridium difficile* infections (Sullivan et al. 2001).

Intestinal microbiota are dominated by anaerobic and facultative anaerobic species which play a major role in the production of volatile fatty acids, however many of which serve as specific targets for antibiotics resulting in destabilizing of the microbiota. As an example, the broad-spectrum antibiotic, clindamycin, has been found in high levels in the bile and fecal matter of treated individuals. However, it has also been associated in decreasing the colonization resistance of the resident microbiota resulting in *C. difficile* infections (Sullivan et al. 2001). This organism is normally present in relatively few numbers in healthy individuals; however, its numbers rapidly increase as a result of antibiotic-induced microbiota disturbances. Other side effects of this antibiotic include diarrhea, gastritis, intestinal pain, and swelling. Another drug of interest is amoxicillin, the use of which has been associated with an increase of antibiotic-resistant bacteria as well as a reduction in the number of gram-positive cocci. Several metagenomic techniques including DGGE have revealed a change in the microbiota constitution with the use of this antibiotic (Brugere et al. 2009). The *Helicobacter pylori* infection treatment protocol consists of the use of a combination of three drugs, namely, omeprazole, metronidazole, and clarithromycin. Dramatic disturbances in the gut microflora have been reported as a result of this treatment regimen with effects lasting for several years after treatment.

4.2 Food Animals

For a long time, the practice of administering antibiotics to animals reared for food, as a means to promote growth or for treatment, has been the norm. However, several studies have highlighted this as a potential source for the transfer of pathogens harboring antibiotic-resistant genes to humans. For example, a study by Levy et al. demonstrated that the use of the antibiotic oxytetracycline to enhance growth in chickens favored the selection of tetracycline-resistant *Escherichia coli* in them, which were subsequently detected in people that consumed them (Levy et al. 1976a, b). Another more recent example relates to the use of avoparcin in animals and its link to the emergence of vancomycin-resistant enterococci (VRE) which are highly pathogenic to humans. Thus, animals used for food, as well as pets and wild animals, can serve as sources of antibiotic-resistant genes. Additionally, several metals and compounds used in the veterinary industry have also been found to induce cross-resistance to various antibiotics. Not limited to animals but animal products such as milk and other dairy items were found to contain lactobacilli and other bacteria that possessed antibiotic-resistant genes.

4.3 Aquaculture

The fairly recent field of aquaculture is another example of the indiscriminate use of antibiotics by direct addition to water or direct administration, to promote growth or treatment of fish. This has resulted in antibiotic-resistant gene pools emerging in fish that are spread throughout the aquatic environment and ultimately to us. Several classes of antibiotics used are related or identical to those used to treat human infections, as a result of which cross-resistance has emerged in intestinal gut microbes. This has been well documented in several *Aeromonas* strains of fish origin that were able to transfer their antibiotic-resistant plasmids to human pathogens in the gut, such as *E. coli* and *Salmonella* (Cabello 2006). Another recent example is the emergence of *Salmonella enterica* serotype Typhimurium DT104 as responsible for outbreaks in the United States and Europe most likely originated from an aquaculture setup (Cabello 2006). The gene encoding a resistance to florfenicol which is widely used in aquaculture was isolated from these pathogens. Another source for the acquisition of antibiotic-resistant genes is through direct consumption of water or fish products that were contaminated.

4.4 Raw Fruits and Fresh Vegetables

Fruits and vegetables could also serve as a source of antibiotic-resistant genes if not washed or cooked adequately before consumption. As an example, *Pseudomonas aeruginosa* strains that were resistant to several commonly used antibiotics such as chloramphenicol, sulfamethoxazole, and ampicillin were isolated from several sources such as lettuce, tomatoes, cucumbers, and carrots (Allydice-Francis and Brown 2012). Another frequently reported source occurs through the consumption of salads and other meals that utilize uncooked vegetable and fruit products.

5 Methods to Study the Gut Resistome

One of the major limitations to study the human gut resistome is the unavailability of techniques to fully characterize resistance genes. In addition, determination and functional characterization of gene sequences providing resistance to antibiotics are challenging. Thus, in order to characterize a gut resistome completely, a combination of methods should be employed.

The simplest method to detect resistance genes present in gut commensals is to isolate strains and characterize them under laboratory conditions. These techniques were used frequently during the 1970s and 1980s to determine antibiotic resistance in gut commensals belonging to the *Bacteroidales* and *Clostridiales* phyla. However, owing to a large number of bacteria present in the gut microbiome, it is practically impossible to culture every gut commensal under laboratory conditions. Therefore, culture-based techniques cannot reveal the complete resistance profile of microbes present in the gut. Furthermore, this approach does not provide information about the mobility of resistance genes. Recent advances in laboratory culture-based techniques have made it possible to perform comparative genome analysis for *Bacteroidales* (Coyne et al. 2014) and *Enterococcaceae* (Palmer et al. 2012; Lebreton et al. 2013). However, this approach is very extensive, and it is unclear whether it can capture the entire gut resistome.

Culture-independent techniques have been found to be fairly useful for the analysis of the gut resistome. Often, PCR and microarray hybridization techniques are used to determine resistance gene reservoirs in the human gut. DNA can be isolated from fecal samples and analyzed for the presence of antibiotic-resistant genes by PCR, using gene-specific primers. Alternatively, isolated DNA samples can be hybridized with known antibiotic-resistant gene sequences (probes) and detected through a technique known as DNA microarray. With the help of quantitative PCR, the relative abundance of specific antibiotic resistance can also be determined (Jernberg et al. 2007; Buelow et al. 2014). The advantage of DNA microarray over PCR-based technique is that large numbers of resistance genes can be detected within a short time period. PCR on the other hand is comparatively slower and yields less information. A major limitation of the above techniques is the detection of known resistance markers as opposed to detection of novel resistance genes (Card et al. 2014; Lu et al. 2014). Furthermore, these techniques do not provide information about the bacterium harboring the resistance gene.

Metagenomic sequencing has proved to be a useful technique for the genetic and functional characterization of the human gut resistome. In this approach, DNA samples are purified from feces and sequenced. With the advent of next-generation sequencing platforms such as the Roche 454 sequencer, the Genome Analyzer of Illumina, and the SOLiD system of Applied Biosystems, sequencing costs have reduced dramatically with considerable increases in throughput. The sequence data sets thus obtained are assembled to form large contiguous DNA fragments and analyzed through bioinformatics tools. This technique not only allows the detection of resistance genes in any sample but also reveals the phylogenetic composition of the microbiota. Additionally, antibiotic-resistant genes in any microbiome can be quantified through these techniques (Forslund et al. 2014). However, even this approach cannot be used to discover new antibiotic resentence genes. On the other hand, functional metagenomics has proved useful in this regard, in which random fragments of metagenomic DNA are cloned in *E. coli* vectors and screened for antibiotic-resistant clones. From this library of antibiotic-resistant clones, gene sequences conferring resistance are determined. Although functional metagenomics is labor intensive, the major advantage of this approach is the identification of novel antibiotic-resistant genes.

6 The Gut Resistome as the Epicenter of Drug-Resistant Genes

Recent advances in metagenomic approaches have dramatically enhanced our knowledge about the gut microbiome (The NIHHMPWG et al. 2009). HGT is a common biological phenomenon in *Enterobacteriaceae* and has also been found to occur between pathogens and gut microflora particularly when the intestinal barrier is altered (Schjorring and Krogfelt 2011; Cremet et al. 2012; Stecher et al. 2012). The gut receives a wide number of bacteria from different sources such as hands, pharyngeal, food, water, beverages, nasal secretions, etc. Neonates acquire environmental microflora rapidly, and in some cases, sepsis occurs due to translocation of new microflora (Tezuka and Ohteki 2010; Das et al. 2011). Surprisingly, in healthy individuals, the gut microflora is stable, and pathogens ingested through food and water are cleared from the gut due to the presence of commensals.

Thus, it is essential to explore the gut microbiota for the presence of antibiotic-resistant genes and subsequently expand our knowledge on mechanism of emergence of multidrug-resistant pathogens. All resistance genes contributed by the gut microbiota have been termed as the "gut resistome." Antibiotic resistance may arise by two mechanisms, namely, "by chance" or "by mutation." The former mechanism describes a gene having a specific role in the host and accidentally neutralizes an antibiotic due to substrate or target similarity. Mutations, on the other hand, generate phenotypic resistance such as target modification, antibiotic inactivation, etc. Thus, antibiotic resistance may naturally be present or can be acquired through the accumulation of mutation. Resistance acquired through any method may be transferred to phylogenetically distant microorganisms sharing a similar ecological niche.

Antibiotics have been the cornerstone of defense against pathogenic infections, particularly after the Second World War. In natural environments, antibiotics provide selective advantage to the producing bacteria and prevent invasive competitors from establishing themselves. Besides, antibiotics may also function as signaling molecules, triggering developmental processes such as biofilm formation (Aminov 2009). Resistance to antibiotics is acquired in a population of susceptible bacteria that accumulate mutations. One such example is the occurrence of point mutation in DNA gyrase which confers resistance to quinolones. Besides this, many bacteria acquire antibiotic-resistant genes that protect microbial cells from the lethal action of antibiotics. Antibiotic-resistant genes confer phenotypic resistance against antibiotics through a variety of mechanisms which include enzymatic inactivation of antibiotics, modification of targets of antibiotics, and pumping antibiotics out of the cells through efflux pumps, thereby preventing accumulation of lethal doses (Martinez 2008, 2014; Allen et al. 2010). Antibiotic-resistant genes have been present in the environment for millennia. For instance, the beta-lactamases have been found to originate about 2 billion years ago (Hall and Barlow 2004). This is evident from the finding that OXA-type beta-lactamases carried on plasmids have been moved between different bacterial phyla for millions of years (Barlow and Hall 2002). Of note, genes conferring antibiotic resistance may have entirely different functions in the original host. For example, 2′-N-acetyltransferase encoded by a gene in a *Gammaproteobacterium*, namely, *Providencia stuartii*, is involved in peptidoglycan modification; however, aminoglycosides are structurally similar to the natural substrate of 2′-N-acetyltransferase and hence are inactivated by the enzyme. Therefore, *Providencia stuartii* are naturally resistant to aminoglycosides. Such genes can be termed as "accidental resistance genes," and when acquired by any pathogen, these genes can provide resistance against aminoglycosides making them antibiotic resistant (Martinez 2008). Emergence of antibiotic resistance among human pathogens has become a major threat to modern medicine. Therefore, identification of niches where microbes acquire antibiotic resistance is of great importance. Further, the mechanisms by which this antibiotic resistance is mobilized to pathogens are also equally important.

There is adequate evidence to suggest that the gut can serve as a reservoir for opportunistic pathogens, which under immunocompromised conditions may cause severe infections. These findings raise a concern particularly for hospitalized patients who undergo antibiotic therapy to prevent infections. Treatment of immunocompromised patients with various antibiotics may provide selective advantage to multidrug-resistant opportunistic pathogens present in the gut.

Since commensals and pathogens present in the gut share similar ecology, therefore, gene transfer is more likely to occur between them (Smillie et al. 2011). Consequently, it becomes very important to understand the nature of the human gut resistome and its role in the spread of antibiotic-resistant genes among members of the gut microflora, particularly between commensals and opportunistic pathogens (Penders et al. 2013).

As previously mentioned, the human gut presents an environment where millions of phylogenetically distinct bacteria may colonize and interact with each other which increases the possibility for the existence of ample antibiotic-resistant genes contributing to the genetic pool. Available facts about the human gut resistome raise concern in light of the emergence of multidrug-resistant opportunistic pathogens. A metagenomic study of 252 fecal samples performed by Forslund et al. revealed the presence of resistance genes for 50 classes of antibiotics accounting for 21 antibiotic-resistant genes per sample collected from different countries (Forslund et al. 2013). In a similar study, 1093 antibiotic-resistant genes were identified in 162 individuals from Denmark, China, and Spain (Hu et al. 2013). The most commonly found resistance genes are *tet32*, *tet40*, *tetO*, *tetQ*, and *tetW* which provide resistance against tetracycline and are found to be present in gut microflora of almost all individuals. Several such genes are ubiquitously present in bacterial genomes such as *ant(6)-Ia*, bacitracin (*bacA*), and glycopeptide vancomycin (*vanRA* and *vanRG*). These genes are thought to confer resistance to aminoglycosides. Similarly, metagenomic screening of gut resistomes of hospitalized patients also revealed many antibiotic-resistant genes in the gut microbiota of patients, which appeared to increase with increasing antibiotic therapy (Perez-Cobas et al. 2013; Buelow et al. 2014). For instance, Buelow et al. reported that genes conferring resistance to aminoglycosides expanded during a hospital stay, especially in intensive care units (Buelow et al. 2014). The observed expansion of the gut resistome was linked with the use of tobramycin which is an aminoglycoside used for selective decontamination of the digestive tract (de Smet et al. 2009). This treatment regime is generally used as a prophylactic measure for patients admitted to intensive care units in order to lower the risk of infection of opportunistic pathogens. Notably, antibiotic use is not always associated with expansion of the gut resistome in patients; rather, in certain cases, resistant genes may be lost during antibiotic treatment (Perez-Cobas et al. 2013; Buelow et al. 2014). This is generally observed during combination therapy in which bacteria carrying resistance against one antibiotic are still susceptible to another one and are thus lost from the microbial population.

7 From the Gut Resistome to Multidrug-Resistant Pathogens

Pathogenic bacterial loads in the gut are generally lower than that of commensal bacteria. Interestingly, the majority of antibiotic-resistant genes are found on the genome of commensals. Despite this, multidrug-resistant pathogens are emerging at an alarming rate posing a serious threat to the future of medicine. Thus, an understanding of the mechanism of transfer of antibiotic resistance from commensals to pathogens is of particular interest. There are two basic mechanisms by which antibiotic-resistant pathogens can emerge: one by opportunity and another by horizontal gene transfer (HGT). In the former, a small population of drug-resistant pathogenic strains exists naturally which, under suitable conditions such as an immunocompromised state of the host, multiply and increase in number. In the second mechanism, a pathogen acquires antibiotic-resistant genes from commensals by HGT.

8 Mobility of Antibiotic-Resistant Genes

Since antibiotic-resistant genes provide acute fitness in various environments, they are often encoded on mobile elements such as conjugative elements, extra chromosomal DNAs, and viral particles. An estimated 1–3 % of people from the developed world undergo antibiotic treatment daily making the gut environment a dynamic niche (Costello et al. 2012). Conjugative elements with resistance genes have been found to be associated with a large number of single-nucleotide polymorphism (SNP) (Schloissnig et al. 2013). Besides point mutation, HGT is also one of the sources of variability. Mobility of resistance genes is further augmented by the tendency of several antibiotics to facilitate gene transfer. For instance, antibiotics which inhibit DNA synthesis are known to induce interspecific transfer of integrating conjugative elements containing multidrug-resistant genes (Beaber et al. 2004). In the mice gut, tetracycline increases the rate of conjugation between *Enterococcus faecalis* and *Listeria monocytogenes* (Doucet-Populaire et al. 1991). Besides conjugation, antibiotics also enhance phage mobility by activating bacterial DNA damage response having cross talk with phage regulation. Treatment with antibiotics has been shown to increase the connectivity between phage and bacterial networks, increasing the accessibility of the microbial genome to phages (Modi et al. 2013). Gut commensals play an important role in host defense against pathogenic microbes by outcompeting for space and nutrients (Brandl et al. 2008; Fukuda et al. 2011; Costello et al. 2012) and inducing host immune responses (Brandl et al. 2008; Fukuda et al. 2011). Antibiotic treatment disrupts the highly organized microbial population structure of the gut, thus exposing a new niche for pathogens and increasing the availability and mobility of resistance genes to virulent species. For instance, methicillin-resistant *Staphylococcus aureus* (MRSA) emerged by obtaining a gene cluster from *Staphylococcus epidermis* which is a skin commensal. This gene cluster improved the colonization of host sites (Diep et al. 2006). There are several other examples of gene cluster movement between different bacteria. For example, a specialized polysaccharide degradation gene cluster which is typically present in marine bacteria was found to be present in the gut microbiota of Japanese individuals. Most likely, the gene cluster was transferred to the gut microbiome due to consumption of seaweeds over a long period of time. Besides this, there is evidence for the exchange of antibiotic-resistant genes between soil bacteria and human pathogens (Forsberg et al. 2012).

Generally, exchange of antibiotic-resistant genes between bacteria from different environments is more frequent than within the same environment, probably because of adaptive advantage provided by the resistance genes (Smillie et al. 2011). However, exchange of antibiotic-resistant genes between gut microbiota and the pathogenic gene pool is yet to be confirmed. Metagenomic analysis of the human gut resistome has revealed that functional sequences of antibiotic-resistant genes present in the gut microbial gene pool share very low homology to those present on known pathogens. Possible uncharacterized barriers hinder the transfer of resistance genes between commensals and pathogens resulting in the compartmentalization of the resistome of gut commensals and pathogens. Although antibiotic treatment has been shown to have significant implications, the mechanisms which govern the flow of genes in vivo have yet to be elucidated. For instance, antibiotic-induced HGT may improve the ability of gut microbiota to withstand a particular stress. One such example is the transfer of carbohydrate-active enzymes across phylogenetically distinct commensal bacteria which in turn helps diverse communities withstand shared challenges encountered in a dynamic gut environment. Antibiotic treatment in mice has been reported to enrich a carbohydrate-active enzyme encoded by phages. This finding suggests that increased gene transfer during antibiotic treatment helps the gut microbiota store and access functional genes that facilitate niche recolonization (Modi et al. 2013).

9 Acquisition of Genes by Pathogens

Human pathogens may acquire antibiotic-resistant genes from the gut reservoir by HGT which can occur by several mechanisms such as transformation, conjugation, and transduction (Fig. 1). In transformation, naked DNA is taken up by competent bacteria. Thus, if the DNA carries an antibiotic-resistant gene, it will be transferred to the recipient bacterium. Transformation is a common phenomenon, and many bacteria have been reported to be naturally competent to take up DNA from different species. Conjugation is a mating process in which exchange of genetic material takes place through a conjugative bridge. During this process, resistance genes can spread from donor cells to recipients. Transduction is mediated by bacteriophages. Several bacteriophages encode antibiotic-resistant genes which can be transferred and integrated into the chromosome or transferred from one cell to another (Furuya and Lowy 2006). Among the three processes, transformation

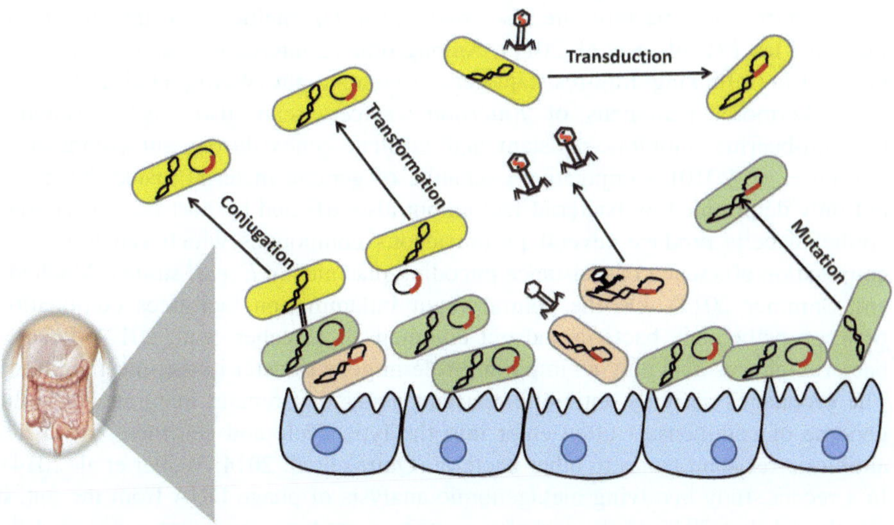

Fig. 1 Spread of drug-resistant genes in the gut microbiome. The figure depicts the mechanisms at play that contribute to the emergence of drug resistance in the gut microbiome. Drug resistance may arise as the mutation of a gene to prevent it from serving as a drug target or an unrelated enzyme whose secondary function serves to inactivate the effect of a drug. Conjugation, transformation, and transduction serve as mechanisms that aid the spread of drug resistance in a population either from pathogens to commensals and vice versa or between the commensals themselves

does not contribute significantly to HGT in the human gut (Nordgard et al. 2012). However, conjugation and transduction play an important role in the exchange of antibiotic genes among gut microflora. Most commonly, gene transfer in the human gut occurs through conjugative plasmids or transposons (Coyne et al. 2014). Genome and plasmid sequence analyses indicate that though conjugation between closely related bacteria is more frequent than remotely related ones, it plays a significant role in the dissemination of antibiotic-resistant genes (Jones et al. 2010; Tamminen et al. 2012). Conjugative transfer of antibiotic resistance from gut commensals to pathogens is evidenced from the finding that the *vanB* gene responsible for vancomycin resistance is present on a transposon of a commensal belonging to the phylum Firmicutes which serves as the source of vancomycin resistance in the nosocomial pathogen *E. faecium* (Stinear et al. 2001; Graham et al. 2008; Howden et al. 2013). Inside the gut, gene transfer takes place in both directions, i.e., from commensals to pathogens and from pathogens to commensals. The most common gram-negative bacteria that play an important role in gene transfer in the gut belong to the phylum *Bacteroidetes* as they are capable of acquiring DNA from a wide range of bacteria including the gram-positives *E. faecalis* and *Clostridium perfringens*. Thus, *Bacteroidetes* can serve as a source of resistance genes for other bacteria within the gut (Coyne et al. 2014). These groups of bacteria contain a conjugative transposon CTnDOT which plays an active role in the spread of erythromycin and tetracycline resistance inside the gut (Waters and Salyers 2013). Besides obligate anaerobes, facultative anaerobic commensals of the gut such as lactic acid bacteria are also involved in channeling gene transfer in the intestinal tract (Ogilvie et al. 2012). Among others, enterococci are exceptionally efficient in trafficking drug-resistant genes in human gut (Werner et al. 2013).

Furthermore, pathogens of *Enterobacteriaceae* can also easily exchange plasmid-bearing antibiotic-resistant and virulent genes during gut colonization (Goren et al. 2010). Conjugative exchange of genetic material inside the gut is not only determined by bacterial factors but also affected by host factors. Human epithelial cells produce several proteinaceous compounds which can lower the conjugation efficiency of resistance-encoding plasmids in *E. coli* strains (Machado and Sommer 2014). On the contrary, gut inflammation facilitates conjugation between pathogenic bacteria and gut commensals (Stecher et al. 2012). Besides bacteria, phages also play an important role in gene transfer in the intestinal tract. The prophages carrying antibiotic-resistant genes that remain integrated into the genome of commensals often enter into the lytic cycle and can thereby transfer antibiotic-resistant genes to other bacteria (Quiros et al. 2014; Waller et al. 2014). In a recent study involving metagenomic analysis of phage DNA from the gut, it was found that 70 % of the samples contained antibiotic-resistant genes such as beta-lactamase resistance gene blaTEM and the quinolone resistance gene *qnrA*. The role of bacteriophages in HGT in the gut has been proved experimentally as well where antibiotic-resistant genes increased considerably in the phage metagenome after treatment with ampicillin or ciprofloxacin (Modi et al. 2013). Phages isolated from mice treated with antibiotics show higher gene transfer rates as compared to those isolated from untreated mice (Modi et al. 2013). Thus, the gut presents a highly dynamic niche where constant gene transfer takes place among different communities by several methods.

10 Geographical Signature of Antibiotic-Resistant Genes

Although the human gut has been found to harbor a large number of resistance genes, there are country-specific variations (although small, approximately 1.5- to twofold) in the gut resistome of individuals. This could be due to differential use of antibiotics in different countries because individuals from countries such as Denmark with restricted use of antibiotics have been found to contain lower levels of antibiotic-resistant genes than the countries like Spain and China where the use of antibiotic is relatively higher. However, the relation between variation of resistome and incidences of antibiotic-resistant infections in different countries still remains to be determined.

Interestingly, analysis of SNPs in the antibiotic-resistant genes isolated from different geographical locations indicates that there is a specific geographical signature present in antibiotic-resistant genes. For example, sequences of antibiotic resistance isolated from Chinese individuals form a distinct cluster than those isolated from Danish and Spanish individuals (Hu et al. 2013). Similar to geographic-level variation, regional-level differences were also observed when resistance genes from different populations of China were analyzed.

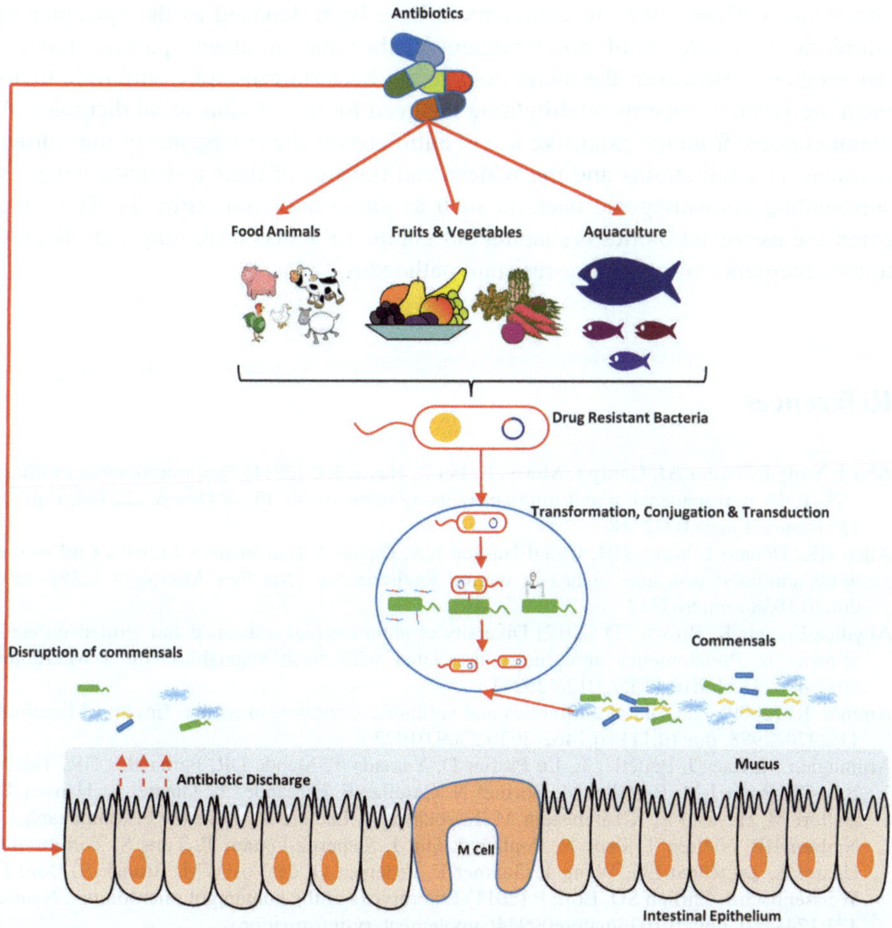

Fig. 2 A summary of the emergence and modes of transmission of drug-resistant genes in the gut microbiome. The figure illustrates the major external and environmental factors contributing to the emergence of drug resistance in the gut microbiome, namely, antibiotic use and the spread of antibiotic-resistant pathogens from various sources such as fruits, vegetables, animals, and fish. The effects of antibiotic use are the destabilization of the gut and selection of drug-resistant strains that in turn serve as sources of drug resistance, thus contributing to the pool of resistance genes in the population

11 Conclusion

The human gut environment presents a dynamic niche for microbes where millions of bacteria belonging to different phyla can colonize and exchange chemical and genetic information. In addition, gut microbes can exchange genetic material with microbes from the external environment through food, water, and beverages. This phenomenon has the result of accumulation of resistance genes in the gut commensals. Hence, the gut commensals have been depicted as the epicenter of antibiotic resistance. Antibiotic treatment has become an integral part of modern-day medicine. However, the disruptions to gut bacteria caused by antibiotic treatment are raising concerns, highlighting the need for new antibacterial therapies. A major concern from the extensive use of antibiotics is the emergence of multidrug-resistant bacterial strains and the widespread transfer of their resistance genes to surrounding nonpathogenic bacteria such as gut commensals (Fig. 2). Thus, the extensive use of antibiotics is shaping our gut microbiota contributing significantly to the emergence of multidrug-resistant pathogens.

References

Ahn J, Yang L, Paster BJ, Ganly I, Morris L, Pei Z, Hayes RB (2011) Oral microbiome profiles: 16S rRNA pyrosequencing and microarray assay comparison. PLoS One 6, e22788. doi:10.1371/journal.pone.0022788

Allen HK, Donato J, Wang HH, Cloud-Hansen KA, Davies J, Handelsman J (2010) Call of the wild: antibiotic resistance genes in natural environments. Nat Rev Microbiol 8:251–259. doi:10.1038/nrmicro2312

Allydice-Francis K, Brown PD (2012) Diversity of antimicrobial resistance and virulence determinants in Pseudomonas aeruginosa associated with fresh vegetables. Int J Microbiol 2012:426241. doi:10.1155/2012/426241

Aminov RI (2009) The role of antibiotics and antibiotic resistance in nature. Environ Microbiol 11:2970–2988. doi:10.1111/j.1462-2920.2009.01972.x

Arumugam M, Raes J, Pelletier E, Le Paslier D, Yamada T, Mende DR, Fernandes GR, Tap J, Bruls T, Batto J-M, Bertalan M, Borruel N, Casellas F, Fernandez L, Gautier L, Hansen T, Hattori M, Hayashi T, Kleerebezem M, Kurokawa K, Leclerc M, Levenez F, Manichanh C, Nielsen HB, Nielsen T, Pons N, Poulain J, Qin J, Sicheritz-Ponten T, Tims S, Torrents D, Ugarte E, Zoetendal EG, Wang J, Guarner F, Pedersen O, de Vos WM, Brunak S, Dore J, Weissenbach J, Ehrlich SD, Bork P (2011) Enterotypes of the human gut microbiome. Nature 473:174–180. doi:10.1038/nature09944(supplementary-information)

Backhed F, Ley RE, Sonnenburg JL, Peterson DA, Gordon JI (2005) Host-bacterial mutualism in the human intestine. Science 307:1915–1920. doi:10.1126/science.1104816

Barlow M, Hall BG (2002) Phylogenetic analysis shows that the OXA beta-lactamase genes have been on plasmids for millions of years. J Mol Evol 55:314–321. doi:10.1007/s00239-002-2328-y

Beaber JW, Hochhut B, Waldor MK (2004) SOS response promotes horizontal dissemination of antibiotic resistance genes. Nature 427:72–74. doi:10.1038/nature02241

Bik EM, Eckburg PB, Gill SR, Nelson KE, Purdom EA, Francois F, Perez-Perez G, Blaser MJ, Relman DA (2006) Molecular analysis of the bacterial microbiota in the human stomach. Proc Natl Acad Sci U S A 103:732–737. doi:10.1073/pnas.0506655103

Brandl K, Plitas G, Mihu CN, Ubeda C, Jia T, Fleisher M, Schnabl B, DeMatteo RP, Pamer EG (2008) Vancomycin-resistant enterococci exploit antibiotic-induced innate immune deficits. Nature 455:804–807. doi:10.1038/nature07250

Brugere JF, Mihajlovski A, Missaoui M, Peyret P (2009) Tools for stools: the challenge of assessing human intestinal microbiota using molecular diagnostics. Expert Rev Mol Diagn 9:353–365. doi:10.1586/erm.09.16

Buelow E, Gonzalez TB, Versluis D, Oostdijk EA, Ogilvie LA, van Mourik MS, Oosterink E, van Passel MW, Smidt H, D'Andrea MM, de Been M, Jones BV, Willems RJ, Bonten MJ, van

Schaik W (2014) Effects of selective digestive decontamination (SDD) on the gut resistome. J Antimicrob Chemother 69:2215–2223. doi:10.1093/jac/dku092

Cabello FC (2006) Heavy use of prophylactic antibiotics in aquaculture: a growing problem for human and animal health and for the environment. Environ Microbiol 8:1137–1144. doi:10.1111/j.1462-2920.2006.01054.x

Card RM, Warburton PJ, MacLaren N, Mullany P, Allan E, Anjum MF (2014) Application of microarray and functional-based screening methods for the detection of antimicrobial resistance genes in the microbiomes of healthy humans. PLoS One 9, e86428. doi:10.1371/journal.pone.0086428

Costello EK, Stagaman K, Dethlefsen L, Bohannan BJ, Relman DA (2012) The application of ecological theory toward an understanding of the human microbiome. Science 336:1255–1262. doi:10.1126/science.1224203

Coyne MJ, Zitomersky NL, McGuire AM, Earl AM, Comstock LE (2014) Evidence of extensive DNA transfer between bacteroidales species within the human gut. mBio 5:e01305–e01314. doi:10.1128/mBio.01305-14

Cremet L, Bourigault C, Lepelletier D, Guillouzouic A, Juvin ME, Reynaud A, Corvec S, Caroff N (2012) Nosocomial outbreak of carbapenem-resistant Enterobacter cloacae highlighting the interspecies transferability of the blaOXA-48 gene in the gut flora. J Antimicrob Chemother 67:1041–1043. doi:10.1093/jac/dkr547

Das P, Singh AK, Pal T, Dasgupta S, Ramamurthy T, Basu S (2011) Colonization of the gut with Gram-negative bacilli, its association with neonatal sepsis and its clinical relevance in a developing country. J Med Microbiol 60:1651–1660. doi:10.1099/jmm.0.033803-0

de Smet AM, Kluytmans JA, Cooper B, Mascini EM, Benus RF, van der Werf TS, van der Hoeven JG, Pickkers P, Bogaers-Hofman D, van der Meer NJ, Bernards AT, Kuijper EJ, Joore JC, Leverstein-van Hall MA, Bindels AJ, Jansz AR, Wesselink RM, de Jongh BM, Dennesen PJ, van Asselt GJ, te Velde LF, Frenay IH, Kaasjager K, Bosch FH, van Iterson M, Thijsen SF, Kluge GH, Pauw W, de Vries JW, Kaan JA, Arends JP, Aarts LP, Sturm PD, Harinck HI, Voss A, Uijtendaal EV, Blok HE, Thieme Groen ES, Pouw ME, Kalkman CJ, Bonten MJ (2009) Decontamination of the digestive tract and oropharynx in ICU patients. N Engl J Med 360:20–31. doi:10.1056/NEJMoa0800394

Dethlefsen L, McFall-Ngai M, Relman DA (2007) An ecological and evolutionary perspective on human-microbe mutualism and disease. Nature 449:811–818. doi:10.1038/nature06245

Diep BA, Gill SR, Chang RF, Phan TH, Chen JH, Davidson MG, Lin F, Lin J, Carleton HA, Mongodin EF, Sensabaugh GF, Perdreau-Remington F (2006) Complete genome sequence of USA300, an epidemic clone of community-acquired meticillin-resistant Staphylococcus aureus. Lancet 367:731–739. doi:10.1016/s0140-6736(06)68231-7

Doucet-Populaire F, Trieu-Cuot P, Dosbaa I, Andremont A, Courvalin P (1991) Inducible transfer of conjugative transposon Tn1545 from Enterococcus faecalis to Listeria monocytogenes in the digestive tracts of gnotobiotic mice. Antimicrob Agents Chemother 35:185–187. doi:10.1128/AAC.35.1.185

Forsberg KJ, Reyes A, Wang B, Selleck EM, Sommer MO, Dantas G (2012) The shared antibiotic resistome of soil bacteria and human pathogens. Science 337:1107–1111. doi:10.1126/science.1220761

Forslund K, Sunagawa S, Kultima JR, Mende DR, Arumugam M, Typas A, Bork P (2013) Country-specific antibiotic use practices impact the human gut resistome. Genome Res 23:1163–1169. doi:10.1101/gr.155465.113

Forslund K, Sunagawa S, Coelho LP, Bork P (2014) Metagenomic insights into the human gut resistome and the forces that shape it. Bioessays 36:316–329. doi:10.1002/bies.201300143

Fukuda S, Toh H, Hase K, Oshima K, Nakanishi Y, Yoshimura K, Tobe T, Clarke JM, Topping DL, Suzuki T, Taylor TD, Itoh K, Kikuchi J, Morita H, Hattori M, Ohno H (2011) Bifidobacteria can protect from enteropathogenic infection through production of acetate. Nature 469:543–547. doi:10.1038/nature09646

Furuya EY, Lowy FD (2006) Antimicrobial-resistant bacteria in the community setting. Nat Rev Microbiol 4:36–45. doi:10.1038/nrmicro1325

Goren MG, Carmeli Y, Schwaber MJ, Chmelnitsky I, Schechner V, Navon-Venezia S (2010) Transfer of carbapenem-resistant plasmid from Klebsiella pneumoniae ST258 to Escherichia coli in patient. Emerg Infect Dis 16:1014–1017. doi:10.3201/eid1606.091671

Graham M, Ballard SA, Grabsch EA, Johnson PD, Grayson ML (2008) High rates of fecal carriage of nonenterococcal vanB in both children and adults. Antimicrob Agents Chemother 52:1195–1197. doi:10.1128/aac.00531-07

Hall BG, Barlow M (2004) Evolution of the serine beta-lactamases: past, present and future. Drug Resist Updat 7:111–123. doi:10.1016/j.drup.2004.02.003

Howden BP, Holt KE, Lam MM, Seemann T, Ballard S, Coombs GW, Tong SY, Grayson ML, Johnson PD, Stinear TP (2013) Genomic insights to control the emergence of vancomycin-resistant enterococci. mBio 4:e00412-13. doi:10.1128/mBio.00412-13

Hu Y, Yang X, Qin J, Lu N, Cheng G, Wu N, Pan Y, Li J, Zhu L, Wang X, Meng Z, Zhao F, Liu D, Ma J, Qin N, Xiang C, Xiao Y, Li L, Yang H, Wang J, Yang R, Gao GF, Wang J, Zhu B (2013) Metagenome-wide analysis of antibiotic resistance genes in a large cohort of human gut microbiota. Nat Commun 4:2151. doi:10.1038/ncomms3151

Jernberg C, Lofmark S, Edlund C, Jansson JK (2007) Long-term ecological impacts of antibiotic administration on the human intestinal microbiota. ISME J 1:56–66. doi:10.1038/ismej.2007.3

Jones BV, Sun F, Marchesi JR (2010) Comparative metagenomic analysis of plasmid encoded functions in the human gut microbiome. BMC Genomics 11:46. doi:10.1186/1471-2164-11-46

Lebreton F, van Schaik W, McGuire AM, Godfrey P, Griggs A, Mazumdar V, Corander J, Cheng L, Saif S, Young S, Zeng Q, Wortman J, Birren B, Willems RJ, Earl AM, Gilmore MS (2013) Emergence of epidemic multidrug-resistant Enterococcus faecium from animal and commensal strains. mBio 4:e00534-13. doi:10.1128/mBio.00534-13

Levy SB, FitzGerald GB, Macone AB (1976a) Changes in intestinal flora of farm personnel after introduction of a tetracycline-supplemented feed on a farm. N Engl J Med 295:583–588. doi:10.1056/nejm197609092951103

Levy SB, FitzGerald GB, Macone AB (1976b) Spread of antibiotic-resistant plasmids from chicken to chicken and from chicken to man. Nature 260:40–42. doi:10.1038/260040a0

Ley RE, Peterson DA, Gordon JI (2006) Ecological and evolutionary forces shaping microbial diversity in the human intestine. Cell 124:837–848. doi:10.1016/j.cell.2006.02.017

Lu N, Hu Y, Zhu L, Yang X, Yin Y, Lei F, Zhu Y, Du Q, Wang X, Meng Z, Zhu B (2014) DNA microarray analysis reveals that antibiotic resistance-gene diversity in human gut microbiota is age related. Sci Rep 4:4302. doi:10.1038/srep04302

Machado AM, Sommer MO (2014) Human intestinal cells modulate conjugational transfer of multidrug resistance plasmids between clinical Escherichia coli isolates. PLoS One 9, e100739. doi:10.1371/journal.pone.0100739

Martinez JL (2008) Antibiotics and antibiotic resistance genes in natural environments. Science 321:365–367. doi:10.1126/science.1159483

Martinez JL (2014) General principles of antibiotic resistance in bacteria. Drug Discov Today Technol 11:33–39. doi:10.1016/j.ddtec.2014.02.001

Modi SR, Lee HH, Spina CS, Collins JJ (2013) Antibiotic treatment expands the resistance reservoir and ecological network of the phage metagenome. Nature 499:219–222. doi:10.1038/nature12212

Mueller S, Saunier K, Hanisch C, Norin E, Alm L, Midtvedt T, Cresci A, Silvi S, Orpianesi C, Verdenelli MC, Clavel T, Koebnick C, Zunft HJ, Dore J, Blaut M (2006) Differences in fecal microbiota in different European study populations in relation to age, gender, and country: a cross-sectional study. Appl Environ Microbiol 72:1027–1033. doi:10.1128/aem.72.2.1027-1033.2006

Nordgard L, Brusetti L, Raddadi N, Traavik T, Averhoff B, Nielsen KM (2012) An investigation of horizontal transfer of feed introduced DNA to the aerobic microbiota of the gastrointestinal tract of rats. BMC Res Notes 5:170. doi:10.1186/1756-0500-5-170

Ogilvie LA, Firouzmand S, Jones BV (2012) Evolutionary, ecological and biotechnological perspectives on plasmids resident in the human gut mobile metagenome. Bioeng Bugs 3:13–31. doi:10.4161/bbug.3.1.17883

Palmer KL, Godfrey P, Griggs A, Kos VN, Zucker J, Desjardins C, Cerqueira G, Gevers D, Walker S, Wortman J, Feldgarden M, Haas B, Birren B, Gilmore MS (2012) Comparative genomics of enterococci: variation in Enterococcus faecalis, clade structure in E. faecium, and defining characteristics of E. gallinarum and E. casseliflavus. mBio 3:e00318-11. doi:10.1128/mBio.00318-11

Penders J, Stobberingh EE, Savelkoul PH, Wolffs PF (2013) The human microbiome as a reservoir of antimicrobial resistance. Front Microbiol 4:87. doi:10.3389/fmicb.2013.00087

Perez-Cobas AE, Artacho A, Knecht H, Ferrus ML, Friedrichs A, Ott SJ, Moya A, Latorre A, Gosalbes MJ (2013) Differential effects of antibiotic therapy on the structure and function of human gut microbiota. PLoS One 8, e80201. doi:10.1371/journal.pone.0080201

Quiros P, Colomer-Lluch M, Martinez-Castillo A, Miro E, Argente M, Jofre J, Navarro F, Muniesa M (2014) Antibiotic resistance genes in the bacteriophage DNA fraction of human fecal samples. Antimicrob Agents Chemother 58:606–609. doi:10.1128/aac.01684-13

Schjorring S, Krogfelt KA (2011) Assessment of bacterial antibiotic resistance transfer in the gut. Int J Microbiol 2011:312956. doi:10.1155/2011/312956

Schloissnig S, Arumugam M, Sunagawa S, Mitreva M, Tap J, Zhu A, Waller A, Mende DR, Kultima JR, Martin J, Kota K, Sunyaev SR, Weinstock GM, Bork P (2013) Genomic variation landscape of the human gut microbiome. Nature 493:45–50. doi:10.1038/nature11711

Sjolund M, Wreiber K, Andersson DI, Blaser MJ, Engstrand L (2003) Long-term persistence of resistant Enterococcus species after antibiotics to eradicate Helicobacter pylori. Ann Intern Med 139:483–487. doi:10.7326/0003-4819-139-6-200309160-00011

Smillie CS, Smith MB, Friedman J, Cordero OX, David LA, Alm EJ (2011) Ecology drives a global network of gene exchange connecting the human microbiome. Nature 480:241–244. doi:10.1038/nature10571

Stecher B, Denzler R, Maier L, Bernet F, Sanders MJ, Pickard DJ, Barthel M, Westendorf AM, Krogfelt KA, Walker AW, Ackermann M, Dobrindt U, Thomson NR, Hardt WD (2012) Gut inflammation can boost horizontal gene transfer between pathogenic and commensal Enterobacteriaceae. Proc Natl Acad Sci U S A 109:1269–1274. doi:10.1073/pnas.1113246109

Stinear TP, Olden DC, Johnson PD, Davies JK, Grayson ML (2001) Enterococcal vanB resistance locus in anaerobic bacteria in human faeces. Lancet 357:855–856. doi:10.1016/s0140-6736(00)04206-9

Sullivan A, Edlund C, Nord CE (2001) Effect of antimicrobial agents on the ecological balance of human microflora. Lancet Infect Dis 1:101–114. doi:10.1016/s1473-3099(01)00066-4

Tamminen M, Virta M, Fani R, Fondi M (2012) Large-scale analysis of plasmid relationships through gene-sharing networks. Mol Biol Evol 29:1225–1240. doi:10.1093/molbev/msr292

Tezuka H, Ohteki T (2010) Regulation of intestinal homeostasis by dendritic cells. Immunol Rev 234:247–258. doi:10.1111/j.0105-2896.2009.00872.x

The NIHHMPWG, Peterson J, Garges S, Giovanni M, McInnes P, Wang L, Schloss JA, Bonazzi V, McEwen JE, Wetterstrand KA, Deal C, Baker CC, Di Francesco V, Howcroft TK, Karp RW, Lunsford RD, Wellington CR, Belachew T, Wright M, Giblin C, David H, Mills M, Salomon R, Mullins C, Akolkar B, Begg L, Davis C, Grandison L, Humble M, Khalsa J, Little AR, Peavy H, Pontzer C, Portnoy M, Sayre MH, Starke-Reed P, Zakhari S, Read J, Watson B, Guyer M (2009) The NIH human microbiome project. Genome Res 19:2317–2323. doi:10.1101/gr.096651.109

Waller AS, Yamada T, Kristensen DM, Kultima JR, Sunagawa S, Koonin EV, Bork P (2014) Classification and quantification of bacteriophage taxa in human gut metagenomes. ISME J 8:1391–1402. doi:10.1038/ismej.2014.30

Waters JL, Salyers AA (2013) Regulation of CTnDOT conjugative transfer is a complex and highly coordinated series of events. MBio 4:e00569-13. doi:10.1128/mBio.00569-13

Werner G, Coque TM, Franz CM, Grohmann E, Hegstad K, Jensen L, van Schaik W, Weaver K (2013) Antibiotic resistant enterococci-tales of a drug resistance gene trafficker. IJMM 303:360–379. doi:10.1016/j.ijmm.2013.03.001

Efflux-Mediated Drug Resistance in *Staphylococcus aureus*

Nitin Pal Kalia

Abstract An intrinsic property of *Staphylococcus aureus* is that it is naturally susceptible to virtually every antibiotic ever developed. Various mechanisms are responsible for multidrug resistance, but efflux pumps play a crucial role in the emergence of multidrug-resistant bacteria by expelling the antimicrobials from inside the cell. These membrane proteins are also involved in a variety of physiological roles. Therefore, there is a need to identify inhibitors and develop strategies to overcome multidrug resistance. One possible option is the use of efflux pump inhibitors (EPIs) in combination with already available antimicrobial agents/antibiotics. Different approaches have been developed for the evaluation of EPIs, ranging from high-throughput screening to bioassay-guided purification. The greatest hurdle to their clinical use is toxicity, as is the case for the existing EPIs, verapamil and reserpine. To date, no single EPI has been approved for use in clinical settings because of uncertainty around their potency and their intolerable adverse effects, particularly inhibition of cytochrome P450. Currently, the use of test active EPIs is limited to epidemiological studies. However, the search for more specific and effective EPIs will continue because of their significant benefits. The development of new chemotherapeutic agents requires contemplation of efflux pump substrate selectivity.

1 Introduction

Staphylococcus aureus, a well-known commensal organism, is considered to be the cause of various life-threatening infections. The incidence of nosocomial and community-associated *S. aureus* infections has increased greatly over the past 20 years because of a rise in antibiotic-resistant strains—specifically, methicillin-resistant *S. aureus* (MRSA) and, more recently, vancomycin-resistant strains

N.P. Kalia (✉)
Lee Kong Chian School of Medicine, Nanyang Technological University, Experimental Medicine Building, 59 Nanyang Drive, Singapore, Singapore
e-mail: kalianpk@ntu.edu.sg

G. Arora et al. (eds.), *Drug Resistance in Bacteria, Fungi, Malaria, and Cancer*,
DOI 10.1007/978-3-319-48683-3_13

(Sievert et al. 2008). Although the factors involved in the evolution of virulent and drug-resistant organisms are not well understood, empirical therapy, abuse of antimicrobials, and the use of substandard pharmaceuticals are clearly contributing factors (Ughachukwu and Unekwe 2012). Bacteria have developed various strategies to escape the action of antibacterial agents, including modification/degradation of the antibiotic (e.g., enzyme-like β-lactamases and transferases for aminoglycosides), permeability modifications (e.g., target accessibility or efflux), and target-site modification. In particular, gram-positive bacteria chromosomes or plasmids are considered to be involved in mediating these resistance mechanisms. Increase in efflux and decrease in uptake are resistance-contributing factors for many antimicrobials, and these factors have recently been recognized as affecting drug concentration within the cell. The development of a multiple-resistance phenotype may be due to nonspecific behavior of these mechanisms. Studies have revealed that efflux pumps are important in more than merely antibiotic export; they may also facilitate the movement of the organism's virulence factors and their survival in respective ecological niches. The literature reveals that efflux pumps help bacteria invade professional phagocytes such as macrophages and even affect the survival of *S. aureus* inside neutrophils (Garzoni and Kelley 2009). These drug transporters become active when they receive a stimuli in the form of stress or some chemical entity and may affect colonization, virulence, and intracellular communication (Kumar and Schweizer 2005; Piddock 2006a; Nishino et al. 2009). However, these effects may belong to only a few transmembrane proteins, which reveals their crucial role in evolution (Krulwich et al. 2005). Inhibitors of one or more efflux pump systems could restore, and would significantly improve, the efficacy of the drug that is a substrate for that particular pump.

2 Efflux Pumps

2.1 Classification of Bacterial Efflux Pumps

Depending on their structural variation, energy source, and mechanisms, efflux pumps have been categorized into five major classes that are of prime importance in bacterial pathogens:

(A) MFS: major facilitator superfamily
(B) MATE family: multidrug and toxic compound extrusion family
(C) SMR family: small multidrug resistance family
(D) RND superfamily: resistance nodulation division superfamily
(E) ABC superfamily: adenosine triphosphate binding (ATP) cassette superfamily.

The pumps classified under the first four categories are known as secondary transporters because of their energy source; they utilize the proton motive force (H^+ antiport) to expel specific chemicals from inside the cell. The Na^+ gradient serves as

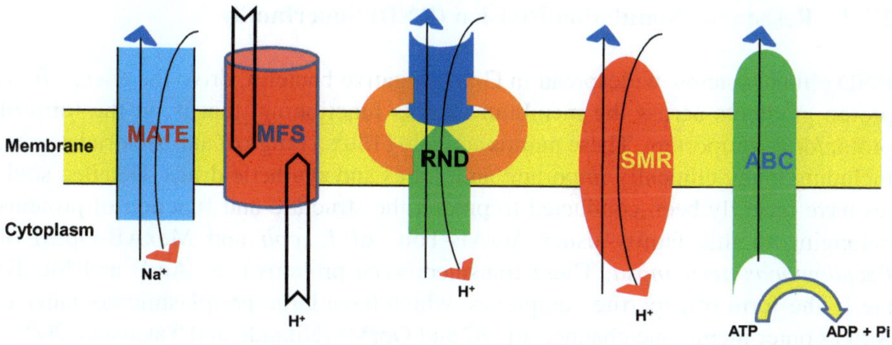

Fig. 1 Drug efflux from inside of bacterial cells is facilitated with the help of energy sources in the form H^+ or Na^+ ions or adenosine triphosphate binding (ATP) hydrolysis

a source of energy for the MATE family. Efflux pumps classified as ABC are considered primary transporters because they use ATP as a source of energy (Fig. 1).

2.1.1 Major Facilitator Superfamily (MFS)

This group of membrane transporters consists of secondary active transporters known for the efflux of multiple substrates, including NorA, QacA, Bmr, Blt, MdfA, and LmrP. These membrane proteins are antiporters and behave as monomers. However, in Gram-negative bacteria, these pumps work as a part of tripartite systems and membrane fusion proteins along with outer membrane channels. This system as a whole enables efflux pumps to effectively remove the drugs across the double membranes. This is in contrast to several other pumps in this group, which can extrude drugs to periplasm only (Li and Nikaido 2004).

2.1.2 Multidrug and Toxic Compound Extrusion (MATE) Family

The group consists of the NorM efflux pumps of *Vibrio parahaemolyticus*, which transport several toxic agents that are cationic (such as fluoroquinolones) as H^+ or Na^+ antiporters. The list of substrates for the MATE family is usually narrower than those for RND transporters. Despite this, to date, only a small number of transporters have been characterized under this group (Kuroda and Tsuchiya 2009); the proteins of this class are present in almost all the kingdoms of life, and genome sequences hold many more examples (Omote et al. 2006). Most of the transporters in this group have been identified by expressing the gene encoding these proteins in antimicrobial-hypersusceptible *Escherichia coli*. Hence, the exact function and contribution of this group of efflux pumps in the actual host bacteria is usually unclear.

2.1.3 Resistance Nodulation Division (RND) Superfamily

RND efflux systems, widespread in Gram-negative bacteria, drive the energy from proton gradients across the membrane for its functioning, that is, in the form of proton/drug antiporters. These pumps actively efflux a range of antibacterial agents, including many clinically important antibiotics and synthetic drugs. Detailed studies have recently been conducted to predict the structure and function of proteins belonging to this family using AcrAB-TolC of *E. coli* and MexAB-OprM of *Pseudomonas aeruginosa*. These transmembrane proteins (i.e., AcrB and MexB) are in the form of tripartite complexes, which have large periplasmic domains as well as outer membrane channels (TolC and OprM) (Nikaido and Takatsuka 2009).

2.1.4 Small Multidrug Resistance (SMR) Family

More than 250 annotated members form this group, which are further sub-categorized into (1) the small multidrug pumps, (2) the paired SMR proteins, and (3) suppressors of *groEL* mutant proteins. These proteins are expressed by genes present on the genomic DNA or on plasmids and are probably associated with integrons. These proteins not only extrude the disinfectants but are also involved in the expulsion of clinically relevant aminoglycosides (Bay et al. 2008; Li et al. 2003). In most cases, these pumps work in conjunction with RND pumps to efflux substrates from inside the cell.

2.1.5 Adenosine Triphosphate Binding cassette (ABC) Superfamily

The ABC family of multidrug efflux pumps is highly conserved throughout the phyla (bacteria to humans) but export a broad range of chemically diverse compounds by using ATP as a source of energy. Better understanding of ABC-mediated drug efflux has been obtained from the Sav1866 multidrug efflux pump of *S. aureus*, which is homologous with human multidrug-resistant (MDR) P-glycoprotein (P-gp) (Dawson and Locher 2006). ATP binding triggers the outward-facing conformation of Sav1866 and hence reflects the ATP-bound state, a central cavity comprising two transmembrane domains in conjunction with two nucleotide-binding domains, assumed to be the path for drug translocation (Davidson and Chen 2004). Separation of adenosine diphosphate (ADP) and phosphate (hydrolysis product of ATP) support the inward-facing conformation and therefore confirms that the substrate binding site is accessible to an inhibitor.

Table 1 Classification of efflux pumps and their respective substrates

Efflux pump	Family	Regulator	Substrates	References
Chromosomally-encoded efflux systems				
NorA	MFS	MgrA, NorG(?)	Hydrophilic fluoroquinolones, quaternary ammonium salts, dyes (e.g., ethidium bromide, rhodamine)	Handzlik et al. (2013)
NorB	MFS	MgrA, NorG	Fluoroquinolones (both hydrophilic and hydrophobic), tetracycline, QACs (e.g., tetraphenylphosphonium, cetrimide), dyes (e.g., ethidium bromide)	
NorB	MFS	MgrA, NorG	Fluoroquinolones (e.g., hydrophilic: ciprofloxacin; hydrophobic: moxifloxacin), dyes (e.g., rhodamine)	
MepA	MATE	MepR	Fluoroquinolones (e.g., hydrophilic: ciprofloxacin, norfloxacin; hydrophobic: moxifloxacin, sparfloxacin), glycylcyclines (e.g., tigecycline), QACs (e.g., tetraphenylphosphonium, cetrimide, benzalkonium chloride), dyes (e.g., ethidium bromide)	
MdeA	MFS	N.I.	Hydrophilic fluoroquinolones (e.g., ciprofloxacin, norfloxacin), virginiamycin, novobiocin, mupirocin, fusidic acid, QACs (e.g., tetraphenylphosphonium, benzalkonium chloride, dequalinium), dyes (e.g., ethidium bromide)	
SepA	N.D.	N.I.	QACs (e.g., benzalkonium chloride), biguanidines (e.g., chlorhexidine), dyes (e.g., acriflavine)	
SdrM	MFS	N.I.	Hydrophilic fluoroquinolones (e.g., norfloxacin), dyes (e.g., ethidium bromide, acriflavine)	
LmrS	MFS	N.I.	Oxazolidinone (linezolid), phenicols (e.g., chloramphenicol, florfenicol), trimethoprim, erythromycin, kanamycin, fusidic acid, QACs (e.g., tetraphenylphosphonium), detergents (e.g., sodium dodecyl sulphate), dyes (e.g., ethidium bromide)	
QacA	MFS	QacR	QACs (e.g., tetraphenylphosphonium, benzalkonium chloride, dequalinium), biguanidines (e.g., chlorhexidine), diamidines (e.g., pentamidine), dyes (e.g., ethidium bromide, rhodamine, acriflavine)	Handzlik et al. (2013)
QacB	MFS	QacR	QACs (e.g., tetraphenylphosphonium, benzalkonium chloride), dyes (e.g., ethidium bromide, rhodamine, acriflavine)	
Smr	SMR	N.I.	QACs (e.g., benzalkonium chloride, cetrimide), dyes (e.g., ethidium bromide)	

(continued)

Table 1 (continued)

Efflux pump	Family	Regulator	Substrates	References
QacG	SMR	N.I.	QACs (e.g., benzalkonium chloride, cetyltrimethylammonium), dyes (e.g., ethidium bromide)	
QacH	SMR	N.I.	QACs (e.g., benzalkonium chloride, cetyltrimethylammonium), dyes (e.g., ethidium bromide)	
QacJ	SMR	N.I.	QACs (e.g., benzalkonium chloride, cetyltrimethylammonium), dyes (e.g., ethidium bromide)	

N.D.: The family of transporters to which SepA belongs is not yet elucidated; N.I.: The transporter has no regulator identified to date
MATE multidrug and toxic compound extrusion family, *MFS* major facilitator superfamily, *QACs* quaternary ammonium compounds, *SMR* small multidrug resistance family

3 Efflux Pumps in *Staphylococcus aureus*

Drug expulsion is one of the most important resistance mechanisms in *S. aureus*. To date, approximately ten such pumps have been found in *S. aureus*. Most are members of the MFS; specifically, efflux pumps NorA, NorB, NorC, MdeA, and SdrM are chromosomally encoded along with some plasmid-encoded QacA/B pumps (Table 1).

4 Why Target Efflux Pumps

4.1 Contribution to Pathogenicity

Bacterial efflux pumps have a broad range of substrates and they export not only antibiotics but also antimicrobial agents produced by the host in response to bacterial entry. This information has revealed the role of these membrane-bound transporters in evading such molecules and allowing bacteria to survive in their ecological niches. Furthermore, recent studies have also proved the importance of efflux pumps in the pathogenicity of the bacteria. Reports have revealed the role of efflux pumps, wherein these pumps efflux not only antibiotics but also virulence-determining factors such as adhesins, toxins, or other proteins that help in the colonization of bacteria and cause infection of human and animal cells. To date, the involvement of RND and MATE family members in pathogenicity has been proven (Piddock 2006b).

4.2 *Efflux Pumps and Biofilm Resistance*

Biofilms are formed by bacteria on a solid surface, and biofilm-associated bacteria are more resistant to antibacterial compounds than are their planktonic counterparts (Mah et al. 2003). This bacterial strategy is most similar to persister cells that close their metabolism rendering targets no longer available to antibacterials that do not kill the cell. Many reasons exist for biofilm-mediated drug resistance, such as slow growth and reduced penetration due to the accumulation of extracellular polysaccharides; however, efflux pumps also play a significant role (Tegos et al. 2011). These drug transporters may influence drug-specific resistance in biofilms, such as in ofloxacin and azithromycin, where expression of the MexAB-OprM pump (Brooun et al. 2000) and the MexCD-OprJ, respectively, serve as a biofilm-specific mechanism for resistance (Gillis et al. 2005).

4.3 *Colonization and Virulence*

The role of MDR pumps (e.g., MFS and RND) in bacterial colonization within the host has been reported. A Salmonella strain, in the absence of all nine major transporters, failed to kill mice. Furthermore, *S. aureus* in which the *norA* gene was deleted did not invade macrophages. On the other hand, invasion increased in a *norA*-overexpressing strain compared with in a wild-type strain (Kalia et al. 2012). Whereas such findings may not indicate direct involvement of pump activity with 'virulence' they do provide evidence and support for the investigation of the role of efflux pumps in transporting virulence factors.

4.4 *Quorum Sensing*

Quorum sensing in bacteria requires the creation and recognition of extracellular signaling molecules called autoinducers, which control cell density and environmental factors, mediating gene expression (Camilli and Bassler 2006). The most thoroughly studied molecules are N-acyl homoserine lactones. Quorum-sensing studies have validated the role of RND pumps, but more data analysis is required to elucidate why some cells in regular batch cultures produce autoinducers (producers), whereas other cells (receivers) respond to these autoinducers. When a specified transporter sends a signal, it is difficult to predict the response because of the conflicting effects on these two types of cells. Furthermore, diffusion of N-acyl homoserine lactones across any membrane is easy because of its lipophilic nature; therefore, it is difficult to conclude whether or not an RND pump is required for its expulsion. Hence, we conclude there is no proof that RND pumps secrete N-acyl homoserine lactones, and the overexpression of these pumps is likely to obstruct autoinducers from entering receiver cells. In fact, overexpression of the MexEF-OprN pump reduces the production of autoinducers from the total population. However, the available literature focuses on the role of RND pumps in quorum sensing without considering these facts, and reports of the myth of RND pump-mediated autoinducer transport continue to appear (Li and Nikaido 2004).

5 Clinical Importance of Efflux Pumps of *S. aureus*

In bacterial pathogens, clinically important efflux-mediated drug resistance can involve both antibiotics and biocides. Antibiotic resistance due to efflux activity may involve (1) efflux pumps responsible for expelling a specific single antibiotic only (e.g., Tet determinants cause resistance to tetracycline) or (2) the expulsion of different groups of antibiotics by efflux pumps such as NorA, which are known for multiple substrate expulsion. Overexpression can lead to MDR pumps promoting resistance to a chemically diverse class of antibiotics but also reducing susceptibility to biocides, resulting in an MDR phenotype. These MDR phenotypes are not only a great risk to therapeutic success but also play a crucial role in co-selection and cross-resistance between efflux-mediated antibiotic and biocide resistance, which is particularly alarming when considering drug-resistant strains such as MRSA (Costa et al. 2013).

6 Strategies to Overcome Efflux-Mediated Resistance

We can reduce the effect of efflux pumps in a number of ways: (1) modify the structure of existing antibiotics to reduce their affinity to binding sites of these proteins; (2) permeabilize membranes to increase the influx of antibiotics, increasing the intracellular drug concentration; (3) suppress efflux pump-encoding genes and/or reduce the activity of active efflux pumps in the bacterial membrane; (4) interfere with the energy source that helps transport the drug; (5) block the assembly of important and active efflux pump parts; (6) place a plug block into membrane channels that transport antibiotics (within pump cavities or at pump exit channels); or (7) create antagonism between the antibiotic and the substrate during transport across the membrane through these pumps (Handzlik et al. 2013).

7 Efflux Pump Inhibitors: Resistance-Modifying Agents

Overexpression of multidrug efflux pumps in clinical isolates results in decreased susceptibility to antibiotics. Therefore, antibiotic resistance in both Gram-positive and Gram-negative pathogens must be considered during antibiotic drug discovery and development. Augmentin (GlaxoSmithKline) is an outcome of the concept whereby clavulanic acid, which inhibits β-lactamase-mediated resistance in a bacterium, may be used with an amoxicillin (β-lactam). Clavulanic acid significantly improved the stability of amoxicillin and inhibited β-lactamases; the combination is used against community-acquired pneumonia or acute bacterial sinusitis caused by β-lactamase-producing bacteria. Chemical entities that play a role in resistance modification may also modulate multidrug resistance mechanisms. Efflux pumps that are present in membranes have a diverse range of substrates and transport antibiotics out from inside the bacterial cell, leaving the cell unable to achieve the concentration required to kill bacteria. An inhibitor combined with the antibiotic substrates of these efflux pumps increases the concentration of the antibiotic inside the cell and therefore helps restore the potency of the drug. These inhibitors may also influence the emergence of antibiotic-resistant variants (Mullin et al. 2004). Prediction and determination of the structure of these transmembrane proteins helps determine the mechanisms of multidrug binding and expulsion as well as the structure-based approaches that can be used to discover efflux pump inhibitors (EPIs).

7.1 Bacterial Efflux Pump Inhibitors (EPIs)

The above discussion and available data confirm the involvement of efflux pumps in the development of resistance and hence their potential as effective antibacterial targets. New effective, safe, and less toxic chemical entities need to be developed as bacterial EPIs (Renau et al. 1999; Li and Nikaido 2004). For molecules to be considered as EPIs, they must be able to decrease the emergence of bacterial resistance to existing antibiotics, modulate resistance in strains with mutations at multiple sites and targets, and reduce the inherent resistance of bacteria to antibiotics (Kriengkauykiat et al. 2005). Different strategies can be used to identify effective EPIs. Modifications in antibiotics that are being effluxed can be used as efflux pumps to stop binding and/or transport. EPIs known and validated against other efflux systems can be used as MDR efflux pumps. Testing of known drugs other than antibiotics (Kristiansen and Amaral 1997) might be cost effective and time saving, as ample data related to their safety index and pharmacokinetics are already available. Another approach would be to screen chemical libraries or biodiverse compounds to identify new possibilities; studies of structure–activity relationships may help in lead optimization. The exact mode of action of these inhibitors is not yet clear, but one proposed mechanism is via binding directly to the pump (Kumar et al. 2013) and thereby blocking it via either a competitive or a non-competitive relationship with the substrates. Interfering directly with and depleting its energy source, either by blocking ATP-binding or by disconnecting the proton gradient across the membrane, can also be considered as another mode of action of such inhibitors. Studies have revealed the affinity of inhibitors for substrates of MDR pumps (Zloh et al. 2004). Therefore, the large size of inhibitor–antibiotic complexes will prevent their expulsion from inside the cell.

7.2 EPIs of Plant Origin

Natural molecules specifically of plant origin have the potential to be lead EPIs and effective antimicrobial agents. Focus on evaluating natural molecules as drug candidates has increased and is promising. Molecules need to satisfy the following criteria to be used as an EPI:

1. They should increase the potential of almost all the substrates of the pump.
2. They should have no effect on antibiotics not being effluxed.
3. They must have no activity for knock strains efflux pumps.
4. They must increase the intracellular concentration of efflux pump substrates and reduce their expulsion.
5. They should not disturb the membrane potential across the inner membrane.

Reserpine, an alkaloid isolated from *Rauwolfia vomitoria* Afz., was initially evaluated for its antihypertensive activity. The Bmr efflux pump of *Bacillus subtilis*, which mediates tetracycline resistance, served as a target to evaluate reserpine as an EPI where it directly interacts with the Bmr protein and binds to the cavity formed by Phe-143, Val-286, and Phe-306. In combination with tetracycline, reserpine reduced the minimum inhibitory concentration (MIC) of tetracycline fourfold against two clinical isolates of MRSA possessing Tet(K) efflux proteins, IS-58 and XU212 (Gibbons and Udo 2000). Reserpine also reversed NorA-conferred multidrug resistance and significantly improved the potential of norfloxacin against *S. aureus*.

Diospyrin (*Diospyros montana*) a naphthoquinone, was tested along with some of its derivatives against *Mycobacterium aurum* A+. Some of the derivatives showed better activity as EPIs in a ciprofloxacin accumulation assay at a concentration of 100 μM and increased ciprofloxacin accumulation (Chakrabarty et al. 2002).

Piperine is a major component of *Piper nigrum* and *Piper longum* and significantly improves the activity of ciprofloxacin against *S. aureus* and MRSA 15187 (Khan et al. 2006). To establish the mechanism of action of piperine, ethidium bromide (EtBr) was tested in combination with piperine. A reduction in MIC was observed against a ciprofloxacin-resistant strain of *S. aureus*, exhibiting an increased MIC against EtBr. Both ciprofloxacin and EtBr are substrates for NorA; therefore, piperine can be considered an inhibitor of this bacterial pump.

Phenylpropanoid ailanthoidiol and two other benzophenanthridine alkaloids, oxychelerythrine and oxynitidine, were isolated from *Zanthoxylum capense* (Thunb.) Harv. (Rutaceae) and evaluated as EPIs against *S. aureus*. Enhanced EtBr accumulation was observed at a concentration of 50 mg/l. Oxychelerythrine and oxynitidine belong to a similar class as berberine, a known inhibitor of bacterial efflux pumps (Stavri et al. 2007).

Daidzein of *Glycine max*, a known P-gp inhibitor (Limtrakul et al. 2005) was evaluated for its EPI activity against *Mycobacterium smegmatis* mc^2 155 ATCC700084, *E. coli*, and *P. aeruginosa* (Lechner et al. 2008). Its effect on the accumulation of EtBr in the presence of daidzein 16 mg/l was promising. Carbonyl cyanide m-chlorophenylhydrazone (CCCP) was used as a positive control in this study.

Cucurbitane-type triterpenes of *Momordica balsamina* were tested as EPIs against *Enterococcus faecalis* and MRSA COL_{OXA} cells. Balsaminagenin B at a concentration of 30 μM showed better EtBr accumulation in *E. faecalis* cells (Ramalhete et al. 2011). Similarly, this compound was also active as a P-gp inhibitor (Ramalhete et al. 2009).

Karavilagenin C and balsaminol F were active in MRSA COL_{OXA} (NorA overexpressing) cells at a concentration of 3 μM when evaluated using an EtBr accumulation assay. None of the triterpenes tested were active when tested against Gram-negative bacteria. Furthermore, no cytotoxicity was observed in an experiment performed against human lymphocytes (Ramalhete et al. 2011).

Capsaicin, a main constituent of *Capsicum annuum* and a p-gp inhibitor, was evaluated as an inhibitor of the NorA efflux pump of *S. aureus*. EtBr accumulation and efflux assay also revealed the potential of capsaicin as an EPI when tested against *S. aureus* 1199B (NorA overproducing). An effect of capsaicin on reducing the macrophage invasion of *S. aureus* was also observed (Kalia et al. 2012).

Phenylpropanoids of *Alpinia galanga* (L.) Wild. (Zingiberaceae) significantly inhibited efflux and accumulation of EtBr in *M. smegmatis* mc^2 155. Specifically, 10-S-10–acetoxy eugenol acetate showed a dose-dependent effect when compared with reserpine for efflux and accumulation (Roy et al. 2012).

The compounds paradol and gingerol from the seeds of *Aframomum melegueta* K. Schum. (Zingiberaceae) have been evaluated as EPIs against *M. smegmatis* mc^2 155 using an EtBr efflux assay (Groeblacher et al. 2012).

A flavonoid, sarothrin of *Alkanna orientalis* inhibited NorA pump *S. aureus* at concentration of 100 μM. Results of EtBr-accumulation studies were similar to those obtained with CCCP (Bame et al. 2013).

Synthetic analogs of bonducellin, a homoisoflavonoid obtained from *Caesalpinia digyna* Rottler (*Leguminosae-Caesalpinioideae*), were tested against *M. smegmatis* mc^2 155; bonducellin and its seven derivatives showed better dose-dependent EtBr efflux and accumulation than did verapamil (Roy et al. 2013a).

Coumarins from *Mesua ferrea* L. (Calophyllaceae) were evaluated for EPI activity using EtBr efflux and accumulation assays. A dose-dependent effect was observed when tested against NorA-overproducing *S. aureus* strains. Furthermore, coumarins were not cytotoxic at a concentration of 200 mg/l (Roy et al. 2013b).

Carnosic acid from *Rosmarinus officinalis* and *Salvia officinalis* is already known for its EPI activity against the NorA-overproducing *S. aureus* 1199B strain. This compound was further evaluated against *E. faecalis* ATCC 29212 and found to be active. Carnosic acid acts by altering the membrane potential without affecting membrane permeability (Ojeda-Sana et al. 2013).

Extract from the leaves of *Baccharoides adoensis* showed EPI activity when tested against *P. aeruginosa* and *S. aureus*. An uptake assay using rhodamine 6G dye was performed against both the bacteria to validate the extract as an EPI (Chitemerere and Mukanganyama 2014).

Extract of *Callistemon citrinus* (Myrtaceae) had EPI activity when tested against *P. aeruginosa* and *S. aureus* using a rhodamine 6G uptake assay. Chemical analysis found α-pinene to be a main constituent of the extract. Furthermore, a diSC3-5 experiment confirmed a role in increasing membrane permeability (Chitemerere and Mukanganyama 2014).

Ursolic acid from of *Eucalyptus tereticornis* and its semi-synthetic esterified analogs performed better than reserpine in an EtBr efflux experiment. A concentration range of 25–50 mg/l was used against an MDR strain of *E. coli* (Dwivedi et al. 2014).

Resveratrol of *Vitis vinifera* has already been evaluated for its potential as an EPI of *M. smegmatis* (Lechner et al. 2008). A dose-dependent effect was observed for EtBr accumulation when tested against *Arcobacter butzleri* LMG 10828 and *Arcobacter cryaerophilus* LMG 10829. The compound also impaired the metabolic activity of these bacteria (Ferreira et al. 2014).

Linoleic acid from *Portulaca oleracea* inhibited EtBr efflux in MRSA strains (64 mg/l) with an effect similar to that of reserpine. MRSA strains tested were erythromycin-resistant and overproducing MsrA efflux pump of the ABC family (Chan et al. 2015).

Extract of *Acer saccharum* Marshall (Sapindaceae) showed potential as an EPI against *E. coli*, *Proteus mirabilis*, and *P. aeruginosa* strains. The extract significantly promoted the accumulation of the EtBr bacteria, similarly to CCCP, a known EPI (Maisuria et al. 2015).

The essential oil of *Eucalyptus grandis* was evaluated as an EPI against *Klebsiella pneumoniae* and *Moraxella catarrhalis*. Observed EPI activity was better than that for berberine (Sewanu et al. 2015).

7.3 From Tactics to Strategies

Plant-based natural compounds play a crucial role in drug discovery particularly in the identification of novel EPIs and effective antimicrobial agents. Over the last decade, antimicrobial drug discovery using natural resources has increased and is considered extremely promising. Fewer natural product libraries exist because of the time-consuming isolation, extraction, characterization, and identification of every single compound required. High-throughput screening for EPIs from natural and synthetic libraries lags behind because of the considerable disadvantages, such as the more specific and precise secondary and tertiary evaluation assays required. Improved and highly effective functional EPI assays for screening, followed by comprehensive secondary validation flowcharts, will help considerably in overcoming these limitations (Tegos et al. 2011).

8 Relevance of EPIs in Clinical Settings

The role of EPIs is clearer, and they are considered capable of restoring the activity of antibiotics that have lost their potential against emerging resistant strains. This combination approach is a promising tool against resistant bacteria (Cabral et al. 2015). Another use of these test-active compounds is in the restoration of the potential of anticancer drugs that have lost their activity due to overexpression of P-gp (ABC transporter) in resistant cancer cells. Semi-synthetically modified lamellarins have been found to be active against P-pg pumps of mammalian cells and the NorA pump of *S. aureus* (Bharate et al. 2015). Hypericin, a natural molecule and inhibitor of ABC transporters, was effective when used alongside light therapy against cancer cells (Kourtesi et al. 2013). The safety profiles and toxicities of these molecules limit their use to research only (Fernandez and Hancock 2012). To be used in antimicrobial regimens, EPIs should be specific only to bacterial efflux pumps (Kourtesi et al. 2013). There is only a single report available for the use of EPIs in the antimicrobial combination MP-601,205, available in an aerosol formulation for respiratory infections due to cystic fibrosis and ventilator-associated pneumonia. Another EPI, piperine, known as an inhibitor of NorA (*S. aureus*) and Rv1258c (*Mycobacterium tuberculosis*), has been formulated at a concentration of 10.0 mg with rifampicin 20.0 mg and isoniazid 30.0 mg; the combination is known as risorine. Piperine plays a crucial role in bioequivalence, and the combination showed similar results to those of available rifampicin 450-mg preparations being marketed in India by Cadila Pharmaceuticals Ltd (Sharma et al. 2014).

9 Conclusion

The potential of *S. aureus* to become resistant to every available drug means it is a supremely important pathogen. This microorganism is capable of developing resistance to antibiotics via different mechanisms, such as modifications at the target site and the inactivation of antibiotics. To develop multidrug resistance, a bacterium must acquire multiple mechanisms, as each mechanism provides resistance only to a single class of compound. Efficient drug efflux is a sophisticated mechanisms where, with the help of drug efflux pumps, bacteria effectively efflux chemically diverse substrates, both antibacterial and non-antibacterial. These efflux pumps are not merely involved in antibiotic efflux; recent studies have also confirmed their involvement in the pathogenicity and survival of the organism. Efflux pumps help bacteria survive within phagocytes such as macrophages and even in neutrophils.

Drug efflux helps in greater selection of resistant mutants with mutations at the target site (Kalia et al. 2012). Overexpression of a broad range of multi-component multidrug efflux systems may cause efflux of intracellular concentrations of such components, which plays a significant role in clinical efficacy. It is also crucial that the frequency of efflux-mediated resistance is higher than the resistance mediated by target alterations (Lomovskaya and Bostian 2006). These are the reasons that validate and support the development of EPIs for adjuvant therapies. Outstanding work has been undertaken in the study of regulatory mechanisms and physiological functions of multidrug efflux pumps in recent years. Structure prediction of efflux pumps, substrates, and inhibitors with X-ray crystallography techniques continue to expand our knowledge of this important aspect of multidrug resistance and are helpful for the development of EPIs. The combination of antibiotics and EPIs is a promising tool with which to fight infections by drug-resistant pathogens. Therefore, drug efflux and related mechanisms should be given prime importance in the development of new antibiotics and in future drug discovery programs. Targeting transcription regulators to prevent the overexpression of efflux genes can be considered an alternative strategy. Furthermore, to enable the clinical application of EPIs, future research must focus on decreasing cytotoxicity, increasing solubility, and exploring new EPI candidates.

References

Bame J, Graf T, Junio H, Bussey R, Jarmusch S, El-Elimat T, Falkinham J, Oberlies N, Cech R, Cech N (2013) Sarothrin from *Alkanna orientalis* is an antimicrobial agent and efflux pump inhibitor. Planta Med 79:327–329. doi:10.1055/s-0032-1328259

Bay DC, Rommens KL, Turner RJ (2008) Small multidrug resistance proteins: a multidrug transporter family that continues to grow. Biochim Biophys Acta 1778:1814–1838. doi:10.1016/j.bbamem.2007.08.015

Bharate J, Singh S, Wani A, Sharma S, Joshi P, Khan I, Kumar A, Vishwakarma R, Bharate S (2015) Discovery of 4-acetyl-3-(4-fluorophenyl)-1-(p-tolyl)-5-methylpyrrole as a dual inhibitor of human P-glycoprotein and *Staphylococcus aureus* Nor A efflux pump. Org Biomol Chem 13:5424–5431. doi:10.1039/c5ob00246j

Brooun A, Liu S, Lewis K (2000) A dose-response study of antibiotic resistance in *Pseudomonas aeruginosa* biofilms. Antimicrob Agents Chemother 44:640–646. doi:10.1128/AAC.44.3.640-646.2000

Cabral V, Luo X, Junqueira E, Costa S, Mulhovo S, Duarte A, Couto I, Viveiros M, Ferreira M (2015) Enhancing activity of antibiotics against *Staphylococcus aureus* : *Zanthoxylum capense* constituents and derivatives. Phytomedicine 22:469–476. doi:10.1016/j.phymed.2015.02.003

Camilli A, Bassler BL (2006) Bacterial small-molecule signaling pathways. Science 311:1113–1116. doi:10.1126/science.1121357

Chakrabarty S, Roy M, Hazra B, Bhattacharya R (2002) Induction of apoptosis in human cancer cell lines by diospyrin, a plant-derived bisnaphthoquinonoid, and its synthetic derivatives. Cancer Lett 188:85–93. doi:10.1016/S0304-3835(02)00494-9

Chan B, Han X, Lui S, Wong C, Wang T, Cheung D, Cheng S, Ip M, Han S, Yang X, Jolivalt C, Lau C, Leung P, Fung K (2015) Combating against methicillin-resistant *Staphylococcus aureus*—two fatty acids from purslane (*Portulaca oleracea* L.) exhibit synergistic effects with erythromycin. J Pharm Pharmacol 67:107–116. doi:10.1111/jphp.12315

Chitemerere T, Mukanganyama S (2014) Evaluation of cell membrane integrity as a potential antimicrobial target for plant products. BMC Complement Altern Med 14:278. doi:10.1186/1472-6882-14-278

Costa SS, Viveiros M, Amaral L, Couto I (2013) Multidrug efflux pumps in *Staphylococcus aureus*: an update. Open Microbiol J 7(Suppl 1-M5):59–71. doi:10.2174/1874285801307010059

Davidson AL, Chen J (2004) ATP-binding cassette transporters in bacteria. Annu Rev Biochem 73:41–268. doi:10.1146/annurev.biochem.73.011303.073626

Dawson RJP, Locher KP (2006) Structure of a bacterial multidrug ABC transporter. Nature 443:180–185. doi:10.1038/nature05155

Dwivedi G, Maurya A, Yadav D, Khan F, Darokar M, Srivastava S (2014) Drug resistance reversal potential of ursolic acid derivatives against nalidixic acid- and multidrug-resistant *Escherichia coli*. Chem Biol Drug Des 86(3):272–283. doi:10.1111/cbdd.12491

Fernandez L, Hancock R (2012) Adaptive and mutational resistance: role of porins and efflux pumps in drug resistance. Clin Microbiol Rev 25:661–681. doi:10.1128/CMR.00043-12

Ferreira S, Silva F, Queiroz J, Oleastro M, Domingues F (2014) Resveratrol against *Arcobacter butzleri* and *Arcobacter cryaerophilus*: activity and effect on cellular functions. Int J Food Microbiol 180:62–68. doi:10.1016/j.ijfoodmicro.2014.04.004

Garzoni C, Kelley WL (2009) *Staphylococcus aureus*: new evidence for intracellular persistence. Trends Microbiol 17:59–65. doi:10.1016/j.tim.2008.11.005

Gibbons S, Udo EE (2000) The effect of reserpine, a modulator of multidrug efflux pumps, on the in vitro activity of tetracycline against clinical isolates of methicillin resistant *Staphylococcus aureus* (MRSA) possessing the tet(K) determinant. Phytother Res 14:139–140. doi:10.1002/(SICI)1099-1573(200003)14:2<139::AID-PTR608>3.0.CO;2-8

Gillis RJ, White KG, Choi KH, Wagner VE, Schweizer HP, Iglewski BH (2005) Molecular basis of azithromycin-resistant *Pseudomonas aeruginosa* biofilms. Antimicrob Agents Chemother 49:3858–3867. doi:10.1128/AAC.49.9.3858-3867.2005

Groeblacher B, Maier V, Kunert O, Bucar F (2012) Putative mycobacterial efflux inhibitors from the seeds of *Aframomum melegueta*. J Nat Prod 75:1393–1399. doi:10.1021/np300375t

Handzlik J, Matys A, Kiec-Kononowicz K (2013) Recent advances in multi-drug resistance (MDR) efflux pump inhibitors of gram-positive bacteria *S. aureus*. Antibiotics 2:28–45. doi:10.3390/antibiotics2010028

Kalia NP, Mahajan P, Mehra R, Nargotra A, Sharma JP, Koul S, Khan IA (2012) Capsaicin, a novel inhibitor of NorA efflux pump reduces the intracellular invasion of *Staphylococcus aureus*. J Antimicrob Chemother 67:2401–2408. doi:10.1093/jac/dks232

Khan IA, Mirza ZM, Kumar A, Verma V, Qazi GN (2006) Piperine, a phytochemical potentiator of ciprofloxacin against *Staphylococcus aureus*. Antimicrob Agents Chemother 50:810–812. doi:10.1128/AAC.50.2.810-812.2006

Kourtesi C, Ball A, Huang Y, Jachak S, Vera D, Khondkar P, Gibbons S, Hamblin M, Tegos G (2013) Microbial efflux systems and inhibitors: approaches to drug discovery and the challenge of clinical implementation. Open Microbiol J 7:34–52. doi:10.2174/1874285801307010034

Kriengkauykiat J, Porter E, Lomovskaya O, Wong-Beringer A (2005) Use of an efflux pump inhibitor to determine the prevalence of efflux pump-mediated fluoroquinolone resistance and multidrug resistance in *Pseudomonas aeruginosa*. Antimicrob Agents Chemother 49:565–570. doi:10.1128/AAC.49.2.565-570.2005

Kristiansen J, Amaral L (1997) The potential management of resistant infections with non-antibiotics. J Antimicrob Chemother 40:319–327. doi:10.1093/jac/40.3.319

Krulwich TA, Lewinson O, Padan E, Bibi E (2005) Do physiological roles foster persistence of drug/multidrug-efflux transporters? A case study. Nat Rev Microbiol 3:566–572. doi:10.1038/nrmicro1181

Kumar A, Schweizer HP (2005) Bacterial resistance to antibiotics: active efflux and reduced uptake. Adv Drug Deliv Rev 57:1486–1513. doi:10.1016/j.addr.2005.04.004

Kumar S, Mukherjee MM, Varela M (2013) Modulation of bacterial multidrug resistance efflux pumps of the major facilitator superfamily. Int J Bacteriol 2013:Article ID 204141, 15 pages. doi:10.1155/2013/204141

Kuroda T, Tsuchiya T (2009) Multidrug efflux transporters in the MATE family. Biochim Biophys Acta 1794(5):763–768. doi:10.1016/j.bbapap.2008.11.012

Lechner D, Gibbon S, Bucar F (2008) Plant phenolic compounds as ethidium bromide efflux inhibitors in *Mycobacterium smegmatis*. J Antimicrob Chemother 62:345–348. doi:10.1093/jac/dkn178

Li XZ, Nikaido H (2004) Efflux-mediated drug resistance in bacteria. Drugs 64:159–204. doi:10.2165/00003495-200464020-00004

Li XZ, Poole K, Nikaido H (2003) Contributions of MexAB-OprM and an EmrE homolog to intrinsic resistance of *Pseudomonas aeruginosa* to aminoglycosides and dyes. Antimicrob Agents Chemother 47:27–33. doi:10.1128/AAC.47.1.27-33.2003

Limtrakul P, Khantamat O, Pintha K (2005) Inhibition of P-glycoprotein function and expression by kaempferol and quercetin. J Chemother 17:86–95. doi:10.1179/joc.2005.17.1.86

Lomovskaya O, Bostian KA (2006) Practical applications and feasibility of efflux pump inhibitors in the clinic—a vision for applied use. Biochem Pharmacol 71:910–918. doi:10.1016/j.bcp.2005.12.008

Mah TF, Pitts B, Pellock B, Walker GC, Stewart PS, O'Toole GA (2003) A genetic basis for *Pseudomonas aeruginosa* biofilm antibiotic resistance. Nature 426:306–310. doi:10.1038/nature02122

Maisuria VB, Hosseinidoust Z, Tufenkji N (2015) Polyphenolic extract from maple syrup potentiates antibiotic susceptibility and reduces biofilm formation of pathogenic bacteria. Appl Environ Microbiol 81:3782–3792. doi:10.1128/AEM.00239-15

Mullin S, Mani N, Grossman TH (2004) Inhibition of antibiotic efflux in bacteria by the novel multidrug resistance inhibitors biricodar (VX-710) and timcodar (VX-853). Antimicrob Agents Chemother 48:4171–4176. doi:10.1128/AAC.48.11.4171-4176.2004

Nikaido H, Takatsuka Y (2009) Mechanisms of RND multidrug efflux pumps. Biochim Biophys Acta 1794:769–781. doi:10.1016/j.bbapap.2008.10.004

Nishino K, Nikaido E, Yamaguchi A (2009) Regulation and physiological function of multidrug efflux pumps in *Escherichia coli* and salmonella. Biochem Biophys Acta 1794(5):834–843. doi:10.1016/j.bbapap.2009.02.002

Ojeda-Sana A, Repetto V, Moreno S (2013) Carnosic acid is an efflux pumps modulator by dissipation of the membrane potential in *Enterococcus faecalis* and *Staphylococcus aureus*. World J Microbiol Biotechnol 29:137–144. doi:10.1007/s11274-012-1166-3

Omote H, Hiasa M, Matsumoto T, Otsuka M, Moriyama Y (2006) The MATE proteins as fundamental transporters of metabolic and xenobiotic organic cations. Trends Pharmacol Sci 27:587–593. doi:10.1016/j.tips.2006.09.001

Piddock LJ (2006a) The AcrAB-TolC efflux system of serovar Typhimurium plays a role in pathogenesis. Cell Microbiol 8:847–856. doi:10.1111/j.1462-5822.2005.00671.x

Piddock LJV (2006b) Multidrug-resistance efflux pumps—not just for resistance. Nat Rev Microbiol 4:629–636. doi:10.1038/nrmicro1464

Ramalhete C, Molnar J, Mulhovo S, Rosario V, Ferreira M (2009) New potent P-glycoprotein modulators with the cucurbitane scaffold and their synergistic interaction with doxorubicin on resistant cancer cells. Bioorg Med Chem 17:6942–6951. doi:10.1016/j.bmc.2009.08.020

Ramalhete C, Spengler G, Martins A, Martins M, Viveiros M, Mulhovo S, Ferreira M, Amaral L (2011) Inhibition of efflux pumps in meticillin-resistant *Staphylococcus aureus* and *Enterococcus faecalis* resistant strains by triterpenoids from *Momordica balsamina*. Int J Antimicrob Agents 37:70–74. doi:10.1016/j.ijantimicag.2010.09.011

Renau TE, Leger R, Flamme EM, Sangalang J, She MW, Yen R, Gannon CL, Griffith D, Chamberland S, Lomovskaya O, Hecker SJ, Lee VJ, Ohta T, Nakayama K (1999) Inhibitors of efflux pumps in *Pseudomonas aeruginosa* potentiate the activity of the fluoroquinolone antibacterial levofloxacin. J Med Chem 42:4928–4931. doi:10.1021/jm9904598

Roy SK, Pahwa S, Nandanwar H, Jachak SM (2012) Phenylpropanoids of *Alpinia galanga* as efflux pump inhibitors in *Mycobacterium smegmatis* mc^2 155. Fitoterapia 83:1248–1255. doi:10.1016/j.fitote.2012.06.008

Roy S, Kumari N, Gupta S, Pahwa S, Nandanwar H, Jachak S (2013a) 7-Hydroxy-(E)-3-phenylmethylene-chroman-4-one analogues as efflux pump inhibitors against *Mycobacterium smegmatis* mc(2) 155. Eur J Med Chem 66:499–507. doi:10.1016/j.ejmech.2013.06.024

Roy S, Kumari N, Pahwa S, Agrahari U, Bhutani K, Jachak S, Newman D (2013b) NorA efflux pump inhibitory activity of coumarins from *Mesua ferrea*. Fitoterapia 90:140–150. doi:10.1016/j.fitote.2013.07.015

Sewanu S, Bongekile M, Folusho O, Adejumobi L, Rowland O (2015) Antimicrobial and efflux pumps inhibitory activities of *Eucalyptus grandis* essential oil against respiratory tract infectious bacteria. J Med Plant Res 9:343–348. doi:10.5897/JMPR2015

Sharma S, Kalia NP, Suden P, Chauhan PS, Kumar M, Ram AB, Khajuria A, Bani S, Khan IA (2014) Protective efficacy of piperine against *Mycobacterium tuberculosis*. Tuberculosis 94:389–396. doi:10.1016/j.tube.2014.04.007

Sievert DM, Rudrik JT, Patel JB, McDonald LC, Wilkins MJ, Hageman JC (2008) Vancomycin-resistant *Staphylococcus aureus* in the United States, 2002–2006. Clin Infect Dis 46:668–674. doi:10.1086/527392

Stavri M, Piddock L, Gibbons S (2007) Bacterial efflux pump inhibitors from natural sources. J Antimicrob Chemother 59:1247–1260. doi:10.1093/jac/dkl460

Tegos GP, Haynes M, Jacob Strouse J, Khan MMT, Bologa CG, Oprea TI, Sklar LA (2011) Microbial efflux pump inhibition: tactics and strategies. Curr Pharm Des 17:1291–1302. doi:10.2174/138161211795703726

Ughachukwu PO, Unekwe PC (2012) Efflux pump mediated resistance in chemotherapy. Ann Med Health Sci Res 2:191–198. doi:10.4103/2141-9248.105671

Zloh M, Kaatz GW, Gibbons S (2004) Inhibitors of multidrug resistance (MDR) have affinity for MDR substrates. Bioorg Med Chem Lett 14:881–885. doi:10.1016/j.bmcl.2003.12.015

Tackling the Antibiotic Resistance: The "Gut" Feeling

Richa Misra, Richa Virmani, Darshan Dhakan, and Abhijit Maji

Abstract Antibiotic resistance is a threat to human health worldwide. The ever-increasing multidrug-resistant (MDR) strains of many bacterial pathogens are paralyzing our efforts to treat many deadly infections. An important measure to deal with the menace is to better understand the process and manage the reservoirs or risk areas. Over the recent past, the importance of gut bacteria in various aspects of human health and physiology has been highlighted. Also, studies are now being carried out to better understand its role in antimicrobial resistance and grave consequences of antibiotic exposure on gut microbiota. This chapter highlights the importance of gut microbiota in better understanding of antibiotic resistance and summarizes the burden imposed by antibiotic use in the healthcare sector. Due to close contact of pathogens with dense human microbiota during the disease progression, gene transfer events might occur frequently. In this context, our microbiome warrants special attention since it can possibly act as one of the most accessible reservoir of antibiotic resistance genes. It seems pertinent to evaluate antimicrobial therapies in the context of this microbial framework, as many life-threatening infections can arise due to antibiotic-associated alterations in the gut microbiota.

1 Introduction

In 1928, when Alexander Fleming discovered penicillin, the first natural chemical compound with antibiotic properties (Fleming 1980), the world rejoiced and called it a "wonder drug." Since then, the discovery of antibiotics has probably been one of

R. Misra (✉)
Department of Zoology, Miranda House, University of Delhi, Delhi, India
e-mail: richa.1403@gmail.com

R. Virmani • A. Maji
CSIR-Institute of Genomics and Integrative Biology (IGIB), Mall Road, Delhi, India

D. Dhakan
Indian Institute of Science Education and Research, Bhopal, India

G. Arora et al. (eds.), *Drug Resistance in Bacteria, Fungi, Malaria, and Cancer*,
DOI 10.1007/978-3-319-48683-3_14

the most successful forms of antimicrobial treatment in the history of medicine. Little did we know that unaccounted use and misuse of many of these wonder drugs would lead to a grave situation of "antimicrobial resistance," where the deadly pathogens would smartly change in ways that make the drugs ineffective! Resistant microorganisms stop responding to the standard drugs, which were originally effective for treatment. This increases the burden of the disease, as it not only lengthens the regular time course of treatment, which in turn increases the risk of spread to others. Moreover, in case of serious pathological conditions, this long treatment exposure also increases the chances of death due to susceptibility to other opportunistic infections.

1.1 Antibiotics

Selman Waksman, the famous soil microbiologist who discovered streptomycin, proposed the term "antibiotic." Although he and his colleagues referred this term to the activity of the compound produced by microbes against other microbes, the term has since then popularly been used for any organic molecule with antibacterial properties (Waksman and Flynn 1973; Davies and Davies 2010). Microbes producing antibiotics are studied largely for exploiting their therapeutic potential; however, these compounds are known to play multifaceted role in microbial community including acting as signaling molecules (Kalia et al. 2007; Kalia and Purohit 2011; Kumar et al. 2013; Sengupta et al. 2013; Kalia 2014; Sajid et al. 2015; Koul et al. 2016). We now understand that antibiotics and antibiotic resistance genes coevolved in the organism to confer a trait, advantageous to the organism in face of selection pressure (Kalia 2013; Arora et al. 2014; Bhaduri et al. 2014; Kalia et al. 2014). So, antibiotic resistance is a naturally evolved phenomenon, without any role of human intervention. However, we also know that the recent upsurge in antibiotic resistance is due to the immense selection pressure created at various levels by anthropogenic activities. The unaccounted use of antibiotics that has resulted in the emergence of drug-resistant strains of *Mycobacterium tuberculosis* is a grim reminder of powerful adaptation of nature (Velayati et al. 2009).

Administration of most antibiotics is associated with adverse effects that range from nausea, fever, allergic reactions, and diarrhea. While antibiotics inhibit pathogenic bacteria, they also exert detrimental effects on the commensal bacterial community that contribute to human health. One such widely studied condition is the antibiotic-associated diarrhea (AAD). Antibiotics affect the human microflora by disrupting the species composition in the gastrointestinal (GI) tract or gut, which leads to overgrowth of pathogenic bacteria, such as *Clostridium difficile*, which is responsible for conditions like diarrhea to pseudomembranous colitis in patients (Cotter et al. 2012). As compared to its presence in healthy individuals, *C. difficile* population is reported to increase in number due to antibiotic-induced disturbances. Most antibiotics have the potential of creating a bacterial imbalance in intestines, and exposure to antibiotics early in life has been implicated to lead to long-term health effects such as development of allergic sensitization and pathogen-induced colitis (Willing et al. 2011). This is most pronounced in case of antibiotics with broad target range. The molecular targets aimed by these broad-spectrum antibiotics (e.g., cell wall components, RNA polymerase, DNA gyrase, etc.) are often highly conserved across many bacterial species, genetically as well as structurally. So, the use of an antibiotic against a pathogenic bacterium most likely also targets the bacterial community sharing the niche with the pathogen. The brazen use of broad-spectrum antibiotics has been pushed in part by the pharmaceutical industry, which focuses largely on the development of broad-spectrum antibiotics, which can be utilized to treat a variety of different infections. Moreover, most of these are available as "over-the-counter" medicines in some countries and are often administered to treat infections of unknown/little known etiology. However, it is only in recent years that the consequence of widespread use of antibiotics has caught our attention. With the help of high-throughput DNA sequencing and related technologies, we are just beginning to comprehend the effect of antibiotic use on the gut microbiota of humans. Over the years, the consumption of antibiotics has led to collateral damage faced by indigenous host-associated microbes, which results in physiological turmoil.

1.2 Microbiome

The collection of microorganisms present in our body is commonly referred as human microbiota/microbiome. It is estimated that microbes constitute 100 trillion cells in our body, almost tenfold the number of human cells (Eckburg et al. 2005; Sears 2005). These microbes colonize mostly all body surfaces like the gut, oral cavity, auditory canal, nares, and skin surfaces (Costello et al. 2009). Of all the microbial population associated with our body, 99 % of the population is bacterial and the majority resides in the gut (Gill et al. 2006; Qin et al. 2010). Of the known phylogenetic categories, bacterial phylotypes, *Bacteroidetes*, and the *Firmicutes* constitute over 90 % of the distal gut microbiota. Populations of *Proteobacteria*, *Actinobacteria*, *Fusobacteria*, and *Verrucobacteria* are present in minor proportions, and change in their relative proportions is indicative of some pathological conditions (Eckburg et al. 2005; Sears 2005; Kalia 2014). It is now well appreciated that these microbes play an important role in the human physiology and health. Some of the crucial processes governed by gut microbes include nutrition, modulation of immune system, and pathogen invasion (Round and Mazmanian 2009; Qin et al. 2010). Since we have been unable to culture most of these bacteria, our understanding of this microbial pool began with the advent of high-throughput sequencing technologies. The microbiome studies have shown substantial diversity of gut microbes between healthy individuals, with lifestyle and diet playing a crucial role in establishing the diversity (Dicksved et al. 2007; Flint et al. 2007; Jernberg et al. 2010). It is now known that the GI tract of a fetus is sterile, and microbiota is acquired primarily during passage through the birth canal in an infant, with the mother's vaginal microbiota being the most common influence. In this context, the mode of infant delivery also governs the microbe diversity and, in turn, susceptibility to various infections (Mändar and Mikelsaar 1996; Huurre et al. 2008). Apart from this, the illness and use of antibiotics are known to cause a drastic change in microbiota or dysbiosis (Dethlefsen and Relman 2011; Faith et al. 2013).

2 Burden of Antibiotic Resistance

Since the first report of antibiotic usage and resistance, the burden of resistance among bacteria has progressively increased and has accelerated within the last decade. In an alarming trend, a survey carried out to account for global consumption of antibiotics from year 2000 to 2010 revealed an increase of 36 % in antibiotic consumption in 71 countries over this period, with India, Brazil, China, Russia, and South Africa accounting for 76 % of this increase. India also emerged as the largest consumer of antibiotics (Van Boeckel et al. 2014). Antibiotic resistance is now established to be a cause of grave concern to human health and welfare. The problem is even intense in developing countries such as India where comprehensive surveillance system is lacking in healthcare sector to monitor the burden of antimicrobial resistance. Due to lack of surveillance data, often the scenario is underrepresented. Depending upon the degree of resistance, most people with antibiotic resistance remain infected for longer. This not only increases the cost of healthcare in medical centers but also increases the chance of death. For example, according to the World Health Organization (WHO), methicillin-resistant *Staphylococcus aureus*-infected patients are 64 % more likely to die as compared to patients infected with a nonresistant form of the bacteria. Similarly, people infected with multidrug-resistant (MDR) and extensively drug-resistant (XDR) strains of *M. tuberculosis* require much longer courses for treatment of tuberculosis (http://apps.who.int/iris/bitstream/10665/112642/1/9789241564748_eng.pdf). According to the US Center for Disease Control and Prevention (CDC), treatment of *C. difficile* infections requires at least $1 billion in excess medical costs per year (CDC, Antibiotic Resistance Threats in the United States, 2013).

Bacterial infections can be broadly classified into healthcare-associated infections (HAIs) or nosocomial infections and community-acquired infections (CAI). CAIs are infections developed outside of a healthcare setting or an infection present on admission, while HAIs are infections that patients contract during the course of receiving treatment for other conditions within a healthcare setting. A large proportion of infections reported in hospitals reveal that HAIs are on a rise, and, usually, strains causing HAIs tend to be naturally more resistant to antibiotics. In a review by Wattal et al., the authors have provided a beautiful summary of the burden of resistance, associated with HAI and CAI, in India (Wattal and Goel 2014). According to the CDC report, at least two million people in the United States acquire bacterial infections that are resistant to one or more of the antibiotics designated to cure those infections, each year (CDC, Antibiotic Resistance Threats in the United States, 2013). The WHO also in its report on resistance surveillance highlighted the serious threat to public health globally due to antibiotic resistance (World Health Organization, "Antimicrobial resistance: Global report on surveillance, 2014").

Some of the key findings from these reports and others are summarized below:

- Resistance to carbapenem antibiotics in *Klebsiella pneumonia*, a common cause of HAI, has spread all over the world.
- One type of HAI—*C. difficile*—is directly associated with administration of antibiotics and may prove to be life-threatening, especially in older adults under medical care.
- Resistance to first-line drugs to treat infections caused by *S. aureus*—a common cause of CAI and HAI—is also widespread.
- MDR and XDR *M. tuberculosis* infections are an increasing threat, globally.
- *Acinetobacter* strains, a cause of pneumonia or bloodstream infections among critically ill patients, are resistant to nearly all antibiotics including carbapenems.
- *Enterococcus* strains have become resistant to vancomycin, making treatment difficult.
- *Salmonella typhi* and non-typhoidal *Salmonella* infections are also reported to be resistant to drugs such as ceftriaxone and ciprofloxacin.
- In 10 countries, millions of people die of gonorrhea, a sexually transmitted disease, because *Neisseria gonorrhoeae* has become resistant to even the last recourse of treatment—the third-generation cephalosporins. An impending danger looms wherein gonorrhea may soon become untreatable, as no alternative treatment measure is currently under development.
- The treatment of urinary tract infections, mainly due to strains of *Escherichia coli*, involves administration of one of the most widely used antibiotic—the fluoroquinolones—resistance to which is very widespread.
- Resistance in *Vibrio cholerae* has also been reported against commonly used antibiotics like furazolidone, co-trimoxazole, and fluoroquinolones (Sharma et al. 2007).
- High prevalence of resistance to first-line drugs is reported in *Shigella* spp., and the recent emergence of ceftriaxone resistance in *Shigella* has led to the use of carbapenems for the treatment of a simple community-acquired diarrheal disease (Taneja et al. 2012).
- Many other infectious agents are increasingly bearing multidrug resistance traits such as *Campylobacter* spp. and *Aeromonas* spp.

3 Source of Antibiotic Resistance in Pathogens

Antibiotic resistance genes (ARGs) have been present in bacterial populations from ancient times (D'Costa et al. 2011), and cycling of ARGs in environmental settings is well documented (Vaz-Moreira et al. 2014; Berendonk et al. 2015). Resistance is known to arise in bacteria either by spontaneous mutations in their genomes (Bagel et al. 1999) or by means of horizontal gene transfer (Frost et al. 2005; Modi et al. 2014) (Fig. 1). Antibiotic resistance genes are often encoded on mobilizable genetic elements, called integrons, in environmental bacteria and commensals and can move between diverse bacteria to disseminate resistance genes, when microbial communities communicate with each other under high selection pressure (Alekshun and Levy 2006). These conditions include communication under highly dense bacterial populations subjected to sub-therapeutic antibiotic concentrations. These interactions can be seen not only in environmental conditions that are subjected to anthropogenic pressure, such as municipal wastewater systems, pharmaceutical manufacturing units, and animal husbandry facilities, but also in hospital settings (Berendonk et al. 2015). The occurrence of New Delhi metallo-beta-lactamase-1 (NDM-1) containing bacteria in water samples in New Delhi is one such stark example of the gravity of the situation (Walsh et al. 2011). The use of sub-therapeutic doses of antibiotics or related compounds in the agricultural industry and animal husbandry to promote animal growth creates a perpetual selective

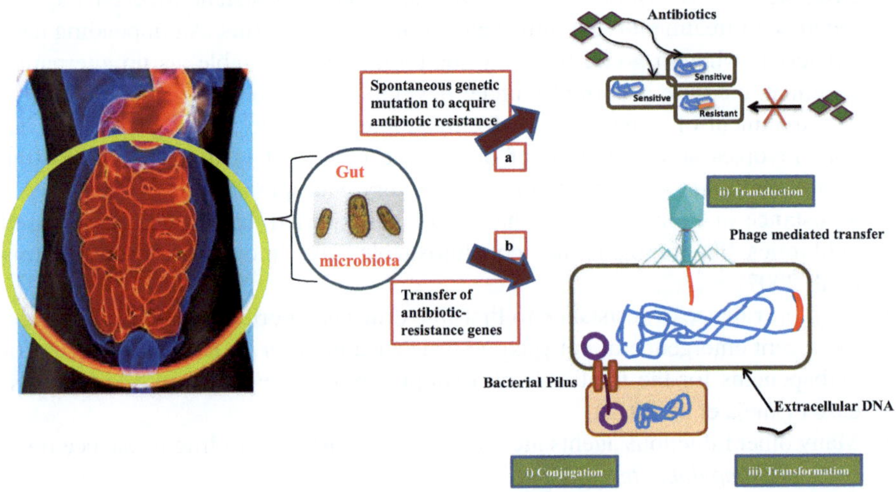

Fig. 1 Acquisition of antibiotic resistance in bacteria. Bacteria are known to acquire resistance to antibiotics by two mechanisms: (a) *spontaneous genetic mutations* which enable bacterium to resist antibiotics either by inactivating them directly or indirectly by modifying the targets/route of entry and (b) acquired resistance genes from another source by means of (i) *conjugation*, which involves direct cell–cell contact with another bacteria; (ii) *transduction*, which is virus-mediated transfer of DNA; and (iii) *transformation*, which is the ability to acquire naked DNA from the environment

pressure which in turn facilitates resistance to develop and may possibly contribute to a larger global resistance reservoir (Heuer and Smalla 2007; Berendonk et al. 2015). In fact, a study reveals that usage of antibiotics in beekeeping in the United States has led to accumulation of extensive tetracycline resistance genes in the microbiota of honeybees (Tian et al. 2012). There is evidence of horizontal transfer of genes encoding carbohydrate-active enzymes between marine bacteria and gut microbiome of Japanese individuals with seaweed-rich diet (Hehemann et al. 2010), which suggests that one such niche can also be the human gut microbiome of healthy individuals, which acts as a reservoir of ARGs, although the direct experimental evidence of in vivo transfer of antibiotic resistance genes within the human microbiome is very limiting (Doucet-Populaire et al. 1991; Ley et al. 2006; Smillie et al. 2011). There is enough alarming evidence to endorse the view that the transfer of ARGs in humans is possible to overcome persistent antibiotic selective pressure, and our urgent attention is warranted to tackle the situation.

4 Effect of Antibiotics on Gut Microbiota

Antibiotics affect the microbial composition in two different ways. Firstly, it decreases the competition for resources for microbes in turn opening up many ecological niches for opportunistic pathogens. In addition, lysis of susceptible bacteria releases carbon sources which can then be utilized by the remaining members of the microflora (Willing et al. 2011). For example, upon disruption of resident microbiota due to antibiotics, two distantly related antibiotic-associated pathogens, *S. enterica* serovar *Typhimurium* and *C. difficile*, utilize the microbiota-liberated sialic acids for their proliferation (Ng et al. 2013). With the help of culture-based approach and next-generation sequencing technologies, many studies have now investigated the effect of administration of antibiotics on persistence of resistance in gut microbiota. These studies have highlighted the potential of gut microbiota as a reservoir accessible to pathogens under extreme antibiotic selection pressure. Some of the key examples from various studies are:

- *Helicobacter pylori*, a gram-negative bacterium colonizing the gastric mucosa, is the main causative agent of peptic ulcer and gastric cancer. Its treatment requires a triple therapy combination with clarithromycin, metronidazole, and omeprazole. It was revealed in a study that apart from the short- and long-term disruption of indigenous microbiota in the throat and in the lower intestine region, there was enrichment of macrolide resistance gene, even after 4 years of treatment (Jakobsson et al. 2010). In an independent study, this treatment regimen was compared with another combination containing amoxicillin and was suggested better for treatment of *H. pylori* since it resulted in emergence of lesser resistant strains, so it can be considered better from an ecological perspective (Adamsson et al. 1999).

- An overgrowth of resistant enterobacterial species was observed on administration of amoxicillin with or without clavulanate (Sullivan et al. 2001).
- Administration of cephalosporin leads to decrease in the abundance of enterobacteria and increased the levels of enterococci, which are known to be intrinsically resistant to this antimicrobial agent (Rafii et al. 2008).
- Another report showed that even short-term exposure to clindamycin treatment led to long-term impacts on the intestinal microbiota and results in dramatic increase in levels of specific resistance genes (Jernberg et al. 2007).
- Karami et al. have demonstrated β-lactamase gene transfer between two *E. coli* strains co-residing in the human gut of a child who was administered ampicillin (Karami et al. 2007).
- Sommer et al. characterized the gut microbes of unrelated healthy individuals who for at least 1 year were not exposed to antibiotics. Surprisingly, with the help of metagenomic as well as culturing approaches, they could decipher many resistance genes that were identical to genes harbored by pathogens, while many were evolutionarily distant from known resistance reservoir. This alarming fact underscored the potential of commensal microbiota of healthy individuals acting as reservoirs of resistance genes, with an imminent contribution to antibiotic resistance (Sommer et al. 2009).

The disturbance in microbiota has also been associated to many disease phenotypes. In many instances, downstream regulation of innate defenses has been associated with antibiotic-induced disruption of microbiota, leading to colonization of other microbes (Brandl et al. 2008; Willing et al. 2011). Antibiotic-associated diarrhea (AAD), a condition characterized by administration of antibiotics and not any other obvious causes, is one of the most common complications arising in hospitals. Although the frequency of AAD can vary between different antibiotics, it is believed that AAD can affect up to 25 % of the patients receiving a particular antibiotic (Young and Schmidt 2004). It is also reported that cephalosporin use during early childhood leads to increased susceptibility to asthma (Kozyrskyj et al. 2007). Kanamycin administration during infancy led to modulation of gut microbiota and was associated with the development of atopic dermatitis-like skin lesions in a mice model (Watanabe et al. 2010). Furthermore, importance of gut microbiota is also emphasized by the fact that fecal microbial transplantation (FMT) has proved to be successful in treatment of recurrent *C. difficile* infection (Seekatz et al. 2014). However, the mechanism and long-term effects of this treatment still remain to be elucidated. In addition, effectiveness of FMT in other diseases still needs to be ascertained.

5 Road to Future

Many steps need to be taken worldwide to tackle the emergence and spread of antibiotic resistance. Firstly, we need to have better understanding about the health and physiology of healthy individuals and the tussle between host and pathogen by modulating various physiological processes (Sachdeva et al. 2010; Maji et al. 2015). Furthermore, detailed understanding about the relationship between the various human-associated and environmental reservoirs that harbor distinct resistance genes need to be attained. With advent of metagenomic approach and advancement in whole-genome sequencing methods, our knowledge about the human microbiome functional diversity and impact of antibiotic treatment on microbial community is improving (Fig. 2). This information needs to be expanded further and applied for effective monitoring of antibiotic load and better formulation of guidelines for therapies. In view of the challenges faced by the community, the WHO focus for the World Health Day in 2011 was antimicrobial resistance, and it also recently concluded the first "Antibiotic Awareness Week" in November 2015. The campaign was aimed to increase awareness in people about the rising antibiotic resistance globally and to encourage good practices among the health practitioners as well as general public to avoid further emergence and spread of drug resistance. However, poor implementation poses a significant challenge toward reaping the benefits of these programs. In order to turn the tide on

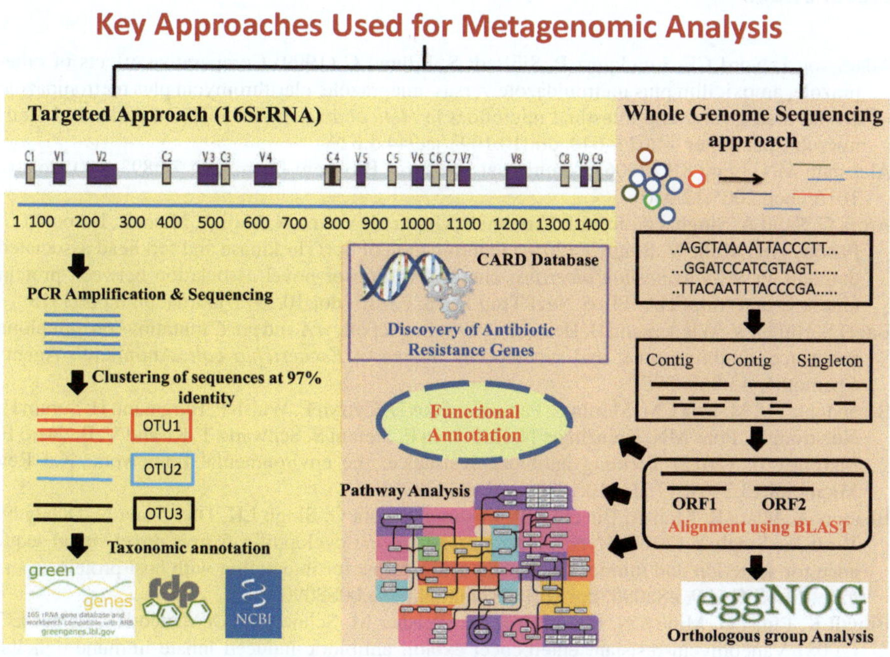

Fig. 2 Schematic representation of metagenomics study

antimicrobial resistance, we need to have a better infrastructural framework, which ensures strict implementation of good practices. Unrestricted use of antibiotics in all sectors, such as healthcare, agricultural, etc., should be curtailed. Examination of antibiotic resistance in environmental samples like sludge, soil, and manure samples should be carried out. Greater importance should be given toward strict implementation of these guidelines for antibiotic therapies. We need to formulate and follow laws against antibiotic dispensing without prescription. The general public should be educated and made aware of the flip side of using antibiotics for any medical problem. Moreover, mandatory refresher courses should be framed for medical practitioners, both in rural and urban settings, to update them about the global medical trends. Improvement in diagnostic facilities, treatment facilities, and overall infrastructural advancement can go a long way in curtailing this deadly menace.

Acknowledgments We thank Prof. Yogendra Singh, the Council of Scientific and Industrial Research-Institute of Genomics & Integrative Biology (CSIR-IGIB), and the University of Delhi for providing us the encouragement for the work and Dr. Vineet K. Sharma and IISER, Bhopal, for insights into the field of metagenomics. RM wishes to thank the principal of Miranda House for the facilities. Financial assistance by the Council of Scientific and Industrial Research (CSIR)-funded project BSC0123 is highly acknowledged.

References

Adamsson I, Nord CE, Lundquist P, Sjöstedt S, Edlund C (1999) Comparative effects of omeprazole, amoxicillin plus metronidazole versus omeprazole, clarithromycin plus metronidazole on the oral, gastric and intestinal microflora in *Helicobacter pylori*-infected patients. J Antimicrob Chemother 44:629–640. doi:10.1093/jac/44.5.629

Alekshun MN, Levy SB (2006) Commensals upon us. Biochem Pharmacol 71:893–900. doi:10.1016/j.bcp.2005.12.040

Arora G, Sajid A, Singhal A, Joshi J, Virmani R, Gupta M, Verma N, Maji A, Misra R, Baronian G, Pandey AK, Molle V, Singh Y (2014) Identification of Ser/Thr kinase and forkhead associated domains in *Mycobacterium ulcerans*: characterization of novel association between protein kinase Q and MupFHA. PLoS Negl Trop Dis 8, e3315. doi:10.1371/journal.pntd.0003315

Bagel S, Hüllen V, Wiedemann B, Heisig P (1999) Impact of *gyrA* and *parC* mutations on quinolone resistance, doubling time, and supercoiling degree of *Escherichia coli*. Antimicrob Agents Chemother 43:868–875

Berendonk TU, Manaia CM, Merlin C, Fatta-Kassinos D, Cytryn E, Walsh F, Bürgmann H, Sørum H, Norström M, Pons MN, Kreuzinger N, Huovinen P, Stefani S, Schwartz T, Kisand V, Baquero F, Martinez JL (2015) Tackling antibiotic resistance: the environmental framework. Nat Rev Microbiol 13:310–317. doi:10.1038/nrmicro3439

Bhaduri A, Misra R, Maji A, Bhetaria PJ, Mishra S, Arora G, Singh LK, Dhasmana N, Dubey N, Virdi JS, Singh Y (2014) *Mycobacterium tuberculosis* cyclophilin A uses novel signal sequence for secretion and mimics eukaryotic cyclophilins for interaction with host protein repertoire. PLoS One 9, e88090. doi:10.1371/journal.pone.0088090

Brandl K, Plitas G, Mihu CN, Ubeda C, Jia T, Fleisher M, Schnabl B, DeMatteo RP, Pamer EG (2008) Vancomycin-resistant enterococci exploit antibiotic-induced innate immune deficits. Nature 455:804–807. doi:10.1038/nature07250

Centers for Disease Control and Prevention (CDC) (2013) Antibiotic resistance threats in the United States, 2013. http://www.cdc.gov/drugresistance/pdf/ar-threats-2013-508.pdf

Costello EK, Lauber CL, Hamady M, Fierer N, Gordon JI, Knight R (2009) Bacterial community variation in human body habitats across space and time. Science 326:1694–1697. doi:10.1126/science.1177486

Cotter PD, Stanton C, Ross RP, Hill C (2012) The impact of antibiotics on the gut microbiota as revealed by high throughput DNA sequencing. Discov Med 13:193–199

Davies J, Davies D (2010) Origins and evolution of antibiotic resistance. Microbiol Mol Biol Rev 74:417–433. doi:10.1128/MMBR.00016-10

D'Costa VM, King CE, Kalan L, Morar M, Sung WW, Schwarz C, Froese D, Zazula G, Calmels F, Debruyne R, Golding GB, Poinar HN, Wright GD (2011) Antibiotic resistance is ancient. Nature 477:457–461. doi:10.1038/nature10388

Dethlefsen L, Relman DA (2011) Incomplete recovery and individualized responses of the human distal gut microbiota to repeated antibiotic perturbation. Proc Natl Acad Sci U S A 108(Suppl 1):4554–4561. doi:10.1073/pnas.1000087107

Dicksved J, Flöistrup H, Bergström A, Rosenquist M, Pershagen G, Scheynius A, Roos S, Alm JS, Engstrand L, Braun-Fahrländer C, von Mutius E, Jansson JK (2007) Molecular fingerprinting of the fecal microbiota of children raised according to different lifestyles. Appl Environ Microbiol 73:2284–2289. doi:10.1128/AEM.02223-06

Doucet-Populaire F, Trieu-Cuot P, Dosbaa I, Andremont A, Courvalin P (1991) Inducible transfer of conjugative transposon Tn1545 from *Enterococcus faecalis* to *Listeria monocytogenes* in the digestive tracts of gnotobiotic mice. Antimicrob Agents Chemother 35:185–187

Eckburg PB, Bik EM, Bernstein CN, Purdom E, Dethlefsen L, Sargent M, Gill SR, Nelson KE, Relman DA (2005) Diversity of the human intestinal microbial flora. Science 308:1635–1638. doi:10.1126/science.1110591

Faith JJ, Guruge JL, Charbonneau M, Subramanian S, Seedorf H, Goodman AL, Clemente JC, Knight R, Heath AC, Leibel RL, Rosenbaum M, Gordon JI (2013) The long-term stability of the human gut microbiota. Science 341:1237439. doi:10.1126/science.1237439

Fleming A (1980) Classics in infectious diseases: on the antibacterial action of cultures of a penicillium, with special reference to their use in the isolation of *B. influenzae* by Alexander Fleming, Reprinted from the British Journal of Experimental Pathology 10:226–236, 1929. Rev Infect Dis 2:129–139

Flint HJ, Duncan SH, Scott KP, Louis P (2007) Interactions and competition within the microbial community of the human colon: links between diet and health. Environ Microbiol 9: 1101–1111. doi:10.1111/j.1462-2920.2007.01281.x

Frost LS, Leplae R, Summers AO, Toussaint A (2005) Mobile genetic elements: the agents of open source evolution. Nat Rev Microbiol 3:722–732. doi:10.1038/nrmicro1235

Gill SR, Pop M, Deboy RT, Eckburg PB, Turnbaugh PJ, Samuel BS, Gordon JI, Relman DA, Fraser-Liggett CM, Nelson KE (2006) Metagenomic analysis of the human distal gut microbiome. Science 312:1355–1359. doi:10.1126/science.1124234

Hehemann JH, Correc G, Barbeyron T, Helbert W, Czjzek M, Michel G (2010) Transfer of carbohydrate-active enzymes from marine bacteria to Japanese gut microbiota. Nature 464: 908–912. doi:10.1038/nature08937

Heuer H, Smalla K (2007) Manure and sulfadiazine synergistically increased bacterial antibiotic resistance in soil over at least two months. Environ Microbiol 9:657–666. doi:10.1111/j.1462-2920.2006.01185.x

Huurre A, Kalliomäki M, Rautava S, Rinne M, Salminen S, Isolauri E (2008) Mode of delivery – effects on gut microbiota and humoral immunity. Neonatology 93:236–240. doi:10.1159/111102

Jakobsson HE, Jernberg C, Andersson AF, Sjölund-Karlsson M, Jansson JK, Engstrand L (2010) Short-term antibiotic treatment has differing long-term impacts on the human throat and gut microbiome. PLoS One 5, e9836. doi:10.1371/journal.pone.0009836

Jernberg C, Löfmark S, Edlund C, Jansson JK (2007) Long-term ecological impacts of antibiotic administration on the human intestinal microbiota. ISME J 1:56–66. doi:10.1038/ismej.2007.3

Jernberg C, Löfmark S, Edlund C, Jansson JK (2010) Long-term impacts of antibiotic exposure on the human intestinal microbiota. Microbiology 156:3216–3223. doi:10.1099/mic.0.040618-0

Kalia VC (2013) Quorum sensing inhibitors: an overview. Biotechnol Adv 31:224–245. doi:10.1016/j.biotechadv.2012.10.004

Kalia VC (2014) Microbes, antimicrobials and resistance: the battle goes on. Indian J Microbiol 54:1–2. doi:10.1007/s12088-013-0443-7

Kalia VC, Purohit HJ (2011) Quenching the quorum sensing system: potential antibacterial drug targets. Crit Rev Microbiol 37:121–140. doi:10.3109/1040841X.2010.532479

Kalia VC, Rani A, Lal S, Cheema S, Raut CP (2007) Combing databases reveals potential antibiotic producers. Expert Opin Drug Discov 2:211–224. doi:10.1517/17460441.2.2.211

Kalia VC, Wood TK, Kumar P (2014) Evolution of resistance to quorum-sensing inhibitors. Microb Ecol 68:13–23. doi:10.1007/s00248-013-0316-y

Karami N, Martner A, Enne VI, Swerkersson S, Adlerberth I, Wold AE (2007) Transfer of an ampicillin resistance gene between two *Escherichia coli* strains in the bowel microbiota of an infant treated with antibiotics. J Antimicrob Chemother 60:1142–1145. doi:10.1093/jac/dkm327

Koul S, Prakash J, Mishra A, Kalia VC (2016) Potential emergence of multi-quorum sensing inhibitor resistant (MQSIR) bacteria. Indian J Microbiol 56:1–18. doi:10.1007/s12088-015-0558-0

Kozyrskyj AL, Ernst P, Becker AB (2007) Increased risk of childhood asthma from antibiotic use in early life. Chest 131:1753–1759. doi:10.1378/chest.06-3008

Kumar P, Patel SK, Lee JK, Kalia VC (2013) Extending the limits of *Bacillus* for novel biotechnological applications. Biotechnol Adv 31:1543–1561. doi:10.1016/j.biotechadv.2013.08.007

Ley RE, Peterson DA, Gordon JI (2006) Ecological and evolutionary forces shaping microbial diversity in the human intestine. Cell 124:837–848. doi:10.1016/j.cell.2006.02.017

Maji A, Misra R, Kumar AM, Kumar D, Bajaj D, Singhal A, Arora G, Bhaduri A, Sajid A, Bhatia S, Singh S, Singh H, Rao V, Dash D, Baby SE, Michael JS, Chaudhary A, Gokhale RS, Singh Y (2015) Expression profiling of lymph nodes in tuberculosis patients reveal inflammatory milieu at site of infection. Sci Rep 5:15214. doi:10.1038/srep15214

Mändar R, Mikelsaar M (1996) Transmission of mother's microflora to the newborn at birth. Biol Neonate 69:30–35. doi:10.1159/000244275

Modi SR, Collins JJ, Relman DA (2014) Antibiotics and the gut microbiota. J Clin Invest 124: 4212–4218. doi:10.1172/JCI72333

Ng KM, Ferreyra JA, Higginbottom SK, Lynch JB, Kashyap PC, Gopinath S, Naidu N, Choudhury B, Weimer BC, Monack DM, Sonnenburg JL (2013) Microbiota-liberated host sugars facilitate post-antibiotic expansion of enteric pathogens. Nature 502:96–99. doi:10.1038/nature12503

Qin J, Li R, Raes J, Arumugam M, Burgdorf KS, Manichanh C, Nielsen T, Pons N, Levenez F, Yamada T, Mende DR, Li J, Xu J, Li S, Li D, Cao J, Wang B, Liang H, Zheng H, Xie Y, Tap J, Lepage P, Bertalan M, Batto JM, Hansen T, Le Paslier D, Linneberg A, Nielsen HB, Pelletier E, Renault P, Sicheritz-Ponten T, Turner K, Zhu H, Yu C, Li S, Jian M, Zhou Y, Li Y, Zhang X, Li S, Qin N, Yang H, Wang J, Brunak S, Doré J, Guarner F, Kristiansen K, Pedersen O, Parkhill J, Weissenbach J, Meta HITC, Bork P, Ehrlich SD, Wang J (2010) A human gut microbial gene catalogue established by metagenomic sequencing. Nature 464: 59–65. doi:10.1038/nature08821

Rafii F, Sutherland JB, Cerniglia CE (2008) Effects of treatment with antimicrobial agents on the human colonic microflora. Ther Clin Risk Manag 4:1343–1358. doi:10.2147/TCRM.S4328

Round JL, Mazmanian SK (2009) The gut microbiota shapes intestinal immune responses during health and disease. Nat Rev Immunol 9:313–323. doi:10.1038/nri2515

Sachdeva P, Misra R, Tyagi AK, Singh Y (2010) The sigma factors of *Mycobacterium tuberculosis*: regulation of the regulators. FEBS J 277:605–626. doi:10.1111/j.1742-4658.2009.07479.x

Sajid A, Arora G, Singhal A, Kalia VC, Singh Y (2015) Protein phosphatases of pathogenic bacteria: role in physiology and virulence. Annu Rev Microbiol 69:527–547. doi:10.1146/annurev-micro-020415-111342

Sears CL (2005) A dynamic partnership: celebrating our gut flora. Anaerobe 11:247–251. doi:10.1016/j.anaerobe.2005.05.001

Seekatz AM, Aas J, Gessert CE, Rubin TA, Saman DM, Bakken JS, Young VB (2014) Recovery of the gut microbiome following fecal microbiota transplantation. mBio 5:e00893-14. doi:10.1128/mBio.00893-14

Sengupta S, Chattopadhyay MK, Grossart HP (2013) The multifaceted roles of antibiotics and antibiotic resistance in nature. Front Microbiol 4:47. doi:10.3389/fmicb.2013.00047

Sharma NC, Mandal PK, Dhillon R, Jain M (2007) Changing profile of *Vibrio cholerae* O1, O139 in Delhi & its periphery (2003–2005). Indian J Med Res 125:633–640

Smillie CS, Smith MB, Friedman J, Cordero OX, David LA, Alm EJ (2011) Ecology drives a global network of gene exchange connecting the human microbiome. Nature 480:241–244. doi:10.1038/nature10571

Sommer MO, Dantas G, Church GM (2009) Functional characterization of the antibiotic resistance reservoir in the human microflora. Science 325:1128–1131. doi:10.1126/science.1176950

Sullivan A, Edlund C, Nord CE (2001) Effect of antimicrobial agents on the ecological balance of human microflora. Lancet Infect Dis 1:101–114. doi:10.1016/S1473-3099(01)00066-4

Taneja N, Mewara A, Kumar A, Verma G, Sharma M (2012) Cephalosporin-resistant *Shigella flexneri* over 9 years (2001–09) in India. J Antimicrob Chemother 67:1347–1353. doi:10.1093/jac/dks061

Tian B, Fadhil NH, Powell JE, Kwong WK, Moran NA (2012) Long-term exposure to antibiotics has caused accumulation of resistance determinants in the gut microbiota of honeybees. mBio 3:e00377-12. doi:10.1128/mBio.00377-12

Van Boeckel TP, Gandra S, Ashok A, Caudron Q, Grenfell BT, Levin SA, Laxminarayan R (2014) Global antibiotic consumption 2000 to 2010: an analysis of national pharmaceutical sales data. Lancet Infect Dis 14:742–750. doi:10.1016/S1473-3099(14)70780-7

Vaz-Moreira I, Nunes OC, Manaia CM (2014) Bacterial diversity and antibiotic resistance in water habitats: searching the links with the human microbiome. FEMS Microbiol Rev 38:761–778. doi:10.1111/1574-6976.12062

Velayati AA, Masjedi MR, Farnia P, Tabarsi P, Ghanavi J, Ziazarifi AH, Hoffner SE (2009) Emergence of new forms of totally drug-resistant tuberculosis bacilli: super extensively drug-resistant tuberculosis or totally drug-resistant strains in Iran. Chest 136:420–425. doi:10.1378/chest.08-2427

Waksman SA, Flynn JE (1973) History of the word 'antibiotic'. J Hist Med Allied Sci 28:284–286. doi:10.1093/jhmas/XXVIII.3.284

Walsh TR, Weeks J, Livermore DM, Toleman MA (2011) Dissemination of NDM-1 positive bacteria in the New Delhi environment and its implications for human health: an environmental point prevalence study. Lancet Infect Dis 11:355–362. doi:10.1016/S1473-3099(11)70059-7

Watanabe J, Fujiwara R, Sasajima N, Ito S, Sonoyama K (2010) Administration of antibiotics during infancy promoted the development of atopic dermatitis-like skin lesions in NC/Nga mice. Biosci Biotechnol Biochem 74:358–363. doi:10.1271/bbb.90709

Wattal C, Goel N (2014) Tackling antibiotic resistance in India. Expert Rev Anti Infect Ther 12:1427–1440. doi:10.1586/14787210.2014.976612

Willing BP, Russell SL, Finlay BB (2011) Shifting the balance: antibiotic effects on host-microbiota mutualism. Nat Rev Microbiol 9:233–243. doi:10.1038/nrmicro2536

World Health Organization (2014) Antimicrobial resistance: global report on surveillance. WHO Press. http://apps.who.int/iris/bitstream/10665/112642/1/9789241564748_eng.pdf

Young VB, Schmidt TM (2004) Antibiotic-associated diarrhea accompanied by large-scale alterations in the composition of the fecal microbiota. J Clin Microbiol 42:1203–1206. doi:10.1128/JCM.42.3.1203-1206.2004

In Silico Analytical Tools for Phylogenetic and Functional Bacterial Genomics

Vipin Chandra Kalia, Ravi Kumar, and Shikha Koul

Abstract Microbial significance in human lives has been gaining importance due to their biotechnological applications and ability to cause diseases. The use of antibiotics to kill them has proved counterproductive. Bacterial resistance to antibiotics has caused huge economic losses. Many bacteria have turned highly drug resistant due to modifications in their genetic reservoirs. It has been recognized that bacteria had another mechanism to circumvent the impact of antibiotics. Bacteria causing infectious diseases form biofilm at high cell density. Biofilm protects bacteria from even extremely high dosages of antibiotics. Under all these conditions, the most important aspect to initiate treatment is to diagnose the organism responsible for the disease. Bacterial identification through the *rrs* gene sequence has been the most prevalent and effective approach. The trouble arises in two main situations: (1) high similarity among gene sequences and (2) the presence of multiple copies of *rrs* gene within a genome. An obvious solution is to employ other highly conserved genes (housekeeping genes), which is uneconomical in terms of time and money. However, a few studies have revealed the presence of latent features within *rrs*. A set of genomic tools allowed identification of organisms up to the species level from their previous status of genus level identity. The most interesting aspect is that the strategy can be extended to all genes from all kinds of organisms.

V.C. Kalia (✉) • S. Koul
Microbial Biotechnology and Genomics, CSIR – Institute of Genomics and Integrative Biology (IGIB), Delhi University Campus, Mall Road, Delhi 110007, India

Academy of Scientific & Innovative Research (AcSIR), 2, Rafi Marg, Anusandhan Bhawan, New Delhi 110001, India
e-mail: vckalia@igib.res.in; vc_kalia@yahoo.co.in

R. Kumar
Microbial Biotechnology and Genomics, CSIR – Institute of Genomics and Integrative Biology (IGIB), Delhi University Campus, Mall Road, Delhi 110007, India

G. Arora et al. (eds.), *Drug Resistance in Bacteria, Fungi, Malaria, and Cancer*,
DOI 10.1007/978-3-319-48683-3_15

1 Introduction

Human interest in microbes has been growing steadily as their applications in our daily life have gained importance (Kalia 2015b; Moroeanu et al. 2015). The role of bacteria in causing diseases has always been studied with the intention of finding mechanisms to kill them. The discovery of antibiotics proved extremely helpful in reducing human misery. Since antibiotics are targeted to kill bacteria (Alipiah et al. 2015), they respond to this stress and undergo genetic changes, which are expressed as functional changes. The immediate impact of these changes has been the evolution of drug resistance in bacteria (Saxena et al. 2014). It was also realized that without undergoing any genetic change, bacteria under certain conditions may show enhanced antibiotic resistance. This acquired antibiotic resistance was attributed to the biofilm formed by bacteria by quorum sensing (QS), a phenomenon which is controlled by cell density. Bacteria under the regulation of QS express virulent metabolic behavior. Under this new regime, the approach is to inhibit QS and prevent bacteria from acquiring resistance to antibiotics (Gui et al. 2014; Kalia 2013a, 2014a, b; Prakasham et al. 2014). Hence, QS system is turning out to be a novel drug target (Kalia and Purohit 2011; Agarwala et al. 2014; Shang et al. 2014; Hema et al. 2015; Kaur et al. 2015; Koul et al. 2015b; Arya and Princy 2016). In all these scenarios, the need is to identify bacteria and provide a rapid diagnosis, which will permit the treatment to proceed.

Initially, phenotypic and biochemical characteristics were used as the basis for their identification and classification. The developments in molecular biology, genomic, and bioinformatics have changed the pace of research in these organisms. The turning point in the new era of genomics came into effect primarily because of the insights provided by Prof. Carl R. Woese (Kalia 2013b; Prakash et al. 2013; Mahale et al. 2014). Microbiologists around the world have used the tools developed in the last three to four decades to identify bacteria: PCR-ribotyping, microarray analysis, restriction endonuclease (RE) digestion, amplified fragment length polymorphism, multi-locus sequence analysis, DNA hybridization, isotope distribution analysis, molecular connectivity, etc. (Sharma et al. 2008; Kapley and Purohit 2009; Nguyen et al. 2013; Prakash et al. 2014; Wang et al. 2014; Yu et al. 2014, 2015; Meza-Lucas et al. 2016; Yagnik et al. 2016).

2 Bacillus

Bacillus is a versatile organism, which has been exploited for a large number of biotechnological applications. This genus has encompassed such a large number of diverse organisms that may be equated with *Pseudomonas*, as the "dumping" ground for gram-positive organisms (Porwal et al. 2009). This genus represents a lot of phenotypic and genotypic heterogeneity, such that an unambiguous identification up to species level has been a tough task (Porwal et al. 2009). Members of

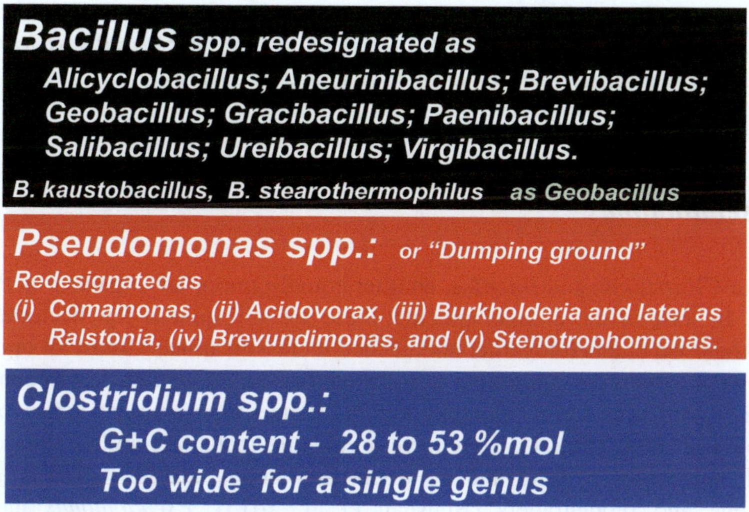

Fig. 1 Reorganization of bacterial systematics: *Bacillus*, *Pseudomonas*, and *Clostridium* (Porwal et al. 2009; Kalia et al. 2011a, b; Bhushan et al. 2013)

Bacillus subtilis group, *B. cereus* group, *B. licheniformis*, and *B. sphaericus* are some of the most notorious trouble spots among *Bacillus* spp. High genomic similarity between *B. subtilis* and *B. halodurans* is recorded for G+C content, genome size, and the characteristics of their ABC transporter genes, ATPases, etc. Information presented by the complete genome of *B. halodurans* show similarity for the enzymes transposases and recombinases, with those recorded among *Anabaena*, *Clostridium*, *Enterococcus*, *Lactococcus*, *Rhodobacter*, *Staphylococcus*, and *Yersinia* species. It clearly hints that *Bacillus* needs further segregation into new genera: *Aneurinibacillus*, *Ureibacillus*, *Virgibacillus*, *Brevibacillus*, and *Paenibacillus*. In fact, *Bacillus stearothermophilus*, *B. thermoleovorans*, *B. kaustophilus*, and *B. thermoglucosidasius* are categorized as *Geobacillus*, whereas *B. pasteurii*, *B. globisporus*, and *B. psychrophilus* are now known as *Sporosarcina* spp. (Fig. 1). The members of *Bacillus marinus* are presently classified as *Marinibacillus marinus* (Yoon et al. 2001).

3 *Clostridium*

The biotechnological importance of *Clostridium* has made researchers to monitor this organism with curiosity and precision. The bacteria is benign on one hand as it can produce solvents, enzymes, biofuels like butanol, ethanol, hydrogen, etc. and is extremely dangerous on the other hand, with the ability to produce deadly toxins (Carere et al. 2008; Bhushan et al. 2015). The heterogeneity of *Clostridium* has been recorded in phenotypic, biochemical, and genotypic characteristics. For quite some time, the issue of accommodating organisms varying in GC content from as low as 24 mol% (*Clostridium perfringens*) to as high as 58 mol% (*Clostridium barkeri*) did not appear justified (Fig. 1) (Kalia et al. 2011a).

4 *Pseudomonas*

Just like *Clostridium* and *Bacillus*, another equally important organism is *Pseudomonas*. In spite of these developments, there are still quite a few bacteria which were otherwise clubbed together and needed reclassification. Almost all organisms which were difficult to categorize were labeled as *Pseudomonas*. It was comprised of phenotypically and biochemically highly diverse organisms and was named the "dumping ground" (Fig. 1). They have a versatile metabolic ability to infect and degrade almost everything and no doubt are among the most widely studied pathogens and biodegraders (Bhushan et al. 2013). *Pseudomonas* has been subjected to repeated taxonomic revisions (Lalucat et al. 2006; Peix et al. 2009). Combined use of housekeeping genes such as*gyrB*, *rpoB*, *rpoD*, *rec*A, *atp*D, and *car*A and the classic gene—*rrs*—was found to be effective in distinguishing different species of *Pseudomonas*: *P. flavescens*, *P. mendocina*, *P. resinovorans*, *P. fluorescens*, *P. chlororaphis*, *P. aeruginosa*, *P. syringae*, *P. putida*, *P. stutzeri*, etc. (Spiers et al. 2000; Hilario et al. 2004; Aremu and Babalola 2015).

5 *Stenotrophomonas*

Another highly versatile organism is *Stenotrophomonas* spp. The phylogenetic diversity of *Stenotrophomonas* spp. is quite interesting as its members were initially grouped under *Pseudomonas* and *Xanthomonas*. Presently, eight recognized *Stenotrophomonas* spp. exist: *S. maltophilia*, *S. nitritireducens*, *S. acidominiphilia*, *S. rhizophila*, *S. koreensis*, *S. terrae*, *S. humi*, and *S. chelatiphaga*. *Stenotrophomonas dokdonensis* has been transferred to *Pseudoxanthomonas* as *P. dokdonensis*. As far as the functional abilities of *Stenotrophomonas* species are concerned, they are able to treat aromatic compounds either individually or in combination with *Bacillus*, *Pseudomonas*, *Flavimonas*, and *Morganella* spp. Ecological and metabolic (genetic and functional) diversity of *S. maltophilia* implies high taxonomic heterogeneity (Anzai et al. 2000; Peix et al. 2007; Tourkya et al. 2009; Verma et al. 2010, 2011; Aremu and Babalola 2015).

6 *Streptococcus*

The genus *Streptococcus* has a big number of species with important clinical relevance. Severe and acute diseases are known to be caused by species such as *S. agalactiae*, *S. pneumoniae,* and *S. pyogenes*. Different analytic methods enable identification to a limited extent and are laborious to apply. The genetic variability among different *Streptococcus* groups is quite low, and distinguishing them is a tough task (Lal et al. 2011; Kalia et al. 2016).

7 *Helicobacter*

Another group of organisms, which have great economic importance and are a cause of worry for health departments, belong to the genus *Helicobacter* (Puri et al. 2016). These organisms cause many diseases in human beings. Among the different species of *Helicobacter*, the following show high genetic variability: *H. cinaedi*, *H. pylori*, *H. felis*, *H. bilis*, and Candidatus *H. heilmannii*. Previously, *Helicobacter* spp. were categorized as *Campylobacter* sp. (Goodwin et al. 1989). Since *H. pylori* is responsible for 50 % of the infections caused by different relatives of *Helicobacter* (Suerbaum and Michetti 2002), it happens to be the most extensively studied species with 450 sequenced genomes. Biochemical assays, including urease test, are cheap but may not be very accurate. Molecular techniques like PCR and MLSA have also not proved to be highly precise (Puri et al. 2016).

8 The Novel Approach to Exploit Hidden Talents of *rrs*

With constant research efforts over the last three decades, *rrs* gene sequencing technique has been simplified to such an extent that almost all research laboratories around the globe have adopted it as routine assay. The RDP database has become a rich referral center, to which the newly sequenced *rrs* gene is subjected and the best match is used for identifying the organism. It must, however, be realized that RDP database can identify the new organism only against what is already known and deposited. The database cannot classify a gene sequence which has not been seen by it so far. It therefore requires a novel overture, wherein we need to first define the taxonomic and phylogenetic limits of each known species and key out the disruptions in the evolutionary scale. In a serial publication of works undertaken in this guidance, extensive genomic analyses were performed to look for any potential characteristics of *rrs* genes, which have not been elucidated so far. In the following text, a few case studies will be described to elucidate the approach, its validity, and significance (Fig. 2) (Kalia 2015a).

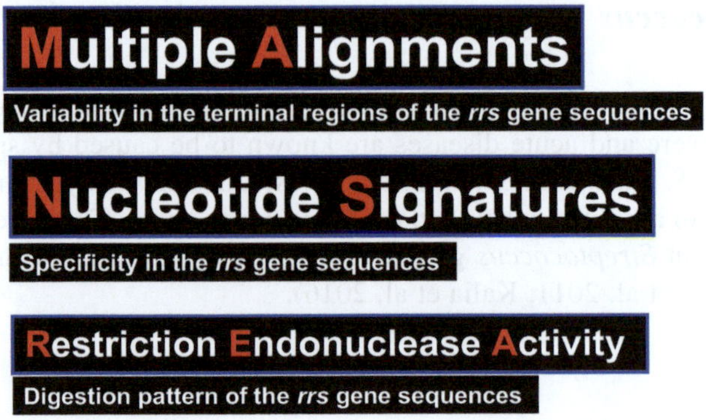

Fig. 2 Novel molecular techniques to explore the microbial taxonomy and phylogeny (Porwal et al. 2009; Kalia et al. 2011a; Lal et al. 2011; Bhushan et al. 2013; Puri et al. 2016)

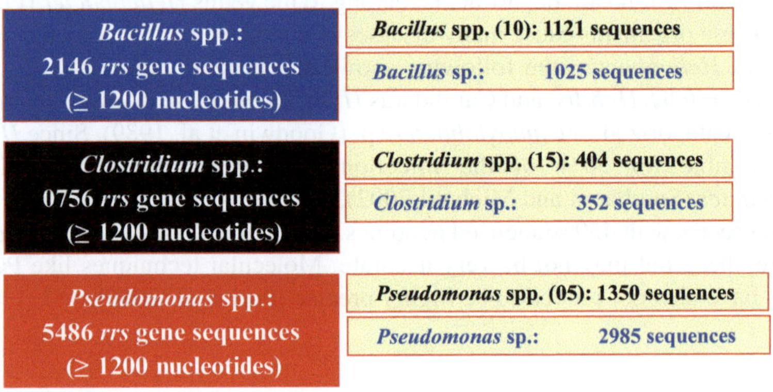

Fig. 3 Number of *rrs* gene sequences of *Bacillus*, *Clostridium*, and *Pseudomonas* used for developing phylogenetic framework—(1) identified up to species level and (2) those identified only up to genus level (Porwal et al. 2009; Kalia et al. 2011a; Bhushan et al. 2013)

9 Bacillus

9.1 Phylogenetic Framework

The first step was to define the phylogenetic boundaries of a species within a given genus. *Bacillus* was the first genus to be studied to evaluate the potential of this approach. Out of the available information in the database (at that time), 1121 *rrs* gene sequences of 10 *Bacillus* species were taken into consideration: *B. thuringiensis*, *B. anthracis*, *B. pumilus*, *B. cereus*, *B. subtilis*, *B. megaterium*, *B. sphaericus*, *B. clausii*, *B. halodurans*, and *B. licheniformis* (36–211 strains) (Figs. 3 and 4) (Porwal et al. 2009). Phylogenetic trees developed on the basis of

Organism	No. of *rrs* gene sequences	No. of aligned groups
Bacillus anthracis	153	04
B. cereus	211	04
B. thuringiensis	108	10
B. subtilis	271	26
B. licheniformis	131	10
B. pumilus	83	12
B. megaterium	47	08
B. sphaericus	42	06
B. clausii	39	06
B. halodurans	36	04
Bacillus sp.		
Total	1121	90

Fig. 4 Segregation of *rrs* gene sequences of *Bacillus* species into multiple sequence alignment groups on the basis of variability in the terminal regions (Adapted from Open Access article: Porwal et al. 2009. doi:10.1371/journal.pone.0004438)

1121 allowed their segregation into 89 clusters. From each cluster, the outermost and innermost *rrs* gene sequence was considered as representative of the limits of the species. In case, there were a large number of sequences in a cluster; additional 1–2 sequences were also taken into account. In all cases, the type strain of the species was also used to develop the phylogenetic framework. On the basis of these criteria, a comprehensive framework consisting of 34 *rrs* sequences representing 10 different *Bacillus* species were established (Fig. 5). With this genomic tool in hand, 305 *Bacillus* strains which were identified only up to genus level could be reclassified as members of these 10 known species (Fig. 6) (Porwal et al. 2009). On the basis of this genomic study, it was proposed to segregate the strains of *B. subtilis* into two species/subspecies. It was revealed in the literature that on the biochemical basis, *B. subtilis* can be divided into subspecies—*subtilis* and *spizizenii* (Nakamura et al. 1999). The study was limited only to 10 species out of around 189 species which are reported to be known today. This indicates that there is a lot of scope to extend this work.

10 Unique Signatures

In order to validate the authenticity of the segregation of *rrs* sequences of strains which could be classified among known *Bacillus* species, 20–30 nucleotide long unique signatures were identified among the 10 known species using MEME program (http://meme.nbcr.net/meme/meme.html). The uniqueness of these

Organism	Accession numbers of Reference *rrs* gene sequence(s)
Bacillus thuringiensis	DQ286308(T)[a], DQ286339, DQ328630, AE017355, DQ286329
B. anthracis[b]	AB190218, AE017334, AE017225
B. cereus[b]	DQ372919, DQ289988
B. subtilis	AB042061(T), DQ420172, AY995569, DQ504376, AY583216, AY881635, AY631853
B. licheniformis	AB039328(T), CP000002, AF234855
B. pumilus	AY260861(T), AY876289, DQ523500
B. megaterium	AJ717381(T), AY373358, AY505510, AY373360
B. sphaericus	AJ310084(T), DQ286309
B. clausii	X76440(T), AB201793, AY960116
B. halodurans	AY423275(T), AY856452
Total	34 strains

Fig. 5 Phylogenetic framework sequences of *rrs* genes of ten *Bacillus* species. [a]Type strain. [b]For *B. cereus* group as a whole, only one type strain was used (Adapted from Open Access article: Porwal et al. 2009. doi:10.1371/journal.pone.0004438)

Bacillus sp.	No.
B. cereus	75
B. thuringiensis	02
B. anthracis	01
B. sphaericus	23
B. licheniformis	21
B. halodurans	07
B. megaterium	69
B. pumilus	32
B. subtilis	44
B. clausii	31
Total	305 / 1025

Pseudomonas sp.	No.
P. aeruginosa	219
P. fluorescens	463
P. putida	347
P. stutzeri	197
P. syringae	141
Total	1367 / 2985

Clostridium sp.	No.
15 species	179
Total	179 / 352

Fig. 6 Number of *rrs* gene sequences of different organisms identified up to species level with the help of genomic frame work (Data on *Bacillus* has been adapted from Open Access article: Porwal et al. 2009. doi:10.1371/journal.pone.0004438) (Porwal et al. 2009; Kalia et al. 2011a; Bhushan et al. 2013)

signature sequences was verified by carrying out a blast search against all the sequences available at NCBI (Fig. 7) (Porwal et al. 2009). The motifs (signature sequences) were reported to be unique to a species, if these were absent from other species. Two to five 29–30 nucleotide long unique signatures were detected for

Bacillus spp. and Nucleotide Signatures	
Bacillus cereus	AAAGTGGAATTCCATGTGTAGCGGTGAAAT
B. thuringiensis	ATAACATTTTGAACTGCATGGTTCGAAATT
B. clausii	AATCCCATAAAGCCATTCTCAGTTCGGATT
B. halodurans	ATAATAAAAAGAACTGCATGGTTCTTTTTT
B. pumilus	AAGGTTTAGCCAATCCCACAAATCTGTTCT
B. megaterium	ATGATTGAAAGATGGTTTCGGCTATCACTT
B. sphaericus	TAAAACTCTGTTGTAAGGGAAGAACAAGTA

Fig. 7 Representative unique nucleotide signatures for *rrs* gene sequences of different *Bacillus* species. No unique signature was detectable for *B. anthracis*, *B. subtilis*, and *B. licheniformis* (Data adapted from Open Access article: Porwal et al. 2009. doi:10.1371/journal.pone.0004438)

Bacillus cereus, *B. thuringiensis*, *B. clausii*, *B. pumilus*, *B. megaterium*, *B. sphaericus*, and *B. halodurans*. In the cases of *B. anthracis*, *B. licheniformis*, and *B. subtilis*, unique signatures were not detectable.

A very interesting observation was made among signatures detected in *rrs* gene sequences of organisms which were identified only as *Bacillus* sp. Their nucleotide signatures indicated that either these organisms belong to those *Bacillus* spp. which have not been used in this study (Porwal et al. 2009) or they represent some other genus/genera. A few of these signatures did show a close resemblance to organisms belonging to *Virgibacillus, Geobacillus, Jeotgalibacillus, Brevibacillus, Marinibacillus, Paenibacillus,* and *Pontibacillus* (Porwal et al. 2009). A survey of published works revealed that some of these organisms (still classified as *Bacillus* sp.) had been reclassified and belong to *Virgibacillus, Geobacillus, Jeotgalibacillus, Brevibacillus, Marinibacillus, Paenibacillus,* and *Pontibacillus* (Heyndrickx et al. 1999; Nazina et al. 2001). It reflected on the strength of the study, which with the help of *in silico* analysis alone provided evidences that these *Bacillus* spp. needed segregation as new genera or species.

11 Restriction Endonuclease Digestion Analysis

Another unique feature to further support the segregation of organisms on the basis of *rrs* gene was the identification of RE, which elucidated a unique digestion pattern. Here the best part was the number of fragments, their size (nucleotides), and the order in which they occur within the gene.

Fourteen type II REs (Table 3) were used: (1) four base pair cutters (*AluI*, *HaeIII*, *DpnII*, *BfaI*, *Tru9I*, and *RsaI*), (2) six base pair cutters (*EcoRI*, *BamHI*, *NruI*, *SmaI*, *HindIII*, *PstI*, and *SacI*), and (3) eight base pair cutter (*NotI*) (rebase.neb.com/rebase/rebase.html). These REs were selected due to the occurrence of highly specific cleavage sites. It was realized that out of these 14 REs, only four base

***Rsa* I**	***Bacillus anthracis***								406	♦	355	♦	146	
	B. thuringiensis								406	♦	355	♦	146	
	B. cereus								406	♦	355	♦	146	
	B. halodurans						171	♦	235	♦	357	♦	146	
	B. megaterium		16	♦	11	♦			406	♦	356	♦	146	♦
	B. clausii	♦	19	♦	11	♦			406	♦	356	♦	146	
	B. licheniformis	♦	19	♦	11	♦			406	♦	501			♦
	B. sphaericus	♦	18	♦	11	♦			405	♦	496			
	B. subtilis	♦	19	♦	11	♦			407	♦	502			♦
	B. pumilus					♦			409	♦	502			

***Hae*III**	***Bacillus anthracis***				22	♦	565	♦			34	♦	456	
	B. thuringiensis			♦	22	♦	568	♦			34	♦	457	♦
	B. cereus			♦	22	♦	565	♦			34	♦	457	
	B. halodurans		78	♦	22	♦			598			♦	459	
	B. megaterium		78	♦	22	♦			578			♦	459	
	B. clausii	♦	78	♦	22	♦	85	♦	216	♦	264	♦		
	B. licheniformis			♦	22	♦			598			♦	457	♦
	B. sphaericus	♦	78	♦	22	♦			596			♦	452	♦
	B. subtilis			♦	22	♦			600			♦	457	
	B. pumilus			♦	22	♦			599			♦	458	♦

Fig. 8 *In silico* restriction endonuclease digestion of *rrs* gene sequences of different *Bacillus* species with *RsaI* and *HaeIII*. Values represent the fragment size (nucleotides). The *filled symbol* represents the RE action site (Data adapted from Open Access article: Porwal et al. 2009. doi:10.1371/journal.pone.0004438)

pair cutter could be exploited as these REs generated 5–9 fragments with sizes, which can be easily distinguished even under experimental conditions (Figs. 8 and 9). RE-*RsaI* digestion of *rrs* of different *Bacillus* spp. resulted in 2–6 fragments ranging in size from 11 to 502 nucleotides. *B. cereus* group members gave similar digestion patterns and were indistinguishable among them. *B. halodurans* and *B. pumilus* were easily distinguished from others based on their unique RE digestion patterns (Fig. 8). *B. clausii* and *B. sphaericus* could be identified as distinct on digestion with RE *HaeIII* (Fig. 8). *In silico* digestion of *rrs* of *B. megaterium*, and *B. pumilus*, gave a unique pattern with RE Tru9I (Fig. 9). The presence of two sets of unique digestion patterns in *rrs* sequences belonging to *B. subtilis* with RE *AluI* (Fig. 9) provided a strong evidence of the potential segregation of this group into two subspecies or as separate species. This observation was made in the phylogenetic framework analysis described above. It may be remarked that certain species segregate together in one RE can be distinguished by analyzing the digestion patterns with another RE.

Enzyme	Species	Fragments
***Tru*9I**	***Bacillus anthracis***	**527 ◆ 270 ◆ 8 ◆ 86 ◆ 134 ◆ 44**
	B. thuringiensis	**527 ◆ 270 ◆ 8 ◆ 86 ◆ 134 ◆ 44 ◆**
	B. cereus	**◆ 527 ◆ 270 ◆ 8 ◆ 85 ◆ 134 ◆ 44**
	B. halodurans	**270 ◆ 8 ◆ 86 ◆ 136**
	B. megaterium	**280 ◆ 86 ◆ 137 ◆**
	B. clausii	**270 ◆ 8 ◆ 86 ◆ 134 ◆**
	B. licheniformis	**◆ 278 ◆ 86 ◆ 134 ◆**
	B. sphaericus	**◆ 278 ◆ 86 ◆ 136**
	B. subtilis	**◆ 278 ◆ 89 ◆ 134 ◆ 181 ◆**
	B. pumilus	**331 ◆ 86 ◆ 134 ◆**
***Alu*1**	***Bacillus anthracis***	**174 ◆ 224 ◆ 599 ◆ 185 ◆ 21 ◆ 58**
	B. thuringiensis	**174 ◆ 225 ◆ 599 ◆ 186 ◆ 21 ◆ 58**
	B. cereus	**174 ◆ 224 ◆ 581 ◆ 42 ◆ 186 ◆ 21 ◆ 58**
	B. halodurans	**173 ◆ 824**
	B. megaterium	**86 ◆ 88 ◆ 615 ◆ 209 ◆ 265 ◆ 123**
	B. clausii	**85 ◆ 88 ◆ 186 ◆ 218 ◆ 419 ◆**
	B. licheniformis	**136 ◆ 33 ◆ 822 ◆ 265**
	B. sphaericus	**175 ◆ 615 ◆ 209 ◆ 265 ◆**
	B. subtilis	**◆ 173 ◆ 186 ◆ 431 ◆ 207 ◆ 265 ◆**
		◆ 164 ◆ 42 ◆ 779 ◆ 374
	B. pumilus	**85 ◆ 88 ◆ 185 ◆ 430 ◆ 207 ◆ 265 ◆**

Fig. 9 *In silico* restriction endonuclease digestion of *rrs* gene sequences of different *Bacillus* species with *Tru9I* and *AluI*. Values represent the fragment size (nucleotides). The *filled symbol* represents the RE action site (Data adapted from Open Access article: Porwal et al. 2009. doi:10.1371/journal.pone.0004438)

12 *Clostridium*

The approach described above for developing genomic tools for phylogenetic analysis were extended to *rrs* gene sequences of *Clostridium* (Kalia et al. 2011a). Here 756 rrs sequences of 110 *Clostridium* species were taken into consideration. *Clostridium botulinum rrs* gene sequences were segregated into four groups. Out of these four groups of *C. botulinum*, 10 *rrs* sequences were selected to represent 128 sequences (Fig. 3). By drawing phylogenetic trees of 15 different *Clostridium* species, 56 *rrs* gene sequences were selected for creating a phylogenetic framework. It defined the phylogenetic limits of the *C. acetobutylicum, C. butyricum, C. beijerinckii, C. perfringens, C. botulinum, C. chauvoei, C. baratii, C. pasteurianum, C. colicanis, C. sardiniense, C. subterminale, C. novyi, C. sporogenes, C. kluyveri*, and *C. tetani*. With this genomic tool in hand, 356 *Clostridium* strains identified only up to genus level could be classified among these 15 known species (Kalia et al. 2011a). A confirmation of this initial reclassification was achieved through nucleotide signature analysis and unique RE digestion patterns. In this case also, REs—*HaeIII*, *AluI*, *RsaI*, *DpnII*, *Tru9I*, and *BfaI*—proved effective in providing relevant information. RE—AluI—was instrumental in

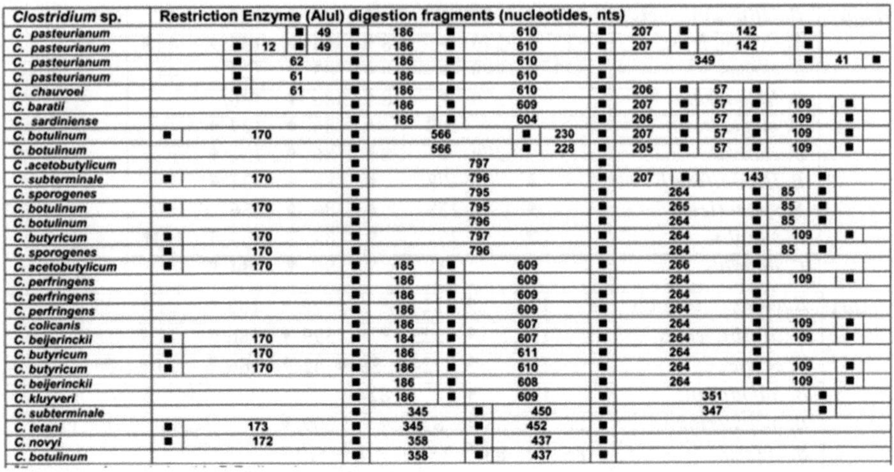

Fig. 10 *In silico* restriction endonuclease digestion of *rrs* gene sequences of different *Clostridium* species with *AluI*. Values represent the fragment size (nucleotides). The *filled symbol* represents the RE action site (Data adapted from Open Access article: Kalia et al. 2011a. doi:10.1186/1471-2-2164-12-18)

clearly segregating *C. chauvoei*, *C. acetobutylicum*, *C. kluyveri*, *C. perfringens*, *C. colicanis*, *C. pasteurianum*, and *C. subterminale* (Fig. 10) (Kalia et al. 2011a).

13 *Pseudomonas*, *Helicobacter*, and *Streptococcus*

Using approaches similar to those defined above for *Bacillus* and *Clostridium* spp., it was found that effective and meaningful phylogenetic and taxonomical information can be retrieved also in the cases of *Stenotrophomonas*, *Streptococcus*, *Pseudomonas*, and *Helicobacter* (Verma et al. 2010, 2011; Lal et al. 2011; Bhushan et al. 2013; Puri et al. 2016).

14 Functional Genomics

In addition to using the genomic tools primarily for bacterial identification, it was realized that these can be extended to derive meaningful information in other genes as well. In attempts to inhibit the virulent behavior of bacteria causing infectious diseases, a search for organisms producing bioactive molecules to act as antibacterial was conducted. Since most infectious diseases are caused by organisms which produce biofilm through QS system, anti-QS molecule producers were targeted. Two enzymes—acyl-homoserine lactone acylase and acyl-homoserine lactone lactonase—have been shown to inhibit QS. Phylogenetic and functional genomic analyses of the genes responsible for the production of these enzymes were carried out. Unique signatures and RE digestion patterns enabled to establish the phenomenon of horizontal gene transfer as well (Kalia et al. 2011b). The unique signatures were proposed as candidates for designing primers for detecting such organisms in unknown samples. On the basis of this functional genomic analysis, three organisms were traced, which possessed genes for both the AHL inhibitory enzymes (Kalia et al. 2011b; Kalia 2014a). The RE digestion of AHL-lactonase with *Tru9I, RsaI*, and *DpnII* could validate the phylogenetic segregation of the organisms based on *rrs* gene (Huma et al. 2011). Diversity analysis of *Citrobacter* species isolated from diverse niches was carried out to check their abilities to degrade aromatic compounds. Nine strains having genes, which coded for aromatic ring-hydroxylating dioxygenases, were analyzed using a diversity of REs—*DpnII*, *RsaI*, and *HaeIII*. Unique signature analysis in combination with RE showed that genomic similarity in a few specific strains supported their closeness in metabolic functions as well (Selvakumaran et al. 2011). Functional genomics of *Stenotrophomonas* diversity in effluent treatment plants was established with precision using RE digestion strategy. This enabled the development of a consortium to be used for bioremediation (Verma et al. 2010, 2011).

More recently, the genomic tool—RE digestion pattern—has been extensively used for identification of organisms, which are economically highly significant for health departments. The primary objective was to find novel markers to be used for diagnostic purposes (Arasu et al. 2015). The use of functional genes was necessitated by the multiple copies of *rrs* genes in different species of *Clostridium, Vibrio, Yersinia, Staphylococcus, Streptococcus*, and *Lactobacillus* (Kalia et al. 2015, 2016; Kalia and Kumar 2015; Kekre et al. 2015; Koul et al. 2015a; Koul and Kalia 2016; Kumar et al. 2016).

15 Opinion

Phylogenetic analysis based on gene sequences is a very handy and effective tool. The gene most widely used for bacterial identification and overall taxonomical purposes is *rrs*. Although it is a widely employed technique, it leads into trouble quite frequently. Often, the most obvious choice is to employ other highly conserved genes (HKGs). It, however, implies higher inputs of time and money. Invariably, additional 7–8 HKGs are required to resolve the issue. To circumvent the efforts needed for identifying bacteria using a single gene—*rrs*—a fresh round of studies were conducted, to develop genomic tools: phylogenetic framework, signatures, and RE digestion patterns. Once again, these tools ran into trouble in case of organisms which possessed multiple copies of *rrs*. The potential solution seems to lie in the genes common to all the species within a genus. Unique gene-RE digestion pattern allowed identification of novel biomarkers. It thus can be envisaged that the use of specific REs-gene combinations can be used for all kinds of phylogenetic and functional genomic analysis.

Acknowledgments We are thankful to the Director of CSIR-Institute of Genomics and Integrative Biology (IGIB) and CSIR project GENESIS (BSC0121) for providing the necessary funds, facilities, and moral support. Authors are also thankful to the Academy of Scientific and Innovative Research (AcSIR), New Delhi.

References

Agarwala M, Choudhury B, Yadav RNS (2014) Comparative study of antibiofilm activity of copper oxide and iron oxide nanoparticles against multidrug resistant biofilm forming uropathogens. Indian J Microbiol 54:365–368. doi:10.1007/s12088-014-0462-z

Alipiah NM, Shamsudin MN, Yusoff FM, Arshad A (2015) Membrane biosynthesis gene disruption in methicillin-resistant *Staphylococcus aureus* (MRSA) as potential mechanism for reducing antibiotic resistance. Indian J Microbiol 54:41–49. doi:10.1007/s12088-014-0488-2

Anzai Y, Kim H, Park JY, Wakabayashi H, Oyaizu H (2000) Phylogenetic affiliation of the pseudomonads based on 16S rRNA sequence. Int J Syst Evol Microbiol 50:1563–1589. doi:10.1099/00207713-50-4-1563

Arasu MV, Al-Dhabi NA, Rejiniemon TS, Lee KD, Huxley VAJ, Kim DH, Duraipandiyan V, Karuppiah P, Choi KC (2015) Identification and characterization of *Lactobacillus brevis* P68 with antifungal, antioxidant and probiotic functional properties. Indian J Microbiol 55:19–28. doi:10.1007/s12088-014-0495-3

Aremu BR, Babalola OO (2015) Classification and taxonomy of vegetable macergens. Front Microbiol 6:1361. doi:10.3389/fmicb.01361

Arya R, Princy SA (2016) Exploration of modulated genetic circuits governing virulence determinants in *Staphylococcus aureus*. Indian J Microbiol 56:19–27. doi:10.1007/s12088-015-0555-3

Bhushan A, Joshi J, Shankar P, Kushwah J, Raju SC, Purohit HJ, Kalia VC (2013) Development of genomic tools for the identification of certain *Pseudomonas* up to species level. Indian J Microbiol 53:253–263. doi:10.1007/s12088-013-0412-1

Bhushan A, Mukherjee T, Joshi J, Shankar P, Kalia VC (2015) Insights into the origin of *Clostridium botulinum* strains: evolution of distinct restriction endonuclease sites in *rrs* (16S rRNA gene). Indian J Microbiol 55:140–150. doi:10.1007/s12088-015-0514-z

Carere CR, Kalia VC, Sparling R, Cicek N, Levin DB (2008) Pyruvate catabolism and hydrogen synthesis pathway genes of *Clostridium thermocellum* ATCC 27405. Indian J Microbiol 48:252–266. doi:10.1007/s12088-008-0036-z

Goodwin CS, Armstrong JA, Chilvers T, Peters M, Collins MD, Sly L, McConnell W, Harper WES (1989) Transfer of *Campylobacter pylori* and *Campylobacter mustelae* to *Helicobacter* gen. nov. as *Helicobacter pylori* comb. nov. and *Helicobacter mustelae* comb. nov. respectively. Int J Syst Bacteriol 39:397–405. doi:10.1099/00207713-39-4-397

Gui Z, Wang H, Ding T, Zhu W, Zhuang X, Chu W (2014) Azithromycin reduces the production of α-hemolysin and biofilm formation in *Staphylococcus aureus*. Indian J Microbiol 54:114–117. doi:10.1007/s12088-013-0438-4

Hema M, Balasubramanian S, Princy SA (2015) Meddling *Vibrio cholerae* murmurs: a neoteric advancement in cholera research. Indian J Microbiol 55:121–130. doi:10.1007/s12088-015-0520-1

Heyndrickx M, Lebbe L, Kersters K, Hoste B, De Wachter R, De Vos P, Forsyth G, Logan NA (1999) Proposal of *Virgibacillus proomii* sp. nov. and emended description of *Virgibacillus pantothenticus* (Proom and Knight 1950) Heyndrickx et al. 1998. Int J Syst Bacteriol 49:1083–1090

Hilario E, Buckley TR, Young JM (2004) Improved resolution on the phylogenetic relationships among *Pseudomonas* by combined analysis of *atpD*, *carA*, *recA* and 16S rDNA. Antonie Van Leeuwenhoek 86:51–64. doi:10.1023/B:ANTO.0000024910.57117.16

Huma N, Shankar P, Kushwah J, Bhushan A, Joshi J, Mukherjee T, Raju S, Purohit HJ, Kalia VC (2011) Diversity and polymorphism in AHL-lactonase gene (aiiA) of *Bacillus*. J Microbiol Biotechnol 21:1001–1011. doi:10.4014/jmb.1105.05056

Kalia VC (2013a) Quorum sensing inhibitors: an overview. Biotechnol Adv 31:224–245. doi:10.1016/j.biotechadv.2012.10.004

Kalia VC (2013b) The visionary: Prof. Carl R. Woese. Indian J Microbiol 53:245–246. doi:10.1007/s12088-013-0417-9

Kalia VC (2014a) In search of versatile organisms for quorum-sensing inhibitors: acyl homoserine lactones (AHL)-acylase and AHL-lactonase. FEMS Microbiol Lett 359:143. doi:10.1111/1574-6968.12585

Kalia VC (2014b) Microbes, antimicrobials and resistance: the battle goes on. Indian J Microbiol 54:1–2. doi:10.1007/s12088-013-0443-7

Kalia VC (2015a) Let's explore the latent features of genes to identify bacteria. J Mol Genet Med 9, e105. doi:10.4172/1747-0862.1000E105

Kalia VC (2015b) Microbes: the most friendly beings? In: Kalia VC (ed) Quorum sensing vs quorum quenching: a battle with no end in sight. Springer, New Delhi, pp 1–5. doi:10.1007/978-81-322-1982-8_1. ISBN 978-81-322-1981-1

Kalia VC, Kumar P (2015) Genome wide search for biomarkers to diagnose *Yersinia* infections. Indian J Microbiol 55:366–374. doi:10.1007/s12088-015-0552-6

Kalia VC, Purohit HJ (2011) Quenching the quorum sensing system: potential antibacterial drug targets. Crit Rev Microbiol 37:121–140. doi:10.3109/1040841X.2010.532479

Kalia VC, Mukherjee T, Bhushan A, Joshi J, Shankar P, Huma N (2011a) Analysis of the unexplored features of *rrs* (16S rDNA) of the genus *Clostridium*. BMC Genomics 12:18. doi:10.1186/1471-2164-12-18

Kalia VC, Raju SC, Purohit HJ (2011b) Genomic analysis reveals versatile organisms for quorum quenching enzymes: acyl-homoserine lactone-acylase and -lactonase. Open Microbiol J 5:1–13. doi:10.2174/1874285801105010000

Kalia VC, Kumar P, Kumar R, Mishra A, Koul S (2015) Genome wide analysis for rapid identification of *Vibrio* species. Indian J Microbiol 55:375–383. doi:10.1007/s12088-015-0553-5

Kalia VC, Kumar R, Kumar P, Koul S (2016) A genome-wide profiling strategy as an aid for searching unique identification biomarkers for *Streptococcus*. Indian J Microbiol 56:46–58. doi:10.1007/s12088-015-0561-5

Kapley A, Purohit HJ (2009) Genomic tools in bioremediation. Indian J Microbiol 49:108–113. doi:10.1007/s12088-009-0012-2

Kaur G, Rajesh S, Princy SA (2015) Plausible drug targets in the *Streptococcus mutans* quorum sensing pathways to combat dental biofilms and associated risks. Indian J Microbiol 55:349–357. doi:10.1007/s12088-015-0534-8

Kekre A, Bhushan A, Kumar P, Kalia VC (2015) Genome wide analysis for searching novel markers to rapidly identify *Clostridium* strains. Indian J Microbiol 55:250–257. doi:10.1007/s12088-015-0535-7

Koul S, Kalia VC (2016) Comparative genomics reveals biomarkers to identify *Lactobacillus* species. Indian J Microbiol 56:253–263. doi:10.1007/s12088-016-0605-5

Koul S, Kumar P, Kalia VC (2015a) A unique genome wide approach to search novel markers for rapid identification of bacterial pathogens. J Mol Genet Med 9:194. doi:10.4172/1747-0862.1000194

Koul S, Prakash J, Mishra A, Kalia VC (2015b) Potential emergence of multi-quorum sensing inhibitor resistant (MQSIR) bacteria. Indian J Microbiol 56:1–18. doi:10.1007/s12088-015-0558-0

Kumar R, Koul S, Kumar P, Kalia VC (2016) Searching biomarkers in the sequenced genomes of *Staphylococcus* for their rapid identification. Indian J Microbiol 56:64–71. doi:10.1007/s12088-016-0565-9

Lal D, Verma M, Lal R (2011) Exploring internal features of 16S rRNA gene for identification of clinically relevant species of the genus *Streptococcus*. Ann Clin Microbiol Antimicrob 10:28. doi:10.1186/1476-0711-10-28

Lalucat J, Bennasar A, Bosch R, Garcia-Valdés E, Palleroni NJ (2006) Biology of *Pseudomonas stutzeri*. Microbiol Mol Biol Rev 70:510–547. doi:10.1128/MMBR.00047-05

Mahale KN, Paranjape PS, Marathe NP, Dhotre DP, Chowdhury S, Shetty SA, Sharma A, Sharma K, Tuteja U, Batra HV, Shouche YS (2014) Draft genome sequences of *Yersinia pestis* strains from the 1994 plague epidemic of Surat and 2002 Shimla outbreak in India. Indian J Microbiol 54:480–482. doi:10.1007/s12088-014-0475-7

Meza-Lucas A, Pérez-Villagómez M, Martínez-López JP, García-Rodea R, Martínez-Castelán MG, Escobar-Gutiérrez A, de-la-Rosa-Arana JL, Villanueva-Zamudio A (2016) Comparison of DOT-ELISA and Standard-ELISA for detection of the *Vibrio cholerae* toxin in culture supernatants of bacteria isolated from human and environmental samples. Indian J Microbiol 56:379–382. doi:10.1007/s12088-016-0596-2

Moroeanu VI, Vamanu E, Paun G, Neagu E, Ungureanu OR, Eremia SAV, Radu GL, Ionescu R, Pelinescu DR (2015) Probiotic strains influence on infant microbiota in the in vitro colonic fermentation model GIS1. Indian J Microbiol 55:423–429. doi:10.1007/s12088-015-0542-8

Nakamura LK, Roberts MS, Cohan FM (1999) Relationship of Bacillus subtilis clades associated with strains 168 and W23: a proposal for Bacillus subtilis subsp. subtilis subsp. nov. and Bacillus subtilis subsp. spizizenii subsp. nov. Int J Syst Bacteriol 49:1211–1215

Nazina TN, Tourova TP, Poltaraus AB, Novikova EV, Grigoryan AA, Ivanova AE, Lysenko AM, Petrunyaka VV, Osipov GA, Belyaev SS, Ivanov MV (2001) Taxonomic study of aerobic thermophilic bacilli: descriptions of *Geobacillus subterraneus* gen. nov., sp nov and *Geobacillus uzenensis* sp nov from petroleum reservoirs and transfer of *Bacillus stearothermophilus, Bacillus thermocatenulatus, Bacillus thermoleovorans, Bacillus kaustophilus, Bacillus thermoglucosidasius* and *Bacillus thermodenitrificans* to *Geobacillus* as the new combinations *G. stearothermophilus, G. thermocatenulatus, G. thermoleovorans, G. kaustophilus, G. thermoglucosidasius* and *G. thermodenitrificans*. Int J Syst Evol Microbiol 51:433–446. doi:10.1099/00207713-51-2-433

Nguyen THK, Doan VTT, Ha LD, Nguyen HN (2013) Molecular cloning, expression of *minD* gene from *Lactobacillus acidophilus* VTCC-B-871 and analyses to identify *Lactobacillus rhamnosus* PN04 from Vietnam *Hottuyniacordata* Thunb. Indian J Microbiol 53:385–390. doi:10.1007/s12088-013-0384-1

Peix A, Valverde A, Rivas R, Igual JM, Ramírez-Bahena MH, Mateos PF, Santa-Regina I, Rodríguez-Barrueco C, Martínez-Molina E, Velázquez E (2007) Reclassification of *Pseudomonas aurantiaca* as a synonym of *Pseudomonas chlororaphis* and proposal of three subspecies, *P. chlororaphis* subsp. chlororaphis subsp. nov., *P. chlororaphis* subsp. aureofaciens subsp. nov., comb. nov. and *P. chlororaphis* subsp. aurantiaca subsp. nov., comb. nov. Int J Syst Evol Microbiol 57:1286–1290. doi:10.1099/ijs.0.000164

Peix A, Ramírez-Bahena MH, Velázquez E (2009) Historical evolution and current status of the taxonomy of genus *Pseudomonas*. Infect Genet Evol 9:1132–1147. doi:10.1016/j.meegid.2009.08.001

Porwal S, Lal S, Cheema S, Kalia VC (2009) Phylogeny in aid of the present and novel microbial lineages: diversity in *Bacillus*. PLoS One 4, e4438. doi:10.1371/journal.pone.0004438

Prakash O, Jangid K, Shouche YS (2013) Carl Woese: from biophysics to evolutionary microbiology. Indian J Microbiol 53:247–252. doi:10.1007/s12088-013-0401-4

Prakash O, Pandey PK, Kulkarni GJ, Mahale KN, Shouche YS (2014) Technicalities and glitches of terminal restriction fragment length polymorphism (T-RFLP). Indian J Microbiol 54:255–261. doi:10.1007/s12088-014-0461-0

Prakasham RS, Kumar BS, Kumar YS, Kumar KP (2014) Production and characterization of protein encapsulated silver nanoparticles by marine isolate *Streptomyces parvulus* SSNP11. Indian J Microbiol 54:329–336. doi:10.1007/s12088-014-0452-1

Puri A, Rai A, Dhanaraj PS, Lal R, Patel DD, Kaicker A, Verma M (2016) An *in silico* approach for identification of the pathogenic species, *Helicobacter pylori* and its relatives. Indian J Microbiol 56:277–286. doi:10.1007/s12088-016-0575-7

Saxena A, Mukherjee M, Kumari R, Singh P, Lal R (2014) Synthetic biology in action: developing a drug against MDR-TB. Indian J Microbiol 54:369–375. doi:10.1007/s12088-014-0498-0

Selvakumaran S, Kapley A, Kashyap SM, Daginawala HF, Kalia VC, Purohit HJ (2011) Diversity of aromatic ring-hydroxylating dioxygenase gene in *Citrobacter*. Bioresour Technol 101:4600–4609. doi:10.1016/j.biortech.2011.01.011

Shang Z, Wang H, Zhou S, Chu W (2014) Characterization of N-acyl-homoserine lactones (AHLs)-deficient clinical isolates of *Pseudomonas aeruginosa*. Indian J Microbiol 54:158–162. doi:10.1007/s12088-014-0449-9

Sharma P, Kumari H, Kumar M, Verma M, Malhotra S, Khurana J, Lal R (2008) From bacterial genomics to metagenomics: concepts, tools and recent advances. Indian J Microbiol 48:173–194. doi:10.1007/s12088-008-0031-4

Spiers AJ, Buckling A, Rainey PB (2000) The causes of *Pseudomonas* diversity. Microbiology 146:2345–2350. doi:10.1099/00221287-146-10-2345

Suerbaum S, Michetti P (2002) *Helicobacter pylori* infection. N Engl J Med 347:1175–1186. doi:10.1056/NEJMra020542

Tourkya B, Boubellouta T, Dufour E, Leriche F (2009) Fluorescence spectroscopy as a promising tool for a polyphasic approach to pseudomonad taxonomy. Curr Microbiol 58:39–46. doi:10.1007/s00284-008-9263-0

Verma V, Raju SC, Kapley A, Kalia VC, Daginawala HF, Purohit HJ (2010) Evaluation of genetic and functional diversity of *Stenotrophomonas* isolates from diverse effluent treatment plants. Bioresour Technol 101:7744–7753. doi:10.1016/j.biortech.2010.05.014

Verma V, Raju SC, Kapley A, Kalia VC, Kanade GS, Daginawala HF, Purohit HJ (2011) Degradative potential of *Stenotrophomonas* strain HPC383 having genes homologous to *dmp* operon. Bioresour Technol 102:3227–3233. doi:10.1016/j.biortech.2010.11.016

Wang R, Fang S, Xiang S, Ling S, Yuan J, Wang S (2014) Generation and characterization of a scFv antibody against T3SS needle of *Vibrio parahaemolyticus*. Indian J Microbiol 54:143–150. doi:10.1007/s12088-013-0428-6

Yagnik B, Patel S, Dave M, Sharma D, Padh H, Desai P (2016) Factors affecting inducible expression of outer membrane protein A (OmpA) of *Shigella dysenteriae* Type-1 in *Lactococcus lactis* using nisin inducible controlled expression (NICE). Indian J Microbiol 56:80–87. doi:10.1007/s12088-015-0556-2

Yoon JH, Weiss N, Lee KC, Lee IS, Kang KH, Park YH (2001) *Jeotgalibacillus alimentarius* gen. nov., sp. nov., a novel bacterium isolated from jeotgal with L-lysine in the cell wall, and reclassification of *Bacillus marinus* Ruger 1983 as *Marinibacillus marinus* gen. nov., comb. nov. Int J Syst Evol Microbiol 51:2087–2093. doi:10.1016/j.biortech.2010.11.016

Yu S, Peng Y, Chen W, Deng Y, Zheng Y (2014) Comparative genomic analysis of two-component signal transduction systems in probiotic *Lactobacillus casei*. Indian J Microbiol 54:293–301. doi:10.1007/s12088-014-0456-x

Yu S, Peng Y, Zheng Y, Chen W (2015) Comparative genome analysis of *Lactobacillus casei*: insights into genomic diversification for niche expansion. Indian J Microbiol 55:102–107. doi:10.1007/s12088-014-0496-2

Exploiting Bacterial Genomes to Develop Biomarkers for Identification

Ravi Kumar, Shikha Koul, and Vipin Chandra Kalia

Abstract Bacteria have unique abilities to adjust itself to diverse environmental conditions. Under adverse conditions, their genetic reservoir provides necessary help. Although bacteria have been perceived as pathogens to most living beings, the most critical are the ones which infect human being. Bacteria also harbour the human gut and skin and have been shown to be helpful. The pathogenic bacteria cause diseases and contribute to ill health. The need is to identify them rapidly, diagnose the disease and initiate the treatment. Most bacteria can be easily identified on the basis of their 16S rRNA (*rrs*) gene. However, in case where multiple copies of *rrs* are present within a bacterial genome, it is difficult to identify them, since they show great homology with other species of a genus. Here, novel approaches have been reviewed, which rely upon certain genes which are common to a large number of species of *Clostridium*, *Lactobacillus*, *Staphylococcus*, *Streptococcus*, *Vibrio* and *Yersinia* and show unique digestion patterns on treatment with restriction endonucleases.

1 Introduction

Bacteria have been bestowed with inherent abilities to withstand extreme environmental conditions and undergo genetic changes to evolve rapidly (Kalia 2010). Their interactions with human beings are extremely varied. The benign type of bacteria persists within the human gut while quite a few exist and survive well on the human body (Yu et al. 2014; Arasu et al. 2015). The pathogenic bacteria are a

R. Kumar
Microbial Biotechnology and Genomics, CSIR – Institute of Genomics and Integrative Biology (IGIB), Delhi University Campus, Mall Road, Delhi 110007, India

S. Koul • V.C. Kalia (✉)
Microbial Biotechnology and Genomics, CSIR – Institute of Genomics and Integrative Biology (IGIB), Delhi University Campus, Mall Road, Delhi 110007, India

Academy of Scientific & Innovative Research (AcSIR), 2, Rafi Marg, Anusandhan Bhawan, New Delhi 110001, India
e-mail: vckalia@igib.res.in; vc_kalia@yahoo.co.in

G. Arora et al. (eds.), *Drug Resistance in Bacteria, Fungi, Malaria, and Cancer*,
DOI 10.1007/978-3-319-48683-3_16

major threat to human health (Gautam et al. 2014; Wang et al. 2014). Bacteria invade the human body and then cause extensive damage, especially in the cases of infectious diseases. The bacterial pathogens may live and survive inside the body for a long term, e.g. tuberculosis, or for a short term as in diarrheas. In these two extreme scenarios, the former are slow growers with an excellent ability to evade the human immune system. In the latter case, the growth is rapid and the expression turns violent. Among the pathogens, which are a worry for Health Departments, notable ones are *Streptococcus*, *Shigella*, *Helicobacter*, *Vibrio*, *Clostridium*, *Yersinia*, *Salmonella*, etc. (Mahale et al. 2014; Kalia et al. 2015, 2016; Kalia and Kumar 2015; Kekre et al. 2015; Koul et al. 2015a, b; Koul and Kalia 2016; Kumar et al. 2016; Meza-Lucas et al. 2016; Puri et al. 2016; Yagnik et al. 2016). The discovery of antibiotics turned out to be a blessing for human beings. However, this arsenal invariably used to get rid of these pathogens didn't live up to its expectations. Bacteria could develop resistance to most of the antibiotics. Few research groups are still showing courage to invest time and money to develop novel antibiotics (Agarwala et al. 2014; Saxena et al. 2014; Prakasham et al. 2014; Alipiah et al. 2015; Sajid et al. 2015). Pharmaceutical companies have completely lost interest in investing in this area. It is now for the governments to look into this societal issue. A mechanism which is adding fuel to the fire is the bacterial ability to become "resistant" to antibiotics, without undergoing any genetic changes. Here, they multiply silently without any evident signs of their presence. Once they reach a threshold population density, they start expressing genes which never get expressed while the bacteria are in low numbers. This phenomenon of quorum sensing is responsible for the expression of pathogenic factors (Kalia and Purohit 2011; Kalia et al. 2011b; Gui et al. 2014; Hema et al. 2015; Kaur et al. 2015; Koul et al. 2015b; Arya and Princy 2016). Among these biofilm formation is one of the most dangerous activities. It provides a shield to the bacteria which thus show "resistance" to antibiotics (Shang et al. 2014). There seems to be a battle which is always going on between bacteria and antibiotics (Kalia 2013a, 2014, 2015b). Under all these circumstances, the need is to rapidly identify the pathogen and diagnose the disease in an unambiguous manner before it becomes unmanageable.

2 The Conventional Ways

A wide range of bacterial identification methods is employed depending upon the need and available facilities: amplified fragment length polymorphism (AFLP), microarray, restriction endonuclease digestion (RE), PCR-ribotyping, multilocus sequence analysis, randomly amplified polymorphic DNA and DNA-DNA re-association (Prakash et al. 2014). One of the most frequently employed molecular markers to identify microbes is the 16S rRNA (*rrs*) gene sequence. The evidence for its popularity among researchers can be gauged from the Ribosomal

Database Project (RDP), which contains around 3.2 million *rrs* entries (https://rdp.cme.msu.edu/). The credit goes to the innovative thinking of Prof. Carl R. Woese (Kalia 2013b; Prakash et al. 2013). In certain cases, the *rrs* gene sequence does not prove helpful, especially if the organisms belong to closely related taxa. As an alternative to *rrs* gene, people resort to the usage of a few housekeeping genes (HKGs), which are also highly conserved throughout the bacterial world—*recA*, *gyrA*, *gyrB*, *rpoB*, etc. These HKGs are used in various combinations and have proved effective in distinguishing closely related organisms (Porwal et al. 2009; Bhushan et al. 2015). Here, invariably up to eight genes are required to achieve meaningful result, which obviously amounts to higher investments of time, resources and money.

3 Exploring the Hidden Potential of *rrs*

At times, full length comparisons of rrs genes from different species or strains don't prove effective enough to distinguish and identify the bacteria unambiguously. Variable regions of *rrs* and their combinations have also been employed for establishing their identity and phylogenetic relations. More recent efforts have revealed the presence of a few features, which have not been exploited for bacterial identification. Molecular markers like unique nucleotide signatures and specific endonuclease restriction digestion patterns were developed for *Bacillus*, *Clostridium* and *Pseudomonas*. Another interesting parameter developed in these studies was the phylogenetic frame work, which could define the phylogenetic limits of a species (Porwal et al. 2009; Kalia et al. 2011a; Bhushan et al. 2013, 2015; Kalia 2015a). It could be extrapolated to identify those bacteria which are yet to be seen by the databases. These genomic tools have been extended to identify members of the genus *Streptococcus* (Lal et al. 2011) and *Helicobacter* (Puri et al. 2016).

4 The Trouble with *rrs* Genes

In the organisms like *Clostridium*, *Lactobacillus*, *Staphylococcus*, *Streptococcus*, *Vibrio* and *Yersinia*, each genome possesses multiple copies (4–13) of the *rrs gene* (Table 1). (Klappenbach et al. 2001; Kalia et al. 2015, 2016; Kalia and Kumar 2015; Kekre et al. 2015; Koul et al. 2015a; Koul and Kalia 2016; Kumar et al. 2016). This is expected to result in overestimation of bacterial populations. The other limitation is a high level of similarity observed between *rrs* copies of genomes from different species, which may lead to mislabelling of the organism.

Table 1 Characteristics of sequenced genomes of certain Gram-positive bacteria: *Lactobacillus*, *Clostridium* and *Staphylococcus* species (www.ncbi.nlm.nih.gov)

	Genome					
Organism	Size (Mb)	%GC	No. of genes	No. of proteins	*rrs* copies	References
Clostridium	2.54–6.00	27.40–32.02	3911–4057	2315–5020	8–13	Kekre et al. (2015)
Lactobacillus	1.37–3.36	33.01–51.50	1434–3148	1298–3004	4–9	Koul and Kalia (2016)
Staphylococcus	2.56–3.07	32.05–37.60	2386–3163	2275–3041	5–6	Kumar et al. (2016)
Streptococcus	1.75–2.38	35.60–41.60	1693–2408	1762–2270	4–7	Kalia et al. (2016)
Vibrio	4.03–6.32	38.37–47.49	3656–5807	3406–5574	7–11	Kalia et al. (2015)
Yersinia	3.60–4.96	46.95–48.05	3219–5596	3079–5498	6–8	Kalia and Kumar (2015)

Table 2 Genes commonly used for identifying different bacterial species

Genus	Genes	Reference
Lactobacillus	*recA*, *pheS*, *pyrG*, *tuf*, *sphI*, *mub*, *fbp*, *bsh*	Naser et al. (2007), Sarmiento-Rubiano et al. (2010), Nguyen et al. (2013), Yu et al. (2015), Koul and Kalia (2016)
Staphylococcus	*coa*, *femA*, *femB*, *gyrA*, *spa*, *mecA*, *atlE*, *tuf*, *ileS*	Brown et al. (2005), Pichon et al. (2012), Roberts (2014), Kumar et al. (2016)
Streptococcus	*cpsA*, *gdh*, *groESL*, *lytA*, *psaA*, *pspA*, *recA*, *recN*, *rpoA*, *rpoB*, *sodA*	Carvalho et al. (2007), Bishop et al. (2009), Abdeldaim et al. (2010), Kalia et al. (2016)
Vibrio	*rpoB*, *hsb60*, *sodB*, *flaE*	Tarr et al. (2007), Taneja et al. (2012), Bhattacharyya and Hou (2013), Kalia and Kumar (2015)

5 The Potential Alternatives to *rrs* Gene

Molecular tools used for distinguishing *Lactobacillus*, *Staphylococcus*, *Streptococcus* and *Vibrio* species involve a wide range of genes (Table 2). The molecular tools like (1) loop-mediated isothermal amplification (LAMP), (2) LAMP combined with lateral flow dipstick (Surasilp et al. 2011; Plaon et al. 2015; Thongkao et al. 2015), (3) a silicon-based optical thin-film biosensor chip (Bai et al. 2010), (4) Fourier transform infrared spectroscopy and (5) matrix-assisted laser desorption/ionization time-of-flight mass spectrometry (Kuhm et al. 2009; Ayyadurai et al. 2010; Stephan et al. 2011) are highly sensitive but costly as well.

Most of these genes had been effective in identifying bacteria up to the species level (Table 2) (Brown et al. 2005; Carvalho et al. 2007; Naser et al. 2007; Tarr et al. 2007; Bishop et al. 2009; Abdeldaim et al. 2010; Kingston et al. 2010;

Sarmiento-Rubiano et al. 2010; Pichon et al. 2012; Taneja et al. 2012; Bhattacharyya and Hou 2013; Liu et al. 2013; Nguyen et al. 2013; Roberts 2014; Illeghems et al. 2015; Kalia and Kumar 2015; Kalia et al. 2015, 2016; Yu et al. 2015; Koul and Kalia 2016; Kumar et al. 2016). A major limitation with the previous studies has been the use of genes which were not present in all the species of a given genus, e.g. *Yersinia* and *Lactobacillus* (Bhagat and Virdi 2007; Kishore et al. 2012; Raftis et al. 2014; Illeghems et al. 2015; Moroeanu et al. 2015; Petrova et al. 2015). In addition, multiple gene analyses have also been employed in many cases. The use of different genes in different studies reflects that in spite of their usage over a long period, no consensus gene has been identified as yet.

6 Screening Genomes for Biomarkers

An innovative approach for searching novel markers in organisms possessing multiple copies of *rrs* has been developed recently (Table 1) (Kalia et al. 2015, 2016; Kalia and Kumar 2015; Kekre et al. 2015; Koul et al. 2015a; Koul and Kalia 2016; Kumar et al. 2016). It involves finding genes, which are common to almost all the species of a genus. From this pool of common genes, 8–34 representative genes (200–4000 nucleotides) were selected (Table 3) (Kalia et al. 2015, 2016). In silico digestion of each of these genes with ten type II restriction endonucleases (REs) (4–6 base cutters) revealed interesting size and sequence of fragments. Unique RE digestion patterns are selected out and can be used for identification of particular organism with high precision (Table 4). Biomarkers for identifying strains have been deduced by using combination of REs (*AluI*, *BfaI*, *BfuCI*, *CviAII*, *HpyCH4V*, *RsaI*, *TaqI* and *Tru9I*) and common genes (Table 4): (a) *Clostridium*—(i) *recN*, *dnaJ*, *secA*, *mutS* and *grpB* (Kekre et al. 2015)(b) *Lactobacillus*—*dnaA*, *dnaJ*, *gyrB*, *pusA*, *recA* and *ruvB* (Koul and Kalia 2016)(c) *Staphylococcus*—*argH*, *argR*, *cysS*, *gyrB*, *purH*, *pyrE* and *recA* (Kumar et al. 2016)(d) *Streptococcus*—*dnaA*, *dnaK*, *gabG*, *mraY*, *purH*, *purK* and *pyrH* (Kalia et al. 2016)(e) *Vibrio*—*dapF*, *hisD*, *ilvH*, *lpxC*, *recF*, *recR*, *rph* and *ruvB* (Kalia et al. 2015)(f) *Yersinia*—*aceE*, *cysJ*, *fadJ*, *gltB*, *gyrB*, *leuD*, *ligA*, *mukB*, *rpoB* and *secA* (Kalia and Kumar 2015)This approach allows detection of bacteria even in a mixed population. It also offers a choice to select a single gene or a combination of genes. Since it does not involve any costly and sophisticated equipment, which demands time, money and highly skilled manpower, the chances of its being exploited on a large scale are high. Thus, the rapidity with which the bacteria can be identified will allow the treatment to initiate quickly.

It was interesting to learn that in the case of the genes which are common to different genera, this approach still has the potential to prove effective in distinguishing them unambiguously (Tables 5 and 6):(1) The *dnaJ* gene in *Clostridium* and *Lactobacillus* and the *gyrB* of *Lactobacillus* and *Yersinia* species were found to provide unique digestion patterns with RE—AluI (Table 5).(2) *purH* gene in *Staphylococcus* and *Streptococcus* showed unique patterns on digestion with REs

Table 3 List of genes common to most species within a genus: digested with different restriction endonucleases[a]

Gene	Genus						Gene	Genus					
	Clostridium	*Lactobacillus*	*Staphylococcus*	*Streptococcus*	*Vibrio*	*Yersinia*[b]		*Clostridium*	*Lactobacillus*	*Staphylococcus*	*Streptococcus*	*Vibrio*	*Yersinia*
aceE	–	–	–	–	+	+	*gyrB*	–	+	+	+	+	+
argG	–	+	+	–	–	–	*ileS*	–	–	–	–	+	+
argH	–	–	+	–	–	–	*ligA*	–	–	–	–	+	+
argR	–	–	+	+	–	–	*mraY*	–	–	+	+	–	–
argS	–	+	+	+	–	–	*murC*	–	–	+	+	–	–
aroB	–	–	+	–	–	–	*nusA*	–	–	–	+	+	+
aroD	–	–	–	+	–	–	*pheS*	–	+	+	–	–	–
asnB	–	–	–	–	+	–	*pheT*	–	–	+	+	+	+
carB	–	+	–	–	+	+	*purH*	–	–	+	+	–	–
clpB	+	–	–	–	–	–	*purK*	–	–	–	+	–	–
clpX	+	–	–	–	–	–	*purR*	–	–	–	+	–	–
cpxA	–	–	–	–	+	–	*pyrB*	–	+	–	–	–	–
cpxP	–	–	–	–	–	+	*pyrE*	–	–	+	–	–	–
cysJ	–	–	–	–	–	+	*pyrG*	–	+	–	+	–	–
cysS	–	+	+	+	–	+	*recA*	+	+	+	+	–	–
dnaA	+	+	+	+	–	–	*recR*	+	–	–	–	+	–
dnaE	–	–	–	–	+	–	*recU*	–	–	+	+	–	–
dnaJ	+	+	–	–	–	–	*rlmL*	–	–	–	–	+	–
dnaK	–	+	+	+	–	+	*rpoB*	–	–	–	–	–	+
fabG	–	–	–	+	–	–	*ruvB*	+	+	+	+	+	+
glyA	–	–	–	+	–	–	*secA*	+	–	–	+	+	+
guaA	–	+	+	–	–	–	*trm D*	–	+	–	+	–	–

References: Kekre et al. (2015), Kalia and Kumar (2015), Kalia et al. (2015, 2016), Koul and Kalia (2016), Kumar et al. (2016)

[a]*AluI*, *BfaI*, *BfuCI*, *CviAII*, *HaeI*, *HinI*, *HpyCH4V*, *RsaI*, *TaqI* and *Tru9I*

[b]The following genes were studies only in the case of *Yersinia: aceK*, *fadJ*, *feoB*, *glmU*, *glpQ*, *gltV*, *gltX*, *hisG*, *lecZ*, *leuD*, *lolD*, *malE*, *metH*, *mltA*, *mukB*, *nagB*, *ribD*, *rnpA*, *rplW* and *thiP* (Kalia and Kumar 2015)

Table 4 Gene–restriction endonuclease combinations with unique digestion patterns

Organisms	Genes	Restriction endonucleases								References
		AluI	*BfaI*	*BfuCI*	*CviAII*	*HpyCH4V*	*RsaI*	*TaqI*	*Tru9I*	
Clostridium	*recN*	+[a]	+	–[b]	–	–	–	–	–	Kekre et al. (2015)
	dnaJ	+	–	–	–	–	–	–	–	
	secA	+	–	–	–	–	–	–	–	
	mutS	–	+	–	–	–	–	–	+	
	grpB	–	–	–	–	–	–	–	+	
Lactobacillus	*dnaA*	+	–	–	–	–	–	–	–	Koul and Kalia (2016)
	dnaJ	+	–	+	+	–	–	–	–	
	gyrB	+	–	–	–	–	–	+		
	pusA	+	–	–	–	–	–	–	+	
	recA	–	–	+	+	+	+	–	–	
	ruvB	+	–	+	–	–	–	+	–	
Staphylococcus	*argH*	–	–	+	–	–	–	–	–	Kumar et al. (2016)
	argR	–	–	–	–	–	–	–	+	
	cysS	+	–	–	–	–	–	–	–	
	gyrB	–	–	–	–	–	+	–	–	
	purH	–	–	+	+	+	–	–	–	
	pyrE	–	–	–	–	–	–	–	+	
	recA	+	–	–	+	+	–	–	+	
Streptococcus	*dnaA*	–	–	–	–	+	–	–	–	Kalia et al. (2016)
	dnaK	–	–	–	–	–	–	+	+	
	fabG	–	–	–	–	–	–	–	+	
	mraY	+	–	–	–	–	–	–	–	
	purH	–	–	+	+	–	–	+	+	
	purK	+	+	–	–	–	–	–	+	
	pyrH	+	–	–	–	+	–	–	+	

(continued)

Table 4 (continued)

Organisms	Genes	Restriction endonucleases								References
		AluI	*BfaI*	*BfuCI*	*CviAII*	*HpyCH4V*	*RsaI*	*TaqI*	*Tru9I*	
Vibrio	*dapF*	–	–	+	+	+	–	+	–	Kalia et al. (2015)
	hisD	+	–	+	–	+	+	+	–	
	ilvH	+	–	–	–	+	–	+	–	
	lpxC	+	–	–	–	–	+	+	+	
	recF	+	–	–	–	+	+	+	–	
	recR	+	–	–	–	+	+	+	–	
	rph	–	–	+	+	+	–	–	+	
	ruvB	–	–	+	+	+	–	+	–	
Yersinia	*aceE*	–	–	+	–	–	–	–	–	Kalia and Kumar (2015)
	cysJ	–	–	–	–	+	–	–	–	
	fadJ	+	–	+	+	+	–	–	–	
	gltB	+	–	+	–	–	+	–	–	
	gyrB	+	–	–	–	–	–	–	–	
	leuD	–	–	–	–	+	–	–	+	
	ligA	+	–	–	–	–	–	–	–	
	mukB	+	–	+	+	+	–	+	–	
	rpoB	–	–	–	–	–	–	–	+	
	secA	+	–	+	–	–	–	+	–	

[a]Unique RE digestion pattern
[b]Not applicable

Table 5 Unique restriction endonuclease digestion patterns in genes common to different genera[a]

Genus	Genes	Restriction endonuclease	References
		AluI	
Clostridium	*dnaJ*	74•33•26•190•491•175•47	Kekre et al. (2015)
Lactobacillus	*dnaJ*	84•268•66•330•109•287•20	Koul and Kalia (2016)
Lactobacillus	*gyrB*	1041•391•280•238	Koul and Kalia (2016)
Yersinia	*gyrB*	884•266•12•775•216•262	Kalia and Kumar (2015)
Staphylococcus	*gyrB*	–	Kumar et al. (2016)
		TaqI	
Lactobacillus	*ruvB*	344•271•184•171•50•27	Koul and Kalia (2016)
Vibrio	*ruvB*	19•121•218•117•30•253•319	Kalia and Kumar (2015)

[a]These patterns are representative of the genus.

Table 6 Unique restriction endonuclease digestion patterns in genes common to different genera[a]

		Restriction endonucleases		
Genus	Genes	*CviAII*	*BfuCI*	*HpyCH4V*
Staphylococcus	*purH*	208•207•33•222•98•8•44•477•63•72•47	234•284•784•108•78	+
Streptococcus	*purH*	15•211•36•21•144•413•8•448•267	117•318•438•261•414	–
Lactobacillus	*recA*	299•232•232•231•80•24	+	296•242•228•202•52•33
Staphylococcus	*recA*	133•57•447•47•204•445•32	–	145•237•153•509

References: Kalia et al. (2016), Koul and Kalia (2016), Kumar et al. (2016)
[a]These patterns are representative of the genus

(*CviAII* and *BfuCI*) and *recA* gene in *Lactobacillus* and *Staphylococcus* with REs—*CviAII* and *HpyCH4V* (Table 6).(3) The *ruvB* gene–*TaqI* combinations in *Lactobacillus* and *Vibrio* species (Table 5).

7 Opinion

Very closely related species and strains possess genes which differ only in a few nucleotides. Efforts to exploit these differences on a global scale don't provide any clues for their being distinct. The search for nucleotide signatures unique to a species along with a unique RE digestion pattern allowed clear-cut distinction to a large extent. However, this approach is likely to fail in case of organisms possessing multiple copies of *rrs*. In either scenario where *rrs* is not able to distinguish organisms, an obvious approach is to go in for exploiting other HKGs. In fact, most studies which used HKGs were successful only on employing information from six to eight genes. Here again, the selection of genes may be limited by the fact that the same genes may not be present in all the strains or species of a genus. Thus, the need is to bank upon genes which are present in most if not all the species of a genus. The approach used in these works relied primarily on genes common to all the sequence genomes. And in case two strains seem quite close in their nucleotide sequences, an additional gene–RE combination can be searched using in silico approach and primers can be developed accordingly. This approach has the potential to be extended to other genes of an organism.

Acknowledgements We are thankful to the Director of CSIR-Institute of Genomics and Integrative Biology (IGIB) and CSIR project GENESIS (BSC0121) for providing the necessary funds, facilities and moral support. Authors are also thankful to Academy of Scientific and Innovative Research (AcSIR), New Delhi.

References

Abdeldaim G, Herrmann B, Mölling P, Holmberg H, Blomberg J, Olcén P, Strålin K (2010) Usefulness of real-time PCR for *lytA*, *ply*, and *Spn9802* on plasma samples for the diagnosis of pneumococcal pneumonia. Clin Microbiol Infect 16:1135–1141. doi:10.1111/j.1469-0691.2009.03069.x

Agarwala M, Choudhury B, Yadav RNS (2014) Comparative study of antibiofilm activity of copper oxide and iron oxide nanoparticles against multidrug resistant biofilm forming uropathogens. Indian J Microbiol 54:365–368. doi:10.1007/s12088-014-0462-z

Alipiah NM, Shamsudin MN, Yusoff FM, Arshad A (2015) Membrane biosynthesis gene disruption in methicillin-resistant *Staphylococcus aureus* (MRSA) as potential mechanism for reducing antibiotic resistance. Indian J Microbiol 54:41–49. doi:10.1007/s12088-014-0488-2

Arasu MV, Al-Dhabi NA, Rejiniemon TS, Lee KD, Huxley VAJ, Kim DH, Duraipandiyan V, Karuppiah P, Choi KC (2015) Identification and characterization of *Lactobacillus brevis* P68 with antifungal, antioxidant and probiotic functional properties. Indian J Microbiol 55:19–28. doi:10.1007/s12088-014-0495-3

Arya R, Princy SA (2016) Exploration of modulated genetic circuits governing virulence determinants in *Staphylococcus aureus*. Indian J Microbiol 56:19–27. doi:10.1007/s12088-015-0555-3

Ayyadurai S, Flaudrops C, Raoult D, Drancourt M (2010) Rapid identification and typing of *Yersinia pestis* and other *Yersinia* species by matrix-assisted laser desorption/ionization time-of flight (MALDI-TOF) mass spectrometry. BMC Microbiol 10:285. doi:10.1186/1471-2180/10/285

Bai S, Zhao J, Zhang Y, Huang W, Xu S, Chen H, Fan L-M, Chen Y, Deng XW (2010) Rapid and reliable detection of 11 foodborne pathogens using thin-film biosensor chips. Appl Microbiol Biotechnol 86:983–990. doi:10.1007/s00253-009-2417-6

Bhagat N, Virdi JS (2007) Distribution of virulence associated genes in *Yersinia enterocolitica* biovar 1A correlates with clonal groups and not the source of isolation. FEMS Microbiol Lett 266:177–183. doi:10.1111/j.1574-6968.2006.00524.x

Bhattacharyya N, Hou A (2013) A pentaplex PCR assay for detection and characterization of *Vibrio vulnificus* and *Vibrio parahaemolyticus* isolates. Lett Appl Microbiol 57:233–240. doi:10.1111/lam.12101

Bhushan A, Joshi J, Shankar P, Kushwah J, Raju SC, Purohit HJ, Kalia VC (2013) Development of genomic tools for the identification of certain *Pseudomonas* up to species level. Indian J Microbiol 53:253–263. doi:10.1007/s12088-013-0412-1

Bhushan A, Mukherjee T, Joshi J, Shankar P, Kalia VC (2015) Insights into the origin of *Clostridium botulinum* strains: evolution of distinct restriction endonuclease sites in *rrs* (16S rRNA gene). Indian J Microbiol 55:140–150. doi:10.1007/s12088-015-0514-z

Bishop CJ, Aanensen DM, Jordan GE, Kilian M, Hanage WP, Spratt BG (2009) Assigning strains to bacterial species via the Internet. BMC Biol 7:3. doi:10.1186/1741-7007-7-3

Brown DF, Edwards DI, Hawkey PM, Morrison D, Ridgway GL, Towner KJ, Wren MW, Joint Working Party of the British Society for Antimicrobial Chemotherapy, Hospital Infection Society, Infection Control Nurses Association (2005) Guidelines for the laboratory diagnosis and susceptibility testing of methicillin-resistant *Staphylococcus aureus* (MRSA). J Antimicrob Chemother 56:1000–1018. doi:10.1093/jac/dki372

Carvalho MG, Tondella ML, McCaustland K, Weidlich L, McGee L, Mayer LW, Steigerwalt A, Whaley M, Facklam RR, Fields B, Carlone G, Ades GW, Dagan R, Sampson JS (2007) Evaluation and improvement of real-time PCR assays targeting *lytA*, ply, and *psaA* genes for detection of pneumococcal DNA. J Clin Microbiol 45:2460–2466. doi:10.1128/JCM.02498-06

Gautam N, Sharma N, Ahlawat OP (2014) Purification and characterization of bacteriocin produced by *Lactobacillus brevis* UN isolated from Dhulliachar: a traditional food product of North East India. Indian J Microbiol 54:185–189. doi:10.1007/s12088-013-0427-7

Gui Z, Wang H, Ding T, Zhu W, Zhuang X, Chu W (2014) Azithromycin reduces the production of α-hemolysin and biofilm formation in *Staphylococcus aureus*. Indian J Microbiol 54:114–117. doi:10.1007/s12088-013-0438-4

Hema M, Balasubramanian S, Princy SA (2015) Meddling *Vibrio cholerae* murmurs: a neoteric advancement in cholera research. Indian J Microbiol 55:121–130. doi:10.1007/s12088-015-0520-1

Illeghems K, De Vuyst L, Weckx S (2015) Comparative genome analysis of the candidate functional starter culture strains *Lactobacillus fermentum* 222 and *Lactobacillus plantarum* 80 for controlled cocoa bean fermentation processes. BMC Genomics 16:766. doi:10.1186/s12864-015-1927-0

Kalia VC (2010) Extending genomic limits through metagenomic exploration. J Cosmol 13:3625–3627

Kalia VC (2013a) Quorum sensing inhibitors: an overview. Biotechnol Adv 31:224–245. doi:10.1016/j.biotechadv.2012.10.004

Kalia VC (2013b) The visionary: Prof. Carl R. Woese. Indian J Microbiol 53:245–246. doi:10.1007/s12088-013-0417-9

Kalia VC (2014) Microbes, antimicrobials and resistance: the battle goes on. Indian J Microbiol 54:1–2. doi:10.1007/s12088-013-0443-7

Kalia VC (2015a) Let's explore the latent features of genes to identify bacteria. J Mol Genet Med 9, e105. doi:10.4172/1747-0862.1000E105

Kalia VC (2015b) Microbes: the most friendly beings? In: Kalia VC (ed) Quorum sensing vs quorum quenching: a battle with no end in sight. Springer India, pp 1–5. ISBN 978-81-322-1981-1. doi:10.1007/978-81-322-1982-8_1

Kalia VC, Kumar P (2015) Genome wide search for biomarkers to diagnose *Yersinia* infections. Indian J Microbiol 55:366–374. doi:10.1007/s12088-015-0552-6

Kalia VC, Purohit HJ (2011) Quenching the quorum sensing system: potential antibacterial drug targets. Crit Rev Microbiol 37:121–140. doi:10.3109/1040841X.2010.532479

Kalia VC, Mukherjee T, Bhushan A, Joshi J, Shankar P, Huma N (2011a) Analysis of the unexplored features of *rrs* (16S rDNA) of the genus *Clostridium*. BMC Genomics 12:18. doi:10.1186/1471-2164-12-18

Kalia VC, Raju SC, Purohit HJ (2011b) Genomic analysis reveals versatile organisms for quorum quenching enzymes: acyl-homoserine lactone-acylase and -lactonase. Open Microbiol J 5:1–13. doi:10.2174/1874285801105010000

Kalia VC, Kumar P, Kumar R, Mishra A, Koul S (2015) Genome wide analysis for rapid identification of *Vibrio* species. Indian J Microbiol 55:375–383. doi:10.1007/s12088-015-0553-5

Kalia VC, Kumar R, Kumar P, Koul S (2016) A genome-wide profiling strategy as an aid for searching unique identification biomarkers for *Streptococcus*. Indian J Microbiol 56:46–58. doi:10.1007/s12088-015-0561-5

Kaur G, Rajesh S, Princy SA (2015) Plausible drug targets in the *Streptococcus mutans* quorum sensing pathways to combat dental biofilms and associated risks. Indian J Microbiol 55:349–357. doi:10.1007/s12088-015-0534-8

Kekre A, Bhushan A, Kumar P, Kalia VC (2015) Genome wide analysis for searching novel markers to rapidly identify *Clostridium* strains. Indian J Microbiol 55:250–257. doi:10.1007/s12088-015-0535-7

Kingston JJ, Radhika M, Roshini PT, Raksha MA, Murali HS, Batra HV (2010) Molecular characterization of lactic acid bacteria recovered from natural fermentation of beet root and carrot Kanji. Indian J Microbiol 50:292–298. doi:10.1007/s12088-010-0022-0

Kishore P, Lalitha KV, Joseph TC, Thampuran N (2012) Biotyping and antibiotic resistance profile of *Yersinia enterocolitica* associated with seafoods from South-west coast of India. Fish Technol 49:64–71

Klappenbach JA, Saxman PR, Cole JR, Schmidt TM (2001) rrndb: the ribosomal RNA operon copy number database. Nucleic Acids Res 29:181–184. doi:10.1093/nar/29.1.181

Koul S, Kalia VC (2016) Comparative genomics reveals biomarkers to identify *Lactobacillus* species. Indian J Microbiol 56:253–263. doi:10.1007/s12088-016-0605-5

Koul S, Kumar P, Kalia VC (2015a) A unique genome wide approach to search novel markers for rapid identification of bacterial pathogens. J Mol Genet Med 9:194. doi:10.4172/1747-0862.1000194

Koul S, Prakash J, Mishra A, Kalia VC (2015b) Potential emergence of multi-quorum sensing inhibitor resistant (MQSIR) bacteria. Indian J Microbiol. doi:10.1007/s12088-015-0558-0

Kuhm AE, Suter D, Felleisen R, Rau J (2009) Identification of *Yersinia enterocolitica* at the species and subspecies levels by Fourier transform infrared spectroscopy. Appl Environ Microbiol 75:5809–5813. doi:10.1128/AEM.00206-09

Kumar R, Koul S, Kumar P, Kalia VC (2016) Searching biomarkers in the sequenced genomes of *Staphylococcus* for their rapid identification. Indian J Microbiol 56:64–71. doi:10.1007/s12088-016-0565-9

Lal D, Verma M, Lal R (2011) Exploring internal features of 16S rRNA gene for identification of clinically relevant species of the genus *Streptococcus*. Ann Clin Microbiol Antimicrob 10:28. doi:10.1186/1476-0711-10-28

Liu Q, Wang S, Zhi J-F, Ming H, Teng D (2013) Efficient production of lactic acid from sweet sorghum juice by a newly isolated *Lactobacillus salivarius* CGMCC 7.75. Indian J Microbiol 53:332–336. doi:10.1007/s12088-013-0377-0

Mahale KN, Paranjape PS, Marathe NP, Dhotre DP, Chowdhury S, Shetty SA, Sharma A, Sharma K, Tuteja U, Batra HV, Shouche YS (2014) Draft genome sequences of *Yersinia pestis* strains from the 1994 plague epidemic of Surat and 2002 Shimla outbreak in India. Indian J Microbiol 54:480–482. doi:10.1007/s12088-014-0475-7

Meza-Lucas A, Pérez-Villagómez M, Martínez-López JP, García-Rodea R, Martínez-Castelán MG, Escobar-Gutiérrez A, de-la-Rosa-Arana JL, Villanueva-Zamudio A (2016) Comparison of DOT-ELISA and Standard-ELISA for detection of the *Vibrio cholerae* toxin in culture supernatants of bacteria isolated from human and environmental samples. Indian J Microbiol 56:379–382. doi:10.1007/s12088-016-0596-2

Moroeanu VI, Vamanu E, Paun G, Neagu E, Ungureanu OR, Eremia SAV, Radu GL, Ionescu R, Pelinescu DR (2015) Probiotic strains influence on infant microbiota in the in vitro colonic fermentation model GIS1. Indian J Microbiol 55:423–429. doi:10.1007/s12088-015-0542-8

Naser SM, Dawyndt P, Hoste B, Gevers D, Vandemeulebroecke K, Cleenwerck I, Vancanneyt M, Swings J (2007) Identification of lactobacilli by *pheS* and *rpoA* gene sequence analysis. Int J Syst Evol Microbiol 57:2777–2789. doi:10.1099/ijs.0.64711-0

Nguyen THK, Doan VTT, Ha LD, Nguyen HN (2013) Molecular cloning, expression of *minD* gene from *Lactobacillus acidophilus* VTCC-B-871 and analyses to identify *Lactobacillus rhamnosus* PN04 from Vietnam *Hottuynia cordata* Thunb. Indian J Microbiol 53:385–390. doi:10.1007/s12088-013-0384-1

Petrova MI, Lievens E, Malik S, Imholz N, Lebeer S (2015) *Lactobacillus* species as biomarkers and agents that can promote various aspects of vaginal health. Front Physiol 6:81. doi:10.3389/fphys.2015.00081

Pichon B, Hill R, Laurent F, Larsen AR, Skov RL, Holmes M, Edwards GF, Teale C, Kearns AM (2012) Development of a real-time quadruplex PCR assay for simultaneous detection of *nuc*, Panton–Valentine leucocidin (PVL), *mecA* and homologue $mecA_{LGA251}$. J Antimicrob Chemother 67:2338–2341. doi:10.1093/jac/dks221

Plaon S, Longyant S, Sithigorngul P, Chaivisuthangkura P (2015) Rapid and sensitive detection of *Vibrio alginolyticus* by loop mediated isothermal amplification combined with a lateral flow dipstick targeted to the rpoX gene. J Aquat Anim Health 27:156–163. doi:10.1080/08997659.2015.1037468

Porwal S, Lal S, Cheema S, Kalia VC (2009) Phylogeny in aid of the present and novel microbial lineages: diversity in *Bacillus*. PLoS One 4, e4438. doi:10.1371/journal.pone.0004438

Prakash O, Jangid K, Shouche YS (2013) Carl Woese: from biophysics to evolutionary microbiology. Indian J Microbiol 53:247–252. doi:10.1007/s12088-013-0401-4

Prakash O, Pandey PK, Kulkarni GJ, Mahale KN, Shouche YS (2014) Technicalities and glitches of terminal restriction fragment length polymorphism (T-RFLP). Indian J Microbiol 54:255–261. doi:10.1007/s12088-014-0461-0

Prakasham RS, Kumar BS, Kumar YS, Kumar KP (2014) Production and characterization of protein encapsulated silver nanoparticles by marine isolate *Streptomyces parvulus* SSNP11. Indian J Microbiol 54:329–336. doi:10.1007/s12088-014-0452-1

Puri A, Rai A, Dhanaraj PS, Lal R, Patel DD, Kaicker A, Verma M (2016) An *in silico* approach for identification of the pathogenic species, *Helicobacter pylori* and its relatives. Indian J Microbiol 56:277–286. doi:10.1007/s12088-016-0575-7

Raftis EJ, Forde BM, Claesson MJ, O'Toole PW (2014) Unusual genome complexity in *Lactobacillus salivarius* JCM1046. BMC Genomics 15:771. doi:10.1186/1471-2164-15-771

Roberts AL (2014) Identification of *Staphylococcus epidermidis* in the clinical microbiology laboratory by molecular methods. Methods Mol Biol 1106:33–53. doi:10.1007/978-1-62703-736-5_3

Sajid A, Arora G, Singhal A, Kalia VC, Singh Y (2015) Protein phosphatases of pathogenic bacteria: role in physiology and virulence. Annu Rev Microbiol 69:527–547. doi:10.1146/annurevmicro-020415-111342

Sarmiento-Rubiano LA, Berger B, Moine D, Zúñiga M, Pérez-Martínez G, Yebra MJ (2010) *Characterization* of a novel *Lactobacillus species* closely related to *Lactobacillus johnsonii* using a combination of molecular and comparative genomics methods. BMC Genomics 11:504. doi:10.1186/1471-2164-11-504

Saxena A, Mukherjee M, Kumari R, Singh P, Lal R (2014) Synthetic biology in action: developing a drug against MDR-TB. Indian J Microbiol 54:369–375. doi:10.1007/s12088-014-0498-0

Shang Z, Wang H, Zhou S, Chu W (2014) Characterization of N-acyl-homoserine lactones (AHLs)-deficient clinical isolates of *Pseudomonas aeruginosa*. Indian J Microbiol 54:158–162. doi:10.1007/s12088-014-0449-9

Stephan R, Cernela N, Ziegler D, Pflueger V, Tonolla M, Ravasi D, Fredriksson-Ahomaa M, Haechler H (2011) Rapid species specific identification and subtyping of *Yersinia enterocolitica* by MALDI-TOF mass spectrometry. J Microbiol Methods 87:150–153. doi:10.1016/j.mimet.2011.08.016

Surasilp T, Longyant S, Rukpratanporn S, Sridulyakul P, Sithigorngul P, Chaivisuthangkura P (2011) Rapid and sensitive detection of *Vibrio vulnificus* by loop-mediated isothermal amplification combined with lateral flow dipstick targeted to *rpoS* gene. Mol Cell Probes 25:158–163. doi:10.1016/j.mcp.2011.04.001

Taneja N, Sangar G, Chowdhury G, Ramamurthy T, Mishra A, Singh M, Sharma M (2012) Molecular epidemiology of *Vibrio cholerae* causing outbreaks & sporadic cholera in northern India. Indian J Med Res 136:656–663

Tarr CL, Patel JS, Puhr ND, Sowers EG, Bopp CA, Strockbine NA (2007) Identification of *Vibrio* isolates by a multiplex PCR assay and *rpoB* sequence determination. J Clin Microbiol 45:134–140. doi:10.1128/JCM.01544-06

Thongkao K, Longyant S, Silprasit K, Sithigorngul P, Chaivisuthangkura P (2015) Rapid and sensitive detection of *Vibrio harveyi* by loop-mediated isothermal amplification combined with lateral flow dipstick targeted to vhhP2 gene. Aquac Res 46:1122–1131. doi:10.1111/are.12266

Wang R, Fang S, Xiang S, Ling S, Yuan J, Wang S (2014) Generation and characterization of a scFv antibody against T3SS needle of *Vibrio parahaemolyticus*. Indian J Microbiol 54:143–150. doi:10.1007/s12088-013-0428-6

Yagnik B, Patel S, Dave M, Sharma D, Padh H, Desai P (2016) Factors affecting inducible expression of outer membrane protein A (OmpA) of *Shigella dysenteriae* Type-1 in *Lactococcus lactis* using nisin inducible controlled expression (NICE). Indian J Microbiol 56:80–87. doi:10.1007/s12088-015-0556-2

Yu S, Peng Y, Chen W, Deng Y, Zheng Y (2014) Comparative genomic analysis of two-component signal transduction systems in probiotic *Lactobacillus casei*. Indian J Microbiol 54:293–301. doi:10.1007/s12088-014-0456-x

Yu S, Peng Y, Zheng Y, Chen W (2015) Comparative genome analysis of *Lactobacillus casei*: insights into genomic diversification for niche expansion. Indian J Microbiol 55:102–107. doi:10.1007/s12088-014-0496-2

Current Antifungal Therapy and Drug Resistance Mechanisms in Dermatophytes

Pawan Kumar, Chitra Latka, and Bhupesh Taneja

Abstract Dermatophytosis is the invasion of keratinized tissue by a group of specialized keratinolytic filamentous fungi called dermatophytes. It is the most common superficial fungal infection affecting millions of people annually worldwide. Dermatophytes have the unique ability to infect immunocompetent people and are associated with considerable morbidity and socioeconomic trauma. The annual estimated burden of treatment of superficial cutaneous fungal infections is nearly $1.7 billion due to direct drug costs in the USA alone.

Among the major current antifungal agents to treat dermatophytosis are polyenes, azoles, griseofulvin and allylamines. While there are many reports of development of drug resistance to azoles in yeasts and molds; dermatophytes usually respond well to antifungal agents. Although treatment of dermatophytosis is usually long term, with several cases of recurrence and numerous side effects, drug resistance in dermatophytes is rare. However, several recent cases of drug resistance in dermatophytes are beginning to emerge. Here, we describe the available therapeutics against fungal infections and some of the resistance mechanisms in fungi and finally highlight the current understanding of drug resistance mechanisms in dermatophytes.

1 Introduction

Fungi are a diverse group of eukaryotic organisms that can exist both in unicellular and multicellular forms in widely diverse environments. As many as 1.5 million species of fungi are known (Hawksworth 1991), many of which cause infections in humans, animals, and plants (Fisher et al. 2012). Fungi have several ecological niches in the human body as well. While the skin harbors many microbes

P. Kumar • C. Latka • B. Taneja (✉)
Genome Informatics and Structural Biology Unit, CSIR – Institute of Genomics and Integrative Biology, Mathura Road, New Delhi 110025, India

Academy of Scientific and Innovative Research, New Delhi 110025, India
e-mail: btaneja@igib.res.in

G. Arora et al. (eds.), *Drug Resistance in Bacteria, Fungi, Malaria, and Cancer*,
DOI 10.1007/978-3-319-48683-3_17

asymptomatically, some of the fungal species may cause infections. As much as 10–20 % of the world population is estimated to carry a cutaneous fungal infection of one type or another during any calendar year (Bickers et al. 2006). Fungal infections in humans may range from superficial common infections, such as dermatophytoses, to deep invasive and disseminated infections (including opportunistic infections), such as candidiasis and aspergillosis. Superficial fungal infections may be caused by dermatophytes or non-dermatophytes. Dermatophytes are a group of highly specialized keratinolytic fungi, belonging to three major fungal genera, namely, *Trichophyton* spp., *Microsporum* spp., and *Epidermophyton* spp. that cause dermatophytosis, commonly referred to as ringworm infections. Dermatophytosis is also referred to as "tinea" infections and named with reference to the area of the infected body part, e.g., tinea barbae (ringworm of facial area with beard and mustache), tinea capitis (ringworm of scalp), tinea corporis (ringworm of the trunk, shoulder, or limbs), tinea cruris (ringworm of groin), tinea manuum (ringworm of the hand), tinea pedis (athlete's foot), tinea unguium (infection of nails) and so on (Weitzman and Summerbell 1995).

Dermatophytosis is the most commonly observed superficial fungal infection that affects millions of people annually worldwide (Weitzman and Summerbell 1995; Brown et al. 2012; Fisher et al. 2012). It is associated with considerable morbidity and socioeconomic trauma. Total annual direct costs attributed to treatment of cutaneous fungal infections is approximately $1.7 billion in the USA alone, and another estimated $282 million are attributed to indirect costs due to lost productivity of the patients (Bickers et al. 2006). Dermatophyte infections can last for months, even when treated, and affect the quality of life of the patient due to associated symptoms such as pain or itching that may occur due to the infections.

Despite significant understanding of the causative pathogens, the mechanistic details for dermatophytes to be confined to the superficial areas of the skin (in immunocompetent people) are poorly understood. Delivery of drugs to these areas at effective therapeutic levels is a major limitation in the clearance of the pathogen. Consequently, intervention methods to treat dermatophyte infections are usually required for long periods of time resulting in several side effects. Poor patient compliance with the long-term therapy as a result of these associated side effects results in several cases of recurrence. The problem is further compounded if they become resistant to currently available therapeutics. Drug resistance in dermatophytes to some of the common antifungal agents is now beginning to emerge (Mukherjee et al. 2003; Alipour and Mozafari 2015; Ghannoum 2016). In the subsequent sections, we briefly describe the currently available clinical antifungal agents and associated resistance mechanisms in yeast and other fungi and finally highlight the current understanding of drug resistance mechanisms in dermatophytes.

2 Fungal Infection and Antifungal Drugs

Antifungal agents in current clinical use are limited to few distinct molecular classes that differ in their efficacies depending on type and site of infections. For instance, azoles, fluoropyrimidine analogs, glucan synthesis inhibitors, and polyenes are generally used to treat systemic fungal infections (Georgopapadakou 1998; Sanglard 2002; Sanglard and Odds 2002), while the use of allylamines, viz., terbinafine, is limited to superficial infections due to adverse side effects during systemic treatments (Vandeputte et al. 2012; Sanglard 2016). All these agents appear to target one of the three distinct fungal metabolic pathways, i.e., synthesis of fungal cell wall component, ergosterol pathway, and nucleotide synthesis. These classes are described below. Representative structures of antifungal agents from each major class described are given in Fig. 1.

2.1 *Fluoropyrimidines*

Molecular Properties and Mechanism of Action Pyrimidine analogs, viz., 5-fluorocytosine and 5-fluorouracil, are synthetic structural analogs of DNA or RNA nucleosides.

5-Fluorocytosine (Fig. 1a) is a prodrug that enters the fungal cell through specific transporters (such as cytosine permeases) and is then converted into 5-fluorouracil by cytosine deaminase. 5-Fluorouracil is then converted into 5-fluorouracil monophosphate (5-FUMP) by uridine phosphoribosyltransferase and finally to its corresponding nucleotide triphosphate, 5-fluorouracil triphosphate

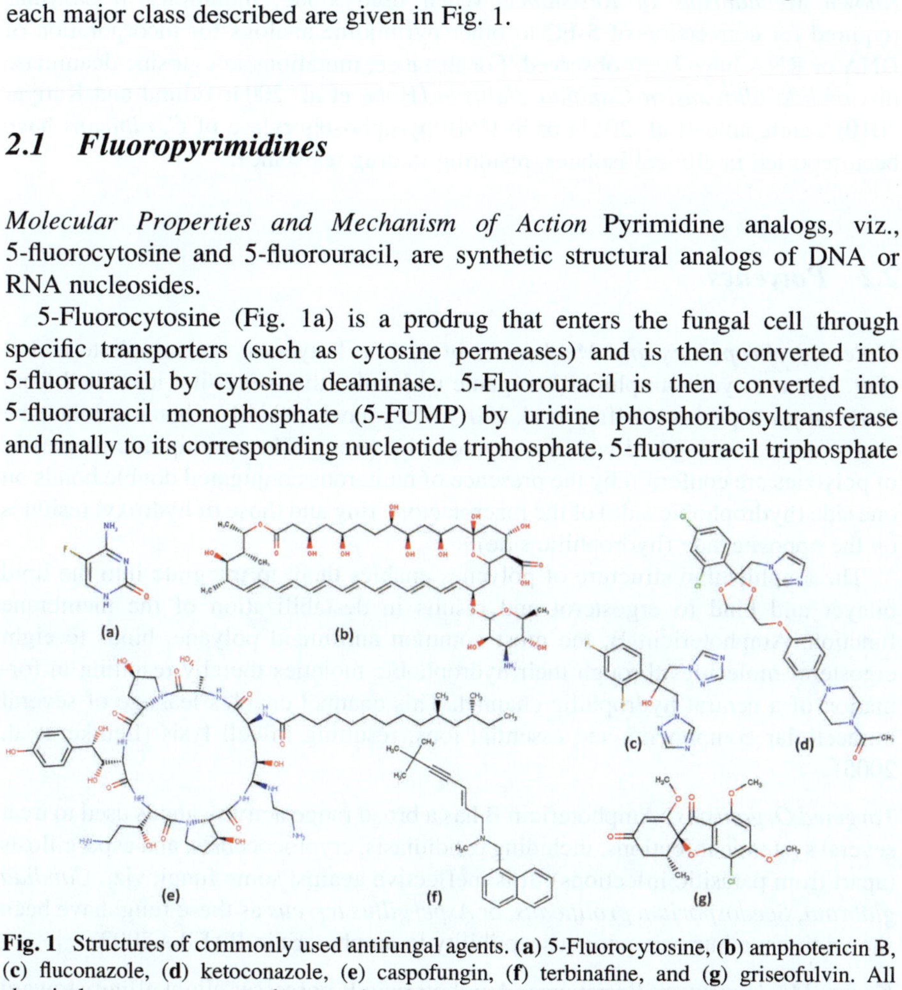

Fig. 1 Structures of commonly used antifungal agents. (**a**) 5-Fluorocytosine, (**b**) amphotericin B, (**c**) fluconazole, (**d**) ketoconazole, (**e**) caspofungin, (**f**) terbinafine, and (**g**) griseofulvin. All chemical structures were drawn using Marvin 16.8.22 (2016), ChemAxon (http://www.chemaxon.com)

(5-FUTP). 5-FUTP may then be incorporated into RNA instead of UTP and inhibit protein synthesis (Bossche et al. 1994; Georgopapadakou 1998). In addition, 5-FUMP may also be converted into the corresponding deoxynucleotide, 5-fluorodeoxyuridine monophosphate, which inhibits thymidylate synthase, thereby inhibiting cell replication and resulting in growth arrest.

Targeted Organisms 5-Fluorocytosine is often used to successfully treat systemic candidiasis and cryptococcal meningitis infections. However, it is not effective against filamentous fungi, viz., *Aspergillus* and dermatophytes. Moreover, 5-FC is rarely used as monotherapy due to rapid emergence of resistance, and its use is more common as a combination therapy with amphotericin B or azoles (Bennett et al. 1979).

Known Mechanisms of Resistance When used alone, mutations in enzymes required for conversion of 5-FC to other pyrimidine analogs for incorporation of DNA or RNA have been observed. For instance, mutations in cytosine deaminase of *Candida albicans* or *Candida glabrata* (Hope et al. 2004; Edlind and Katiyar 2010; Vandeputte et al. 2011) or in UMP pyrophosphorylase of *C. albicans* have been reported in clinical isolates, resulting in drug resistance.

2.2 Polyenes

Molecular Properties and Mechanism of Action Polyenes, viz., amphotericin B (Fig. 1b), are cyclic amphiphilic organic molecules that were first identified from *Streptomyces* species (Caffrey et al. 2001). They have a 20–40 carbon macrolactone ring structure conjugated with a D-mycosamine group. The amphiphilic properties of polyenes are conferred by the presence of numerous conjugated double bonds on one side (hydrophobic side) of the macrolactone ring and those of hydroxyl residues on the opposite side (hydrophilic side).

The amphipathic structure of polyenes enables them to integrate into the lipid bilayer and bind to ergosterol and results in destabilization of the membrane function. Amphotericin B, the most common antifungal polyene, binds to eight ergosterol molecules through their hydrophobic moieties thereby resulting in formation of a central hydrophilic channel. This channel enables leakage of several intracellular components and essential ions, resulting in cell lysis (Lemke et al. 2005).

Targeted Organisms Amphotericin B has a broad range activity and is used to treat several systemic infections, including candidiasis, cryptococcosis, and aspergillosis (apart from parasitic infections). It is ineffective against some fungi, viz., *Candida glabrata*, *Scedosporium prolificans*, or *Aspergillus terreus* as these fungi have been shown to have intrinsic poor susceptibility to amphotericin B (Ellis 2002).

Known Mechanisms of Resistance Amphotericin B possesses slight affinity toward cholesterol, leading to several side effects. It also exerts intrinsic hepato- and nephrotoxicity, which may be overcome with liposomal formulations to minimize the side effects (Barrett et al. 2003). Intrinsic clinical resistance in the targeted pathogens, viz., *Candida* and *Aspergillus*, is less common, and toxicity remains a bigger challenge. The few reported cases of drug resistance have been found to be associated with alteration in membrane lipids. Some of these changes have been found to be low levels of ergosterol in the mutants or disturbance in the levels and composition of phospholipids (Sokol-Anderson et al. 1988). Mutations in *ERG2*, *ERG3*, and *ERG11* genes, involved in the ergosterol biosynthetic pathway, have been reported. Treatment with polyenes induces oxidative stress in fungal cells; resistant strains, hence, may exhibit alterations in the enzymes involved in oxidative stress response (Warn et al. 2004; Cuenca-Estrella 2014).

2.3 Azoles

Molecular Properties and Mechanism of Action Azoles are synthetic organic cyclic molecules that are now widely used as antifungal agents. They are classified as imidazoles, viz., miconazole and ketoconazole (Fig. 1d), or triazoles, viz., itraconazole, fluconazole (Fig. 1c), and voriconazole, on the basis of whether they have two or three nitrogen atoms in the five-membered azole rings.

Azoles act on ergosterol biosynthesis at the C-14 demethylation step by inhibiting ERG11, cytochrome P450 lanosterol 14-alpha demethylase, thereby resulting in depletion of ergosterol as well as accumulation of lanosterol and other intermediates in the fungi, and eventually disrupting the integrity of the plasma membrane in the fungi (Georgopapadakou 1998; Vandeputte et al. 2012).

Targeted Organisms Azoles are fungistatic with a broad-spectrum activity against several yeast as well as filamentous fungi, including dermatophytes.

Known Mechanisms of Resistance Although azoles of both classes have been popular in antifungal therapy despite associated side effects, drug resistance mechanisms have been reported for azoles in several fungi, leading to search of new generation triazoles. Drug resistance against azoles has been reported in yeast, *Candida*, and *Aspergillus* and typically arises through multiple distinct mechanisms, including mutations in the target *ERG11*, mutations in the promoter of *ERG11*, buffering of mutations by Hsp90 chaperone, increase in multidrug transporters, changes in membrane composition, or alterations in sterol biosynthesis (Cowen and Lindquist 2005; Vandeputte et al. 2012; Abdolrasouli et al. 2015; Sanglard 2016).

2.4 Glucan Synthesis Inhibitors

Fungal cell wall majorly consists of a β-(1, 3)-D-glucan polysaccharide which are homopolymers of β-(1, 3)-linked residues with occasional side chains involving β-(1, 6)-linkages. The biosynthesis of β-glucans is essential for fungal viability and is brought about by β-(1, 3)-glucan synthase. The catalytic subunit of glucan synthase is hence an important target for antifungal agents (Georgopapadakou 1998).

Molecular Properties and Mechanism of Action Echinocandins were originally identified as naturally produced lipopeptides produced by several fungi. Echinocandins, viz., micafungin, caspofungin (Fig. 1e), and anidulafungin, are synthetic derivatives of these lipopeptides that act as noncompetitive inhibitors of glucan synthase. Echinocandins thus target polysaccharides of cell wall, thereby weakening the rigidity of the cell wall and eventually leading to clearance of the fungal infection (Georgopapadakou and Tkacz 1995; Stone et al. 2002).

Targeted Organisms Echinocandins show a wide range of effectiveness against different fungi. They are fungicidal against *Candida* spp. and fungistatic in *Aspergillus*. However, they are ineffective against certain *Fusarium* spp., *Cryptococcus neoformans*, or dermatophytes, viz., *Trichophyton* spp.

Known Mechanisms of Resistance Molecular resistance to echinocandins is rare. The poor efficacy of echinocandins in *C. neoformans* has been attributed to unique polysaccharide composition in this species, rather than mutations in *FKS1* or *FKS2* (that encode different isoforms of glucan synthase) (Maligie and Selitrennikoff 2005).

2.5 Allylamines

Molecular Properties and Mechanism of Action Allylamines are a class of synthetic organic molecules that act to inhibit components of the ergosterol biosynthetic pathway. Terbinafine (Fig. 1f), the most widely used allylamine, is a noncompetitive inhibitor of ERG1, squalene epoxidase, an enzyme involved in cyclization of squalene to lanosterol in the ergosterol biosynthesis pathway. The resulting depletion of ergosterol and accumulation of squalene inside the cell affects membrane structure and nutrient uptake (Ryder 1991; Georgopapadakou and Bertasso 1992; Georgopapadakou 1998).

Targeted Organisms Terbinafine shows a wide spectrum of activity in vitro against *Aspergillus* spp. and other filamentous fungi, including dermatophytes, but is not effective against *Candida* spp. However, pharmacokinetic studies have limited the clinical efficacy of allylamines primarily to topical treatment, and oral usage is limited due to associated side effects and adverse drug reactions. Terbinafine is hence the primary drug of choice for treatment of superficial dermatophyte infections (Elewski 1998).

Known Mechanisms of Resistance Clinical drug resistance against terbinafine is largely unreported. There have been a few reported cases of relapses but were attributed to the inability of the drug to penetrate the site of infection (viz., nail plate) rather than acquired drug resistance. The few reported cases of clinical drug resistance of dermatophytes to terbinafine are discussed in Sect. 3.2.

2.6 Microtubules Assembly Synthesis Inhibitors

Molecular Properties and Mechanism of Action Griseofulvin (Fig. 1g) is one of the oldest molecules being used for selective inhibitory activity against fungi. It acts by disrupting microtubules inside the fungal cell.

Microtubules form highly organized structures and important components of cytoskeleton in all eukaryotic cells. Microtubules are long, dynamic polymers made up of α- and β-tubulin dimers. Griseofulvin is active against growing hyphae and arrests fungal cell mitosis in metaphase through inhibition of microtubule assembly (Elewski 1998; Georgopapadakou 1998).

Targeted Organisms With the availability of several other antifungal agents of comparative higher efficacy, griseofulvin currently finds limited use primarily to clear some dermatophyte infections.

Known Mechanisms of Resistance There have been only few reported cases of treatment failures to griseofulvin and have mostly been attributed to host factors. Some of the dermatophyte-related reported cases are discussed in the specific section later.

In summary, there are several albeit limited classes of therapeutic agents available against fungi. The molecular details of inhibition by these agents and their molecular targets are now beginning to be understood and are summarized in Fig. 2. However, not all of these agents are effective against all manifestations of fungal infections, and the pathogen may develop different mechanisms to evade clearance. Although an exhaustive description of the molecular details of these drug resistance mechanisms is beyond the scope of this current review, the most well-understood or common mechanisms have been briefly described with the specific therapeutic agent. Drug resistance in dermatophytes, on the other hand, is rare. However, new cases are beginning to emerge. We next describe the current understanding of this upcoming and emerging area of interest.

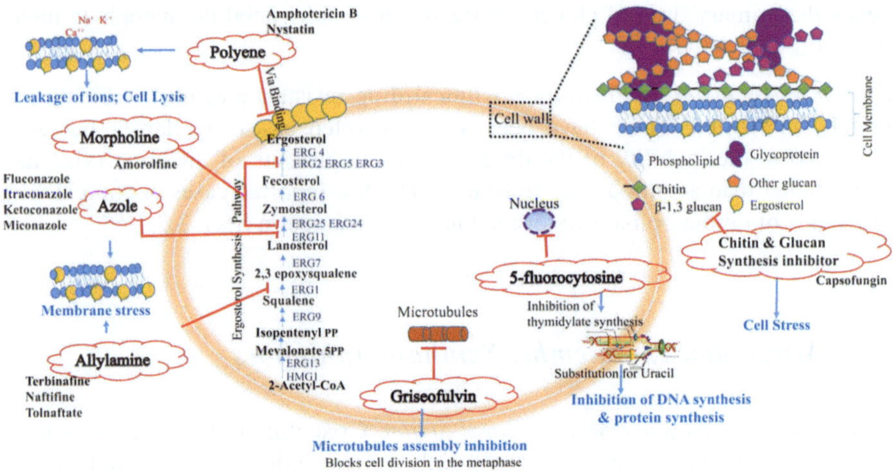

Fig. 2 Diagrammatic representation of antimycotic agents and their molecular targets

3 Dermatophytosis: Treatment and Evasion by the Pathogen

3.1 Available Therapy and Limitations of Antifungal Agents in Dermatophytes

Dermatophytes have the unique ability to infect immunocompetent people, resulting in superficial infections, as well as immunocompromised people, resulting in deep-seated systemic infections. Current antifungal medications include polyenes, viz., amphotericin B; azoles, viz., ketoconazole and itraconazole; and allylamines, viz., terbinafine and griseofulvin. While griseofulvin is active against growing hyphae and arrests fungal cell mitosis in metaphase through inhibition of microtubule assembly, all other currently prescribed antifungal agents act to inhibit specific enzymes of the ergosterol biosynthesis pathway. Early or mild cases of infection by dermatophytes are usually dealt to appropriate levels with topical treatments. However, effective clearance of superficial and often deep-seated infections with currently available antifungal agents presents several limitations in clinical treatment.

Limitations in Dermatophyte Treatments Topical treatment of dermatophytosis is often ineffective in treating certain cases of dermatophytosis, viz., onychomycosis, when the drug fails to penetrate the nail unit and hence is unable to completely eradicate the infection. Systemic therapy is, hence, often necessary to treat persistent or severe infections. Most antifungal agents interact with many medications and have other associated problems. For instance, terbinafine interacts with several agents and medications, most notably, caffeine and cimetidine, while ketoconazole is less preferred due to several associated instances of hepatotoxicity and has largely been replaced by the more effective triazole derivatives, namely, fluconazole and itraconazole (reviewed in Elewski 1998). Moreover, none of the antifungals are generally advised during pregnancy.

As part of therapy, intervention methods to treat dermatophyte infections are usually long term due to limited delivery of drug at therapeutic levels at the site of infection, with several cases of recurrence and numerous side effects. For instance, certain *T. rubrum* infections have been found to persist within villous hair shafts and follicles (Weitzman and Summerbell 1995), leading to chronic recurrences of the infection. In a few other instances of systemic treatment of onychomycosis with pulse itraconazole or continuous or intermittent terbinafine, a high relapse rate of approximately 20 % in patients followed up to 3 years after treatment was observed (Tosti et al. 1998). Similarly, unsuccessful treatment of dermatophytes with the fungistatic griseofulvin is common and reported in several cases (Artis et al. 1981; Robertson et al. 1982). However, there have been only limited reports of nonresponder dermatophyte strains exhibiting generalized resistance to most antifungal agents.

3.2 Drug Resistance in Dermatophytes

The strong biological variability of dermatophytes has prevented the emergence and/or availability of a single drug regimen that may be effective against all manifestations of dermatophytoses. Owing to the limited therapeutics and poorly defined molecular targets in many cases, the molecular aspects involved in the resistance of dermatophytes to available antifungals remain obscure. Only few cases of drug resistance in dermatophytes are known so far (Mukherjee et al. 2003; Alipour and Mozafari 2015; Ghannoum 2016). The major biochemical mechanisms that may contribute to drug resistance phenotype in dermatophytes are hence thought to be similar to those observed earlier in different bacterial and fungal pathogens, namely, (1) modification of delivery pathways, (2) metabolism or degradation of the drug by the pathogen, (3) sequestration of the drug in intracellular compartments, (4) alterations in interaction of drug and target, (5) mutations/genic deletions in molecular target, (6) overexpression of target, and (7) increased incidences of efflux. While one or more of these mechanisms may come into play in drug-resistant dermatophytes, only a few drug-resistant cases have been observed against currently used antifungal agents so far (Martinez-Rossi et al. 2008; Peres et al. 2010) and are next described.

Griseofulvin One of the first reported studies of clinical drug resistance against griseofulvin was in the dermatophyte *Microsporum gypseum* (Lenhart 1970). Two different unlinked loci, grf-1 and grf-2, were identified that were found to be associated with drug resistance in this pathogen. However, further genetic studies were restricted as most of the other griseofulvin-resistant mutants did not cross with wild type. Although in vitro development of drug resistance has been reported in a few cases (Martinez-Rossi et al. 2008), clinical resistance to griseofulvin remains rare, possibly due to its limited use.

Terbinafine and Efflux Pumps The first reported case of terbinafine resistance in dermatophytes was reported in 2003 (Mukherjee et al. 2003). The resistance in the clinical strain was ascribed to mutations leading to amino acid substitutions in conserved residues, (L393F) (Osborne et al. 2005) and (F397L) (Osborne et al. 2006) of *ergA*, encoding squalene epoxidase. These remain the only reported cases of drug resistance to terbinafine in dermatophytes till date. However, apart from mutations in the target genes, alternate mechanisms may also be utilized by the fungi to evade drug response. For instance, in *T. rubrum*, ATP-binding cassette (ABC) transporters involved in drug efflux, TruMDR1 and TruMDR2, have also been shown to aid drug resistance in some mutant strains (Cervelatti et al. 2006; Fachin et al. 2006; Maranhao et al. 2007). Molecular details in most of these limited cases of drug resistance, however, remain obscure as clinical resistance to antifungal agents in dermatophytes is largely unknown or unreported. This is somewhat surprising as spontaneous *T. rubrum* mutants to terbinafine and amorolfine were generated in a recent in vitro study, although at a low frequency of 10^{-9} (Ghelardi et al. 2014). However, in a recent analysis of in vitro antifungal susceptibility with previously isolated clinical isolates from Iran, several *Trichophyton mentagrophytes* exhibited in vitro resistance (Alipour and Mozafari 2015). Further studies will be required to identify the molecular basis of resistance. Nevertheless, these recent reports further endorse urgent attention in this area.

Azoles and Other Drugs Drug resistance to azoles in dermatophytes is now beginning to emerge through several reports across the world (Goh et al. 1994; Ghannoum et al. 2006; Manzano-Gayosso et al. 2008). In a recent study to investigate the in vitro response to different drugs, including ketoconazole, in *T. rubrum* (Yu et al. 2007), transcriptional profiles of the response to these drugs were studied by microarray analysis. Changes in expression of several genes involved in lipid biosynthesis of the cell, viz., lipid and sterol metabolism, were observed, suggesting expression of several alternate pathways and compensatory genes to overcome the effect of the drug. Clinical drug resistance to azoles in dermatophytes and their clinical implications have been recently reviewed (Ghannoum 2016), though the molecular basis of drug resistance is yet to be established.

3.3 Future Perspectives in Dermatophyte Research

Dermatophytes present their own unique problems of persistence and/or inaccessibility to topical therapeutic agents. Hence, molecular studies specific to dermatophytes are required to obtain a detailed understanding of pathogen-specific response and to unravel the molecular basis of emerging drug resistance. However, the absence of efficient molecular tools for genetic manipulations of dermatophytes, poor efficiency of transformation of exogenous DNA, and low observed rates of homologous recombination (White et al. 2008) pose further hurdles in detailed investigations. Whole genome sequences of several dermatophytes have become available in recent years (see Table 1 for summary of genome sequence data of dermatophytes, Burmester et al. 2011; Martinez et al. 2012; Latka et al. 2015) and offer an opportunity to understand genomic context of dermatophyte variability, pathogenicity, as well as drug resistance.

4 Conclusion

Combating the drug resistance problem in fungi, including dermatophytes, is extremely important for effective clearance of these pathogens. While there are many reports of the development of drug resistance to antifungal agents in yeasts and molds, drug resistance cases in dermatophytes are beginning to emerge only recently. It is now becoming clear that commonly used antimycotic agents used against dermatophytes, namely, azoles and terbinafine, have a potential for inducing drug resistance in dermatophytes and resistance mutants could be generated in the lab at a frequency of 10^{-7} or 10^{-9}, respectively (Ghelardi et al. 2014). Clinical cases of drug resistance in dermatophytes to these agents are now being reported from all across the world. With the availability of whole genome sequences of several dermatophyte strains from different locations across the world, integrated genome sequence-based studies will help understand fungal pathogenesis and associated drug resistance in greater details.

5 Opinion

Both primary and opportunistic fungal pathogens are a major cause of morbidity and/or mortality in immunocompromised as well as immunocompetent people. However, our understanding of pathogenesis of pathogens of eukaryote origin, viz., protozoa or fungi, lags far behind that of bacteria. Treatment modes of fungal pathogens, especially dermatophytes, are limited, and research in this area is scarce. One of the reasons is the technical challenges encountered in dermatophyte research, viz., slow growth of dermatophytes in the lab or limited tools for efficient genetic manipulations in dermatophytes. Next-generation sequencing (NGS)-based methods, hence, offer an important key to address the emerging problem of drug resistance in dermatophytes. Whole genome sequences of at least eight different dermatophyte strains have been published in recent years and many more are available in NCBI. Detailed comparative analysis of whole genomes of these organisms with drug-resistant strains will help understand the genomic context of drug resistance through identification of mutations either in genic or promoter regions of key genes and drug targets. Such interdisciplinary approaches that utilize the strength of modern sequencing-based technologies in a clinical context of drug resistance may hold the key to tackle drug resistance problems in the future.

Table 1 Sequencing and gene prediction statistics of dermatophyte genome sequences

Dermatophyte	*Arthroderma benhamiae* CBS 112371[a]	*Microsporum canis* CBS 113480[b]	*Microsporum gypseum* CBS 118893[b]	*Trichophyton equinum* CBS 127.97[b]	*Trichophyton tonsurans* CBS 112818[b]	*Trichophyton rubrum* CBS 118892[b]	*Trichophyton rubrum IGIB-SBL-CI1*[c]	*Trichophyton verrucosum* HKI 0517[a]
Epidemiological group	Zoophilic	Zoophilic	Geophilic	Zoophilic	Anthropophilic	Anthropophilic	Anthropophilic	Zoophilic
Location	Switzerland	Germany	Germany	Finland	Canada	Germany	India	Germany
Size of genome assembly (Mb)	22.2	23.3	23.3	24.2	23.0	22.5	22.5	22.5
Predicted protein coding genes (CDS)	7980	8915	8907	8679	8523	8707	8265	8024
Mean length, CDS (bp)	1482	1459	1436	1371	1409	1393	1463	1458
tRNA	80	82	83	85	82	82	91	77

[a]From Burmester et al. (2011)
[b]From Martinez et al. (2012)
[c]From Latka et al. (2015)

Acknowledgments The authors wish to thank Dr. Rajesh Gokhale for support, Sanchita Sanchaya Dey for discussions, and BSC0302 Project Grant of CSIR for providing necessary funds and facilities. PK acknowledges CSIR-GATE for Junior Research Fellowship.

References

Abdolrasouli A, Rhodes J, Beale MA, Hagen F, Rogers TR, Chowdhary A, Meis JF, Armstrong-James D, Fisher MC (2015) Genomic context of azole resistance mutations in *Aspergillus fumigatus* determined using whole-genome sequencing. MBio 6, e00536. doi:10.1128/mBio.00536-15

Alipour M, Mozafari NA (2015) Terbinafine susceptibility and genotypic heterogeneity in clinical isolates of *Trichophyton mentagrophytes* by random amplified polymorphic DNA (RAPD). J Mycol Med 25:e1–e9. doi:10.1016/j.mycmed.2014.09.001

Artis WM, Odle BM, Jones HE (1981) Griseofulvin-resistant dermatophytosis correlates with in vitro resistance. Arch Dermatol 117:16–19. doi:10.1001/archderm.1981.01650010022016

Barrett JP, Vardulaki KA, Conlon C, Cooke J, Daza-Ramirez P, Evans EG, Hawkey PM, Herbrecht R, Marks DI, Moraleda JM, Park GR, Senn SJ, Viscoli C, Amphotericin B Systematic Review Study Group (2003) A systematic review of the antifungal effectiveness and tolerability of amphotericin B formulations. Clin Ther 25:1295–1320. doi:10.1016/S0149-2918(03)80125-X

Bennett JE, Dismukes WE, Duma RJ, Medoff G, Sande MA, Gallis H, Leonard J, Fields BT, Bradshaw M, Haywood H, McGee ZA, Cate TR, Cobbs CG, Warner JF, Alling DW (1979) A comparison of amphotericin B alone and combined with flucytosine in the treatment of cryptoccal meningitis. N Engl J Med 301:126–131. doi:10.1056/NEJM197907193010303

Bickers DR, Lim HW, Margolis D, Weinstock MA, Goodman C, Faulkner E, Gould C, Gemmen E, Dall T, American Academy of Dermatology Association, Society for Investigative Dermatology (2006) The burden of skin diseases: 2004 a joint project of the American Academy of Dermatology Association and the Society for Investigative Dermatology. J Am Acad Dermatol 55:490–500. doi:10.1016/j.jaad.2006.05.048

Bossche HV, Marichal P, Odds FC (1994) Molecular mechanisms of drug resistance in fungi. Trends Microbiol 2:393–400. doi:10.1016/0966-842X(94)90618-1

Brown GD, Denning DW, Gow NA, Levitz SM, Netea MG, White TC (2012) Hidden killers: human fungal infections. Sci Transl Med 4:165rv13. doi:10.1126/scitranslmed.3004404

Burmester A, Shelest E, Glöckner G, Heddergott C, Schindler S, Staib P, Heidel A, Felder M, Petzold A, Szafranski K, Feuermann M, Pedruzzi I, Priebe S, Groth M, Winkler R, Li W, Kniemeyer O, Schroeckh V, Hertweck C, Hube B, White TC, Platzer M, Guthke R, Heitman J, Wöstemeyer J, Zipfel PF, Monod M, Brakhage AA (2011) Comparative and functional genomics provide insights into the pathogenicity of dermatophytic fungi. Genome Biol 12: R7. doi:10.1186/gb-2011-12-1-r7

Caffrey P, Lynch S, Flood E, Finnan S, Oliynyk M (2001) Amphotericin biosynthesis in *Streptomyces nodosus*: deductions from analysis of polyketide synthase and late genes. Chem Biol 8:713–723. doi:10.1016/S1074-5521(01)00046-1

Cervelatti EP, Fachin AL, Ferreira-Nozawa MS, Martinez-Rossi NM (2006) Molecular cloning and characterization of a novel ABC transporter gene in the human pathogen *Trichophyton rubrum*. Med Mycol 44:141–147. doi:10.1080/13693780500220449

Cowen LE, Lindquist S (2005) Hsp90 potentiates the rapid evolution of new traits: drug resistance in diverse fungi. Science 309:2185–2189. doi:10.1126/science.1118370

Cuenca-Estrella M (2014) Antifungal drug resistance mechanisms in pathogenic fungi: from bench to bedside. Clin Microbiol Infect 6(Suppl):54–59. doi:10.1111/1469-0691.12495

Edlind TD, Katiyar SK (2010) Mutational analysis of flucytosine resistance in *Candida glabrata*. Antimicrob Agents Chemother 54:4733–4738. doi:10.1128/AAC.00605-10

Elewski BE (1998) Onychomycosis: pathogenesis, diagnosis and management. Clin Microbiol Rev 11:415–429

Ellis D (2002) Amphotericin B: spectrum and resistance. J Antimicrob Chemother 49(Suppl 1):7–10. doi:10.1093/jac/49.suppl_1.7

Fachin AL, Ferreira-Nozawa MS, Maccheroni W Jr, Martinez-Rossi NM (2006) Role of the ABC transporter TruMDR2 in terbinafine, 4-nitroquinoline N-oxide and ethidium bromide susceptibility in *Trichophyton rubrum*. J Med Microbiol 55:1093–1099. doi:10.1099/jmm.0.46522-0Ghelardi

Fisher MC, Henk DA, Briggs CJ, Brownstein JS, Madoff LC, McCraw SL, Gurr SJ (2012) Emerging fungal threats to animal, plant and ecosystem health. Nature 484:186–194. doi:10.1038/nature10947

Georgopapadakou NH (1998) Antifungals: mechanism of action and resistance, established and novel drugs. Curr Opin Microbiol 1:547–557. doi:10.1016/S1369-5274(98)80087-8

Georgopapadakou NH, Bertasso A (1992) Effects of squalene epoxidase inhibitors on *Candida albicans*. Antimicrob Agents Chemother 36:1779–1781. doi:10.1128/AAC.36.8.1779

Georgopapadakou NH, Tkacz JS (1995) The fungal cell wall as a drug target. Trends Microbiol 3:98–104. doi:10.1016/S0966-842X(00)88890-3

Ghannoum M (2016) Azole resistance in dermatophytes: prevalence and mechanism of action. J Am Podiatr Med Assoc 106:79–86. doi:10.7547/14-109

Ghannoum M, Isham N, Sheehan D (2006) Voriconazole susceptibilities of dermatophyte isolates obtained from a worldwide tinea capitis clinical trial. J Clin Microbiol 44:2579–2580. doi:10.1128/JCM.00818-06

Ghelardi E, Celandroni F, Gueye SA, Salvetti S, Senesi S, Bulgheroni A, Mailland F (2014) Potential of Ergosterol synthesis inhibitors to cause resistance or cross-resistance in Trichophyton rubrum. Antimicrob Agents Chemother 58:2825–2829. doi:10.1128/AAC.02382-13

Goh CL, Tay YK, Bin Ali K, Koh MT, Seow CS (1994) *In vitro* evaluation of griseofulvin, ketoconazole, and itraconazole against various dermatophytes in Singapore. Int J Dermatol 33:733–737. doi:10.1111/j.1365-4362.1994.tb01523.x

Hawksworth DL (1991) The fungal dimension of biodiversity: magnitude, significance, and conservation. Mycol Res 95:641–655. doi:10.1016/S0953-7562(09)80810-1

Hope WW, Tabernero L, Denning DW, Anderson MJ (2004) Molecular mechanisms of primary resistance to flucytosine in *Candida albicans*. Antimicrob Agents Chemother 48:4377–4386. doi:10.1128/AAC.48.11.4377-4386.2004

Latka C, Dey SS, Mahajan S, Prabu R, Jangir PK, Gupta C, Das S, Ramachandran VG, Bhattacharya SN, Pandey R, Sharma R, Ramachandran S, Taneja B (2015) Genome sequence of a clinical isolate of dermatophyte, *Trichophyton rubrum* from India. FEMS Microbiol Lett 362:fnv039. doi:10.1093/femsle/fnv039

Lemke A, Kiderlen AF, Kayser O (2005) Amphotericin B. Appl Microbiol Biotechnol 68:151–162. doi:10.1007/s00253-005-1955-9

Lenhart K (1970) Griseofulvin resistance in dermatophytes. Experientia 26:109–110. doi:10.1007/BF01900425

Maligie MA, Selitrennikoff CP (2005) *Cryptococcus neoformans* resistance to echinocandins: (1,3)beta-glucan synthase activity is sensitive to echinocandins. Antimicrob Agents Chemother 49:2851–2856. doi:10.1128/AAC.49.7.2851-2856.2005

Manzano-Gayosso P, Mendez-Tovar LJ, Hernandez-Hernandez F, López-Martínez R (2008) Antifungal resistance: an emerging problem in Mexico. Gac Med Mex 144:23–26

Maranhao FC, Paiao FG, Martinez-Rossi NM (2007) Isolation of transcripts over-expressed in human pathogen *Trichophyton rubrum* during growth in keratin. Microb Pathog 43:166–172. doi:10.1016/j.micpath.2007.05.006

Martinez DA, Oliver BG, Gräser Y, Goldberg JM, Li W, Martinez-Rossi NM, Monod M, Shelest E, Barton RC, Birch E, Brakhage AA, Chen Z, Gurr SJ, Heiman D, Heitman J, Kosti I, Rossi A, Saif S, Samalova M, Saunders CW, Shea T, Summerbell RC, Xu J,

Young S, Zeng Q, Birren BW, Cuomo CA, White TC (2012) Comparative genome analysis of *Trichophyton rubrum* and related dermatophytes reveals candidate genes involved in infection. MBio 3:e00259-12. doi:10.1128/mBio.00259-12

Martinez-Rossi NM, Peres NT, Rossi A (2008) Antifungal resistance mechanisms in dermatophytes. Mycopathologia 166:369–383. doi:10.1007/s11046-008-9110-7

Mukherjee PK, Leidich SD, Isham N, Leitner I, Ryder NS, Ghannoum MA (2003) Clinical *Trichophyton rubrum* strain exhibiting primary resistance to terbinafine. Antimicrob Agents Chemother 47:82–86. doi:10.1128/AAC.47.1.82-86.2003

Osborne CS, Leitner I, Favre B, Ryder NS (2005) Amino acid substitution in *Trichophyton rubrum* squalene epoxidase associated with resistance to terbinafine. Antimicrob Agents Chemother 49:2840–2844. doi:10.1128/AAC.49.7.2840-2844.2005

Osborne CS, Leitner I, Hofbauer B, Fielding CA, Favre B, Ryder NS (2006) Biological, biochemical, and molecular characterization of a new clinical *Trichophyton rubrum* isolate resistant to terbinafine. Antimicrob Agents Chemother 50:2234–2236. doi:10.1128/AAC.01600-05

Peres NT, Maranhão FC, Rossi A, Martinez-Rossi NM (2010) Dermatophytes: host-pathogen interaction and antifungal resistance. An Bras Dermatol 85:657–667. doi:10.1590/S0365-05962010000500009

Robertson MH, Rich P, Parker F, Hanifin JM (1982) Ketoconazole in griseofulvin-resistant dermatophytosis. J Am Acad Dermatol 6:224–229. doi:10.1016/S0190-9622(82)70015-5

Ryder NS (1991) Squalene epoxidase as a target for the allylamines. Biochem Soc Trans 19:774–777. doi:10.1042/bst0190774

Sanglard D (2002) Resistance of human fungal pathogens to antifungal drugs. Curr Opin Microbiol 5:379–385. doi:10.1016/S1369-5274(02)00344-2

Sanglard D (2016) Emerging threats in antifungal-resistant fungal pathogens. Front Med (Lausanne) 3:11. doi:10.3389/fmed.2016.00011

Sanglard D, Odds FC (2002) Resistance of *Candida* species to antifungal agents: molecular mechanisms and clinical consequences. Lancet Infect Dis 2:73–85. doi:10.1016/S1473-3099(02)00181-0

Sokol-Anderson M, Sligh JE Jr, Elberg S, Brajtburg J, Kobayashi GS, Medoff G (1988) Role of cell defense against oxidative damage in the resistance of Candida albicans to the killing effect of amphotericin B. Antimicrob Agents Chemother 32:702–705. doi:10.1128/AAC.32.5.702

Stone EA, Fung HB, Kirschenbaum HL (2002) Caspofungin: an echinocandin antifungal agent. Clin Ther 24:351–377. doi:10.1016/S0149-2918(02)85039-1

Tosti A, Piraccini BM, Stinchi C, Colombo MD (1998) Relapses of onychomycosis after successful treatment with systemic antifungals: a three-year follow-up. Dermatology 197:162–166. doi:10.1159/000017990

Vandeputte P, Pineau L, Larcher G, Noel T, Brèthes D, Chabasse D, Bouchara JP (2011) Molecular mechanisms of resistance to 5-fluorocytosine in laboratory mutants of *Candida glabrata*. Mycopathologia 171:11–21. doi:10.1007/s11046-010-9342-1

Vandeputte P, Ferrari S, Coste AT (2012) Antifungal resistance and new strategies to control fungal infections. Int J Microbiol 2012:713687. doi:10.1155/2012/713687

Warn PA, Sharp A, Guinea J, Denning DW (2004) Effect of hypoxic conditions on in vitro susceptibility testing of amphotericin B, itraconazole and micafungin against *Aspergillus* and *Candida*. J Antimicrob Chemother 53:743–749. doi:10.1093/jac/dkh153

Weitzman I, Summerbell RC (1995) The dermatophytes. Clin Microbiol Rev 8:240–259. doi:10.1002/9780470688618.taw0140

White TC, Oliver BG, Gräser Y, Henn MR (2008) Generating and testing molecular hypotheses in the dermatophytes. Eukaryot Cell 7:1238–1245. doi:10.1128/EC.00100-08

Yu L, Zhang W, Wang L, Yang J, Liu T, Peng J, Leng W, Chen L, Li R, Jin Q (2007) Transcriptional profiles of the response to ketoconazole and amphotericin B in *Trichophyton rubrum*. Antimicrob Agents Chemother 51:144–153. doi:10.1128/AAC.00755-06

Human Fungal Pathogens and Drug Resistance Against Azole Drugs

Preetida J. Bhetariya, Neha Sharma, Pragati Singh, Priyanka Tripathi, Santosh K. Upadhyay, and Poonam Gautam

Abstract Pathogenic fungi causing severe infections in humans with immunocompromised immune system have been the major reasons of deaths in the world. *Candida albicans*, *Cryptococcus neoformans*, and *Aspergillus fumigatus* are among the most prevalent human fungal pathogens. The most widely used therapy used for the invasive fungal infections is the treatment with azole antifungal drugs; however, drug resistance against azole drugs is a major limitation in treatment of fungal infections. High-throughput techniques such as genomics and proteomics have been applied to understand the molecular mechanisms involved in drug resistance against azole drugs in human pathogenic fungi. These studies could be useful to prevent the increase in drug resistance and better response to antifungals. Here, we focus on the incidences of drug resistance against azole antifungal drugs in human fungal pathogens, molecular mechanisms of drug resistance, and new strategies for combating drug resistance to improve clinical treatment of invasive fungal infections.

P.J. Bhetariya
Department of Infectious Disease, University of Utah School of Medicine, Salt Lake City, UT 84132, USA

N. Sharma
Department of Biosciences, Jamia Milia Islamia University, New Delhi 110025, India

P. Singh • S.K. Upadhyay
Department of Biotechnology, Kumaun University, Nainital, Bhimtal Campus, Bhimtal 263136, India

P. Tripathi • P. Gautam, PhD (✉)
Tumor Biology Group, National Institute of Pathology (NIOP), Indian Council of Medical Research (ICMR), Post Box No. 4909, Safdarjung Hospital Campus, New Delhi 110029, India
e-mail: gautam.poonam@gmail.com; gautamp@icmr.org.in

G. Arora et al. (eds.), *Drug Resistance in Bacteria, Fungi, Malaria, and Cancer*,
DOI 10.1007/978-3-319-48683-3_18

1 Introduction

Opportunistic fungal infections are among the most difficult diseases to manage in patients. These infections frequently endanger the success of cancer treatments, transplant and surgery complications, autoimmune disease therapies, and also intensive care. In spite of several antifungal drugs and prophylaxis available, there is an increase in reported invasive fungal infections (IFIs). Worldwide mortality due to candidiasis, by *Candida albicans* (*C. albicans*) and related species, was 46–75 %; aspergillosis, by *Aspergillus fumigatus* (*A. fumigatus*) and related species, was 30–95 %; and mortality rate of cryptococcosis mainly caused by *Cryptococcus neoformans* (*C. neoformans*) worldwide was 20–70 % (d'Enfert 2009).

Fungal infections are established by fungi that are ubiquitous in nature and routinely inhaled by us in daily life. However, only individuals with compromised or diminished immunity are susceptible to invasive mycoses. Pathogenic fungi mainly *C. albicans*, *A. fumigatus*, and *C. neoformans* are leading pathogens among IFIs in humans. *C. albicans* is common microflora in human that may cause systemic infections in individuals with immunocompromised immune system leading to high mortality rates approaching up to 40 % (d'Enfert 2009). Other *Candida* species colonizes specific locations in the hosts, mainly in the gastrointestinal tract, genital tract, and the skin, whereas *C. glabrata* and *C. parapsilosis* also cause IFIs. Opportunistic pathogens, such as *A. fumigatus* and *C. neoformans*, are ubiquitously present in the soil. Normally host innate immunity, especially alveolar macrophages, manages these fungal spores, yeasts, or mycelial fragments. However, inhaled fungal spores can survive for extended period in macrophages and can establish infection when host immunity is weakened. The common causative agent of invasive aspergillosis is the filamentous mold *A. fumigatus*, with mortality rates up to 40–90 % (Dagenais and Keller 2009). The other opportunistic fungal pathogen known to cause infections in immunocompromised individuals is *C. neoformans* that leads to complications in central nervous system (CNS) such as cryptococcal meningitis among acquired immune deficiency syndrome (AIDS) patients (Sloan and Parris 2014). *Pneumocystis* spp. among other pathogenic fungi causes pneumonia in immunocompromised hosts.

These fungi have developed molecular mechanisms to combat the defense system in immunocompromised hosts. Detailed understanding of the complex interactions between genetic variations and its contributions to the disease phenotype is lacking and essential to study. Insight into host-pathogen interactions with respect to host genetic susceptibility can improve the identification of novel therapeutic targets and the design of better antifungal prophylaxis strategies.

It is a challenge to develop new techniques for early diagnosis of IFIs and effective treatment considering the increased burden of the disease and death in patients with their vulnerable immune status (Panackal et al. 2006). Diagnosis is extremely challenging, due to lack of sensitive or specific diagnostic methods, and results from currently used test are often available too late to be clinically useful.

Considering the population at risk for fungal infections and the presence of wide range of fungal pathogens in the environment niche, these opportunistic fungi present significant challenge for diagnostics and therapeutics. In comparison to the bacterial antibiotics, few antifungal drugs have been reported with minimum risk of side effects and the occurrence of resistance.

Resistance to first-line drugs in most of the pathogenic fungi causing invasive infections ranges from 0 to almost 100 %. Though the evaluation of the overall impact of drug resistance in health is difficult, it is quite clear that morbidity and mortality due to resistant pathogens have increased in view less effective treatment [WHO global strategy for containment of antimicrobial resistance, 2001]. Resistance to various antifungal drugs, azoles in particular, has been reported in pathogenic fungi (Warris et al. 2002). Rates of resistance to widely used azole and fluconazole (FLC) have been reported between 10 % and 25 % in invasive candidiasis. Cross-resistance to other azole agents like voriconazole and itraconazole has also been reported among some of these FLC-resistant isolates (Cuenca-Estrella et al. 2006; Arendrup et al. 2013). Resistance rate to triazoles in *Aspergillus* spp. is less common and assumed to be below 5 % in most of countries (Snelders et al. 2008; Alastruey-Izquierdo et al. 2013). However, resistance in *A. fumigatus* seems to increase majorly due to evolution of resistant isolates in response to azole fungicides that is being widely used in Europe for crop protection (Snelders et al. 2008). Several factors have been reported to be associated to drug resistance against azoles in pathogenic fungi including upregulation of functional genes controlling drug efflux, alterations in sterol synthesis, mitochondrial dysfunctioning, decreased affinity for the cellular target and chromosomal abnormalities, and high-osmolarity glycerol (HOG) pathway. The major factor for resistance to azoles involves alteration in target enzyme, 14α-demethylase, and specific drug efflux pumps in terms of the quantity or quality (Ghannoum and Rice 1999; Nascimento et al. 2003), mainly two classes of efflux transporters, class ATP-binding cassette (ABC) and major facilitator superfamily (MFS) (Slaven et al. 2002; da Silva Ferreira et al. 2006).

Here, we have focused on incidences of drug resistance and molecular mechanism in three of the major human pathogenic fungi, *Candida*, *Aspergillus*, and *Cryptococcus*, against azole drugs.

2 Pathogenic Fungi and Invasive Fungal Infections

Pathogenic fungal infections are categorized as primary or opportunistic. Primary infections may develop in immunocompetent hosts, while opportunistic fungal infections are frequent in "high-risk" populations of immunocompromised individuals undergoing treatment for cancer, organ transplant cases, and autoimmune disease; in patients who are at risk of infection after prosthetic surgery or under the treatment of broad-spectrum antibiotics leading to changes in the normal flora; and in patients with HIV with immune deficiency (Nanjappa and Klein 2014). Opportunistic fungal infections could be superficial or systemic. IFIs have

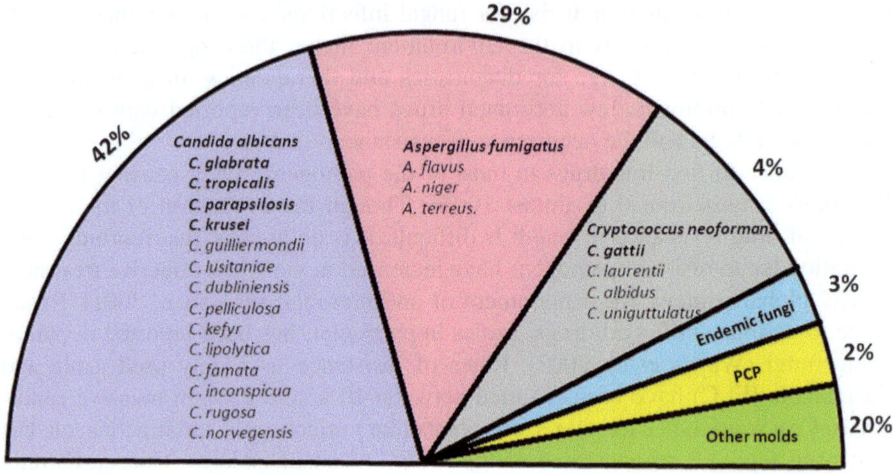

Fig. 1 Pathogenic fungi causing invasive fungal infections among transplant recipients. *Candida* spp. predominates among pathogenic fungi followed by *Aspergillus* spp. and *Cryptococcus* spp. as reported by Pfaller et al. (2009)

increased in the last two decades with increased use of steroids. Opportunistic fungal pathogens include *Candida*, *Aspergillus*, and *Cryptococcus* species. Figure 1 shows pathogenic fungi belonging to these genus and percentage of infection caused by these fungi.

2.1 Candida *spp.*

Candida species are *Ascomycota* group of fungi of the order Saccharomycotina, normally commensal organisms in mucous membranes, mainly gastrointestinal tract. There are a total of 150 *Candida* spp., and only 15 of them are commonly associated with humans as colonizers or opportunistic pathogens (Kam and Xu 2002). Of these, *Candida albicans* is the frequently isolated human commensal and pathogen (Krcmery and Barnes 2002). *C. albicans*, *Candida tropicalis*, *and Candida glabrata* are more virulent species, while *Candida parapsilosis* and *Candida krusei* are less virulent spp. Infections due to *C. parapsilosis* are more prevalent among children, and *C. glabrata* are frequently encountered among older adults (Yapar 2014; Kullberg and Arendrup 2015). These species have the differences in virulence as well as susceptibility to the azoles and the echinocandin drugs.

C. albicans is a dimorphic fungus and has ability to transform into different morphologies such as yeast, hyphae, and pseudohyphae upon perception of environmental signals (van der Meer et al. 2010). Most healthy individuals carry *C. albicans*, which harmlessly colonizes mucous membranes in different anatomical sites. *Candida* spp. are predominant constituents of the vaginal microflora

(Seed 2014). However, it is an efficient invasive pathogen for establishing infections in individuals with ineffective adaptive cellular immunity and in patients lacking neutrophils. *Candida* spp. cause invasive candidiasis which is mostly reported among immunocompromised patients. Candidemia, also known as blood-borne systemic candidiasis, generally develops in neutropenic patients or in patients exposed to contaminated indwelling catheters during surgery (Yapar 2014). For this, *Candida* is majorly the cause of Candidemia associated with healthcare in the USA (Magill et al. 2014). Despite of proper antifungal therapy being given to the patients, mortality rate has still been reported as high as 40 %. Non-albicans *Candida* species is also gaining attentions as it is emerging resistance to antifungal drugs (Kullberg and Arendrup 2015).

C. albicans infections are usually prevented by our defense system by mucosal tissues and the peripheral circulation. Polymorphic nuclear leukocytes (PMNL) are the first-line defense against blood-borne *Candida* infections. Neutrophil defects, decrease in neutrophil counts, and dysregulation in Th-cell reactivity are main risk factors which contribute to severe *Candida* infections (Romani 2004; Netea et al. 2004). *C. albicans* induces immunosuppression and leads to production of CD4+CD25+ T-regulatory cells. Further, the presence of specific antibodies also protects against fungal infections (Polonelli et al. 2000).

2.2 Aspergillus *spp.*

The *Aspergillus* species are *Ascomycota* group of fungi of the order *Eurotiales*, which grow on high osmotic concentrations and on carbon-rich sources. *Aspergillus* has 339 identified species (Samson et al. 2014); however, only few of them, namely, *A. fumigatus*, *A. flavus*, *A. niger*, and *A. terreus*, have been considered to be pathogenic to humans. *A. fumigatus* is the common etiological agent of human aspergillosis accounted for ~90 % of cases; *A. flavus*, *A. niger*, and *A. terreus* are secondary agents. *Aspergillus* species causes extensive spectrum of diseases with clinical manifestations that ranges from colonization of fungi in the organs leading to asthma, allergic bronchopulmonary aspergillosis (ABPA), and invasive aspergillosis (IA). The major site of *Aspergillus* infection is the lung (Ellis et al. 2009). *A. fumigatus* is a ubiquitous, saprophytic mold that releases airborne conidia which are inhaled by humans everyday (Latge 1999). *A. fumigatus* causes IA which primarily occurs in transplant recipients and hematological malignancy patients. The smaller (~2 μm) conidial size of *A. fumigatus* gives the fungus advantage to remain airborne for long periods and enter human alveoli. Conidia are coated with hydrophobic proteins and with the chemoprotectant melanin (Pihet et al. 2009) to withstand harsh environment in air and in vivo. On the onset of their germination, hyphae are recognized by innate immune cells in paranasal sinus or the lung. The airway mucus serves as a physical, chemical, and biological barrier secreting fluids that contain glycoproteins, proteoglycans, lipids, etc. and lead to clearance of fungal conidia (McCormack and Whitsett 2002).

A. fumigatus also produces various metabolites, e.g., gliotoxin, fumagillin, and helvolic acid which damages epithelium and can have inhibitory effects on ciliary movement (Amitani et al. 1995). Host proteins such as pattern recognition receptors (PRRs), lung surfactant proteins A and D (SP-A and SP-D), mannan-binding lectin (MBL), and toll-like receptors (TLRs) (Walsh et al. 2005; Johnson et al. 2014) are significant for the host defense against *Aspergillus*, and any alterations in these molecules may affect susceptibility to *Aspergillus* infections in the individuals (Johnson et al. 2014).

2.3 Cryptococcus *and Other Pathogenic Fungi*

Cryptococcus causes fatal infections in patients with weak immune status such as those associated with T-lymphocyte deficiency, common in AIDS patients. These infections are frequently reported in patients with stem cell recipients and malignancy, though 15–40 % of *Cryptococcus* cases are reported in HIV-negative patients (Speed and Dunt 1995). Glucuronoxylomannan (GXM) capsule of *Cryptococcus* is a well-known virulence factor that suppresses the host inflammatory response and prevents phagocytosis of the fungus (Speed and Dunt 1995).

The use of antifungal prophylaxis has become more prevalent for common and also uncommon fungal pathogens thus complicating clinical management. The other fungal pathogens include *Fusarium* spp., zygomycete, and the opportunistic yeastlike fungi such as *Histoplasma capsulatum* and *Pneumocystis* spp.; *Fusarium* species causes onychomycosis and fungal keratitis. *Fusarium* infections in the lungs majorly cause allergic bronchopulmonary fusariosis and hypersensitivity pneumonitis. Although radiological indications for fusariosis is similar to invasive aspergillosis, frequent occurrence of disseminated nodular skin lesions and blood culture positivity is the classical marker of *Fusarium* infection which is to differentiate it from aspergillosis (Nucci et al. 2015). Members of *Fusarium* genus may affect humans and cause mycotoxicosis by ingestion of toxin-contaminated food (Bennett and Klich 2003). Mucormycosis is another fungal infections usually found in spreading pneumonia due to invasion in blood vessels (Saxena et al. 2015).

3 Current Diagnosis and Clinical Treatment

3.1 Diagnosis

Early diagnosis of the infection along with species differentiation are of great importance for improved treatment of IFIs. Standard criteria for systemic fungal infections are histopathologic examination for the presence of fungus in the tissue or culture and isolation of etiologic agent from clinical sterile specimens (blood, sputum, urine, cerebrospinal fluid, or tissue biopsy). Radiological tests are used for diagnosis of patients where invasive procedures for sample collection are very difficult and risky due to low immune status. Pulmonary lesions or nodules, infiltration, and halo signs are indicative of pulmonary fungal infections such as aspergillosis, fusariosis, scedosporiosis, or zygomycosis (Greene et al. 2007; Godoy et al. 2012). Among non-culture methods, serological tests are considered as standard practice for the diagnosis of fungal infections. Commercially available ELISA kits are available for *Candida* and *Aspergillus* antigens which detect mannan and galactomannan, respectively, and demonstrate good specificity but variable sensitivity (Musher et al. 2004). Detection of fungal DNA signatures in body fluids using polymerase chain reaction (PCR) assay in combination with serological test in high-risk patients is also viable diagnostic option. Advances in the qualitative methods, such as panfungal PCR, for fungal DNA in human blood samples, tissues, bronchoalveolar lavage, and other body fluids, are now in clinical practice as reliable test for fungal infections (Orsi et al. 2015).

3.2 Treatment

Standard antifungal drugs in use are polyenes, azoles, and echinocandins. Amphotericin B (AMB) deoxycholate, AMB-D and lipid formulations of AMB, and L-AMB are polyene drugs in use. Fluconazole and voriconazole are mainline azole compounds in use. Caspofungin, micafungin, and anidulafungin are recommended antifungal echinocandins. Fluconazole remains the choice of drug for invasive candidiasis, while for *Candida* species known to be susceptible to fluconazole, other compounds like echinocandins and AMB or AMB-D are recommended (Pappas et al. 2009a). Voriconazole is recommended for the primary treatment of invasive aspergillosis in most patients including infections resistant to AMB (also with *Aspergillus terreus* and *Aspergillus nidulans*) (Walsh et al. 2008). Recommended antifungal therapy for major fungal infections in adults is discussed in Table 1.

4 Azoles and Other Antifungal Drugs

The use of potassium iodide (KI) for treating sporotrichosis was the first successful chemotherapy in 1903. After this, nystatin, the polyene compound, was successfully used as antifungal followed by amphotericin B in 1956, which is still the standard antifungal drug to evaluate new systemic antifungals (Al-Mohsen and Hughes 1998). The antifungal drugs include the natural products (polyenes, griseofulvin, and echinocandins) and the synthetic chemicals (azoles, allylamines, flucytosine, and phenylmorpholines). The available antifungal drugs are broadly

Table 1 Summary of recommendations for the treatment of major IFIs in adult

Fungal infection	Primary antifungal therapy	Alternative antifungal therapy	References
Invasive pulmonary aspergillosis, extrapulmonary aspergillosis, *Aspergillus* infections of the heart, eyes, or cutaneous	Voriconazole IV or orally	L-AMB, ABLC, caspofungin, micafungin, posaconazole, itraconazole	Walsh et al. (2008)
Allergic bronchopulmonary aspergillosis	Itraconazole	Oral voriconazole or posaconazole	
Candidemia, nonneutropenic adults	Fluconazole or an echinocandin	L-AMB or AMB or voriconazole	Pappas et al. (2009a)
Candidemia, neutropenic patients	Echinocandin or L-AMB	Fluconazole or voriconazole	
CNS and disseminated cryptococcosis	Induction therapy, AMB-D and 5-FC (B); consolidation therapy, fluconazole; suppressive therapy, fluconazole	Induction therapy, AMB-D Third line, 5-FC plus fluconazole	Thursky et al. (2008)
Mucormycosis	L-AMB	Posaconazole	Chang et al. (2014)
Fusarium species infections	Voriconazole or L-AMB	Posaconazole	
Scedosporium	Commence voriconazole with terbinafine	–	
Rare and emerging fungal infections such as *Paecilomyces* spp., *Phaeohyphomycosis* spp.	Voriconazole, posaconazole, and itraconazole	–	

AMB amphotericin B, *L-AMB* liposomal amphotericin B, *AMB-D* amphotericin B deoxycholate, *5-FC* flucytosine

classified into azoles, polyenes, nucleoside analogues, and echinocandins based on their mechanism of action.

4.1 Azoles

Azole antifungal agents are now considered to be important for therapeutics of systemic fungal infections. Fluconazole (FLC), itraconazole (ITC), voriconazole (VRC), posaconazole (POS), and ketoconazole (KTC) are some of the most widely used antifungal agents (Gallagher et al. 2003). Triazole drugs are more effective against many fungal pathogens with no severe nephrotoxic effects in comparison to AMB. Among azoles, FLC remains an effective and low-cost drug for the treatment of candidiasis and cryptococcosis. It is the first line of drug used for *Candida* (except *Candida krusei* and *glabrata*) and cryptococcal infections including non-meningeal coccidioidal infections. This drug is safe; however, it is not effective against filamentous fungi. VRC has been effective for the treatment of aspergillosis; however, its use is limited by significant drug interactions. It is effective against most of the *Candida* and *Aspergillus* spp. ITC is effective in vitro and in vivo against *A. fumigatus*, *Candida* spp., *C. neoformans*, and the dimorphic fungi. Finally, POS is the latest addition to the azole drugs and is effective against the zygomycetes. POS is effective against most *Candida* and *Aspergillus* species in vitro. It is also active against *C. neoformans* and *Fusarium* spp. (Zonios and Bennett 2008). KTC was the azole that was available for oral usage and was found with constant levels in blood; however, its use is limited due to hepatotoxicity and resistance reported in patients with candidiasis or AIDS and esophageal and oropharyngeal candidiasis.

Azoles inhibit fungal cytochrome P_{450} demethylase (an enzyme encoded by *CYP51* or ERG11 gene) which converts lanosterol to ergosterol. This leads to reduced amount of ergosterol in the cell membrane of the fungus (Andriole and Bodey 1994; Georgopapadakou and Walsh 1996; Andriole 1998, 1999). In order to understand the molecular targets of azole drugs, various high-throughput studies have been carried out. Altered expression of *Saccharomyces cerevisiae* with different classes of antifungal compounds (KTC, amphotericin B, caspofungin, and 5-fluorocytosine) was carried out to identify altered gene expression specific to each drug. Exposure of KTC led to altered levels of genes involved in ergosterol biosynthesis pathway as well as sterol uptake, while exposure to caspofungin led to altered levels of genes belonging to cell wall integrity, and exposure to 5-FC led to altered levels of genes involved in DNA damage repair, DNA synthesis, protein synthesis, and regulation of cell cycle. On the other hand, exposure to AMB led to altered levels of genes involved in cell stress, membrane reorganization, cell wall integrity, and transport (Agarwal et al. 2003). In order to understand the molecular targets of antifungal azoles, VRC, and ITC, high-throughput techniques have been used. Using microarray hybridization, 2271 genes were found as differentially expressed in wild-type strain which describes decrease in biosynthesis of genes involved in ergosterol biosynthesis and increased mRNA level of gene-encoding transporters (da Silva Ferreira et al. 2006). Proteomic profiling of *A. fumigatus* on exposure to ITC (ITC) using 2-DE followed by mass spectrometric analysis led to identification of 54 differentially expressed proteins including proteins related to cell stress, carbohydrate metabolism, and amino acid metabolism.

4.2 *Other Antifungal Drugs*

The polyene antifungal agents are fungicidal with large spectrum of antifungal activity than any other antifungal agents. Amphotericin B, the polyene compound, has been proposed to interact with ergosterol that leads to the production of aqueous pores (Holz 1974). These pores result in dysregulated permeability and leakage of necessary cytoplasmic components leading to the killing of the organism (Kerridge

1980; Kerridge 1985). Intravenous amphotericin B has been recommended for severe and for invasive aspergillosis, *Candida* infections of CNS, blastomycosis, coccidioidomycosis, and mucormycosis (Andriole and Kravetz 1962; Andriole and Bodey 1994; Andriole 1998, 1999). Liposomal amphotericin B has reduced nephrotoxicity than conventional AMB (Hiemenz and Walsh 1996; Groll et al. 1998).

Flucytosine 5-FC is a fluorinated pyrimidine deregulating RNA and protein synthesis in the fungal cell mainly targeting pyrimidine metabolism (Andriole 1998, 1999; Groll et al. 1998). Flucytosine activity in vitro has been reported against *Aspergillus* spp., *Candida* spp., *C. glabrata*, and *C. neoformans*. However, if treated with it alone, it is less effective and has increased chances of developing fungal resistance in *Candida* and cryptococcal organisms. Echinocandins are cyclic lipopeptide agents with fungicidal activity through inhibition of cell wall enzyme, 1,3-β-D-glucan synthase, not expressed in mammalian cells, and hence form a potential antifungal target (Georgopapadakou and Walsh 1996; Groll et al. 1998). The investigational compound, LY 303366, an inhibitor of β-1, 3-beta-D-glucan synthase, is found to be active against *Candida* and *Aspergillus* organisms (Andriole 1998, 1999).

5 Drug Resistance to Azoles in Human Fungal Pathogens

Antifungal drug resistance is among the major causes of therapeutic failure in invasive fungal infections other than low bioavailability of the antifungal drug, weakened immune function, or a higher metabolism of the drug. Primary resistance occurs in organisms which are not at all exposed to the specific drug, while secondary or acquired resistance is due to exposure of an organism to the drug. Clinical resistance is reversion of infection due to therapeutic failure in an organism and is not linked to in vitro resistance (Rex et al. 1997).

For effective treatment of IFI, we need to be familiar with the susceptibility of the resistant fungal isolates to the antifungal drugs. The Clinical Laboratory Standard Institute (CLSI) in the USA and the European Committee on Antimicrobial Susceptibility Testing (EUCAST) in Europe have recommended protocols for antifungal susceptibility testing. The minimum inhibitory concentration (MIC) of the drug is calculated as the threshold levels of drugs leading to in vitro growth inhibition. The CLSI has recommended antifungal MIC breakpoints to separate susceptible and resistant population for azoles and echinocandins by analyzing the in vitro susceptibility data, in vitro outcome, and pharmacokinetics/pharmacodynamic studies, while EUCAST has defined the breakpoint derived from MIC as the epidemiological cutoff value (ECV) to avoid confusion with clinical breakpoints. ECVs are the MIC values that capture >95 % of the observed population. In this chapter, we discuss antifungal drug susceptibility for three major pathogenic fungal spp. against azole drugs (Fig. 2).

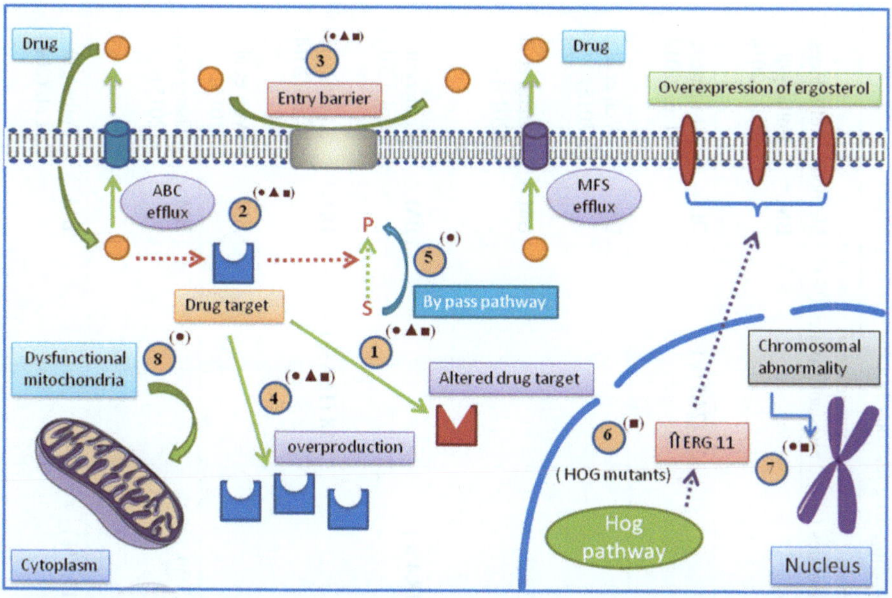

Fig. 2 Molecular mechanisms of drug resistance to azoles in human pathogenic fungi. Various mechanisms involved in fungal drug resistance are numbered from 1 to 8. Pathogenic fungi are shown with sign *closed circle* (*Candida* spp.), *closed triangle* (*Aspergillus* spp.), and closed square (*Cryptococcus* spp.). (1) Mutations in the target enzyme (lanosterol 14α-demethylase) result in the complete inhibition of the binding of the azole drug to its target. (2) The two different drug efflux systems in fungi, i.e., the ATP-binding cassette (ABC) superfamily and the major facilitator superfamily (MFS), contribute to azole drug resistance by pumping out the azole drugs out of the fungal cell leading to their less accumulation. (3) Modifications in the composition of plasma membrane affect the membrane asymmetry which alters the uptake of the drug. (4) Increased levels of target enzyme (lanosterol 14α-demethylase) may overwhelm the drug entering the cell, thereby resulting in increased level of resistance. Overproduction of the enzyme also results in cross-resistance between azoles. (5) The cell has a bypass pathway that balances for the loss-of-function inhibition due to the drug activity. (6) The HOG pathway negatively regulates the expression of ergosterol biosynthetic genes. Thus HOG mutants exhibit a decreased sensitivity toward azoles. (7) Chromosomal abnormalities have been linked with azole resistance in *Candida* and *Cryptococcus* species. *Candida* spp. acquire azole resistance by increasing the copy number of ERG11 (disomy) present on chromosome 5, whereas in *Cryptococcus* azole resistance is associated with chromosomes 1 and 4. (8) Loss of mitochondrial genome and changes in mitochondria membrane structures lead to the potential activation of drug resistance pathway in *Candida* spp.

5.1 Candida *spp.*

In vitro antifungal drug susceptibility of *Candida* spp. against different azole drugs (FLC, ITC, VRC, KTC) has been carried out each year for an update on the trends on resistant clinical isolates. Table 2 provides the ECV/MIC data compiled from various studies carried out in the last 2 years for five major *Candida* species. The comparative analysis of resistant isolates for four azole drugs ITC, FLC, VRC, and

Table 2 Antifungal drug susceptibility against azole drugs in *Candida* clinical isolates revealing drug resistant isolates

Source	Species/Total no. of isolates tested	Fluconazole resistant isolates (%)	Itraconazole resistant isolates (%)	Voriconazole resistant isolates (%)	Ketoconazole resistant isolates (%)	Method	References
NA	*C. albicans* *FLC (1196); VRC (593)*	**68 (5.7 %)**	–	27 (4.6%)	–	BM	Fothergill et al. (2014)
VVI	*C. albicans* *FLC (93; ITC (93); KTC (93)*	**83 (89.24 %)**	**8 (8.6 %)**	–	**23 (24.7 %)**	DD	Al-mamari et al. (2014)
BSI	*C. albicans* *FLC (183); VRC (183)*	0/183	–	0/183	–	BM	Won et al. (2015)
S,B,U	*C. albicans* *FLC (625); ITC (625)* *VRC (625)*	27/625 (4.3%)	8/625 (1.3%)	13/625 (2.1%)	–	BM	Zhang et al. (2015)
NA	*C. glabrata* *FLC (882); VRC (522)*	**70/882 (7.9 %)**	–	**96/522 (18.4 %)**	–	BM	Fothergill et al. (2014)
VVI	*C. glabrata* *FLC (18); ITC (18)* *KTC (18)*	**17/18 (94.4 %)**	0/18	–	**3/18 (16.6 %)**	DD	Al-mamari et al. (2014)
BSI	*C. glabrata* *FLC (58); VRC (58)*	6/58 (10.3%)	–	0/58	–	BM	Won et al. (2015)
S,B,U	*C. glabrata* *FLC (193); ITC (193)* *VRC (193)*	12/193 (6.2%)	4/193 (2.1%)	9/193 (4.7%)	–	BM	Zhang et al. (2015)
NA	*C. tropicalis* *FLC (327); VRC (205)*	32/327 (9.9%)	–	**36/205 (17.6 %)**	–	BM	Fothergill et al. (2014)

VVI	*C. tropicalis* *FLC (24); ITC (24)* *KTC (24)*	4/24 (17%)	0/24	–	1/24 (4.2%)	DD	Al-mamari et al. (2014)
BSI	*C. tropicalis* *FLC (82); VRC (82)*	2/82 (2.44%)	–	0/82	–	BM	Won et al. (2015)
S,B,U	*C. tropicalis* *FLC (84); ITC (84)* *VRC (84)*	**9/84 (10.7%)**	**4/84 (4.8%)**	6/84 (7.1%)	–	BM	Zhang et al. (2015)
NA	*C. krusei* *FLC (NA); VRC (98)*	NA	–	7/98 (12.2%)	–	BM	Fothergill et al. (2014)
VVI	*C. krusei* *FLC (6); ITC (6)* *KTC (6)*	**5/6 (83.3%)**	0/6	–	0/6	DD	Al-mamari et al. (2014)
BSI	*C. krusei* *FLC (4); VRC (4)*	**4/4 (100%)**	–	0/4	–	BM	Won et al. (2015)
S,B,U	*C. krusei* *FLC (29); ITC (29)* *VRC (29)*	22/29 (75.9%)	1/29 (3.4%)	2/29 (6.9%)	–	BM	Zhang et al. (2015)
NA	*C. parapsilosis* *FLC (497); VRC (298)*	11/497 (2.2%)	–	6/298 (2%)	–	BM	Fothergill et al. (2014)
BSI	*C. parapsilosis* *FLC (101); VRC (101)*	1/101 (0.99%)	–	0/101	–	BM	Won et al. (2015)
S,B,U	*C. parapsilosis* *FLC (13); ITC (13)* *VRC (13)*	**2/13 (15.4%)**	0/13	0/13	–	BM	Zhang et al. (2015)

BM broth microdilution; *DD* disk diffusion; *ECV* epidemiologic cutoff value for susceptible, susceptible dose-dependent and resistant isolates; *MIC* minimum inhibitory concentration at which more than 50% of the total isolates are inhibited in growth; *VVI* vulvovaginal isolates; *BSI* bloodstream infection isolates; *S*, *B*, *U* sputum, blood, urine; *NA* not available; (–) experiment was not performed for the corresponding azole and ECV/MIC values were not available

KTC showed maximum number of resistant isolates to FLC in *C. albicans*, i.e., 68 isolates (5.7 %) and 83 isolates (89.24 %); *C. glabrata*, 70 isolates (7.9 %) and 17 isolates (94.4 %); to VRC in *C. glabrata*, i.e., 96 isolates (18.4 %); *C. tropicalis*, 36 isolates (17.6 %); to ITC in *C. albicans* i.e. 8 isolates (8.6 %); *C. tropicalis* i.e. 4 isolates (4.8 %); to KTC in *C. albicans* i.e. 23 isolates (24.7 %); *C. glabrata i.e.* 3 isolates (16.6 %) (Fothergill et al. 2014; Al-mamari et al. 2014; Zhang et al. 2015; Won et al. 2015; 83 isolates (Al-mamari et al. 2014); *C. glabrata*, 111 isolates (Fothergill et al. 2014); to VRC in *C. albicans*, i.e., 82 isolates (Fothergill et al. 2014); *C. glabrata*, 258 isolates (Fothergill et al. 2014); and *C. tropicalis*, 94 isolates (Fothergill et al. 2014) (refer Table 2). The trend of increasing resistance of FLC leads to therapeutic failure and raises the need of new effective azoles for treatment. Reports of low resistance rates in VRC suggest that VRC may be used with confidence for the treatment of candidiasis (98 % success rates of VRC and 95 % in case of FLC) against FLC-resistant isolates (Ally et al. 2001).

There are several reports suggesting correlation of resistant isolates against different azoles and mutations in ERG11 gene in *Candida* spp. ERG11 (CnCYP51) gene encodes lanosterol 14α-demethylase enzyme which appears to be important target for azole antifungal drug. In a detailed study, ERG11 genes from 17 clinical isolates of *C. albicans* were analyzed for FLC resistance. These strains were observed to have 27 point mutations in ERG11 gene. Out of these, five mutations mainly Y132H, A114S, Y257H, G464S, and F72S substitutions were most prevalent in resistance spp., while two novel substitutions, T285A and S457P, in hotspot regions were relevant (Wang et al. 2015a). In the clinical isolates of *C. albicans*, the correlation between naturally occurring mutations in a gene for multidrug resistance regulator 2 (MRR2) and FLC resistance was analyzed. In a group of 20 FLC-resistant *C. albicans* isolates, 12 isolates showed overexpression of *Candida* drug resistance 1 (CDR1) gene. Of these, only three FLC-resistant isolates showed 11 identical missense mutations in MRR2 gene, 6 of which were among azole-resistant isolates. In addition, the role of MRR2 mutations in CDR1 overexpression and thus to FLC resistance was verified using recombinant strains with mutated MRR2 gene (Wang et al. 2015b). Hence, increasing rates of azole resistance in these species emphasize on the development of antifungal strategies with greater efficacy toward azole class of antifungals.

5.2 Aspergillus *spp.*

Appearance of azole resistance in strains of *A. fumigatus* has become a serious public health problem. Antifungal drug susceptibility in *Aspergillus* against different azoles has been performed by several groups. We looked for the data on antifungal drug susceptibility of *Aspergillus* spp. against different azole drugs from major studies in the last 10 years in literature. The comparative analysis of resistant isolates for three azole drugs in *A. fumigatus*, *A. flavus*, and *A. niger* showed maximum number of resistant isolates to VRC in *A. fumigatus* [3.1 % (Espinel-Ingroff et al. 2010), 0.8 % (Pfaller et al. 2009), and 2.2 % (Rodriguez-Tudela et al. 2008)], to ITC in *A. flavus* (5.6 %), and to POS in *A. niger* (8.8 %) (Espinel-Ingroff et al. 2010) (see Table 3). The data suggest increased resistance to VRC, first line of therapy for invasive aspergillosis, in *A. fumigatus* (Mikulska et al. 2012). This is in line with the observation that VRC treatment is failing due to resistant isolates in the patients as mentioned earlier (Verweij et al. 2007).

There are several reports suggesting correlation of resistant isolates against different azoles and mutations in CYP51A1 gene in *A. fumigatus* (Denning et al. 1997). In a study (Bader et al. 2013), 527 clinical isolates of *A. fumigatus* were analyzed and found that 17 (3.2 %) strains showed increased MIC_0 values. Out of these 17 isolates, 14 were found to have resistance to ITC ($MIC_0 > 32$ mg/l) and 1 was highly resistant to *POS* [$MIC_0 > 32$ mg/l]. All the resistant isolates showed mutations in CYP51A1, and most common mutation was TR/L98H. Recently, ITC resistance also has been found to be increased due to TR/L98H mutations in CYP51A1 (Snelders et al. 2008). In another study, 38 clinical isolates were analyzed, where 3 isolates showed multi-azole resistance to ITC, VRC, and *POS*. It was found that all three isolates also showed exclusively TR_{34}/L98H mutations in *CYP51A1* gene, not present in the remaining 35 azole-susceptible isolates (Wu et al. 2015).

5.3 Cryptococcus *spp.*

The antifungal drug susceptibility for azoles in *C. neoformans* and *C. gatti* is provided in Table 4. Molecular typing for *Cryptococcus* spp. done using molecular methodologies like AFLP and PCR fingerprinting identified eight molecular types of *Cryptococcus*. An in vitro antifungal susceptibility test was done for the abovementioned molecular types of *Cryptococcus* species using the CLSI broth dilution method (Espinel-Ingroff et al. 2012). In *C. neoformans* strains with high FLC MICs ($\geq$32 μg/ml), it was observed that a point mutation (involving the glycine to serine substitution at 484th position) in the ERG11 (CnCYP51) gene resulted in an alteration in the binding site of the target enzyme (Rodero et al. 2003). Earlier work done by Sanguinetti et al. showed a reduction in concentration of FLC in *C. neoformans* due to overexpression of the gene *C. neoformans* Antifungal Resistance 1 (Cn*AFR1*) which codes for an ABC transporter, a membrane efflux pump (Sanguinetti et al. 2006).

Cross-resistance among pathogenic fungi against azoles is expected as their target of action is similar, but azole cross-resistance is rarely seen among the *Cryptococcal* strains. In *C. neoformans*, cross-resistance between ITC and FLC is not observed due to the dual targets of ITC (the lanosterol 14α-demethylase and the 3-ketosteroid reductase). In *C. neoformans* isolates, the cross-resistance is also reported between FLC (MICs >64 μg/ml) and VRC (MICs $\geq$ 2 μg/ml), but no cross-resistance was seen between FLC and ITC. The fluconazole-voriconazole

Table 3 Antifungal drug susceptibility against azole drugs in *Aspergillus* clinical isolates revealing resistant isolates

Aspergillus spp./Total number of isolates tested	Itraconazole		Posaconazole		Voriconazole		Method	References
	WT (%)	NWT (%)	WT (%)	NWT (%)	WT (%)	NWT (%)		
A. fumigatus *ITC (2554)* *POS (1647)* *VRC (2778)*	2488 (98.8)	68 (2.6)	1611 (97.8)	37 (2.2)	2692 (96.9)	**88 (3.1)**	CLSI M38-A2 microdilution method	Espinel-Ingroff et al. (2010)
A. fumigatus *ITC (637)* *POS (637)* *VRC (637)*	636 (99.8)	1 (0.2)	636 (99.8)	1 (0.2)	632 (99.2)	**5 (0.8)**	EUCAST broth microdilution (BMD) method	Pfaller et al. (2009)
A. fumigatus *ITC (393)* *POS (393)* *VRC (393)*	635 (99.7)	2 (0.3)	387 (98.6)	6 (1.4)	384 (97.8)	**9 (2.2)**	CLSI M38-A2 microdilution method	Rodriguez-Tudela et al. (2008)
A. flavus *ITC (536)* *POS (321)* *VRC (590)*	532 (99.3)	41 (0.7)	303 (94.7)	**18 (5.6)**	578 (98)	12 (2)	CLSI M38-A2 microdilution method	Espinel-Ingroff et al. (2010)
A. niger *ITC (427)* *POS (325)* *VRC (479)*	389 (91.2)	**41 (8.8)**	308 (94.8)	19 (5.2)	474 (99)	5 (1)	CLSI M38-A2 microdilution method	Espinel-Ingroff et al. (2010)

Maximum percentage (%) of non-wild-type isolates of *Aspergillus* spp. against specific azole drug is shown in bold
WT wild type, *NWT* non-wild type

Table 4 Antifungal drug susceptibility against azoles for *Cryptococcus* species revealing resistant isolates

Species/total no of isolates used in the study	Genotype	Antifungal agents: Fluconazole WT/Non-WT [% of isolates with MICs > ECV]	Itraconazole WT/Non-WT [% of isolates with MICs > ECV]	Posaconazole WT/Non-WT [% of isolates with MICs > ECV]	Voriconazole WT/Non-WT [% of isolates with MICs > ECV]	References
C. neoformans FLC (5637) *ITC (3875)* *PSC (2879)* *VRC (4665)*	Non genotyped	4370/76 [1.7]	2700/30 [1.1]	1999/121 [5.7]	3351/122 [3.5]	Espinel-Ingroff et al. (2012)
	VNI	1104/33 [2.9]	1132/13 [1.1]	757/2 [0.3]	1063/26 [2.4]	
	VNIII	53/1 [1.9]	–	–	50/1 [2]	
	VNIV	–	–	–	50/3 [5.7]	
C. neoformans *FLC (285)* *PSC (285)* *VRC (285)*	Non genotyped	276/9 [3.1]	–	275/10 [3.5]	271/14 [4.9]	Pfaller et al. (2011)
C. gatt *FLC (827)* *ITC (504* *PSC (250)* *VRC (649)*	Non genotyped	124/13 [9.5]	70/0 [0]	67/1 [1.5]	93/5 [5.1]	Espinel-Ingroff et al. (2012)
	VGI	257/3 [1.2]	256/1 [0.4]	181/1 [0.5]	258/0 [0]	
	VGII[b]	94/7 [6.9]	46/1 [2.1]	–	92/4 [4.1]	

(continued)

Table 4 (continued)

Species/total no of isolates used in the study	Genotype	Antifungal agents				References
		Fluconazole	Itraconazole	Posaconazole	Voriconazole	
		WT/Non-WT [% of isolates with MICs > ECV]	WT/Non-WT [% of isolates with MICs > ECV]	WT/Non-WT [% of isolates with MICs > ECV]	WT/Non-WT [% of isolates with MICs > ECV]	
	VGIIa	192/8 [4]	–	–	192/5 [2.5]	
	VGIII	42/1 [2.3]	44/0 [0]	–	–	
	VGIV	82/4 [4.7]	86/0 [0]	–	–	

[b]Isolates identified as VGII molecular type and not being one of the VGIIa subtypes examined as a separate group

(–) Data not available

cross-resistance was attributed to a missense mutation [involving tyrosine to phenylalanine (Y145F) substitution] in the *ERG11* gene (Espinel-Ingroff et al. 2012). In addition, the level of heteroresistance to FLC was found to be more in *C. gattii* than in *C. neoformans* isolates (Kwon-Chung and Rhodes 1986). Heteroresistance is the expression of different resistance profiles in subpopulations of a strain (Nunes et al. 2007).

6 Molecular Mechanism of Drug Resistance to Azoles

Resistance to first line of therapy such as FLC and VRC has increased in recent years. Several high-throughput studies with clinical isolates and in vitro-developed resistant isolates of *Candida* spp. have been carried out in order to understand the genes/proteins and pathways involved in drug resistance. Table 5 represents high-throughput studies using transcriptomic, proteomic, and lipidomic analysis to understand molecular pathways involved in drug resistance. Here, we discuss the factors associated with drug resistance against azoles in major pathogenic fungi.

6.1 Alteration in Ergosterol Biosynthesis Pathway Enzyme

As discussed earlier, azoles generally target lanosterol 14α-demethylase which is a cytochrome P450-dependent enzyme encoded by *CYP51* or ERG11 gene. This is an oxidative process which involves the removal of 14α-methyl group from lanosterol. Azole binds to the ferric ion moiety of the heme-binding site and blocks the enzyme's natural substrate lanosterol, disrupting the biosynthetic pathway (Odds et al. 2003). Amino acid substitutions in the drug target that inhibit drug binding are common azole drug resistance mechanisms in fungi.

Azole resistance in *Candida* species is attributed to the point mutation in ERG11 gene resulting in the alteration of drug target that inhibits the azole drug from binding to its target (Vandeputte et al. 2012). In a study on *Candida*, five different mutations in CYP51A1 gene (G129A, Y132H, S405F, G464S, and R467K) were observed. In the clinical isolates of *C. albicans*, combined mutations in CYP51A1 gene resulted in a greater decrease in binding affinity of the drugs and then single mutations (Sanglard et al. 1998). Studies on *C. albicans* and *C. glabrata* also found that overproduction of drug target (lanosterol demethylase) which resulted from genome rearrangement and chromosome duplication was another factor for an increased azole resistance (Marichal et al. 1997; Selmecki et al. 2008). In another study, a zinc transcription factor, CaUpc2, was essential for controlling the regulation of ERG genes in the subsistence of ergosterol biosynthesis inhibitors (Vandeputte et al. 2012). Further, in other study the binding capacity of CaUpc2 to the promoter region of ERG11 in *C. albicans* was confirmed (MacPherson et al. 2005).

Table 5 Various studies carried out to understand the molecular mechanisms of azole resistance in *Candida* Species

Candida species/strains	Azoles tested	Techniques used	No. of differentially expressed proteins/genes	Verifications/functional studies	References
Transcriptomic studies					
C. albicans *(FLC susceptible and resistant isogenic strains showing overexpression of CDR1 and CDR2)*	FLC	RNA sequencing	228 genes	Transcription factor encoding gene, CZF1, overexpressed in FLC resistant isolate, was further studied and its role in drug resistance was shown in *C. albicans* and proposed as a potential target for FLC resistant isolates of *C. albicans*	Dhamgaye et al. (2012)
C. albicans *(Azole-resistant clinical strains and strains with disrupted TFs encoding genes)*	FLC (MIC for resistance—64 μg/ml)	Microarray and qRT-PCR	209 genes were upregulated in response to FLC	To analyze the role of CAS5 transcriptonal regulator in azole resistant isolates of *C. albicans*, CAS5 mutants (exhibiting the resistance mutations in TAC1, MRR1 and ERG11) were generated and it was shown that there was reduction of both MIC and MFCs values in these isolates. These data suggests that in absence of CAS5 regulator, susceptibility to FLC was increased in resistant isoates having specific resistance mutations whereas in the presence of CAS5, resistance was persisted in these isolates. Further, genome-wide transcriptional analysis revealed the role of CAS5 in cell wall organization, iron transport and homeostasis	Vasicek et al. (2014)

C. glabrata *(Seven pairs of susceptible and resistant isogenic isolates)*	FLC (MIC range for resistance—128–512 μg/ml)	DNA micro-array and qRT-PCR	Expression of 45 genes was significantly altered in at least one clinical pairs. Of these 19 were upregulated in majority of the resistant isolates	All the resistant isolates had acquired mutation in *C. glabrata* pleiotropic drug resistance (CgPDR1) open reading frame. Transcript analysis showed twofold upregulation of CgPDR1 and its known target genes in all the seven resistant isolates. The study showed that gain-of-function mutations in CgPDR1were associated with azole resistance	Tsai et al. (2010)
C. glabrata *(FLC susceptible and resistant isogenic strains)*	FLC (MIC for resistance—> 256 μg/ml)	Microarray and qRT-PCR	379 genes	qRT-PCR analysis of FLC susceptible and resistant isogenic strains showed the resistant isolate exhibited mitochondrial dysfunction and upregulation of the ABC transporter genes, C. glabrata CDR1 (CgCDR1), CgCDR2, and CgSNQ2, involved in drug resistance. Further, the resistant isolate showed increased virulence in vivo in both systemic and vaginal murine infection models. Microarray analysis of FLC susceptible and resistant isogenic strains further confirmed the overexpression of ABC transporter genes, *C. glabrata* CDR1 (CgCDR1), CgCDR2, and CgSNQ2 and cell wall proteins, GPI anchored proteins, yapsins (CgYps1, 3, 5, 8-11). Overall, this study showed that mitochondrial dysfunction was one of the factor in *C. glabrata* virulence and may serves a foundation for identification of virulence factors in *C. glabrata*	Ferrari et al. (2011)

(continued)

Table 5 (continued)

Candida species/strains	Azoles tested	Techniques used	No. of differentially expressed proteins/genes	Verifications/functional studies	References
C. parapsilosis (FLC resistant,VRC resistant, PSC resistant strains) (*invitro* study)	FLC (MIC for resistance —≥64 μg/ml), VRC (MIC for resistance —≥4 μg/ml), POS (MIC for susceptibility —≤1 μg/ml)	cDNA microarray and qRT-PCR	1128 genes (in FLC resistant), 210 genes (in VRC resistant), 598 genes (in PSC resistant strain)	Resistant *C. parapsilosis* strains were obtained after constant exposure to VRC, FLC, POS. Microarray analysis showed increased expression of MDR1 and other efflux pump members, MFS, and transcription factor MRR1, involved in regulation of MDR, to be associated with VRC and FLC resistance in *C. parapsilosis* resistant strains, while increased expression of ergosterol biosynthesis genes and transcription factors, UPC2 and NDT80, involved in regulation of ergosterol biosynthesis genes,were associated with PSC resistance. Some of the genes involved in ergostrol biosynthesis, ERG5, 11, MDR1, MRR1 and NTD80, were further confirmed by qRT-PCR	Silva et al. (2011)
Proteomic studies					
C. albicans (FLC resistant and sensitive strains, clinical isolates)	FLC (MIC for resistance—≥64 μg/ml)	2D-PAGE and MALDI-TOF MS	15 proteins	The comparative proteomic analysis showed altered expression of proteins was majorly involved in energy metabolism and amino acid biosynthesis in resistant strain. Proteomic analysis revealed upregulation of alcohol dehydrogenase (Adh1p), involved in biofilm formation and interaction with host, in FLC resistant strain, which was earlier reported to be associated with FLC resistance in *C. albicans* using differential display-PCR technique	Wang et al. (2012)

C. glabrata (FLC resistant and susceptible strains, clinical isolates)	FLC (MIC for resistance—64 μg/ml)	2D SDS-PAGE and LC-MS/MS	65 proteins 39 intracellular and 26 membrane proteins	The study analyzed membrane and cellular proteins differentially expressed in FLC resistant isolate. Resistant strains showed upregulation of membrane proteins while cellular proteins were downregulated. Heat shock protein and stress response proteins were observed to be upregulated in membrane fractions of resistant strains	Yoo et al. (2012)
C. glabrata (VRC resistant,susceptible, susceptible dose-dependent strains, clinical isolates)	VRC (MIC for resistance—4 μg/ml)	2D SDS-PAGE and LC-MS/MS	46 proteins 15 intracellular and 31 membrane proteins	The study analyzed membrane and cellular proteins differentially expressed in VRC resistant isolate. This study showed that there was overexpression of heat shock protein 70 (Hsp70) isoforms in intracellular fraction and decreased expression of nine Hsp70 protein isoforms in membrane fractions of susceptible, susceptible dose-dependent and resistant *C. glabrata* strains suggesting that this protein may be associated with VRC resistance in resistant *C.glabrata* strains	Yoo et al. (2013)
C. glabrata (FLC resistant strain) (in vitro study)	FLC (MIC for resistance—> 256 μg/ml)	2D-PAGE, MALDI-TOF MS, qRT-PCR	25 proteins	FLC resistant strains of *C. glabrata* were generated and eight of the selected mutants with large (n=4) and small colonies (n=4) analyzed by CHEF showed four genotypes for these mutants. Two of the randomly selected stable FLC resistant mutants further analyzed by proteomic analysis showed	Samaranayake et al. (2013)

(continued)

Table 5 (continued)

Candida species/strains	Azoles tested	Techniques used	No. of differentially expressed proteins/genes	Verifications/functional studies	References
				a total of 25 proteins to be differentially expressed in resistant mutants, some of them, ERG11, CDR1, CDR2, MFS, MTI, TPR, VPS and EFT2, were further confirmed by qRT-PCR analysis. Interestingly, FLC resistant isolates also showed resistance to other antifungal azole-ITC (MIC for resistance—>32 μg/ml), Ketoconazole (MIC for resistance—>32 μg/ml) and VRC (MIC for resistance—>32 μg/ml). The study also showed increased bud formation of yeast and metallothionein production in resistant mutants and proposed these phenotypes to be associated with drug resistance in *C. glabrata*	
Lipidomics studies					
C. albicans (*Eight pairs of FLC susceptible and resistant clinical isolates*)	FLC (MIC range for resistance—16 to 128 μg/ml)	ESI-MS/MS	>200 lipids species[a]	This study performed comparative lipidomic analysis of FLC susceptible and resistant clinical isolates (with over-expression of an ABC transporter encoding gene CaCDR1 or MFS encoding gene, CaMDR1) by mass spectrometry. Molecular lipid species ranging from monounsaturated to	Singh and Prasad (2011)

				polyunsaturated fatty acid-containing phosphoglycerides were identified to be among the commonalities in the lipid profiles of these pairs. While fluctuation in phosphatidyl serine, mannosylinositol phosphorylceramides, and sterol esters levels indicated their compensatory role in maintaining lipid homeostasis among most AR isolates. This sudy establishes the versatility of lipid metabolism in handling azole stress among various matched azole resistant isolates	
C. albicans (Six isolates collected from 3 different time points of FLC therapy from HIV patients)	FLC (MIC for resistance—> 125 μM)	ESI-MS/MS,RT-PCR and HP-TLC	242 lipids species[a]	The study compared the lipidome profiling of FLC suceptible, intermediate and resistant isogenic isolates. They observed differential levels of plasma membrane microdomain specific lipids such as mannosylinositolphosphorylceramides and ergosterol, and in a mitochondrial specific phosphoglyceride, phosphatidyl glycerol in resistant isolates. They further confirmed the expression of key genes involved in lipid metabolism by RT-PCR. Further, the lipid contents were observed to be altered in resistant isolates. This study revealed correlation of development of FLC resistance	Singh et al. (2012)

(continued)

Table 5 (continued)

Candida species/strains	Azoles tested	Techniques used	No. of differentially expressed proteins/genes	Verifications/functional studies	References
				with lipid remodelling in *C. albicans*. Further, mitochondrial dysfunction and defective cell wall was observed in clinical azole resistant isolates of *C. albicans* providing an evidence of a cross-talk between mitochondrial lipid homeostatis,cell wall integrity and azole resistance	

MIC minimum inhibitory concentration, *FLC* fluconazole, *ITC* itraconazole, *VRC* voriconazole, *PSC* posaconazole, *qRT-PCR* quantitative real-time polymerase chain reaction, *ABC* ATP-binding cassette transporter, *2D-PAGE* two-dimensional polyacrylamide gel electrophoresis, *MALDI-TOF MS* matrix-assisted laser desorption ionization-time of flight mass spectrometry, *SDS* sodium dodecyl sulfate, *LC-MS/MS* liquid chromatography tandem mass spectrometry, *ESI-MS/MS* electron spray ionization tandem mass spectrometry, *HP-TLC* high performance thin layer chromatography, *RT-PCR* reverse transcriptase polymerase chain reaction, *MDR* multidrug resistance, *CHEF* contour-clamped homogeneous electrophoretic field analysis

The azole resistance in *A. fumigatus* is majorly mediated by mutations in gene CYP51A1. Target gene, cyp51 (ERG11), has two copies, i.e., CYP51A1 and CYP51B, each encoding a different protein (Mellado et al. 2001); however, mutations in CYP51A1 gene are reported as major cause of resistance in *A. fumigatus* (Odds et al. 2003; Diaz-Guerra et al. 2003; Warrilow et al. 2010). In *A. fumigatus* the first mutation to be identified was the glycine 54 (G54) point mutation detected in clinical isolates resistant to ITC and POS (Mellado et al. 2007). A high-throughput multiplex RT-PCR has been developed for detecting mutations in *A. fumigatus* CYP51A1 that leads to ITC resistance (Balashov et al. 2005). In another study, clinical isolates of *A. fumigatus* with reduced susceptibility were identified, and it was seen that the resistance developed among these species may be linked to two factors: firstly due to a point mutation involving substitution of leucine by histidine at 98th position (L98H) in CYP51A1 gene and secondly due to the presence of two copies of a 34-bp tandem repeats (TR) in the CYP51A1 promoter region (Snelders et al. 2010). Methionine 220 (M220) and glycine 138 (G138) are the less common mutations in *A. fumigatus* azole resistance strains (Slaven et al. 2002). TR/L98H genotype is the most ubiquitous mutation responsible for resistance mechanism observed for azole-resistant strain (Willger et al. 2008). A recently identified mechanism in the CYP51A1 gene that decreases the susceptibility of *A. fumigatus* against voriconazole consists single polymorphisms mainly substitutions in tandem repeat of 46-bp in the promoter region (Dirr et al. 2010).

But a limited number of mutations in *ERG11* has been reported in *C. neoformans*, including Y145F (Espinel-Ingroff et al. 2012) and G484S (Rodero et al. 2003). Study by Espinel-Ingroff et al. showed by sequencing the ERG11 gene from clinical isolates of MRL862 (*C. neoformans* strain isolated from a FLC-treated patient) that the strain MRL862 contained five unique mutations compared to reference strain H99. Triazole susceptibility coupled with the molecular changes in the ERG11 gene revealed that a single missense Y145F, tyrosine replaced by phenylalanine, mutation resulted in high FLC resistance of the strain (Espinel-Ingroff et al. 2012). The study by Rodero et al. analyzed five *C. neoformans* isolates that were sequentially isolated from an AIDS patient with frequent meningitis and found that out of five isolates, four were FLC susceptible while FLC resistance was seen to be developed in the fifth isolate. The analysis further revealed that a point mutation G484S (involving the glycine to serine substitution at 484th position) in the ERG11 gene was the key to the development of FLC resistance in the fifth isolate (Rodero et al. 2003). Studies by various groups have demonstrated that this substitution confers an orientation change in the P450 heme-binding domain that leads to a decrease in the binding affinity of azole drug as well as a decrease in the enzyme catalytic efficiency (Sanglard et al. 1998; Kelly et al. 1999).

In *C. albicans*, point mutations in ERG3 gene result in the alteration of C5 sterol desaturase enzyme; therefore, 14α-methyl-3,6-diol (toxic sterol) cannot be synthesized. However, in the presence of azoles, ergosterol is replaced by sterol species resulting in functional fungal cell membrane (Sanguinetti et al. 2015).

6.2 Drug Efflux

There are two different drug efflux systems in fungi that contribute to azole drug resistance. They belong to the superfamily ATP-binding cassette (ABC) and the major facilitator superfamily (MFS). ABC proteins are the ATP-dependent transporters, usually arranged in duplicates, comprise of two transmembrane span (TMS) domains and the two cytoplasmic nucleotide-binding domains (NBDs) which facilitate the ATP hydrolysis. In *C. albicans* drug resistance isolates, it was reported that CDR1 and CDR2 are main contributors of azole resistance which belongs to ABC transporters superfamily (Vandeputte et al. 2012). Studies on regulation of CDR1 and CDR2 showed cis-acting regulatory elements: a basal expression element (BEE), a drug-responsive element (DRE), two steroid responsive elements (SREs), and a negative regulatory element (NRE) in CDR1, while the CDR2 promoter contains only a DRE motif (de Micheli et al. 2002; Karnani et al. 2004; Gaur et al. 2005). Among these DRE was reported to be the exclusively responsible element for overexpression/upregulation of both CDR1 and CDR2. In azole-resistant clinical isolates of *C. glabrata*, three transporters mainly CgCDR1, CgCDR2, and ABC transporter co-regulated with CgCDR1 and CgCDR2, called SNQ2, were found to be upregulated and involved in azole resistance (Torelli et al. 2008). In *A. fumigatus*, atrF and AfuMDR4 were found to be upregulated in itraconazole-resistant strains (Vandeputte et al. 2012). In *C. albicans*, a gene encoding a protein CaNdt80p that was found to regulate CDR1 participating in drug resistance mechanism was reported (Chen et al. 2004). In a study, it was demonstrated that mutations lead to hyperactive alleles in *C. albicans* and consequent loss of heterozygosity (LOH) at the transcriptional activator of CDR gene (TAC1) and multidrug resistance 1 (MRR1) loci (Coste et al. 2009).

In azole resistance clinical isolates of *C. albicans*, multidrug resistance 1 (MDR1) transporter was reported as MFS transporter (Ben-Yaacov et al. 1994). Among FLC-resistant isolates of *C. albicans*, nucleotide region, called MDR1 drug resistance element (MDRE), was found to be responsible for the overexpression of MDR1 (Riggle and Kumamoto 2006; Rognon et al. 2006). A study by Hiller et al. reported three different cis-activating regions (regions 1, 2, and 3) in MDR1, and regions 1 and 3, close to the MDRE region, were reported to be necessary for controlling the expression of MDR1 in an azole-resistant isolate (Vandeputte et al. 2012). Studies of MRR1 gene deletion in azole-resistant strains were shown to diminish the overexpression of MDR1 suggesting MRR1p identification as a main controller of MDR1. The CdMDR1 and CtMDR1 are homologues of MDR1 in *C. dubliniensis* and *C. tropicalis*, respectively, which are found to be upregulated in azole-resistant strains (Pinjon et al. 2003, 2005; Vandeputte et al. 2005).

Drug efflux pumps are also responsible for resistance in *A. fumigatus* as they mediate reduced accumulation of intracellular drugs (Coleman and Mylonakis 2009). A probe derived from CDR1 gene, ABC transporter genes for drug efflux in *C. albicans*, was used to clone atrF gene from *A. fumigatus* with characteristics of multidrug resistance motifs. In *A. fumigatus* isolate, AF72 has approximately fivefold higher levels of expression of atrF compare to the susceptible isolates AF10 and H06-03 with sub-MIC levels of ITC (Slaven et al. 2002). In *A. fumigatus*, overexpression of multidrug resistance 3 (AfuMDR3) or AfuMDR4 is linked to high-level resistance against itraconazole due to the mutations at the drug target site (Denning et al. 1997; Nascimento et al. 2003). In the biofilms of azole resistance of *A. fumigatus* efflux pump, AfuMDR4 pump was seen to be upregulated that is responsible for resistance to VRC (Rajendran et al. 2011).

In *A. nidulans*, high levels and differential expression of AtrA to AtrD genes were reported in the presence of drugs such as camptothecin, imazalil, ITC, hygromycin, and 4-nitroquinoline oxide using real-time RT-PCR. Further, AtrA to AtrD expression levels were verified in the *A. nidulans* imazalil-resistant mutants (Semighini et al. 2002). One of the four ABC-type transporter genes, abcD, was reported to have two- to sixfold increased mRNA level expression following exposure to miconazole, camptothecin, methotrexate, and ethidium bromide (do Nascimento et al. 2002).

Many ABC proteins are found in *C. neoformans* in comparison to other pathogenic fungi (Lamping et al. 2010), but only a few are linked to azole drug resistance. The upregulation of the ABC transporter-encoding gene *AFR1* in *C. neoformans* is contributing to the in vitro resistance to FLC (Coleman and Mylonakis 2009) and corelated to less azole accumulation. Sanguinetti et al. generated a set of recombinant strains of *C. neoformans* BPY22.17 (FLC-resistant mutant strain), BPY444 (*afr1* mutant strain), and BPY445 (AFR1-overexpressing mutants), all derived from a FLC-susceptible isolate of *C. neoformans* strain BPY22. In infectious mice model exposed to these strains showed, strain BPY445 was more virulent than BPY22 and displayed enhanced intracellular survival due to upregulation of *AFR1* (Sanguinetti et al. 2006). In spite of several studies on the role of ABC multidrug efflux transporter drug resistance, little is known about the contribution of the Drug:H^+ Antiporter (DHA) family in azole resistance. There are nine DHA1 and seven DHA2 transporters present in *C. neoformans* which play role in antifungal drug resistance, but their role in azole drug resistance has yet not been studied (Costa et al. 2014).

Influx of Drugs into the Cell Import of azoles occurs via facilitated diffusion mediated by transporters in fungi such as *C. albicans* and *C. neoformans*, and mutations in the transporter may greatly influence resistance. A study by Mansfield et al. demonstrated that azole compounds utilized the same mechanism for incorporating drugs inside the cell membrane which was carried out by a transporter. The mutation in that putative transporter resulted in the azole cross-resistance. Among 35 studied clinical isolates of *C. albicans*, 4 isolates showed overexpression of genes; ERG11, MDR1, CDR1, and/or CDR2. A mutation is reported in ERG11 which significantly alters [3H]-FLC import suggesting that modification of azole import mediated by mutated transporter may be associated with antifungal resistance (Mansfield et al. 2010). Composition alterations in plasma membrane are some other factor that affects fluidity and asymmetry of the membrane that leads to a decreased drug uptake (Parks and Casey 1996).

6.3 Chromosomal Abnormalities: Loss of Heterozygosity and Aneuploidy

Azole drug resistance has been associated with multiple genomic alterations, including loss of heterozygosity (LOH) of specific genomic regions (Coste et al. 2007; Selmecki et al. 2010), increase in the chromosome copy number, as well as segmental or chromosomal aneuploidies.

Comparative genome hybridization (CGH) analysis showed that there were 37 aneuploid chromosomes in *C. albicans* strains (FLC-resistant and FLC-sensitive strains). This aneuploidy was frequent on chromosome 5, while segmental aneuploidy was present in eight FLC-resistant strains with extra copies of chr5L which was confirmed by Southern blot analysis (Selmecki et al. 2006). In *C. albicans*, events of LOH were observed antifungal drug-resistant strains (Sanguinetti et al. 2015). It was shown in *C. albicans* that LOH events were increased during in vitro exposure to heat stress (onefold to 40-fold), oxidative stress (threefold to 72-fold), and treatment with antifungal agents (FLC, 285-fold) (Forche et al. 2011) and led to the development of antifungal drug resistance. Further, it was demonstrated that increasing extremity of stress was associated with increased rates of LOH. Chromosomal rearrangements and duplications also have a role in increased resistance toward azoles.

Numerous analyses of azole-resistant and azole-sensitive strains obtained by comparative genome hybridization (CGH) from clinical and laboratory sources reflects a clear link between "heteroresistance" (an azole-associated acquisition of aneuploidy) and azole resistance (Kwon-Chung and Chang 2012). Among the screened isolates of *C. neoformans* and *C. gatti*, it was showed that both the species displayed intrinsic heteroresistance phenotype to FLC (Sionov et al. 2009), which was observed to diminish upon release of drug stress. It was observed that the strain resistance to FLC was always disomic for Chr1, and further elevation in drug level resulted in disomy of Chr4. Infrequent doubling of chromosome 1 has been reported in *erg11* or *afr1* mutants suggesting their role during times of stress (Sionov et al. 2010). Since both *ERG11* and *AFR1* were present on Chr1 in *C. neoformans*, it was seen that the duplication of the chromosome has been an advantage during the development of a resistant strain during FLC stress.

6.4 Mitochondrial Dysfunction

Mitochondrial dysfunction contributes greatly toward the virulence as well as drug tolerance of fungal pathogens. Loss of mitochondrial genome and changes in mitochondria membrane structures lead to the potential activation of drug resistance pathway. In a study, FLC-susceptible and FLC-resistant strains of *C. glabrata* showed upregulation of the ABC transporter genes, CgCDR1, CgCDR2, and CgSNQ2 (Ferrari et al. 2011). A study by Vazquez et al. showed that deletion of the mitochondrial inner membrane translocase gene, OXA1, essential gene for the assembly of mitochondrial-encoded subunits, activates the drug resistance pathway in *C. glabrata*. In addition, *C. glabrata* modification of phosphatidylglycerol synthase gene CgPGS1 responsible for synthesis of the phospholipids in mitochondria has been associated with drug resistance (Shingu-Vazquez and Traven 2011).

6.5 High-Osmolarity Glycerol Pathway

The HOG pathway is often reported to play an important role in controlling various activities like stress response, biosynthesis of ergosterol, production of virulence factor, and differentiation in pathogenic fungi. In *C. neoformans*, transcriptome analysis of the HOG pathway discovered Hrk1 (gene regulated by Hog1), encoding a putative protein kinase. This gene plays a role in stress, virulence of the fungus in Hog1-dependent and Hog1-independent manner, and thus antifungal drug susceptibility (Kim et al. 2011). The HOG pathway controls expression levels of ergosterol biosynthesis genes and in turn affects azole drug susceptibility. Consequently, the HOG mutants exhibit a decrease in drug sensitivity toward azoles such as FLC and KTC (Ko et al. 2009).

6.6 Other Mechanisms

Stress response can regulate azole resistance in *A. fumigatus* but very little is known about it. Sterol regulatory element-binding protein, SrbA, a highly conserved transcription factor, was found to be mediating stress responses under hypoxic conditions. It was further reported to be implicated in maintenance of sterol biosynthesis, hyphal morphology, and specifically in azole resistance, such as FLC and VRC (Willger et al. 2008). Another study involving mutants of MAP kinase kinase 2 (Mkk2), which is generally associated with positive regulation of calcium-mediated signaling, cellular response to salt stress, and cell wall integrity pathway, showed increased sensitivity to both *POS* and VRC (Dirr et al. 2010). The inhabitation of Hsp90 increases the efficiency of VRC in *A. fumigatus* suggesting that it may have significant role in the azole resistance in *A. fumigatus* (Cowen 2009). There are several other mutations in protein-coding regions which were found to favor azole resistance in *A. fumigatus*. HapE gene encodes for CCAAT-binding transcription factor; mutation in this gene that substitutes proline to leucine is one such example (Camps et al. 2012). However, the mechanism of resistance is not clearly understood.

7 Novel Strategies to Combat Drug Resistance

Due to the limited antifungal treatments available, antifungal agents with broad spectrum of fungicidal activity using versatile mechanism of action are often required. Specific treatment of resistant organisms with less toxicity is of paramount importance (Messer et al. 2009). Novel strategies as discussed below may be considered for combating drug resistance.

Understanding the key molecular mechanisms in host-fungal interactions in vitro and in vivo is the key to finding novel therapies. These may also uncover the mechanisms behind the development of drug resistance. The present antifungal agents in use can be altered in a way which can target fungal virulent genes, enzyme inhibitors, and fungal metabolic pathways. Genome-wide analysis studies can be performed to identify the genes responsible in fungal survival and can serve as efficient targets. The catalogue of essential genes for *C. albicans* and *Aspergillus fumigatus* has been studied by molecular techniques such as gene replacement and conditional expression (GRACE) or conditional promoter replacement (CPR) (Roemer et al. 2003; Hu et al. 2007). Such analyses provide specific drug targets for designing effective antifungal drugs. Combinatorial antifungal therapy is used as an alternate approach for primary and salvage treatment of fungal infections, e.g., the use of amphotericin B deoxycholate and flucytosine is highly recommended for cryptococcal meningitis treatment (Pappas et al. 2009b). However, the use of combination therapy for in invasive candidiasis and in aspergillosis has mixed review and still awaits studies, which may come in near future.

The use of multifunctional approach using computational modeling and biochemical studies can be employed to screen the large libraries of chemical compounds effective against broad range of fungal pathogen. The use of in vivo model of *Caenorhabditis elegans–Candida albicans* killing assay presents rapid and economical approach for antifungal discovery (Breger et al. 2007) and presents limited and definite compounds to be tested for in vivo models or in human clinical trials. Compounds from natural sources (plants, sea, microorganisms) and chemical sources should be screened for their antifungal properties and used as another novel approach to overcome current drug resistance. A cysteine-rich antifungal protein (AFP) isolated from *Aspergillus giganteus* is reported to exert potent antifungal activity against pathogenic fungi without altering the viability of host mammalian cells (Meyer 2008); however, its bioactivity is restricted to filamentous fungi. A chitosan, a polymer isolated from crustacean exoskeletons, is another well-researched compound known to be effective against *C. neoformans* biofilms formed on indwelling surgical devices (Martinez et al. 2010). A compound named as 25-azalanosterol inhibits the growth of *C. albicans* in vitro and has potential as a new class of anti-*Candida* agents with no toxic side effects in the mammalian host (Wang and Wu 2008). Although it will be extremely useful to have new and improved antifungal drugs, the immediate focus should be on improvisation of strategies to use the antifungal treatment we possess at present.

8 Conclusions

There has been an increase in incidences of IFIs and emergence of new human pathogenic fungal species. One of the major factors for increased resistance to antifungal drugs in these pathogenic fungi is due to exposure to the antifungal drugs. In view of this, species identification and antifungal susceptibility testing to identify resistant isolate needs are urgent needs for better treatment management of IFIs. Clinical breakpoints have been defined for *Candida* species; however, more studies are required for *Aspergillus* and *Cryptococcus* species. Various studies to understand the molecular mechanisms of resistance in fungi have been carried out majorly in *C. albicans*. However, such studies need to be performed in *Aspergillus* and *Cryptococcus* species. A combinatorial therapy targeting multiple pathways involved in drug resistance might help to combat drug-resistant isolates. Further, new antifungals targeting survival genes would be helpful to in combating drug resistance and improved treatment management of IFIs. Novel strategies in search for antimycotic compounds that should be cost-effective along with high performance and limited toxicity are required.

9 Opinion

The improved diagnostic method based on molecular and serologic techniques is an urgent need for early diagnosis of IFI. Further, development of diagnostic platform detecting drug-resistant isolates from clinical samples is imperative for better IFI treatment. To improve the detection of antifungal resistance, novel assays identifying fungal isolates, which possess mutations known to be associated with antifungal resistance, should be developed and included in diagnostic platform. A reproducible and clinically relevant antifungal susceptibility testing and its use in routine clinical practice are needed. In view of the increasing resistance to antifungal drugs leading to persistent fungal infections, focused research on understanding molecular mechanisms involved in drug resistance should be promoted. Research outcomes of such studies may be beneficial for designing new antifungal drugs to combat these life-threatening infections and develop tools enabling the rapid detection of resistance in fungal population of diverse origins. Novel antifungal agents with better efficacy are required due to relative shortage of antifungal agents for treating invasive mycoses diseases.

References

Agarwal AK, Rogers PD, Baerson SR, Jacob MR, Barker KS, Cleary JD, Walker LA, Nagle DG, Clark AM (2003) Genome-wide expression profiling of the response to polyene, pyrimidine, azole, and echinocandin antifungal agents in Saccharomyces cerevisiae. J Biol Chem 278:34998–35015. doi:10.1074/jbc.M306291200

Alastruey-Izquierdo A, Mellado E, Peláez T, Pemán J, Zapico S, Alvarez M, Rodríguez-Tudela JL, Cuenca-Estrella M (2013) Population-based survey of filamentous fungi and antifungal resistance in Spain (FILPOP Study). Antimicrob Agents Chemother 57:3380–3387. doi:10.1128/AAC.00383-13

Ally R, Schürmann D, Kreisel W, Carosi G, Aguirrebengoa K, Dupont B, Hodges M, Troke P, Romero AJ, Esophageal Candidiasis Study Group (2001) A randomized, double-blind, double-dummy, multicenter trial of voriconazole and fluconazole in the treatment of esophageal candidiasis in immunocompromised patients. Clin Infect Dis 33:1447–1454. doi:10.1086/322653

Al-mamari A, Al-buryhi M, Al-heggami MA, Al-hag S (2014) Identify and sensitivity to antifungal drugs of Candida species causing vaginitis isolated from vulvovaginal infected patients in Sana'a city. Der Pharma Chemica 6:336–342

Al-Mohsen I, Hughes WT (1998) Systemic antifungal therapy: past, present and future. Ann Saudi Med 18:28–38

Amitani R, Taylor G, Elezis EN, Llewellyn-Jones C, Mitchell J, Kuze F, Cole PJ, Wilson R (1995) Purification and characterization of factors produced by *Aspergillus fumigatus* which affect human ciliated respiratory epithelium. Infect Immun 63:3266–3271

Andriole VT (1998) Current and future therapy of invasive fungal infections. In: Remington J, Swartz M (eds) Current clinical topics in infections diseases. Blackwell Sciences, Malden, pp 19–36

Andriole VT (1999) The 1998 Garrod Plecture. Current and future antifungal therapy:new targets for antifungal agents. J Antimicrob Chemother 44:151–162. doi:10.1093/jac/44.2.151

Andriole VT, Bodey GP (1994) Pocket guide to systemic antifungal therapy. Scientific Therapeutics, Madison, NJ, 148 p

Andriole VT, Kravetz HM (1962) The use of amphotericin B in man. JAMA 180:269–272

Arendrup MC, Dzajic E, Jensen RH, Johansen HK, Kjaeldgaard P, Knudsen JD, Kristensen L, Leitz C, Lemming LE, Nielsen L, Olesen B, Rosenvinge FS, Røder BL, Schønheyder HC (2013) Epidemiological changes with potential implication for antifungal prescription recommendations for fungaemia: data from a nationwide fungaemia surveillance programme. Clin Microbiol Infect 19:E343–E353. doi:10.1111/1469-0691.12212

Bader O, Weig M, Reichard U, Lugert R, Kuhns M, Christner M, Held J, Peter S, Schumacher U, Buchheidt D, Tintelnot K, Groß U, Partners MLN-D (2013) cyp51A-Based mechanisms of *Aspergillus fumigatus* azole drug resistance present in clinical samples from Germany. Antimicrob Agents Chemother 57:3513–3517. doi:10.1128/AAC.00167-13

Balashov SV, Gardiner R, Park S, Perlin DS (2005) Rapid, high-throughput, multiplex, real-time PCR for identification of mutations in the cyp51A gene of *Aspergillus fumigatus* that confer resistance to itraconazole. J Clin Microbiol 43:214–222. doi:10.1128/JCM.43.1.214-222.2005

Bennett JW, Klich M (2003) Mycotoxins. Clin Microbiol Rev 16:497–516. doi:10.1128/CMR.16.3.497-516.2003

Ben-Yaacov R, Knoller S, Caldwell GA, Becker JM, Koltin Y (1994) *Candida albicans* gene encoding resistance to benomyl and methotrexate is a multidrug resistance gene. Antimicrob Agents Chemother 38:648–652. doi:10.1128/AAC.38.4.648

Breger J, Fuchs BB, Aperis G, Moy TI, Ausubel FM, Mylonakis E (2007) Antifungal chemical compounds identified using a C. elegans pathogenicity assay. PLoS Pathog 3, e18. doi:10.1371/journal.ppat.0030018

Camps SM, Dutilh BE, Arendrup MC, Rijs AJ, Snelders E, Huynen MA, Verweij PE, Melchers WJ (2012) Discovery of a HapE mutation that causes azole resistance in *Aspergillus fumigatus* through whole genome sequencing and sexual crossing. PLoS One 7, e50034. doi:10.1371/journal

Chang CC, Ananda-Rajah M, Belcastro A, McMullan B, Reid A, Dempsey K, Athan E, Cheng AC, Slavin MA (2014) Consensus guidelines for implementation of quality processes to prevent invasive fungal disease and enhanced surveillance measures during hospital building works. Intern Med J 44:1389–1397. doi:10.1111/imj.12601

Chen CG, Yang YL, Shih HI, Su CL, Lo HJ (2004) CaNdt80 is involved in drug resistance in *Candida albicans* by regulating CDR1. Antimicrob Agents Chemother 48:4505–4512. doi:10.1128/AAC.48.12.4505-4512.2004

Coleman JJ, Mylonakis E (2009) Efflux in fungi:la pièce de résistance. PLoS Pathog 5, e1000486. doi:10.1371/journal.ppat.1000486

Costa C, Dias PJ, Sá-Correia I, Teixeira MC (2014) MFS multidrug transporters in pathogenic fungi: do they have real clinical impact? Front Physiol 5:197. doi:10.3389/fphys.2014.00197

Coste A, Selmecki A, Forche A, Diogo D, Bougnoux ME, d'Enfert C, Berman J, Sanglard D (2007) Genotypic evolution of azole resistance mechanisms in sequential *Candida albicans* isolates. Eukaryot Cell 6:1889–1904. doi:10.1128/EC.00151-07

Coste AT, Crittin J, Bauser C, Rohde B, Sanglard D (2009) Functional analysis of cis-and trans-acting elements of the *Candida albicans* CDR2 promoter with a novel promoter reporter system. Eukaryot Cell 8:1250–1267. doi:10.1128/EC.00069-09

Cowen LE (2009) Hsp90 orchestrates stress response signaling governing fungal drug resistance. PLoS Pathog 8, e1000471. doi:10.1371/journal.ppat.1000471

Cuenca-Estrella M, Gomez-Lopez A, Mellado E, Buitrago MJ, Monzon A, Rodriguez-Tudela JL (2006) Head-to-head comparison of the activities of currently available antifungal agents against 3,378 Spanish clinical isolates of yeasts and filamentous fungi. Antimicrob Agents Chemother 50:917–921. doi:10.1128/AAC.50.3.917-921.2006

d'Enfert C (2009) Hidden killers: persistence of opportunistic fungal pathogens in the human host. Curr Opin Microbiol 12:358–364. doi:10.1016/j.mib.2009.05.008

da Silva Ferreira ME, Malavazi I, Savoldi M, Brakhage AA, Goldman MH, Kim HS, Nierman WC, Goldman GH (2006) Transcriptome analysis of *Aspergillus fumigatus* exposed to voriconazole. Curr Genet 50:32–44

Dagenais TR, Keller NP (2009) Pathogenesis of *Aspergillus fumigatus* in invasive aspergillosis. Clin Microbiol Rev 22:447–465. doi:10.1128/CMR.00055-08

de Micheli M, Bille J, Schueller C, Sanglard D (2002) A common drug-responsive element mediates the upregulation of the *Candida albicans* ABC transporters CDR1 and CDR2, two genes involved in antifungal drug resistance. Mol Microbiol 43:1197–1214. doi:10.1046/j.1365-2958.2002.02814.x

Denning DW, Venkateswarlu K, Oakley KL, Anderson MJ, Manning NJ, Stevens DA, Warnock DW, Kelly SL (1997) Itraconazole resistance in *Aspergillus fumigatus*. Antimicrob Agents Chemother 41:1364–1368

Dhamgaye S, Bernard M, Lelandais G, Sismeiro O, Lemoine S, Coppée JY, Le Crom S, Prasad R, Devaux F (2012) RNA sequencing revealed novel actors of the acquisition of drug resistance in *Candida albicans*. BMC Genomics 13:396. doi:10.1186/1471-2164-13-396

Diaz-Guerra TM, Mellado E, Cuenca-Estrella M, Rodriguez-Tudela JL (2003) A point mutation in the 14-alpha-sterol demethylase gene *cyp51A* contributes to itraconazole resistance in *A. fumigatus*. Antimicrob Agents Chemother 47:1120–1124. doi:10.1128/AAC.47.3.1120-1124.2003

Dirr F, Echtenacher B, Heesemann J, Hoffmann P, Ebel F, Wagener J (2010) AfMkk2 is required for cell wall integrity signaling, adhesion, and full virulence of the human pathogen *Aspergillus fumigatus*. Int J Med Microbiol 300:496–502. doi:10.1016/j.ijmm.2010.03.001

do Nascimento AM, Goldman MH, Goldman GH (2002) Molecular characterization of ABC transporter-encoding genes in Aspergillus nidulans. Genet Mol Res 1:337–349

Ellis M, Al-Ramadi B, Bernsen R, Kristensen J, Alizadeh H, Hedstrom U (2009) Prospective evaluation of mannan and anti-mannan antibodies for diagnosis of invasive Candida infections in patients with neutropenic fever. J Med Microbiol 58:606–615. doi:10.1099/jmm.0.006452-0

Espinel-Ingroff A, Diekema DJ, Fothergill A, Johnson E, Pelaez T, Pfaller MA, Rinaldi MG, Canton E, Turnidge J (2010) Wild-type MIC distributions and epidemiological cutoff values for the triazoles and six Aspergillus spp. for the CLSI broth microdilution method (M38-A2 document). J Clin Microbiol 48:3251–3257. doi:10.1128/JCM.00536-10

Espinel-Ingroff A, Aller AI, Canton E, Castañón-Olivares LR, Chowdhary A, Cordoba S, Cuenca-Estrella M, Fothergill A, Fuller J, Govender N, Hagen F, Illnait-Zaragozi MT, Johnson E, Kidd S, Lass-Flörl C, Lockhart SR, Martins MA, Meis JF, Melhem MS, Ostrosky-Zeichner L, Pelaez T, Pfaller MA, Schell WA, St-Germain G, Trilles L, Turnidge J (2012) Cryptococcus neoformans-Cryptococcus gattii species complex: an international study of wild-type susceptibility endpoint distributions and epidemiological cutoff values for fluconazole, itraconazole, posaconazole, and voriconazole. Antimicrob Agents Chemother 56:5898–5906. doi:10.1128/AAC.01115-12

Ferrari S, Sanguinetti M, De Bernardis F, Torelli R, Posteraro B, Vandeputte P, Sanglard D (2011) Loss of mitochondrial functions associated with azole resistance in Candida glabrata results in enhanced virulence in mice. Antimicrob Agents Chemother 55:1852–1860. doi:10.1128/AAC.01271-10

Forche A, Abbey D, Pisithkul T, Weinzierl MA, Ringstrom T, Bruck D, Petersen K, Berman J (2011) Stress alters rates and types of loss of heterozygosity in *Candida albicans*. MBio 2: e00129-11. doi:10.1128/mBio.00129-11

Fothergill AW, Sutton DA, McCarthy DI, Wiederhold NP (2014) Impact of new antifungal breakpoints on antifungal resistance in *Candida* species. J Clin Microbiol 52:994–997. doi:10.1128/JCM.03044-13

Gallagher JC, Dodds Ashley ES, Drew RH, Perfect JR (2003) Antifungal pharmacotherapy for invasive mould infections. Expert Opin Pharmacother 4:147–164. doi:10.1517/14656566.4.2.147

Gaur NA, Manoharlal R, Saini P, Prasad T, Mukhopadhyay G, Hoefer M, Morschhäuser J, Prasad R (2005) Expression of the CDR1 efflux pump in clinical *Candida albicans* isolates is controlled by a negative regulatory element. Biochem Biophys Res Commun 332:206–214. doi:10.1016/j.bbrc.2005.04.113

Georgopapadakou NH, Walsh TJ (1996) Antifungal agents: chemotherapeutic targets and immunologic strategies. Antimicrob Agents Chemother 40:279–291

Ghannoum MA, Rice LB (1999) Antifungal agents: mode of action, mechanisms of resistance, and correlation of these mechanisms with bacterial resistance. Clin Microbiol Rev 12:501–517

Godoy MC, Viswanathan C, Marchiori E, Truong MT, Benveniste MF, Rossi S, Marom EM (2012) The reversed halo sign: update and differential diagnosis. Br J Radiol 85:1226–1235. doi:10.1259/bjr/54532316

Greene RE, Schlamm HT, Oestmann JW, Stark P, Durand C, Lortholary O, Wingard JR, Herbrecht R, Ribaud P, Patterson TF, Troke PF, Denning DW, Bennett JE, de Pauw BE, Rubin RH (2007) Imaging findings in acute invasive pulmonary aspergillosis: clinical significance of the halo sign. Clin Infect Dis 44:373–379. doi:10.1086/509917

Groll AH, Piscitelli SC, Walsh TJ (1998) Clinical pharmacology of systemic antifungal agents: a comprehensive review of agents in clinical use, current investigational compounds, and putative targets for antifungal drug development. Adv Pharmacol 44:343–500

Hiemenz JW, Walsh TJ (1996) Lipid formulations of amphotericin B: recent progress and future directions. Clin Infect Dis 22(Suppl 2):S133–S144. doi:10.1093/clinids/22.Supplement_2.S13

Holz RW (1974) The effects of the polyene antibiotics nystatin and amphotericin B on thin lipid membranes. Ann N Y Acad Sci 235:469–479. doi:10.1111/j.1749-6632.1974.tb43284.x

Hu W, Sillaots S, Lemieux S, Davison J, Kauffman S, Breton A, Linteau A, Xin C, Bowman J, Becker J, Jiang B, Roemer T (2007) Essential gene identification and drug target prioritization in *Aspergillus fumigatus*. PLoS Pathog 3, e24. doi:10.1371/journal.ppat.0030024

Johnson G, Ferrini A, Dolan SK, Nolan T, Agrawal S, Doyle S, Bustin SA (2014) Biomarkers for invasive aspergillosis: the challenges continue. Biomark Med 8:429–451. doi:10.2217/bmm.13.129

Kam AP, Xu J (2002) Diversity of commensal yeasts within and among healthy hosts. Diagn Microbiol Infect Dis 43:19–28. doi:10.1016/S0732-8893(02)00364-4

Karnani N, Gaur NA, Jha S, Puri N, Krishnamurthy S, Goswami SK, Mukhopadhyay G, Prasad R (2004) SRE1 and SRE2 are two specific steroid-responsive modules of Candida drug resistance gene 1 (CDR1) promoter. Yeast 21:219–239. doi:10.1002/yea.1067

Kelly SL, Lamb DC, Loeffler J, Einsele H, Kelly DE (1999) The G464S amino acid substitution in *Candida albicans* sterol 14alpha-demethylase causes fluconazole resistance in the clinic through reduced affinity. Biochem Biophys Res Commun 262:174–179. doi:10.1006/bbrc.1999.1136

Kerridge D (1980) The plasma membrane of *Candida albicans* and its role in the action of antifungal drugs. In: Gooday GW, Lloyd D, Trinci APJ (eds) The eukaryotic microbial cell. Cambridge University Press, Cambridge, 103 p

Kerridge D (1985) The protoplast membrane and antifungal drugs. In: Peberdy JF, Ferenczy L (eds) Fungal protoplasts: applications in biochemistry and genetics. Marcel Dekker Inc., New York, NY, 135 p

Kim SY, Ko YJ, Jung KW, Strain A, Nielsen K, Bahn YS (2011) Hrk1 plays both Hog1-dependent and -independent roles in controlling stress response and antifungal drug resistance in Cryptococcus neoformans. PLoS One 6, e18769. doi:10.1371/journal.pone.0018769

Ko YJ, Yu YM, Kim GB, Lee GW, Maeng PJ, Kim S, Floyd A, Heitman J, Bahn YS (2009) Remodeling of global transcription patterns of Cryptococcus neoformans genes mediated by the stress-activated HOG signaling pathways. Eukaryot Cell 8:1197–1217. doi:10.1128/EC.00120-09

Krcmery V, Barnes AJ (2002) Non-albicans Candida spp. causing fungaemia: pathogenicity and antifungal resistance. J Hosp Infect 50:243–260. doi:10.1053/jhin.2001.1151

Kullberg BJ, Arendrup MC (2015) Invasive Candidiasis. N Engl J Med 373:1445–1456. doi:10.1056/NEJMra1315399

Kwon-Chung KJ, Chang YC (2012) Aneuploidy and drug resistance in pathogenic fungi. PLoS Pathog 8, e1003022. doi:10.1371/journal.ppat.1003022

Kwon-Chung KJ, Rhodes JC (1986) Encapsulation and melanin formation as indicators of virulence in Cryptococcus neoformans. Infect Immun 51:218–223

Lamping E, Baret PV, Holmes AR, Monk BC, Goffeau A, Cannon RD (2010) Fungal PDR transporters: phylogeny, topology, motifs and function. Fungal Genet Biol 47:127–142. doi:10.1016/j.fgb.2009.10.007

Latge JP (1999) *Aspergillus fumigatus* and aspergillosis. Clin Microbiol Rev 12:310–350

MacPherson S, Akache B, Weber S, De Deken X, Raymond M, Turcotte B (2005) *Candida albicans* zinc cluster protein Upc2p confers resistance to antifungal drugs and is an activator of ergosterol biosynthetic genes. Antimicrob Agents Chemother 49:1745–1752. doi:10.1128/AAC.49.5.1745-1752.2005

Magill SS, Edwards JR, Bamberg W, Beldavs ZG, Dumyati G, Kainer MA, Lynfield R, Maloney M, McAllister-Hollod L, Nadle J, Ray SM, Thompson DL, Wilson LE, Fridkin SK (2014) Multistate point-prevalence survey of health care-associated infections. N Engl J Med 370:1198–1208. doi:10.1056/NEJMoa1306801

Mansfield BE, Oltean HN, Oliver BG, Hoot SJ, Leyde SE, Hedstrom L, White TC (2010) Azole drugs are imported by facilitated diffusion in *Candida albicans* and other pathogenic fungi. PLoS Pathog 6, e1001126. doi:10.1371/journal.ppat.1001126

Marichal P, Vanden Bossche H, Odds FC, Nobels G, Warnock DW, Timmerman V, Van Broeckhoven C, Fay S, Mose-Larsen P (1997) Molecular biological characterization of an azole-resistant Candida glabrata isolate. Antimicrob Agents Chemother 41:2229–2237

Martinez LR, Mihu MR, Han G, Frases S, Cordero RJ, Casadevall A, Friedman AJ, Friedman JM, Nosanchuk JD (2010) The use of chitosan to damage Cryptococcus neoformans biofilms. Biomaterials 31:669–679. doi:10.1016/j.biomaterials.2009.09.087

McCormack FX, Whitsett JA (2002) The pulmonary collectins, SP-A and SP-D, orchestrate innate immunity in the lung. J Clin Invest 109:707–712. doi:10.1172/jci15293

Mellado E, Diaz-Guerra TM, Cuenca-Estrella M, Rodriguez-Tudela JL (2001) Identification of two different 14-alpha sterol demethylase related genes (cyp51A and cyp51B) in *A. fumigatus* and other *A.* species. J Clin Microbiol 39:2431–2438. doi:10.1128/JCM.39.7.2431-2438.2001

Mellado E, Garcia-Effron G, Alcázar-Fuoli L, Melchers WJ, Verweij PE, Cuenca-Estrella M, Rodríguez-Tudela JL (2007) A new *Aspergillus fumigatus* resistance mechanism conferring in vitro cross-resistance to azole antifungals involves a combination of cyp51A alterations. Antimicrob Agents Chemother 51:1897–1904. doi:10.1128/AAC.01092-06

Messer SA, Moet GJ, Kirby JT, Jones RN (2009) Activity of contemporary antifungal agents, including the novel echinocandin anidulafungin, tested against Candida spp., Cryptococcus spp., and Aspergillus spp.: report from the SENTRY Antimicrobial Surveillance Program (2006 to 2007). J Clin Microbiol 47:1942–1946. doi:10.1128/JCM.02434-08

Meyer V (2008) A small protein that fights fungi: AFP as a new promising antifungal agent of biotechnological value. Appl Microbiol Biotechnol 78:17–28

Mikulska M, Novelli A, Aversa F, Cesaro S, de Rosa FG, Girmenia C, Micozzi A, Sanguinetti M, Viscoli C (2012) Voriconazole in clinical practice. J Chemother 24:311–327. doi:10.1179/1973947812Y.0000000051

Musher B, Fredricks D, Leisenring W, Balajee SA, Smith C, Marr KA (2004) *Aspergillus* galactomannan enzyme immunoassay and quantitative PCR for diagnosis of invasive aspergillosis with bronchoalveolar lavage fluid. J Clin Microbiol 42:5517–5522. doi:10.1128/jcm.42.12.5517-5522.2004

Nanjappa SG, Klein BS (2014) Vaccine immunity against fungal infections. Curr Opin Immunol 28:27–33. doi:10.1016/j.coi.2014.01.014

Nascimento AM, Goldman GH, Park S, Marras SA, Delmas G, Oza U, Lolans K, Dudley MN, Mann PA, Perlin DS (2003) Multiple resistance mechanisms among *Aspergillus fumigatus* mutants with high-level resistance to itraconazole. Antimicrob Agents Chemother 47:1719–1726. doi:10.1128/AAC.47.5.1719-1726.2003

Netea MG, Sutmuller R, Hermann C, van der Graaf CA, van der Meer JW, van Krieken JH, Hartung T, Adema G, Kullberg BJ (2004) Toll-like receptor 2 suppresses immunity against *Candida albicans* through induction of IL-10 and regulatory T cells. J Immunol (Baltimore, Md: 1950) 172:3712–3718. doi:10.4049/jimmunol.172.6.3712

Nucci F, Nouer SA, Capone D, Anaissie E, Nucci M (2015) Fusariosis. Semin Respir Crit Care Med 36:706–714. doi:10.1055/s-0035-1562897

Nunes AP, Schuenck RP, Bastos CC, Magnanini MM, Long JB, Iorio NL, Santos KR (2007) Heterogeneous resistance to vancomycin and teicoplanin among Staphylococcus spp. isolated from bacteremia. Braz J Infect Dis 11:345–350. doi:10.1590/S1413-86702007000300009

Odds FC, Brown AJ, Gow NA (2003) Antifungal agents: mechanisms of action. Trends Microbiol 11:272–279. doi:10.1016/S0966-842X(03)00117-3

Orsi CF, Bettua C, Pini P, Venturelli C, La Regina A, Morace G, Luppi M, Forghieri F, Bigliardi S, Luppi F, Codeluppi M, Girardis M, Blasi E (2015) Detection of *Pneumocystis jirovecii* and *Aspergillus* spp. DNA in bronchoalveolar lavage fluids by commercial real-time PCR assays: comparison with conventional diagnostic tests. New Microbiol 38:75–84

Panackal AA, Imhof A, Hanley EW, Marr KA (2006) *Aspergillus* ustus infections among transplant recipients. Emerg Infect Dis 12:403–408. doi:10.3201/eid1205.050670

Pappas PG, Kauffman CA, Andes D, Benjamin DK Jr, Calandra TF, Edwards JE Jr, Filler SG, Fisher JF, Kullberg BJ, Ostrosky-Zeichner L, Reboli AC, Rex JH, Walsh TJ, Sobel JD, Infectious Diseases Society of America (2009a) Clinical practice guidelines for the management of candidiasis: 2009 update by the Infectious Diseases Society of America. Clin Infect Dis 48:503–535. doi:10.1086/596757

Pappas PG, Chetchotisakd P, Larsen RA, Manosuthi W, Morris MI, Anekthananon T, Sungkanuparph S, Supparatpinyo K, Nolen TL, Zimmer LO, Kendrick AS, Johnson P, Sobel JD, Filler SG (2009b) A phase II randomized trial of amphotericin B alone or combined with fluconazole in the treatment of HIV-associated cryptococcal meningitis. Clin Infect Dis 48:1775–1783. doi:10.1086/599112

Parks LW, Casey WM (1996) Fungal sterols. In: Prasad R, Ghannoum MA (eds) Lipids of pathogenic fungi. CRC Press, Boca Raton, FL, pp 63–82

Pfaller MA, Diekema DJ, Ghannoum MA, Rex JH, Alexander BD, Andes D, Brown SD, Chaturvedi V, Espinel-Ingroff A, Fowler CL, Johnson EM, Knapp CC, Motyl MR, Ostrosky-Zeichner L, Sheehan DJ, Walsh TJ, Clinical and Laboratory Standards Institute Antifungal Testing Subcommittee (2009) Wild-type MIC distribution and epidemiological cutoff values for *Aspergillus fumigatus* and three triazoles as determined by the Clinical and Laboratory Standards Institute broth microdilution methods. J Clin Microbiol 47:3142–3146. doi:10.1128/JCM.00940-09

Pihet M, Vandeputte P, Tronchin G, Renier G, Saulnier P, Georgeault S, Mallet R, Chabasse D, Symoens F, Bouchara JP (2009) Melanin is an essential component for the integrity of the cell wall of *Aspergillus fumigatus* conidia. BMC Microbiol 9:177. doi:10.1186/1471-2180-9-177

Pinjon E, Moran GP, Jackson CJ, Kelly SL, Sanglard D, Coleman DC, Sullivan DJ (2003) Molecular mechanisms of itraconazole resistance in Candida dubliniensis. Antimicrob Agents Chemother 47:2424–2437. doi:10.1128/AAC.47.8.2424-2437.2003

Pinjon E, Jackson CJ, Kelly SL, Sanglard D, Moran G, Coleman DC, Sullivan DJ (2005) Reduced azole susceptibility in genotype 3 Candida dubliniensis isolates associated with increased CdCDR1 and CdCDR2 expression. Antimicrob Agents Chemother 49:1312–1318. doi:10.1128/AAC.49.4.1312-1318.2005

Polonelli L, Casadevall A, Han Y, Bernardis F, Kirkland TN, Matthews RC, Adriani D, Boccanera M, Burnie JP, Cassone A, Conti S, Cutler JE, Frazzi R, Gregory C, Hodgetts S, Illidge C, Magliani W, Rigg G, Santoni G (2000) The efficacy of acquired humoral and cellular immunity in the prevention and therapy of experimental fungal infections. Med Mycol 38:281–292

Rajendran R, Mowat E, McCulloch E, Lappin DF, Jones B, Lang S, Majithiya JB, Warn P, Williams C, Ramage G (2011) Azole resistance of *Aspergillus fumigatus* biofilms is partly associated with efflux pump activity. Antimicrob Agents Chemother 55:2092–2097. doi:10.1128/AAC.01189-10

Rex JH, Pfaller MA, Galgiani JN, Bartlett MS, Espinel-Ingroff A, Ghannoum MA, Lancaster M, Odds FC, Rinaldi MG, Walsh TJ, Barry AL (1997) Development of interpretive breakpoints for antifungal susceptibility testing: conceptual framework and analysis of in vitro-in vivo correlation data for fluconazole, itraconazole, and candida infections. Subcommittee on Antifungal Susceptibility Testing of the National Committee for Clinical Laboratory Standards. Clin Infect Dis 24:235–247. doi:10.1093/clinids/24.2.235

Riggle P, Kumamoto CA (2006) Transcriptional regulation of MDR1, encoding a drug efflux determinant, in fluconazole-resistant *Candida albicans* strains through an Mcm1p binding site. Eukaryot Cell 5:1957–1968. doi:10.1128/EC.00243-06

Rodero L, Mellado E, Rodriguez AC, Salve A, Guelfand L, Cahn P, Rodriguez-Tudela JL (2003) G484S amino acid substitution in lanosterol 14-α demethylase (*ERG11*) is related to fluconazole resistance in a recurrent Cryptococcus neoformans clinical isolate. Antimicrob Agents Chemother 47:3653–3656. doi:10.1128/AAC.47.11.3653-3656.2003

Rodriguez-Tudela JL, Alcazar-Fuoli L, Mellado E, Alastruey-Izquierdo A, Monzon A, Cuenca-Estrella M (2008) Epidemiological cutoffs and cross-resistance to azole drugs in *Aspergillus fumigatus*. Antimicrob Agents Chemother 52:2468–2472. doi:10.1128/AAC.00156-08

Roemer T, Jiang B, Davison J, Ketela T, Veillette K, Breton A, Tandia F, Linteau A, Sillaots S, Marta C, Martel N, Veronneau S, Lemieux S, Kauffman S, Becker J, Storms R, Boone C, Bussey H (2003) Large-scale essential gene identification in *Candida albicans* and applications to antifungal drug discovery. Mol Microbiol 50:167–181. doi:10.1046/j.1365-2958.2003.03697.x

Rognon B, Kozovska Z, Coste AT, Pardini G, Sanglard D (2006) Identification of promoter elements responsible for the regulation of MDR1 from *Candida albicans*, a major facilitator transporter involved in azole resistance. Microbiology 152:3701–3722. doi:10.1099/mic.0.29277-0

Romani L (2004) Immunity to fungal infections. Nat Rev Immunol 4:1–23. doi:10.1038/nri1255
Samaranayake YH, Cheung BP, Wang Y, Yau JY, Yeung KW, Samaranayake LP (2013) Fluconazole resistance in *Candida glabrata* is associated with increased bud formation and metallothionein production. J Med Microbiol 62:303–318. doi:10.1099/jmm.0.044123-0
Samson RA, Visagie CM, Houbraken J, Hong SB, Hubka V, Klaassen CH, Perrone G, Seifert KA, Susca A, Tanney JB, Varga J, Kocsubé S, Szigeti G, Yaguchi T, Frisvad JC (2014) Phylogeny, identification and nomenclature of the genus Aspergillus. Stud Mycol 78:141–173. doi:10.1016/j.simyco.2014.07.004
Sanglard D, Ischer F, Koymans L, Bille J (1998) Amino acid substitutions in the cytochrome P-450 lanosterol 14alpha-demethylase (CYP51A1) from azole-resistant *Candida albicans* clinical isolates contribute to resistance to azole antifungal agents. Antimicrob Agents Chemother 42:241–253
Sanguinetti M, Posteraro B, La Sorda M, Torelli R, Fiori B, Santangelo R, Delogu G, Fadda G (2006) Role of AFR1, an ABC transporter-encoding gene, in the in vivo response to fluconazole and virulence of Cryptococcus neoformans. Infect Immun 74:1352–1359. doi:10.1128/IAI.74.2.1352-1359.2006
Sanguinetti M, Posteraro B, Lass-Flörl C (2015) Antifungal drug resistance among Candida species: mechanisms and clinical impact. Mycoses 58(Suppl 2):2–13. doi:10.1111/myc.12330
Saxena P, Shen SH, Morrissey O, Gooi JH (2015) Challenges in the management of invasive pulmonary zygomycosis: the Alfred experience. ANZ J Surg 85:700–701. doi:10.1111/ans.13264
Seed PC (2014) The human mycobiome. Cold Spring Harb Perspect Med 5:a019810. doi:10.1101/cshperspect.a019810
Selmecki A, Forche A, Berman J (2006) Aneuploidy and isochromosome formation in drug-resistant *Candida albicans*. Science 313:367–370. doi:10.1126/science.1128242
Selmecki A, Gerami-Nejad M, Paulson C, Forche A, Berman J (2008) An isochromosome confers drug resistance in vivo by amplification of two genes, ERG11 and TAC1. Mol Microbiol 68:624–641. doi:10.1111/j.1365-2958.2008.06176.x
Selmecki A, Forche A, Berman J (2010) Genomic plasticity of the human fungal pathogen *Candida albicans*. Eukaryot Cell 9:991–1008. doi:10.1128/EC.00060-10
Semighini CP, Marins M, Goldman MH, Goldman GH (2002) Quantitative analysis of the relative transcript levels of ABC transporter Atr genes in Aspergillus nidulans by real-time reverse transcription-PCR assay. Appl Environ Microbiol 68:1351–1357. doi:10.1128/AEM.68.3.1351-1357.2002
Shingu-Vazquez M, Traven A (2011) Mitochondria and fungal pathogenesis: drug tolerance, virulence, and potential for antifungal therapy. Eukaryot Cell 10:1376–1383. doi:10.1128/EC.05184-11
Silva AP, Miranda IM, Guida A, Synnott J, Rocha R, Silva R, Amorim A, Pina-Vaz C, Butler G, Rodrigues AG (2011) Transcriptional profiling of azole resistant *Candida parapsilosis* strains. Antimicrob Agents Chemother 55:3546–3556. doi:10.1128/AAC.01127-10
Singh A, Prasad R (2011) Comparative lipidomics of azole sensitive and resistant clinical isolates of *Candida albicans* reveals unexpected diversity in molecular lipid imprints. PLoS One 6:e19266. doi:10.1371/journal.pone.0019266
Singh A, Yadav V, Prasad R (2012) Comparative lipidomics in clinical isolates of *Candida albicans* reveal crosstalk between mitochondria, cell wall integrity and azole resistance. PLoS One 7:e39812. doi:10.1371/journal.pone.0039812
Sionov E, Chang YC, Garraffo HM, Kwon-Chung KJ (2009) Heteroresistance tofluconazole in Cryptococcus neoformans is intrinsic and associated with virulence. Antimicrob Agents Chemother 53:2804–2815. doi:10.1128/AAC.00295-09
Sionov E, Lee H, Chang YC, Kwon-Chung KJ (2010) Cryptococcus neoformans overcomes stress of azole drugs by formation of disomy in specific multiple chromosomes. PLoS Pathog 6, e1000848. doi:10.1371/journal.ppat.1000848

Slaven JW, Anderson MJ, Sanglard D, Dixon GK, Bille J, Roberts IS, Denning DW (2002) Increased expression of a novel Aspergillus fumigatus ABC transporter gene, atrF, in the presence of itraconazole in an itraconazole resistant clinical isolate. Fungal Genet Biol 36:199–206. doi:10.1016/S1087-1845(02)00016-6

Sloan DJ, Parris V (2014) Cryptococcal meningitis: epidemiology and therapeutic options. Clin Epidemiol 6:169–182. doi:10.2147/CLEP.S38850

Snelders E, van der Lee HA, Kuijpers J, Rijs AJ, Varga J, Samson RA, Mellado E, Donders AR, Melchers WJ, Verweij PE (2008) Emergence of azole resistance in *Aspergillus fumigatus* and spread of a single resistance mechanism. PLoS Med 5, e219. doi:10.1371/journal.pmed.0050219

Snelders E, Karawajczyk A, Schaftenaar G, Verweij PE, Melchers WJ (2010) Azole resistance profile of amino acid changes in *A. fumigates* cyp51A based on protein homology modeling. Antimicrob Agents Chemother 54:2425–2430. doi:10.1128/AAC.01599-09

Speed B, Dunt D (1995) Clinical and host differences between infections with the two varieties of Cryptococcus neoformans. Clin Infect Dis 21:28–34. doi:10.1093/clinids/21.1.28

Thursky KA, Playford EG, Seymour JF, Sorrell TC, Ellis DH, Guy SD, Gilroy N, Chu J, Shaw DR (2008) Recommendations for the treatment of established fungal infections. Intern Med J 38:496–520. doi:10.1111/j.1445-5994.2008.01725.x

Torelli R, Posteraro B, Ferrari S, La Sorda M, Fadda G, Sanglard D, Sanguinetti M (2008) The ATP-binding cassette transporter-encoding gene CgSNQ2 is contributing to the CgPDR1-dependent azole resistance of Candida glabrata. Mol Microbiol 68:186–201. doi:10.1111/j.1365-2958.2008.06143.x

Tsai HF, Sammons LR, Zhang X, Suffis SD, Su Q, Myers TG, Marr KA, Bennett JE (2010) Microarray and molecular analyses of the azole resistance mechanism in *Candida glabrata* oropharyngeal isolates. Antimicrob Agents Chemother 54:3308–3317. doi:10.1128/AAC.00535-10

van der Meer JW, van de Veerdonk FL, Joosten LA, Kullberg BJ, Netea MG (2010) Severe Candida spp. infections: new insights into natural immunity. Int J Antimicrob Agents 36:S58–S62. doi:10.1016/j.ijantimicag.2010.11.013

Vandeputte P, Larcher G, Bergès T, Renier G, Chabasse D, Bouchara JP (2005) Mechanisms of azole resistance in a clinical isolate of Candida tropicalis. Antimicrob Agents Chemother 49:4608–4615. doi:10.1128/AAC.49.11.4608-4615.2005

Vandeputte P, Ferrari S, Coste AT (2012) Antifungal resistance and new strategies to control fungal infections. Int J Microbiol 713687. doi:10.1155/2012/713687

Vasicek EM, Berkow EL, Bruno VM, Mitchell AP, Wiederhold NP, Barker KS, Rogers PD (2014) Disruption of the transcriptional regulator Cas5 results in enhanced killing of *Candida albicans* by Fluconazole. Antimicrob Agents Chemother 58:6807–6818. doi:10.1128/AAC.00064-14

Verweij PE, Mellado E, Melchers WJ (2007) Multiple-triazole-resistant aspergillosis. N Engl J Med 356:1481–1483. doi:10.1056/NEJMc061720

Walsh TJ, Roilides E, Cortez K, Kottilil S, Bailey J, Lyman CA (2005) Control, immunoregulation, and expression of innate pulmonary host defenses against *Aspergillus fumigatus*. Med Mycol 43:S165–S172. doi:10.1080/13693780500064672

Walsh TJ, Anaissie EJ, Denning DW, Herbrecht R, Kontoyiannis DP, Marr KA, Morrison VA, Segal BH, Steinbach WJ, Stevens DA, van Burik JA, Wingard JR, Patterson TF (2008) Treatment of aspergillosis: clinical practice guidelines of the Infectious Diseases Society of America. Clin Infect Dis 46:327–360. doi:10.1086/525258

Wang J, Wu J (2008) Antifungal activity of 25-azalanosterol against Candida species. Eur J Clin Microbiol Infect Dis 27:1131–1136. doi:10.1007/s10096-008-0554-y

Wang Z, Zhang G, Zhang X, Wu S, Yin X, Zhang H (2012) Proteomic analysis of fluconazole resistance in *Candida albicans*. Afr J Pharm Pharmacol 6:1226–1230. doi:10.5897/AJPP11.801

Wang B, Huang L-H, Zhao J-X, Wei M, Fang H, Wang D-Y, Wang H-F, Yin J-G, Xiang M (2015a) ERG11 mutations associated with azole resistance in *Candida albicans* isolates from vulvovaginal candidiasis patients. Asian Pac J Trop Biomed 5:909–914. doi:10.1016/j.apjtb.2015.08.002

Wang Y, Liu JY, Shi C, Li WJ, Zhao Y, Yan L, Xiang MJ (2015b) Mutations in transcription factor Mrr2p contribute to fluconazole resistance in clinical isolates of *Candida albicans*. Int J Antimicrob Agents 46:552–559. doi:10.1016/j.ijantimicag.2015.08.001

Warrilow AG, Melo N, Martel CM, Parker JE, Nes WD, Kelly SL, Kelly DE (2010) Expression, purification, and characterization of *Aspergillus fumigatus* sterol 14-alpha demethylase (CYP51) isoenzymes A and B. Antimicrob Agents Chemother 54:4225–4234. doi:10.1128/AAC.00316-10

Warris A, Weemaes CM, Verweij PE (2002) Multidrug resistance in *Aspergillus fumigatus*. N Engl J Med 347:2173–2174. doi:10.1056/NEJM200212263472618

Willger SD, Puttikamonkul S, Kim KH, Burritt JB, Grahl N, Metzler LJ, Barbuch R, Bard M, Lawrence CB, Cramer RA Jr (2008) A sterol-regulatory element binding protein is required for cell polarity, hypoxia adaptation, azole drug resistance, and virulence in *Aspergillus fumigatus*. PLoS Pathog 4, e1000200. doi:10.1371/journal

Won EJ, Shin JH, Choi MJ, Lee WG, Park YJ, Uh Y, Kim SY, Lee MK, Kim SH, Shin MG, Suh SP, Ryang DW (2015) Antifungal susceptibilities of bloodstream isolates of Candida species from nine hospitals in Korea: application of new antifungal breakpoints and relationship to antifungal usage. PLoS One 10, e0118770. doi:10.1371/journal.pone.0118770

Wu CJ, Wang HC, Lee JC, Lo HJ, Dai CT, Chou PH, Ko WC, Chen YC (2015) Azole-resistant *Aspergillus fumigatus* isolates carrying TR_{34}/L98H mutations in Taiwan. Mycoses 58:544–549. doi:10.1111/myc.12354

Yapar N (2014) Epidemiology and risk factors for invasive candidiasis. Therap Clin Risk Manag 10:95–105. doi:10.2147/tcrm.s40160

Yoo JI, Choi CW, Kim HS, Yoo JS, Jeong YH, Lee YS (2012) Proteomic analysis of cellular and membrane proteins in fluconazole-resistant *Candida glabrata*. Osong Public Health Res Perspect 3:74–78. doi:10.1016/j.phrp.2012.04.001

Yoo JI, Kim HS, Choi CW, Yoo JS, Yu JY, Lee YS (2013) Proteomic analysis of intracellular and membrane proteins from voriconazole-resistant *Candida glabrata*. Osong Public Health Res Perspect 4:293–300. doi:10.1016/j.phrp.2013.10.001

Zhang L, Zhou S, Pan A, Li J, Liu B (2015) Surveillance of antifungal susceptibilities in clinical isolates of *Candida* species at 36 hospitals in China from 2009 to 2013. Int J Infect Dis 33:1–4. doi:10.1016/j.ijid.2014.12.033

Zonios DI, Bennett JE (2008) Update on azole antifungals. Semin Respir Crit Care Med 29:198–210. doi:10.1055/s-2008-1063858

Drug Resistance in Malaria

Santosh C.M. Kumar

Abstract Malaria is one of the most prevalent and most fatal parasitic disease humans with 1–2 million fatalities every year. Although intense efforts to eradicate malaria are progressing with several treatment regimens, emergence of drug-resistant parasites impedes these efforts. Resistance has emerged to all the mainstream antimalarial drugs including the artemisinin and therefore is responsible for an immense increase in malaria-related mortality worldwide, particularly in Africa. Fortunately, inherent emergence of resistance has been effectively prevented by the combinatorial antimalarial drugs such as the artemisinin derivative combinations. Therefore, sustained improvements in antimalarial medicines with comprehensive understanding of the mechanism of action and resistance are essential to treat and control malaria. This chapter presents the current understanding on the mechanism of the evolution of drug resistance and the strategies followed to delay or curtail it.

1 Introduction

The members of *Plasmodium* genera, constituting the hematoprotozoan parasites, cause human malarias that are transmitted through a certain species of the anopheline mosquito vectors (Bray and Garnham 1982; Petersen et al. 2011). Among the five species affecting humans, namely, *Plasmodium falciparum*, *P. vivax*, *P. ovale*, *P. malariae*, and *P. knowlesi*, the *P. falciparum* accounts for the majority of instances of morbidity and mortality causing approximately 225 million infections and resulting in nearly one million deaths every year in the suburban Africa (White 2004; Cox-Singh et al. 2008; Petersen et al. 2011). Furthermore, *P. vivax* is the second most common species that can cause a relapsing form of malaria principally in the populations of Asia and South America (Prince et al. 2007). Each year, an estimate of 300–500 million clinical cases of malaria occur that include 1.5–2.0

S.C.M. Kumar (✉)
Laboratory of Structural Biology (LSB), National Centre for Cell Science (NCCS), Pune 411007, MH, India
e-mail: cms@nccs.res.in

G. Arora et al. (eds.), *Drug Resistance in Bacteria, Fungi, Malaria, and Cancer*,
DOI 10.1007/978-3-319-48683-3_19

million deaths, making malaria one of the most common infectious diseases worldwide (World Malaria Report 2014). In addition, malaria occurs in over 97 countries worldwide, wherein 36 % of the global population live in the risk areas of malaria transmission, 7 % reside in areas with limited control on the transmission, and 29 % live in areas where the transmission has been reestablished (World Malaria Report 2014). Remarkably, the increase in the financing and coverage of malaria control programs worldwide has resulted in significant reduction in malaria incidence and mortality rates. Interestingly, global estimates for malaria case incidence rates fell by 30 % between 2000 and 2013, while estimated mortality rates fell by 47 % (Fig. 1). Despite this tremendous progress, malaria still remains an important public health concern in countries where transmission occurs regularly and thus a global estimate of 3.3 billion people in 97 countries are still at risk of malaria with 1.2 billion people at high risk (>1 case of malaria per 1000 population each year) (World Malaria Report 2014). Owing to its importance in the world's clinical manifestations, this year's Nobel Prize has been awarded to the Chinese scientist Prof. Youyou Tu for discovering artemisinin for the treatment of malaria.

The classical discovery that mosquito mediates malarial transmission initiated the malaria control measures that largely included protection from mosquito bites and minimizing mosquito reproduction, which successfully limited disease transmission and eliminated the disease from more than ten countries between 1900 and 1946 (Hay et al. 2004). Following this, the World Health Organization launched the "Global Malaria Eradication Programme" in 1955, wherein chloroquine chemotherapy was implemented, which successfully increased the number of malaria-free countries to 27 by the end of this program in 1969. However, despite the impressive initial success, eradication in many countries failed due to technical, operational, and socioeconomic difficulties, thereby leading to resurgence of malaria in many parts of the world. Paradoxically, the initial malarial control has relied largely on the inexpensive and easily available antimalarials (mainly chloroquine and antifolate drugs). The extensive deployment of these drugs has imposed tremendous selection pressure on human malaria parasites, particularly in *P. falciparum*, to evolve the mechanisms of resistance, which has been a major contributor to the global resurgence of malaria in the last three decades (Marsh 1998) and doubling of malaria-attributable child mortality in southeast Africa (Korenromp et al. 2003). These control programs, therefore, have been hampered by the spread of drug resistance in the parasites and insecticide resistance in the mosquito vectors (Ballou et al. 1987).The causative agents of malaria, the plasmodial parasites, are unicellular protozoans with a complex life cycle that involves sexual and asexual reproductive stages (Fig. 2). Parasites reproduce asexually while in human hosts, whereas the sexual stage takes place inside the mosquito vector. During the sexual stage, the recombination of parasitic genetic material occurs in the mosquito, which increases the chances of resistance-causing mutations in proportion to the number of sexual reproductions and the number of mosquitoes that participate in the transmission of the parasite.

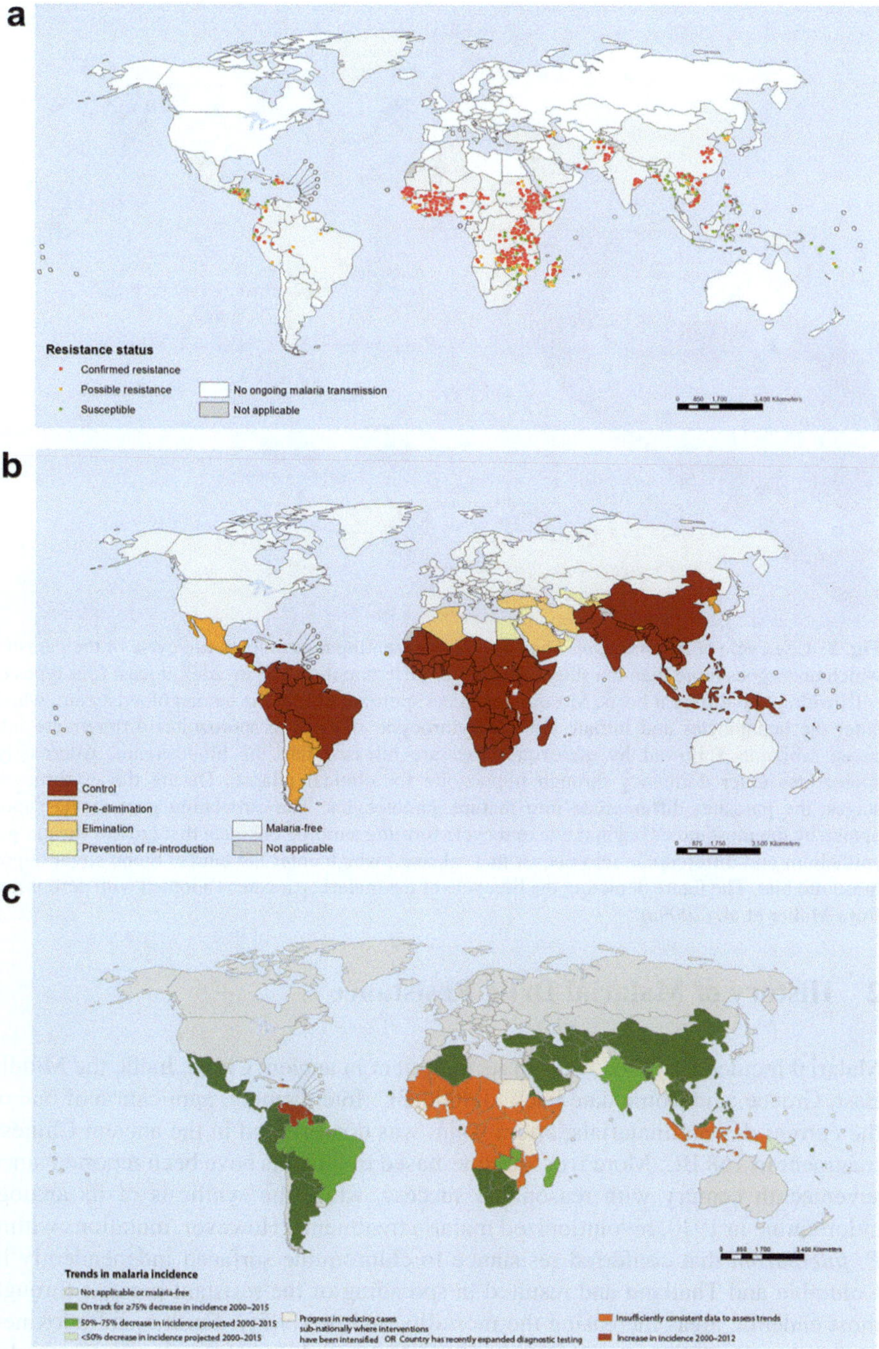

Fig. 1 Incidence and spread of malaria across the globe. The maps depicting insecticide susceptibility status for malaria vectors across the globe (**a**), classification of countries by stage of malaria elimination (**b**), and trends in reported malaria incidence since 2000–2012 (**c**). The maps were adopted from the World Malaria Report (2013)

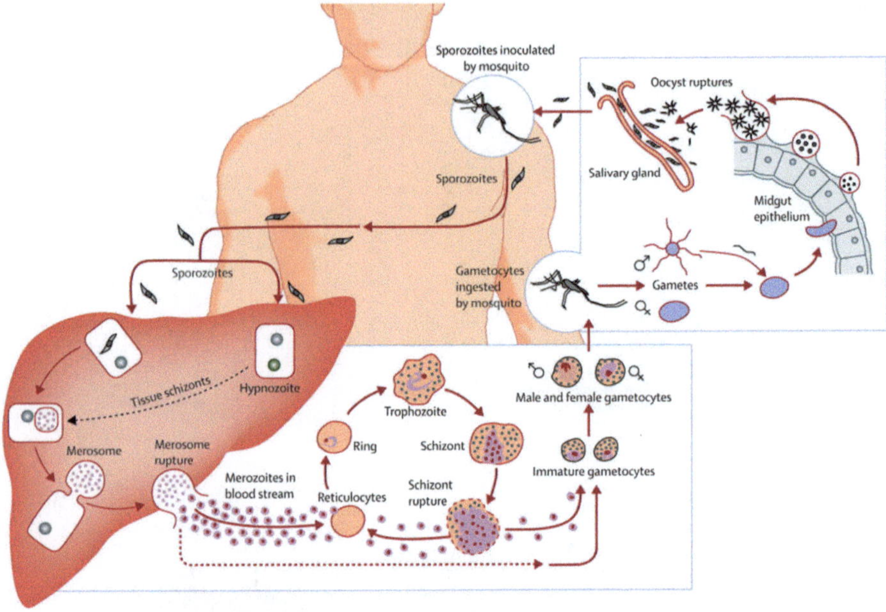

Fig. 2 Life cycle of *Plasmodium*. The cartoon representing the complex life cycle of the parasite, which undergoes more than ten stages of cellular differentiation and invades at least four types of cells within two different hosts. Mosquitoes inject sporozoites into the human bloodstream, which enter the hepatocytes and initiate the exoerythrocytic stage. The sporozoites differentiate into tissue schizonts followed by merozoites that are released into the bloodstream. Alternately, sporozoites enter dormancy through hypnozoite for clinical relapse. During the erythrocytic stages, the parasites differentiate into mature gametocytes. The circulating gametocytes upon uptake by the mosquitoes begin the sexual cycle forming a motile ookinete that crosses the midgut epithelium and differentiate into oocyst that releases, which enter the human blood stream upon mosquito bite. The figure depicting the lifecycle of the malaria parasite is adopted with permission from Müller et al. (2009a)

2 History of Malarial Drug Resistance

Malarial incidents and the ways of its treatment in ancient China, India, the Middle East, Greece, and Rome date back to 4000 BC. Interestingly, application of one of the current day antimalarials, artemisinin, was documented in the ancient Chinese treatment in 168 BC. Moreover, quinine-based treatments have been reported since seventeenth century with reasonable success, while the synthesis of its analog, chloroquine in 1930, revolutionized malaria treatments. However, mutations within *P. falciparum* that conferred resistance to chloroquine surfaced independently in Columbia and Thailand and resulted in spreading of the resistant parasite through most endemic areas increasing the mortality and morbidity. Further, sulfadoxine–pyrimethamine (SP), a combination of two drugs, replaced chloroquine effectively, and resistance to SP evolved rapidly and now occurs at high frequency in major malarious regions (Laxminarayan 2004). Therefore, alternative drugs such as the artemisinin-based combination therapies have been deployed currently. Although the development of other novel therapeutic agents is in progress, however, higher production costs limit their application in major endemic areas. On the other hand, the evolution of resistance against affordable drugs incurs an enormous societal cost for combating the disease. Therefore, efforts on the understanding the routes of the emergence of malarial drug resistance are in progress.

3 General Mechanisms of Drug Resistance in Malarial Parasites

Understanding the mechanisms leading to the drug resistance will facilitate the clinicians and health workers to devise ways to combat the drug-resistant varieties of the parasite and, thereby, contain the disease progression. The important factors for the transmission of malarial disease progression are defined largely by either the geographically specific variables including the rates of transmission and migration and host immunity or evolutionary-genetic mode of resistance, which includes the number and fitness effects of mutated genes and their interactions. Understandably, several mechanisms of drug resistance are observed for malarial parasites that are believed to have emerged by the inadequate drug exposure, improper dosing, poor pharmacokinetic properties, fake drugs, and reinfections during the drug elimination phase (Müller 2011). However, current understanding on molecular events leading to this complex phenomenon is limited, which severely limits efforts to analyze and evaluate treatment models. For the practical purposes, however, the factors governing resistance can be classified in the following: (a) the rate of mutations in the parasite, (b) the fitness costs associated with the resistance mutations, (c) the parasite load, (d) the drug selection vigor, and (e) the treatment compliance. Understandably, the mutation rate of the parasite has a direct influence on the frequency at which resistance can emerge. Therefore, attempts to measure the rate of spontaneous mutations in *P. falciparum* revealed a relatively low rate of 2.5×10^{-9} mutations/replication in the dihydrofolate reductase gene (Paget-McNicoL and Saul 2001). This means if an infected host carries 10^{11} parasites in the body, at least 100 of them will be drug resistant, and the ratio of drug-resistant varieties will be increasing because of the selective advantage of these varieties over the drug-sensitive wild-type strains. However, upon changing the drug selection pressures, the parasite encounters increased rate of mutations, which results in the "accelerated resistance to multiple drug" (ARMD) phenotype observed in the isolates from Southeast Asia (Rathod et al. 1997). Additionally, since the mutations inducing drug resistance often impart a fitness cost compelling the parasite to balance between the selective advantage of drug resistance and the biological cost owing to the mutated protein (Petersen et al. 2011), the parasite generally mitigates this conundrum by the acquisition of compensatory mutations (Levin et al. 2000), thereby eliminating the deleterious mutants while retaining the isolates with compensatory mutations, as observed in the high parasitemias (Hastings 2004). Additionally, in the high-transmission regions, reinfection after a treatment regime often results in the exposure of parasites to subtherapeutic drug concentrations and consequent selection of the drug-tolerant parasites with moderate to full resistance (Hastings and Watkins 2006; Stepniewska and White 2008). Agonizingly, the drug-resistant parasites that emerge due to such reinfections result in polyclonal infection with up to seven clones coexisting within one host (Färnert et al. 2009). Different treatment regimens that have been used to combat the parasite effectively resulted in the emergence of resistant varieties over the time, thereby necessitating the need for the development of sustainable treatment regimens that delay the emergence and spread of drug resistance.

4 Antimalarial Drugs and the Resistance Mechanisms

As noted before, chemotherapy, principally comprising the quinine derivatives, has traditionally played an important role in the treatment and control of malaria. Several structural variants of quinines, such as chloroquine and mefloquine, have been employed in the treatment and containment of the malaria. However, successful implementation of antimalarial drugs requires quick antimalarial action, affordability, tolerability, and safety. Glaringly, drug-resistant varieties for several individual first-line drugs have emerged making treatment regimen complicated and thus leading to the combinatorial therapy of almost all the antimalarials. Although the genetic basis of the emergence of drug resistance follows a typical two-step path, the de novo selection of resistance followed by the spread of resistance, with each drug targeting distinct mechanisms within the parasite, the mechanism of action of these drugs and modes of the resistance follow the drug-specific fashion (White 2004).

4.1 Quinine

The arylamino alcohol, quinine, is one of the oldest antimalarial agents and has been used by the Quenchua of Peru in the form of "tonic water," a mixture of minced bark of cinchona trees and sweetened water, to halt shivering due to low temperatures—the typical symptoms of malaria (Butler et al. 2010). Although the first use to treat malaria was reported in Rome in 1631, the active alkaloid, quinine, was first isolated from the cinchona bark in 1820 and the artificially synthesized in 1944 (Woodward and Doering 1945). Quinine has been widely used till 1940 to treat severe cases of malaria but currently used as a second-line treatment in combination with antibiotics such as tetracycline/doxycycline to treat resistant malaria (Farooq and Mahajan 2004). Although molecular mechanism by which quinine acts is only partly understood, its accumulation in the digestive food vacuoles of the parasite (Fitch 2004) and the multiple genes influencing its response encode its transporters (Cooper et al. 2002; Sidhu et al. 2005; Cooper et al. 2007; Nkrumah et al. 2009) suggest its action is dependent on the intra-parasitic concentration. Biochemically, quinine is believed to act similar to its synthetic variant, chloroquine, by inhibiting hemozoin crystallization during the heme detoxification pathway thereby causing the accumulation of cytotoxic heme in the parasites (Egan 2008). In addition to the dose-dependent action, its short half-life of 8–10 h (Meng et al. 2010; Briolant et al. 2011) and widespread administration (Farooq and Mahajan 2004) are believed to have contributed to the emergence of antibiotic resistance. The first case of quinine resistance was reported in South America in the early 1900s and later in the Thai–Cambodian border in the mid-1960s (Wongsrichanalai et al. 2002). Further emergence of clinically relevant resistance to quinine followed a sporadic geographical pattern. It has been widespread in Thailand (Wernsdorfer 1994) but noticed sporadically in Southeast Asia and western Oceania and less frequently in South America (Zalis et al. 1998) and Africa (Jelinek et al. 2001).

4.2 Chloroquine

Owing to the expensive synthetic process of quinine (Woodward and Doering 1945) and its side effects on the pregnant women, children (McGready et al. 2001), and HIV-coinfected patients, an easily synthesizable derivative called chloroquine has been introduced in the late 1940s and has been extensively used on a massive scale for malaria treatment since then. Moreover, its long half-life of 60 days, efficacy, affordability, and safety made chloroquine a better successor to quinine and in the treatment of malaria for many years that followed (AlKadi 2007). Although the 60-day-long half-life of chloroquine is advantageous in providing a chemoprophylactic effect during the drug elimination phase, it adversely exposes the parasites for longer period of time after the drug concentrations fall below therapeutic range thereby contributing to the selection for drug-resistant parasites (Stepniewska and White 2008). Chloroquine-resistant parasites emerged approximately 10 years after its introduction, simultaneously along the Thai–Cambodian border in Southeast Asia (Spencer 1985; Wernsdorfer and Payne 1991) and Colombia in South America (Young and Moore 1961) in the late 1950s (Mita et al. 2009). Since then, chloroquine resistance has been reported from all parts of the world where malaria is endemic except in the Central America, Caribbean Hispaniola Island, parts of the Middle East, and Central Asia (Hastings and D'Alessandro 2000). Genetic epidemiological data suggests that chloroquine-resistant *P. falciparum* strains have spread in all endemic areas of South America by 1970, Asia and Oceania by 1989, and Africa by late 1970s (Peters 1987). Furthermore, chloroquine resistance spread from eastern Africa to Central and Southern Africa, followed by the Western Africa by 1983 and sub-Saharan Africa by 1989 (Wongsrichanalai et al. 2002) and emerged independently in several parts of

Asia, including Papua New Guinea and the Philippines (Mita et al. 2009). Although chloroquine-resistant *P. falciparum* is widespread, clinical efficacy of the drug is retained for the premune patients that have acquired partial immunity through the reinfection (Wellems and Plowe 2001), and therefore it remains as the first-line treatment of *P. falciparum* and *P. vivax* infections (Petersen et al. 2011).

Understanding the mechanism of action of chloroquine has been of great research and medical interest for the past few decades. Although the exact mechanism is not understood, since most of the drug targets are localized in the acidic food vacuole of the parasite (Geary et al. 1986; Krogstad et al. 1987), the accepted model proposes inhibition of heme polymerization into the cytotoxic hemozoin by chloroquine followed by the accumulation of cytotoxic heme within the parasite (Egan 2008). Moreover, as a weak base with pK_a values of 8.1 and 10.2, chloroquine remains neutral at the physiological pH and thus diffuses readily across membranes (Martin et al. 2009). However, in the acidic food vacuole, chloroquine becomes diprotonated and thus unable to transverse across the membrane, thus accumulates inside the vacuoles, binds to hematin, prevents detoxification of the heme, and ultimately leads to the death of the parasite. Therefore, the mechanism of resistance, understandably, relies on increased capacity of the parasite to expel chloroquine faster than the drug could reach optimal concentration for its action (Foley and Tilley 1997). Moreover, the observations that resistant varieties have been demonstrated to efflux chloroquine 40–50 times faster than the wild-type parasites (Krogstad et al. 1987) and the chemicals which inhibit the efflux pumps could effectively reverse the resistance (Martin et al. 1987) further support the presumption on efflux-mediated resistance.

Therefore, the genes associated with the resistance essentially encode the transport proteins; for example, the gene encoding multidrug-resistant transporter, *pfmdr*-1 (located on chromosome 5), and the gene encoding a more specific transporter, chloroquine-resistant transporter, *pfcrt* (located on chromosome 7), have been implicated in the resistance phenotype (Fidock et al. 2004). Polymorphisms in the chloroquine-resistant transporter (*PfCRT*) have been demonstrated to be the principal determinants of chloroquine resistance (Sidhu et al. 2002), while mutations in the multidrug-resistant transporter (PfMDR1) have been demonstrated to modulate chloroquine resistance (Barnes et al. 1992). The point mutation in *pfmdr*-1 at Asp-Tyr (D86Y) has been demonstrated with chloroquine resistance (von Seidlein et al. 1997; Póvoa et al. 1998; Price et al. 1999; Babiker et al. 2001), while several other polymorphisms at F184, C1034, D042, and Y1246 have been implicated in chloroquine resistance (Djimdé et al. 2001). Likewise, the point mutation in *pfcrt* at Thr-Lys (T76K) has been associated with the resistance mechanisms (Djimdé et al. 2001; Durand et al. 2001). Interestingly, the linkage disequilibrium between PfMDR1 and PfCRT alleles in chloroquine-resistant parasites isolated from the Southeast Asia and African patients suggests a functional interaction of both proteins (Hastings 2006; Osman et al. 2007).

4.3 Mefloquine

Mefloquine drug introduced in the mid-1970s by the United States Department of Defense (Trenholme et al. 1975) is a racemic mixture of two 4-methanolquinoline enantiomers: (+)-anti-mefloquine and (−)-anti-mefloquine (Rastelli and Coltart 2015) with a reasonably longer half-life of 14–18 days (Stepniewska and White 2008). Unlike the parental quinine, mefloquine is widely used for the treatment of malaria in pregnancy (González et al. 2014) but is inefficient to act on the parasites in the hepatocellular phase of the disease and thus administered in combination with primaquines for the *P. vivax* patients (Schlagenhauf et al. 2010). Similar to chloroquine, the resistance to mefloquine was first observed in the Southeast Asia along the Thai–Cambodian border during the late 1980s (Shanks 1994; Wongsrichanalai et al. 2001). Since then, resistance has been reported in several other parts of Southeast Asia, in the Amazon region of South America, and sporadically in Africa (Muckenhaupt 1995). Since the mechanism of action of mefloquine is similar to its quinine synthetic analogs (Fitch et al. 1979; Eastman and Fidock 2009), resistance to mefloquine is implicated to the copy number and polymorphism of *pfmdr*-1 gene (Cowman et al. 1994). However, different resistant isolates of the parasites showed geographically distinct behavior; for example, while resistance in the Thai isolates correlated to the higher copy number of *pfmdr*-1 (Price et al. 1999), the resistance in the Brazilian (Zalis et al. 1998) and African (Basco et al. 1995) isolates was insensitive to the copy number. In addition, functional studies on the mefloquine resistance employing recombinant parasites have suggested a primary mode of action outside of the food vacuoles (von Seidlein et al. 1997), probably targeting the transport function of PfMDR-1 (Martin et al. 1987). Notably, the D86Y mutation in *pfmdr*-1 that confers resistance to quinine has been demonstrated to induce sensitivity to mefloquine, suggesting an inverse relationship between mefloquine and chloroquine resistance phenotype (Prince et al. 2001; Duraisingh et al. 2000). On the other hand, point mutations at Ser-1034, Asn-1042, and Asp-1246 have been demonstrated to induce mefloquine resistance (Reed et al. 2000), indicating a unique, yet unexplored, relation of *pfmdr*-1 and mefloquine resistance.

4.4 The Antifolate Drugs

The antifolate drugs used for malarial therapy interfere with folate metabolism, a pathway essential to the survival of the parasite. These synthetic antimicrobial agents that contain the sulfonamide group and thus are also known as the sulfa drugs comprise essentially two components: (a) sulfadoxine or dapsone that competitively inhibits folate biosynthetic enzyme, the dihydropteroate synthetase enzyme (PfDHPS) by competing with p-aminobenzoic acid (Bruce-Chwatt 1985), and (b) pyrimethamine or proguanil, which principally inhibits the protozoan dihydrofolate reductase (PfDHFR) activity of the bifunctional dihydrofolate reductase/thymidylate synthase enzyme thereby preventing the synthesis of the active form of folate, the tetrahydrofolate. These two drugs, therefore, are generally administered in combination since they inhibit two different steps in the biosynthesis of tetrahydrofolate. The drug combination of sulfadoxine–pyrimethamine (SP), which possesses a half-life of approximately 4–5 days, was introduced in the late 1970s as a drug of choice to treat chloroquine-resistant malaria (Winstanley 2001) which turned out as an effective, affordable, and well-tolerated single-dose drug. Unfortunately, resistance to SP emerged rapidly (Uhlemann and Krishna 2005) with the first report from the Thai–Cambodian border (Björkman and Phillips-Howard 1990) followed by reports from parts of Southeast Asia, Southern China, and Amazon basin in the early 1980s (Aramburu et al. 1999; Vasconcelos et al. 2000). In addition, low degree of resistance has been reported from the west coastal South America, southern Asia including Iran, and western Oceania (Bloland and World Health Organization 2001). Notably, SP resistance occurred much later in Africa with the first case in the 1980s, followed by an alarmingly rapid spreading to the eastern Africa (Rønn et al. 1996). Owing to the emergence and fast-spreading resistance, SP is currently limited to intermittent preventative malaria treatment during pregnancy.

Unlike the quinine-derived drugs, the mechanisms of action of and resistance to SP are well understood. Since these antifolate compounds inhibit the action of DHFR and DHPS, plausibly, the resistance mapped to the point mutations in these target enzymes. Five-point mutations in the *dhps* gene at S436A/P, A437G, K540E/G, A581G, and A613S/T that are known to confer resistance by lowering enzyme affinity have been identified from various parts of the world including Indonesia (Nagesha et al. 2001), Malawi, Bolivia, Kenya (Plowe et al. 1997), Gabon (Mawili-Mboumba et al. 2001), and South America (Urdaneta et al. 1999). Likewise, five-point mutations in the *dhfr* gene at A16V, N51I, C59R, S108N, and T164L/I that are known to elicit pyrimethamine resistance by reduction in drug-binding affinity of the enzyme have been identified the isolates from Thailand (Farooq and Mahajan 2004). Although the precise relation between these mutations and SP resistance is unclear (Peters 1987), the DHFR enzymes from the resistant strains have been demonstrated to bind pyrimethamine with 400- to 800-fold lower affinity than the enzymes from the drug-sensitive strains (Ferone 1970), indicating a direct relation between the mutations and resistance. On the other hand, the other antifolate drug combinations of dapsone–proguanil resulted in dapsone-driven hemolysis in G6PD-deficient patients and have been immediately discontinued (Luzzatto 2010).

4.5 Artemisinins

Artemisinin (aka qinghaosu), a sesquiterpene lactone containing an unusual peroxide bridge that is required for antimalarial activity (Eastman and Fidock 2009), and its semisynthetic endoperoxide derivatives are currently used in the treatment of *P. falciparum* malaria owing to its rapid action against the parasite (White 1997). Since all the mainstream drugs in the treatment of malaria have resulted in the emergence of resistant parasites, the artemisinin combination therapies (ACT) are now used as the linchpin in antimalarial chemotherapy worldwide, and consequently artemisinin discovery has been awarded with this year's Nobel Prize in Medicine. Current ACTs include artemether–lumefantrine, artesunate–mefloquine, artesunate–amodiaquine, artesunate–sulfadoxine–pyrimethamine, dihydroartemisinin–piperaquine, and artesunate–pyronaridine (World Malaria Report 2014). Artemisinin, as a natural product, is isolated from the Chinese herbal plant, *Artemisia annua* (Chinese sweet wormwood), and the precursor compound has been produced in yeast using recombinant DNA technology. Since low bioavailability, high production cost, poor pharmacokinetic properties, short half-life at 0.5–1.4 h (Bloland and World Health Organization 2001), and low water solubility limit the use of the parental artemisinin, several semisynthetic derivatives such as artemether, artesunate, and dihydroartemisinin have been developed for clinical use (Bray et al. 2005). Conspicuously, since the use of the drug as monotherapy is ineffective, therapies that combine artemisinin or its derivatives with other antimalarial drugs have been effective (Fig. 3).

Although initial studies could not identify any clinically significant artemisinin resistance except in an animal model, agonizingly, recent studies have identified certain resistance isolates of the parasites. Initial widespread administration of artemisinin derivatives as monotherapy in Southeast Asia along the Thai–Cambodian border is believed to have resulted in the emergence of resistance as reported

Drug	Main limitations
Chloroquine Quinine Amodiaquine Mefloquine Halofantrine	Resistance Compliance/safety/resistance Safety/resistance (Safety)/resistance/(cost) Safety/resistance/cost
Artemisinins (artemether, arteether, artesunate)	Compliance/(safety)/(GMP)/(cost)
Sulphadoxine–pyrimethamine	Resistance
Atovaquone–proguanil	Resistance potential/cost
Lumefantrine–artemether	(Compliance)/resistance potential/(cost)

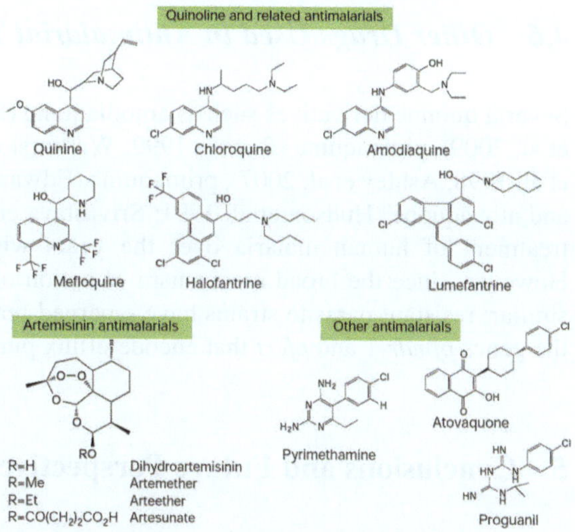

Fig. 3 Principal antimalarial drugs. The biochemical structures and clinical limitations of the principal drugs used currently in the treatment of human malarias are presented. The figure is adopted with permission from Ridley (2002)

sporadically from this region in the year 2008 (Noedl et al. 2008; Dondorp et al. 2009; Eastman and Fidock 2009). Following this, Thailand, Northern and North-eastern Cambodia, Vietnam, Southern Laos, and Central and Eastern Myanmar have reported cases of clinically significant artemisinin resistance (Ashley et al. 2014). Although the mechanism of action of artemisinin drugs is unclear, several lines of evidence suggest that generation of oxygen free radicals through the action of the endoperoxide bridge, which follows a series of biochemical events, ultimately acts against the parasite (Ginsburg and Atamna 1994). Moreover, although mutations in the gene encoding the Ca^{2+}-transporting ATPase 6 (*pfatp6*) (Jambou et al. 2005) and polymorphisms in the gene encoding a deubiquitination enzyme (*ubp1*) (Hunt et al. 2007) have been associated with increased artesunate resistance, phenotypic association of these two candidates, however, with the parasites requiring longer clearance time has not been correlated with the in vivo phenotype of resistant parasites (Imwong et al. 2010). In addition to the other drugs, overexpression of *pfmdr1* has been demonstrated to significantly reduce parasite susceptibility to artemisinin (Sidhu et al. 2005; Chavchich et al. 2010). In addition, to avoid the artemisinin drug selection pressure, the parasites have been demonstrated to enter quiescence during the intraerythrocytic development stage (Witkowski et al. 2010). Surprisingly, genetic analyses have reported that the reduction in artemisinin susceptibility bears a heritable component (Anderson et al. 2010). Therefore, artemisinin derivatives are deployed in combination with other antimalarial compound to prolong the lifespan of artemisinin drugs by reducing the emergence of drug-resistant parasites (Müller et al. 2009b).

4.6 Other Drugs Used in Antimalarial Therapy

Several quinine derivatives such as amodiaquine (Stepniewska and White 2008; Sá et al. 2009), piperaquine (Raynes 1999; Warhurst et al. 2007), lumefantrine (Ezzet et al. 1998; Ashley et al. 2007), primaquine (Edwards et al. 1993; Wells et al. 2010), and atovaquone (Hudson et al. 1991; Srivastava et al. 1997) have been used in the treatment of human malaria over the years with different ranges of success. However, since the broad mechanism of action of these quinine-derived drugs is similar, resistant parasite strains have emerged readily with the polymorphisms in the genes *pfmdr*-1 and *pfcrt* that encode efflux pumps (Gil et al. 2003).

5 Conclusions and Future Perspectives

With the emergence of drug resistance to several mainstream drug regimens including the artemisinin derivatives—the primary stronghold in malaria chemotherapeutical treatment—a novel and advanced therapeutic strategy needs to be implemented to contain and curtail the spread of resistance. Applying advanced genomics, empirical and theoretical research in population genetics in combination with the progress in drug discovery and development would be able to present a larger choice of effective, appropriate, and affordable antimalarial drugs and design promising strategies to delay the evolution of resistance against the newly introduced antimalarial drugs. Since the current choice of drugs is limited, it is difficult to balance the widespread administration of the effective drugs against the reducing the emergence of resistance through controlled administration. Therefore, currently the combinatorial therapies are applied to maximize the life span of the individual antimalarials and to reduce the emergence and spread of parasites resistant to any one particular antimalarial (Boni et al. 2008). In addition, public awareness, training of the clinicians, adherence to the full treatment regimen, and the vector control strategies are essential to pare down the emergence of drug-resistant parasites and ensure adequate treatment of individual infections. Therefore, combined with effective and rational drug treatment policies, the prevailing tools should be able to further reduce morbidity and mortality. However, in addition to the thorough understanding of the genetic mechanisms and pharmacological factors that could impact treatment, successful curtailment of drug-resistant malaria would require sustained financial support and political cooperation to monitor and curtail the emergence of drug-resistant parasites. In conclusion, the reduction in the global malaria burden and control of drug-resistant malaria require reducing the overall drug pressure through more selective drug combinations (possibly vaccines) that either are inherently less likely to foster resistance or have properties that do not facilitate development or spread of resistant parasites.

Acknowledgments This work was supported by grants from the Department of Biotechnology, India (BT/PR3260/BRB/10/967/2011). We thank Kalyani PV for assistance with proof reading the manuscript.

References

AlKadi HO (2007) Antimalarial drug toxicity: a review. Chemotherapy 53:385–391. doi:10.1159/000109767

Anderson TJ, Nair S, Nkhoma S, Williams JT, Imwong M, Yi P, Socheat D, Das D, Chotivanich K, Day NP, White NJ, Dondorp AM (2010) High heritability of malaria parasite clearance rate indicates a genetic basis for artemisinin resistance in western Cambodia. J Infect Dis 201:1326–1330. doi:10.1086/651562

Aramburú Guarda J, Ramal Asayag C, Witzig R (1999) Malaria reemergence in the Peruvian Amazon region. Emerg Infect Dis 5:209–215. doi:10.3201/eid0502.990204

Ashley EA, Stepniewska K, Lindegårdh N, Annerberg A, Kham A, Brockman A, Singhasivanon P, White NJ, Nosten F (2007) How much fat is necessary to optimize lumefantrine oral bioavailability? Trop Med Int Health 12:195–200. doi:10.1111/j.1365-3156.2006.01784.x

Ashley EA, Dhorda M, Fairhurst RM, Amaratunga C, Lim P, Suon S, Sreng S, Anderson JM, Mao S, Sam B, Sopha C, Chuor CM, Nguon C, Sovannaroth S, Pukrittayakamee S, Jittamala P, Chotivanich K, Chutasmit K, Suchatsoonthorn C, Runcharoen R, Hien TT, Thuy-Nhien NT, Thanh NV, Phu NH, Htut Y, Han KT, Aye KH, Mokuolu OA, Olaosebikan RR, Folaranmi OO, Mayxay M, Khanthavong M, Hongvanthong B, Newton PN, Onyamboko MA, Fanello CI,

Tshefu AK, Mishra N, Valecha N, Phyo AP, Nosten F, Yi P, Tripura R, Borrmann S, Bashraheil M, Peshu J, Faiz MA, Ghose A, Hossain MA, Samad R, Rahman MR, Hasan MM, Islam A, Miotto O, Amato R, MacInnis B, Stalker J, Kwiatkowski DP, Bozdech Z, Jeeyapant A, Cheah PY, Sakulthaew T, Chalk J, Intharabut B, Silamut K, Lee SJ, Vihokhern B, Kunasol C, Imwong M, Tarning J, Taylor WJ, Yeung S, Woodrow CJ, Flegg JA, Das D, Smith J, Venkatesan M, Plowe CV, Stepniewska K, Guerin PJ, Dondorp AM, Day NP, White NJ, (TRAC) TRtAC (2014) Spread of artemisinin resistance in *Plasmodium falciparum* malaria. N Engl J Med 371:411–423. doi:10.1056/NEJMoa1314981

Babiker HA, Pringle SJ, Abdel-Muhsin A, Mackinnon M, Hunt P, Walliker D (2001) High-level chloroquine resistance in Sudanese isolates of *Plasmodium falciparum* is associated with mutations in the chloroquine resistance transporter gene *pfcrt* and the multidrug resistance Gene *pfmdr1*. J Infect Dis 183:1535–1538. doi:10.1086/320195

Ballou WR, Hoffman SL, Sherwood JA, Hollingdale MR, Neva FA, Hockmeyer WT, Gordon DM, Schneider I, Wirtz RA, Young JF (1987) Safety and efficacy of a recombinant DNA *Plasmodium falciparum* sporozoite vaccine. Lancet 1:1277–1281

Barnes DA, Foote SJ, Galatis D, Kemp DJ, Cowman AF (1992) Selection for high-level chloroquine resistance results in deamplification of the *pfmdr1* gene and increased sensitivity to mefloquine in *Plasmodium falciparum*. EMBO J 11:3067–3075

Basco LK, Ramiliarisoa O, Le Bras J (1995) In vitro activity of atovaquone against the African isolates and clones of *Plasmodium falciparum*. Am J Trop Med Hyg 53:388–391

Björkman A, Phillips-Howard PA (1990) The epidemiology of drug-resistant malaria. Trans R Soc Trop Med Hyg 84:177–180

Bloland PB, World Health Organization (2001) Drug resistance in malaria. World Health Organization, Geneva

Boni MF, Smith DL, Laxminarayan R (2008) Benefits of using multiple first-line therapies against malaria. Proc Natl Acad Sci U S A 105:14216–14221. doi:10.1073/pnas.0804628105

Bray RS, Garnham PC (1982) The life-cycle of primate malaria parasites. Br Med Bull 38:117–122

Bray PG, Ward SA, O'Neill PM (2005) Quinolines and artemisinin: chemistry, biology and history. Curr Top Microbiol Immunol 295:3–38

Briolant S, Pelleau S, Bogreau H, Hovette P, Zettor A, Castello J, Baret E, Amalvict R, Rogier C, Pradines B (2011) In vitro susceptibility to quinine and microsatellite variations of the *Plasmodium falciparum* Na+/H+ exchanger (*Pfnhe*-1) gene: the absence of association in clinical isolates from the Republic of Congo. Malar J 10:37. doi:10.1186/1475-2875-10-37

Bruce-Chwatt LJ (1985) Essential malariology. Wiley/Heinemann Medical Books, New York/London

Butler AR, Khan S, Ferguson E (2010) A brief history of malaria chemotherapy. J R Coll Physicians Edinb 40:172–177

Chavchich M, Gerena L, Peters J, Chen N, Cheng Q, Kyle DE (2010) Role of pfmdr1 amplification and expression in induction of resistance to artemisinin derivatives in *Plasmodium falciparum*. Antimicrob Agents Chemother 54:2455–2464. doi:10.1128/AAC.00947-09

Cooper RA, Ferdig MT, Su X-Z, Ursos LMB, Mu J, Nomura T, Fujioka H, Fidock DA, Roepe PD, Wellems TE (2002) Alternative mutations at position 76 of the vacuolar transmembrane protein PfCRT are associated with chloroquine resistance and unique stereospecific quinine and quinidine responses in *Plasmodium falciparum*. Mol Pharmacol 61:35–42. doi:10.1124/mol.61.1.35

Cooper RA, Lane KD, Deng B, Mu J, Patel JJ, Wellems TE, Su X, Ferdig MT (2007) Mutations in transmembrane domains 1, 4 and 9 of the *Plasmodium falciparum* chloroquine resistance transporter alter susceptibility to chloroquine, quinine and quinidine. Mol Microbiol 63:270–282. doi:10.1111/j.1365-2958.2006.05511.x

Cowman AF, Galatis D, Thompson JK (1994) Selection for mefloquine resistance in *Plasmodium falciparum* is linked to amplification of the *pfmdr1* gene and cross-resistance to halofantrine and quinine. Proc Natl Acad Sci U S A 91:1143–1147

Cox-Singh J, Davis TM, Lee KS, Shamsul SS, Matusop A, Ratnam S, Rahman HA, Conway DJ, Singh B (2008) *Plasmodium knowlesi* malaria in humans is widely distributed and potentially life threatening. Clin Infect Dis 46:165–171. doi:10.1086/524888

Djimdé A, Doumbo OK, Cortese JF, Kayentao K, Doumbo S, Diourté Y, Coulibaly D, Dicko A, Su XZ, Nomura T, Fidock DA, Wellems TE, Plowe CV (2001) A molecular marker for chloroquine-resistant *falciparum* malaria. N Engl J Med 344:257–263. doi:10.1056/NEJM200101253440403

Dondorp AM, Nosten F, Yi P, Das D, Phyo AP, Tarning J, Lwin KM, Ariey F, Hanpithakpong W, Lee SJ, Ringwald P, Silamut K, Imwong M, Chotivanich K, Lim P, Herdman T, An SS, Yeung S, Singhasivanon P, Day NP, Lindegardh N, Socheat D, White NJ (2009) Artemisinin resistance in *Plasmodium falciparum* malaria. N Engl J Med 361:455–467. doi:10.1056/NEJMoa0808859

Duraisingh MT, Roper C, Walliker D, Warhurst DC (2000) Increased sensitivity to the antimalarials mefloquine and artemisinin is conferred by mutations in the *pfmdr1* gene of *Plasmodium falciparum*. Mol Microbiol 36:955–961

Durand R, Jafari S, Vauzelle J, Delabre JF, Jesic Z, Le Bras J (2001) Analysis of *pfcrt* point mutations and chloroquine susceptibility in isolates of *Plasmodium falciparum*. Mol Biochem Parasitol 114:95–102

Eastman RT, Fidock DA (2009) Artemisinin-based combination therapies: a vital tool in efforts to eliminate malaria. Nat Rev Microbiol 7:864–874. doi:10.1038/nrmicro2239

Edwards G, McGrath CS, Ward SA, Supanaranond W, Pukrittayakamee S, Davis TM, White NJ (1993) Interactions among primaquine, malaria infection and other antimalarials in Thai subjects. Br J Clin Pharmacol 35:193–198

Egan TJ (2008) Haemozoin formation. Mol Biochem Parasitol 157:127–136. doi:10.1016/j.molbiopara.2007.11.005

Ezzet F, Mull R, Karbwang J (1998) Population pharmacokinetics and therapeutic response of CGP 56697 (artemether + benflumetol) in malaria patients. Br J Clin Pharmacol 46:553–561

Färnert A, Williams TN, Mwangi TW, Ehlin A, Fegan G, Macharia A, Lowe BS, Montgomery SM, Marsh K (2009) Transmission-dependent tolerance to multiclonal *Plasmodium falciparum* infection. J Infect Dis 200:1166–1175. doi:10.1086/605652

Farooq U, Mahajan RC (2004) Drug resistance in malaria. J Vector Borne Dis 41:45–53

Ferone R (1970) Dihydrofolate reductase from pyrimethamine-resistant *Plasmodium berghei*. J Biol Chem 245:850–854

Fidock DA, Rosenthal PJ, Croft SL, Brun R, Nwaka S (2004) Antimalarial drug discovery: efficacy models for compound screening. Nat Rev Drug Discov 3:509–520. doi:10.1038/nrd1416

Fitch CD (2004) Ferriprotoporphyrin IX, phospholipids, and the antimalarial actions of quinoline drugs. Life Sci 74:1957–1972. doi:10.1016/j.lfs.2003.10.003

Fitch CD, Chan RL, Chevli R (1979) Chloroquine resistance in malaria: accessibility of drug receptors to mefloquine. Antimicrob Agents Chemother 15:258–262

Foley M, Tilley L (1997) Quinoline antimalarials: mechanisms of action and resistance. Int J Parasitol 27:231–240

Geary TG, Jensen JB, Ginsburg H (1986) Uptake of [3H] chloroquine by drug-sensitive and -resistant strains of the human malaria parasite *Plasmodium falciparum*. Biochem Pharmacol 35:3805–3812

Gil JP, Nogueira F, Strömberg-Nörklit J, Lindberg J, Carrolo M, Casimiro C, Lopes D, Arez AP, Cravo PV, Rosário VE (2003) Detection of atovaquone and Malarone resistance conferring mutations in *Plasmodium falciparum* cytochrome b gene (*cytb*). Mol Cell Probes 17:85–89

Ginsburg H, Atamna H (1994) The redox status of malaria-infected erythrocytes: an overview with an emphasis on unresolved problems. Parasite 1:5–13

González R, Mombo-Ngoma G, Ouédraogo S, Kakolwa MA, Abdulla S, Accrombessi M, Aponte JJ, Akerey-Diop D, Basra A, Briand V, Capan M, Cot M, Kabanywanyi AM, Kleine C, Kremsner PG, Macete E, Mackanga JR, Massougbodgi A, Mayor A, Nhacolo A,

Pahlavan G, Ramharter M, Rupérez M, Sevene E, Vala A, Zoleko-Manego R, Menéndez C (2014) Intermittent preventive treatment of malaria in pregnancy with mefloquine in HIV-negative women: a multicentre randomized controlled trial. PLoS Med 11, e1001733. doi:10.1371/journal.pmed.1001733

Hastings IM (2004) The origins of antimalarial drug resistance. Trends Parasitol 20:512–518. doi:10.1016/j.pt.2004.08.006

Hastings IM (2006) Gametocytocidal activity in antimalarial drugs speeds the spread of drug resistance. Trop Med Int Health 11:1206–1217. doi:10.1111/j.1365-3156.2006.01668.x

Hastings IM, D'Alessandro U (2000) Modelling a predictable disaster: the rise and spread of drug-resistant malaria. Parasitol Today 16:340–347

Hastings IM, Watkins WM (2006) Tolerance is the key to understanding antimalarial drug resistance. Trends Parasitol 22:71–77. doi:10.1016/j.pt.2005.12.011

Hay SI, Guerra CA, Tatem AJ, Noor AM, Snow RW (2004) The global distribution and population at risk of malaria: past, present, and future. Lancet Infect Dis 4:327–336. doi:10.1016/S1473-3099(04)01043-6

Hudson AT, Dickins M, Ginger CD, Gutteridge WE, Holdich T, Hutchinson DB, Pudney M, Randall AW, Latter VS (1991) 566C80: a potent broad spectrum anti-infective agent with activity against malaria and opportunistic infections in AIDS patients. Drugs Exp Clin Res 17:427–435

Hunt P, Afonso A, Creasey A, Culleton R, Sidhu AB, Logan J, Valderramos SG, McNae I, Cheesman S, do Rosario V, Carter R, Fidock DA, Cravo P (2007) Gene encoding a deubiquitinating enzyme is mutated in artesunate- and chloroquine-resistant rodent malaria parasites. Mol Microbiol 65:27–40. doi:10.1111/j.1365-2958.2007.05753.x

Imwong M, Dondorp AM, Nosten F, Yi P, Mungthin M, Hanchana S, Das D, Phyo AP, Lwin KM, Pukrittayakamee S, Lee SJ, Saisung S, Koecharoen K, Nguon C, Day NP, Socheat D, White NJ (2010) Exploring the contribution of candidate genes to artemisinin resistance in *Plasmodium falciparum*. Antimicrob Agents Chemother 54:2886–2892. doi:10.1128/AAC.00032-10

Jambou R, Legrand E, Niang M, Khim N, Lim P, Volney B, Ekala MT, Bouchier C, Esterre P, Fandeur T, Mercereau-Puijalon O (2005) Resistance of *Plasmodium falciparum* field isolates to in-vitro artemether and point mutations of the SERCA-type *PfATPase6*. Lancet 366:1960–1963. doi:10.1016/S0140-6736(05)67787-2

Jelinek T, Grobusch MP, Löscher T (2001) Patterns of *Plasmodium falciparum* drug resistance in nonimmune travellers to Africa. Eur J Clin Microbiol Infect Dis 20:284–286

Korenromp EL, Williams BG, Gouws E, Dye C, Snow RW (2003) Measurement of trends in childhood malaria mortality in Africa: an assessment of progress toward targets based on verbal autopsy. Lancet Infect Dis 3:349–358

Krogstad DJ, Gluzman IY, Kyle DE, Oduola AM, Martin SK, Milhous WK, Schlesinger PH (1987) Efflux of chloroquine from *Plasmodium falciparum*: mechanism of chloroquine resistance. Science 238:1283–1285

Laxminarayan R (2004) Act now or later? Economics of malaria resistance. Am J Trop Med Hyg 71:187–195

Levin BR, Perrot V, Walker N (2000) Compensatory mutations, antibiotic resistance and the population genetics of adaptive evolution in bacteria. Genetics 154:985–997

Luzzatto L (2010) The rise and fall of the antimalarial Lapdap: a lesson in pharmacogenetics. Lancet 376:739–741. doi:10.1016/S0140-6736(10)60396-0

Marsh K (1998) Malaria disaster in Africa. Lancet 352:924. doi:10.1016/S0140-6736(05)61510-3

Martin SK, Oduola AM, Milhous WK (1987) Reversal of chloroquine resistance in *Plasmodium falciparum* by verapamil. Science 235:899–901. doi:10.1126/science.3544220

Martin RE, Ginsburg H, Kirk K (2009) Membrane transport proteins of the malaria parasite. Mol Microbiol 74:519–528. doi:10.1111/j.1365-2958.2009.06863.x

Mawili-Mboumba DP, Ekala MT, Lekoulou F, Ntoumi F (2001) Molecular analysis of DHFR and DHPS genes in *P. falciparum* clinical isolates from the Haut—Ogooué region in Gabon. Acta Trop 78:231–240

McGready R, Cho T, Samuel, Villegas L, Brockman A, van Vugt M, Looareesuwan S, White NJ, Nosten F (2001) Randomized comparison of quinine-clindamycin versus artesunate in the treatment of falciparum malaria in pregnancy. Trans R Soc Trop Med Hyg 95:651–656

Meng H, Zhang R, Yang H, Fan Q, Su X, Miao J, Cui L, Yang Z (2010) In vitro sensitivity of *Plasmodium falciparum* clinical isolates from the China-Myanmar border area to quinine and association with polymorphism in the Na+/H+ exchanger. Antimicrob Agents Chemother 54:4306–4313. doi:10.1128/AAC.00321-10

Mita T, Tanabe K, Kita K (2009) Spread and evolution of *Plasmodium falciparum* drug resistance. Parasitol Int 58:201–209. doi:10.1016/j.parint.2009.04.004

Muckenhaupt FP (1995) Mefloquine resistance in *Plasmodium falciparum*. Parasitol Today 11:248–253

Müller O (2011) Malaria in Africa: challenges for control and elimination in the 21st century. Peter Lang, Frankfurt am Main

Müller I, Galinski MR, Baird JK, Carlton JM, Kochar DK, Alonso PL, del Portillo HA (2009a) Key gaps in the knowledge of *Plasmodium vivax*, a neglected human malaria parasite. Lancet Infect Dis 9:555–566. doi:10.1016/S1473-3099(09)70177-X

Müller I, Genton B, Rare L, Kiniboro B, Kastens W, Zimmerman P, Kazura J, Alpers M, Smith TA (2009b) Three different *Plasmodium* species show similar patterns of clinical tolerance of malaria infection. Malar J 8:158. doi:10.1186/1475-2875-8-158

Nagesha HS, Din-Syafruddin, Casey GJ, Susanti AI, Fryauff DJ, Reeder JC, Cowman AF (2001) Mutations in the *pfmdr1*, *dhfr* and *dhps* genes of *Plasmodium falciparum* are associated with in-vivo drug resistance in West Papua, Indonesia. Trans R Soc Trop Med Hyg 95:43–49

Nkrumah LJ, Riegelhaupt PM, Moura P, Johnson DJ, Patel J, Hayton K, Ferdig MT, Wellems TE, Akabas MH, Fidock DA (2009) Probing the multifactorial basis of *Plasmodium falciparum* quinine resistance: evidence for a strain-specific contribution of the sodium-proton exchanger PfNHE. Mol Biochem Parasitol 165:122–131. doi:10.1016/j.molbiopara.2009.01.011

Noedl H, Se Y, Schaecher K, Smith BL, Socheat D, Fukuda MM, Consortium ARiCAS (2008) Evidence of artemisinin-resistant malaria in western Cambodia. N Engl J Med 359:2619–2620. doi:10.1056/NEJMc0805011

Osman ME, Mockenhaupt FP, Bienzle U, Elbashir MI, Giha HA (2007) Field-based evidence for linkage of mutations associated with chloroquine (*pfcrt*/*pfmdr1*) and sulfadoxine-pyrimethamine (*pfdhfr*/*pfdhps*) resistance and for the fitness cost of multiple mutations in *P. falciparum*. Infect Genet Evol 7:52–59. doi:10.1016/j.meegid.2006.03.008

Paget-McNicol S, Saul A (2001) Mutation rates in the dihydrofolate reductase gene of *Plasmodium falciparum*. Parasitology 122:497–505

Peters W (1987) Chemotherapy and drug resistance in malaria. Academic Press, London

Petersen I, Eastman R, Lanzer M (2011) Drug-resistant malaria: molecular mechanisms and implications for public health. FEBS Lett 585:1551–1562. doi:10.1016/j.febslet.2011.04.042

Plowe CV, Cortese JF, Djimde A, Nwanyanwu OC, Watkins WM, Winstanley PA, Estrada-Franco JG, Mollinedo RE, Avila JC, Cespedes JL, Carter D, Doumbo OK (1997) Mutations in *Plasmodium falciparum* dihydrofolate reductase and dihydropteroate synthase and epidemiologic patterns of pyrimethamine-sulfadoxine use and resistance. J Infect Dis 176:1590–1596

Póvoa MM, Adagu IS, Oliveira SG, Machado RL, Miles MA, Warhurst DC (1998) *Pfmdr1* Asn1042Asp and Asp1246Tyr polymorphisms, thought to be associated with chloroquine resistance, are present in chloroquine-resistant and -sensitive Brazilian field isolates of *Plasmodium falciparum*. Exp Parasitol 88:64–68. doi:10.1006/expr.1998.4195

Price RN, Cassar C, Brockman A, Duraisingh M, van Vugt M, White NJ, Nosten F, Krishna S (1999) The *pfmdr1* gene is associated with a multidrug-resistant phenotype in *Plasmodium falciparum* from the western border of Thailand. Antimicrob Agents Chemother 43:2943–2949

Prince RJ, Geissler PW, Nokes K, Maende JO, Okatcha F, Gringorenko E, Sternberg R (2001) Knowledge of herbal and pharmaceutical medicines among Luo children in western Kenya. Anthropol Med 8:211–235. doi:10.1080/13648470020011814

Rastelli EJ, Coltart DM (2015) A concise and highly enantioselective total synthesis of (+)-anti- and (-)-syn-Mefloquine hydrochloride: definitive absolute stereochemical assignment of the mefloquines. Angew Chem Int Ed Engl 127:14276–14280. doi:10.1002/anie.201507304

Rathod PK, McErlean T, Lee PC (1997) Variations in frequencies of drug resistance in *Plasmodium falciparum*. Proc Natl Acad Sci U S A 94:9389–9393

Raynes K (1999) Bisquinoline antimalarials: their role in malaria chemotherapy. Int J Parasitol 29:367–379

Reed MB, Saliba KJ, Caruana SR, Kirk K, Cowman AF (2000) *Pgh1* modulates sensitivity and resistance to multiple antimalarials in *Plasmodium falciparum*. Nature 403:906–909. doi:10.1038/35002615

Ridley RG (2002) Medical need, scientific opportunity and the drive for antimalarial drugs. Nature 415:686–693. doi:10.1038/415686a

Rønn AM, Msangeni HA, Mhina J, Wernsdorfer WH, Bygbjerg IC (1996) High level of resistance of *Plasmodium falciparum* to sulfadoxine-pyrimethamine in children in Tanzania. Trans R Soc Trop Med Hyg 90:179–181

Sá JM, Twu O, Hayton K, Reyes S, Fay MP, Ringwald P, Wellems TE (2009) Geographic patterns of *Plasmodium falciparum* drug resistance distinguished by differential responses to amodiaquine and chloroquine. Proc Natl Acad Sci U S A 106:18883–18889. doi:10.1073/pnas.0911317106

Schlagenhauf P, Adamcova M, Regep L, Schaerer MT, Rhein HG (2010) The position of mefloquine as a 21st century malaria chemoprophylaxis. Malar J 9:357. doi:10.1186/1475-2875-9-357

Shanks GD (1994) Drugs for prophylaxis and treatment of malaria. J Travel Med 1:40–47

Sidhu AB, Verdier-Pinard D, Fidock DA (2002) Chloroquine resistance in *Plasmodium falciparum* malaria parasites conferred by *pfcrt* mutations. Science 298:210–213. doi:10.1126/science.1074045

Sidhu AB, Valderramos SG, Fidock DA (2005) *pfmdr1* mutations contribute to quinine resistance and enhance mefloquine and artemisinin sensitivity in *Plasmodium falciparum*. Mol Microbiol 57:913–926. doi:10.1111/j.1365-2958.2005.04729.x

Spencer HC (1985) Drug-resistant malaria–changing patterns mean difficult decisions. Trans R Soc Trop Med Hyg 79:748–758

Srivastava IK, Rottenberg H, Vaidya AB (1997) Atovaquone, a broad spectrum antiparasitic drug, collapses mitochondrial membrane potential in a malarial parasite. J Biol Chem 272:3961–3966

Stepniewska K, White NJ (2008) Pharmacokinetic determinants of the window of selection for antimalarial drug resistance. Antimicrob Agents Chemother 52:1589–1596. doi:10.1128/AAC.00903-07

Trenholme CM, Williams RL, Desjardins RE, Frischer H, Carson PE, Rieckmann KH, Canfield CJ (1975) Mefloquine (WR 142,490) in the treatment of human malaria. Science 190:792–794. doi:10.1126/science.1105787

Uhlemann AC, Krishna S (2005) Antimalarial multi-drug resistance in Asia: mechanisms and assessment. Curr Top Microbiol Immunol 295:39–53

Urdaneta L, Plowe C, Goldman I, Lal AA (1999) Point mutations in dihydrofolate reductase and dihydropteroate synthase genes of *Plasmodium falciparum* isolates from Venezuela. Am J Trop Med Hyg 61:457–462

Vasconcelos KF, Plowe CV, Fontes CJ, Kyle D, Wirth DF, Pereira da Silva LH, Zalis MG (2000) Mutations in *Plasmodium falciparum* dihydrofolate reductase and dihydropteroate synthase of isolates from the Amazon region of Brazil. Mem Inst Oswaldo Cruz 95:721–728

von Seidlein L, Duraisingh MT, Drakeley CJ, Bailey R, Greenwood BM, Pinder M (1997) Polymorphism of the *Pfmdr1* gene and chloroquine resistance in *Plasmodium falciparum* in The Gambia. Trans R Soc Trop Med Hyg 91:450–453

Warhurst DC, Craig JC, Adagu IS, Guy RK, Madrid PB, Fivelman QL (2007) Activity of piperaquine and other 4-aminoquinoline antiplasmodial drugs against chloroquine-sensitive

and resistant blood-stages of *Plasmodium falciparum*. Role of beta-haematin inhibition and drug concentration in vacuolar water- and lipid-phases. Biochem Pharmacol 73:1910–1926. doi:10.1016/j.bcp.2007.03.011
Wellems TE, Plowe CV (2001) Chloroquine-resistant malaria. J Infect Dis 184:770–776. doi:10.1086/322858
Wells TN, Burrows JN, Baird JK (2010) Targeting the hypnozoite reservoir of *Plasmodium vivax*: the hidden obstacle to malaria elimination. Trends Parasitol 26:145–151. doi:10.1016/j.pt.2009.12.005
Wernsdorfer WH (1994) Epidemiology of drug resistance in malaria. Acta Trop 56:143–156
Wernsdorfer WH, Payne D (1991) The dynamics of drug resistance in *Plasmodium falciparum*. Pharmacol Ther 50:95–121
White NJ (1997) Assessment of the pharmacodynamic properties of antimalarial drugs in vivo. Antimicrob Agents Chemother 41:1413–1422
White NJ (2004) Antimalarial drug resistance. J Clin Invest 113:1084–1092. doi:10.1172/JCI21682
Winstanley P (2001) Modern chemotherapeutic options for malaria. Lancet Infect Dis 1:242–250. doi:10.1016/S1473-3099(01)00119-0
Witkowski B, Lelièvre J, Barragán MJ, Laurent V, Su XZ, Berry A, Benoit-Vical F (2010) Increased tolerance to artemisinin in *Plasmodium falciparum* is mediated by a quiescence mechanism. Antimicrob Agents Chemother 54:1872–1877. doi:10.1128/AAC.01636-09
Wongsrichanalai C, Sirichaisinthop J, Karwacki JJ, Congpuong K, Miller RS, Pang L, Thimasarn K (2001) Drug resistant malaria on the Thai-Myanmar and Thai-Cambodian borders. Southeast Asian J Trop Med Public Health 32:41–49
Wongsrichanalai C, Pickard AL, Wernsdorfer WH, Meshnick SR (2002) Epidemiology of drug-resistant malaria. Lancet Infect Dis 2:209–218
Woodward RB, Doering WE (1945) The total synthesis of quinine. J Am Chem Soc 67:860–874. doi:10.1021/ja01221a051
World Malaria Report (2013) Accessed at http://www.who.int/malaria/publications/world_malaria_report_2013/report/en/
World Malaria Report (2014) Accessed at http://www.who.int/malaria/publications/world_malaria_report_2013/report/en/
Young MD, Moore DV (1961) Chloroquine resistance in *Plasmodium falciparum*. Am J Trop Med Hyg 10:317–320
Zalis MG, Pang L, Silveira MS, Milhous WK, Wirth DF (1998) Characterization of *Plasmodium falciparum* isolated from the Amazon region of Brazil: evidence for quinine resistance. Am J Trop Med Hyg 58:630–637

Drug Resistance in Cancer

Santosh K. Upadhyay, Ramesh C. Rai, Rekha Gehtori, Ashutosh Paliwal, Poonam Gautam, and Penny Joshi

Abstract Cancer is currently a major public health problem worldwide. Recent data estimates that approximately 40 % of men and women will develop some form of cancer, at some point during their lifetime. Major challenges in treating cancer include delayed diagnosis, emergence of resistance toward existing drugs and metastasis, etc. In majority of advance-stage cancers, the conventional therapies are mostly ineffective. Targeted therapies raise new hope in several difficult-to-treat cases, but their application is restricted by variable efficacies of these drugs in different cancers (tissue backgrounds) and genetic backgrounds. Efflux of drugs, inactivation of drugs, modification of drug targets, inhibition of cell death, induction of epithelial–mesenchymal transition (EMT), and enhanced DNA damage repair activity are the most common mechanisms of drug resistance in cancer. Here, we present an account of these drug resistance mechanisms in cancer and recent breakthroughs in drug development, which have been successful in deceiving most of these mechanisms of drug resistance. Additionally, prognostic biomarkers and approaches for diagnosis of drug resistance have also been discussed.

S.K. Upadhyay (✉) • R. Gehtori • A. Paliwal
Department of Biotechnology, Kumaun University Nainital, Bhimtal Campus, Bhimtal 263136, Uttarakhand, India
e-mail: upadhyaysk97@gmail.com

R.C. Rai
Immunology Group, International Centre for Genetic Engineering and Biotechnology (ICGEB), New Delhi 110067, India

P. Gautam
Tumor Biology Lab, National Institute of Pathology, Safdarjung Hospital Campus, New Delhi 110029, India

P. Joshi
Department of Chemistry, Kumaun University Nainital, D.S.B. Campus, Nainital 263001, Uttarakhand, India

G. Arora et al. (eds.), *Drug Resistance in Bacteria, Fungi, Malaria, and Cancer*,
DOI 10.1007/978-3-319-48683-3_20

1 Introduction

Cancer is the unrestrained proliferation of cells, which can invade and spread to distant sites in the body. Usually, in a normal tissue, the growth of the cells and their division proceed according to the need of the body and these processes are stringently controlled. However, sometimes because of intrinsic or extrinsic reasons, certain cells lose responsiveness to the growth control mechanisms. These cells expand their population and produce a tumor or neoplasm. A tumor that does not invade the surrounding tissue and does not spread is "benign," whereas tumor that attains the characteristic of continuous growth and invasiveness is considered "malignant" or cancerous. In the process of being malignant, tumors acquire characteristics of invasion and metastasis. During metastasis, cancerous cells get dislodged from the original tumor and disperse and invade through the lymph and blood vessels to the other tissues. These metastatic cells may also form secondary tumors in the other tissues by continuous proliferation (NCI website: http://www.who.int/cancer/en/; Goldsby et al. 2002; Mukherjee 2010). Cancers are classified mainly in three groups, viz., carcinomas, leukemias/lymphomas, and sarcomas, based on their tissue origin during embryonic development. About 90 % of the cancers in humans are malignancies of epithelial cells, the carcinomas. Leukemias and lymphomas arise from the hematopoietic cells of bone marrow and are responsible for roughly 8 % of human cancers. Whereas leukemias tend to proliferate as single cells, lymphomas grow as aggregates of tumor cells. Sarcomas, which are relatively rare in humans (approximately 1 % of total cancer cases), are solid tumors of connective tissues (Cooper 2000; Goldsby et al. 2002).

1.1 Transformation and Hallmarks of Cancer

Cancer cells differ from normal cells in various ways that allow them to grow indefinitely and become metastatic. Sometimes, normal cells in culture alter their morphology and growth properties when treated with chemical carcinogens, DNA damage-inducing irradiation, or certain oncogenic viruses, and occasionally this may also impart tumor-producing ability to these cells, when injected into animals. The process is referred to as "transformation" or "malignant transformation," and these cells exhibit very similar in vitro properties to those of cancer cells. Therefore, malignant transformation has been used as a model to study the process of cancer induction (Goldsby et al. 2002). Development of a cancer from its normal progenitor cell is a multistage process, and interactions of the person's genetic factors with that of the environmental factors, such as physical, chemical, and biological carcinogens, are considered responsible for this (Stewart and Wild 2014). There is accumulation of these factors over the age which puts elderly people at the higher risk of developing cancer. Infection with certain viruses, viz., hepatitis B and C viruses, Kaposi's sarcoma herpesvirus (KSHV, also known as *human herpesvirus 8*),

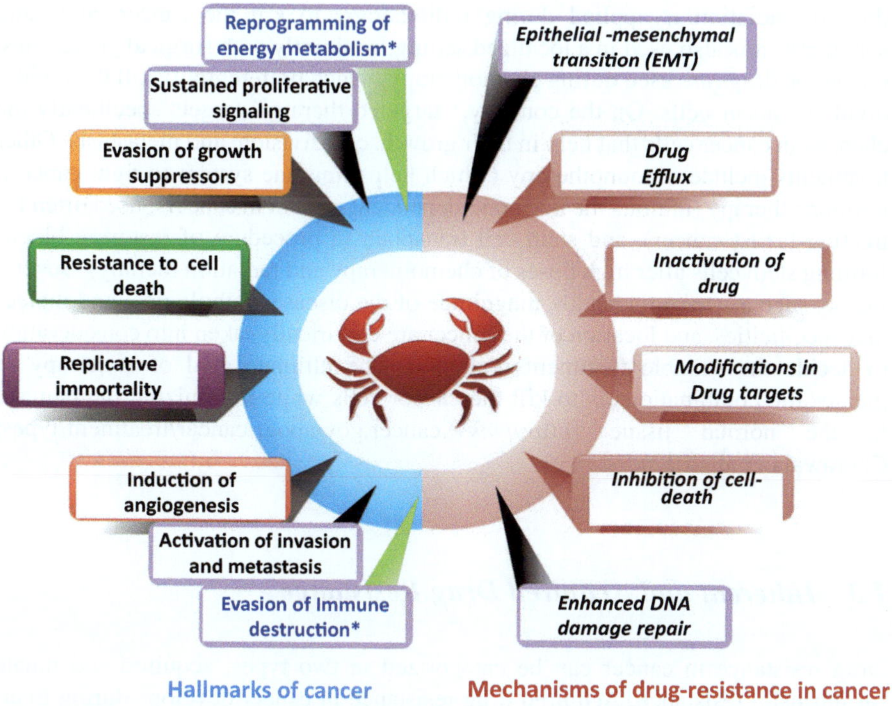

Fig. 1 Hallmarks and emerging hallmarks* of cancer and mechanisms of drug resistance in cancer

Epstein–Barr virus (EBV, also known as *human herpesvirus 4*), and certain papilloma viruses, also causes cancer (Cai et al. 2012; de Martel et al. 2012; Upadhyay et al. 2012; Jha et al. 2013).

Hanahan and Weinberg (2000) established an organizing principle to explain the complexities of neoplastic disease and proposed the six lead changes which drive the malignancy. These changes include sustained signaling for proliferation, evasion from the influence of growth inhibitors, defiance to cell death, everlasting replication ability, enhanced angiogenesis, and induction of invasion and metastasis. Based on new progress in oncology research made during 2000–2010, they also proposed two emerging hallmarks of cancer, which include ability of cancer cells to reprogram energy metabolism and evade destruction by immune system (Hanahan and Weinberg 2011) (Fig. 1).

1.2 Various Treatment Regimes of Cancer

The cancer management mainly involves surgery, chemotherapy, radiotherapy, and targeted therapy. Surgery is the preferred method for "early-stage" cancers. A high dose of radiation is applied during radiotherapy to kill the cancer cells, and sometimes it is also used in a localized setting, conjointly with surgical procedures. Cytotoxic drugs are used during chemotherapy, which preferentially kill the rapidly dividing cancer cells. On the contrary, "targeted therapy" targets specifically the changes in cancer cells that help in their growth, cell division, and metastasis. Other treatments include immunotherapy (which helps immune system to fight cancer), hormone therapy (inhibits the hormone-dependent growth in cancers, used often for treating breast cancer), and stem cell transplant (a procedure of restoring blood-forming stem cells after high doses of chemotherapy and radiation therapy). Several factors such as patient's health, magnitude of the disease, pathological and molecular specificities, and location of the cancer are empirically taken into consideration to decide the suitable treatment regime(s). The ultimate goal of a therapy or therapeutic combinations is to kill the cancer cells while minimizing the damage to the normal tissue (http://www.cancer.gov/about-cancer/treatment/types; Chorawala et al. 2012).

1.3 Inherent and Acquired Drug Resistance

Drug resistance in cancer can be categorized in two types, acquired and innate (or intrinsic) resistance. Acquired drug resistance in cancer develops during treatment, usually in response to therapy, while innate resistance preexists in cancer cells, even before starting the treatment. Acquired resistance arises during the course of treatment and could be due to mutations or adaptive responses to the treatment regimens, including activation of alternative compensatory signaling or upregulation of the therapeutic target(s) (Holohan et al. 2013). The drug resistance can also be a result of therapy-induced selection of a small subgroup of resistant cells present in tumors, which usually tend to display high degree of molecular heterogeneity (Swanton 2012).

2 Mechanism(s) of Drug Resistance in Cancer

2.1 Drug Efflux

This mechanism relies on reducing accumulation of drug inside the cell by enhancing its efflux. ATP-binding cassette (ABC) transporters are most well-studied proteins carrying out this function. More than 100 transporters from this evolutionarily conserved protein family are distributed from prokaryotes to humans and are known to move substrates in (influx) or out (efflux) of cells, or carry out their transport across intracellular membranes. Substrates for ABC transporters may be endogenous, such as metal ions, inorganic anions, amino acids, peptides, sugars, hydrophobic compounds, and metabolites, or exogenous substrates, such as drugs and other xenobiotic compounds (Choi 2005; Vasiliou et al. 2009). ABC transporters have two discrete domains—a nucleotide-binding domain (NBD) and a transmembrane domain. Substrate binding at the transmembrane domain of protein induces hydrolysis of ATP molecule at the nucleotide-binding site, which drives a conformational change facilitating efflux of drugs. This efflux is important for preventing the accumulation of drugs inside the cell. Expression of ABC transporters is very high in the epithelial cells of the intestine and liver, where they carry out efflux of drugs besides other harmful substrates into the intestinal lumen and bile duct. Additionally, these proteins have crucial role(s) in the maintenance of blood–brain barrier (Housman et al. 2014). The human genome codes for 49 ABC transporters, from which at least 14 genes are implicated in human diseases, including cystic fibrosis, drug-resistant tumors, ataxia, adrenoleukodystrophy, Dubin–Johnson syndrome, Stargardt's disease, Byler's disease, X-linked sideroblastic anemia, progressive familiar intrahepatic cholestasis, and persistent hyperinsulinemic hypoglycemia in children (Choi 2005; Vasiliou et al. 2009).

"Multidrug resistance (MDR)" is an important mechanism implicated in the development of resistance toward chemotherapeutic agents in many cancers. MDR impacts cancers ranging from blood cancers to a variety of solid tumors (Persidis 1999). In human cancers, elevation of three ABC transporters, P-glycoprotein (expressed from MDR1 gene), multidrug resistance-associated protein 1 (MRP1), and mitoxantrone resistance protein (MXR, *aliases*: ABC-P, BCRP) have been found to be linked with resistance to various antineoplastic drugs (Gottesman 2002; Zahreddine and Borden 2013). Work on a doxorubicin-resistant lung cancer cell line revealed that instead of P-glycoprotein, this cell line depends on another protein, namely, MRP, for drug resistance. MRP was also identified as a member of the ABC transmembrane transporter superfamily and a drug efflux pump (Persidis 1999). P-Glycoprotein is the best studied ABC transporter involved in drug resistance, whereas majority of multidrug transporters belong to the multidrug resistance protein (MRP) family (alias ABCC family). Increased expression of MDR1 has been recorded in cancerous tissue of the colon, liver, and kidney, whereas in lung and breast tissues and prostate cells which lacks MDR1, upregulation of MRP1 or MXR has been associated with drug resistance (Housman et al. 2014). Increased expression of P-glycoprotein imparts resistance against vinblastine, vincristine, doxorubicin, teniposide, daunorubicin, etoposide, and taxol, whereas overexpressed MXR is known to provide resistance against topoisomerase I inhibitors, viz., anthracyclines and mitoxantrone (Gottesman 2002; Chorawala et al. 2012; Zahreddine and Borden 2013). Yanase et al. have succeeded in reversing BCRP (MXR)-mediated drug resistance in breast cancer cells by using gefitinib, a tyrosine kinase inhibitor that blocks this transporter (Yanase et al. 2004). MRP1 is ubiquitously expressed in normal tissues and human cancers (Chorawala et al. 2012). This protein displays a broad spectrum of resistance for anticancer drug and is involved in the transportation of glutathione (GSH)-/glucuronic acid-/sulfate-conjugated drugs and natural product drugs which are negatively charged. Additionally, cotransport of glutathione with drugs which are positively charged, e.g., vinblastine, has also been observed in certain cases (Gottesman 2002; Zahreddine and Borden 2013). In a study of neuroblastoma, high levels of MRP1 have also been associated with poor prognosis (Haber et al. 2006). In addition to MRP1, MRPs 2, 3, and 5 have also been implicated in drug resistance in cancers (Chorawala et al. 2012).

2.2 Inactivation of Drug

A number of drugs turn into their cytotoxic variants, after metabolic activation/alteration by enzymes. Various researchers have shown that certain drug-resistant cancer cells display reduced drug activation under the influence of selection pressure exerted by the cytotoxic drugs (Kufe and Spriggs 1985; Bardenheuer et al. 2005). This is usually achieved by abrogating the activity of enzyme [by downregulation of expression or mutation(s) in the coding gene(s)] involved in drug activation (Sampath et al. 2006). Cytarabine (also called AraC) is a nucleoside analogue used for treating acute myelogenous leukemia (Sampath et al. 2006; Zahreddine and Borden 2013). Multiple phosphorylations convert AraC to AraC-triphosphate, which is a substrate for human DNA polymerases. Downregulation at expression level, or mutation in the genes of the pathway, has been identified as the common mechanism facilitating AraC drug resistance (Sampath et al. 2006; Zahreddine and Borden 2013). Additionally, inactivation of drugs is also used as a mechanism of development of resistance. One example is conjugation of the drugs to glutathione, which is a potent cellular antioxidant (Wilson et al. 2006). Conjugation of GSH to platinum drugs such as oxaliplatin and cisplatin converts them into better substrates for ABC transporters resulting in their efflux outside the cells (Zahreddine and Borden 2013). Elevation of GST expression in cancer cells thereby leads to increased rate of detoxification of the anticancer drugs, resulting into reduced cytotoxic damage to the cells (Peters and Roelofs 1997; Holohan et al. 2013).

Irinotecan is a topoisomerase I inhibitor that is inactivated by UGT1A1-mediated glucuronidation. Expression of UGT1A1 is usually suppressed by promoter methylation, and induction of UGT1A1 expression by epigenetic means also facilitates resistance to Irinotecan (Gagnon et al. 2006). Additionally, CYP450 is also known to inactivate irinotecan during phase I metabolism (Xu and Villalona-Calero 2002; Holohan et al. 2013; Zahreddine and Borden 2013).

2.3 Modification in Drug Target

Target alterations also lead to drug resistance in cancers. Signaling kinases are among the most important drug targets for combating cancer. Constitutive activation of many of these kinases transmit signal for continuous cell division. Important components of this signaling include proteins belonging to the family of epidermal growth factor receptor (EGFR) and the molecules involved in the downstream signaling, e.g., Ras, Src, Raf, MEK, etc. Mutations are usually responsible for the overactivation of these kinases; however, gene overexpression has also been found to be responsible for similar functional outcome. In non-small cell lung carcinoma (NSCLC), mutations which activate EGFR have been identified as key drivers of carcinogenesis in 10–15 % of European patients and approximately 30 % of East Asian patients (Rosell et al. 2009). Mutation of exon 21 (L858R) and exon 19 (delE746-A750) are two most common mutations which activate EGFR. Patients with these EGFR mutations usually show excellent responses to therapy with erlotinib or gefitinib, which are first-generation reversible inhibitors of EGFR (Mok et al. 2009; Fukuoka et al. 2011; Rosell et al. 2012). However, despite of significant initial response to treatment by erlotinib or gefitinib, there have been reports of relapse in roughly 60 % of patients after 9–14 months of therapy. This is due to occurrence of a different point mutation at the amino acid position 790 (T790M) (Pao et al. 2005; Sharma et al. 2007; Sequist et al. 2011; Yu et al. 2013). By creating steric hindrance, the T790M mutation prevents binding of the inhibitor in the ATP-binding pocket, ultimately leading to drug resistance (Kobayashi et al. 2005; Kwak et al. 2005; Pao et al. 2005; Yun et al. 2008; Walter et al. 2013). Few second-generation irreversible HER family tyrosine kinase inhibitors, viz., dacomitinib (PF299804) and afatinib (BIBW2992), revealed significant efficacy in treating T790M mutants in vitro; however, response to these drugs was not convincing in patients who have failed first-generation TKIs (Miller et al. 2012). PKC412 (midostaurin) is an indolocarbazole compound identified as a potential inhibitor of T790M NSCLC (Lee et al. 2013); however, its clinical application is also debatable due to its broad kinase inhibition profile (Karaman et al. 2008; Walter et al. 2013). CO-1686 is a 2,4-disubstituted pyrimidine compound that has recently been found to irreversibly and selectively inhibit drug resistance due to T790M mutation, in NSCLC models. When CO-1686 is administered orally in transgenic mouse model of T790M mutation and in patient-derived xenograft models, it resulted in significant tumor regression (Walter et al. 2013). Phase I clinical trial of CO-1686 (Sequist et al. 2014) and another T790M-targeting drug AZD9291 (Janne et al. 2014) in patients harboring T790M mutation, whose disease aggravated after initial response to EGFR-targeted therapy, has shown promising results. Remarkably, in around 60 % of T790M-positive patients who received CO-1686 and close to 50 % of those who received AZD9291, shrinkage in the tumor was observed. As the drugs CO-1686 and AZD9291 specifically target mutant *EGFR*, they only insignificantly affect normal EGFR in the body, resulting in fewer debilitating adverse effects (Masters et al. 2015). CO-1686 also received the US FDA's "breakthrough therapy" designation in May 2014 (Chernew 2014).

Inhibitors of topoisomerase II stabilize the (otherwise transient) interaction of this enzyme on substrate (DNA), leading to DNA damage and ultimately inhibition of cell division. Mutation in the gene coding for topoisomerase II has been identified as one of the mechanisms by which various cell lines attain resistant to topoisomerase II inhibitors (Hinds et al. 1991; Holohan et al. 2013; Housman et al.

2014). Resistance to paclitaxel and other taxanes is another example of attaining drug resistance (in ovarian cancers) by target alteration, by mutations in β-tubulin, and by other means (Holohan et al. 2013; Housman et al. 2014).

2.4 Inhibition of Cell Death

Resistance to apoptosis is considered one of the six hallmarks of cancer and could be a contributor toward drug resistance. However, surprisingly not many of the anti-apoptotic proteins have actually been implicated in the drug resistance (Letai 2008). The proteins which are known to have role in imparting resistance to apoptosis include anti-apoptotic proteins of BCl-2 family, FLIP, a known inhibitor of caspase 8 and inhibitor of apoptosis proteins (IAPs). Enhancement in activities of these proteins by any means (including mutation, chromosomal translocation, overexpression, etc.) has been affiliated with several malignancies, as well as to resistance against chemotherapy or targeted therapies. Expression of these anti-apoptotic proteins is induced by transcription factors which are involved in pro-survival signaling, e.g., STAT3 and NF-κB. The pro-survival signaling is activated during tumorigenesis usually via oncogenic mutations in kinases that in turn induces anti-apoptotic signaling, thereby bridging together two important hallmarks of cancer (Holohan et al. 2013).

BCL-2 family members are among the most well-studied anti-apoptotic proteins, in regulating responses to chemotherapy. In leukemic cells and mouse thymocytes, overexpression of BCL2 imparts resistance to cytotoxic chemotherapeutic drugs (Sentman et al. 1991; Miyashita and Reed 1992). Other BCL-2 family proteins which regulate apoptosis induced by chemotherapy include BCL2 family proteins which are anti-apoptotic in nature, e.g., BCL-XL and MCL1, and pro-apoptotic family members, e.g., BAX, BAD, and BAK, and also BH3-only proteins which antagonize functions of the anti-apoptotic BCL-2 family members (Kitada et al. 1998; Wang et al. 2001; Holohan et al. 2013). Recently a method called BH3 profiling has been developed that measures the level of "priming" of a cell for undergoing apoptosis, and a good correlation between mitochondrial priming level and clinical response to chemotherapy has been shown for many cancer types (Ni Chonghaile et al. 2011). This shows the importance of BCL-2 family proteins in deciding cell fate after chemotherapy (Holohan et al. 2013; Housman et al. 2014). Bcl-2-like protein 11, also known as BIM, has also been shown to have important role in gefitinib- and erlotinib-induced apoptosis in EGFR-mutated NSCLC and in imatinib-induced apoptosis in CML (Kuribara et al. 2004; Costa et al. 2007; Gong et al. 2007). Interestingly, in East Asian patients, a deletion in the BIM gene has shown resistance to growth inhibition by TKI therapies in EGFR mutant lung cancer and chronic phase CML (Ng et al. 2012). BIM levels have also been implicated as prognostic marker for predicting clinical responsiveness to EGFR inhibitors, ERBB2 (alias HER2) or PI3K (Faber et al. 2011). Recently, a number of pharmacological inhibitors have been discovered for targeting proteins of BCL-2 family for cancer treatment, and most notable of them is ABT-737 and its orally bioavailable derivative ABT-263 (or "navitoclax"). This drug antagonizes the function of anti-apoptotic proteins BCL-2, BCL-XL, and BCL-W and promotes the pro-apoptotic function of BAX and BAK (van Delft et al. 2006). ABT-737/ABT-263 has revealed its effectiveness as a single agent against various tumor types (Oltersdorf et al. 2005; Konopleva et al. 2006) and in combination with chemotherapies or radiation as well (Oltersdorf et al. 2005). However, the efficacy of these drugs is abrogated by the action of anti-apoptotic protein MCL1, which is a member of BCL2 family. MCL1 has been shown to bind with ABT-737, and higher expression level of MCL1 has been associated with resistance to these agents (Konopleva et al. 2006; van Delft et al. 2006; Lin et al. 2007; Holohan et al. 2013; Housman et al. 2014). Apoptosis signaling is also induced through the extrinsic signaling pathway via death receptors present on the cell surface. Various recombinant forms of ligands that induce apoptosis (such as TRAIL) and agonistic antibodies have been developed which recognize either death receptor 4 (known as DR4 or TRAILR1) or DR5 (TRAILR2) (Holohan et al. 2013). However, despite of profound antitumor activity of these compounds in xenograft and in vitro models, the use of recombinant TRAIL or agonistic antibodies targeting TRAIL receptor as monotherapies has not been successful in clinical trials. Translation of TRAIL to the clinic has also been largely unsuccessful due to its short half-life, insufficient delivery methods, and presence of the TRAIL-resistant cancer cell populations (Stuckey and Shah 2013; Housman et al. 2014).

Higher level of expression of IAPs has been found to be linked with chemoresistance in cancer treatment (Hunter et al. 2007). To tackle with this chemoresistance, researchers are exploring application of small molecule inhibitors of IAPs, those are inspired by a tetrapeptide motif (AVPI) which is present in SMAC, the endogenous antagonist of IAP. Therefore, SMAC-mimetic drugs are being designed which act by inhibiting XIAP (an inhibitor of caspases 3, 7, and 9) and by facilitating degradation of IAPs, 1 (BIRC3) and 2 (BIRC2) via the ubiquitin–proteasome system. Therefore, the SMAC-mimetic drugs can stimulate the activation of caspases 3, 7, 8, and 9 and could also sensitize various tumors for the chemotherapeutic treatment or to the TRAIL (Chen and Huerta 2009). Furthermore, some SMAC mimetics have been evaluated and few others are under examination, in patient-derived xenograft models (Benetatos et al. 2014), or clinical trials [LCL against multiple myeloma, clinical trial identifier: NCT01955434; birinapant against solid tumors or lymphoma (Holohan et al. 2013; Amaravadi et al. 2015)].

2.5 *Epithelial to Mesenchymal Transition*

Recent studies suggest that during their progression, tumors undergo epithelial to mesenchymal transition (EMT). During this transition, epithelial cells give up their differentiation features such as cell–cell adherence, cell polarity, and immotile nature and instead gain mesenchymal characteristics which include invasiveness, motility, and increased resistance to apoptosis (Polyak and Weinberg 2009). EMT is not only implicated in metastasis, but owing to involvement of a number of common transcription factors during EMT and induction of drug resistance, it is also now considered responsible for acquired drug resistance (Holohan et al. 2013). Numerous studies are indicative of correlation between MDR and cancer invasiveness. Recently, Saxena et al. have shown co-expression of various ABC transporters with markers of EMT in invasive breast cancer cells, during treatment with chemotherapeutic drugs. Additionally, induction of EMT phenotype in immortalized, noninvasive cell lines was found sufficient to increase the expression of ABC transporters, drug resistance, migration, and invasion. On the other hand, reversal of EMT by knocking down EMT-promoting transcription factors reduced the expression of ABC transporters, their chemoresistance, and invasion properties. The promoter region of ABC transporters contains multiple binding sites for the EMT-promoting transcription factors (Saxena et al. 2011). Promoter activity assay showed that overexpression of FOXC2, Twist, and Snail could increase the expression of ABC transporters. Chromatin immunoprecipitation studies further unveiled the binding of Twist to E-box elements of the ABC transporters. This study thus provides the molecular justifications for the correlation between invasiveness and MDR and suggests EMT-promoting transcription factors as molecular target for simultaneously curbing the metastasis as well as the associated drug resistance in cancer (Saxena et al. 2011).

A number of cell lines undergoing EMT show resistance to EGFR inhibitors (Fuchs et al. 2008; Yao et al. 2010). In NSCLC cell lines, resistance to erlotinib and gefitinib is also associated with EMT (Thomson et al. 2005; Yauch et al. 2005; Witta et al. 2006). Erlotinib sensitivity of NSCLC cells harboring normal (wild-type) EGFR depends on whether they express CDH1 (epithelial marker) or VIM (mesenchymal marker) (Thomson et al. 2005). A recently developed 76-gene EMT signature could characterize EMT in NSCLC. This signature has also been validated to successfully predict the efficacy of NSCLC cells toward the EGFR and PI3K inhibitors (Byers et al. 2013). EMT has also been observed in clinical samples of NSCLC, which showed resistance to EGFR inhibitors (Holohan et al. 2013). More recent studies reveal the roles of EMT in conferring drug resistance to the EGFR inhibitors and also in response to other mutations. These mutations are present in K-Ras, EGFR (T790M), Alk, MEK, and BRAF and are associated with EMT by various degrees and combinedly represent more than 70 % cases of lung adenocarcinoma (Seo et al. 2012). Receptor tyrosine kinase AXL has been recognized as a potential therapeutic target to overcome the EMT-associated resistance to EGFR inhibitors (Zhang et al. 2012; Byers et al. 2013). Additionally, an siRNA screen identified MED12 as determinants of tumor sensitivity to inhibitors of ALK and EGFR. MED12 is a component of the mediator transcription complex which is often mutated in cancers. MED12 loss has been shown to activate the signaling via transforming growth factor-β receptor (TGF-βR), which is a known inducer of EMT phenotype. Inhibition of TGF-βR signaling resulted in restoration of drug sensitivity in MED12-depleted cells (Huang et al. 2012; Holohan et al. 2013).

Interestingly, EMT-induced resistance provides protection to cancer cells not only against targeted therapies but against chemotherapy and radiation therapy as well (Huang et al. 2012; Gomez-Casal et al. 2013). Arumugam et al. characterized patterns of resistance of pancreatic cancer cells to three common chemotherapeutic agents with different mechanisms of action: gemcitabine, 5-fluorouracil (5-FU), and cisplatin. Gene expression profiling shows that the cell line which is either sensitive or resistant to these agents differed in expression of EMT-related genes. E-Cadherin and Zeb-1 (transcriptional suppressor of E-cadherin) mostly show inverse correlation in their expression patterns. Moreover, immunohistochemical analysis also confirmed this inverse correlation between these two proteins, in primary tumors. Silencing of Zeb-1 in the mesenchymal cells resulted in upregulation of E-cadherin expression and restoration of drug sensitivity (Arumugam et al. 2009). In a recent study, Meidhof et al. used HDAC inhibitor mocetinostat to reverse ZEB1-associated EMT and drug resistance in cancer cells. In this study, miR-203 was identified as a major drug sensitizer, and restoration of its expression by mocetinostat successfully sensitized the cancer cells against chemotherapy (Meidhof et al. 2015). EMT has also been implicated in radioresistance, in prostate cancer, which was found to be associated with enhanced cancer stem cell (CSC) phenotypes via activation of the PI3K/Akt/mTOR signaling. A PI3K/mTOR inhibitor, BEZ235, effectively sensitized radioresistant prostate cancer cells for treatment with radiation (Chang et al. 2013). In another study, hypoxia or TGF-β-induced EMT was shown to reduce the expression of epithelial markers and enhance the expression of mesenchymal markers in MCF7, A549, and NMuMG epithelial cells. Transition to mesenchymal phenotype and E-cadherin loss were associated with radioresistance in these cells (Theys et al. 2011). Therefore, inhibition of EMT is now seen as a promising approach of inhibiting tumor progression by inhibiting metastasis as well as drug/radiation resistance (Holohan et al. 2013). Even the drug-resistant tumors could be resensitized for radiation or chemotherapy by using EMT-inhibiting molecules (Kang et al. 2013; Wilson et al. 2014). EMT has been linked to the generation of cancer stem cells (Mani et al. 2008; Lan et al. 2013). These cells are multi-/pluripotent cells which have the ability to initiate tumors and self-renew. They divide asymmetrically and generate differentiated progenitors (Singh and Settleman 2010). As expression of ABC transporters and anti-apoptotic proteins is high in cancer stem cells and these cells also show enhanced DNA damage repair and aldehyde dehydrogenase (ALDH) activities, they tend to display good degree of drug resistance (Holohan et al. 2013) (Fig. 2).

2.5.1 Inducers of Epithelial–Mesenchymal Transition

Inflammation, hypoxia, and certain mutations (Table 1) are well-known inducers of EMT. Inflammation is the body's response to injury, infection, or perturbation in its homeostasis by any other means. Although it is a protective response, however uncontrolled inflammation is associated with the several disease pathologies

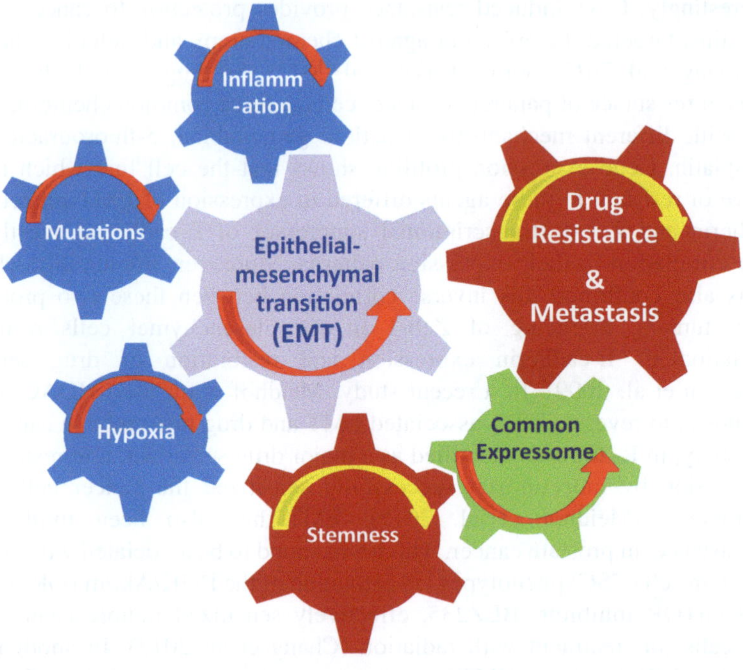

Fig. 2 EMT is driven by various factors and plays an important role in inducing drug resistance and metastasis

including cancer. In cancer, pro-inflammatory cytokine interleukin-6 (IL-6) can promote EMT through induced expression of SNAIL1 by Janus kinase (JAK)-STAT3 signaling which leads to drug-resistant phenotype (Sullivan et al. 2009). Hypoxia also guides EMT in cancerous condition by stabilizing hypoxia-inducible factor-1 (HIF-1) transcription complex. HIF-1 binds to the hypoxia response element in the proximal promoter of TWIST and induces its expression. Elevated expression of TWIST has been shown to promote EMT in tumor environment (Yang et al. 2008) (Fig. 2).

2.6 Enhanced DNA Damage Repair

Numerous chemotherapeutic agents work by damaging DNA. Additionally, many others affect the DNA damage repair system by targeting topoisomerases. In response to DNA damage, the cellular machinery is activated to repair it; however, if repair is not achievable, cell death pathway is turned on to terminate the cell lineage with damaged DNA (Holohan et al. 2013). p53 is a protein sometimes referred to as "guardian of the genome" because of its important role in conserving stability by preventing genome mutation (Lane 1992). Various cell cycle

Table 1 Contribution of EMT in drug/treatment resistance

Mutations implicated in drug resistance		
K-Ras	Tumors with K-Ras-activating mutations are among most difficult to treat; interestingly, it is the only mutation that shows statistically significant association with mesenchymal phenotype in NSCLC tumors	Byers et al. (2013)
EGFR, Alk, MEK, BRAF	MED12 loss confers resistance to Alk, EGFR, BRAF, and MEK inhibitors in NSCLC and various other cancer types by activation of TGF-β signaling (or by EMT)	Huang et al. (2012)
EGFR (T790M)	Analysis of transcriptome of transgenic mice with T790M mutation shows prominent expression of mesenchymal signatures in lung tissue	GEO dataset: GSE17373 Regales et al. (2009)
Cancer therapies		
Chemotherapy	*Cisplatin*: reduced expression of MED12 (i.e., increased TGF-β signaling or EMT) confers resistance to chemotherapy drug cisplatin	Huang et al. (2012)
	Docetaxel: inhibition of zeb1 reverses EMT and chemoresistance in docetaxel-resistant human lung adenocarcinoma cell line	Ren et al. (2013)
	5-Fluorouracil (5-FU): reduced expression of MED12 (i.e., increased TGF-β signaling or EMT) confers resistance to chemotherapy drug 5-FU	Huang et al. (2012)
Radiation therapy	Non-small cell lung cancer cells that survived ionizing radiation treatment display cancer stem cell and epithelial–mesenchymal transition phenotypes	Gomez-Casal et al. (2013)

checkpoints are regulated by p53 and mutations in this gene can interfere with the mechanism of cell cycle arrest induced by DNA damage. Another p53 function is to induce apoptosis in case of failure of DNA damage repair, and therefore mutation in p53 can also contribute to drug resistance by inhibition of apoptosis (Enoch and Norbury 1995; Holohan et al. 2013). DNA damage response (DDR) mechanisms are able to overturn the drug-induced damage and thereby can impart the drug resistance characteristics to cancers. Therefore, damaging DNA by chemotherapy along with targeting DDR pathways makes an efficient therapeutic combination that could sensitize cancer cells and enhance the therapeutic efficacy (Housman et al. 2014).

Cancer cells have many mutations accumulated in their genomes. Sometimes mutation(s) in key gene(s) renders DDR pathways nonfunctional. This makes the cell dependent on another pathway for the DDR function. Therefore, if one pathway is inactivated by mutations, this provides opportunity to selectively kill cancer cells by targeting another pathway by the use of specific inhibitors (Kaelin 2005; Chan and Giaccia 2011). This concept is known as "synthetic lethality" and identification of targetable synthetic lethal interactions always presents new opportunity of anticancer drug discovery. For targeted anticancer therapies, a number of synthetic lethal interactions are at the focus of many preclinical and clinical studies (Chan and Giaccia 2011). Inhibitors of the poly(ADP–ribose) polymerase 1 (PARP1), an enzyme involved in the repair of the single-strand DNA breaks, are recommended in ovarian and breast tumors harboring mutations in the BRCA1 or BRCA2 genes (Farmer et al. 2005). However, when BRCA2-mutant tumors are treated with PARP inhibitors, due to an in-frame deletion in BRCA2, its DNA repair function is partially restored and these tumor cells survive treatment by chemotherapeutic agents (Edwards et al. 2008; Sakai et al. 2008; Holohan et al. 2013). Similarly, mismatch repair (MMR) deficiency has been associated with resistance to various cytotoxic chemotherapies, such as resistance to carboplatin and cisplatin (Holohan et al. 2013). In MSH2-deficient cells, accumulation of oxidative lesions, e.g., 8-oxoguanine (8-oxoG), is caused by methotrexate, which induces apoptosis of cancer cells (Martin et al. 2009; Holohan et al. 2013). Nucleotide excision repair is an important signaling pathway involved in the repair of damaged DNA (Kirschner and Melton 2010). Expression level of ERCC1 is an important determinant for proper functioning of NER pathway, and its high expression has been linked to poor chemotherapeutic outcome in many cancers, including NSCLC, ovarian cancer, and gastric cancer (Lord et al. 2002; Kwon et al. 2007). Testicular cancers usually have very low levels of ERCC1 and have been found to be very sensitive to cisplatin treatment (Usanova et al. 2010; Holohan et al. 2013).

Some chemotherapy drugs induce guanine O6 alkylation. The O6-methylguanine DNA methyltransferase (MGMT), repairs and revert it back to guanine. Therefore, many tumors which have high levels of MGMT expression show resistance to alkylating agents (Blanc et al. 2004). Inhibiting this repair system, therefore, could sensitize cancers to alkylating agents. However, due to their lack of cancer specificity, many of the MGMT-targeting drugs show significant toxicity (Hegi et al. 2005; Rabik et al. 2008; Curtin 2012). Taking clue from glioma patients whose disease-free and overall survival rates have increased because of epigenetically silenced MGMT genes, recommendations are made for individualizing dosage of O6-guanine alkylating agent by measuring MGMT-promoter-CpG-methylation levels as a biomarker of sensitivity (Housman et al. 2014).

3 Prognostic Markers and Diagnosis of Drug Resistance

Numerous evidences suggest that the drug sensitivity of tumors is remarkably affected by variations in the cancer genome. In chronic myeloid leukemia (CML), selective targeting of BCR-ABL translocation with drugs has revolutionized the treatment, and 5-year survival rate of 90 % has been achieved in treated patients (Druker et al. 2006). Similarly, in NSCLC, a mutation which activates EGFR is predictive of marked sensitivity to EGFR-TKIs. L858R point mutation and exon 19 deletions of EGFR are linked with the sensitivity to EGFR TKIs (Janne and Johnson 2006). In patients with tumors harboring these mutations, response rates of $>70\,\%$ have been observed, when treated with either erlotinib or gefitinib (Jackman et al. 2009; West et al. 2009). However, NSCLC patients with KRAS mutations in their tumors exhibited high degree of resistance to erlotinib and gefitinib, with a response rate of $\leq 5\,\%$. Few other *EGFR* mutations, including T790M and exon 20 insertions, are also associated with resistance to TKIs (Janne and Johnson 2006). Clinical and histological features, up to some extent, may correlate with selected genetic alterations; however, for precise identification of the mutations associated with drug sensitivity or resistance, molecular testing needs to be done. In order to link drug activity with the complexity associated with cancer genomes, Garnett et al. screened several cancer cell lines representing most of the tissue-type and genetic backgrounds of human cancers. Approximately 130 drugs already under clinical and preclinical studies were tested in this screen. In addition to classic oncogene addiction paradigms, the drug activity was influenced by tissue-specific and/or expression biomarkers. On the other hand, sensitivity to a wide range of therapeutic agents has been correlated well with certain frequently mutated genes. The pharmacogenomic profiling studies in well-characterized cancer cell lines and clinical samples therefore provide guide to better rationalize cancer therapeutic strategies by utilizing biomarker discovery platform (Garnett et al. 2012).

Protein biomarkers of cancers have attracted attention of researchers earlier than that at the level of nucleic acids (Lippert et al. 2011). The obvious focus has been on cancer-derived and cancer-specific molecules present in the blood, and a number of markers, viz., carcinoembryonic antigen (CEA) and alpha-fetoprotein (AFP), were among the initial molecules assessed. Most of the markers explored for this purpose turned out to be nonspecific to tumors. Tumor markers, those found their way to clinical trials, include prostate-specific antigen (PSA) for prostate cancer, thyroglobulin for thyroid cancer, CA 125 for ovarian cancer, and human chorionic gonadotropin (HCG) for chorionepithelioma. These markers have also been useful in examining the effectiveness of treatment (Lippert et al. 2011). Recently we've proposed a subset of 13 proteins, including annexins A1, A2, and A3, reported to be differentially expressed at tissue level, as circulatory markers of gall bladder carcinoma (GBC), for clinical applications (Singh et al. 2015).

Cells and fragments of DNA shed by tumors into the bloodstream can now be used for noninvasive screening of early-stage cancers as well as for monitoring responses to treatment (Brock et al. 2015). Circulating tumor cells, cell-free DNA (cfDNA), and exosomes (harboring DNA, RNA, proteins) are the important constituents of these liquid biopsies. Liquid biopsies can be used for molecular characterization of the tumor, and their noninvasive nature allows repeated sampling to monitor the changes in molecular/cellular markers over time (Brock et al. 2015; Karachaliou et al. 2015). In certain cases, it becomes inevitable to examine the drug resistance in actual tumor sample. For in vitro diagnosis of drug resistance, a test based on 3H-thymidine uptake into cultured cells from tumor biopsies, in the presence of various drugs, has been developed by IMPATH (Los Angeles, Calif., USA). An algorithm is applied to the data obtained from this experiment to determine the probabilities of tumor responding to various drugs. This test has been found to be very accurate in predicting the drug-resistant in vivo (Kern and Weisenthal 1990; Chorawala et al. 2012). Positron emission tomography (PET) is another important method of diagnosis of drug resistance during cancer treatment. PET has been in clinical use for many years for examining and localization of cancer localization. Now it can be utilized to determine the metabolic activity of tumors (Lippert et al. 2011).

4 Conclusion

Traditional therapeutic modalities for cancer, including radiation and chemotherapy, have been the backbone of cancer therapy in the past and are still playing major role in its treatment. However, systemic toxicity, relapse after treatment, and non-applicability/reduced efficacy in advance stages are the major drawbacks of these treatments. Genetic diversity among cancers, as well as in the constituent cells, presents a major complication in devising successful, specific drugs for cancer treatment. Variability in the tissue background, added by that in the tumor microenvironment and in the epigenome, further makes the designing of specific drugs cumbersome. As the classical hallmarks of cancer have been postulated to understand components and progression of cancer, in a similar fashion, key target pathways and molecules are being identified by researchers, and new compounds are being developed, to specifically target these pathway/molecules. In the recent past, new promising drugs have been approved and many of them are entering clinical trials, though there is still very limited coverage of cancer-specific treatments available across the myriad of neoplasms. There is an immense need of expediting the drug discovery to widen the coverage of specific treatments and to outpace the rate of evolution of drug resistance at the individual (cancer) cell level to ensure the availability of drugs for intervening the continuously evolving resistance mechanisms.

5 Opinion

With the establishment of the fact that genetic makeup (mutational status) and resulting expressome of cancers constitute an important determinant in predicting their sensitivity toward anticancer agents, a bigger challenge for researchers is to develop new agents to cover treatments for diversity of tumor expressome(s). Furthermore, there should be increased awareness in clinicians and patients across the world, regarding carrying out genetic testing of cancers. A rigorous campaigning should also be carried out for registration of patients in clinical trials, particularly those with advance-stage cancers. Policy makers from across the world should facilitate conduction of these trials at broader geographical and ethnic levels. Adopting a comprehensive approach at the level of drug development, such as finding newer agents for nano-targeting, EMT-inhibition, etc., would provide more efficient treatment options in the future.

Acknowledgments The authors wish to thank the Vice Chancellor, Kumaun University, Nainital, and University Grant Commission [UGC-BSR-Research startup grant, Letter No. F.30-74/2014 (BSR)], Government of India, for providing necessary facilities and funds.

References

Amaravadi RK, Schilder RJ, Martin LP, Levin M, Graham MA, Weng DE, Adjei AA (2015) A phase I study of the SMAC-MIMETIC BIRINAPANT IN ADULTS WITH REFRACTORY SOLID TUMORS OR LYMPHOMA. Mol Cancer Ther 14:2569–2575. doi:10.1158/1535-7163.MCT-15-0475

Arumugam T, Ramachandran V, Fournier KF, Wang H, Marquis L, Abbruzzese JL, Gallick GE, Logsdon CD, McConkey DJ, Choi W (2009) Epithelial to mesenchymal transition contributes to drug resistance in pancreatic cancer. Cancer Res 69:5820–5828. doi:10.1158/0008-5472.CAN-08-2819

Bardenheuer W, Lehmberg K, Rattmann I, Brueckner A, Schneider A, Sorg UR, Seeber S, Moritz T, Flasshove M (2005) Resistance to cytarabine and gemcitabine and in vitro selection of transduced cells after retroviral expression of cytidine deaminase in human hematopoietic progenitor cells. Leukemia 19:2281–2288. doi:10.1038/sj.leu.2403977

Benetatos CA, Mitsuuchi Y, Burns JM, Neiman EM, Condon SM, Yu G, Seipel ME, Kapoor GS, Laporte MG, Rippin SR, Deng Y, Hendi MS, Tirunahari PK, Lee YH, Haimowitz T, Alexander MD, Graham MA, Weng D, Shi Y, McKinlay MA, Chunduru SK (2014) Birinapant (TL32711), a bivalent SMAC mimetic, targets TRAF2-associated cIAPs, abrogates TNF-induced NF-kappaB activation, and is active in patient-derived xenograft models. Mol Cancer Ther 13:867–879. doi:10.1158/1535-7163.MCT-13-0798

Blanc JL, Wager M, Guilhot J, Kusy S, Bataille B, Chantereau T, Lapierre F, Larsen CJ, Karayan-Tapon L (2004) Correlation of clinical features and methylation status of MGMT gene promoter in glioblastomas. J Neurooncol 68:275–283. doi:10.1023/B:NEON.0000033385.37098.85

Brock G, Castellanos-Rizaldos E, Hu L, Coticchia C, Skog J (2015) Liquid biopsy for cancer screening, patient stratification and monitoring. Transl Cancer Res 4:280–290. doi:10.3978/j.issn.2218-676X.2015.06.05

Byers LA, Diao L, Wang J, Saintigny P, Girard L, Peyton M, Shen L, Fan Y, Giri U, Tumula PK, Nilsson MB, Gudikote J, Tran H, Cardnell RJ, Bearss DJ, Warner SL, Foulks JM, Kanner SB, Gandhi V, Krett N, Rosen ST, Kim ES, Herbst RS, Blumenschein GR, Lee JJ, Lippman SM, Ang KK, Mills GB, Hong WK, Weinstein JN, Wistuba II, Coombes KR, Minna JD, Heymach JV (2013) An epithelial-mesenchymal transition gene signature predicts resistance to EGFR and PI3K inhibitors and identifies Axl as a therapeutic target for overcoming EGFR inhibitor resistance. Clin Cancer Res 19:279–290. doi:10.1158/1078-0432.CCR-12-1558

Cai Q, Xiao B, Si H, Cervini A, Gao J, Lu J, Upadhyay SK, Verma SC, Robertson ES (2012) Kaposi's sarcoma herpesvirus upregulates Aurora A expression to promote p53 phosphorylation and ubiquitylation. PLoS Pathog 8(3), e1002566. doi:10.1371/journal.ppat.1002566

Cancer Information web-page of National Cancer Institute (NCI). http://www.cancer.gov/about-cancer/treatment/types

Chan DA, Giaccia AJ (2011) Harnessing synthetic lethal interactions in anticancer drug discovery. Nat Rev Drug Discov 10:351–364. doi:10.1038/nrd3374

Chang L, Graham PH, Hao J, Ni J, Bucci J, Cozzi PJ, Kearsley JH, Li Y (2013) Acquisition of epithelial-mesenchymal transition and cancer stem cell phenotypes is associated with activation of the PI3K/Akt/mTOR pathway in prostate cancer radioresistance. Cell Death Dis 4, e875. doi:10.1038/cddis.2013.407

Chen DJ, Huerta S (2009) Smac mimetics as new cancer therapeutics. Anticancer Drugs 20:646–658. doi:10.1097/CAD.0b013e32832ced78

Chernew ME (2014) The American Society of Clinical Oncology Annual Meeting. Am J of Managed Care SP354–SP355.

Choi CH (2005) ABC transporters as multidrug resistance mechanisms and the development of chemosensitizers for their reversal. Cancer Cell Int 5:30. doi:10.1186/1475-2867-5-30

Chorawala MR, Oza PM, Shah GB (2012) Mechanisms of anticancer drugs resistance: an overview. Int J Pharmaceut Sci Drug Res 4:1–10. doi:10.3389/fphar.2013.00028

Cooper G (2000) The cell: a molecular approach. Sinauer Associates Inc., Sunderland, MA

Costa DB, Halmos B, Kumar A, Schumer ST, Huberman MS, Boggon TJ, Tenen DG, Kobayashi S (2007) BIM mediates EGFR tyrosine kinase inhibitor-induced apoptosis in lung cancers with oncogenic EGFR mutations. PLoS Med 4:1669–1679. doi:10.1371/journal.pmed.0040315, discussion 1680

Curtin NJ (2012) DNA repair dysregulation from cancer driver to therapeutic target. Nat Rev Cancer 12:801–817. doi:10.1038/nrc3399

de Martel C, Ferlay J, Franceschi S, Vignat J, Bray F, Forman D, Plummer M (2012) Global burden of cancers attributable to infections in 2008: a review and synthetic analysis. Lancet Oncol 13:607–615. doi:10.1016/S1470-2045(12)70137-7

Druker BJ, Guilhot F, O'Brien SG, Gathmann I, Kantarjian H, Gattermann N, Deininger MW, Silver RT, Goldman JM, Stone RM, Cervantes F, Hochhaus A, Powell BL, Gabrilove JL, Rousselot P, Reiffers J, Cornelissen JJ, Hughes T, Agis H, Fischer T, Verhoef G, Shepherd J, Saglio G, Gratwohl A, Nielsen JL, Radich JP, Simonsson B, Taylor K, Baccarani M, So C, Letvak L, Larson RA (2006) Five-year follow-up of patients receiving imatinib for chronic myeloid leukemia. N Engl J Med 355:2408–2417. doi:10.1056/NEJMoa062867

Edwards SL, Brough R, Lord CJ, Natrajan R, Vatcheva R, Levine DA, Boyd J, Reis-Filho JS, Ashworth A (2008) Resistance to therapy caused by intragenic deletion in BRCA2. Nature 451:1111–1115. doi:10.1038/nature06548

Enoch T, Norbury C (1995) Cellular responses to DNA damage: cell-cycle checkpoints, apoptosis and the roles of p53 and ATM. Trends Biochem Sci 20:426–430. doi:10.1016/S0968-0004(00)89093-3

Faber AC, Corcoran RB, Ebi H, Sequist LV, Waltman BA, Chung E, Incio J, Digumarthy SR, Pollack SF, Song Y, Muzikansky A, Lifshits E, Roberge S, Coffman EJ, Benes CH, Gomez HL, Baselga J, Arteaga CL, Rivera MN, Dias-Santagata D, Jain RK, Engelman JA (2011) BIM expression in treatment-naive cancers predicts responsiveness to kinase inhibitors. Cancer Discov 1(4):352–365. doi:10.1158/2159-8290.CD-11-0106

Farmer H, McCabe N, Lord CJ, Tutt AN, Johnson DA, Richardson TB, Santarosa M, Dillon KJ, Hickson I, Knights C, Martin NM, Jackson SP, Smith GC, Ashworth A (2005) Targeting the DNA repair defect in BRCA mutant cells as a therapeutic strategy. Nature 434:917–921. doi:10.1038/nature03445

Fuchs BC, Fujii T, Dorfman JD, Goodwin JM, Zhu AX, Lanuti M, Tanabe KK (2008) Epithelial-to-mesenchymal transition and integrin-linked kinase mediate sensitivity to epidermal growth factor receptor inhibition in human hepatoma cells. Cancer Res 68:2391–2399. doi:10.1158/0008-5472.CAN-07-2460

Fukuoka M, Wu YL, Thongprasert S, Sunpaweravong P, Leong SS, Sriuranpong V, Chao TY, Nakagawa K, Chu DT, Saijo N, Duffield EL, Rukazenkov Y, Speake G, Jiang H, Armour AA, To KF, Yang JC, Mok TS (2011) Biomarker analyses and final overall survival results from a phase III, randomized, open-label, first-line study of gefitinib versus carboplatin/paclitaxel in clinically selected patients with advanced non-small-cell lung cancer in Asia (IPASS). J Clin Oncol 29:2866–2874. doi:10.1200/JCO.2010.33.4235

Gagnon JF, Bernard O, Villeneuve L, Tetu B, Guillemette C (2006) Irinotecan inactivation is modulated by epigenetic silencing of UGT1A1 in colon cancer. Clin Cancer Res 12:1850–1858. doi:10.1158/1078-0432.CCR-05-2130

Garnett MJ, Edelman EJ, Heidorn SJ, Greenman CD, Dastur A, Lau KW, Greninger P, Thompson IR, Luo X, Soares J, Liu Q, Iorio F, Surdez D, Chen L, Milano RJ, Bignell GR, Tam AT, Davies H, Stevenson JA, Barthorpe S, Lutz SR, Kogera F, Lawrence K, McLaren-Douglas A, Mitropoulos X, Mironenko T, Thi H, Richardson L, Zhou W, Jewitt F, Zhang T, O'Brien P, Boisvert JL, Price S, Hur W, Yang W, Deng X, Butler A, Choi HG, Chang JW, Baselga J, Stamenkovic I, Engelman JA, Sharma SV, Delattre O, Saez-Rodriguez J, Gray NS, Settleman J, Futreal PA, Haber DA, Stratton MR, Ramaswamy S, McDermott U, Benes CH (2012) Systematic identification of genomic markers of drug sensitivity in cancer cells. Nature 483:570–575. doi:10.1038/nature11005

Goldsby RA, Kindt TJ, Kube J, Osborne BA (2002) Immunology. W. H. Freeman & Company, New York , NY

Gomez-Casal R, Bhattacharya C, Ganesh N, Bailey L, Basse P, Gibson M, Epperly M, Levina V (2013) Non-small cell lung cancer cells survived ionizing radiation treatment display cancer stem cell and epithelial-mesenchymal transition phenotypes. Mol Cancer 12:94. doi:10.1186/1476-4598-12-94

Gong Y, Somwar R, Politi K, Balak M, Chmielecki J, Jiang X, Pao W (2007) Induction of BIM is essential for apoptosis triggered by EGFR kinase inhibitors in mutant EGFR-dependent lung adenocarcinomas. PLoS Med 4, e294. doi:10.1371/journal.pmed.0040294

Gottesman MM (2002) Mechanisms of cancer drug resistance. Annu Rev Med 53:615–627. doi:10.1146/annurev.med.53.082901.103929

Haber M, Smith J, Bordow SB, Flemming C, Cohn SL, London WB, Marshall GM, Norris MD (2006) Association of high-level MRP1 expression with poor clinical outcome in a large prospective study of primary neuroblastoma. J Clin Oncol 24:1546–1553. doi:10.1200/JCO.2005.01.6196

Hanahan D, Weinberg RA (2000) The hallmarks of cancer. Cell 100:57–70. doi:10.1016/S0092-8674(00)81683-9

Hanahan D, Weinberg RA (2011) Hallmarks of cancer: the next generation. Cell 144:646–674. doi:10.1016/j.cell.2011.02.013

Hegi ME, Diserens AC, Gorlia T, Hamou MF, de Tribolet N, Weller M, Kros JM, Hainfellner JA, Mason W, Mariani L, Bromberg JE, Hau P, Mirimanoff RO, Cairncross JG, Janzer RC, Stupp R (2005) MGMT gene silencing and benefit from temozolomide in glioblastoma. N Engl J Med 352:997–1003. doi:10.1056/NEJMoa043331

Hinds M, Deisseroth K, Mayes J, Altschuler E, Jansen R, Ledley FD, Zwelling LA (1991) Identification of a point mutation in the topoisomerase II gene from a human leukemia cell line containing an amsacrine-resistant form of topoisomerase II. Cancer Res 51:4729–4731

Holohan C, Van Schaeybroeck S, Longley DB, Johnston PG (2013) Cancer drug resistance: an evolving paradigm. Nat Rev Cancer 13:714–726. doi:10.1038/nrc3599

Housman G, Byler S, Heerboth S, Lapinska K, Longacre M, Snyder N, Sarkar S (2014) Drug resistance in cancer: an overview. Cancers (Basel) 6:1769–1792. doi:10.3390/cancers6031769

Huang S, Holzel M, Knijnenburg T, Schlicker A, Roepman P, McDermott U, Garnett M, Grernrum W, Sun C, Prahallad A, Groenendijk FH, Mittempergher L, Nijkamp W, Neefjes J, Salazar R, Ten Dijke P, Uramoto H, Tanaka F, Beijersbergen RL, Wessels LF, Bernards R (2012) MED12 controls the response to multiple cancer drugs through regulation of TGF-beta receptor signaling. Cell 151:937–950. doi:10.1016/j.cell.2012.10.035

Hunter AM, LaCasse EC, Korneluk RG (2007) The inhibitors of apoptosis (IAPs) as cancer targets. Apoptosis 12:1543–1568. doi:10.1007/s10495-007-0087-3

Jackman DM, Miller VA, Cioffredi LA, Yeap BY, Janne PA, Riely GJ, Ruiz MG, Giaccone G, Sequist LV, Johnson BE (2009) Impact of epidermal growth factor receptor and KRAS mutations on clinical outcomes in previously untreated non-small cell lung cancer patients: results of an online tumor registry of clinical trials. Clin Cancer Res 15:5267–5273. doi:10.1158/1078-0432.CCR-09-0888

Janne PA, Johnson BE (2006) Effect of epidermal growth factor receptor tyrosine kinase domain mutations on the outcome of patients with non-small cell lung cancer treated with epidermal growth factor receptor tyrosine kinase inhibitors. Clin Cancer Res 12:4416s–4420s. doi:10.1158/1078-0432.CCR-06-0555

Janne PA, Ramalingam SS, Yang JC (2014) Clinical activity of the mutant-selective EGFR inhibitor AZD9291 in patients with EGFR inhibitor-resistant non-small cell lung cancer. J Clin Oncol 32(Suppl 15s):abstr 8009

Jha HC, Upadhyay SK, AJP M, Lu J, Cai Q, Saha A, Robertson ES (2013) H2AX phosphorylation is important for LANA-mediated Kaposi's sarcoma-associated herpesvirus episome persistence. J Virol 87(9):5255–5269. doi:10.1128/JVI.03575-12

Kaelin WG (2005) The concept of synthetic lethality in the context of anticancer therapy. Nat Rev Cancer 5:689–698. doi:10.1038/nrc1691

Kang J, Kim E, Kim W, Seong KM, Youn H, Kim JW, Kim J, Youn B (2013) Rhamnetin and cirsiliol induce radiosensitization and inhibition of epithelial-mesenchymal transition (EMT) by miR-34a-mediated suppression of Notch-1 expression in non-small cell lung cancer cell lines. J Biol Chem 288:27343–27357. doi:10.1074/jbc.M113.490482

Karachaliou N, Mayo-de-Las-Casas C, Molina-Vila MA, Rosell R (2015) Real-time liquid biopsies become a reality in cancer treatment. Ann Transl Med 3:36. doi:10.3978/j.issn.2305-5839.2015.01.16

Karaman MW, Herrgard S, Treiber DK, Gallant P, Atteridge CE, Campbell BT, Chan KW, Ciceri P, Davis MI, Edeen PT, Faraoni R, Floyd M, Hunt JP, Lockhart DJ, Milanov ZV, Morrison MJ, Pallares G, Patel HK, Pritchard S, Wodicka LM, Zarrinkar PP (2008) A quantitative analysis of kinase inhibitor selectivity. Nat Biotechnol 26:127–132. doi:10.1038/nbt1358

Kern DH, Weisenthal LM (1990) Highly specific prediction of antineoplastic drug resistance with an in vitro assay using suprapharmacologic drug exposures. J Natl Cancer Inst 82:582–588

Kirschner K, Melton DW (2010) Multiple roles of the ERCC1-XPF endonuclease in DNA repair and resistance to anticancer drugs. Anticancer Res 30:3223–3232

Kitada S, Andersen J, Akar S, Zapata JM, Takayama S, Krajewski S, Wang HG, Zhang X, Bullrich F, Croce CM, Rai K, Hines J, Reed JC (1998) Expression of apoptosis-regulating proteins in chronic lymphocytic leukemia: correlations with in vitro and in vivo chemoresponses. Blood 91:3379–3389

Kobayashi S, Boggon TJ, Dayaram T, Janne PA, Kocher O, Meyerson M, Johnson BE, Eck MJ, Tenen DG, Halmos B (2005) EGFR mutation and resistance of non-small-cell lung cancer to gefitinib. N Engl J Med 352:786–792. doi:10.1056/NEJMoa044238

Konopleva M, Contractor R, Tsao T, Samudio I, Ruvolo PP, Kitada S, Deng X, Zhai D, Shi YX, Sneed T, Verhaegen M, Soengas M, Ruvolo VR, McQueen T, Schober WD, Watt JC, Jiffar T, Ling X, Marini FC, Harris D, Dietrich M, Estrov Z, McCubrey J, May WS, Reed JC, Andreeff M (2006) Mechanisms of apoptosis sensitivity and resistance to the BH3 mimetic ABT-737 in acute myeloid leukemia. Cancer Cell 10:375–388. doi:10.1016/j.ccr.2006.10.006

Kufe DW, Spriggs DR (1985) Biochemical and cellular pharmacology of cytosine arabinoside. Semin Oncol 12:34–48

Kuribara R, Honda H, Matsui H, Shinjyo T, Inukai T, Sugita K, Nakazawa S, Hirai H, Ozawa K, Inaba T (2004) Roles of Bim in apoptosis of normal and Bcr-Abl-expressing hematopoietic progenitors. Mol Cell Biol 24:6172–6183. doi:10.1128/MCB.24.14.6172-6183.2004

Kwak EL, Sordella R, Bell DW, Godin-Heymann N, Okimoto RA, Brannigan BW, Harris PL, Driscoll DR, Fidias P, Lynch TJ, Rabindran SK, McGinnis JP, Wissner A, Sharma SV, Isselbacher KJ, Settleman J, Haber DA (2005) Irreversible inhibitors of the EGF receptor may circumvent acquired resistance to gefitinib. Proc Natl Acad Sci U S A 102:7665–7670. doi:10.1073/pnas.0502860102

Kwon HC, Roh MS, Oh SY, Kim SH, Kim MC, Kim JS, Kim HJ (2007) Prognostic value of expression of ERCC1, thymidylate synthase, and glutathione S-transferase P1 for 5-fluorouracil/oxaliplatin chemotherapy in advanced gastric cancer. Ann Oncol 18:504–509. doi:10.1093/annonc/mdl430

Lan L, Luo Y, Cui D, Shi BY, Deng W, Huo LL, Chen HL, Zhang GY, Deng LL (2013) Epithelial-mesenchymal transition triggers cancer stem cell generation in human thyroid cancer cells. Int J Oncol 43:113–120. doi:10.3892/ijo.2013.1913

Lane DP (1992) Cancer. p53, guardian of the genome. Nature 358:15–16. doi:10.1038/358015a0

Lee HJ, Schaefer G, Heffron TP, Shao L, Ye X, Sideris S, Malek S, Chan E, Merchant M, La H, Ubhayakar S, Yauch RL, Pirazzoli V, Politi K, Settleman J (2013) Noncovalent wild-type-sparing inhibitors of EGFR T790M. Cancer Discov 3:168–181. doi:10.1158/2159-8290.CD-12-0357

Letai AG (2008) Diagnosing and exploiting cancer's addiction to blocks in apoptosis. Nat Rev Cancer 8:121–132. doi:10.1038/nrc2297

Lin X, Morgan-Lappe S, Huang X, Li L, Zakula DM, Vernetti LA, Fesik SW, Shen Y (2007) 'Seed' analysis of off-target siRNAs reveals an essential role of Mcl-1 in resistance to the small-molecule Bcl-2/Bcl-XL inhibitor ABT-737. Oncogene 26:3972–3979. doi:10.1038/sj.onc.1210166

Lippert TH, Ruoff HJ, Volm M (2011) Current status of methods to assess cancer drug resistance. Int J Med Sci 8:245–253. doi:10.7150/ijms.8.245

Lord RV, Brabender J, Gandara D, Alberola V, Camps C, Domine M, Cardenal F, Sanchez JM, Gumerlock PH, Taron M, Sanchez JJ, Danenberg KD, Danenberg PV, Rosell R (2002) Low ERCC1 expression correlates with prolonged survival after cisplatin plus gemcitabine chemotherapy in non-small cell lung cancer. Clin Cancer Res 8:2286–2291

Mani SA, Guo W, Liao MJ, Eaton EN, Ayyanan A, Zhou AY, Brooks M, Reinhard F, Zhang CC, Shipitsin M, Campbell LL, Polyak K, Brisken C, Yang J, Weinberg RA (2008) The epithelial-mesenchymal transition generates cells with properties of stem cells. Cell 133:704–715. doi:10.1016/j.cell.2008.03.027

Martin SA, McCarthy A, Barber LJ, Burgess DJ, Parry S, Lord CJ, Ashworth A (2009) Methotrexate induces oxidative DNA damage and is selectively lethal to tumour cells with defects in the DNA mismatch repair gene MSH2. EMBO Mol Med 1:323–337. doi:10.1002/emmm.200900040

Masters GA, Krilov L, Bailey HH, Brose MS, Burstein H, Diller LR, Dizon DS, Fine HA, Kalemkerian GP, Moasser M, Neuss MN, O'Day SJ, Odenike O, Ryan CJ, Schilsky RL, Schwartz GK, Venook AP, Wong SL, Patel JD (2015) Clinical cancer advances 2015: annual report on progress against cancer from the American Society of Clinical Oncology. J Clin Oncol 33:786–809. doi:10.1200/JCO.2014.59.9746

Meidhof S, Brabletz S, Lehmann W, Preca BT, Mock K, Ruh M, Schuler J, Berthold M, Weber A, Burk U, Lubbert M, Puhr M, Culig Z, Wellner U, Keck T, Bronsert P, Kusters S, Hopt UT, Stemmler MP, Brabletz T (2015) ZEB1-associated drug resistance in cancer cells is reversed by the class I HDAC inhibitor mocetinostat. EMBO Mol Med 7:831–847. doi:10.15252/emmm.201404396

Miller VA, Hirsh V, Cadranel J, Chen YM, Park K, Kim SW, Zhou C, Su WC, Wang M, Sun Y, Heo DS, Crino L, Tan EH, Chao TY, Shahidi M, Cong XJ, Lorence RM, Yang JC (2012) Afatinib versus placebo for patients with advanced, metastatic non-small-cell lung cancer after failure of erlotinib, gefitinib, or both, and one or two lines of chemotherapy (LUX-Lung 1): a phase 2b/3 randomised trial. Lancet Oncol 13:528–538. doi:10.1016/S1470-2045(12)70087-6

Miyashita T, Reed JC (1992) bcl-2 gene transfer increases relative resistance of S49.1 and WEHI7.2 lymphoid cells to cell death and DNA fragmentation induced by glucocorticoids and multiple chemotherapeutic drugs. Cancer Res 52:5407–5411

Mok TS, Wu YL, Thongprasert S, Yang CH, Chu DT, Saijo N, Sunpaweravong P, Han B, Margono B, Ichinose Y, Nishiwaki Y, Ohe Y, Yang JJ, Chewaskulyong B, Jiang H, Duffield EL, Watkins CL, Armour AA, Fukuoka M (2009) Gefitinib or carboplatin-paclitaxel in pulmonary adenocarcinoma. N Engl J Med 361:947–957. doi:10.1056/NEJMoa0810699

Mukherjee S (2010) The emperor of all maladies: a biography of cancer by Siddhartha Mukherjee. Scribner, New York , NY

National Cancer Institute (NCI) website. http://www.who.int/cancer/en/

Ng KP, Hillmer AM, Chuah CT, Juan WC, Ko TK, Teo AS, Ariyaratne PN, Takahashi N, Sawada K, Fei Y, Soh S, Lee WH, Huang JW, Allen JC Jr, Woo XY, Nagarajan N, Kumar V, Thalamuthu A, Poh WT, Ang AL, Mya HT, How GF, Yang LY, Koh LP, Chowbay B, Chang CT, Nadarajan VS, Chng WJ, Than H, Lim LC, Goh YT, Zhang S, Poh D, Tan P, Seet JE, Ang MK, Chau NM, Ng QS, Tan DS, Soda M, Isobe K, Nothen MM, Wong TY, Shahab A, Ruan X, Cacheux-Rataboul V, Sung WK, Tan EH, Yatabe Y, Mano H, Soo RA, Chin TM, Lim WT, Ruan Y, Ong ST (2012) A common BIM deletion polymorphism mediates intrinsic resistance and inferior responses to tyrosine kinase inhibitors in cancer. Nat Med 18:521–528. doi:10.1038/nm.2713

Ni Chonghaile T, Sarosiek KA, Vo TT, Ryan JA, Tammareddi A, Moore Vdel G, Deng J, Anderson KC, Richardson P, Tai YT, Mitsiades CS, Matulonis UA, Drapkin R, Stone R, Deangelo DJ, McConkey DJ, Sallan SE, Silverman L, Hirsch MS, Carrasco DR, Letai A (2011) Pretreatment mitochondrial priming correlates with clinical response to cytotoxic chemotherapy. Science 334:1129–1133. doi:10.1126/science.1206727

Oltersdorf T, Elmore SW, Shoemaker AR, Armstrong RC, Augeri DJ, Belli BA, Bruncko M, Deckwerth TL, Dinges J, Hajduk PJ, Joseph MK, Kitada S, Korsmeyer SJ, Kunzer AR, Letai A, Li C, Mitten MJ, Nettesheim DG, Ng S, Nimmer PM, O'Connor JM, Oleksijew A, Petros AM, Reed JC, Shen W, Tahir SK, Thompson CB, Tomaselli KJ, Wang B, Wendt MD, Zhang H, Fesik SW, Rosenberg SH (2005) An inhibitor of Bcl-2 family proteins induces regression of solid tumours. Nature 435:677–681. doi:10.1038/nature03579

Pao W, Miller VA, Politi KA, Riely GJ, Somwar R, Zakowski MF, Kris MG, Varmus H (2005) Acquired resistance of lung adenocarcinomas to gefitinib or erlotinib is associated with a second mutation in the EGFR kinase domain. PLoS Med 2, e73. doi:10.1371/journal.pmed.0020073

Persidis A (1999) Cancer multidrug resistance. Nat Biotechnol 17(1):94–95. doi:10.1038/5289

Peters WH, Roelofs HM (1997) A mutation in exon 7 of the human cytochrome P-4501A1 gene as marker for sensitivity to anti-cancer drugs? Br J Cancer 75:1397

Polyak K, Weinberg RA (2009) Transitions between epithelial and mesenchymal states: acquisition of malignant and stem cell traits. Nat Rev Cancer 9(4):265–273. doi:10.1038/nrc2620

Rabik CA, Fishel ML, Holleran JL, Kasza K, Kelley MR, Egorin MJ, Dolan ME (2008) Enhancement of cisplatin [cis-diammine dichloroplatinum (II)] cytotoxicity by O6-benzylguanine involves endoplasmic reticulum stress. J Pharmacol Exp Ther 327:442–452. doi:10.1124/jpet.108.141291

Regales L, Gong Y, Shen R, de Stanchina E, Vivanco I, Goel A, Koutcher JA, Spassova M, Ouerfelli O, Mellinghoff IK, Zakowski MF, Politi KA, Pao W (2009) Dual targeting of EGFR can overcome a major drug resistance mutation in mouse models of EGFR mutant lung cancer. J Clin Invest 119:3000–3010. doi:10.1172/JCI38746

Ren J, Chen Y, Song H, Chen L, Wang R (2013) Inhibition of ZEB1 reverses EMT and chemoresistance in docetaxel-resistant human lung adenocarcinoma cell line. J Cell Biochem 114:1395–1403. doi:10.1002/jcb.24481

Rosell R, Moran T, Queralt C, Porta R, Cardenal F, Camps C, Majem M, Lopez-Vivanco G, Isla D, Provencio M, Insa A, Massuti B, Gonzalez-Larriba JL, Paz-Ares L, Bover I, Garcia-Campelo-R, Moreno MA, Catot S, Rolfo C, Reguart N, Palmero R, Sanchez JM, Bastus R, Mayo C, Bertran-Alamillo J, Molina MA, Sanchez JJ, Taron M (2009) Screening for epidermal growth factor receptor mutations in lung cancer. N Engl J Med 361:958–967. doi:10.1056/NEJMoa0904554

Rosell R, Carcereny E, Gervais R, Vergnenegre A, Massuti B, Felip E, Palmero R, Garcia-Gomez R, Pallares C, Sanchez JM, Porta R, Cobo M, Garrido P, Longo F, Moran T, Insa A, De Marinis F, Corre R, Bover I, Illiano A, Dansin E, de Castro J, Milella M, Reguart N, Altavilla G, Jimenez U, Provencio M, Moreno MA, Terrasa J, Munoz-Langa J, Valdivia J, Isla D, Domine M, Molinier O, Mazieres J, Baize N, Garcia-Campelo R, Robinet G, Rodriguez-Abreu D, Lopez-Vivanco G, Gebbia V, Ferrera-Delgado L, Bombaron P, Bernabe R, Bearz A, Artal A, Cortesi E, Rolfo C, Sanchez-Ronco M, Drozdowskyj A,

Queralt C, de Aguirre I, Ramirez JL, Sanchez JJ, Molina MA, Taron M, Paz-Ares L (2012) Erlotinib versus standard chemotherapy as first-line treatment for European patients with advanced EGFR mutation-positive non-small-cell lung cancer (EURTAC): a multicentre, open-label, randomised phase 3 trial. Lancet Oncol 13:239–246. doi:10.1016/S1470-2045(11)70393-X

Sakai W, Swisher EM, Karlan BY, Agarwal MK, Higgins J, Friedman C, Villegas E, Jacquemont C, Farrugia DJ, Couch FJ, Urban N, Taniguchi T (2008) Secondary mutations as a mechanism of cisplatin resistance in BRCA2-mutated cancers. Nature 451:1116–1120. doi:10.1038/nature06633

Sampath D, Cortes J, Estrov Z, Du M, Shi Z, Andreeff M, Gandhi V, Plunkett W (2006) Pharmacodynamics of cytarabine alone and in combination with 7-hydroxystaurosporine (UCN-01) in AML blasts in vitro and during a clinical trial. Blood 107:2517–2524. doi:10.1182/blood-2005-08-3351

Saxena M, Stephens MA, Pathak H, Rangarajan A (2011) Transcription factors that mediate epithelial-mesenchymal transition lead to multidrug resistance by upregulating ABC transporters. Cell Death Dis 2, e179. doi:10.1038/cddis.2011.61

Sentman CL, Shutter JR, Hockenbery D, Kanagawa O, Korsmeyer SJ (1991) bcl-2 inhibits multiple forms of apoptosis but not negative selection in thymocytes. Cell 67:879–888. doi:10.1016/0092-8674(91)90361-2

Seo JS, Ju YS, Lee WC, Shin JY, Lee JK, Bleazard T, Lee J, Jung YJ, Kim JO, Yu SB, Kim J, Lee ER, Kang CH, Park IK, Rhee H, Lee SH, Kim JI, Kang JH, Kim YT (2012) The transcriptional landscape and mutational profile of lung adenocarcinoma. Genome Res 22:2109–2119. doi:10.1101/gr.145144.112

Sequist LV, Waltman BA, Dias-Santagata D, Digumarthy S, Turke AB, Fidias P, Bergethon K, Shaw AT, Gettinger S, Cosper AK, Akhavanfard S, Heist RS, Temel J, Christensen JG, Wain JC, Lynch TJ, Vernovsky K, Mark EJ, Lanuti M, Iafrate AJ, Mino-Kenudson M, Engelman JA (2011) Genotypic and histological evolution of lung cancers acquiring resistance to EGFR inhibitors. Sci Transl Med 3:75ra26. doi:10.1126/scitranslmed.3002003

Sequist LV, Soria JC, Gadgeel SM (2014) First-in-human evaluation of CO-1686, an irreversible, highly selective tyrosine kinase inhibitor of mutations of EGFR (activating and T790M). J Clin Oncol 32(Suppl 15s):abstr 8010

Sharma SV, Bell DW, Settleman J, Haber DA (2007) Epidermal growth factor receptor mutations in lung cancer. Nat Rev Cancer 7(3):169–181. doi:10.1038/nrc2088

Singh A, Settleman J (2010) EMT, cancer stem cells and drug resistance: an emerging axis of evil in the war on cancer. Oncogene 29:4741–4751. doi:10.1038/onc.2010.215

Singh P, Upadhyay SK, Gautam P (2015) Protein-based candidate biomarkers for gallbladder carcinoma (manuscript in communication)

Stewart BW, Wild CP (eds) (2014) World cancer report 2014. International Agency for Research on Cancer (IARC)

Stuckey DW, Shah K (2013) TRAIL on trial: preclinical advances in cancer therapy. Trends Mol Med 19:685–694. doi:10.1016/j.molmed.2013.08.007

Sullivan NJ, Sasser AK, Axel AE, Vesuna F, Raman V, Ramirez N, Oberyszyn TM, Hall BM (2009) Interleukin-6 induces an epithelial-mesenchymal transition phenotype in human breast cancer cells. Oncogene 28:2940–2947. doi:10.1038/onc.2009.180

Swanton C (2012) Intratumor heterogeneity: evolution through space and time. Cancer Res 72:4875–4882. doi:10.1158/0008-5472.CAN-12-2217

Theys J, Jutten B, Habets R, Paesmans K, Groot AJ, Lambin P, Wouters BG, Lammering G, Vooijs M (2011) E-Cadherin loss associated with EMT promotes radioresistance in human tumor cells. Radiother Oncol 99:392–397. doi:10.1016/j.radonc.2011.05.044

Thomson S, Buck E, Petti F, Griffin G, Brown E, Ramnarine N, Iwata KK, Gibson N, Haley JD (2005) Epithelial to mesenchymal transition is a determinant of sensitivity of non-small-cell lung carcinoma cell lines and xenografts to epidermal growth factor receptor inhibition. Cancer Res 65:9455–9462. doi:10.1158/0008-5472.CAN-05-1058

Upadhyay SK, Jha HC, Saha A, Robertson ES (2012) Lymphocryptoviruses: EBV and its role in human cancer. In: Robertson ES (ed) Cancer associated viruses. Springer, New York, NY. doi:10.1007/978-1-4614-0016-5. ISBN 978-1-4419-9999-3

Usanova S, Piee-Staffa A, Sied U, Thomale J, Schneider A, Kaina B, Koberle B (2010) Cisplatin sensitivity of testis tumour cells is due to deficiency in interstrand-crosslink repair and low ERCC1-XPF expression. Mol Cancer 9:248. doi:10.1186/1476-4598-9-248

van Delft MF, Wei AH, Mason KD, Vandenberg CJ, Chen L, Czabotar PE, Willis SN, Scott CL, Day CL, Cory S, Adams JM, Roberts AW, Huang DC (2006) The BH3 mimetic ABT-737 targets selective Bcl-2 proteins and efficiently induces apoptosis via Bak/Bax if Mcl-1 is neutralized. Cancer Cell 10:389–399. doi:10.1016/j.ccr.2006.08.027

Vasiliou V, Vasiliou K, Nebert DW (2009) Human ATP-binding cassette (ABC) transporter family. Hum Genomics 3:281–290. doi:10.1186/1479-7364-3-3-281

Walter AO, Sjin RT, Haringsma HJ, Ohashi K, Sun J, Lee K, Dubrovskiy A, Labenski M, Zhu Z, Wang Z, Sheets M, St Martin T, Karp R, van Kalken D, Chaturvedi P, Niu D, Nacht M, Petter RC, Westlin W, Lin K, Jaw-Tsai S, Raponi M, Van Dyke T, Etter J, Weaver Z, Pao W, Singh J, Simmons AD, Harding TC, Allen A (2013) Discovery of a mutant-selective covalent inhibitor of EGFR that overcomes T790M-mediated resistance in NSCLC. Cancer Discov 3:1404–1415. doi:10.1158/2159-8290.CD-13-0314

Wang GQ, Gastman BR, Wieckowski E, Goldstein LA, Gambotto A, Kim TH, Fang B, Rabinovitz A, Yin XM, Rabinowich H (2001) A role for mitochondrial Bak in apoptotic response to anticancer drugs. J Biol Chem 276:34307–34317. doi:10.1074/jbc.M103526200

West H, Lilenbaum R, Harpole D, Wozniak A, Sequist L (2009) Molecular analysis-based treatment strategies for the management of non-small cell lung cancer. J Thorac Oncol 4: S1029–S1039, quiz S1041–1022. doi:10.1097/JTO.0b013e3181b27170

Wilson TR, Longley DB, Johnston PG (2006) Chemoresistance in solid tumours. Ann Oncol 17 (Suppl 10):x315–x324. doi:10.1093/annonc/mdl280

Wilson C, Ye X, Pham T, Lin E, Chan S, McNamara E, Neve RM, Belmont L, Koeppen H, Yauch RL, Ashkenazi A, Settleman J (2014) AXL inhibition sensitizes mesenchymal cancer cells to antimitotic drugs. Cancer Res 74:5878–5890. doi:10.1158/0008-5472.CAN-14-1009

Witta SE, Gemmill RM, Hirsch FR, Coldren CD, Hedman K, Ravdel L, Helfrich B, Dziadziuszko R, Chan DC, Sugita M, Chan Z, Baron A, Franklin W, Drabkin HA, Girard L, Gazdar AF, Minna JD, Bunn PA Jr (2006) Restoring E-cadherin expression increases sensitivity to epidermal growth factor receptor inhibitors in lung cancer cell lines. Cancer Res 66:944–950. doi:10.1158/0008-5472.CAN-05-1988

Xu Y, Villalona-Calero MA (2002) Irinotecan: mechanisms of tumor resistance and novel strategies for modulating its activity. Ann Oncol 13:1841–1851. doi:10.1093/annonc/mdf337

Yanase K, Tsukahara S, Asada S, Ishikawa E, Imai Y, Sugimoto Y (2004) Gefitinib reverses breast cancer resistance protein-mediated drug resistance. Mol Cancer Ther 3:1119–1125

Yang MH, Wu MZ, Chiou SH, Chen PM, Chang SY, Liu CJ, Teng SC, Wu KJ (2008) Direct regulation of TWIST by HIF-1alpha promotes metastasis. Nat Cell Biol 10:295–305. doi:10.1038/ncb1691

Yao Z, Fenoglio S, Gao DC, Camiolo M, Stiles B, Lindsted T, Schlederer M, Johns C, Altorki N, Mittal V, Kenner L, Sordella R (2010) TGF-beta IL-6 axis mediates selective and adaptive mechanisms of resistance to molecular targeted therapy in lung cancer. Proc Natl Acad Sci U S A 107:15535–15540. doi:10.1073/pnas.1009472107

Yauch RL, Januario T, Eberhard DA, Cavet G, Zhu W, Fu L, Pham TQ, Soriano R, Stinson J, Seshagiri S, Modrusan Z, Lin CY, O'Neill V, Amler LC (2005) Epithelial versus mesenchymal phenotype determines in vitro sensitivity and predicts clinical activity of erlotinib in lung cancer patients. Clin Cancer Res 11:8686–8698. doi:10.1158/1078-0432.CCR-05-1492

Yu HA, Arcila ME, Rekhtman N, Sima CS, Zakowski MF, Pao W, Kris MG, Miller VA, Ladanyi M, Riely GJ (2013) Analysis of tumor specimens at the time of acquired resistance to EGFR-TKI therapy in 155 patients with EGFR-mutant lung cancers. Clin Cancer Res 19:2240–2247. doi:10.1158/1078-0432.CCR-12-2246

Yun CH, Mengwasser KE, Toms AV, Woo MS, Greulich H, Wong KK, Meyerson M, Eck MJ (2008) The T790M mutation in EGFR kinase causes drug resistance by increasing the affinity for ATP. Proc Natl Acad Sci U S A 105:2070–2075. doi:10.1073/pnas.0709662105
Zahreddine H, Borden KL (2013) Mechanisms and insights into drug resistance in cancer. Front Pharmacol 4:28. doi:10.3389/fphar.2013.00028
Zhang Z, Lee JC, Lin L, Olivas V, Au V, LaFramboise T, Abdel-Rahman M, Wang X, Levine AD, Rho JK, Choi YJ, Choi CM, Kim SW, Jang SJ, Park YS, Kim WS, Lee DH, Lee JS, Miller VA, Arcila M, Ladanyi M, Moonsamy P, Sawyers C, Boggon TJ, Ma PC, Costa C, Taron M, Rosell R, Halmos B, Bivona TG (2012) Activation of the AXL kinase causes resistance to EGFR-targeted therapy in lung cancer. Nat Genet 44:852–860. doi:10.1038/ng.2330

BCL-2 Proteins and their Role in Cancer Resistance

Hamida Thakur and Abid R. Mattoo

Abstract BCl2 family proteins are major players in regulating the mitochondrial or intrinsic pathway of cell death. The BCL2 proteins are divided into prosurvival, proapoptotic, and BH3-only proteins. The balance between the three classes of proteins regulates the intrinsic or mitochondrial pathway of apoptosis. Defects in the expression or regulation of these proteins have been reported in different cancers. In fact the overexpression of prosurvival members of the BCL2 family has been associated with the resistance of various cancers to current therapies. Currently inhibitors are being developed for prosurvival members of the BCL2 proteins, some of which like ABT-199 have shown a very good response in clinical trials. The early promise shown by these inhibitors in clinical trials has opened avenues for therapeutic intervention of a number of highly resistant cancers, alone or in combination with other currently available therapies.

1 Introduction

Apoptosis is the cellular process of programed cell death to remove unwanted cells and maintain tissue homeostasis in multicellular organisms. The cell death is triggered by a subset of proteases, called caspases which cleave several hundred cellular substrates. There are two pathways of caspase activation which have been identified so far. The extrinsic pathway is activated by involvement of death receptors on the cell surface while as the intrinsic pathway or the mitochondrial pathway involves BCL2 proteins. Various inputs like DNA damage, problems with cell signaling, metabolic stress, and hypoxia can trigger apoptosis by activation of BCL2 proteins. The overexpression of BCL2 family proteins has been associated with

H. Thakur
Department of Biochemistry, Faculty of Medicine, Jawaharlal Nehru Medical College, Aligarh Muslim University, Aligarh, India

A.R. Mattoo (✉)
Department of Radiation Oncology, Methodist Hospital Research Institute, Houston, TX, USA
e-mail: abidmattoo@gmail.com

G. Arora et al. (eds.), *Drug Resistance in Bacteria, Fungi, Malaria, and Cancer*,
DOI 10.1007/978-3-319-48683-3_21

different cancers and is one of the important reasons for cancer resistance to current therapeutic drugs (Yip and Reed 2008). BCL2 family proteins are important players involved in the regulation of cell death and have been shown to be involved in varied cell death mechanisms that include apoptosis, necrosis, and autophagy (Adams et al. 2005; Yip and Reed 2008). Changes in the expression and function of BCL2 family proteins are responsible for the induction and progression of human cancers.

2 Family of BCL2 Proteins

BCL2 family members have been clustered into three classes (Fig. 1). The first class inhibits apoptosis (BCL2, BCL-XL, BCLW, MCL1, BCL2A1), while a second class supports apoptosis (BAX, BAK, and BOK). A third diverse class includes BH3-only proteins like BAD, NOXA, BID, BIK, BMF, BIM, and PUMA. These BH3-only proteins have a conserved BH3 domain that can bind and regulate the BCL2 proteins to support apoptosis. In response to cytoplasmic stress signals, BH3-only proteins regulate the proapoptotic function both at the transcriptional or posttranslational level, either by negating the activity of prosurvival BCL2 family proteins or by directly binding to and activating the proapoptotic proteins BAX and BAK. Antiapoptotic or prosurvival proteins can bind directly to the BH3 domains of active BAK and BAX and thereby inhibit their proapoptotic activity. All of the prosurvival proteins appear to bind BAX, but BAK appears to be inhibited preferably by BCL-XL, A1, and MCL-1 (Shamas-Din et al. 2013). The affinity of BH3-only proteins is variable toward prosurvival proteins. Some like Puma, Bim, and Bid can interact with all the prosurvival molecules, whereas others have a more defined affinity (Chen et al. 2005). Bad binds only to BCL2, BCL-XL, and BCLW but does not bind to MCL-1 or A1. However, BH3-only proteins like Noxa

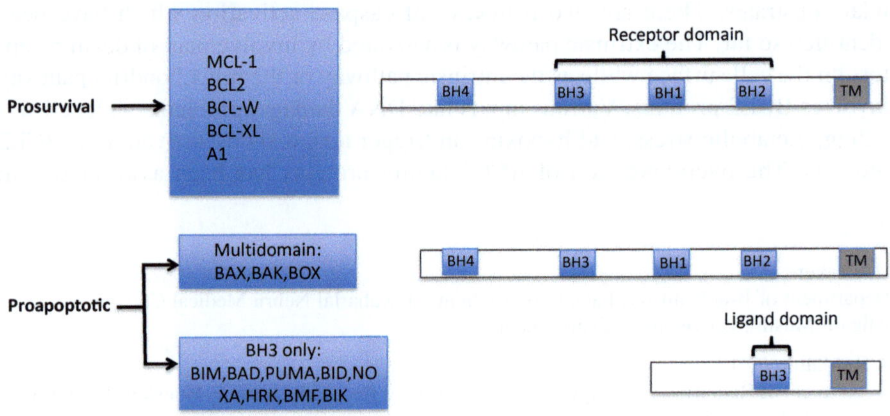

Fig. 1 Classification of BCL2 family proteins

show binding affinity only toward MCL-1 and A1 but not to other prosurvival proteins. The BH3 domain has been shown to be involved in this interaction, and it is through this interaction that BH3-only proteins regulate apoptosis. Moreover, BH3-only proteins like Bid, Puma, and Bim were also shown to interact with BAX (Marani et al. 2002; Willis et al. 2005). BH3 domains of proteins like Bid, Bmf, and Bad achieve the active conformation upon binding to a prosurvival protein (Czabotar et al. 2014).

The three classes of these BCL2 proteins collaborate with each other to regulate the process of apoptosis. There are different models based on the interaction of these proteins which have been put forth to explain how proapoptotic proteins BAK/BAX are regulated and how these proteins regulate apoptosis. The direct activation model is based on the affinities of binding of BH3 proteins to the BCl2 family proteins and based on their affinities they have been characterized into activators or sensitizers (Letai et al. 2002). The BH3 activator proteins—Bim, t-Bid, and Puma—can bind to both proapoptotic and prosurvival proteins. The BH3 sensitizer proteins—Bad, Noxa, Bix, Bmf, Bnip3, and Hrk—bind to prosurvival proteins and release activator BH3 proteins which in turn activate proapoptotic proteins and promote membrane permeabilization. According to this model, the prosurvival proteins bind to both activator and sensitizer BH3 proteins, but they do not form complex with BAK and BAX. Hence, for a cell to protect from apoptosis, prosurvival proteins must sequester the supply of BH3 proteins to prevent BAK/BAX activation and apoptosis.

According to the indirect (displacement) model, BAX and BAK remain blocked by antiapoptotic members of the family, which are displaced by BH3-only proteins upon apoptosis initiation. Among BH3-only proteins, Bid, Bim, and PUMA are believed to be more efficient because of their ability to bind all the prosurvival members of the family (Shamas-Din et al. 2013).

The BAK and BAX proteins once released mediate pro-death function at mitochondrial outer membrane (MOM). They bind to MOM resulting in the permeabilization of the membrane and release of proteins from the mitochondrial intermembrane space like cytochrome c, SMAC/DIABLO, endonuclease G, etc. Cytochrome *c* binds to apoptotic protease-activating factor 1 (APAF1) and activates it, which in turn induces the formation of apoptosome. The apoptosome triggers the recruitment and activation of Caspase 9, which is the initiator caspase. Furthermore, Caspase 9 cleaves and activates caspase 3 and caspase 7 which are considered as the executioner caspases. The release of second mitochondria-derived activator of caspase (SMAC) from mitochondria blocks X-linked inhibitor of apoptosis protein (XIAP)-mediated inhibition of caspase activity (Dewson and Kluck 2009). The difference between these two proapoptotic proteins lies in the fact that BAK has a high affinity for the antiapoptotic proteins, MCL-1 and Bcl-XL (Willis et al. 2005; Shamas-Din et al. 2013). Another difference is that BAK is found constitutively bound to the MOM, whereas BAX is primarily cytosolic but shifts to the MOM after activation of apoptosis. The summary of events of the intrinsic pathway is indicated in Fig. 2.

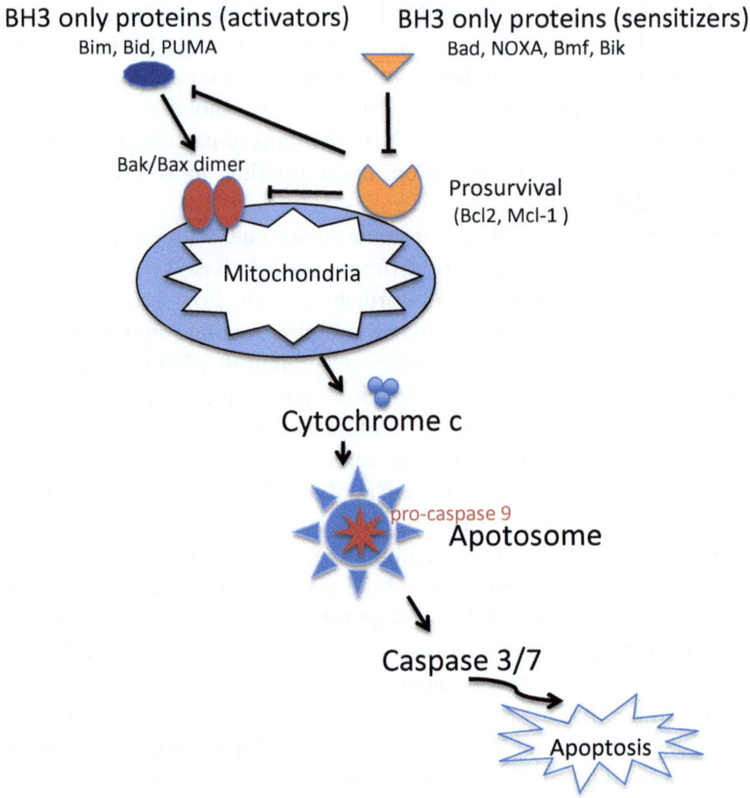

Fig. 2 Mitochondrial pathway of apoptosis

3 Role of Antiapoptotic BCL2 Proteins in Cancer Resistance

There are a number of instances where regulation of genes encoding pro- and anti-apoptotic are altered in cancer. The discovery of BCl2 protein is related to its role in translocations between chromosomes 14 and 18, observed in non-Hodgkin's lymphomas. In this translocation, the BCl2 gene comes under the direct control of enhancer of immunoglobulin gene on chromosome 14, thereby upregulating the expression of BCl2 gene. MCL1 and BCL-XL are frequently amplified or overexpressed in a number of hematological as well as solid tumors (Reed et al. 1996). In addition to chromosomal translocations, other mechanisms which contribute to overexpression of prosurvival BCL2 proteins include gene amplification, loss of endogenous microRNA that repress BCL2 gene expression, and changes in epigenetic regulation of BCl2 genes. Moreover, posttranslational mechanisms that negatively regulate protein degradation pathways may also contribute to elevated expression of prosurvival BCL2 family proteins (Yip and Reed 2008; Hata et al.

2015). For example, genetic inactivation of the ubiquitin ligase complex protein FBW7 can result in overexpression of MCL1 due to enhanced stability of this protein. Overexpression of BCl2 and other prosurvival proteins have been shown to prevent cell death induced by external stimuli like oxidative stress, hypoxia, and deprivation of growth factors. The most important outcome of this overexpression of BCL2 proteins in cancers is that it makes the cancer cells resistant to cell killing by cytotoxic anticancer drugs. This property to prevent apoptosis makes the BCL2 proteins important targets for cancer drug discovery.

4 Therapeutic Potential of BCl2 Family Protein Inhibitors

With the fact that BCl2 family proteins have been shown to mediate the apoptotic response to cancer therapies, these proteins have the potential to serve as biomarkers in predicting treatment response, particularly in cancers which overexpress BCl2 proteins. Moreover, as discussed above the apoptotic response of the cell is governed by relative balance of pro- and antiapoptotic BCl2 proteins. Therefore, for the treatment of the cancers, which overexpress this family of proteins, direct inhibition of BCL2 proteins alone or in combination with other drugs may be useful. The protein inhibitors currently in clinical trials inhibit BCl2 proteins by binding within the BH3-binding groove of BCL2 proteins and disrupting their interaction with BH3 proteins and are thus termed "BH3 mimetics." The BCL2 inhibitors currently being developed are either pan-BCl2 inhibitors or inhibitors for BCL2/BCL-XL/BCLW, BCL2 alone, or MCL-1 alone (Table 1). The most successful among these inhibitors (BH3 mimetics) in terms of clinical utility target BCl2, BCL-XL&BCLW, or BCl2 only. ABT-737 and its clinical analogue ABT-263 are small molecules that target antiapoptotic BCL2 family proteins (BCL2, BCL-XL, and BCLW). These two BH3 mimetics act by sequestering proapoptotic BH3 domain proteins, supporting BAX and BAK oligomerization and ultimately resulting in programmed cell death of cancerous cells (Oltersdorf et al. 2005). The preclinical activity of ABT-737 or ABT-263 as a single agent or in combination with various other drugs has shown promising activity against SCLC, multiple myeloma, lymphoma, CLL, and acute lymphoblastic leukemia (Li et al. 2009; Reuland et al. 2012; Billard 2013; Mattoo and FitzGerald 2013; Mattoo et al. 2013, 2014). Although recent clinical trials of ABT-263 have demonstrated activity in chronic lymphocytic leukemia (CLL), the efficacy of single-agent BCL2/BCL-XL inhibitors in SCLC has been disappointing (Rudin et al. 2012). Moreover, since BCL-XL is important for survival of platelets, the administration of ABT-263 to patients results in platelet killing or thrombocytopenia. ABT-199 a selective BCl2 inhibitor does not cause platelet killing and has shown very good response in malignancies which overexpress BCL2 like CLL and acute myeloid leukemia (AML) (Souers et al. 2013). In fact a phase 1 clinical trial of ABT-199 in patients with relapsed/refractory CLL including del(17p) and fludarabine refractory disease has shown an overall objective response rate of 80 % (Hata et al. 2015).

Table 1 BCL2 protein inhibitors in clinical development

Active agent	Sponsor	Target	Stage
ABT-199/GDC-0199	Abbvie	BCL2	Phase III
ABT-737/ABT-263	Abbott	BCL2, BCL-XL, BCLW	Phase I and Phase II
Apogossypol	Burnham Institute/NCI	BCL2, BCL-XL, MCL-1	Preclinical
Gossypol (AT-101)	Ascenta	BCL2, MCL-1, BCLW, BCL-XL	Phase I and Phase II
GX-15-070 (Obatoclox)	Gemin X	BCL2, BCL-XL, BCLW, MCL-1	Phase I and Phase II
Genasense (oblimersen sodium)	Genta	BCL2	Phase III
Antimycin A	Univ. of Washington	BCL2/BCL-XL	Preclinical
TW-37	Univ. of Michigan	MCL-1	Preclinical
MIM-1	Harvard Med. School	MCL-1	Preclinical

Inhibition of MCL-1 alone or in combination with ABT-737/ABT-263/ABT-199 has shown very good response in preclinical models for malignancies, which are addicted to more than one BCl2 protein like SCLC, mesothelioma, pancreatic cancers, etc. (Mattoo and FitzGerald 2013; Mattoo et al. 2013). But none of the MCL-1 inhibitors developed so far have entered clinical trials because of the lack of specificity toward MCL-1 protein for these inhibitors.

5 Perspective

Overexpression of BCL2 family proteins is one of the major factors responsible for resistance of cancer cells to current therapies. The development of recent inhibitors has shown promise in treatment of hematological cancers like CLL and AML, as a single agent or in combination with other drugs. Since some cancers show overexpression and addiction toward more than one BCL2 protein, lack of clinically relevant inhibitors for BCL2 proteins like MCL-1 is still an area which needs to be addressed. The rational use of these BCL2 inhibitors alone or in combination has a potential in overcoming the resistance shown by current therapies and providing novel treatment options for wide range of cancers.

Acknowledgments Abid Mattoo acknowledges the support from Intramural research program of the National Cancer Institute at NIH. Abid Mattoo also thanks Dr. Tej Pandita, Director, Department for Radiation Oncology, Methodist hospital Research institute for research support.

References

Adams JM, Huang DC, Strasser A, Willis S, Chen L, Wei A, van Delft M, Fletcher JI, Puthalakath H, Kuroda J, Michalak EM, Kelly PN, Bouillet P, Villunger A, O'Reilly L, Bath ML, Smith DP, Egle A, Harris AW, Hinds M, Colman P, Cory S (2005) Subversion of the BCL2 life/death switch in cancer development and therapy. Cold Spring Harb Symp Quant Biol 70:469–477. doi:10.1101/sqb.2005.70.009

Billard C (2013) BH3 mimetics: status of the field and new developments. Mol Cancer Ther 12: 1691–1700. doi:10.1158/1535-7163.MCT-13-0058

Chen L, Willis SN, Wei A, Smith BJ, Fletcher JI, Hinds MG, Colman PM, Day CL, Adams JM, Huang DC (2005) Differential targeting of prosurvival BCL2 proteins by their BH3-only ligands allows complementary apoptotic function. Mol Cell 17:393–403. doi:10.1016/j.molcel.2004.12.030

Czabotar PE, Lessene G, Strasser A, Adams JM (2014) Control of apoptosis by the BCL2 protein family: implications for physiology and therapy. Nat Rev Mol Cell Biol 15:49–63. doi:10.1038/nrm3722

Dewson G, Kluck RM (2009) Mechanisms by which BAK and BAX permeabilise mitochondria during apoptosis. J Cell Sci 122:2801–2808

Hata AN, Engelman JA, Faber AC (2015) The BCL2 family: key mediators of the apoptotic response to targeted anticancer therapeutics. Cancer Discov 5:475–487. doi:10.1158/2159-8290.CD-15-0011

Letai A, Bassik MC, Walensky LD, Sorcinelli MD, Weiler S, Korsmeyer SJ (2002) Distinct BH3 domains either sensitize or activate mitochondrial apoptosis, serving as prototype cancer therapeutics. Cancer Cell 2:183–192. doi:10.1016/S1535-6108(02)00127-7

Li R, Zang Y, Li C, Patel NS, Grandis JR, Johnson DE (2009) ABT-737 synergizes with chemotherapy to kill head and neck squamous cell carcinoma cells via a Noxa-mediated pathway. Mol Pharmacol 75:1231–1239. doi:10.1124/mol.108.052969

Marani M, Tenev T, Hancock D, Downward J, Lemoine NR (2002) Identification of novel isoforms of the BH3 domain protein Bim which directly activate BAX to trigger apoptosis. Mol Cell Biol 22:3577–3589. doi:10.1128/MCB.22.11.3577-3589.2002

Mattoo AR, FitzGerald DJ (2013) Combination treatments with ABT-263 and an immunotoxin produce synergistic killing of ABT-263-resistant small cell lung cancer cell lines. Int J Cancer 132:978–987. doi:10.1002/ijc.27732

Mattoo AR, Pastan I, Fitzgerald D (2013) Combination treatments with the PKC inhibitor, enzastaurin, enhance the cytotoxicity of the anti-mesothelin immunotoxin, SS1P. PLoS One 8, e75576. doi:10.1371/journal.pone.0075576

Mattoo AR, Zhang J, Espinoza LA, Jessup JM (2014) Inhibition of NANOG/NANOGP8 down-regulates MCL-1 in colorectal cancer cells and enhances the therapeutic efficacy of BH3 mimetics. Clin Cancer Res 20:5446–5455. doi:10.1158/1078-0432.CCR-14-1134

Oltersdorf T, Elmore SW, Shoemaker AR, Armstrong RC, Augeri DJ, Belli BA, Bruncko M, Deckwerth TL, Dinges J, Hajduk PJ, Joseph MK, Kitada S, Korsmeyer SJ, Kunzer AR, Letai A, Li C, Mitten MJ, Nettesheim DG, Ng S, Nimmer PM, O'Connor JM, Oleksijew A, Petros AM, Reed JC, Shen W, Tahir SK, Thompson CB, Tomaselli KJ, Wang B, Wendt MD, Zhang H, Fesik SW, Rosenberg SH (2005) An inhibitor of BCL2 family proteins induces regression of solid tumours. Nature 435:677–681. doi:10.1038/nature03579

Reed JC, Miyashita T, Krajewski S, Takayama S, Aime-Sempe C, Kitada S, Sato T, Wang HG, Harigai M, Hanada M, Krajewska M, Kochel K, Millan J, Kobayashi H (1996) BCL2 family proteins and the regulation of programmed cell death in leukemia and lymphoma. Cancer Treat Res 84:31–72, http://link.springer.com/chapter/10.1007/978-1-4613-1261-1_3

Reuland SN, Goldstein NB, Partyka KA, Smith S, Luo Y, Fujita M, Gonzalez R, Lewis K, Norris DA, Shellman YG (2012) ABT-737 synergizes with Bortezomib to kill melanoma cells. Biol Open 1:92–100. doi:10.1242/bio.2011035

Rudin CM, Hann CL, Garon EB, Ribeiro de Oliveira M, Bonomi PD, Camidge DR, Chu Q, Giaccone G, Khaira D, Ramalingam SS, Ranson MR, Dive C, McKeegan EM, Chyla BJ, Dowell BL, Chakravartty A, Nolan CE, Rudersdorf N, Busman TA, Mabry MH, Krivoshik AP, Humerickhouse RA, Shapiro GI, Gandhi L (2012) Phase II study of single-agent navitoclax (ABT-263) and biomarker correlates in patients with relapsed small cell lung cancer. Clin Cancer Res 18:3163–3169. doi:10.1158/1078-0432.CCR-11-3090

Shamas-Din A, Kale J, Leber B, Andrews DW (2013) Mechanisms of action of BCL2 family proteins. Cold Spring Harb Perspect Biol 5:a008714. doi:10.1101/cshperspect.a008714

Souers AJ, Leverson JD, Boghaert ER, Ackler SL, Catron ND, Chen J, Dayton BD, Ding H, Enschede SH, Fairbrother WJ, Huang DC, Hymowitz SG, Jin S, Khaw SL, Kovar PJ, Lam LT, Lee J, Maecker HL, Marsh KC, Mason KD, Mitten MJ, Nimmer PM, Oleksijew A, Park CH, Park CM, Phillips DC, Roberts AW, Sampath D, Seymour JF, Smith ML, Sullivan GM, Tahir SK, Tse C, Wendt MD, Xiao Y, Xue JC, Zhang H, Humerickhouse RA, Rosenberg SH, Elmore SW (2013) ABT-199, a potent and selective BCL2 inhibitor, achieves antitumor activity while sparing platelets. Nat Med 19:202–208. doi:10.1038/nm.3048

Willis SN, Chen L, Dewson G, Wei A, Naik E, Fletcher JI, Adams JM, Huang DC (2005) Proapoptotic BAK is sequestered by MCL-1 and Bcl-xL, but not BCL2, until displaced by BH3-only proteins. Genes Dev 19:1294–1305. doi:10.1101/gad.1304105

Yip KW, Reed JC (2008) BCL2 family proteins and cancer. Oncogene 27:6398–6406. doi:10.1038/onc.2008.307

Perturbed Signaling and Role of Posttranslational Modifications in Cancer Drug Resistance

Suruchi Aggarwal, Manu Kandpal, Shailendra Asthana, and Amit Kumar Yadav

Abstract Cancer is a complex disease in which erratic cellular signaling leads to uncontrolled growth and proliferation. Several drugs and therapies have been developed to control these signaling perturbations so as to kill the tumor cells. Despite these advances, cancer is a compounding global health problem made severe by the ever-increasing drug resistance. The number of new drugs approved is hopelessly outpaced by the instances of drug resistance and relapses. Posttranslational modifications (PTMs) are emerging as a hidden regulatory layer controlling metabolism and homeostasis. Drugs usually target PTMs to kill tumor cells. PTMs are also exploited by cancer cells to maintain their growth and survival by rewiring survival signaling pathways that can introduce drug resistance, both intrinsic and acquired. In this chapter, we discuss major known resistance mechanisms in cancer, exemplify how PTMs are involved in those, and attract the attention of drug discovery community toward this regulatory mechanism. A thorough understanding of the role of PTMs in these signaling changes can play a significant role in solving the drug resistance problem. We believe that combination therapies exploiting the PTMs may have a better chance of treating cancer and averting the intractable problems of drug resistance and cancer relapse.

1 Introduction

Cancer is a heterogeneous disease claiming 13 % patient lives each year (2008 figures) and increasing (Ferlay et al. 2010; Singh and Settleman 2010). Every year, hundreds of drugs are designed to combat widespread cancer in human population, but the emergence of drug resistance toward most of these drugs is a persistent

S. Aggarwal • M. Kandpal • S. Asthana • A.K. Yadav, PhD (✉)
Drug Discovery Research Center (DDRC), Translational Health Science and Technology Institute (THSTI), NCR Biotech Science Cluster, 3rd Milestone, Faridabad-Gurgaon Expressway, Faridabad 121001, Haryana, India
e-mail: amit.yadav@thsti.res.in

G. Arora et al. (eds.), *Drug Resistance in Bacteria, Fungi, Malaria, and Cancer*,
DOI 10.1007/978-3-319-48683-3_22

hurdle in the progression of drug development. Despite the success of currently available drugs, the decrease in the mortality rate of cancer patients is only 1 % (Karimi et al. 2014). This is because of the rapid development of resistance against drugs in cancer cells. Drug resistance is the leading cause of cancer treatment failures across the globe. It can be either intrinsic, which is present in cancer cells before treatment, or acquired, where resistance develops in response to treatment (Holohan et al. 2013). In either case, the failure of drugs to alleviate the disease condition is a critical issue that leads to reduced number of viable drug molecules approved for cancer treatment.

Cancer cells have mechanisms to evade the harmful effects of chemotherapy, targeted therapy or radiotherapy. Cancer cells survive the effect of harmful drugs by accumulating genetic and epigenetic changes that bypass the molecular mechanisms targeted by the drug molecules. Many of these mechanisms are linked to mutations, amplifications, chromosomal translocations, and differential expression of genes as reviewed previously (Goldie 2001; Gottesman 2002; Gillet and Gottesman 2010; Holohan et al. 2013; Housman et al. 2014).

For decades, it has been known that posttranslational modifications are required for critical signaling events for the cancer cells to undergo neoplastic transformation (Krueger and Srivastava 2006). While most of these changes are well characterized from the genetic mutation perspective, the role of PTMs has largely remained hidden or underappreciated. Though there are many amino acid substitutions/mutations occurring in cancer cells, not all mutations have a direct steric effect on drug binding. Some of the mutations can lead to altered protein function due to loss or gain of a PTM site or PTM refactoring (changing one PTM site to another), making the cells either sensitive or resistant to the chemotherapy. It is also possible that PTMs may not have any effect on the chemosensitivity of the cells but they still cause system wide disruption and deregulation of important signaling cascades by creating heterogeneity in metabolism which is a precursor to adaptive resistance.

In this chapter, we enumerate some examples and the emergent generic mechanisms of PTM-aided drug resistance in cancer. Although there are other mechanisms of resistance like stress/environmental effects, horizontal gene transfer, faster mutation rates, etc. to name a few, our major focus is to discuss the emergent role of PTMs in inducing drug resistance. While major signaling events highlighted in most of the mechanisms here are highly dominated by phosphorylation, since most studies have been conducted on kinases and mechanisms are well understood, it is worth noting that phosphorylation is not the only PTM involved. PTM crosstalk is now recognized as a global regulatory phenomenon although techniques to successfully tease out the crosstalk are still in infancy but developing at a high pace. Next-generation proteomics advances have made such studies feasible and will act as the much needed required shot in the arm for PTM crosstalk knowledge-driven cancer therapeutic development.

2 Mechanisms of Resistance

The basic mechanisms of drug resistance can either be intrinsic or acquired. The role of PTMs may be to disrupt or enhance a specific binding or modulation of a signaling pathway and could be either direct (disruption of binding site) or indirect (upstream changes leading to pathway blockage). The role of PTMs is discussed with reference to the classically known mechanisms like drug penetration and efflux, inactivation/alteration in drugs or their targets, DNA damage repair, apoptosis, epigenetics, cancer cell heterogeneity, etc. Also, newer mechanisms like *oncogenic bypass* or "kinome reprogramming" and tumor microenvironment are discussed. Although all of these mechanisms are not protein PTMs per se, broadly chemical modifications or modifying enzymes are at play, and we are discussing these for coherence and completeness. The role of PTMs in drug resistance is also summarized in Table 1 with respect to the discussed mechanisms.

2.1 Drug Inactivation

Certain drugs require metabolic activation after reaching the tumor cell. The activation includes interaction of the drug with other molecules in vivo that could modify or partially degrade (by enzymes) or form a complex with the drug molecules (by small molecule ligands) leading to their activation. This mechanism is not regulated directly by modifications of proteins but by modifying enzymes that act directly on the drug. The absence of any mechanism that could lead to drug activation will render it nonfunctional and eventually removed from the cells (Fig. 1a). Shutting down drug activation mechanisms or denigrating them is one of the abilities of cancer cells to protect themselves from the effects of the drugs. For example, the administration of *cytarabine* (Ara-C) drug in the treatment of acute myelogenous leukemia (AML) requires multiple phosphorylation events that convert Ara-C to Ara-C triphosphate by deoxycytidine kinase (DCK), the activated form of drug. Disruption or deactivation of the enzyme that leads to its multiple phosphorylations can induce drug resistance in leukemia cells (Sampath et al. 2006; Zahreddine and Borden 2013). *Capecitabine* is also administered in its pro-form, but once it reaches inside the cancer cells, it is converted to 5-FU (*5-fluorouracil*) by thymidine phosphorylase (TYMP). The absence of the phosphorylase enzyme causes the inactivation of drug and thus resistance (Malet-Martino and Martino 2002). This theme has been covered in detail elsewhere and beyond the scope of this review (Sampath et al. 2006; Michael and Doherty 2005; Shen et al. 2007).

Table 1 The PTMs involved in the various mechanisms of cancer resistance and the pathways targeted by the drugs

Mechanism	PTMs involved	Drugs targeted	Pathways targeted	Proteins/genes	Effect	References
Drug inactivation	Phosphorylation	*Cytarabine, capecitabine*	DNA damage repair	Deoxycytidine kinase (DCK), thymidine phosphorylase	Addition or removal of phosphorylation to the drugs	Malet-Martino and Martino (2002), Sampath et al. (2006), Zahreddine and Borden (2013)
Alteration in drug target	Phosphorylation	*Gefitinib, erlotinib, crizotinib*	EGFR receptor tyrosine kinase signaling	EGFR	Drug blocks ATP-binding site, blocking phosphorylation	Kobayashi et al. (2005), Kosaka et al. (2011)
Drug efflux	Glycosylation, ubiquitination, phosphorylation	*Mitoxantrone, topotecan, flavopiridol*		MRP1/ABCC1, MDR1/P-glycoprotein, BCRP/ABC-P	Ubiquitin degrades protein, phosphorylation protects it from ubiquitin, and glycosylation helps in maturation and membrane relocation	Katayama et al. (2014)
DNA damage repair	Acetylation, sumoylation, ubiquitination, phosphorylation	*Bendamustine, camptothecin, etoposide, bleomycin, enediyne, gemcitabine, cisplatin*	p53-mediated DNA repair, base excision repair, homologous recombination, nonhomologous end joining	p53, MDM2, BRCA1, CHK1, CHK2, GADD45, KU70, Ku80, CBP, PCAF, PARP	Phosphorylation of p53 prevents its ubiquitination, inhibiting PARP activates BRCA1, CHK1-/CHK2-mediated DNA repair by phosphorylating p53. Deacetylation of KU70 initiates NHEJ	Enoch and Norbury (1995), Martin et al. (2008), Bergink and Jentsch (2009), Fattah et al. (2010), Bouwman and Jonkers (2012), Lord and Ashworth (2012), Woods and Turchi (2013)
Cell death inhibition	Ubiquitination, phosphorylation, and lipidation	*Rapamycin*	mTOR-mediated phosphorylation of ATG-13,	ATG1-17, MCL1, VPS34, BAX,	Multi-phosphorylation of ATG13 prevents autophagy,	Brooks and Gu (2003), Schwickart et al. (2010)

			mitochondrial outer membrane permeability (MOMP)	BAK, NOXA, PUMA	deubiquitination of MCL1 helps in blocking BAX and BAK for MOMP	
EMT and metastasis	Glycosylation, ubiquitination, phosphorylation	*Gefitinib, U0126*	Epithelial–mesenchymal transitions	N-cadherins, vimentin, fibronectin, E-cadherins, SNAIL, FBXO11, PKD1, GSK-3β	Blocking phosphorylation of SNAIL by PKD1 leads to accumulation of SNAIL and repression of E-cadherin production	Zhou et al. (2004), Yook et al. (2005), Buonato and Lazzara (2014)
Cancer cell heterogeneity	Phosphorylation	*Crizotinib*	Signaling cascades	Oct4, POU5F1, ABCG2, AKT, BCR, ABL, ERK	Phosphorylation of Oct4 induces stem cell-like properties	Linn et al. (2010), Zhao et al. (2015)
Epigenetic regulation	Acetylation, sumoylation, methylation, ubiquitination, phosphorylation	*Vorinostat, romidepsin, panobinostat*	Transcriptional regulation of genes through histone modifications	ABC transporter family, tumor suppressors, and oncogenes	Histone modifications like acetylations help in unwinding of the DNA for transcription, deacetylation helps in rewinding	Kelly et al. (2010), Henrique et al. (2013), Easwaran et al. (2014)
Kinome reprogramming	Phosphorylation	*Sorafenib, erlotinib, gefitinib*	Signaling cascades	EGFR, HGFR, MET, RAS, PI3K, STAT, AKT, MAPK, ERK	Blocking of one receptor tyrosine kinase (RTK) by the drug activates the other RTK to carry out tyrosine phosphorylation signaling	Graves et al. (2013)
Tumor microenvironment	Glycosylation, phosphorylation	*Rituximab, trastuzumab, tamoxifen*	Hippo signaling pathway, MET-mediated	HGF, MCP-1, PDGF, TGF-β, VEGFA, MET,	Provides cytokines, growth signals to cancer cells to	Mufson (1997); Ciravolo et al. (2012), Pontiggia

(continued)

Table 1 (continued)

Mechanism	PTMs involved	Drugs targeted	Pathways targeted	Proteins/genes	Effect	References
			signaling, JAK/STAT pathway, p38/MAPK signaling pathway	EGFR, STAT, NOTCH, JAK, SOCS	promote tyrosine phosphorylation signaling	et al. (2012), Straussman et al. (2012), Mao et al. (2013), Kise et al. (2015), Mumenthaler et al. (2015)

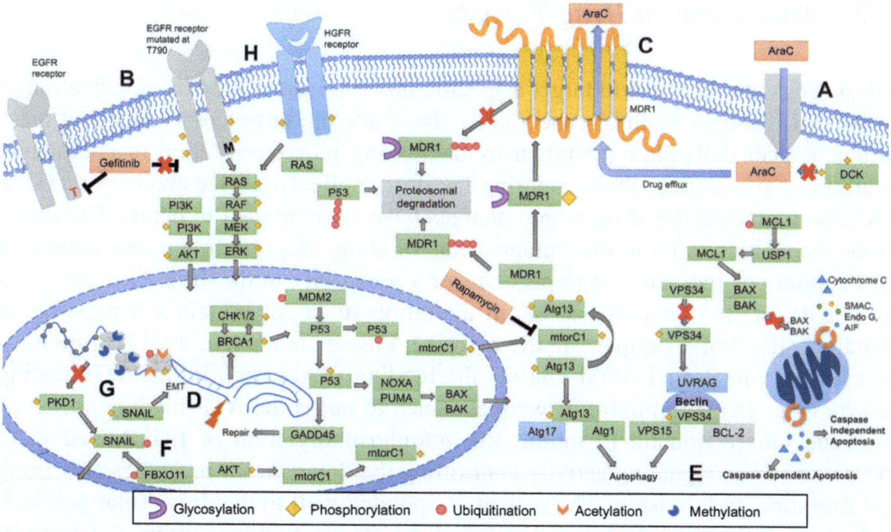

Fig. 1 The resistance mechanisms in cancer cells with the role of PTMs are broadly shown. (**a**) The inactivation of drug Ara-C in tumor by the absence of phosphorylation by deoxycytidine kinase (DCK) due to which Ara-C is effluxed out of the cell by MDR1 protein. (**b**) The alteration of drug (*gefitinib*) target by mutation in the EGFR receptor at the gatekeeper mutation T790 (threonine) to M (methionine). Gatekeeper residue guides the entry of ATP as well as drug in the binding pocket of the tyrosine kinase domain. In the case of mutated gatekeeper residue, ATP can bind but the drug cannot, which causes resistance. (**c**) The phosphorylation and glycosylation of MDR1 protein help its maturation and relocation to cell membrane. In normal cells, MDR1 production is controlled by ubiquitination at the motif of phosphorylation, blocking phosphorylation and leading the protein toward proteasomal degradation. (**d**) The DNA damage response in the cancer cells can also cause resistance. In the case of DNA breaks, BRCA1 and CHK1/CHK2 get phosphorylated and in turn phosphorylate p53 which otherwise gets proteasomally degraded by its interaction with MDM2. Once p53 is stabilized, it helps in GADD45 production that starts DNA repair. (**e**) The regulation of cell death in cancer cells. When p53 gets stabilized, it may also trigger the production of BAX and BAK that after translocation to cytoplasm initiate mitochondrial outer membrane permeability (MOMP) to release cytochrome C, EndoG, SMAC, and AIF proteins from intramembrane compartment that initiates caspase-dependent and caspase-independent cell apoptosis. In cancer cells, once BAK and BAX are released in cytoplasm, MCL1 gets stabilized by interacting with USP1 (DUB) and blocks BAK and BAX to oligomerize and form a channel in mitochondria. To induce autophagy, *rapamycin* blocks mTORC1 from phosphorylating ATG-13 and thus forms a complex with ATG-1 and ATG-17. The cancer cells block the phosphorylation of Vps34 thus blocking the activation of autophagy. (**f**) The protein SNAIL activates epithelial to mesenchymal transition (EMT) due to its phosphorylation by PKD1 and ubiquitination by FXBO11. In cancer cells, PKD1 is blocked epigenetically, and in the absence of phosphorylation at SNAIL, ubiquitination cannot occur, leading to repression of E-cadherin by SNAIL and initiation of EMT. (**g**) Represents the epigenetic control of gene expression by histone modifications. Methylations wind up the DNA and stop transcription, while acetylation opens up the DNA to activate transcription. Phosphorylation of H2AX helps in initiating DNA damage response. (**h**) Represents the oncogenic bypass or kinome reprogramming in cancer cells. In cases where EGFR receptor is blocked by the drug, HGFR receptor gets activated to initiate tyrosine signaling cascade via RAS and PI3K

2.2 Alterations in Drug Targets

Most drugs are designed to target a specific protein, inactivation or modification of which will lead to death of cancer cells. To evade the unwelcome effects of these drugs, cancer cells alter the target by decreasing its concentration or completely stalling its production, thus rendering the drug ineffective (Housman et al. 2014). Genetic mutations for drug target changes have been studied in detail. The mutations can either aid or discourage drug binding, depending on the nature of mutational change, and can impact cancer's sensitivity to the drug. Genetic effect is not direct. It is manifested as an alteration in target protein's expression or binding site. For example, in the tyrosine kinase domain of EGFR protein, a gatekeeper mutation T790M inhibits the binding of *erlotinib* in the ATP-binding cavity (Fig. 1b). The inhibitors were designed to target the ATP-binding pocket of tyrosine kinase domain to inhibit the autophosphorylation of EGFR leading to decreased or no signaling activity controlling the downstream processes governing proliferation and division. The patients responded well to the drugs (like *gefitinib*, *erlotinib*, *crizotinib*, etc.) but developed resistance within a year of treatment (Kobayashi et al. 2005). These drugs have a high affinity for the ATP-binding pocket. Moreover, the gatekeeper residue threonine in wild type is supposed to enhance the binding of drug by creating hydrogen bonds with the drug molecule. But in the cases where the threonine 790 was mutated to methionine, the drug molecule could not enter the binding pocket because of the steric hindrance introduced by methionine. Also, the methionine group lacks a hydroxyl group for hydrogen bonding, thus inhibiting the binding of drug, but the binding of ATP can still occur just as freely as in wild type, thus phosphorylating the activity of EGFR inducing the downstream signaling cascade for functioning (Kobayashi et al. 2005; Kosaka et al. 2011; Cortot and Janne 2014).

2.3 Drug Efflux

Drug resistance has also been linked to drug efflux mechanisms. As soon as the drug enters the tumor cells, it is pumped out by a family of ATP-binding cassette (ABC) membrane transporters protecting the cells. Till now, 49 members of this family have been identified, but only three have been linked to drug resistance (Holohan et al. 2013). These three proteins—MRP1/ABCC1, MDR1/P-glycoprotein, and BCRP/ABC-P—are known to induce multidrug resistance in cancer tissues owing to their regulation by PTMs. In many tumors, MDR1 protein is known to be overexpressed leading to intrinsic drug resistance (Fig. 1c). Cancer stem cells also exhibit a high level of drug efflux proteins portraying an intrinsically drug-resistant phenotype toward many drugs (Gottesman et al. 2002).

The levels of ABC transporters (majorly p-glycoproteins) are modulated by ubiquitin-mediated proteasomal degradation. However, in cancer cells the transporters evade this pathway by phosphorylating serine 683 that hinders with the ubiquitination of the protein, thus preventing its degradation. After phosphorylation, the sequence motif allows glycosylation which matures the p-glycoprotein and helps in its transport to the cell surface. The serine/threonine protein kinase, Pim1, phosphorylates P-glycoprotein. It is a transcription factor that controls expression of genes to suppress apoptosis and promotes cell migration, cell cycle progression, and protein translation. It also phosphorylates threonine at position 362 in BCRP protein that helps in its dimerization at cell surface. In cancer cells, the overexpression of ABC transporters has been positively correlated with increased expression of PIM kinases (Katayama et al. 2014). Phosphorylation, glycosylation, and ubiquitination are the important PTMs regulating the activity and dimerization of transporter proteins that help in drug efflux.

2.4 DNA Damage Repair

Chemotherapeutic agents are designed to cause damage to the DNA either directly or indirectly. DNA damage response (DDR) can either lead to cell cycle arrest, apoptosis of the cell, or can repair the DNA to protect the cell. So, evading DNA damage and inducing repair becomes a necessity for the cancer cell survival in response to such chemotherapeutics. DDR induces a rapid but faulty repair mechanism that can introduce mutations in the repaired DNA. This is a known vehicle for rapid evolution to help cancer cells evade the harmful effects of drug-induced damages. There have been studies that show how protein posttranslational modifications like phosphorylation, ubiquitination, and sumoylation help in regulating the DNA damage response (DDR) proteins (Bergink and Jentsch 2009). Targeting the proteins governing the DNA damage response pathway can aid in extermination of cancer cells completely since the impaired DNA repair can activate p53-mediated apoptosis or cell cycle arrest of cancer cells. There are certain checkpoints present in the cell which govern the fate of the cell during cell division (Enoch and Norbury 1995). These checkpoints can halt the replication fork during DNA replication by activating p53-mediated DNA damage response, thus preventing the cells with faulty mutations to multiply (Bouwman and Jonkers 2012). In cancer cells, p53 is either suppressed or mutated to render its function as tumor suppressor ineffective, thus increasing DNA repair response (Brooks and Gu 2003). There are many mechanisms in a cell to activate DNA repair based on the damage caused. There are base excision repairs for single-stranded breaks, homologous recombination for double-stranded breaks, mismatch repair for correcting the wrong base pair addition, nucleotide excision repair for removal of bulky groups added, and direct reversal (Martin et al. 2008). The details of the biochemical pathways activating these responses are discussed in depth in several articles (Tutt et al. 2001; Lord and Ashworth 2008, 2012, 2013; Postel-Vinay et al. 2013). Usually the cancer cells are targeted with DNA-damaging alkaloids in combination with PARP inhibitors, topoisomerase inhibitors, or platinum-based drugs. But the DNA damage responses are buffered in response to drugs. When the single-stranded break excision repair mechanism is blocked using PARP1 inhibitors, double-stranded break (DSB) excision repair mechanisms get activated. DSB is detected through the sensors in the cell-like DNA-dependent protein kinase (DNA-PK) which in turn phosphorylates checkpoint proteins CHK1 and CHK2 which act as a relay of signaling cascade to activate p53 or BRCA1 or Nbs1 or cdc25C for DNA repair by homologous recombination (Chen and Sanchez 2004; Smith et al. 2010).

Usually in normal cells, p53 is suppressed with the help of MDM2 protein which is an E3 ligase that polyubiquitinates p53 and degrades it using the proteasomes. However, in cells experiencing DNA damage, p53 is phosphorylated, and the interaction with MDM2 gets blocked thus stabilizing p53. Subsequently, the phosphorylated p53 forms complexes with other proteins to result in transcriptional activation of GADD45 (Fig. 1d), which initiates DNA repair and thus imparts resistance to cancer cells from DNA-damaging drugs (Geske et al. 2000). This is a very important example where specific protein–protein interaction (PPI) between MDM2 and p53 is mediated by ubiquitination and the disruption of this PPI brings about corrective DDR and the ensuing resistance to drugs.

Other way in which DNA repair gets activated is through nonhomologous end joining, where the KU70 gets deacetylated due to activity of histone deacetylases (HDAC) and interacts with FLIP to initiate DNA repair mechanism and thus desensitizes cells to the DNA damage elements (Kerr et al. 2012). In normal cells, KU70 with its interacting partner KU80 is acetylated by HATs (histone acetyltransferases) to reduce its efficiency to bind to DNA and initiate DNA repair. This in turn leads to activation of apoptotic signals in cells. However, when DNA-damaging elements are introduced, HDAC activity reduces KU70 acetylation, and it forms complex with KU80, CBP, and PCAF which bind to DNA and repair it through nonhomologous end joining method. Phosphorylation and its dynamic interactions with acetylation, sumoylation, and ubiquitination thus drive the DDR-related drug resistance in tumor cells.

2.5 Cell Death Inhibition (Apoptosis and Autophagy Regulation)

Evading apoptosis and autophagy is an important regulatory event for cancer cells to survive. The processes of apoptosis and autophagy are antagonistic to each other. Though the stimuli to activate either pathway might be similar, but the activation sometimes depends on mutual inhibition in either pathway. Deregulation of apoptotic pathway for survival has been marked as a hallmark of cancer (Hanahan and Weinberg 2000). The cancer cells are addicted to a small number of anti-apoptotic proteins required for the deregulation of apoptotic pathway and subsequent survival (Holohan et al. 2013). BCL-2 family proteins, inhibitor of apoptosis proteins (IAPs) and FLIP, the caspase-8 inhibitor, are the most prominent proteins targeted for drug development. One of the examples of this type of drug resistance is found in MCL1 protein (a proapoptotic survival protein of BCL-2 family). Generally, cancer cells induce cell survival via deregulation of apoptosis by increasing the expression of pro-survival proteins. But in the case of MCL1, it can be achieved by altered ubiquitination posttranslationally. In normal cells, the MCL1 protein gets proteasomally degraded due to its ubiquitination. However, in cancer cells it interacts with USP9x which is a deubiquitinating enzyme (DUB) that helps in removal of ubiquitin from MCL1 and thus its overexpression leading to cancer cell survival (Schwickart et al. 2010).

Autophagy is the eradication of cancer cells via lysosomal degradation pathway. Even though this mechanism acts as a tumor suppressor pathway by degrading tumor cells, this also doubles up as a drug resistance pathway, by degrading drug molecule as soon as it enters the cells (Holohan et al. 2013). Activation of autophagy depends on the stimulus induced by stress, ionizing radiations, and chemotherapeutic drugs that damage the DNA. In normal cells this pathway is kept in check through multiple phosphorylations of ATG-13 (autophagy protein-13) by mTOR Ser/Thr kinase. To treat cancer cells, *rapamycin* is given that inhibits phosphorylation of ATG-13 by mTOR, causing autophagy induction. In this condition, ATG-13 remains dephosphorylated allowing it to bind to ATG-1 and ATG-17 to form an active complex that initiates autophagy. To counterattack this process, resistant cancer cells can inhibit the formation of type III PI3K complex by beclin-1 with VPS34, by dephosphorylating it. The Beclin 1-VPS34 protein complex drives the autophagy pathway by autophagy induction and autophagosome maturation. If this type III PI3K complex is not in place, autophagy cannot take place (Maiuri et al. 2007). The phosphoactivation of class III PI3K and ULK complexes expands the autophagosomal membrane by recruiting two ubiquitin-like conjugation systems. Atg7 (E1-like enzyme) and Atg10 (E2-like enzyme) mediated the formation of Atg-5-Atg12-Atg-16L complex that binds to phagophore after self-oligomerization (He and Klionsky 2009; He et al. 2009) (Fig. 1e). Atg8 (or LC3) is conjugated to phosphatidylethanolamine (PE), and delipidation of LC3-PE is carried out by Atg4 to increase free LC3 pool. Ubiquitinated proteins bind to p62/SQSTM1 accumulated under oxidation stress and are delivered to autophagosomes followed by lysosomal degradation (Komatsu et al. 2007; Kongara and Karantza 2012). Both autophagy and apoptosis depend on mitochondrial outer membrane permeabilization (MOMP) for their functioning. MOMP is the intrinsic pathway for cell death. During DNA damage response when p53 is stabilized by phosphorylation, it regulates the transcriptional activity of BAX or BAK which oligomerizes to mitochondrial outer membrane to form a channel between the cytoplasm and intermembrane space of the mitochondria, initiating the release of cytochrome C, Omi, EndoG, and SMAC in the cytoplasm that leads to cell death. Cytochrome C mediates caspase-dependent cell death, while the mode of cell death by the other three proteins mediates caspase-independent cell death (Kroemer and Martin 2005). But all these pathways are regulated by BCL-2 family proteins which act as sentinels to cell stress and regulate these pathways via posttranslational modifications (Maiuri et al. 2007). In cancer cells, this pathway is blocked via three blocks.

In class-A block, p53 is not activated, which in turn will not activate NOXA or PUMA to activate and express BAX or BAK. Class-B block involves the deletion of DNA segment for BAX or BAK, since without BAX and BAK, MOMP cannot be triggered. Class-C blocks involve BCL-2 family proteins like MCL-1 to inactivate BAX or BAK by binding to their BH3 binding cleft, thus blocking them for MOMP. For BCL-2 family proteins to act, they need to be deubiquitinated which is facilitated by the deubiquitinating enzymes (DUBs) (Letai 2008; van de Kooij et al. 2013).

2.6 Epithelial–Mesenchymal Transition and Metastasis

The epithelial to mesenchymal transition is a process by which cancer cells transform themselves to allow their movement from the site of origin to other sites where the tumor can progress. This process is known as metastasis. Due to the molecular and conformational changes occurring in the tumor cells, the drugs targeted at specific proteins (mainly receptors) are rendered nonspecific and or ineffective thus causing resistance. One of the major factors responsible for EMT induction and metastasis in cancer cells is the tumor microenvironment (discussed in detail later) that consists of extracellular matrix (ECM) and other soluble factors. Also linked closely to cancer cell heterogeneity (CCH), this represents an intricate interplay between these mechanisms although subtle but noteworthy differences warrant their discussion under separate sections. Metastasis introduces heterogeneity in tumor cell population by introduction of the metastatic stem cells (MSCs) which are the major reason for tumor relapse after chemotherapy. Since these cells are different from the targeted epithelial cells, they evade the effect of the drugs quite easily (Voulgari and Pintzas 2009).

In EMT, the cancer cells lose apical–basal polarity and cell adhesion phenotype to become mesenchymal in nature. During EMT, tumor cells experience a functional loss of E-cadherin and gain of N-cadherins, vimentin, and fibronectin. E-cadherins are glycoproteins (proteins modified with glycosylations) that help in epithelial cell anchorage at the adherent junctions between cells that work in concert along with β-catenins; N-cadherins have been suggested to destabilize the adherent junctions that facilitate the motility even in the presence of E-cadherins. The mesenchymal cells also express R-cadherin and cadherin-11. Moreover, claudins, connexins, occludins, and zonula occluden proteins that are localized in tight junctions are also perpetrated to be involved in EMT (Voulgari and Pintzas 2009). To reduce the expression of E-cadherins, cancer cells express SNAIL, a transcription factor that induces EMT. In tumor cells, SNAIL is produced in low quantities which are required to inhibit fructose-1,6-biphosphatase (FBP1) expression, to increase macromolecular biosynthesis, glucose uptake, and maintenance of ATP production in hypoxic conditions. For maintaining the desired low level, SNAIL is constantly being degraded by the ubiquitin proteasomal system in the cells with the help of FBXO11 E3 ligase. But FBXO11 cannot interact with SNAIL until it has been phosphorylated, particularly at serine 11 by PKD1, which is a serine/threonine protein kinase. The SNAIL protein is exported out of the nucleus after the formation of phosphorylated SNAIL and FBXO11 for ubiquitination. When EMT inducers are activated in tumor cells, the expression of PKD1 is halted by epigenetically controlling its expression. If PKD1 is not present to phosphorylate SNAIL, it will not form a complex with FBXO11 and in turn will repress the expression of E-cadherin in tumor cells leading to metastasis (Fig. 1f).

Another mode in which SNAIL evades phosphorylation by PKD1 is by mutating its serine at position 11. In the absence of a mutable serine at the particular position, phosphorylation will not be possible, but the functioning of SNAIL is not affected by it. Thus it will bind to the promoter of E-cadherin and thereby will repress its activation (Zheng et al. 2014).

In a study on behavior of SNAIL in prostate cancer, it was found that phosphorylation of SNAIL at serine 246 leads to its activation by accumulation in nucleus and thus repressing E-cadherin. Another kinase GSK-3β (glycogen synthase kinase—3β) phosphorylates SNAIL and its subsequent ubiquitination exporting it from nucleus for proteasomal degradation in cytosol (Zhou et al. 2004; Yook et al. 2005; Smith and Odero-Marah 2012). EMT is a highly dynamic process controlled by phosphorylation, glycosylation, and ubiquitination posttranslational modifications also assisted by the epigenetic modifications of methylation and acetylation along with several other histone modifications.

2.7 *Cancer Cell Heterogeneity*

The cancer cells are not made up of the same type of cells due to higher rate of evolution; certain cells are inherently resistance to drugs. A tumor consists of heterogeneous cell types like cancer cells and cancer stem cells (CSCs). There are two modes in which heterogeneity develops in cancer—the cancer stem cell model and clonal evolution model.

Cancer stem cell model states that the cancer population contains few cells with the capacity of self-renewal. Termed cancer stem cells (CSCs), these cells can form all other cell types present in a tumor. These stem cells may give rise to different type of cancer cells which is one of the reasons of heterogeneity. Due to this heterogeneity, some cells may be intrinsically resistant to drugs owing to their molecular phenotype. As most examples about the role of PTMs in drug resistance overlap across the different mechanisms discussed in this chapter, we provide a passing reference for such examples, since these are discussed in detail elsewhere. As we have already discussed, CCH is also related to EMT and also with tumor microenvironment (discussed in Sect. 2.10).

In chemoresistant hepatocellular carcinoma, overexpression of Oct4 helps in persistence of stem cell-like properties (Wang et al. 2010). It is an important transcription factor that maintains the self-renewal and pluripotency in mouse and human embryonic stem cells (Kellner and Kikyo 2010). Oct4 is encoded by

POU5F1 gene that controls multiple drug-resistant ABCG2 gene expression. Oct4 is activated by phosphorylation mediated through the central AKT kinase. Oct4-TCL1-AKT signaling pathway can affect cell survival and the sensitivity to drugs by regulating the efflux pump, ABCG2, a protective efflux pump, thereby imparting resistance. Phosphorylation of Oct4 also plays an important role in its stabilization and resistance to apoptosis in cancer cells. Oct4 is an important substrate of AKT for maintaining the self-renewal of stem-like cancer cells (Zhao et al. 2015). The phosphorylation at threonine 235 (pT235) on Oct4 was not only detected in human glioblastoma but also in liver cancer specimens (Zhao et al. 2015).

Cancer cell heterogeneity may also arise from clonal selection wherein the tumor starts from a single mutated cell with the ability to mutate further. Each accumulation step can give rise to subpopulations of heterogeneously mutated cells, ultimately resulting in cancer cell heterogeneity (Fig. 2). Every subpopulation has a different mutation which is the basis for a variable response to different therapies. Some cells may intrinsically be resistant owing to their mutations, while others may develop acquired resistance by subsequent selection pressure. *Imatinib* targets the

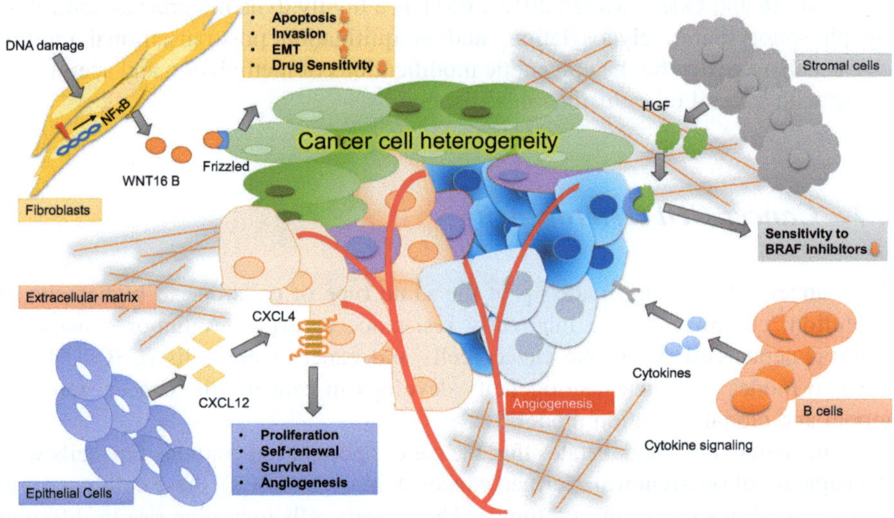

Fig. 2 Tumor microenvironment and cancer cell heterogeneity. Cancer-associated fibroblasts (CAFs), stromal cells, tumor-associated endothelial cells (TECs), vasculature, extracellular matrix (ECM), exosomes (not shown), and immune cells (B-cells) make up the tumor microenvironment and regulate cancer growth, sensitivity to drugs, and even drug resistance. Fibroblast cells in response to DNA damage induce NF-κB-mediated signaling process leading to WNT16B-facilitated activation of frizzled receptor increasing drug resistance. Epithelial cells are involved in CXCL12-based signaling increasing tumor cell survival, proliferation, and angiogenesis. B-cells are involved in the cytokine-based signaling. Stromal cell HGF secretion increases tumor sensitivity to BRAF inhibitors. Cancer cell heterogeneity also adds to the resistance wherein the intrinsically resistant cells are selected on drug application. *Different colors* and *sizes* of cells in tumor depict cancer cell heterogeneity, where all *light-colored cells* are drug-sensitive and dark-colored cells are resistant. Cells in *purple color* are cancer stem cells which on metastasis can give rise to any type of cancer cell at the new tumor foci.

signaling mediated by the fusion protein BCR-ABL (Hantschel et al. 2008). Mutation in the kinase domain of ABL in BCR-ABL fusion protein inhibits drug binding to the kinase domain and causes resistance in chronic myeloid leukemia (CML) patients (Soverini et al. 2011). Mutations in ABL kinase domain are divided into two groups—(1) Mutation in the residues, which directly interact with the drug (e.g., T315, F317, F395), that block drug binding without affecting the ATP binding. (2) Mutation in the distinct residues is responsible for conformational changes that favor drug binding. For example, deformation of ATP phosphate-binding loop (P loop) by mutations at residues G250, Q252, Y253, and E255 creates a hydrophobic cage that hinders the binding of drug.

The resistance to *imatinib* allows normal dimerization of BCR-ABL and activates autophosphorylation, which in turn activates the kinase activity and causes conformational changes allowing attachment of GRB2. BCR-ABL signaling activates many such pathways that enhance the survival of cancer cells, inhibit apoptosis, and change cell adhesion and migration.

In glioblastomas (GBMs), cancer cell heterogeneity manifests itself in the form of copy number variations in two receptor tyrosine kinases (RTKs), EGFR and PDGFRA that drive the signaling through phosphorylation. A minute fraction of cells have equal amplification of the genes, while most cells have mutually exclusive amplification, apart from the spatial distribution of the amplification as shown by PDGFRA-amplified cells which are present close to the endothelial cells. These differentially amplified kinases provide heterogeneous response to drugs and possibly transient targets for phosphorylation (Little et al. 2012; Szerlip et al. 2012; Sun and Yu 2015).

Another mechanism that can contribute to heterogeneous protein modification is that some signals are transduced in oscillatory manner impacting different cells in a pulse-like manner with different cells at different phases. For example, ERK activation acts in such a pulsed manner. It has been hypothesized that the heterogeneity of the signal transductions which are thus possible could be functional, some of which may be transient while others stable. As evolutionary selection operates on both equally, some may acquire resistance.

Heterogeneity of cells has also been observed in EGFR mutations discussed above (alterations in drug targets, Sect. 2.2). Same phenomenon has been observed in resistance to anaplastic lymphoma. Mutation in the ALK tyrosine kinase domain results in resistance against *crizotinib*. There are several mutations leading to more than one resistance mechanism. Thus the tumor cells in this case depict heterogeneity in resistance mechanism (Katayama et al. 2012).

Gatenby et al. propose "evolutionary double-blind therapy" in which two therapies will be given to cancer patients consecutively—one making tumor cell adaptive in a specific way while making it vulnerable for the second therapy (Cunningham et al. 2011). Improved knowledge of tumor heterogeneity can be highly relevant for the drug development strategies in oncology and can provide therapeutic advancement. Identification of driving cause, loss or gain of which may lead to heterogeneity in cancer cells, will provide better modes of designing therapies.

2.8 Epigenetics

DNA methylation and histone modifications are the two main types of epigenetic changes that influence growth and proliferation of tumor cells by controlling gene activity and chromatin architecture. Of these, the histone modifications are post-translational modifications, and the complex interplay involving crosstalk of these modifications to control gene expression in specific ways has been termed as the "histone code" (Jenuwein and Allis 2001). While DNA methylation has been studied in great detail, the histone PTMs have been understudied as their role was underappreciated for a long time. DNA methylation keeps DNA in heterochromatin form and inaccessible to transcription factors, some of which express the tumor suppressor genes. DNA methylation and histone modifications both mediate cell growth by different modes of regulation. Histones get modified posttranslationally to bind to or unwind DNA for transcription. DNA methylation silences the genes that could regulate other proteins in cell survival. For example, in the case of silencing of cytokine signaling-3 (SOCS-3) gene, the DNA is hypermethylated, promoting the phosphorylation of STAT3 by JAK in cytokine signaling cascade. The SOCS-3 protein is the inhibitor of JAK/STAT phosphorylation, but inhibition of SOCS-3 expression promotes phosphorylation signaling in cancer cells.

Posttranslational modifications in histones influence chromatin structure by recruiting enzyme pairs for reversible acetylation (histone acetyltransferases or HATs and histone deacetylases or HDACs), methylation (histone methyl-transferases or HMTs and histone demethylases or HDMs), phosphorylation (kinases and phosphatases), ubiquitination (ubiquitination ligases, conjugases, and deubiquitinases), etc., to add or remove functional groups (or PTMs). While HATs help in opening up the chromatin by acetylation and thus induce gene expression, the HDACs on the opposite hand help making it inaccessible for transcription by deacetylating the histones. Other pairs of enzymes also facilitate such reversible functions. These PTMs expand the functional repertoire of the histones by creating docking sites or desired binding regions for the specific interaction of proteins or even for formation of protein complexes. The proteins that add, remove, or interpret these modifications on histones are thus called as chromatin "writers", "erasers" or "readers" respectively (Marsh et al. 2014). The combination of different modifications at specific positions in the histone proteins is known to regulate the open or closed configuration of the chromatin for transcription. For example, mono-methylated histone 3 at lysine 79 (written as H3K79me), di- and trimethylated histone 3 at lysine 4 (H3K4me2 and H3K4me3), monoubiquitinated of histone 2B at lysine 120 (H2BK120ub), and several histone acetylations mark the open chromatin and active transcription, whereas monomethylated histone 3 at lysines 9 and 27(H3K9me and H3K27me) and of lysine 20 in histone 4 (H4K20me) mark the closed configuration of the chromatin (Barski et al. 2007; Dawson and Kouzarides 2012; Easwaran et al. 2014) (Fig. 1g).

In cancer cells, few proteins of the ABC transporter family were found to be overexpressed leading to reduced cellular permeability to drugs and higher drug efflux—the two basic mechanisms for cancer cells to develop inherent drug resistance. Initially it was thought to be associated with promoter demethylation of genes coding for ABC transporter proteins. But in 2007, it was observed that the histones binding the genes of ABCG2 are hypoacetylated in normal cells. But in cancer cells, hyperacetylation of H3K9 and K14 and increase in trimethylation of H3K4 and decreased trimethylation of H3K9 increase its production and exhibit resistance to the drugs like *mitoxantrone*, *topotecan*, and *flavopiridol* (To et al. 2008). Similar mechanism of regulation for protein MDR1 (P-glycoprotein) was also observed. In cancer cells, the hyperacetylation in histone 3 and demethylation at lysine 4 of histone 3 caused the enhanced expression of MDR1 protein leading to drug resistance (Henrique et al. 2013).

During DNA damage response, phosphorylation of histone 2 (H2AX) by PI3K in cancer cells promotes retention and accumulation of repair proteins at the damaged site of the chromatin. This phosphorylation also mediates the association of HATs at the damaged site, where it leads to relaxing of chromatin threads for the repair proteins to bind and initiate repair mechanisms instead of signaling the apoptosis (Sawan and Herceg 2010).

Epigenetic regulation of genes through histone modifications like phosphorylation, methylation, acetylation, ubiquitination, sumoylation, and their crosstalk thus plays a very active role in regulation of mechanisms inducing resistance in cancer cells.

2.9 Kinome Reprogramming or Oncogenic Bypass

Protein phosphorylation is the most widely observed modification on proteins for controlling signaling changes and expansion of functions and interactions through conformational changes. It is dynamically regulated in the cell with the help of kinases and phosphatases, which add and remove phosphate group, respectively, from specific amino acids on the protein. Protein kinases regulate most of the signal transduction pathways in eukaryotes by phosphorylating other proteins (Graves et al. 2013). There are about 518 protein kinases (Manning et al. 2002) and 156 phosphatases known till now in human proteome. Since the kinases act as the key signaling nodes for cancer cells to function, shutting them will affect cell metabolism, survival, and growth (Barouch-Bentov and Sauer 2011). When these key nodes are inhibited by the drugs, cancer cells reprogram their cellular machinery to activate alternate signaling pathways to compensate for the inhibited kinase. This could also be termed as "adaptive molecular reprogramming" or "kinome reprogramming."

Molecular targeted therapeutics is gaining much popularity recently with more than 25 oncology drugs approved as kinase inhibitors and hundreds in various phases of clinical trials (Gross et al. 2015). But the targeting of single kinases tends to trigger resistance quickly in cancer cells by activating other kinases through mutation or rewiring of signaling networks. In malignant melanomas, small inhibitors like *sorafenib* were designed to target B-Raf kinases to inhibit MAPK signaling cascade. But a mutation in B-Raf at valine 600 to glutamic acid helps in chronic activation of all three Rafs (A, B, and C), which after phosphorylation get activated and subsequently activate ERK1/ERK2 by phosphorylation, leading to enhanced cancer cell proliferation or drug refractory tumors (Hatzivassiliou et al. 2010; Osborne et al. 2012). In an already discussed example, EGFR inhibition by *erlotinib* and *gefitinib* in tumors where the gatekeeper mutation does not occur, amplification of MET is observed. MET or HGFR is another tyrosine kinase that gets phosphorylated in cases where EGFR is blocked; and through RAS, PI3K, and STAT; reactivates signaling cascades to promote tumor growth and survival (Fig. 1h). As discussed in TME (Sect. 2.10), HGFR is contributed not directly by tumor cells but by the microenvironment which helps in the rewiring or reprogramming of the signaling cascades to bring about drug resistance effects.

In light of the above observations, instead of single targeted therapy, combination therapies are being designed to overcome oncogenic resistance in these cases, and several success stories are known. In combination therapies instead of targeting a single kinase, many kinases in a single pathway are targeted to overcome the kinome rewiring in tumor cells (Gross et al. 2015). But to be alert, cancer cells may still rewire their signaling, and a deeper understanding of the kinase signaling networks can throw light on the possibilities of resistance and probably the counteracting mechanisms to prevent it.

2.10 Tumor Microenvironment (TME)

Cancer cells do not develop in isolation but in a complex microenvironment that facilitates its survival, growth, and proliferation. Tumor microenvironment has been shown to have an important supportive role in cancer development and progression (Sung et al. 2007; Mao et al. 2013; Kise et al. 2015). Of late, its contribution to therapeutic resistance is also starting to be understood (Shain et al. 2000; Sung et al. 2007; Andre et al. 2010; Mao et al. 2013; Kise et al. 2015; Mumenthaler et al. 2015; Sun 2015). TME can contribute toward decreased drug penetration and helps in cancer cell proliferation and bypassing apoptotic pathways. These processes do not require the cancer cells to undergo mutations or any epigenetic changes themselves. Figure 2 shows a schematic representation of TME—tumor crosstalk.

Malignant cancer cells and benign stromal cells coevolve, and their intertwined crosstalks are being recognized as a hallmark of cancer (Hanahan and Weinberg 2011). Most resistance mechanisms recognized in cancer revolve majorly around mutational or epigenetic changes to bring about lowered permeability, efflux, modification/degradation of drugs, metabolic alterations, strengthening DNA damage machinery, enhancing pro-apoptotic pathways, etc. (Sun 2015). TME can contribute not only to the development of therapeutic resistance via altered signaling pathways of cancer cells, modulating the major kinases through phosphorylation cascades by cytokines, but also may provide ways to treat the malignancy by exploiting the growth factor treatment with antibodies disrupting the tumor cell-TME crosstalk.

2.10.1 Stromal Cells

Hepatocyte growth factor (HGF) secretion by stromal cells has been shown to be responsible for cancer resistance in *BRAF*-mutant melanomas (Straussman et al. 2012; Masuda and Izpisua Belmonte 2013; Lohr et al. 2014). Targeted therapy of human lung adenocarcinoma with RAF, ALK, or EGFR kinase inhibitors leads to therapy-induced signaling changes driven by TME which leads to drug resistance. AKT pathway is hyperactivated, and dual inhibition of RAF and the PI3K/AKT/mTOR intracellular signaling helps check the growth of drug-resistant cell population, mitigating the TME secretome effects. This combination therapy can work against relapse due to promotion of resistant clones by tumor microenvironment secretome (Obenauf et al. 2015).

HGF also induces EMT by modulating the Hippo signaling pathway which controls the tumor proliferation and apoptosis (Farrell et al. 2014). HGF not only activates EMT-related c-MET pathway but also provides resistance to conventional EGFR inhibitors in lung cancers (Wang et al. 2009). TME stromal contribution seems to be the most powerful neutralizing factor against personalized cancer therapy. For example, in prostate cancer, chemotherapy resistance develops after stromal cells exposed to DNA damage induce NF-κB-mediated Wnt-16B expression. This promotes signaling changes that lead to tumor cell survival and enhances treatment resistance (Sun et al. 2012).

2.10.2 Cancer-Associated Fibroblasts (CAFs)

The TME orchestrates disease progression and dominates therapeutic responses with active help from fibroblasts. The normal cancer fibroblasts can turn into cancer-associated fibroblasts (CAFs) on stimulation by fibroblast growth factor (FGF), monocyte chemotactic protein 1 (MCP-1), platelet-derived growth factor (PDGF), tissue inhibitor of metalloproteinase 1 (TIMP-1), and tumor transforming growth factor β (TGF-β) (Quail and Joyce 2013; Song et al. 2015). Cancer-associated fibroblasts proliferate aggressively, show enhanced ECM deposition, and produce a secretome that triggers chemoresistance and consists of interleukins and growth factors (HGF, IL-6, PDGF, SDF-1, VEGF, etc.) that modulate the cancer tissue signaling toward aggressive growth and proliferation. These factors promote angiogenesis and vascular permeability and are actively driven by phosphorylation cascades.

2.10.3 Vasculature System

Tumor vasculature develops from new vessels and modification of old vessels by differentiation of endothelial precursors from the bone marrow. The vasculature develops by remodeling and is the major source of oxygen and nutrient transport to the tumor tissue. Despite this vital task, the poorly developed and loosely formed vasculature may lead to hypoxic conditions and limitations of growth factors. This may form an infiltration gradient from tumor foci to the vasculature. Mesenchymal stem cells (MSCs), tumor-associated macrophages (TAMs), and CAFs jointly contribute to tumor vascularization and secrete several angiogenic growth factors like vascular endothelial growth factor A (VEGFA) into the TME. VEGFA has been inversely correlated with worsening prognosis in lung, renal, and colorectal carcinomas (Kise et al. 2015; Mumenthaler et al. 2015; Sun 2015).

2.10.4 Tumor-Associated Endothelial Cells (TECs)

Tumor-associated endothelial cells (TECs) overexpress chemokine CXC motif ligand receptor (CXCR7) that promotes angiogenesis through activating the ERK1/ERK2 phosphorylation pathway. The ligand (CXCL12) for CXCR7 is absent in normal endothelial cells. Since this CXCL12–CXCR7 autocrine loop is involved in TEC-associated pro-angiogenesis, tumor growth, lung metastasis, and resistance, it is a viable target for anti-angiogenesis therapies that are targeted to disrupt the formation of tumor blood vessels (Sun 2015).

2.10.5 Extracellular Matrix

All cell types within a cancer microenvironment contribute to the formation of extracellular matrix (ECM). ECM provides the basic structural support for the tumor and serves as the framework for signal integration, movement for tumor cells, proliferation, and progression. ECM also drives the cell adhesion-mediated resistance to drugs constituting the integrins like fibronectin, collagen, and laminin, which can attenuate activities of RTKs like EGFR (Bishop et al. 1995; Byron et al. 2011; Pontiggia et al. 2012).

2.10.6 TME-Derived Exosomes

Exosomes carry a large pool of bioactive compounds. Exosomes mediate growth and metastasis by activating the receptor tyrosine kinase, MET-mediated signaling by phosphorylation (Peinado et al. 2012). Therapeutic antibodies like *rituximab* and *trastuzumab* are trapped by exosomes and rendered ineffective leading to resistance (Aung et al. 2011; Ciravolo et al. 2012). Fibroblast-secreted exosomes induce Wnt signaling-dependent enhanced motility and protrusion of tumor cells. Upregulated STAT1 and NOTCH3 signaling are also known to be associated with enhancing the resistance of breast cancer subpopulations (Boelens et al. 2014). Tumor-derived microvesicles mediate human breast cancer invasion through hyper-glycosylated extracellular matrix metalloproteinase inducer (EMMPRIN). This modified EMMPRIN brings about a pro-invasive effect by activating the p38/MAPK signaling pathway in tumor cells (Menck et al. 2015).

2.10.7 Inflammatory/Immune Cells

Cells of the innate immune system provide a generic defense mechanism as well as regulation of cellular homeostasis and wound healing, which are essential components for tumor survival. Tumor cells are nurtured by tumor-associated macrophages (TAMs) by regulating signaling changes via the paracrine loop, that involves EGF/CSF-1 signaling between TAMs and these two and also through the activation of WNT signaling in these macrophages (Sun 2015). TAMs contribute cysteine cathepsins to drive tumor progression by regulation of angiogenesis and tumor growth. TAMs also secrete matrix metallopeptidases (MMPs) that not only degrade the ECM but also increase the availability of ECM-bound factors such as VEGFA, required for signaling changes leading to angiogenesis (Shain et al. 2000; Quail and Joyce 2013).

2.10.8 Cytokines and Growth Factors

Cytokines or interleukins are the small soluble molecules prominently found in immunological or hematopoietic cells for signal transduction (Wormald and Hilton 2004). When cytokines bind to their receptors, the receptors dimerize and cause the transphosphorylation of Janus kinase (JAK) at the tyrosine residues. Once JAK is phosphorylated, it phosphorylates the tyrosine residues in the cytoplasmic domain of the receptor, forming recognition sites for the signaling cascade proteins (Mufson 1997).

One of the downstream processes mediated by cytokines/interleukins is the activation of transcription factors like STAT through JAK phosphorylation. Once STAT is phosphorylated, it dimerizes and translocates to the nucleus to regulate transcription (Mufson 1997).

In tumor cells, this pathway is very important as STAT regulates the transcription of BCL-2 family proteins, which are actively involved in regulation of apoptosis. In normal cells, SOCS protein inhibits the activation of JAK/STAT pathway to regulate normal apoptosis. But in cancer cells, hypermethylation of gene coding for SOCS has been observed that causes the silencing of transcription of SOCS. In the absence of silencing proteins, STAT is phosphorylated, and BCL-2 family proteins are transcribed (Krebs and Hilton 2000).

3 Conclusion

With the appreciation of existence of PTM codes (Creixell and Linding 2012) governing the regulation of cellular signaling, growth, metabolism, and survival (Karve and Cheema 2011), there has been an increase in studies identifying and validating PTMs as the players involved in cancer cell resistance. In cancer cells, modulations of posttranslational modifications can lead to drug resistance as exemplified throughout the text. The recent therapeutic techniques that target the kinome of a cell to induce cell death or reduce cellular signaling are displaying great results with more than 25 drugs approved (Gross et al. 2015). This chapter provides a brief understanding on how PTMs regulate various mechanisms of drug resistance.

Due to their central role in signaling and our deep understanding of phosphorylation (kinase signaling networks) as compared to other PTMs, kinase inhibition has been the most powerful therapy till now for the treatment of various cancers (Radivojac et al. 2008; Li et al. 2010). Most therapeutics aiming to overcome the resistance problem are dependent on harnessing combination therapy, where multiple checkpoints are controlled instead of one. With an increased understanding of the role of other PTMs and their crosstalk along with phosphorylation signaling, better therapeutic interventions can be developed.

Acknowledgement AKY is supported by the Innovative Young Biotechnologist Award (IYBA) and MK is supported by the IYBA Junior Research Fellowship (IYBA-JRF) from the Department of Biotechnology, India. Suruchi Aggarwal is supported by the Senior Research Fellowship (SRF) from the Indian Council of Medical Research (ICMR), India. AKY and Shailendra Asthana also acknowledge DDRC-SFC grant from the Department of Biotechnology, India. Authors thank Dr. Kanury V.S. Rao for critical comments on the manuscript.

References

Andre F, Berrada N, Desmedt C (2010) Implication of tumor microenvironment in the resistance to chemotherapy in breast cancer patients. Curr Opin Oncol 22:547–551. doi:10.1097/CCO.0b013e32833fb384

Aung T, Chapuy B, Vogel D, Wenzel D, Oppermann M, Lahmann M, Weinhage T, Menck K, Hupfeld T, Koch R, Trumper L, Wulf GG (2011) Exosomal evasion of humoral immunotherapy in aggressive B-cell lymphoma modulated by ATP-binding cassette transporter A3. Proc Natl Acad Sci USA 108:15336–15341. doi:10.1073/pnas.1102855108

Barouch-Bentov R, Sauer K (2011) Mechanisms of drug resistance in kinases. Expert Opin Investig Drugs 20:153–208. doi:10.1517/13543784.2011.546344

Barski A, Cuddapah S, Cui K, Roh TY, Schones DE, Wang Z, Wei G, Chepelev I, Zhao K (2007) High-resolution profiling of histone methylations in the human genome. Cell 129:823–837. doi:10.1016/j.cell.2007.05.009

Bergink S, Jentsch S (2009) Principles of ubiquitin and SUMO modifications in DNA repair. Nature 458:461–467. doi:10.1038/nature07963

Bishop LA, Rahman D, Pappin DJ, Watt FM (1995) Identification of an 80kD protein associated with the alpha 3 beta 1 integrin as a proteolytic fragment of the alpha 3 subunit: studies with human keratinocytes. Cell Adhes Commun 3:243–255. doi:10.3109/15419069509081290

Boelens MC, Wu TJ, Nabet BY, Xu B, Qiu Y, Yoon T, Azzam DJ, Twyman-Saint VC, Wiemann BZ, Ishwaran H, Ter Brugge PJ, Jonkers J, Slingerland J, Minn AJ (2014) Exosome transfer from stromal to breast cancer cells regulates therapy resistance pathways. Cell 159:499–513. doi:10.1016/j.cell.2014.09.051

Bouwman P, Jonkers J (2012) The effects of deregulated DNA damage signalling on cancer chemotherapy response and resistance. Nat Rev Cancer 12:587–598. doi:10.1038/nrc3342

Brooks CL, Gu W (2003) Ubiquitination, phosphorylation and acetylation: the molecular basis for p53 regulation. Curr Opin Cell Biol 15:164–171. doi:10.1016/S0955-0674(03)00003-6

Buonato JM, Lazzara MJ (2014) ERK1/2 blockade prevents epithelial-mesenchymal transition in lung cancer cells and promotes their sensitivity to EGFR inhibition. Cancer Res 74:309–319. doi:10.1158/0008-5472.CAN-12-4721

Byron A, Humphries JD, Bass MD, Knight D, Humphries MJ (2011) Proteomic analysis of integrin adhesion complexes. Sci Signal 4:t2. doi:10.1126/scisignal.2001827

Chen Y, Sanchez Y (2004) Chk1 in the DNA damage response: conserved roles from yeasts to mammals. DNA Repair (Amst) 3:1025–1032. doi:10.1016/j.dnarep.2004.03.003

Ciravolo V, Huber V, Ghedini GC, Venturelli E, Bianchi F, Campiglio M, Morelli D, Villa A, Della MP, Menard S, Filipazzi P, Rivoltini L, Tagliabue E, Pupa SM (2012) Potential role of HER2-overexpressing exosomes in countering trastuzumab-based therapy. J Cell Physiol 227: 658–667. doi:10.1002/jcp.22773

Cortot AB, Janne PA (2014) Molecular mechanisms of resistance in epidermal growth factor receptor-mutant lung adenocarcinomas. Eur Respir Rev 23:356–366. doi:10.1183/09059180.00004614

Creixell P, Linding R (2012) Cells, shared memory and breaking the PTM code. Mol Syst Biol 8: 598. doi:10.1038/msb.2012.33

Cunningham JJ, Gatenby RA, Brown JS (2011) Evolutionary dynamics in cancer therapy. Mol Pharm 8:2094–2100. doi:10.1021/mp2002279

Dawson MA, Kouzarides T (2012) Cancer epigenetics: from mechanism to therapy. Cell 150: 12–27. doi:10.1016/j.cell.2012.06.013

Easwaran H, Tsai HC, Baylin SB (2014) Cancer epigenetics: tumor heterogeneity, plasticity of stem-like states, and drug resistance. Mol Cell 54:716–727. doi:10.1016/j.molcel.2014.05.015

Enoch T, Norbury C (1995) Cellular responses to DNA damage: cell-cycle checkpoints, apoptosis and the roles of p53 and ATM. Trends Biochem Sci 20:426–430. doi:10.1016/S0968-0004(00)89093-3

Farrell J, Kelly C, Rauch J, Kida K, Garcia-Munoz A, Monsefi N, Turriziani B, Doherty C, Mehta JP, Matallanas D, Simpson JC, Kolch W, von Kriegsheim A (2014) HGF induces epithelial-to-mesenchymal transition by modulating the mammalian hippo/MST2 and ISG15 pathways. J Proteome Res 13:2874–2886. doi:10.1021/pr5000285

Fattah F, Lee EH, Weisensel N, Wang Y, Lichter N, Hendrickson EA (2010) Ku regulates the non-homologous end joining pathway choice of DNA double-strand break repair in human somatic cells. PLoS Genet 6:e1000855. doi:10.1371/journal.pgen.1000855

Ferlay J, Shin HR, Bray F, Forman D, Mathers C, Parkin DM (2010) Estimates of worldwide burden of cancer in 2008: GLOBOCAN 2008. Int J Cancer 127:2893–2917. doi:10.1002/ijc.25516

Geske FJ, Nelson AC, Lieberman R, Strange R, Sun T, Gerschenson LE (2000) DNA repair is activated in early stages of p53-induced apoptosis. Cell Death Differ 7:393–401. doi:10.1038/sj.cdd.4400663

Gillet JP, Gottesman MM (2010) Mechanisms of multidrug resistance in cancer. Methods Mol Biol 596:47–76. doi:10.1007/978-1-60761-416-6_4

Goldie JH (2001) Drug resistance in cancer: a perspective. Cancer Metastasis Rev 20:63–68

Gottesman MM (2002) Mechanisms of cancer drug resistance. Annu Rev Med 53:615–627. doi:10.1146/annurev.med.53.082901.103929

Gottesman MM, Fojo T, Bates SE (2002) Multidrug resistance in cancer: role of ATP-dependent transporters. Nat Rev Cancer 2:48–58. doi:10.1038/nrc706

Graves LM, Duncan JS, Whittle MC, Johnson GL (2013) The dynamic nature of the kinome. Biochem J 450:1–8. doi:10.1042/BJ20121456
Gross S, Rahal R, Stransky N, Lengauer C, Hoeflich KP (2015) Targeting cancer with kinase inhibitors. J Clin Invest 125:1780–1789. doi:10.1172/JCI76094
Hanahan D, Weinberg RA (2000) The hallmarks of cancer. Cell 100:57–70. doi:10.1016/S0092-8674(00)81683-9
Hanahan D, Weinberg RA (2011) Hallmarks of cancer: the next generation. Cell 144:646–674. doi:10.1016/j.cell.2011.02.013
Hantschel O, Rix U, Superti-Furga G (2008) Target spectrum of the BCR-ABL inhibitors imatinib, nilotinib and dasatinib. Leuk Lymphoma 49:615–619. doi:10.1080/10428190801896103
Hatzivassiliou G, Song K, Yen I, Brandhuber BJ, Anderson DJ, Alvarado R, Ludlam MJ, Stokoe D, Gloor SL, Vigers G, Morales T, Aliagas I, Liu B, Sideris S, Hoeflich KP, Jaiswal BS, Seshagiri S, Koeppen H, Belvin M, Friedman LS, Malek S (2010) RAF inhibitors prime wild-type RAF to activate the MAPK pathway and enhance growth. Nature 464:431–435. doi:10.1038/nature08833
He C, Klionsky DJ (2009) Regulation mechanisms and signaling pathways of autophagy. Annu Rev Genet 43:67–93. doi:10.1146/annurev-genet-102808-114910
He C, Baba M, Klionsky DJ (2009) Double duty of Atg9 self-association in autophagosome biogenesis. Autophagy 5:385–387. doi:10.4161/auto.5.3.7699
Henrique R, Oliveira AI, Costa VL, Baptista T, Martins AT, Morais A, Oliveira J, Jeronimo C (2013) Epigenetic regulation of MDR1 gene through post-translational histone modifications in prostate cancer. BMC Genomics 14:898. doi:10.1186/1471-2164-14-898
Holohan C, Van SS, Longley DB, Johnston PG (2013) Cancer drug resistance: an evolving paradigm. Nat Rev Cancer 13:714–726. doi:10.1038/nrc3599
Housman G, Byler S, Heerboth S, Lapinska K, Longacre M, Snyder N, Sarkar S (2014) Drug resistance in cancer: an overview. Cancers (Basel) 6:1769–1792. doi:10.3390/cancers6031769
Jenuwein T, Allis CD (2001) Translating the histone code. Science 293:1074–1080. doi:10.1126/science.1063127
Karimi P, Shahrokni A, Ranjbar MR (2014) Implementation of proteomics for cancer research: past, present, and future. Asian Pac J Cancer Prev 15:2433–2438
Karve TM, Cheema AK (2011) Small changes huge impact: the role of protein posttranslational modifications in cellular homeostasis and disease. J Amino Acids 2011:1–13. doi:10.4061/2011/207691
Katayama R, Shaw AT, Khan TM, Mino-Kenudson M, Solomon BJ, Halmos B, Jessop NA, Wain JC, Yeo AT, Benes C, Drew L, Saeh JC, Crosby K, Sequist LV, Iafrate AJ, Engelman JA (2012) Mechanisms of acquired crizotinib resistance in ALK-rearranged lung Cancers. Sci Transl Med 4:120ra17. doi:10.1126/scitranslmed.3003316
Katayama K, Noguchi K, Sugimoto Y (2014) Regulations of P-Glycoprotein/ABCB1/MDR1 in human cancer cells. New J Sci 2014:1–10. doi:10.1155/2014/476974
Kellner S, Kikyo N (2010) Transcriptional regulation of the Oct4 gene, a master gene for pluripotency. Histol Histopathol 25:405–412
Kelly TK, De Carvalho DD, Jones PA (2010) Epigenetic modifications as therapeutic targets. Nat Biotechnol 28:1069–1078. doi:10.1038/nbt.1678
Kerr E, Holohan C, McLaughlin KM, Majkut J, Dolan S, Redmond K, Riley J, McLaughlin K, Stasik I, Crudden M, Van SS, Fenning C, O'Connor R, Kiely P, Sgobba M, Haigh D, Johnston PG, Longley DB (2012) Identification of an acetylation-dependant Ku70/FLIP complex that regulates FLIP expression and HDAC inhibitor-induced apoptosis. Cell Death Differ 19: 1317–1327. doi:10.1038/cdd.2012.8
Kise K, Kinugasa-Katayama Y, Takakura N (2015) Tumor microenvironment for cancer stem cells. Adv Drug Deliv Rev. doi:10.1016/j.addr.2015.08.005
Kobayashi S, Boggon TJ, Dayaram T, Janne PA, Kocher O, Meyerson M, Johnson BE, Eck MJ, Tenen DG, Halmos B (2005) EGFR mutation and resistance of non-small-cell lung cancer to gefitinib. N Engl J Med 352:786–792. doi:10.1056/NEJMoa044238

Komatsu M, Waguri S, Koike M, Sou YS, Ueno T, Hara T, Mizushima N, Iwata J, Ezaki J, Murata S, Hamazaki J, Nishito Y, Iemura S, Natsume T, Yanagawa T, Uwayama J, Warabi E, Yoshida H, Ishii T, Kobayashi A, Yamamoto M, Yue Z, Uchiyama Y, Kominami E, Tanaka K (2007) Homeostatic levels of p62 control cytoplasmic inclusion body formation in autophagy-deficient mice. Cell 131:1149–1163. doi:10.1016/j.cell.2007.10.035

Kongara S, Karantza V (2012) The interplay between autophagy and ROS in tumorigenesis. Front Oncol 2:171. doi:10.3389/fonc.2012.00171

Kosaka T, Yamaki E, Mogi A, Kuwano H (2011) Mechanisms of resistance to EGFR TKIs and development of a new generation of drugs in non-small-cell lung cancer. J Biomed Biotechnol 2011:165214. doi:10.1155/2011/165214

Krebs DL, Hilton DJ (2000) SOCS: physiological suppressors of cytokine signaling. J Cell Sci 113:2813–2819

Kroemer G, Martin SJ (2005) Caspase-independent cell death. Nat Med 11:725–730. doi:10.1038/nm1263

Krueger KE, Srivastava S (2006) Posttranslational protein modifications: current implications for cancer detection, prevention, and therapeutics. Mol Cell Proteomics 5:1799–1810. doi:10.1074/mcp.R600009-MCP200

Letai AG (2008) Diagnosing and exploiting cancer's addiction to blocks in apoptosis. Nat Rev Cancer 8:121–132. doi:10.1038/nrc2297

Li S, Iakoucheva LM, Mooney SD, Radivojac P (2010) Loss of post-translational modification sites in disease. Pac Symp Biocomput 2010:337–347. doi:10.1142/9789814295291_0036

Linn DE, Yang X, Sun F, Xie Y, Chen H, Jiang R, Chen H, Chumsri S, Burger AM, Qiu Y (2010) A role for OCT4 in tumor initiation of drug-resistant prostate cancer cells. Genes Cancer 1: 908–916. doi:10.1177/1947601910388271

Little SE, Popov S, Jury A, Bax DA, Doey L, Al-Sarraj S, Jurgensmeier JM, Jones C (2012) Receptor tyrosine kinase genes amplified in glioblastoma exhibit a mutual exclusivity in variable proportions reflective of individual tumor heterogeneity. Cancer Res 72:1614–1620. doi:10.1158/0008-5472.CAN-11-4069

Lohr JG, Stojanov P, Carter SL, Cruz-Gordillo P, Lawrence MS, Auclair D, Sougnez C, Knoechel B, Gould J, Saksena G, Cibulskis K, McKenna A, Chapman MA, Straussman R, Levy J, Perkins LM, Keats JJ, Schumacher SE, Rosenberg M, Getz G, Golub TR (2014) Widespread genetic heterogeneity in multiple myeloma: implications for targeted therapy. Cancer Cell 25:91–101. doi:10.1016/j.ccr.2013.12.015

Lord CJ, Ashworth A (2008) Targeted therapy for cancer using PARP inhibitors. Curr Opin Pharmacol 8:363–369. doi:10.1016/j.coph.2008.06.016

Lord CJ, Ashworth A (2012) The DNA damage response and cancer therapy. Nature 481:287–294. doi:10.1038/nature10760

Lord CJ, Ashworth A (2013) Mechanisms of resistance to therapies targeting BRCA-mutant cancers. Nat Med 19:1381–1388. doi:10.1038/nm.3369

Maiuri MC, Zalckvar E, Kimchi A, Kroemer G (2007) Self-eating and self-killing: crosstalk between autophagy and apoptosis. Nat Rev Mol Cell Biol 8:741–752. doi:10.1038/nrm2239

Malet-Martino M, Martino R (2002) Clinical studies of three oral prodrugs of 5-fluorouracil (capecitabine, UFT, S-1): a review. Oncologist 7:288–323. doi:10.1634/theoncologist.7-4-288

Manning G, Whyte DB, Martinez R, Hunter T, Sudarsanam S (2002) The protein kinase complement of the human genome. Science 298:1912–1934. doi:10.1126/science.1075762;298/5600/1912

Mao Y, Keller ET, Garfield DH, Shen K, Wang J (2013) Stromal cells in tumor microenvironment and breast cancer. Cancer Metastasis Rev 32:303–315. doi:10.1007/s10555-012-9415-3

Marsh DJ, Shah JS, Cole AJ (2014) Histones and their modifications in ovarian cancer - drivers of disease and therapeutic targets. Front Oncol 4:144. doi:10.3389/fonc.2014.00144

Martin LP, Hamilton TC, Schilder RJ (2008) Platinum resistance: the role of DNA repair pathways. Clin Cancer Res 14:1291–1295. doi:10.1158/1078-0432.CCR-07-2238

Masuda S, Izpisua Belmonte JC (2013) The microenvironment and resistance to personalized cancer therapy. Nat Rev Clin Oncol 10:79. doi:10.1038/nrclinonc.2012.127-c1

Menck K, Scharf C, Bleckmann A, Dyck L, Rost U, Wenzel D, Dhople VM, Siam L, Pukrop T, Binder C, Klemm F (2015) Tumor-derived microvesicles mediate human breast cancer invasion through differentially glycosylated EMMPRIN. J Mol Cell Biol 7:143–153. doi:10.1093/jmcb/mju047

Michael M, Doherty MM (2005) Tumoral drug metabolism: overview and its implications for cancer therapy. J Clin Oncol 23:205–229. doi:10.1200/JCO.2005.02.120

Mufson RA (1997) The role of serine/threonine phosphorylation in hematopoietic cytokine receptor signal transduction. FASEB J 11:37–44

Mumenthaler SM, Foo J, Choi NC, Heise N, Leder K, Agus DB, Pao W, Michor F, Mallick P (2015) The impact of microenvironmental heterogeneity on the evolution of drug resistance in cancer cells. Cancer Inform 14:19–31. doi:10.4137/CIN.S19338

Obenauf AC, Zou Y, Ji AL, Vanharanta S, Shu W, Shi H, Kong X, Bosenberg MC, Wiesner T, Rosen N, Lo RS, Massague J (2015) Therapy-induced tumour secretomes promote resistance and tumour progression. Nature 520:368–372. doi:10.1038/nature14336

Osborne JK, Zaganjor E, Cobb MH (2012) Signal control through Raf: in sickness and in health. Cell Res 22:14–22. doi:10.1038/cr.2011.193

Peinado H, Aleckovic M, Lavotshkin S, Matei I, Costa-Silva B, Moreno-Bueno G, Hergueta-Redondo M, Williams C, Garcia-Santos G, Ghajar C, Nitadori-Hoshino A, Hoffman C, Badal K, Garcia BA, Callahan MK, Yuan J, Martins VR, Skog J, Kaplan RN, Brady MS, Wolchok JD, Chapman PB, Kang Y, Bromberg J, Lyden D (2012) Melanoma exosomes educate bone marrow progenitor cells toward a pro-metastatic phenotype through MET. Nat Med 18:883–891. doi:10.1038/nm.2753

Pontiggia O, Sampayo R, Raffo D, Motter A, Xu R, Bissell MJ, Joffe EB, Simian M (2012) The tumor microenvironment modulates tamoxifen resistance in breast cancer: a role for soluble stromal factors and fibronectin through beta1 integrin. Breast Cancer Res Treat 133: 459–471. doi:10.1007/s10549-011-1766-x

Postel-Vinay S, Bajrami I, Friboulet L, Elliott R, Fontebasso Y, Dorvault N, Olaussen KA, Andre F, Soria JC, Lord CJ, Ashworth A (2013) A high-throughput screen identifies PARP1/2 inhibitors as a potential therapy for ERCC1-deficient non-small cell lung cancer. Oncogene 32:5377–5387. doi:10.1038/onc.2013.311

Quail DF, Joyce JA (2013) Microenvironmental regulation of tumor progression and metastasis. Nat Med 19:1423–1437. doi:10.1038/nm.3394

Radivojac P, Baenziger PH, Kann MG, Mort ME, Hahn MW, Mooney SD (2008) Gain and loss of phosphorylation sites in human cancer. Bioinformatics 24:i241–i247. doi:10.1093/bioinformatics/btn267

Sampath D, Cortes J, Estrov Z, Du M, Shi Z, Andreeff M, Gandhi V, Plunkett W (2006) Pharmacodynamics of cytarabine alone and in combination with 7-hydroxystaurosporine (UCN-01) in AML blasts in vitro and during a clinical trial. Blood 107:2517–2524. doi:10.1182/blood-2005-08-3351

Sawan C, Herceg Z (2010) Histone modifications and cancer. Adv Genet 70:57–85. doi:10.1016/B978-0-12-380866-0.60003-4

Schwickart M, Huang X, Lill JR, Liu J, Ferrando R, French DM, Maecker H, O'Rourke K, Bazan F, Eastham-Anderson J, Yue P, Dornan D, Huang DC, Dixit VM (2010) Deubiquitinase USP9X stabilizes MCL1 and promotes tumour cell survival. Nature 463:103–107. doi:10.1038/nature08646

Shain KH, Landowski TH, Dalton WS (2000) The tumor microenvironment as a determinant of cancer cell survival: a possible mechanism for de novo drug resistance. Curr Opin Oncol 12: 557–563

Shen H, He MM, Liu H, Wrighton SA, Wang L, Guo B, Li C (2007) Comparative metabolic capabilities and inhibitory profiles of CYP2D6.1, CYP2D6.10, and CYP2D6.17. Drug Metab Dispos 35: 1292–1300. doi:10.1124/dmd.107.015354

Singh A, Settleman J (2010) EMT, cancer stem cells and drug resistance: an emerging axis of evil in the war on cancer. Oncogene 29:4741–4751. doi:10.1038/onc.2010.215

Smith BN, Odero-Marah VA (2012) The role of Snail in prostate cancer. Cell Adh Migr 6: 433–441. doi:10.4161/cam.21687

Smith J, Tho LM, Xu N, Gillespie DA (2010) The ATM-Chk2 and ATR-Chk1 pathways in DNA damage signaling and cancer. Adv Cancer Res 108:73–112. doi:10.1016/B978-0-12-380888-2.00003-0

Song T, Dou C, Jia Y, Tu K, Zheng X (2015) TIMP-1 activated carcinoma-associated fibroblasts inhibit tumor apoptosis by activating SDF1/CXCR4 signaling in hepatocellular carcinoma. Oncotarget 6:12061-12079. doi: 10.18632/oncotarget.3616

Soverini S, Hochhaus A, Nicolini FE, Gruber F, Lange T, Saglio G, Pane F, Muller MC, Ernst T, Rosti G, Porkka K, Baccarani M, Cross NC, Martinelli G (2011) BCR-ABL kinase domain mutation analysis in chronic myeloid leukemia patients treated with tyrosine kinase inhibitors: recommendations from an expert panel on behalf of European LeukemiaNet. Blood 118: 1208–1215. doi:10.1182/blood-2010-12-326405

Straussman R, Morikawa T, Shee K, Barzily-Rokni M, Qian ZR, Du J, Davis A, Mongare MM, Gould J, Frederick DT, Cooper ZA, Chapman PB, Solit DB, Ribas A, Lo RS, Flaherty KT, Ogino S, Wargo JA, Golub TR (2012) Tumour micro-environment elicits innate resistance to RAF inhibitors through HGF secretion. Nature 487:500–504. doi:10.1038/nature11183

Sun Y (2015) Tumor microenvironment and cancer therapy resistance. Cancer Lett. doi:10.1016/j.canlet.2015.07.044

Sun XX, Yu Q (2015) Intra-tumor heterogeneity of cancer cells and its implications for cancer treatment. Acta Pharmacol Sin 36:1219–1227. doi:10.1038/aps.2015.92

Sun Y, Campisi J, Higano C, Beer TM, Porter P, Coleman I, True L, Nelson PS (2012) Treatment-induced damage to the tumor microenvironment promotes prostate cancer therapy resistance through WNT16B. Nat Med 18:1359–1368. doi:10.1038/nm.2890

Sung SY, Hsieh CL, Wu D, Chung LW, Johnstone PA (2007) Tumor microenvironment promotes cancer progression, metastasis, and therapeutic resistance. Curr Probl Cancer 31:36–100. doi:10.1016/j.currproblcancer.2006.12.002

Szerlip NJ, Pedraza A, Chakravarty D, Azim M, McGuire J, Fang Y, Ozawa T, Holland EC, Huse JT, Jhanwar S, Leversha MA, Mikkelsen T, Brennan CW (2012) Intratumoral heterogeneity of receptor tyrosine kinases EGFR and PDGFRA amplification in glioblastoma defines subpopulations with distinct growth factor response. Proc Natl Acad Sci USA 109:3041–3046. doi:10.1073/pnas.1114033109

To KK, Polgar O, Huff LM, Morisaki K, Bates SE (2008) Histone modifications at the ABCG2 promoter following treatment with histone deacetylase inhibitor mirror those in multidrug-resistant cells. Mol Cancer Res 6:151–164. doi:10.1158/1541-7786.MCR-07-0175

Tutt A, Bertwistle D, Valentine J, Gabriel A, Swift S, Ross G, Griffin C, Thacker J, Ashworth A (2001) Mutation in Brca2 stimulates error-prone homology-directed repair of DNA double-strand breaks occurring between repeated sequences. EMBO J 20:4704–4716. doi:10.1093/emboj/20.17.4704

van de Kooij B, Rooswinkel RW, Kok F, Herrebout M, de Vries E, Paauwe M, Janssen GM, van Veelen PA, Borst J (2013) Polyubiquitination and proteasomal turnover controls the anti-apoptotic activity of Bcl-B. Oncogene 32:5439–5448. doi:10.1038/onc.2013.99

Voulgari A, Pintzas A (2009) Epithelial-mesenchymal transition in cancer metastasis: mechanisms, markers and strategies to overcome drug resistance in the clinic. Biochim Biophys Acta 1796:75–90. doi:10.1016/j.bbcan.2009.03.002

Wang W, Li Q, Yamada T, Matsumoto K, Matsumoto I, Oda M, Watanabe G, Kayano Y, Nishioka Y, Sone S, Yano S (2009) Crosstalk to stromal fibroblasts induces resistance of lung cancer to epidermal growth factor receptor tyrosine kinase inhibitors. Clin Cancer Res 15: 6630–6638. doi:10.1158/1078-0432.CCR-09-1001

Wang XQ, Ongkeko WM, Chen L, Yang ZF, Lu P, Chen KK, Lopez JP, Poon RT, Fan ST (2010) Octamer 4 (Oct4) mediates chemotherapeutic drug resistance in liver cancer cells through a potential Oct4-AKT-ATP-binding cassette G2 pathway. Hepatology 52:528–539. doi:10.1002/hep.23692

Woods D, Turchi JJ (2013) Chemotherapy induced DNA damage response: convergence of drugs and pathways. Cancer Biol Ther 14:379–389. doi:10.4161/cbt.23761

Wormald S, Hilton DJ (2004) Inhibitors of cytokine signal transduction. J Biol Chem 279: 821–824. doi:10.1074/jbc.R300030200

Yook JI, Li XY, Ota I, Fearon ER, Weiss SJ (2005) Wnt-dependent regulation of the E-cadherin repressor snail. J Biol Chem 280:11740–11748. doi:10.1074/jbc.M413878200

Zahreddine H, Borden KL (2013) Mechanisms and insights into drug resistance in cancer. Front Pharmacol 4:28. doi:10.3389/fphar.2013.00028

Zhao QW, Zhou YW, Li WX, Kang B, Zhang XQ, Yang Y, Cheng J, Yin SY, Tong Y, He JQ, Yao HP, Zheng M, Wang YJ (2015) Akt mediated phosphorylation of Oct4 is associated with the proliferation of stemlike cancer cells. Oncol Rep 33:1621–1629. doi:10.3892/or.2015.3752

Zheng H, Shen M, Zha YL, Li W, Wei Y, Blanco MA, Ren G, Zhou T, Storz P, Wang HY, Kang Y (2014) PKD1 phosphorylation-dependent degradation of SNAIL by SCF-FBXO11 regulates epithelial-mesenchymal transition and metastasis. Cancer Cell 26:358–373. doi:10.1016/j.ccr. 2014.07.022

Zhou BP, Deng J, Xia W, Xu J, Li YM, Gunduz M, Hung MC (2004) Dual regulation of Snail by GSK-3 beta-mediated phosphorylation in control of epithelial-mesenchymal transition. Nat Cell Biol 6:931–940. doi:10.1038/ncb1173

Ovarian Cancer and Resistance to Therapies: Clinical and Laboratory Perspectives

Riyaz Basha, Zainab Mohiuddin, Abdul Rahim, and Sarfraz Ahmad

Abstract The cancer that originates in ovary is one of the most deadly gynecological malignancies. Despite advances in surgical and therapeutic options, patient survival remains poor in ovarian cancer. Debulking surgery is the primary option to manage patients with this malignancy. Radiation has been used in adjuvant therapy for ovarian cancer patients but has largely been replaced with platinum-based chemotherapy. Response of chemotherapy is impulsive in some patients, and long-term analyses showed recurrence of disease in approximately half of the patients. Several combinations or regimen were tested for achieving optimal response and increasing ovarian cancer patient survival. Morbidity associated with intensive therapy and resistance to widely used chemotherapy are the major concerns in treating ovarian cancer patients. Therefore, it is important for developing effective strategies to sensitize malignant cells to standard therapy. Such strategies have the potential in achieving improved response with minimal side effects. Platinum-based drugs are abundantly used for treating initial malignancy and recurrent disease, especially, improving the response of platinum-based chemotherapy and addressing drug resistance is highly beneficial to treat relapsed patients or the patients with advanced-stage disease. This chapter summarizes research findings in this area with the support of current peer-reviewed literature and elaborates future directions for improving the therapeutic response in ovarian cancer.

R. Basha
Preclinical Services, Molecular & Medical Genetics, Institute for Cancer Research, University of North Texas Health Science Center, Fort Worth, TX 76107, USA

Z. Mohiuddin • A. Rahim
Department of Gynecologic Oncology, Florida Hospital Cancer Institute, Orlando, FL 32804, USA

S. Ahmad, Ph.D., F.A.B.A.P., F.A.C.B. (✉)
Department of Gynecologic Oncology, Florida Hospital Cancer Institute, Orlando, FL 32804, USA

UCF & FSU Colleges of Medicine, 2501 N. Orange Ave., Suite 786, Orlando, FL 32804, USA
e-mail: sarfraz.ahmad@flhosp.org

G. Arora et al. (eds.), *Drug Resistance in Bacteria, Fungi, Malaria, and Cancer*,
DOI 10.1007/978-3-319-48683-3_23

1 Introduction

Ovarian cancer typically occurs in the ovary, fallopian tubes, and primary peritoneal cavity. It is the deadliest form of gynecologic malignancies. According to recent estimates, in the United States, 21,290 new cases and 14,180 deaths are expected to occur during the year 2015 (Siegel et al. 2015). Age is an important factor for the incidences of this malignancy. The median age of ovarian cancer patients is 65 years, and about 88 % of patients are aged 45 years and older. Unfortunately, due to the unsystematic nature of the disease at early stage and limitations in effective screening, >75 % of patients are diagnosed at advanced stages (Das and Bast 2008).

The risk factors for this malignancy are not fully understood. Some of the common risk factors include start of menstrual cycle before 12 years of age, late menopause (after 52 years of age), first child at the older age than 30 years, family history of ovarian or other cancers such as breast and colon, and infertility and/or use of fertility medicines. Ashkenazi Jewish women's heredity and mutations in genes BRCA1 and BRCA2 are also among the common risk factors for this malignancy (Nelson et al. 1993; Carlson et al. 1994; Satagopan et al. 2002; Carcangiu et al. 2006).

To date, platinum- and taxane-based combination therapy has been the prominent therapy for the effective treatment and management of patients with ovarian, fallopian tube, and the primary peritoneal cancers. Unfortunately, a sizable number of patients experience drug-related toxicity, and some of the side effects are so serious that cause the discontinuation of therapy. Despite advances in therapeutic options for ovarian cancer treatment, the disease prognosis and 5-year survival rate are still relatively poor, and patients often develop resistance to therapy.

2 Biology of the Disease

2.1 *Occurrence and Subtypes*

Ovarian cancer consists of numerous related, albeit distinct, tumors. Depending on the origin site and type of cells, ovarian cancer tumors are mainly of three types: (1) epithelial, (2) germ cell, and (3) stromal. Transformation of surface epithelium seems to contribute epithelial ovarian cancer. The exact mechanism associated with this transformation is not yet fully uncovered. It has been implicated that pro-inflammatory state may contribute to the ovarian carcinogenesis. Epithelial cell type of ovarian malignancies is the predominant form and is responsible for about 90 % of the incidences (and the majority of deaths) due to this cancer. Histologically, epithelial ovarian tumors are further divided in subtypes: (1) serous, (2) mucinous, (3) endometrioid, (4) clear cell, and (5) transitional. Ovarian germ cell tumors develop from the cells that produce the egg (or ova). Stromal tumors are typically rare in ovarian cancer. These tumors originate from the cells of connective tissue, which support ovary and secreted hormones (Kalli et al. 2004; Rosen et al. 2009; Romero and Bast 2012). Similar to other cancers, patients with ovarian cancer respond well if the disease is diagnosed at early stage. Since this disease is mainly asymptomatic in the early stages, most patients are diagnosed at advanced stage. Unfortunately, in such instances, especially due to metastasis, the treatment options become somewhat less effective.

2.2 *Hypercoagulability in Ovarian Cancer*

Platelet activation and systemic coagulation often present in patients that characterize the thrombotic form of ovarian cancer (Wang et al. 2005; Holmes et al. 2009). Due to thromboembolic episodes, ovarian cancer patients are more likely to develop blood clots. Factors underlying the hypercoagulable state in ovarian cancer have been implicated with tumorigenesis and metastasis. Immunohistochemical studies have shown that ovarian cancer cells can generate a complete coagulation pathway. They are able to generate thrombin and activate platelets, causing both inflammation and thrombosis (Amirkhosravi et al. 2013; Chen et al. 2013).

Correlation of hypercoagulability and high levels of activation markers is associated in patients with recurrent and at advanced stage of the disease. Thus, identification of hypercoagulable patients would be important not only for effective thromboprophylaxis but also for possible future treatment strategies for histologically proven patients with ovarian cancer. Crucial laboratory markers/tests including D-dimer, hemoglobin, platelet and leukocyte count, selectins (e.g., soluble P-selection), tissue factor, factor VIII, and C-reactive protein can potentially aid and/or predict the hypercoagulability in ovarian cancer (Amirkhosravi et al. 2013; Tas et al. 2013).

2.3 *Detection and Diagnostic Usefulness of Tumor Biomarkers*

It is well known that the lack of specific biomarkers makes it difficult to detect the malignancy at a relatively early stage of the disease. This is mainly due to the fact of complex classification of this malignancy. This classification of the disease is based on clinical presentation, tumorigenesis, and the analysis of gene expression profiles. Due to such complexity, it is challenging to identify precise biomarkers for the disease prognosis. Following are some of the well-known biomarkers and their clinical applications in ovarian cancer.

CA125 The glycoprotein antigen, cancer antigen 125 (CA125), also known as mucin 16 (MUC16), typically presents on the cell surface of tissues derived from coelomic epithelia. Ovary and fallopian tube tissue expresses CA125. The sensitivity of CA125 is about 50 % in stage I ovarian cancer patients, but its association is highly sensitive (~90 %) for the patients with advanced-stage disease (van Nagell et al. 2007). It has been approved by the Food & Drug Administration (FDA) as a biomarker for determining residual or recurrent epithelial ovarian cancer (EOC) patients, especially for the cases after their first-line therapy. Since approximately 80 % of ovarian cancer patients show elevated levels of CA125 (Skates and Singer 1991), monitoring its levels is routinely used for determining the effectiveness of the treatment. One of the limitations is the lack of specificity. Elevated levels of CA125 can also be found in some benign conditions and other cancer patients including breast, endometrial, colon, and pancreatic cancers. The levels of CA125 and sonographic analysis of the adnexal mass are used for determining the risk stratification. The CA125 concentration, menopausal status (MS), and ultrasound score (US) are used to measure the risk of malignancy index (RMI) (Anton et al. 2012). To calculate the RMI, the concentration of CA125 is multiplied by MS and US (Anton et al. 2012).

HE4 Human epididymis protein 4 (HE4) is a member of whey acidic four disulfide core (WFDC) family proteins. It is overexpressed in all endometrioid ovarian carcinomas and some other non-mucinous carcinomas such as serous (93 %) and clear cell tumors (50 %). HE4 was approved by the FDA in June 2008 to monitor the progressive or recurrent disease in EOC patients. Notably, it is not approved as a screening test for ovarian cancer in asymptomatic women. The biomarker is not widely available on automated immunoassay platforms. Physicians often use both HE4 and CA125 for assessing risk of ovarian cancer at the pelvic mass surgery in women.

The use of ROMA (Risk of Ovarian Malignancy Algorithm) has recently been proposed, which accounts for the concentration of the two analytes along with the menopausal status of a given subject leading to the generation of score (at 0–10 scale). Based on this score, the low or high likelihood of the malignancy can be determined utilizing appropriate cutoff values. Obviously, a subject with ROMA scale higher than the cutoff level should be referred to gynecologic oncologist to rule out the possibility of higher risk for the disease (along with other clinical assessments/judgments, including the women's pre- or postmenopausal status).

OVA1 This multi-analyte assay is relatively newer, which was FDA cleared in 2009 as an aid in assessing the risk of malignancy in subjects presenting with an adnexal mass (when independent clinical/radiological evaluation for malignancy is indeterminate). This assay measures serum levels of CA125, transthyretin (prealbumin), apolipoprotein A1, β2-microglobulin, and transferrin together. The assay utilizes two immunoassay platforms, i.e., Roche Elecsys (for CA125) and Siemens BNII (for rest of the four analytes). The OvaCalc (company's proprietary software) uses the values of these analytes along with the menopausal status of the subject in order to calculate the score. While a score of $\geq$5.0 in premenopausal subject may suggest high probability of malignancy, a score of $\geq$4.4 in postmenopausal subject. Notably, OVA1 score above the cutoff is not a diagnosis of cancer,

Table 1 Useful clinical applications of key serum biomarkers for the detection of ovarian cancer

Biomarkers			
Useful applications	CA125	HE4	OVA1
Screening	No	No	No
Discrimination of pelvic masses	Yes[a]	Yes[b]	Yes
Monitoring treatment	Yes	Yes	No
Detection of disease recurrence	Yes	Yes	No

[a]As a component of the RMI or ROMA score
[b]As a component of the ROMA score
CA125: Cancer antigen 125
HE4: Human epididymis protein 4
OVA1: A registered biomarker blood test measuring 5 candidates: (1) CA-125, (2) apolipoprotein A1, (3) beta-2 microglobulin, (4) transthyretin, and (5) transferrin
RMI: Risk of malignancy index
ROMA: Risk of ovarian malignancy algorithm

but indicative of increased risk of malignancy. One must recognize that this test does not replace currently available other methods of risk assessment (or screening). We summarize the clinical applications and usefulness of these three widely available serum biomarkers for ovarian cancer testing in Table 1.

3 Treatment Options for Ovarian Cancer

3.1 Standard of Care

Similar to many malignancies, surgery and chemotherapy are used as the standard options for treating patients with ovarian cancer. After the debulking surgery, chemotherapy involving platinum-based drugs and taxane is given to patients (Rueda et al. 2010; Schorge et al. 2010; Raja et al. 2012). Even though the initial therapy is effective in most of the cases, almost all patients develop chemoresistance, and the disease can relapse in a majority of patients.

The biological behavior of ovarian cancer suggests that the malignant portion is confined to peritoneal cavity and hence chemotherapy is often delivered via intraperitoneal (IP) administration. The IP chemotherapy has limited systemic toxicity and exposes higher drug concentrations to malignant area and highly effective for improving the survival of ovarian cancer patients (Markman et al. 2001; Armstrong et al. 2006; Frenel et al. 2011). While surgery and chemotherapy are part of the standard of care, rarely radiation is used for the treatment of this malignancy. Despite multiple clinical trials and attempts to improve the therapeutic efficacy, still the outcomes remain poor. The current therapeutic options are often intensive and cause severe side effects. The morbidities associated with current intensive therapies support the development of alternative therapeutic strategies, which may enhance the effectiveness of current treatment regimens.

Radiation therapy is commonly used as an adjuvant therapy (depending on the tumor size and related histopathological features/assessments) in most cancers, including ovarian cancer. In some cases, surgery and radiation therapy seem to be more effective than just chemotherapy or surgery followed by chemotherapy (Dembo 1992; Thomas 1993). Whole abdominal and pelvic irradiation (WART) was used to treat certain subsets of ovarian cancer; however, now it is largely switched to chemotherapy (Goldberg and Peschel 1988; Franchin et al. 1991; Fyles et al. 1997; Firat et al. 2003). Recently, IP chemotherapy is showing higher response when compared to intravenous (IV) administration (Markman 2003). Use of radiation is primarily not encouraged due to its toxicity—and lack of proper clinical trials also made it somewhat impossible to assess the benefits of radiation in combination with current chemotherapeutic options.

3.2 Chemoresponse Assays

Chemoresponse assay, often referred to as chemotherapy sensitivity and resistance assay (CSRA), reports a panel of markers characterizing a tumor's response to multiple chemotherapy agents (Brower et al. 2008). Different types of CSRAs are used in testing. Adenosine triphosphate (ATP) and methylthiazolyl-diphenyl-tetrazolium bromide (MTT) assays are used for measuring cell growth inhibition. Other assays include human tumor cloning assay (HTCA) and extreme drug resistance (EDR) assay. These assays provide tumor response information aimed at aiding in the selection of effective, individualized treatment regimens. The assays are generally based on phenotypic rather than molecular characterization. The advantages and disadvantages of various CSRA methods are debatable and require further validation.

Currently, two best validated assays are commercially available for solid tumors in the United States, i.e., the Microculture-Kinetic (MiCK) assay (DiaTech Oncology, Franklin, TN) and the ChemoFx® assay (Precision Therapeutics, Inc., Pittsburgh, PA). MiCK assay is based on drug-induced apoptosis. Validation of MiCK assay is used for determining the overall survival of naive patients diagnosed at advanced-stage primary ovarian cancer. This assay can serve as an independent predictor for the survival of such patients. The ChemoFx® assay characterizes both the sensitivity and resistance of the tumor, quantifying the chemotherapeutic effect by direct visualization and enumeration of live cells following exposure to the treatments (Grendys et al. 2014). The assay process is highly automated; hence the process strongly contributes to the high throughput and reproducibility of the assay requiring minimum specimens. Enormous peer-reviewed literature exists about the analytical performance and clinical validation of ChemoFx® assay. It is relatively inexpensive and rather more effective for predicting the treatment outcomes. Thus, the inclusion of either chemoresponse assay results, in concert with other clinical assessments/factors/biomarkers, proves to be quite helpful toward the management of ovarian cancer patients.

3.3 *Role of Single Nucleotide Polymorphisms*

Single nucleotide polymorphisms (SNPs) are employed to assess the risk of adverse events in chemotherapy in ovarian cancer patients. SNPs in genes associated with platinum and taxane metabolism/detoxification have been correlated to increased risk of severe adverse events (AE) when patients receive these drugs (Moxley et al. 2013). In addition, SNPs in genes for drug resistance proteins and antiapoptotic proteins have been associated with cardiotoxicity in adriamycin/doxorubicin-treated patients and hematological toxicities in gemcitabine-treated patients (Okazaki et al. 2010; Volkova and Russell 2011; Blanco et al. 2012). The combination of platinum- and taxane-based regimen is used to treat most of the patients. While initial response rates are extremely high (>90 %), some patients experience severe AE which can lead to discontinuation of the therapy. The Gynecologic Oncology Group (GOG) and paclitaxel package insert include treatment guidelines that involve a decreased dose regimen, if AEs are encountered. Hence, clinical validation of the genetic differences in these genes as biomarkers for severe AEs would be helpful strategy for the treating physicians to alter dosing, thereby increasing the time a patient could remain on the drug while decreasing side effects and unnecessary morbidity.

Another clinical utility of these genetic differences in ovarian cancer patient care is the identification of such patients who may not benefit from IP chemotherapy or dose-dense chemotherapy. While IP chemotherapy has been shown to improve the patient outcomes, the side effects are much more frequent and severe due to the high dose (Wenzel et al. 2007). Testing the patients prior to the treatment for predictive genotypes may factor into a treating physician's decision to forego IP chemotherapy in favor of the standard IV delivery.

Dose-dense chemotherapy has demonstrated improvements in outcome but also some increases in side effects (Katsumata et al. 2009; Glaze et al. 2013). In both therapy regimens, the ability to stratify patients based on the risks of toxicities associated with treatment may lead to a greater benefit of IP or dose-dense therapy while minimizing side effects and associated healthcare costs. Furthermore, in the recurrent setting, both doxorubicin and gemcitabine are treatment options referenced in the guidelines of the National Comprehensive Cancer Network (NCCN). Using genetic risk factors to personalize drug selection may minimize side effects in the recurrent setting and improve drug response and outcomes.

It is well established that the expression of certain proteins such as p53, glutathione S-transferase (GST), and Excision Repair Cross-Complementation Group1 (ERCC1) potentially impacts the response to platinum-based treatment in ovarian cancer patients (Hirazono et al. 1995; Ferrandina et al. 1999; Bali et al. 2004; Howells et al. 2004; Steffensen et al. 2008, 2009; Scheil-Bertram et al. 2010; Milovic-Kovacevic et al. 2011). When SNP in the genes that encode these proteins were evaluated, a correlation to AE in response to platinum-based therapies was made (Khrunin et al. 2010; Sakano et al. 2010). Mutations that affected activity of ERCC1 were associated with nephrotoxicity. Mutations in the GST family members were associated with neutropenia (GSTA1), neuropathy (GSTM3 and GSTP1), or anemia and thrombocytopenia (GSTM3). Mutations in p53 were associated with neutropenia. In these studies, mutations in XPD and XRCC1 were also found to be predictive of neutropenia (both XPD and XRCC1) and anemia (XPD only). Hence, validating additional SNPs in patients with ovarian cancer may further predict adverse events when treated with therapies that include platinum, taxanes, doxorubicin, and/or gemcitabine. Using this technology (determination of genotypes for SNP in blood specimens) may stratify patients to different dosing regimens and routes of administration or, in recurrent cancer to aid in drug selection, may improve outcomes and potentially reduce the costs associated with the management of drug-related side effects while not changing the standard of care.

3.4 Cellular and Immune Therapies

New treatments improve short-term and median survivals, but not long-term survival or cure. Hence, the use of IP cellular therapy in combination with cytokines to treat the disease has been advocated as a novel combination of enhanced immune therapy. IP chemotherapy has been used in the United States since the late 1980s and has resulted in an improvement in time-to-progression and overall survival (Armstrong et al. 2006). Intra-tumoral T-cells are correlated with improvement in survival indicating an important role of the immune system (Zhang et al. 2003a). In addition, cytokines are currently being used to treat a variety of malignancies; therefore, the combination of IP cellular therapy with known biologically active cytokines is a rational hypothesis for an innovative approach to this localized disease.

Cytokine-induced killer cells (CIK) are currently being investigated as effector cells in cellular therapy treatment regimens. CIK are a heterogeneous population of cells that exhibit nonmajor histocompatibility complex antitumor effect (Li et al. 2012). However, tumors are known to produce an immunosuppressive environment, which limits the immune system's ability to fight the tumor and is a major obstacle in the development of successful immunotherapeutic treatment regimens. Myeloid-derived suppressor cells (MDSC) and T regulatory cells (Tregs) have been identified in a majority of cancer types and are known to inhibit both innate and adaptive immunity (Ostrand-Rosenberg and Sinha 2009).

Immunotherapy with stimulated immune cells also provided some evidence for effectively treating ovarian cancer. Research from Zhang et al. (2003a) group studied the overall survival of advanced-stage (III/IV) patients and reported that infiltration of CD+ T-cells was associated with improved survival. The 5-year overall survival was 4.5 % and 38 %, respectively, for the patients with low T-cell infiltration and high T-cell infiltration. This report suggested a strong association of immune system response in the survival of ovarian cancer patients (Zhang et al.

2003a). It was also proposed that the patients with higher immune response possibly will have a relatively longer disease-free survival (Nelson 2008).

Laboratory experimental observations demonstrated that healthy immune effector cells induce cytotoxicity in ovarian cancer cells. Preclinical studies also showed that cytokines (IL-2: interleukin-2; IFNα-2b: interferon α-2b) in combination with peripheral blood mononuclear cells (PBMC) caused higher inhibition of ovarian cancer cell proliferation and tumor growth in mouse model (xenografts) for ovarian cancer. The animal studies also suggested that this combination was not toxic and did not result in overt hazard effects (Ingersoll et al. 2009, 2011, 2012). More recently, using the xenograft mouse model, we further demonstrated that treatment with healthy PBMC (IP) and IL-2 (IP) results in improved survival compared to "only IL-2"-treated mice (Ingersoll et al. 2015). These preclinical results provided the required evidence and facilitated to potentially initiate a phase 1 clinical trial for the treatment of ovarian cancer.

3.5 Emerging Role of HIPEC in Ovarian Cancer

Use of hyperthermic intraperitoneal chemotherapy (HIPEC) procedure is a relatively newer form of IP therapy. It is an alternative locoregional treatment strategy employed for treating the advanced-stage ovarian cancer patients (Helm 2009). This strategy is used instantly after the surgery and tumor resection. It is designed to increase the distribution and penetration of chemotherapy for improved results. Currently, there are several ongoing prospective clinical trials evaluating the HIPEC approach to ascertain its definitive role in treating patients with ovarian cancer.

The HIPEC delivery technique requires intraoperative perfusion machines, elaborate logistics, and a high degree of organizational effort. Potential advantages that make it a promising therapeutic option as part of a multimodality treatment are: (1) a high volume of chemotherapy can be delivered, and a homogenous distribution can be achieved, (2) there is no interval between cytoreduction and chemotherapy, (3) hyperthermia has a pharmacokinetic benefit, including tumor penetration of cisplatin and DNA cross-linking, and (4) low systemic exposure to chemotherapy and higher distribution in IP compartments can be achieved (in a single intraoperative treatment) (Oseledchyk and Zivanovic 2015).

It is to be cautioned, based on the currently available scientific evidence, that HIPEC should not be considered yet a standard therapeutic option after optimal cytoreduction in advanced ovarian cancer, nor should it be offered outside of a clinical trial. One of the limitations with HIPEC approach is "small disease volume" requirement; hence the fundamental question remains whether the hyperthermia is worth the risk of renal toxicity in the absence of greater insight into platinum resistance and its reversal by hyperthermia. Indeed, as noted above, the ongoing randomized clinical trials in ovarian cancer population may address some of these concerns (Bakrin et al. 2014; Oseledchyk and Zivanovic 2015).

4 Drug Resistance and Treatment Options

In ovarian cancer treatment, platinum-based drugs and taxane are used as the first-line treatment (Parmar et al. 2003) while cisplatin, doxorubicin, gemcitabine, and taxane derivatives are used as the second line of therapy (Pfisterer et al. 2006; Ferrandina et al. 2008; Sehouli et al. 2008). Platinum-based therapy is effective in majority of the patients with epithelial ovarian cancer, but relapse is common among the advanced-staged disease patients. Cisplatin causes cytotoxicity via inducing apoptosis through interacting with DNA. Drug resistance is responsible for majority of deaths in these patients, and the underlying mechanisms associated with drug resistance are not fully uncovered. In the patients treated at advanced-stage disease (III and/or IV), even though they seem to respond well initially, relapse is common in the majority of cases. Cisplatin is a commonly used drug in both the first- and second-line chemotherapy (Januchowski et al. 2013).

Even though a few distinct markers are established, still sensitive biomarkers for the early diagnosis of the disease and assessing the prognosis are lacking. There are certain effective approaches that are being extensively tested, and a few are currently under investigation to identify sensitive markers and effective pathways/candidates to target ovarian cancer treatment.

4.1 Modulators of Apoptosis

Bcl2 Family Apoptosis is biologically controlled cell death that is primarily mediated by two important families, Bcl2 and inhibitor of apoptosis proteins (IAP) (Oto et al. 2007). Bcl2 members include both pro- and anti-apoptotic properties. Bcl-2 was first discovered and immensely studied as anti-death gene in cellular processes (Yip and Reed 2008). Since Bcl-2 family proteins mediate cell death, there are multiple studies that clearly elucidated the involvement of alterations in antiapoptotic and/or proapoptotic members of the Bcl-2 family proteins in ovarian cancer (Witham et al. 2007; Chaudhry et al. 2010, 2012; Yasmeen et al. 2011; Aust et al. 2013; Zhou et al. 2015). Inhibitors of Bcl-2 are tested for evaluating their efficacy in ovarian cancer treatment and also on overcoming drug resistance. Bcl-2 family proteins include both death agonists and antagonists and regulate apoptosis in cells. Analysis of clinical specimens revealed that high expression of death antagonists, Bcl-2 and Bcl-XL, is associated with low survival rates, whereas high expression of death agonist, Bax, is known to a relatively longer survival in epithelial ovarian cancer patients. It has also been established that upregulation of Bcl-2 family proteins is associated with resistance to chemotherapy in some cancers (Adams and Cory 2007; Vogler et al. 2009). The NCI panel of 60 cell lines representing several malignancies showed a negative correlation of Bcl-XL and drug sensitivity (Amundson et al. 2000; Vogler et al. 2009). Overall, as presented in Fig. 1, the overexpression of death agonists, Bcl-2, Bcl-XL, and Mcl-1,

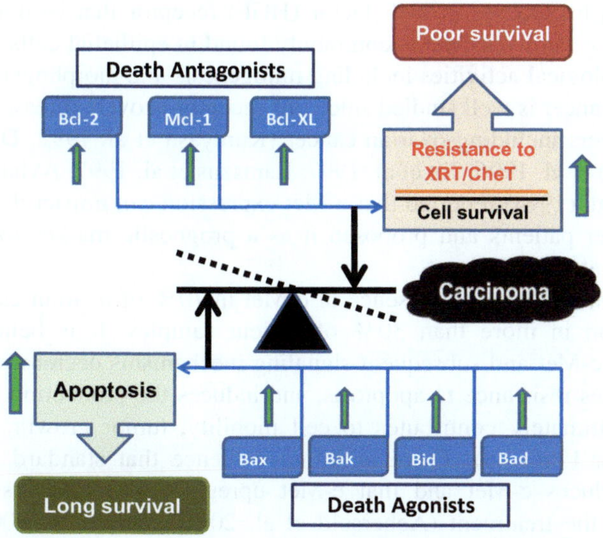

Fig. 1 Ovarian cancer: Distinct functions of Bcl-2 family proteins. The Bcl-2 family proteins have both death antagonists and death agonists. Bcl-2, Bcl-XL, and Mcl-1 are classified as death antagonists. Bax, Bak, Bid, and Bad are death agonists. Overexpression of Bcl-2, Bcl-XL, and Mcl-1 suppresses apoptosis, induces resistance to radiation/chemotherapy, and causes poor survival. The upregulation of death agonists (Bax, Bak, Bid, and Bad) increases apoptosis, enhances the response to therapy, and helps to achieve longer survival. The imbalance between the death agonists and death antagonists impacts the distinct biological features of carcinoma

suppresses apoptosis, increases resistance to radiation/chemotherapy, and is associated with poor survival, while the upregulation of death agonists, Bax, Bak, Bid, and Bad, induces apoptosis, potentiates response to the treatments, and thereby helps for longer survival (Wang et al. 2015). The imbalance between these two categories of Bcl-2 family proteins impacts the biological features of carcinoma (Giovannetti et al. 2006). Furthermore, screening of Bcl-2 inhibitors (e.g., TW-37) demonstrated the anticancer activity and potentiation of the effect of chemotherapeutic agent, cisplatin, in preclinical models of ovarian cancer (Zeitlin et al. 2008; Wang et al. 2015).

Survivin The IAP family has eight members, which play roles in cell survival. Specifically, DNA damage and activation of effector caspases are countered by the members of the IAP family. Survivin is one of the best studied IAPs in cancer, which mediates several biochemical and cellular functions and impacts the cell survival. Studies have shown the functional roles of survivin in both the cell division and apoptosis control (Deveraux and Reed 1999; Altieri 2006). Survivin is overexpressed in several malignancies, and its upregulation is associated with a relatively poor prognosis. It is also evident that the expression of survivin is elevated in radiation-resistant cell lines, and inhibiting survivin causes sensitization to radiation or chemotherapy (Lu et al. 2004; Shinohara et al. 2004; Rodel et al. 2005; Pennati et al. 2007; Konduri et al. 2009).

c-Met It is a hepatocyte growth factor (HGF) receptor that belongs to tyrosine kinase receptor family. c-Met is commonly found in epithelial cells and arbitrated to several biological activities including mitogenesis and morphogenesis. Its association with cancer is well studied since this receptor is overexpressed/amplified in multiple cancers including ovarian cancer (Kuniyasu et al. 1992; Di Renzo et al. 1994; Medico et al. 1996; Jin et al. 1997; Lamszus et al. 1997; Ayhan et al. 2005). Di Renzo et al. (1994) reported that c-Met expression can impact the prognosis of ovarian cancer patients and proposed it as a prognostic marker for this disease (Di Renzo et al. 1994).

Research has shown the presence of c-Met in 70 % of ovarian carcinomas and overexpression in more than 30 % of patient samples. It is believed that the activation of c-Met and subsequent signaling mechanisms decreases cell proliferation, increases resistance to apoptosis, and induces the production of serine proteases and ultimately contributes to cell mobility, tumor growth, and invasion (Jeffers et al. 1996a, b). There is strong evidence that standard therapy (e.g., radiation) induces c-Met and that c-Met upregulation is also associated with resistance to the treatment (Aebersold et al. 2001; Qian et al. 2003; Lal et al. 2005; Chu et al. 2006; Bhardwaj et al. 2012). De Bacco et al. (2011) showed that radiation upregulates c-Met, and induction of Met causes resistance to radiation and invasive growth of cancer cells (De Bacco et al. 2011). The study suggested that radiation upregulates c-Met and subsequent signaling cascades that triggers pro-survival activity, which diffuses the response of the standard treatment (De Bacco et al. 2011). The activation of c-Met in cancer cells is also believed to protect from DNA damage and c-Met inhibitor-caused radiosensitization (Welsh et al. 2009). These studies strongly supported the role of c-Met in inducing the cell survival and resistance to the standard treatment options and suggested that the strategies to suppress its activation will be a viable option to enhance the response of cancer treatment.

4.2 Targeting Sp Transcription Factors

The Specificity protein (Sp)-family of transcription factors have been shown to regulate a variety of genes involved in critical processes (ranging from cell cycle, proliferation, cell differentiation, and apoptosis). The specific Sp transcription factors (i.e., Sp1, Sp3, and Sp4) bind to GC-rich promoter sites and regulate key sets of genes associated with cancer. Hence, the role of Sp proteins is obvious in the development of various cancers. The Sp1 expression correlates with the aggressive disease and poor prognosis. Sp proteins also regulate vascular endothelial growth factor (VEGF) and play key role(s) in tumorigenesis (Abdelrahim et al. 2006; Bermudez et al. 2007).

An interesting link exists between Sp proteins and the other two candidates (c-Met and survivin), as described above. While multiple candidates are involved in the aggressive disease and poor prognosis, Sp1 and Sp3 transcription factors mediate the expression of both c-Met and survivin. High relevance of Sp proteins have been implicated in the signaling cascade associated with c-Met activation. Both c-Met and survivin contain GC-rich promoters, and specific Sp proteins (e.g., Sp1 and Sp3) mediate their expression (Zhang et al. 2003b; Papineni et al. 2009; Basha et al. 2011; Colon et al. 2011; Sankpal et al. 2012). Recent preclinical studies emphasize that downregulation of c-Met and survivin is well correlated with the decrease in Sp1 and Sp3 expression (Papineni et al. 2009; Basha et al. 2011; Colon et al. 2011).

Understanding the key players associated with the aggressive disease and inducing resistance to therapy is critical in treating cancer. It is clear that both c-Met and survivin are responsible for cancer cell survival, aggressive disease, and poor prognosis and also induce resistance to radiation and chemotherapy. Research focusing on the strategies to target either c-Met or survivin to induce sensitivity to irradiation and/or chemotherapy has been attempted previously. It is plausible that clinical response may be achievable using a strategy to target these two candidates through transcriptional regulation. Targeting Sp1 and Sp3 potentially modulates c-Met, survivin, and downstream cascades, thereby inhibiting tumor development, and also induces radiosensitization.

Mithramycin A (Mit-A) is a known Sp protein inhibitor, which has been tested in several cancers, including ovarian cancer (Blume et al. 1991; Previdi et al. 2010). Despite its efficacy for inhibiting cancer cells and tumor growth, it has not been able to gain popularity in cancer treatment due to apparent toxicity and side effects (Parsons et al. 1971; Kofman et al. 1973; Margileth et al. 1973). The analogs of Mit-A were introduced so as to enhance therapeutic efficacy and alleviate the issues related to side effects, which are currently under preclinical testing (Previdi et al. 2010; Fernandez-Guizan et al. 2014, 2015; Vizcaino et al. 2014).

There is growing evidence to support the use of nonsteroidal anti-inflammatory drugs (NSAIDs) as anticancer agents (Jacoby et al. 2000; Tarnawski and Jones 2003; Gately and Li 2004). Abdelrahim et al. (2006) screened the ability of various structural classes of NSAIDs to better identify effective candidates with the ability to decrease the levels of Sp proteins associated with cancer using the orthotopic animal model for pancreatic cancer. This study identified a fenamate, tolfenamic acid (TA), as a potent agent (also an NSAID) for inhibiting pancreatic cancer cell proliferation and tumor growth in mice (Abdelrahim et al. 2006). TA activates the degradation of specific Sp transcription factors (such as Sp1, Sp3, and Sp4), thereby reduces the VEGF expression and decreases tumor growth and metastasis (Abdelrahim et al. 2006). Recent studies from our laboratories and research collaborations further confirmed the relationship of Sp proteins in various cancers models. These studies revealed that targeting Sp1 with TA serve as a promising approach for cancer therapy (Abdelrahim et al. 2006; Konduri et al. 2009; Papineni et al. 2009; Basha et al. 2011; Eslin et al. 2013; Sankpal et al. 2012; 2016). The primary advantage of suppressing Sp proteins is to regulate their downstream targets such as survivin, c-Met, and the proteins associated with cell cycle. Inhibition of survivin and c-Met potentially increases apoptosis and drug sensitivity. Modulating the expression of candidates required to pass various phases (G0/G1, S,

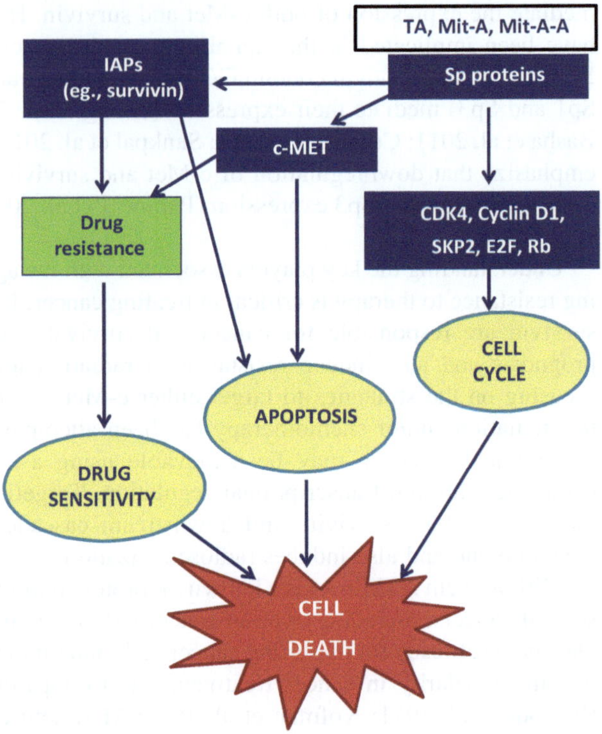

Fig. 2 Inhibition of Sp proteins for inducing anticancer activity. Sp proteins regulate the expression of survivin, c-Met, and key proteins that regulate cell cycle phase distribution. Inhibition of survivin and c-Met induces cell death via increasing apoptosis and sensitivity to standard therapy. In a complementary mechanism, the inhibition of Sp proteins can also modulate the expression of CDK4, cyclin D1, SKP2, E2F, and Rb, which are required for cell cycle phase distribution and ceases the cell division thereby inducing cell death in cancer cells

and G2) of cell cycle such as CDK4, cyclin D1, SKP2, E2F, and Rb can cause cell cycle arrest, cease the cell division, and lead to cell growth inhibition. By targeting the candidates involved in cell survival, drug resistance, and cell cycle phase distribution, inhibitors of Sp proteins can effectively enhance the cancer cell death in cancer cells (Fig. 2).

4.3 Multiple Drug Resistance

There are several mechanisms proposed to understand the drug resistance. The ability of cancer cells to eliminate the drug from the cellular compartments via transport proteins is an important mechanism known to cause drug resistance (Stavrovskaya 2000; Kruh 2003). This may facilitate cancer cells to develop insensitivity leading to multiple drug resistance (MDR). It is especially detrimental since MDR in cancer cells may indiscriminately develop resistance to pharmacological agents along with anticancer (cytotoxic) agents (Januchowski et al. 2014). Extensive research revealed the involvement of certain transmembrane proteins that belong to ABC family for inducing MDR (Leonard et al. 2003). Glycoprotein P (P-gp), MDR-related protein 1 (MRP1), and MRP2 are the important members of the ABC family proteins, which are implicated in the MDR (Januchowski et al. 2014). The expression of P-gp is often upregulated by chemotherapy leading the cancer cells to acquire MDR. Overexpression of P-gp induces the removal of multiple (approximately 20) anticancer agents including doxorubicin, paclitaxel, and vincristine. MRP1 and MRP2 are associated with developing resistance to cisplatin in ovarian cancer (Cole et al. 1992; Leonard et al. 2003; Surowiak et al. 2006; Januchowski et al. 2014). Januchowski et al. (2013) developed ovarian cancer cell lines resistant to chemotherapeutic drugs such as cisplatin, doxorubicin, methotrexate, paclitaxel, topotecan, and vincristine. These cells were tested for the signatures of drug resistance and found to be correlated with P-gp overexpression and resistance to doxorubicin, paclitaxel, and vincristine. The MRP1, MRP2, P-gp, and breast cancer resistance protein (BCRP) are ABC family proteins, which are encoded by ABCC1, ABCC2, ABCB1, and ABCG2, respectively. The non-ABC family protein, lung resistance-related protein (LRP)/major vault protein (MVP) is Vault 1 gene. All of these proteins are implicated in developing the resistance to such chemotherapeutic drugs as doxorubicin, vincristine, cisplatin, paclitaxel, topotecan, and etoposide. Drugs and the proteins that are associated with developing resistance are presented schematically in Fig. 3.

5 Economic Impact

It is generally recognized that ovarian cancer diagnosis could be a major event in a person who is diagnosed with the disease and her loved ones also. Given the complexity and often prolonged nature of the disease, the life of patient with ovarian cancer and potentially the family members face multiple challenges, including financial toxicity. A recent economic analysis of caregiving in advanced ovarian cancer patients demonstrated that, in addition to the patient, there are significant burdens that cancer treatment puts on the patients' families and the society at large (Angioli et al. 2015). Both direct and indirect medical costs are attributable to the patients' cancer leading to adverse economic impact (burden), particularly to the elderly patients. The geographic location of the patient and caregiving facilities also impacts the overall surgical and treatment economics. As the number of new ovarian cancer patients is expected to increase in the future due to life expectancy (despite trends in the overall improvement in the quality of life and survival), the cost of cancer treatment is obviously expected go higher. Likewise, this will have ripple effect on the families and caregivers with regard to financial burden. Thus, healthcare professionals should identify the potential risk and significant burden of the financial toxicities associated with the disease and be considerate while treating/managing ovarian cancer patients.

Over the years, the clinical validation of chemoresponse assays has demonstrated a significant increase in the overall survival of patients with recurrent ovarian cancer that were treated with such therapies to which their tumor was sensitive/resistant in the assay system. In a most recent study, cost-effectiveness of

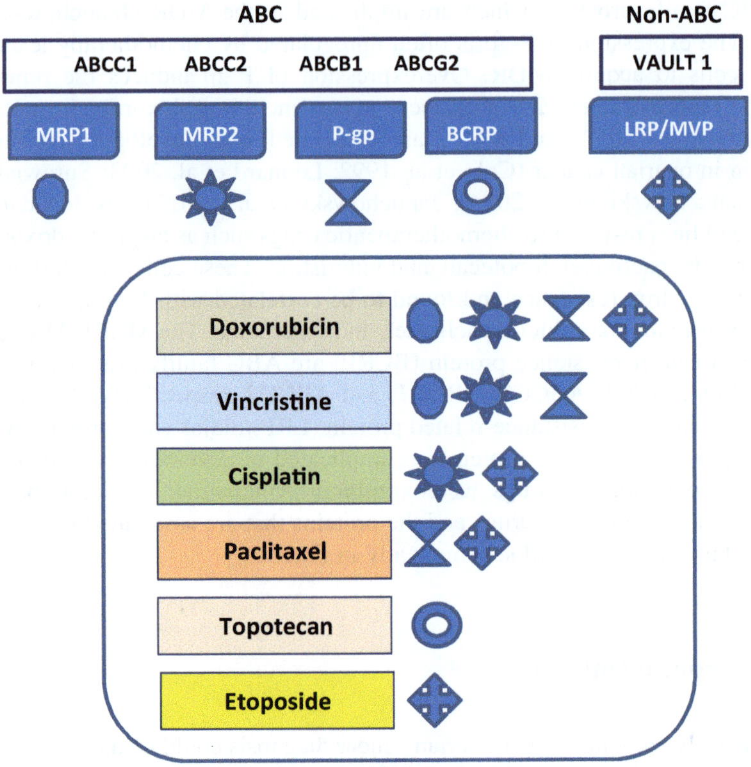

Fig. 3 Proteins implicated in multiple drug resistance (MDR). The ABC family proteins are implicated in MDR. The MDR-related protein 1 (MRP1), MRP2, glycoprotein P (P-gp), breast cancer resistance protein (BCRP) and non-ABC family protein, and lung resistance-related protein (LRP)/major vault protein (MVP) are the ABC family proteins that play active role in developing resistance to chemotherapeutic drugs. MRP1, MRP2, P-gp, BCRP, and LRP/MVP proteins are encoded by ABCC1, ABCC2, ABCB1, ABCG2, and VAULT1 genes, respectively. Each protein is given a symbol, and specific protein(s) associated with developing resistance to each drug is indicated with the corresponding symbols

using the assay during ovarian cancer recurrence (from the payer's perspective) has been reported (Plamadeala et al. 2015). The authors found that the cost-effectiveness was associated with platinum-sensitive and platinum-resistant subject treated with assay-sensitive therapies, which also concluded that the use of a chemoresponse assay for treatment decisions in recurrent ovarian cancer patients has the potential for cost-effectiveness in both platinum-sensitive as well as platinum-resistant subjects. Early palliative care intervention has also been shown to have the potential for reducing the costs associated with the end-of-life care for recurrent and platinum-resistant ovarian cancer patients (Lowery et al. 2013).

6 Future Perspectives

Traditionally, chemotherapy has been utilized successfully to treat ovarian cancer; however, peritoneal carcinomatosis is associated with a relatively poor prognosis and survival. Surgery alone is inadequate at the microscopic level, and systemic chemotherapy is of limited value because of the peritoneum–plasma barrier. Multiple alternate regimens have been utilized, most of them based on the platinum and taxane standard, but augmented with additional therapies and/or altered sequencing. For patients with recurrent, persistent, or progressive ovarian cancer, chemotherapy choice is generally based on the duration and type of response to the initial therapy. The greatest potential for ovarian cancer cure appears to be aggressive cytoreductive surgery to no residual disease followed by IP chemotherapy.

Despite significant advances, there is urgent need and potential for developing newer strategies to kill ovarian tumor cells because of aggressive relapsed disease and drug-resistant cancers. Predictive biomarkers and its measurements in tumor or blood specimens that can help to determine how well currently approved and investigational agents work will be highly advantageous. Studying the compounds' detailed mechanism(s) of action, their potential resistance mechanism(s), and how to combine them with other therapeutic agents (and targets) will be promising approach. Part of the success for combining drugs is not only to have better therapeutic effects on the ovarian tumors but also to delay or overcome any kind of resistance mechanisms. Studies are attempted to unraveling precise markers associated with resistance to platinum drugs and finding new candidates including ATP11B (Moreno-Smith et al. 2013) and microRNAs (e.g., let-7 family) (Cai et al. 2013). Apart from the compounds to modify Bcl2 family proteins (Witham et al. 2007; Wang et al. 2015), new agents for targeting DNA repair and inhibiting poly (ADP-ribose) polymerase (Mukhopadhyay et al. 2010; Ratner et al. 2012; Wiggans et al. 2015) are under rigorous screening in laboratory experiments (Fig. 4). Small molecules such as buthionine sulfoximine, triethylenetetramine, colchicine, genistein, and rapamycin are also under preclinical testing for using to reverse the resistance to cisplatin (Yellepeddi et al. 2012). Research is much needed in this direction to identify effective drugs for preventing tumors to acquire resistance (Fig. 4). Silencing gene expression is an emerging technology for the applications in cancer treatment. It took more than a decade since its inception to improve this tool for using in therapeutic application. Precise knockdown of oncogenes using short interfering RNAs (siRNAs) is under testing in several malignancies, including ovarian cancer (Fig. 4).

Relative to the traditional treatment strategies using standard chemotherapeutic agents, there are potentially excellent benefits for using the new technologies. Research on understanding drug resistance including multiple drug resistance and the specific markers associated with resistance to commonly used chemotherapeutic drugs is currently underway. Targeted therapy using novel drug delivery systems and microRNAs for diagnosis is gaining momentum. Gene silencing is also currently utilized as an effective tool in ovarian cancer research. The important

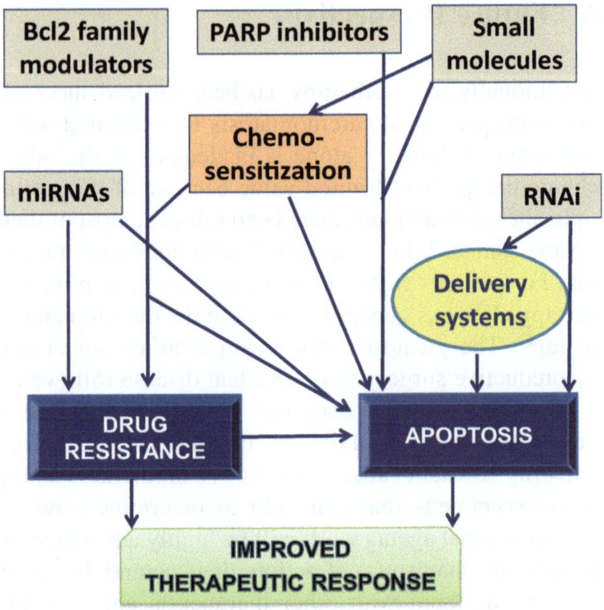

Fig. 4 Strategies for improving therapeutic response in ovarian cancer. Multiple strategies are currently under testing for improving the therapeutic response in patients with ovarian cancer. Modulators of Bcl2 family proteins, PARP inhibitors, small molecule microRNAs, and RNAi technology are used to induce apoptosis, while some of these candidates can also induce chemosensitization or work against drug resistance. Effective drug delivery systems are also under development for specifically introducing the agents to the target site(s)

advantages of gene silencing include the following: it requires less quantity (dose), specifically targets the genes of interest, and is effective against complex druggable targets (Bumcrot et al. 2006; Goldberg 2013). But the major limitations are stability and delivering the compound to the target site (Pecot et al. 2011; Wu et al. 2014). Future studies should focus on overcoming such constraints and bringing more agents for clinical testing. Especially, improving the stability and developing effective drug delivery systems could pave the way for utilizing the application of personalized care for the treatment of ovarian cancer.

7 Conclusions

Ovarian cancer is the second most important malignancy occurring among women. Despite significant advances in surgical and therapeutic methods, still serious concerns exist for treating the patients with recurrent disease and the issues related to drug resistance. Combination treatments using multiple chemotherapeutic agents and involving other compounds such as small molecules are under screening and some of these strategies are at various stages of clinical testing. Recently, more sophisticated tools are becoming available to surgeons and medical oncologists to apply in surgery and clinical interventions. Novel agents including the modulators of Bcl2 family proteins, inhibitor of apoptosis family proteins, and small molecules as chemo-sensitizers are showing promising results, at least in preclinical investigations. Cellular and immune therapy coupled with genomic analysis and genetic tastings are especially adding further strengths for diagnosis in personalized care to cancer patients. These tools are significant milestones in cancer therapy even though optimal treatment options are not yet available for all patients. Recent rapid advances in both diagnosis and therapy are expected to unfold existing obstacles and provide effective personalized care for ovarian cancer patients to improve quality of life and survival and may completely cure this devastating disease someday.

Conflict of Interest Statement We declare that there are no potential conflicts of interest pertaining to this article.

Acknowledgments RB is supported by the Institute for Cancer Research and Department of Pediatrics, University of North Texas Health Science Center, Fort Worth, TX.

References

Abdelrahim M, Baker CH, Abbruzzese JL, Safe S (2006) Tolfenamic acid and pancreatic cancer growth, angiogenesis, and Sp protein degradation. J Natl Cancer Inst 98:855–868. doi:10.1093/jnci/djj232

Adams JM, Cory S (2007) The Bcl-2 apoptotic switch in cancer development and therapy. Oncogene 26:1324–1337. doi:10.1038/sj.onc.1210220

Aebersold DM, Kollar A, Beer KT, Laissue J, Greiner RH, Djonov V (2001) Involvement of the hepatocyte growth factor/scatter factor receptor c-met and of Bcl-xL in the resistance of oropharyngeal cancer to ionizing radiation. Int J Cancer 96:41–54

Altieri DC (2006) The case for survivin as a regulator of microtubule dynamics and cell-death decisions. Curr Opin Cell Biol 18:609–615. doi:10.1016/j.ceb.2006.08.015

Amirkhosravi A, Bigsby G, Desai H, Rivera-Amaya M, Coll E, Robles-Carrillo L, Faust P, Waters A, Meyer T, Reyes E, Langer F, Francis JL (2013) Blood clotting activation analysis for preoperative differentiation of benign versus malignant ovarian masses. Blood Coagul Fibrinolysis 24:510–517. doi:10.1097/MBC.0b013e32835e63b7

Amundson SA, Myers TG, Scudiero D, Kitada S, Reed JC, Fornace AJ Jr (2000) An informatics approach identifying markers of chemosensitivity in human cancer cell lines. Cancer Res 60:6101–6110

Angioli R, Capriglione S, Aloisi A, Miranda A, de Cicco NC, Terranova C, Adrower R, Plotti F (2015) Economic impact among family caregivers of patients with advanced ovarian cancer. Int J Gynecol Cancer 25:1541–1546. doi:10.1097/IGC.0000000000000512

Anton C, Carvalho FM, Oliveira EI, Maciel GA, Baracat EC, Carvalho JP (2012) A comparison of CA125, HE4, risk ovarian malignancy algorithm (ROMA), and risk malignancy index (RMI) for the classification of ovarian masses. Clinics (Sao Paulo) 67:437–441

Armstrong DK, Bundy B, Wenzel L, Huang HQ, Baergen R, Lele S, Copeland LJ, Walker JL, Burger RA (2006) Intraperitoneal cisplatin and paclitaxel in ovarian cancer. N Engl J Med 354:34–43. doi:10.1056/NEJMoa052985

Aust S, Pils S, Polterauer S, Horvat R, Cacsire Castillo-Tong D, Pils D, Dudek G, Schmid B, Speiser P, Reinthaller A, Grimm C (2013) Expression of Bcl-2 and the antiapoptotic BAG family proteins in ovarian cancer. Appl Immunohistochem Mol Morphol 21:518–524. doi:10.1097/PAI.0b013e318284a053

Ayhan A, Ertunc D, Tok EC (2005) Expression of the c-Met in advanced epithelial ovarian cancer and its prognostic significance. Int J Gynecol Cancer 15:618–623. doi:10.1111/j.1525-1438.2005.00117.x

Bakrin N, Classe JM, Pomel C, Gouy S, Chene G, Glehen O (2014) Hyperthermic intraperitoneal chemotherapy (HIPEC) in ovarian cancer. J Visc Surg 151:347–353. doi:10.1016/j.jviscsurg.2014.07.005

Bali A, O'Brien PM, Edwards LS, Sutherland RL, Hacker NF, Henshall SM (2004) Cyclin D1, p53, and p21Waf1/Cip1 expression is predictive of poor clinical outcome in serous epithelial ovarian cancer. Clin Cancer Res 10:5168–5177. doi:10.1158/1078-0432.CCR-03-0751

Basha R, Ingersoll SB, Sankpal UT, Ahmad S, Baker CH, Edwards JR, Holloway RW, Kaja S, Abdelrahim M (2011) Tolfenamic acid inhibits ovarian cancer cell growth and decreases the expression of c-Met and survivin through suppressing specificity protein transcription factors. Gynecol Oncol 122:163–170. doi:10.1016/j.ygyno.2011.03.014

Bermudez Y, Yang H, Saunders BO, Cheng JQ, Nicosia SV, Kruk PA (2007) VEGF- and LPA-induced telomerase in human ovarian cancer cells is Sp1-dependent. Gynecol Oncol 106:526–537. doi:10.1016/j.ygyno.2007.05.005

Bhardwaj V, Zhan Y, Cortez MA, Ang KK, Molkentine D, Munshi A, Raju U, Komaki R, Heymach JV, Welsh J (2012) C-Met inhibitor MK-8003 radiosensitizes c-Met-expressing non-small-cell lung cancer cells with radiation-induced c-Met-expression. J Thorac Oncol 7:1211–1217. doi:10.1097/JTO.0b013e318257cc89

Blanco JG, Sun CL, Landier W, Chen L, Esparza-Duran D, Leisenring W, Mays A, Friedman DL, Ginsberg JP, Hudson MM, Neglia JP, Oeffinger KC, Ritchey AK, Villaluna D, Relling MV, Bhatia S (2012) Anthracycline-related cardiomyopathy after childhood cancer: role of polymorphisms in carbonyl reductase genes--a report from the Children's Oncology Group. J Clin Oncol 30:1415–1421. doi:10.1200/JCO.2011.34.8987

Blume SW, Snyder RC, Ray R, Thomas S, Koller CA, Miller DM (1991) Mithramycin inhibits SP1 binding and selectively inhibits transcriptional activity of the dihydrofolate reductase gene in vitro and in vivo. J Clin Invest 88:1613–1621. doi:10.1172/JCI115474

Brower SL, Fensterer JE, Bush JE (2008) The ChemoFx assay: an ex vivo chemosensitivity and resistance assay for predicting patient response to cancer chemotherapy. Methods Mol Biol 414:57–78

Bumcrot D, Manoharan M, Koteliansky V, Sah DW (2006) RNAi therapeutics: a potential new class of pharmaceutical drugs. Nat Chem Biol 2:711–719. doi:10.1038/nchembio839

Cai J, Yang C, Yang Q, Ding H, Jia J, Guo J, Wang J, Wang Z (2013) Deregulation of let-7e in epithelial ovarian cancer promotes the development of resistance to cisplatin. Oncogenesis 2: e75. doi:10.1038/oncsis.2013.39

Carcangiu ML, Peissel B, Pasini B, Spatti G, Radice P, Manoukian S (2006) Incidental carcinomas in prophylactic specimens in BRCA1 and BRCA2 germ-line mutation carriers, with emphasis on fallopian tube lesions: report of 6 cases and review of the literature. Am J Surg Pathol 30:1222–1230. doi:10.1097/01.pas.0000202161.80739.ac

Carlson KJ, Skates SJ, Singer DE (1994) Screening for ovarian cancer. Ann Intern Med 121:124–132

Chaudhry P, Srinivasan R, Patel FD (2010) Expression of the major fas family and Bcl-2 family of proteins in epithelial ovarian cancer (EOC) and their correlation to chemotherapeutic response and outcome. Oncol Res 18:549–559. doi:10.3727/096504010X12767359113884

Chaudhry P, Srinivasan R, Patel FD (2012) Differential expression of Fas family members and Bcl-2 family members in benign versus malignant epithelial ovarian cancer (EOC) in North Indian population. Mol Cell Biochem 368:119–126. doi:10.1007/s11010-012-1350-7

Chen LM, Wang W, Lee JC, Chiu FH, Wu CT, Tai CJ, Wang CK, Tai CJ, Huang MT, Chang YJ (2013) Thrombomodulin mediates the progression of epithelial ovarian cancer cells. Tumour Biol 34:3743–3751. doi:10.1007/s13277-013-0958-x

Chu SH, Zhu ZA, Yuan XH, Li ZQ, Jiang PC (2006) In vitro and in vivo potentiating the cytotoxic effect of radiation on human U251 gliomas by the c-Met antisense oligodeoxynucleotides. J Neurooncol 80:143–149. doi:10.1007/s11060-006-9174-5

Cole SP, Bhardwaj G, Gerlach JH, Mackie JE, Grant CE, Almquist KC, Stewart AJ, Kurz EU, Duncan AM, Deeley RG (1992) Overexpression of a transporter gene in a multidrug-resistant human lung cancer cell line. Science 258:1650–1654

Colon J, Basha MR, Madero-Visbal R, Konduri S, Baker CH, Herrera LJ, Safe S, Sheikh-Hamad-D, Abudayyeh A, Alvarado B, Abdelrahim M (2011) Tolfenamic acid decreases c-Met expression through Sp proteins degradation and inhibits lung cancer cells growth and tumor formation in orthotopic mice. Invest New Drugs 29:41–51. doi:10.1007/s10637-009-9331-8

Das PM, Bast RC Jr (2008) Early detection of ovarian cancer. Biomark Med 2:291–303. doi:10.2217/17520363.2.3.291

De Bacco F, Luraghi P, Medico E, Reato G, Girolami F, Perera T, Gabriele P, Comoglio PM, Boccaccio C (2011) Induction of MET by ionizing radiation and its role in radioresistance and invasive growth of cancer. J Natl Cancer Inst 103:645–661. doi:10.1093/jnci/djr093

Dembo AJ (1992) Epithelial ovarian cancer: the role of radiotherapy. Int J Radiat Oncol Biol Phys 22:835–845

Deveraux QL, Reed JC (1999) IAP family proteins--suppressors of apoptosis. Genes Dev 13:239–252

Di Renzo MF, Olivero M, Katsaros D, Crepaldi T, Gaglia P, Zola P, Sismondi P, Comoglio PM (1994) Overexpression of the Met/HGF receptor in ovarian cancer. Int J Cancer 58:658–662

Eslin D, Sankpal UT, Lee C, Sutphin RM, Maliakal P, Currier E, Sholler G, Khan M, Basha R (2013) Tolfenamic acid inhibits neuroblastoma cell proliferation and induces apoptosis: a novel therapeutic agent for neuroblastoma. Mol Carcinog 52:377–386. doi:10.1002/mc.21866

Fernandez-Guizan A, Mansilla S, Barcelo F, Vizcaino C, Nunez LE, Moris F, Gonzalez S, Portugal J (2014) The activity of a novel mithramycin analog is related to its binding to DNA, cellular accumulation, and inhibition of Sp1-driven gene transcription. Chem Biol Interact 219:123–132. doi:10.1016/j.cbi.2014.05.019

Fernandez-Guizan A, Lopez-Soto A, Acebes-Huerta A, Huergo-Zapico L, Villa-Alvarez M, Nunez LE, Moris F, Gonzalez S (2015) Pleiotropic anti-angiogenic and anti-oncogenic activities of the novel Mithralog Demycarosyl-3D-ss-D-Digitoxosyl-Mithramycin SK (EC-8042). PLoS One 10:e0140786. doi:10.1371/journal.pone.0140786

Ferrandina G, Fagotti A, Salerno MG, Natali PG, Mottolese M, Maneschi F, De Pasqua A, Benedetti-Panici P, Mancuso S, Scambia G (1999) p53 overexpression is associated with cytoreduction and response to chemotherapy in ovarian cancer. Br J Cancer 81:733–740. doi:10.1038/sj.bjc.6690756

Ferrandina G, Ludovisi M, Lorusso D, Pignata S, Breda E, Savarese A, Del Medico P, Scaltriti L, Katsaros D, Priolo D, Scambia G (2008) Phase III trial of gemcitabine compared with pegylated liposomal doxorubicin in progressive or recurrent ovarian cancer. J Clin Oncol 26:890–896. doi:10.1200/JCO.2007.13.6606

Firat S, Murray K, Erickson B (2003) High-dose whole abdominal and pelvic irradiation for treatment of ovarian carcinoma: long-term toxicity and outcomes. Int J Radiat Oncol Biol Phys 57:201–207

Franchin G, Tumolo S, Scarabelli C, De Paoli A, Boz G, Crivellari D, Arcicasa M, Bortolus R, Gobitti C, Minatel E et al (1991) Whole abdomen radiation therapy after a short chemotherapy course and second-look laparotomy in advanced ovarian cancer. Gynecol Oncol 41:206–211

Frenel JS, Leux C, Pouplin L, Ferron G, Berton Rigaud D, Bourbouloux E, Dravet F, Jaffre I, Classe JM (2011) Oxaliplatin-based hyperthermic intraperitoneal chemotherapy in primary or recurrent epithelial ovarian cancer: a pilot study of 31 patients. J Surg Oncol 103:10–16. doi:10.1002/jso.21732

Fyles A, Bolis G, Ferraris C, Parazzini F, Bolla M (1997) Is abdomino-pelvic radiation therapy the optimal treatment for completely resected stage I and II high risk ovarian cancer? Eur J Cancer 33:12–19

Gately S, Li WW (2004) Multiple roles of COX-2 in tumor angiogenesis: a target for antiangiogenic therapy. Semin Oncol 31:2–11

Giovannetti E, Mey V, Nannizzi S, Pasqualetti G, Del Tacca M, Danesi R (2006) Pharmacogenetics of anticancer drug sensitivity in pancreatic cancer. Mol Cancer Ther 5:1387–1395. doi:10.1158/1535-7163.MCT-06-0004

Glaze S, Teitelbaum L, Chu P, Ghatage P, Nation J, Nelson G (2013) Dose-dense Paclitaxel with Carboplatin for advanced ovarian cancer: a feasible treatment alternative. J Obstet Gynaecol Can 35:61–67

Goldberg MS (2013) siRNA delivery for the treatment of ovarian cancer. Methods 63:95–100. doi:10.1016/j.ymeth.2013.01.007

Goldberg N, Peschel RE (1988) Postoperative abdominopelvic radiation therapy for ovarian cancer. Int J Radiat Oncol Biol Phys 14:425–429

Grendys EC Jr, Fiorica JV, Orr JW Jr, Holloway R, Wang D, Tian C, Chan JK, Herzog TJ (2014) Overview of a chemoresponse assay in ovarian cancer. Clin Transl Oncol 16:761–769. doi:10.1007/s12094-014-1192-8

Helm CW (2009) The role of hyperthermic intraperitoneal chemotherapy (HIPEC) in ovarian cancer. Oncologist 14:683–694. doi:10.1634/theoncologist.2008-0275

Hirazono K, Shinozuka T, Kuroshima Y, Itoh H, Kawai K (1995) Immunohistochemical expression of glutathione S-transferase pi (GST-pi) and chemotherapy response in malignant ovarian tumors. J Obstet Gynaecol Res (Tokyo) 21:305–312

Holmes CE, Levis JE, Ornstein DL (2009) Activated platelets enhance ovarian cancer cell invasion in a cellular model of metastasis. Clin Exp Metastasis 26:653–661. doi:10.1007/s10585-009-9264-9

Howells RE, Dhar KK, Hoban PR, Jones PW, Fryer AA, Redman CW, Strange RC (2004) Association between glutathione-S-transferase GSTP1 genotypes, GSTP1 over-expression, and outcome in epithelial ovarian cancer. Int J Gynecol Cancer 14:242–250. doi:10.1111/j.1048-891X.2004.014207.x

Ingersoll SB, Patel S, Caballero L, Ahmad S, Edwards D, Holloway RW, Edwards JR (2009) Synergistic cytotoxicity of interferon alpha-2b and interleukin-2 in combination with PBMC against ovarian cancer: development of an experimental model for cellular therapy. Gynecol Oncol 112:192–198. doi:10.1016/j.ygyno.2008.09.028

Ingersoll SB, Ahmad S, Stoltzfus GP, Patel S, Radi MJ, Finkler NJ, Edwards JR, Holloway RW (2011) Functional characterization of a fluorescent highly tumorigenic ovarian cancer line to test cellular therapy in experimental models. Int J Gynecol Cancer 21:457–465. doi:10.1097/IGC.0b013e31820f4ef0

Ingersoll SB, Ahmad S, Finkler NJ, Edwards JR, Holloway RW (2012) Cellular therapy for ovarian cancer: experimental and clinical perspectives. Curr Med Chem 19:3787–3793

Ingersoll SB, Ahmad S, McGann HC, Banks RK, Stavitzski NM, Srivastava M, Ali G, Finkler NJ, Edwards JR, Holloway RW (2015) Cellular therapy in combination with cytokines improves survival in a xenograft mouse model of ovarian cancer. Mol Cell Biochem 407:281–287. doi:10.1007/s11010-015-2475-2

Jacoby RF, Seibert K, Cole CE, Kelloff G, Lubet RA (2000) The cyclooxygenase-2 inhibitor celecoxib is a potent preventive and therapeutic agent in the min mouse model of adenomatous polyposis. Cancer Res 60:5040–5044

Januchowski R, Wojtowicz K, Andrzejewska M, Zabel M (2014) Expression of MDR1 and MDR3 gene products in paclitaxel-, doxorubicin- and vincristine-resistant cell lines. Biomed Pharmacother 68:111–117. doi:10.1016/j.biopha.2013.09.004

Januchowski R, Wojtowicz K, Sujka-Kordowska P, Andrzejewska M, Zabel M (2013) MDR gene expression analysis of six drug-resistant ovarian cancer cell lines. Biomed Res Int 2013:241763. doi:10.1155/2013/241763

Jeffers M, Rong S, Vande Woude GF (1996a) Enhanced tumorigenicity and invasion-metastasis by hepatocyte growth factor/scatter factor-met signalling in human cells concomitant with induction of the urokinase proteolysis network. Mol Cell Biol 16:1115–1125

Jeffers M, Rong S, Woude GF (1996b) Hepatocyte growth factor/scatter factor-Met signaling in tumorigenicity and invasion/metastasis. J Mol Med 74:505–513

Jin L, Fuchs A, Schnitt SJ, Yao Y, Joseph A, Lamszus K, Park M, Goldberg ID, Rosen EM (1997) Expression of scatter factor and c-met receptor in benign and malignant breast tissue. Cancer 79:749–760. doi:10.1002/(SICI)1097-0142(19970215)79:4<749::AID-CNCR12>3.0.CO;2-#

Kalli KR, Bradley SV, Fuchshuber S, Conover CA (2004) Estrogen receptor-positive human epithelial ovarian carcinoma cells respond to the antitumor drug suramin with increased proliferation: possible insight into ER and epidermal growth factor signaling interactions in ovarian cancer. Gynecol Oncol 94:705–712. doi:10.1016/j.ygyno.2004.05.059

Katsumata N, Yasuda M, Takahashi F, Isonishi S, Jobo T, Aoki D, Tsuda H, Sugiyama T, Kodama S, Kimura E, Ochiai K, Noda K, Japanese Gynecologic Oncology Group (2009) Dose-dense paclitaxel once a week in combination with carboplatin every 3 weeks for advanced ovarian cancer: a phase 3, open-label, randomised controlled trial. Lancet 374:1331–1338. doi:10.1016/S0140-6736(09)61157-0

Khrunin AV, Moisseev A, Gorbunova V, Limborska S (2010) Genetic polymorphisms and the efficacy and toxicity of cisplatin-based chemotherapy in ovarian cancer patients. Pharmacogenomics J 10:54–61. doi:10.1038/tpj.2009.45

Kofman S, Perlia CP, Economou SG (1973) Mithramycin in the treatment of metastatic Ewing's sarcoma. Cancer 31:889–893

Konduri S, Colon J, Baker CH, Safe S, Abbruzzese JL, Abudayyeh A, Basha MR, Abdelrahim M (2009) Tolfenamic acid enhances pancreatic cancer cell and tumor response to radiation therapy by inhibiting survivin protein expression. Mol Cancer Ther 8:533–542. doi:10.1158/1535-7163.MCT-08-0405

Kruh GD (2003) Introduction to resistance to anticancer agents. Oncogene 22:7262–7264. doi:10.1038/sj.onc.1206932

Kuniyasu H, Yasui W, Kitadai Y, Yokozaki H, Ito H, Tahara E (1992) Frequent amplification of the c-met gene in scirrhous type stomach cancer. Biochem Biophys Res Commun 189:227–232. doi:10.1016/0006-291X(92)91548-5

Lal B, Xia S, Abounader R, Laterra J (2005) Targeting the c-Met pathway potentiates glioblastoma responses to gamma-radiation. Clin Cancer Res 11:4479–4486. doi:10.1158/1078-0432.CCR-05-0166

Lamszus K, Jin L, Fuchs A, Shi E, Chowdhury S, Yao Y, Polverini PJ, Laterra J, Goldberg ID, Rosen EM (1997) Scatter factor stimulates tumor growth and tumor angiogenesis in human breast cancers in the mammary fat pads of nude mice. Lab Invest 76:339–353

Leonard GD, Fojo T, Bates SE (2003) The role of ABC transporters in clinical practice. Oncologist 8:411–424

Li JJ, Gu MF, Pan K, Liu LZ, Zhang H, Shen WX, Xia JC (2012) Autologous cytokine-induced killer cell transfusion in combination with gemcitabine plus cisplatin regimen chemotherapy for metastatic nasopharyngeal carcinoma. J Immunother 35:189–195. doi:10.1097/CJI.0b013e318241d9de

Lowery WJ, Lowery AW, Barnett JC, Lopez-Acevedo M, Lee PS, Secord AA, Havrilesky L (2013) Cost-effectiveness of early palliative care intervention in recurrent platinum-resistant ovarian cancer. Gynecol Oncol 130:426–430. doi:10.1016/j.ygyno.2013.06.011

Lu B, Mu Y, Cao C, Zeng F, Schneider S, Tan J, Price J, Chen J, Freeman M, Hallahan DE (2004) Survivin as a therapeutic target for radiation sensitization in lung cancer. Cancer Res 64:2840–2845

Margileth DA, Smith FE, Lane M (1973) Sudden arterial occlusion associated with mithramycin therapy. Cancer 31:708–712

Markman M (2003) Role of intraperitoneal chemotherapy in the front-line setting. J Clin Oncol 21:145s–148s. doi:10.1200/JCO.2003.01.067

Markman M, Bundy BN, Alberts DS, Fowler JM, Clark-Pearson DL, Carson LF, Wadler S, Sickel J (2001) Phase III trial of standard-dose intravenous cisplatin plus paclitaxel versus moderately high-dose carboplatin followed by intravenous paclitaxel and intraperitoneal cisplatin in small-volume stage III ovarian carcinoma: an intergroup study of the Gynecologic Oncology Group, Southwestern Oncology Group, and Eastern Cooperative Oncology Group. J Clin Oncol 19:1001–1007

Medico E, Mongiovi AM, Huff J, Jelinek MA, Follenzi A, Gaudino G, Parsons JT, Comoglio PM (1996) The tyrosine kinase receptors Ron and Sea control "scattering" and morphogenesis of liver progenitor cells in vitro. Mol Biol Cell 7:495–504

Milovic-Kovacevic M, Srdic-Rajic T, Radulovic S, Bjelogrlic S, Gavrilovic D (2011) Expression of ERCC1 protein in biopsy specimen predicts survival in advanced ovarian cancer patients treated with platinum-based chemotherapy. J Buon 16:708–714

Moreno-Smith M, Halder JB, Meltzer PS, Gonda TA, Mangala LS, Rupaimoole R, Lu C, Nagaraja AS, Gharpure KM, Kang Y, Rodriguez-Aguayo C, Vivas-Mejia PE, Zand B, Schmandt R, Wang H, Langley RR, Jennings NB, Ivan C, Coffin JE, Armaiz GN, Bottsford-Miller J, Kim SB, Halleck MS, Hendrix MJ, Bornman W, Bar-Eli M, Lee JS, Siddik ZH, Lopez-Berestein G, Sood AK (2013) ATP11B mediates platinum resistance in ovarian cancer. J Clin Invest 123:2119–2130. doi:10.1172/JCI65425

Moxley KM, Benbrook DM, Queimado L, Zuna RE, Thompson D, McCumber M, Premkumar P, Thavathiru E, Hines L, Moore KN (2013) The role of single nucleotide polymorphisms of the ERCC1 and MMS19 genes in predicting platinum-sensitivity, progression-free and overall survival in advanced epithelial ovarian cancer. Gynecol Oncol 130:377–382. doi:10.1016/j.ygyno.2013.04.054

Mukhopadhyay A, Elattar A, Cerbinskaite A, Wilkinson SJ, Drew Y, Kyle S, Los G, Hostomsky Z, Edmondson RJ, Curtin NJ (2010) Development of a functional assay for homologous recombination status in primary cultures of epithelial ovarian tumor and correlation with sensitivity to poly(ADP-ribose) polymerase inhibitors. Clin Cancer Res 16:2344–2351. doi:10.1158/1078-0432.CCR-09-2758

Nelson BH (2008) The impact of T-cell immunity on ovarian cancer outcomes. Immunol Rev 222:101–116. doi:10.1111/j.1600-065X.2008.00614.x

Nelson CL, Sellers TA, Rich SS, Potter JD, McGovern PG, Kushi LH (1993) Familial clustering of colon, breast, uterine, and ovarian cancers as assessed by family history. Genet Epidemiol 10:235–244. doi:10.1002/gepi.1370100404

Okazaki T, Javle M, Tanaka M, Abbruzzese JL, Li D (2010) Single nucleotide polymorphisms of gemcitabine metabolic genes and pancreatic cancer survival and drug toxicity. Clin Cancer Res 16:320–329. doi:10.1158/1078-0432.CCR-09-1555

Oseledchyk A, Zivanovic O (2015) Intraoperative hyperthermic intraperitoneal chemotherapy in patients with advanced ovarian cancer. Oncology (Williston Park) 29:695–701

Ostrand-Rosenberg S, Sinha P (2009) Myeloid-derived suppressor cells: linking inflammation and cancer. J Immunol 182:4499–4506. doi:10.4049/jimmunol.0802740

Oto OA, Paydas S, Tanriverdi K, Seydaoglu G, Yavuz S, Disel U (2007) Survivin and EPR-1 expression in acute leukemias: prognostic significance and review of the literature. Leuk Res 31:1495–1501. doi:10.1016/j.leukres.2007.01.005

Papineni S, Chintharlapalli S, Abdelrahim M, Lee SO, Burghardt R, Abudayyeh A, Baker C, Herrera L, Safe S (2009) Tolfenamic acid inhibits esophageal cancer through repression of specificity proteins and c-Met. Carcinogenesis 30:1193–1201. doi:10.1093/carcin/bgp092

Parmar MK, Ledermann JA, Colombo N, du Bois A, Delaloye JF, Kristensen GB, Wheeler S, Swart AM, Qian W, Torri V, Floriani I, Jayson G, Lamont A, Trope C, Icon CAGO (2003) Paclitaxel plus platinum-based chemotherapy versus conventional platinum-based chemotherapy in women with relapsed ovarian cancer: the ICON4/AGO-OVAR-2.2 trial. Lancet 361:2099–2106

Parsons V, Scott G, Baum M, Molland E, Makin J (1971) Mithramycin treatment of hypercalcaemia and renal failure in a patient with paratesticular embryonic sarcoma. Br J Cancer 25:306–310

Pecot CV, Calin GA, Coleman RL, Lopez-Berestein G, Sood AK (2011) RNA interference in the clinic: challenges and future directions. Nat Rev Cancer 11:59–67. doi:10.1038/nrc2966

Pennati M, Folini M, Zaffaroni N (2007) Targeting survivin in cancer therapy: fulfilled promises and open questions. Carcinogenesis 28:1133–1139. doi:10.1093/carcin/bgm047

Pfisterer J, Plante M, Vergote I, du Bois A, Hirte H, Lacave AJ, Wagner U, Stahle A, Stuart G, Kimmig R, Olbricht S, Le T, Emerich J, Kuhn W, Bentley J, Jackisch C, Luck HJ, Rochon J, Zimmermann AH, Eisenhauer E, Ago O, Ncic CTG, Eortc GCG (2006) Gemcitabine plus carboplatin compared with carboplatin in patients with platinum-sensitive recurrent ovarian cancer: an intergroup trial of the AGO-OVAR, the NCIC CTG, and the EORTC GCG. J Clin Oncol 24:4699–4707. doi:10.1200/JCO.2006.06.0913

Plamadeala V, Kelley JL, Chan JK, Krivak TC, Gabrin MJ, Brower SL, Powell MA, Rutherford TJ, Coleman RL (2015) A cost-effectiveness analysis of a chemoresponse assay for treatment of patients with recurrent epithelial ovarian cancer. Gynecol Oncol 136:94–98. doi:10.1016/j.ygyno.2014.11.019

Previdi S, Malek A, Albertini V, Riva C, Capella C, Broggini M, Carbone GM, Rohr J, Catapano CV (2010) Inhibition of Sp1-dependent transcription and antitumor activity of the new aureolic acid analogues mithramycin SDK and SK in human ovarian cancer xenografts. Gynecol Oncol 118:182–188. doi:10.1016/j.ygyno.2010.03.020

Qian LW, Mizumoto K, Inadome N, Nagai E, Sato N, Matsumoto K, Nakamura T, Tanaka M (2003) Radiation stimulates HGF receptor/c-Met expression that leads to amplifying cellular response to HGF stimulation via upregulated receptor tyrosine phosphorylation and MAP kinase activity in pancreatic cancer cells. Int J Cancer 104:542–549. doi:10.1002/ijc.10997

Raja FA, Chopra N, Ledermann JA (2012) Optimal first-line treatment in ovarian cancer. Ann Oncol 23(Suppl 10):x118–x127. doi:10.1093/annonc/mds315

Ratner ES, Sartorelli AC, Lin ZP (2012) Poly (ADP-ribose) polymerase inhibitors: on the horizon of tailored and personalized therapies for epithelial ovarian cancer. Curr Opin Oncol 24:564–571. doi:10.1097/CCO.0b013e3283564230

Rodel F, Hoffmann J, Distel L, Herrmann M, Noisternig T, Papadopoulos T, Sauer R, Rodel C (2005) Survivin as a radioresistance factor, and prognostic and therapeutic target for radiotherapy in rectal cancer. Cancer Res 65:4881–4887. doi:10.1158/0008-5472.CAN-04-3028

Romero I, Bast RC Jr (2012) Minireview: Human ovarian cancer: biology, current management, and paths to personalizing therapy. Endocrinology 153:1593–1602. doi:10.1210/en.2011-2123

Rosen DG, Yang G, Liu G, Mercado-Uribe I, Chang B, Xiao XS, Zheng J, Xue FX, Liu J (2009) Ovarian cancer: pathology, biology, and disease models. Front Biosci (Landmark Ed) 14:2089–2102

Rueda BR, Tilly JL, Schorge JO (2010) Surgical debulking before or after chemotherapy: stemming the tide on ovarian cancer recurrence. Onkologie 33:286–287. doi:10.1159/000314637

Sakano S, Hinoda Y, Sasaki M, Wada T, Matsumoto H, Eguchi S, Shinohara A, Kawai Y, Hara T, Nagao K, Naito K, Matsuyama H (2010) Nucleotide excision repair gene polymorphisms may predict acute toxicity in patients treated with chemoradiotherapy for bladder cancer. Pharmacogenomics 11:1377–1387. doi:10.2217/pgs.10.106

Sankpal UT, Abdelrahim M, Connelly SF, Lee CM, Madero-Visbal R, Colon J, Smith J, Safe S, Maliakal P, Basha R (2012) Small molecule tolfenamic acid inhibits PC-3 cell proliferation and invasion in vitro, and tumor growth in orthotopic mouse model for prostate cancer. Prostate 72:1648–1658. doi:10.1002/pros.22518

Sankpal UT, Ingersoll SB, Ahmad S, Holloway RW, Bhat VB, Simecka JW, Daniel L, Kariali E, Vishwanatha JK, Basha R (2016) Association of Sp1 and survivin in epithelial ovarian cancer: Sp1 inhibitor and cisplatin, a novel combination for inhibiting epithelial ovarian cancer cell proliferation. Tumour Biol 37:14259–14269

Satagopan JM, Boyd J, Kauff ND, Robson M, Scheuer L, Narod S, Offit K (2002) Ovarian cancer risk in Ashkenazi Jewish carriers of BRCA1 and BRCA2 mutations. Clin Cancer Res 8:3776–3781

Scheil-Bertram S, Tylus-Schaaf P, du Bois A, Harter P, Oppitz M, Ewald-Riegler N, Fisseler-Eckhoff A (2010) Excision repair cross-complementation group 1 protein overexpression as a

predictor of poor survival for high-grade serous ovarian adenocarcinoma. Gynecol Oncol 119:325–331. doi:10.1016/j.ygyno.2010.07.018
Schorge JO, McCann C, Del Carmen MG (2010) Surgical debulking of ovarian cancer: what difference does it make? Rev Obstet Gynecol 3:111–117
Sehouli J, Stengel D, Oskay-Oezcelik G, Zeimet AG, Sommer H, Klare P, Stauch M, Paulenz A, Camara O, Keil E, Lichtenegger W (2008) Nonplatinum topotecan combinations versus topotecan alone for recurrent ovarian cancer: results of a phase III study of the North-Eastern German Society of Gynecological Oncology Ovarian Cancer Study Group. J Clin Oncol 26:3176–3182. doi:10.1200/JCO.2007.15.1258
Shinohara ET, Hallahan DE, Lu B (2004) The use of antisense oligonucleotides in evaluating survivin as a therapeutic target for radiation sensitization in lung cancer. Biol Proced Online 6:250–256. doi:10.1251/bpo95
Siegel RL, Miller KD, Jemal A (2015) Cancer statistics, 2015. CA Cancer J Clin 65:5–29. doi:10.3322/caac.21254
Skates SJ, Singer DE (1991) Quantifying the potential benefit of CA 125 screening for ovarian cancer. J Clin Epidemiol 44:365–380
Stavrovskaya AA (2000) Cellular mechanisms of multidrug resistance of tumor cells. Biochemistry (Mosc) 65:95–106
Steffensen KD, Waldstrom M, Jeppesen U, Brandslund I, Jakobsen A (2008) Prediction of response to chemotherapy by ERCC1 immunohistochemistry and ERCC1 polymorphism in ovarian cancer. Int J Gynecol Cancer 18:702–710. doi:10.1111/j.1525-1438.2007.01068.x
Steffensen KD, Waldstrom M, Jakobsen A (2009) The relationship of platinum resistance and ERCC1 protein expression in epithelial ovarian cancer. Int J Gynecol Cancer 19:820–825. doi:10.1111/IGC.0b013e3181a12e09
Surowiak P, Materna V, Kaplenko I, Spaczynski M, Dolinska-Krajewska B, Gebarowska E, Dietel M, Zabel M, Lage H (2006) ABCC2 (MRP2, cMOAT) can be localized in the nuclear membrane of ovarian carcinomas and correlates with resistance to cisplatin and clinical outcome. Clin Cancer Res 12:7149–7158. doi:10.1158/1078-0432.CCR-06-0564
Tarnawski AS, Jones MK (2003) Inhibition of angiogenesis by NSAIDs: molecular mechanisms and clinical implications. J Mol Med 81:627–636
Tas F, Kilic L, Bilgin E, Keskin S, Sen F, Ciftci R, Yildiz I, Yasasever V (2013) Clinical and prognostic significance of coagulation assays in advanced epithelial ovarian cancer. Int J Gynecol Cancer 23:276–281. doi:10.1097/IGC.0b013e31827b8796
Thomas GM (1993) Is there a role for consolidation or salvage radiotherapy after chemotherapy in advanced epithelial ovarian cancer? Gynecol Oncol 51:97–103. doi:10.1006/gyno.1993.1253
van Nagell JR Jr, DePriest PD, Ueland FR, DeSimone CP, Cooper AL, McDonald JM, Pavlik EJ, Kryscio RJ (2007) Ovarian cancer screening with annual transvaginal sonography: findings of 25,000 women screened. Cancer 109:1887–1896. doi:10.1002/cncr.22594
Vizcaino C, Nunez LE, Moris F, Portugal J (2014) Genome-wide modulation of gene transcription in ovarian carcinoma cells by a new mithramycin analogue. PLoS One 9:e104687. doi:10.1371/journal.pone.0104687
Vogler M, Dinsdale D, Dyer MJ, Cohen GM (2009) Bcl-2 inhibitors: small molecules with a big impact on cancer therapy. Cell Death Differ 16:360–367. doi:10.1038/cdd.2008.137
Volkova M, Russell R 3rd (2011) Anthracycline cardiotoxicity: prevalence, pathogenesis and treatment. Curr Cardiol Rev 7:214–220
Wang H, Zhang Z, Wei X, Dai R (2015) Small-molecule inhibitor of Bcl-2 (TW-37) suppresses growth and enhances cisplatin-induced apoptosis in ovarian cancer cells. J Ovarian Res 8:3. doi:10.1186/s13048-015-0130-x
Wang X, Wang E, Kavanagh JJ, Freedman RS (2005) Ovarian cancer, the coagulation pathway, and inflammation. J Transl Med 3:25. doi:10.1186/1479-5876-3-25
Welsh JW, Mahadevan D, Ellsworth R, Cooke L, Bearss D, Stea B (2009) The c-Met receptor tyrosine kinase inhibitor MP470 radiosensitizes glioblastoma cells. Radiat Oncol 4:69. doi:10.1186/1748-717X-4-69

Wenzel LB, Huang HQ, Armstrong DK, Walker JL, Cella D (2007) Health-related quality of life during and after intraperitoneal versus intravenous chemotherapy for optimally debulked ovarian cancer: a Gynecologic Oncology Group Study. J Clin Oncol 25:437–443. doi:10.1200/JCO.2006.07.3494

Wiggans AJ, Cass GK, Bryant A, Lawrie TA, Morrison J (2015) Poly(ADP-ribose) polymerase (PARP) inhibitors for the treatment of ovarian cancer. Cochrane Database Syst Rev 5: CD007929. doi:10.1002/14651858.CD007929.pub3

Witham J, Valenti MR, De-Haven-Brandon AK, Vidot S, Eccles SA, Kaye SB, Richardson A (2007) The Bcl-2/Bcl-XL family inhibitor ABT-737 sensitizes ovarian cancer cells to carboplatin. Clin Cancer Res 13:7191–7198. doi:10.1158/1078-0432.CCR-07-0362

Wu SY, Yang X, Gharpure KM, Hatakeyama H, Egli M, McGuire MH, Nagaraja AS, Miyake TM, Rupaimoole R, Pecot CV, Taylor M, Pradeep S, Sierant M, Rodriguez-Aguayo C, Choi HJ, Previs RA, Armaiz-Pena GN, Huang L, Martinez C, Hassell T, Ivan C, Sehgal V, Singhania R, Han HD, Su C, Kim JH, Dalton HJ, Kovvali C, Keyomarsi K, McMillan NA, Overwijk WW, Liu J, Lee JS, Baggerly KA, Lopez-Berestein G, Ram PT, Nawrot B, Sood AK (2014) 2′-OMe-phosphorodithioate-modified siRNAs show increased loading into the RISC complex and enhanced anti-tumour activity. Nat Commun 5:3459. doi:10.1038/ncomms4459

Yasmeen A, Beauchamp MC, Piura E, Segal E, Pollak M, Gotlieb WH (2011) Induction of apoptosis by metformin in epithelial ovarian cancer: involvement of the Bcl-2 family proteins. Gynecol Oncol 121:492–498. doi:10.1016/j.ygyno.2011.02.021

Yellepeddi VK, Vangara KK, Kumar A, Palakurthi S (2012) Comparative evaluation of small-molecule chemosensitizers in reversal of cisplatin resistance in ovarian cancer cells. Anticancer Res 32:3651–3658

Yip KW, Reed JC (2008) Bcl-2 family proteins and cancer. Oncogene 27:6398–6406. doi:10.1038/onc.2008.307

Zeitlin BD, Zeitlin IJ, Nor JE (2008) Expanding circle of inhibition: small-molecule inhibitors of Bcl-2 as anticancer cell and antiangiogenic agents. J Clin Oncol 26:4180–4188. doi:10.1200/JCO.2007.15.7693

Zhang L, Conejo-Garcia JR, Katsaros D, Gimotty PA, Massobrio M, Regnani G, Makrigiannakis A, Gray H, Schlienger K, Liebman MN, Rubin SC, Coukos G (2003a) Intratumoral T cells, recurrence, and survival in epithelial ovarian cancer. N Engl J Med 348:203–213. doi:10.1056/NEJMoa020177

Zhang X, Li Y, Dai C, Yang J, Mundel P, Liu Y (2003b) Sp1 and Sp3 transcription factors synergistically regulate HGF receptor gene expression in kidney. Am J Physiol 284:F82–F94

Zhou Y, Liu QH, Liu CL, Lin L (2015) Calycosin induces apoptosis in human ovarian cancer SKOV3 cells by activating caspases and Bcl-2 family proteins. Tumour Biol 36:5333–5339. doi:10.1007/s13277-015-3194-8

Initiatives Across Countries to Reduce Antibiotic Utilisation and Resistance Patterns: Impact and Implications

Brian Godman, Joseph Fadare, Dan Kibuule, Lyna Irawati, Mwangana Mubita, Olayinka Ogunleye, Margaret Oluka, Bene D. Anand Paramadhas, Juliana de Oliveira Costa, Lívia Lovato Pires de Lemos, Augusto Afonso Guerra Júnior, Alian A. Alrasheedy, Mohamed Azmi Hassali, Fahad Saleem, Thu Huong, and Ilse Truter

Abstract Introduction: Greater accessibility to antibiotics has resulted in their excessive use, leading to increasing antimicrobial resistance (AMR) and strains on healthcare systems, with only a limited number of patients in ambulatory

---

B. Godman (✉)
Department of Laboratory Medicine, Division of Clinical Pharmacology, KarolinskaInstitutet, Karolinska University Hospital Huddinge, 141 86 Stockholm, Sweden

Strathclyde Institute of Pharmacy and Biomedical Sciences, University of Strathclyde, Glasgow, UK
e-mail: Brian.Godman@ki.se; Brian.godman@strath.ac.uk

J. Fadare
Departments of Pharmacology and Medicine, Ekiti State University/Ekiti State University Teaching Hospital, Ado-Ekiti, Nigeria

D. Kibuule
School of Pharmacy, University of Namibia, Windhoek, Namibia

L. Irawati • M.A. Hassali
Discipline of Social and Administrative Pharmacy, School of Pharmaceutical Sciences, University Sains Malaysia, 11800 Minden, Penang, Malaysia

M. Mubita
School of Pharmacy, University of Namibia, Windhoek, Namibia

School of Pharmacy, Queen's University Belfast, Belfast, Northern Ireland, United Kingdom

O. Ogunleye
Department of Pharmacology and Medicine, Lagos State University College of Medicine and Teaching Hospital, Ikeja, Lagos, Nigeria

Clinical Pharmacology Unit, Department of Medicine, Lagos State, University Teaching Hospital, Ikeja, Lagos, Nigeria

M. Oluka
Department of Pharmacology and Pharmacognosy, School of Pharmacy, University of Nairobi, Nairobi, Kenya

G. Arora et al. (eds.), *Drug Resistance in Bacteria, Fungi, Malaria, and Cancer*,
DOI 10.1007/978-3-319-48683-3_24

care treated according to guidelines. High rates of AMR are now seen across countries and continents, resulting in AMR becoming one of the most critical issues facing healthcare systems. It is estimated that AMR could potentially cause over 10 million deaths per year by 2050 unless addressed, resulting in appreciable economic consequences. There are also concerns with under-treatment especially if patients are forced to fund more expensive antibiotics as a result of AMR to first-line antibiotics and do not have available funds. Overprescribing of antibiotics is not helped by patient pressure even when physicians are aware of the issues. There is also extensive dispensing of antibiotics without a prescription, although this is now being addressed in some countries. Aim: Review interventions that have been instigated across continents and countries to reduce inappropriate antibiotic prescribing and dispensing, and associated AMR, to provide future guidance. Method: Narrative case history approach. Findings: A number of successful activities have been instigated to reduce inappropriate prescribing and dispensing of antibiotics across sectors. These include the instigation of quality indicators, suggested activities of pharmacists as well as single and multiple interventions among all key stakeholder groups. Multiple interlinking strategies are typically needed to enhance appropriate antibiotic prescribing and dispensing. The impact of ongoing activities need to be continually analysed to provide future direction if AMR rates, and their impact on subsequent morbidity, mortality and costs, are to be reduced.

B.D.A. Paramadhas
Department of Pharmacy, Nyangabwe Referral Hospital, Francistown, Botswana

J.d.O. Costa • L.L.P. de Lemos • A.A.G. Júnior
SUS Collaborating Centre – Health Technology Assessment & Excellence in Health, Department of Social Pharmacy, College of Pharmacy, Federal University of Minas Gerais, Belo Horizonte, Brazil

A.A. Alrasheedy
Department of Pharmacy Practice, College of Pharmacy, Qassim University, Buraydah, Saudi Arabia

F. Saleem
Faculty of Pharmacy & Health Sciences, University of Balochistan, Quetta 87300, Pakistan

T. Huong
Pharmacy Department, E Hospital, Hanoi, Vietnam

Administration Medical Service, Ministry of Health, Hanoi, Vietnam

I. Truter
Drug Utilization Research Unit (DURU), Department of Pharmacy, Nelson Mandela Metropolitan University, Port Elizabeth 6031, South Africa

1 Introduction

1.1 Extent and Threat of Antimicrobial Resistance and Its Impact on Future Healthcare and Costs

Before the discovery of antibiotics, infectious diseases were the principal cause of morbidity and mortality (Md Rezal et al. 2015). This changed with their advent (Bosch and Roschi 2008; Alharbi et al. 2014; HMG 2014). Antibiotics have now become the cornerstone of treatment for bacterial infections across healthcare sectors (van de Sande-Bruinsma et al. 2008; Holloway et al. 2011; WHO Europe 2014; Laxminarayan et al. 2016) and are seen as an essential part of modern medical practice (Hwang et al. 2015). As a result, they are now among the most prescribed and dispensed medicines across healthcare settings (Larsson et al. 2000; Jande et al. 2012; Kho et al. 2013; Almeman et al. 2014; Md Rezal et al. 2015; Podolsky et al. 2015; Truter 2015). Antibiotic utilisation increased 36 % globally between 2000 and 2010, with Brazil, China, India, South Africa and Russia accounting for 76 % of this increase (Van Boeckel et al. 2014; Laxminarayan et al. 2016). This is set to continue unless addressed.

Greater accessibility to antibiotics has resulted in their irrational and excessive use, leading to increasing antimicrobial resistance (AMR) and a concomitant strain on healthcare systems (Goossens et al. 2005; de Jong et al. 2008; van de Sande-Bruinsma et al. 2008; Davies and Davies 2010; Kesselheim and Outterson 2010; de Kraker et al. 2013; Barnett and Linder 2014; Versporten et al. 2014; Truter 2015; Dyar et al. 2016). For instance in Pakistan, the average number of antibiotics prescribed per physician encounter is 1.1 in ambulatory care and 2.4 for inpatient care (Riaz et al. 2015). In Botswana, only 31 % of children were correctly prescribed an antibiotic, and 40.3 % of children who did not require an antibiotic left the health facility with a prescription for one (IMCI 2004). In Kenya, antibiotics were prescribed in two thirds of patients seeing a physician for diarrhoea (Brooks et al. 2006). In Kampala, Uganda, there was also appreciable self-purchasing of antibiotics for acute respiratory tract infections (RTIs) including common colds, mainly amoxicillin and co-trimoxazole, before patients or their families sought medical help (Massele et al. 2015; Kibuule et al. 2016). Self-purchasing of antibiotics is also common in many other countries despite being illegal in most (Chatterjee and Fleck 2011; Godman et al. 2014; WHO Europe 2014; Massele et al. 2015; WHO 2015; Kalungia et al. 2016) and is strongly correlated with increasing AMR rates in low-income and middle-income countries (LMICs) (Alsan et al. 2015).

The consumption patterns of antibiotics among humans and its use in agriculture, especially in animal husbandry, have been found globally to correlate with the development of AMR (Oduyebo et al. 2008; Goossens 2009; Meyer et al. 2013; Adesokan et al. 2015; Velickovic-Radovanovic et al. 2015; Dyar et al. 2016). Resistant strains of bacteria have been isolated from food animals, plant source and dairy products in different parts of the world (Brooks et al. 2006; Goossens

2009; de Kraker et al. 2013; WHO 2015). The HIV/AIDS epidemic, especially on the African continent, coupled with the use of co-trimoxazole prophylaxis, has also increased AMR rates (Cotton et al. 2008; Marwa et al. 2015).

In addition, only a limited number of ambulatory care patients with infections are currently treated according to guidelines. Over 40 % of prescriptions for antibiotics are considered inappropriate particularly for upper respiratory tract infections (URTIs), which are typically viral in origin (Gonzales et al. 2001; Holloway et al. 2011; Little et al. 2013; Barnett and Linder 2014; Hassali et al. 2015; Dyar et al. 2016). For instance, a study in the UK concluded that the complication rate after URTIs is very low and antibiotic therapy is ineffective. As a result, the authors estimated 4000 patients or more with URTIs need to be treated with antibiotics to prevent a single episode of pneumonia (Petersen et al. 2007).

Overall, it is estimated that resistance rates of $\geq$50 % are seen worldwide to common bacteria such as *Escherichia coli*, *Klebsiella pneumoniae* and *Staphylococcus aureus* (WHO 2014). In 15 European countries, more than 10 % of bloodstream *S. aureus* infections are due to *methicillin-resistant S. aureus* (MRSA), with seven of these countries having resistance rates above 25 % (ECDPC 2015; HMG 2014). High rates of antibiotic resistance are seen across Africa, the Middle East and Asia (Table 1), with five out of six WHO regions reporting high resistance rates leading to increased prescribing of second-line antibiotics (WHO 2014).

For instance, in Tanzania in children aged between 0 and 7 years with septicaemia, AMR was a major risk factor for a poor outcome (Blomberg et al. 2007). In Uganda in patients with surgical site infections, extended spectrum beta-lactamase (ESBL)-producing *Enterobacteriaceae* and MRSA organisms were also major risk factors for a poor outcome (Seni et al. 2013). AMR is of particular concern in community settings where infections are common and can easily be transmitted among the population (NCCID 2010; WHO 2014). However, the true burden of bacterial infections in many African countries, including South Africa, remains unknown (Laxminarayan et al. 2016; Mendelson and Matoso 2015).

MRSA is also a common problem worldwide. New community-acquired strains of MRSA (CA-MRSA) are also now emerging causing concern (FIP 2008).

Increasing AMR rates is now seen as one of the most critical problems facing healthcare systems (WHO 2001; Sumpradit et al. 2012; Llor and Bjerrum 2014; Barlam and Gupta 2015; Hoffman and Outterson 2015; Jinks et al. 2016). It is estimated that AMR infections currently cause approximately 50,000 deaths a year in Europe and the USA alone (HMG 2014). This increases to several hundred thousand deaths each year when other countries are included (HMG 2014). The continual rise in AMR could result in it becoming a leading cause of death worldwide by 2050 with over 10 million deaths per year, potentially reducing GDP by 2 % to 3.5 % and costing up to US$100 trillion (HMG 2014; Md Rezal et al. 2015). Other authors have suggested lower costs; however, their estimates typically fail to consider the costs for patients or health authorities after patients are discharged from hospitals (Gandra et al. 2014).

As a result of increasing AMR rates, common infections are becoming more difficult to treat, causing life-threatening illnesses and potentially death (Costelloe et al. 2010). This combination of overuse of antibiotics, misuse, stopping treatment

Table 1 Antibiotic resistance rates among African, Middle East and Asian countries

Country	Antibiotic resistance patterns
Asia	• High prevalence of AMR across Asian countries (Jean and Hsueh 2011; Kim et al. 2012; Kang and Song 2013; Lai et al. 2014) •The prevalence of *Streptococcus pneumonia* resistant to erythromycin is 80.7 % in Vietnam, 84.9 % in Taiwan and 96.4 % in China (Song et al. 2004; Kim et al. 2012; Kang and Song 2013; Lai et al. 2014)
Botswana	• MRSA in 22.6 % of isolates from skin and soft tissue infections from hospitalised children and adults (Wood et al. 2009) • Hospital antibiograms in 2013 and 2014 in a tertiary care hospital in northern Botswana showed *Klebsiella pneumoniae* is resistant to most beta-lactam antibiotics with less than 50 % sensitivities (Massele et al. 2015)
India and Pakistan	• Up to 95 % of adults have bacteria that are resistant to β-lactam antibiotics. This includes the carbapenems (Reardon 2014)
Kenya	• Multidrug-resistant (MDR) *non-typhi Salmonella* was 42 % in 2003 (Kariuki et al. 2005), and the estimated incidence of community-acquired *non-typhi Salmonella* was 166/100,000 people per year for children <5 years (Berkley et al. 2005a). 35 % of all *non-typhi Salmonella* cases in newborns resulted in death in a national referral hospital (Kariuki et al. 2006) • MDR *S. typhi* was over 75 % of all *S. typhi* among private clinics and the main referral hospital in Nairobi (Kariuki et al. 2010), with 43 % of *S. pneumoniae* isolates in Nairobi resistant to penicillin (Kariuki et al. 2003). The prevalence of MRSA was 33 % of *S. aureus* isolates at a national referral hospital (Ngumi 2006) • *E. coli* isolated from among hospital patients were highly resistant to commonly used antibiotics including ampicillin, amoxycillin/clavulanic acid and trimethoprim/sulphamethoxazole, with less resistance to ciprofloxacin and third-generation cephalosporins (Kiiru et al. 2012). In one region, resistance to *E. coli* was 85 % to co-trimoxazole, 78 % to amoxicillin and 42 % to chloramphenicol (Bejon et al. 2005) • 50 % of isolates of *Haemophilus influenzae type B* from children with severe pneumonia were resistant to penicillin (Berkley et al. 2005b) • 65 % of the *N. gonorrhoeae* isolates were resistant to penicillin and plasmid-mediated tetracycline resistance was 97 %. In 2007, quinolone-resistant *N. gonorrhoeae* first appeared, increasing to 50 % in 2009 (Mehta et al. 2011)
Malaysia	• The National Surveillance of Antibiotic Resistance in 2014 reported an increase in antibiotic resistance rates among common strains of bacteria such as *Streptococcus pneumonia*, *Enterobacter cloacae* and *Salmonella* spp. (Ahmad 2014) • Hospital-acquired pneumonia associated with *Acinetobacter* spp. showed a very high rate of resistance to imipenem (86.7 %) (Kang and Song 2013)

(continued)

Table 1 (continued)

Country	Antibiotic resistance patterns
	• In a multicentre surveillance study, 60.6 % of *Streptococcus pneumococcus* isolates were resistant to erythromycin (Kim et al. 2012)
Nigeria	• The susceptibility of antibiotics commonly used as empirical treatment for many infections in hospitals—especially those of the urinary tract, ear infections and post-operative wound infections—has declined considerably, e.g. as many as 88 % of *Staphylococcus aureus* infections are resistant to methicillin (Reardon 2014) • Studies have reported susceptibility rates to empiric antibiotics below 60 % in urinary tract isolates and over 98 % resistance to β-lactam antibiotics in post-operative wound infections(Okesola and Aroundegbe 2011; Muoneke et al. 2012; Dibua et al. 2014) • Studies have also reported susceptibility rates below 50 % for antibiotics used for empirical treatment of patients with community-acquired pneumonia (Iroezindu et al. 2014)
South Africa (Mendelson and Matoso 2015)	• 50 % of all hospital-acquired *S. aureus* in public hospitals in 2010 were MRSA, with MRSA accounting for 75 % of all hospital-acquired *S. aureus* infections in a large tertiary-level paediatric hospital • *Enterococcus faecium* bloodstream isolates from the private sector showed variable sensitivity to vancomycin, ranging between 33 and 100 % depending on the geographical location • Up to 75 % of *K. pneumonia* isolated from hospitalised patients were ESBL-producing bacteria • 16 % of carbapenem-susceptible *Enterobacteriaceae* in the private sector contained a carpapenemase-producing gene, and carpapenemase-producing *Enterobacteriaceae* are widespread among public hospitals in South Africa
Vietnam (Hoa et al. 2011; Van Nguyen et al. 2013; Van et al. 2014)	• Pneumococcal penicillin resistance rates are typically the highest in Asia, with carbapenem-resistant bacteria (notably NDM-1) recently emerging • *Streptococcus pneumoniae* penicillin resistance rates increased from 8 to 75 % from 1999 to 2007 in Ba Vi, not helped by high rates of self-purchasing of antibiotics (Table 3) • More than 90 % of isolates from principally patients in an intensive care unit in Hanoi tested for resistance to *A. baumannii* were resistant to tested β-lactamase inhibitors/β-lactamase, carbapenems, cephalosporins, fluoroquinolones and trimethoprim/sulfamethoxazole. Overall, 25.4 % of isolates were resistant to all tested aminoglycosides, β-lactams and quinolones • There has been a significant increase in resistance of *Streptococcus suis* to tetracycline and chloramphenicol in isolates between 1997 and 2008, concurrent with an increase in multidrug-resistant organisms

before courses are finished, cultural differences and underuse due to a lack of access and financial support is seen as key driver of AMR (Llor and Bjerrum 2014; WHO Europe 2014; Klein et al. 2015; Laxminarayan et al. 2016; Md Rezal et al. 2015; Dyar et al. 2016). These factors lead to the phenomenon which has been termed selective pressure (WHO 2001). As mentioned, increasing antibiotic resistance poses a threat to health and healthcare systems across countries as it can lead to high associated costs (Van Nguyen et al. 2013), for example, forcing a shift to more expensive and more broad-spectrum antibiotics (Laxminarayan and Heymann 2012; Md Rezal et al. 2015).

Improving the rational use of antibiotics is one of the best ways to slow down the development and spread of AMR (van de Sande-Bruinsma et al. 2008; Sumpradit et al. 2012; Earnshaw et al. 2013; Llor and Bjerrum 2014). This means addressing issues such as physicians' lack of adherence to treatment guidelines and their lack of knowledge and training regarding antibiotics, the lack of diagnostic facilities as well as uncertainty over the diagnosis, pressures from patients and the pharmaceutical industry and finally fear of clinical failure (Little et al. 2013; Van Nguyen et al. 2013; Hassali et al. 2015; Md Rezal et al. 2015; Riaz et al. 2015). In LMIC countries and others, this also includes implementing and enforcing regulations surrounding the dispensing of antibiotics including self-purchasing where this is a concern (Radyowijati and Haak 2003; Kotwani et al. 2010; Li et al. 2012; Van Nguyen et al. 2013; Holloway and Henry 2014; Holloway et al. 2015; Kalungia et al. 2016).

Adoption of antimicrobial resistance strategies in countries including South Africa and Vietnam as well as the WHO Europe strategic action plan on AMR, the recent WHO global report on antimicrobial resistance documenting alarming levels of AMR in many countries and the endorsement of the global action plan on antimicrobial resistance in May 2015 are seen as important steps to help reduce AMR (WHO Europe 2011; Department of Health RSA 2014; WHO 2014; Abdula et al. 2015; Cluzeau and Manandhar 2015; Mendelson and Matoso 2015).

1.2 Physician Attitudes Towards Antibiotics

A recent systematic review showed that physicians still have limited knowledge and misconceptions about antibiotics and their prescribing (Md Rezal et al. 2015). In addition, some physicians still prescribed antibiotics despite knowing they are generally of limited benefit in a number of situations (Md Rezal et al. 2015). For instance in Brazil, 38.2 % of interviewed patients declared that they had taken antibiotics in the previous 6 months. For patients who had received an antibiotic prescription for oral/dental infections, only 9.3 % presented with a fever, indicating overprescribing of antibiotics by physicians (Del Fiol et al. 2010).

Several factors influence physician prescribing of antibiotics. These include the severity and duration of infections, expectations of patients, uncertainty over diagnosis, the risk of losing patients and pharmaceutical company influence (Md Rezal et al. 2015). Inadequate knowledge regarding the prescribing of antibiotics appears prevalent among physicians across countries. However, many physicians are interested in addressing this to help reduce AMR rates (Md Rezal et al. 2015). Qualitative research undertaken by Thatoyaone Kenaope and colleagues also found that among pharmacists in South Africa, the socio-economic status of patients, patient satisfaction, their knowledge of antibiotic indications and the professional relationships between healthcare professionals also influenced physician prescribing behaviour (Massele et al. 2015).

To be able to successfully reduce AMR rates, physicians need to be knowledgeable about the prevailing epidemiology of infections and current antimicrobial sensitivity status in their location as well as strictly comply with evidence-based treatment guidelines. However, a cross-national study conducted among physicians in England and France found only 31 % and 26 % of physicians, respectively, knew the correct prevalence of antibiotic misuse and of MRSA in their hospitals (Pulcini et al. 2011). Among medical doctors and students in the Congo Democratic Republic, there was very limited knowledge of local antimicrobial resistance patterns (Thriemer et al. 2013). These and other studies demonstrate the need to tackle all key stakeholder groups to enhance appropriate prescribing and dispensing of antibiotics (Jinks et al. 2016).

1.3 Antibiotic Prescribing and Dispensing in Pharmacies

There is a large body of literature discussing pharmacists' antibiotic dispensing practices across countries, e.g. in Kenya 25 % of the population go to retail pharmacies first before seeking outpatient care (Sharma et al. 2008; Thoithi and Okalebo 2009). This is illustrated by 24 % of patients with symptoms of acute RTIs in rural Kenya already purchasing an antibiotic from pharmacies before seeing a physician (Bigogo et al. 2010).

As mentioned, dispensing of antibiotics without a prescription, albeit unlawful, is common. Countries where this happens include Albania (Hoxha et al. 2015), Brazil (Rauber et al. 2009), Egypt (Dooling et al. 2014; Sabry et al. 2014), Greece (Contopoulos-Ioannidis et al. 2001; Plachouras et al. 2010), India (Kotwani et al. 2012; Salunkhe et al. 2013), Iraq (Mikhael 2014), Jordan (Almaaytah et al. 2015), Lebanon (Farah et al. 2015), Portugal (Roque et al. 2013,2014), Saudi Arabia (Abdulhak et al. 2011; Emeka et al. 2012; Al-Mohamadi et al. 2013), Serbia (Godman et al. 2014), Spain (Zapata-Cachafeiro et al. 2014), Syria (Bahnassi 2015), the UAE (Abasaeed et al. 2013), and Zambia (Kalungia et al. 2016). Table 2 illustrates the extent of self-purchasing across countries.

However in some countries such as Thailand, pharmacists are allowed to dispense antibiotics without a prescription (Saengcharoen et al. 2008). Among the wider European countries including former Soviet Union countries, it is also currently possible and legal to purchase antibiotics over-the-counter (OTC) without a prescription in 19 out of the 44 countries (WHO Europe 2014), and this is also possible outside of pharmacies (12 out of 44 countries). In 5 out of the 44 countries surveyed, it is also possible to purchase antibiotics over the Internet without a

Table 2 Examples of the extent of dispensing antibiotics without a prescription

Authors	Country	Extent of dispensing antibiotics without a prescription (%)
Hoxha et al. (2015)	Albania	80.0
Farah et al. (2015)	Beirut, Lebanon	32.0
Volpato et al. (2005)[a]	Brazil	74.0
Almaaytah et al. (2015)	Jordan	74.3
Salunkhe et al. (2013)	Pune, India	94.7
Abdulhak et al. (2011)	Riyadh, Saudi Arabia	77.6

[a]NB: Studies conducted before prescription requirements in November 2010 among private pharmacies

prescription (WHO Europe 2014). The situation is different in many high-income countries where regulations are typically strictly implemented, e.g. New Zealand (Dameh et al. 2012).

Among many Latin American countries, including Brazil and Mexico, restrictions on OTC sales of antibiotics were implemented in 2010 (Santa-Ana-Tellez et al. 2015). This included the requirement for a prescription for antibiotics to be dispensed even in private pharmacies. A prescription for antibiotics was always a requirement among public pharmacies in Brazil. A recent study showed a reduction in inappropriate penicillin use (as proxy for antibiotics use) among private pharmacies in Mexico after this government initiative (Santa-Ana-Tellez et al. 2015), although there was a limited reduction in Brazil. However, another study showed a significant decrease in antimicrobial sales among private pharmacies in Brazil following the restrictions in November 2010 (Moura et al. 2015). The impact of the restrictions was greater in regions with overall higher socio-economic status (Moura et al. 2015).

In a systematic review of published work on antibiotic self-medication in developing countries, the overall prevalence of antimicrobial self-medication was 38.8 %. Identified determinants of self-medication were citizens' age, level of education, sex, income, severity of their disease condition and history of previous successful use of antibiotics (Ocan et al. 2015). Table 3 contains additional data on the extent of self-purchasing.

Several studies have reported that pharmacists have dispensed antibiotics for inappropriate conditions (Kotwani et al. 2012; Salunkhe et al. 2013; Dooling et al. 2014; Mikhael 2014; Sabry et al. 2014; Kalungia et al. 2016), e.g. in Iraq, 45 % of community pharmacists dispensed antibiotics for a common cold. Community pharmacists also dispensed several types of antibiotics including amoxicillin, amoxicillin/clavulanate, azithromycin and ciprofloxacin (Mikhael 2014). In India, azithromycin was commonly dispensed for a sore throat in 51.2 % of cases. Furthermore, antibiotics for a sore throat were only dispensed in correct doses and durations in 64.2 % of cases (Salunkhe et al. 2013). In Zambia; pharmacy personnel also dispensed antibiotics without justifiable causes (Kalungia et al. 2016).

Consequently, pharmacists in many countries play a key role with enhancing the appropriate use of antibiotics given the extent of self-purchasing (WHO Europe

Table 3 Extent of self-purchasing of antibiotics across continents and countries (building on Table 2)

Countries	Extent of self-purchasing
Brazil[a]	• In a study conducted in Tubarão, 85.0 % of pharmacists dispensed an antibiotic without a prescription, mainly for the treatment of respiratory (62.8 %) and urinary (12.0 %) tract disorders. In most cases, this was for adults (64.0 %), but also for children (27.8 %), the elderly (3.3 %) and pregnant women (1.6 %) (Rauber et al. 2009) • In Jataí, antibiotic self-medication was seen in 9.1 % of participants; however, 9.1 % used antibiotics as recommended by pharmacists in the last month. 20.5 % of responders also traditionally recommend the use of antimicrobials to family and friends (Braoios et al. 2013) • In a population-based study in Bambuí, self-medication in the last 90 days was reported for 28 % of the 1221 residents and antibiotics and chemotherapeutics accounted for 6.2 % of non-prescribed drugs (Loyola Filho et al. 2002)
Ethiopia (Gebeyehu et al. 2015)	• Antibiotic self-medication was common among community dwellers with RTIs, diarrhoea and physical injury
Greece (Mitsi et al. 2005)	• 74.6 % of the general public reported taking non-prescribed antibiotics
Northern Israel (Raz et al. 2005)	• 18.7 % of the general public reported taking antibiotics without seeking medical advice
Kenya (Karambu 2011)	• 70 % of sampled pharmacies did not ask patients for a prescription before dispensing antibiotics as required by the law • In addition, only 9 % of pharmacies asked for a prescription, and 18 % declined to sell the antibiotics after a prescription was presented
Malaysia (Islahudin et al. 2014)	• 45.1 % of patients did not consult a physician before taking antibiotics
Middle East/Jordan	• Specifically in Jordan among Middle East countries, self-medication with antibiotics is high (39.5 %). This is significantly associated with age, education and income (Al-Azzam et al. 2007) • A high level of self-medication (40.7 %) was also reported in another Jordanian study (Sawair et al. 2009)
Nigeria (Enato et al. 2011).	• In a community-based study conducted in the southern part of Nigeria, over 93.6 % of community dwellers who reported they were ill self-medicated with antibiotics
Serbia	• Illegal self-purchasing of antibiotics increased their utilisation by 115–128 % in recent years compared with reimbursed utilisation (Godman et al. 2014) • This resulted in Serbia having the third highest utilisation for cephalosporins in 2007 among European countries, highest for penicillins, second highest for macrolides and third highest for quinolones (Bajcetic et al. 2012) • Following this, there has been tightening of the regulations regarding self -purchasing in pharmacies, leading to a decrease in overall utilisation in recent years (Godman et al. 2014)

(continued)

Table 3 (continued)

Countries	Extent of self-purchasing
Vietnam (Van Nguyen et al. 2013; Cluzeau and Manandhar 2015)	• The Pharmaceutical Law (2005) made antibiotics prescription-only medicines; however, there are currently no sanctions. In addition, no pharmacist has been penalised for dispensing antibiotics without a prescription (as of March 2013) • Self-medication persists as it avoids lengthy and costly consultations with physicians and others in the formal healthcare systems • 91 % of children in Ba Vi, Vietnam, in 1991 with symptoms of acute RTIs were treated with broad-spectrum penicillins, with 78 % self-medicating. This corresponded to 75 % of all children in this particular community • This is likely to change with the implementation of the National Plan to reduce antimicrobial resistance rates in Vietnam

[a]NB: All studies conducted before the changes in the law in Brazil

2014) (Tables 2 and 3). This, coupled with concerns with appropriate prescribing and rising AMR rates, led FIP (International Pharmaceutical Federation) to urge pharmacists to undertake a number of activities to reduce this, supported by suggested activities among health authorities and governments (Box 1).

Box 1: Suggested Activities of Pharmacists to Reduce AMR [Adapted from FIP (2008)]

(A) *FIP urges pharmacists to:*

- Patients given proper counselling as well as provided with written information that is appropriate when dispensing antibiotics. In addition, encourage patients to comply with the full prescribed regimen. If not possible, ask them to dispose of any unused antibiotics appropriately.
- Work with physicians to ensure patients complete their course and correct antibiotics and doses are prescribed by providing up-to-date information, similarly for other healthcare professionals influencing or administering antibiotics.
- Monitor the supply of antibiotics.
- Recommend treatments other than antibiotics for minor conditions such as a common cold.
- Become actively involved regarding hygiene and infection control across healthcare settings.

(continued)

Box 1 (continued)

(B) *FIP urges governments and health authorities to*:

- Implement antimicrobial surveillance plans nationally including monitoring of antibiotic utilisation in agriculture, humans and veterinary medicine.
- Implement measures to enhance the appropriate use of antibiotics as well as stop their sale/dispensing without a prescription or order from a qualified healthcare professional. This includes measures to strengthen regulatory and legislative controls regarding the supply of antibiotics including the enforcement of any statutes and regulations to improve the rational use and dispensing of antibiotics.
- Impose restrictions on the prescribing of selected antibiotics, the objective being to limit development of AMR.
- Conduct education campaigns among all key groups to help promote the appropriate use of antibiotics. In addition, collaborate with all health professional societies to develop and help implement educational and behavioural interventions to improve appropriate antibiotic prescribing.
- Help with establishing infection control programmes to reduce AMR.
- Instigate methods to dispose of antibiotics where necessary that are environmentally sound.

The WHO in Europe has also produced guidance encouraging the prudent use of antibiotics including the role of pharmacists (WHO Europe 2014). This is alongside documenting seven key areas on potential ways to reduce AMR rates across Europe (Box 2).

Box 2: Seven Key Action Areas Identified by WHO Europe to Address AMR [Adapted from WHO Europe (2014)]

- Strengthen co-ordination across healthcare sectors and personnel.
- Strengthen the surveillance of AMR rates as well as antibiotic utilisation patterns among the countries in WHO Europe.
- Instigate measures to enhance the rational use of antibiotics (building on existing programmes).
- Strengthen infection control across all healthcare settings.
- Instigate measures to help prevent emerging antibiotic resistance strains in food and veterinary sectors.
- Help instigate appropriate measures that promote research into new antibiotics given their current scarcity.
- Improve awareness including partnerships to promote patient safety and reduce AMR rates across Europe.

Other authors and organisations have also endorsed the need for greater education of pharmacists regarding antibiotics during their training (Ahmad et al. 2015).

1.4 Public Knowledge and Attitudes Towards Antibiotics

Published studies have shown that there is a general lack of knowledge about antibiotics among the general public (Eng et al. 2003; McNulty et al. 2007a; Grigoryan et al. 2008; You et al. 2008; André et al. 2010; Oh et al. 2011; Al-Haddad 2012; Chan et al. 2012; Lim and Teh 2012; Napolitano et al. 2013; Van Nguyen et al. 2013; Fatokun 2014; Islahudin et al. 2014; Gualano et al. 2015; WHO 2015). Even in some countries in which public awareness campaigns on antibiotics had been conducted, there was still widespread belief that antibiotics are effective against viral infections (WHO 2014). Further inappropriate antibiotic prescribing, including unnecessary prescribing of antibiotics, has heightened public perceptions that these medicines are first-line treatment for common infections such as URTIs. These perceptions do lead patients to pressurise physicians to prescribe antibiotics regardless of the indication (Hassali et al. 2015; Md Rezal et al. 2015). These pressures are enhanced by patients' perception of antibiotics as powerful therapies, with the ability to quickly stop symptoms (Sharma et al. 2008).

Gualano et al. (2015) recently conducted a systematic review regarding public knowledge and attitudes towards antibiotics. The authors demonstrated that 53.9 % of the general public believed antibiotics can be used to treat viral infections, with 50.9 % perceiving antibiotics can be used to reduce inflammation with 33.7 % unaware that antibiotics should only be used to treat bacterial infections.

There are also concerns that the public do not take their antibiotics according to the instructions on the label, take incomplete courses, take leftover antibiotics from previous treatments and share their antibiotics with others, particularly in LMICs (Zarb and Goossens 2012). For instance in Bangladesh, 51.4 % of the general public did not take their antibiotics according to the label instructions (Sutradhar et al. 2014), and in Brazil, 30 % of those interviewed could have had their antibiotic treatment compromised by ignoring their current diagnosis (the indication for which an antibiotic was prescribed), for not understanding the dosage and administration schedule or both (Nicolini et al. 2008).

In Asia, Europe, North America and Oceania, 47.1 % of patients stopped taking antibiotics when their symptoms disappeared (Gualano et al. 2015). In Tetovo in the Republic of Macedonia, 60.8 % of parents kept leftover antibiotics at home for future use (Alili-Idrizi et al. 2014), whilst 25 % of the general population in Belgium, Colombia, France, Italy, Morocco, Spain, Thailand, Turkey and the UK also saved leftover antibiotics for future use (Pechere 2001). In Taiwan, 13 % of the general public shared their antibiotics with their family and friends (Chen et al. 2005). This compares to 8 % in similar studies in Hong Kong and Malaysia (You et al. 2008; Al-Haddad 2012).

Several studies have shown a strong association between patient expectations and demands with excessive prescribing of antibiotics (Hamm et al. 1996; Macfarlane et al. 1997; Mangione-Smith et al. 1999; Gonzales 2005; Mustafa et al. 2014). In Malaysia, 73.8 % of the general population expect physicians to prescribe antibiotics for a common cold (Lim and Teh 2012), whilst in Oklahoma, USA, approximately 65 % of patients with respiratory infections expected to be prescribed with antibiotics (Hamm et al. 1996). This may be because public awareness of antibiotic resistance is low across countries (Hawkings et al. 2007; Brooks et al. 2008; Brookes-Howell et al. 2012; Wun et al. 2013; Gualano et al. 2015; WHO 2015).

Several studies have also explored public knowledge and perceptions of antibiotic resistance in Europe (Brookes-Howell et al. 2012), South India (Chandy et al. 2013) and the UK (Hawkings et al. 2007; Brooks et al. 2008). Most of the public had heard of antibiotic resistance but only a minority had the correct understanding of this term. The public often understood the term 'antibiotic resistance' as the body getting used to or becoming immune to antibiotics. Other members perceived antibiotic resistance as the body becoming incompatible with antibiotics, the illness becoming too strong for antibiotics to manage and a hereditary illness (Brookes-Howell et al. 2012).

Most of the general public are also unaware of the causes and consequences of antibiotic resistance. In addition, they attributed antibiotic resistance to external factors such as overprescribing of antibiotics by physicians as opposed to internal factors, e.g. non-adherence to antibiotic therapy. As a result, they typically do not perceive themselves as being responsible for preventing the misuse of antibiotics (Hawkings et al. 2007; Brooks et al. 2008).

1.5 The Need for Public Engagement to Combat Antibiotic Resistance

In view of these misconceptions, public campaigns and engagement are needed to increase patients' knowledge about antibiotics and antibiotic resistance. Public engagement has been broadly defined as 'involving the general public in decision-making and in the planning, design, governance and delivery of initiatives' (O'Mara-Eves et al. 2013).

In this context, the School of Pharmaceutical Sciences of Universiti Sains Malaysia, in collaboration with Action on Antibiotic Resistance (ReAct), has taken the initiative to engage Yayasan Bina Ilmu (YBI) in addressing antibiotic resistance at the community level in Jelutong District, Penang, Malaysia. The school offers support and materials to the organisation to undertake relevant work on antibiotic resistance, in line with their focus areas. Coordination, communication and commitment among all stakeholders play a key role in making this approach a success, with the findings used to support the development and implementation of a national policy to manage antibiotic resistance at the community level in Malaysia (Irawati et al. 2015).

2 Aims

Given the global importance of AMR, there is a need to review potential interventions and measures that have been instigated across continents, countries and regions to improve antibiotic utilisation. Consequently, the aim of this chapter is to review interventions that have been introduced to reduce inappropriate prescribing and dispensing of antibiotics, and associated reduction in AMR, and their effectiveness to provide future guidance. This is seen as a conservatism strategy (Hoffman and Outterson 2015).

It is recognised that alongside this, strategies must be instigated to enhance the development of new antibiotics to address current unmet need as well as limit their utilisation to much targeted populations once launched (Hoffman and Ottersen 2015; Hoffman and Outterson 2015; Outterson et al. 2015). This will potentially involve new methods of funding research (Brogan and Mossialos 2013; Balasegaram et al. 2015; Hwang et al. 2015; Outterson et al. 2015). There is also a need for restrictions on the use of antibiotics in animals with, for instance, the utilisation of antibiotics in animals in the USA at least three times greater than the overall use in humans (So et al. 2015). However, new funding models for financing antibiotic research as well as programmes to reduce antibiotic use in animals are outside the scope of this chapter.

3 Methods

A narrative review of publications, including case histories, is known to the co-authors to provide future guidance. This is because systematic reviews of potential approaches to improve the utilisation of antibiotics have already been undertaken by the co-authors and others (Huttner et al. 2010; Dar et al. 2016; Md Rezal et al. 2015; Dyar et al. 2016).

We have undertaken similar approaches when reviewing measures to enhance the prescribing and dispensing of generics as well as initiatives to optimise the use of new medicines (Godman et al. 2012; Dylst et al. 2013; Godman et al. 2013; Godman et al. 2014; Godman et al. 2015).

4 Findings

The case histories will be divided into hospitals and ambulatory care. This will include the findings of meta-analyses as well as country and regional studies. There will also be a review of the impact of educational interventions on public knowledge and attitudes towards antibiotics and resistance. Typically multifaceted interventios are needed to change prescribing (Llor and Bjerrum 2014; Dyar et al. 2016).

4.1 Hospitals

Meta-analysis (Cochrane Review) (Davey et al. 2013)

- This meta-analysis included 95 interventions principally targeting antibiotic prescribing. This included antibiotic choices as well as the timing of first dose and the route of administration. There was reliable data from 76 interventions including persuasive (majority), restrictive and structural interventions. Interventions that were restrictive in nature had a significantly greater impact on subsequent outcomes at 1 month and 6 months. However, there were no significant differences between the different interventions on outcomes at 12 or 24 months.

  There was a reduction in *Clostridium difficile* infections as well as a reduction in the colonisation or infections with resistant bacteria, including cephalosporin-resistant and aminoglycoside-resistant gram-negative bacteria, vancomycin-resistant *Enterococcus faecalis* and MRSA, with interventions that sought to reduce excessive antibiotic prescribing. There was also a significant reduction in mortality in interventions instigated to improve antibiotic prescribing in patients with pneumonia.

China (Zou et al. 2014)

- There have been considerable concerns regarding the overuse of antibiotics in China, driven by incentive systems for hospitals and physicians encouraging the use of IV versus oral antibiotics as well as their overuse (Reynolds and McKee 2009, 2011). However, antibiotic prescribing is changing following a nationwide campaign instigating antibiotic stewardship programmes (ASPs) to address their overuse and AMR rates. A recent study assessing the impact of this nationwide campaign showed the following between 2011 and 2012:

  - Decreasing antibiotic use (26.54 instead of 39.37 DDDs/100 inpatient days)
  - Decreasing % of antibiotics among outpatient prescriptions and their use in hospital inpatients over the 2 years
  - Correlation between subsequent antibiotic utilisation and the type and size of specialised hospitals, however not with the region

Scotland

- The Scottish Antimicrobial Prescribing Group (SAPG) was established in Scotland in 2008 to coordinate a national antimicrobial stewardship programme (Nathwani et al. 2011). The collection of prevalence data in 2009 led to the development of two quality prescribing indicators: (1) the extent of compliance with antibiotic policies among acute admission units and (2) surgical prophylaxis duration (Malcolm et al. 2013). Studies showed compliance with current antibiotic policies (81.0 %) was similar to other European countries; however, the duration of surgical prophylaxis <24hr (68.6 %) was higher than generally seen across Europe.

  Following the implementation of prescribing indicators, there was an improvement in the indication noted in the patient records of ≥90 %. Compliance with antibiotic prescribing policies also increased to 90 % (Malcolm et al. 2013), with the mean proportion of patients receiving single-dose prophylaxis following colorectal surgery exceeding 95 % (the target). Overall, SAPG has shown that the implementation of national prescribing indicators, which are acceptable to clinicians and regularly audited, can improve subsequent antibiotic utilisation (Malcolm et al. 2013).

Zambia (Mubita et al. 2013; Massele et al. 2015)

- Mubita et al. conducted a prospective quasi-experimental clinical audit project among internal medicine wards at a University Teaching Hospital in Zambia. The findings showed that the implementation of an antimicrobial prescribing care bundle was associated with an improvement in the quality of antibiotic prescribing in terms of compliance with its care elements.

  A pilot study was undertaken to assess the implementation of an adapted antimicrobial prescribing care bundle. Subsequently, compare compliance of antimicrobial prescribing with the care elements (standards) of the antimicrobial prescribing care bundle before and after implementation. Implementation of the bundle involved educational programmes on antimicrobial stewardship. Topics covered included the antimicrobial prescribing care bundle, standard treatment guidelines on infections and the hospital antibiogram. Feedback on the control phase results was communicated to the prescribers. Posters of the antimicrobial stewardship treatment algorithm were displayed within the medical wards. Pocket-sized cards of the prescribers' checklist as an aide—memoire—were distributed to all prescribers in internal medicine.

Outcome measures were compliance with the care bundle's care elements as audit standards as well as:

- Presence of clinical evidence of bacterial infection
- Documentation of the clinical indication for antibiotics, the duration or review date as well as the route and dose of antibiotics prescribed
- Collection of appropriate culture specimens
- Appropriate empirical selection of antibiotics
- Documentation of appropriate prescribing decision option by 48 h of antimicrobial therapy

Table 4 Findings of the clinical audit project conducted in a teaching hospital in Zambia

Care element (audit standard)	Compliance level *n* (%) Control phase (*n* = 128)	Intervention Phase (*n* = 128)	*p*
Clinical evidence of bacterial infection	112 (87.5)	116 (90.6)	0.022
Documentation of the clinical indication, the duration or review date as well as the route and dose of antibiotics	9 (7)	58 (45.3)	0.522
Culture specimens collected according to standard treatment guidelines	84 (65.6)	92 (71.9)	<0.001
Appropriate empirical selection of antibiotics, i.e. prescribing according to standard treatment guidelines at initiation of antibiotics	73 (57)	80 (62.5)	0.333
Documentation of appropriate prescribing decision option by 48 h of antimicrobial therapy	64 (50)	83 (64.8)	<0.001
Overall compliance	7 (5.5)	46 (35.9)	0.044

Table 4 shows a summary of the findings of the clinical audit project.

This study recommended implementation of a formal antimicrobial stewardship programme among internal medicine wards in Zambia. This would involve establishing multidisciplinary antimicrobial stewardship teams in hospitals with core membership comprising the following: an infectious diseases physician, a clinical microbiologist and a clinical pharmacist with expertise in infectious diseases.

4.2 *Ambulatory Care (Settings, Regions and Countries)*

4.2.1 Pharmacy Dispensing

Section 1.3 together with Boxes 1 and 2 describes potential ways forward for pharmacists and others to reduce inappropriate dispensing and use of antibiotics. We are already seeing a number of countries tighten illegal self-purchasing of antibiotics, e.g. Brazil (private sector), Mexico and Serbia, and this will grow (Godman et al. 2014; Moura et al. 2015; Santa-Ana-Tellez et al. 2015).

4.2.2 European Surveillance of Antimicrobial Consumption (ESAC)

ESAC have suggested a number of aggregated and patient-level quality indicators to improve future use of antibiotics. These include:

Aggregated indexes (in defined daily doses/1000 inhabitants/day—DIDs) (Coenen et al. 2007):

- % combination penicillins including β-lactamase-sensitive penicillins vs. all antibiotics
- % of third- and fourth-generation cephalosporins vs. first- and second-generation cephalosporins (and all antibiotics)
- % fluoroquinolones vs. all antibiotics
- % broad vs. narrow penicillins, cephalosporins and macrolides

Quality indicators where patient level data is available include (Adriaenssens et al. 2011):

- % of patients between 18 and 75 years with acute bronchitis prescribed antibiotics (acceptable range 0–30 %); within this % those receiving recommended antibiotics (acceptable range 80–100 %) or fluoroquinolones (acceptable range 0–5 %)
- % of patients older than 1 year with upper respiratory tract infections (URTIs) prescribed antibiotics for systemic use (acceptable range 0–20 %); within this % receiving recommended antibiotics (acceptable range 80–100 %) or fluoroquinolones (acceptable range 0–5 %)

4.2.3 Internet-Based Training on Antibiotic Prescribing Rates for Patients with Acute RTIs (Little et al. 2013)

Primary care practices among six European countries were randomised to (1) usual care, (2) training in the use of a C-reactive protein (CRP) tests at the point of care, (3) training in enhanced communication skills, or (4) both training in CRP and enhanced communication via the Internet.

The study showed that antibiotic prescribing rates were lower with training in CRP tests and lower in practices with enhanced-communication training than without such. The greatest reduction in subsequent antibiotic prescribing was seen when the interventions were combined.

The authors concluded that Internet training did achieve important reductions in antibiotic prescribing for RTIs across languages and cultures, consequently a potential way forward to enhance appropriate antibiotic prescribing across countries where resources are limited.

4.2.4 Belgium (WHO Europe 2011)

Since 2000, national campaigns among GPs and communities on the prudent use of antibiotics have been organised by the Belgian Antibiotic Policy Coordination Committee, alongside this, establishing surveillance systems across healthcare sectors, improving legislation on hospital hygiene and disseminating guidelines on the prevention and treatment of bacterial infections. These multifaceted campaigns resulted in a steady decrease (6.2 % per year) in the use of antibiotics in ambulatory care in recent years. In addition, the correct use of antibiotics is also increasing across sectors and resistance to streptococcal infections steadily declining.

Similar successful campaigns have been initiated in other countries including France, Spain, Poland and the UK.

4.2.5 France (Sabuncu et al. 2009)

Among European countries, France typically has the highest rate of antibiotic consumption and beta-lactam resistance in *Streptococcus* spp. This resulted in the French government instigating in 2001 a programme entitled 'Keep Antibiotics Working', the main component being 'Les antibiotiques c'est pas automatique' ('Antibiotics are not automatic').

Compared to the pre-intervention period (2000–2002):

- Decrease by 226.5 % over 5 years in the number of antibiotic prescriptions, adjusted for flu-like syndrome frequency during the winter season.
- This decline in antibiotic utilisation occurred throughout France, which affected all antibiotic classes except quinolones.
- The greatest decrease in antibiotic utilisation was seen among young children (6–15 years of age).

The authors concluded that this multifaceted campaign was associated with an appreciable reduction in unnecessary antibiotic prescribing. This was particularly the case in children.

4.2.6 Italy (Formoso et al. 2013)

Interventions to try and reduce antibiotic prescribing among the public included posters, brochures and adverts in the local media. There was also a newsletter on local antibiotic resistance targeting physicians and pharmacists. The design of the materials and messages was facilitated by GPs and paediatricians working in the intervention locality.

Antibiotic prescribing was significantly reduced in the intervention locality compared with controls outside the locality. There was a greater decrease for penicillins resistant to beta-lactamase, as well as a decrease in the prescribing of penicillins susceptible to beta-lactamase, consistent with the content of the newsletter. However, there was no difference with respect to knowledge and attitudes regarding the correct use of antibiotics among the public in the two groups.

Table 5 Some of the activities undertaken in Slovenia to enhance the appropriate use of antibiotics (adapted from Fürst et al. 2015)

Activity	Institution (organiser)	Targeted public	Introduction/ frequency
Two-day symposium on antibiotics once a year	Department of Infectious Diseases of the Ljubljana UMC	GPs	1995/every year
Prescribing restrictions for amoxicillin/clavulanic acid and the fluoroquinolones	National Health Insurance	All physicians	2000/ permanent
Workshops in primary health centres	Primary health centres, National Health Insurance	GPs	2001/ sporadically
Informative budget targets for prescribed drugs	National Health Insurance	All physicians	2001/ permanent
Guidelines on treatment of infectious diseases	Medical professionals	GPs	2002
Audits	National Health Insurance	All physicians	2002/ regularly
Workshop on rational prescribing of antibiotics	Faculty of Medicine, University of Ljubljana	Specialising GPs	2004/every year
Prescribing restrictions for cephalosporins	National Health Insurance	All physicians	2005/ permanent
Booklet 'My Child Has a Fever'	National Health Insurance, medical professionals	Parents	2007/always available
Workshops in regions with the highest utilisation of antibiotics	National Committee for the Rational Use of Antimicrobials	GPs	2007/once a year
Antibiotic Awareness Day	Ministry of Health and the National Committee for the Rational Use of Antimicrobials	Lay public and GPs	2008/every year
Prescribing restrictions for the macrolides	National Health Insurance	All physicians	2009/ permanent
Workshop on rational prescribing of antibiotics	Slovenian Society of Chemotherapy	Young physicians	2010/every year
Flyer 'Get well without antibiotics'	National Health Insurance, medical professionals	Lay public	2010/always available
Quality indicators including antibiotics	National Health Insurance	GPs	2011

4.2.7 Slovenia (Fürst et al. 2015)

Multifaceted interventions among all key stakeholder groups in Slovenia decreased antibiotic utilisation by 2–9 % per year from 1999 to 2012, overall by 31 %. This followed a 24 % increase in antibiotic utilisation at the end of the 1990s in Slovenia. There were also appreciable changes in the prescribing of different antibiotics.

Table 5 contains some of the multiple initiatives and policies undertaken in Slovenia in recent years to reduce overall antibiotic utilisation as well as significantly reduce the utilisation of seven out of ten antibiotics.

Streptococcus resistance to penicillin decreased, mirroring decreasing utilisation. However, *Streptococcus* resistance to macrolides increased despite their utilisation halving, and *E. coli* resistance to fluoroquinolones doubled despite their utilisation decreasing by a third.

4.2.8 Thailand (Sumpradit et al. 2012; Antibiotic Smart use, Thailand, 2016)

The Antibiotics Smart Use programme was initiated to promote rational antibiotic use and reduce AMR rates. Multifaceted interventions at individual and organisational levels were implemented to increase public knowledge regarding antibiotics and change antibiotic prescription practices among physicians. Interventions at the network and policy levels were used to maintain behaviour change and scale up the various programmes. Key healthcare professionals, including physicians and pharmacists, as well as community leaders, were trained to promote rational antibiotic use in their healthcare settings and communities. Educational materials were also provided for display or distribution to patients.

The findings showed that the combined interventions increased public knowledge regarding antibiotics and changed their attitudes towards them. Antibiotic utilisation decreased by 18–46 % after the programme, and the percentage of patients who were not prescribed with antibiotics for the three targeted illnesses, i.e. acute diarrhoea, URTIs and simple wounds, increased by 29.1 %. Furthermore, almost all patients who were not prescribed antibiotics fully recovered within 7–10 days after the medical visits demonstrating that antibiotics were not necessary.

4.2.9 UK: Scotland (Nathwani et al. 2011; Colligan et al. 2015)

We have previously documented the initiation and activities of SAPG in Scotland (Sect. 4.1), which significantly reduced *Clostridium difficile* infection rates. An integrated approach to antimicrobial stewardship is being achieved through engagement with key stakeholder groups at all levels, aided by implementation of data management systems and training materials on antimicrobial stewardship. Improving the treatment of infections such as community-acquired pneumonia was also helped by quality improvement methodologies.

A self-reported survey in 2014 evaluating stewardship activities by the regional antimicrobial management teams (AMTs) as part of the SAPG programme demonstrated good compliance with nine of the ten key European indicators. 50 % of the AMTs achieved all nine indicators and 100 % achieved at least six out of the nine indicators (67 %). The authors concluded that collaborative working between SAPG and AMTs, together with central funding, has been a key to achieving good success, providing direction to other countries and regions.

4.3 Impact of Educational Interventions on Public Knowledge and Attitudes Towards Antibiotics and Antibiotic Resistance

Numerous public education campaigns have been conducted, particularly in high-income countries, to improve antibiotic utilisation (Finch et al. 2004; Rodis et al. 2004; Gonzales et al. 2005; Curry et al. 2006; Huttner et al. 2010; Greene et al. 2011; Holloway 2011; McKay et al. 2011; Fürst et al. 2015), some of which have already been described earlier. All campaigns have tried to convey that AMR is a major public health problem and, in addition, that the misuse of antibiotics contributes to this. Furthermore, most campaigns have tried to educate the public that most RTIs are viral in origin and consequently should not be treated with antibiotics (Finch et al. 2004; Huttner et al. 2010).

The campaigns have used various means of communication. The most common has been printed educational materials such as brochures, leaflets and posters. Some campaigns have used mass media such as newspapers, radio, television, billboards and public transport advertisements, whilst others have conducted public seminars to provide opportunities for interactive education and behavioural change (Freimuth et al. 2000; Finch et al. 2004; Huttner et al. 2010; Holloway 2011; Fürst et al. 2015). Nearly all campaigns have used the Internet in some way (Finch et al. 2004; Huttner et al. 2010).

Table 6 contains details of a number of the campaigns and their impact, building on the experiences of Thailand and other countries described above.

Shehadeh et al. recently conducted a study to assess the impact of a pharmacist-initiated educational intervention on public knowledge of antibiotics and antibiotic resistance in Jordan (Shehadeh et al. 2016). The mean knowledge score for the pre- and post-education was 59.4 % and 65.9 %, respectively. In addition, participants within poor and adequate knowledge categories were significantly shifted to the good knowledge category after the educational intervention. However, the authors believed the improvement in participants' knowledge may not always translate into a change in subsequent antibiotic-seeking behaviour.

Despite these initiatives, it appears difficult in practice to educate the public about the differences between bacterial and viral infections (Finch et al. 2004; Goossens et al. 2006; Huttner et al. 2010). Telephone surveys among the public in Canada did not show any impact of a campaign educating them of the bacterial or viral nature of certain infections. Prior to the campaign, 54 % of the public agreed that antibiotics are effective for the treatment of viral infections. After the campaign, 53 % still did not know that antibiotics do not work against viruses (NIPA 2013). In New Zealand, a post-education survey demonstrated that the general public still had misconceptions that antibiotics are needed to treat viral infections (Curry et al. 2006), and in France 54 % of the public still remained unaware that most URTIs are caused by viruses and do not require antibiotic treatment despite successive campaigns over 5 years (Huttner B et al. 2010). However, this should not dissuade health authorities from instigating programmes in the future to try and

Table 6 Influence of educational activities among patients

Country	Programmes and their influence
Belgium (Goossens et al. 2006)	• The majority of the public (79 %) who recalled the national antibiotic campaign recalled information from the television rather than other means of communication • Compared with the pre-campaign, patient expectations to be prescribed an antibiotic decreased significantly for acute bronchitis, cold, flu, diarrhoea and sore throats
Israel (Hemo et al. 2009)	• Exposure to media campaigns was associated with positive attitudes by patients towards antibiotics • Parents exposed to the media campaign were more likely to agree with current standards of the appropriate prescribing of antibiotics and less likely to expect physicians to prescribe antibiotics for RTIs, otitis media and pharyngitis than parents not exposed to the campaign
USA—North Carolina (Greene et al. 2011)	• A pharmacy student-driven education programme raised public awareness of the threat of AMR and the appropriate use of antibiotics • After participating in the programme, 71 % of patients stated that they would not now be requesting antibiotics from their physicians
USA—Ohio (Rodis et al. 2004)	• Patients' knowledge about the appropriate indication for prescribing antibiotics improved following a pharmacist-initiated educational intervention • The post-intervention survey demonstrated a significant increase in the % of patients agreeing that antibiotics should not be used for a cold as well as for influenza combined with cough and body aches

educate patients regarding the appropriate use and requests for antibiotics given the impact of inappropriate antibiotic use on increasing AMR rates.

Increasing knowledge of antibiotics among the public can result in a paradoxical effect (Huttner et al. 2010). In the UK, studies have shown that public education campaigns are associated with increased knowledge of antibiotics. However, they also resulted in increased likelihood of self-medication, sharing antibiotics with someone else and keeping leftover antibiotics (McNulty et al. 2007a; McNulty et al. 2007b). This needs to be factored into any future campaign.

Care is also needed when translating programmes from one country to another where the health systems and context can be different (Holloway 2011). Typically, educational approaches will only be effective if they are tailored to the local context and able to address local educational needs and organisational barriers (Siddiqi et al. 2005; Sumpradit et al. 2012).

5 Conclusion

The threat of AMR with its impact on morbidity, mortality and costs is growing and, as a result, becoming one of the most critical problems facing healthcare systems worldwide. A number of interlinking strategies are needed to address this. These include enhancing access where co-payments are an issue, reducing inappropriate prescribing (conservatism) where this is a problem as well as encouraging the development of new antibiotics given the current scarcity (Hoffman and Outterson 2015; O'Neill 2015). Knowledge regarding current utilisation and resistance patterns in a country or region is a pre-requisite to instigating future interventions to improve future rational prescribing and dispensing of antibiotics (WHO Europe 2014).

We have shown that multiple approaches, including all key stakeholder groups such as health authorities, physicians, pharmacists and patients, are needed to reduce inappropriate antibiotic prescribing. Within this, physicians and pharmacists are particularly important with FIP urging a number of activities among pharmacists and governments (Box 1). This is endorsed by the WHO in Europe (Box 2). Activities need to be continually analysed and sustained for maximum impact given the misconceptions that can still remain coupled with people's short memories from one year to the next.

We will continue to monitor the impact of the interventions across regions and countries to reduce future inappropriate prescribing and dispensing of antibiotics to provide direction to others to address this critical area.

Acknowledgements and Financial Interests There are no conflicts of interest from any author and no writing assistance was employed in the writing of this chapter. Each author contributed to and approved the final manuscript.

This work was in part supported by a VR-Link grant from Swedish Research Council (VR-Link 2013-6710). The write-up was in part supported by a grant from the Newton Fellowship.

References

Abasaeed AE, Vlcek J, Abuelkhair MA, Andrajati R, Elnour AA (2013) A comparative study between prescribed and over-the-counter antibiotics. Saudi Med J 34:1048–1054

Abdula N, Macharia J, Motsoaledi A, Swaminathan S, Vijay Raghavan K (2015) National action for global gains in antimicrobial resistance. Lancet 10014:e3–e25. doi:10.1016/S0140-6736(15)00668-6

Abdulhak A, Al-Tannir M, Almansor M, Almohaya M, Onazi A, Marei M, Aldossary O, Obeidat S, Obeidat M, Riaz M, Tleyjeh I (2011) Non prescribed sale of antibiotics in Riyadh, Saudi Arabia: a cross sectional study. BMC Pub Health 11:538. doi:10.1186/1471-2458-11-538

Adesokan HK, Akanbi IO, Akanbi IM, Obaweda RA (2015) Pattern of antimicrobial usage in livestock animals in south-western Nigeria: the need for alternative plans. Onderstepoort J Vet Res 82:816. doi:10.4102/ojvr.v82i1.816

Adriaenssens N, Coenen S, Tonkin-Crine S, Verheij TJ, Little P, Goossens H (2011) European Surveillance of Antimicrobial Consumption (ESAC): disease-specific quality indicators for outpatient antibiotic prescribing. BMJ Qual Saf 20:764–772. doi:10.1136/bmjqs.2010.049049

Ahmad N (2014) Summary of national surveillance of antibiotic resistance report, 2014. http://www.imr.gov.my/en/component/content/article/75-english-content/national-collabration/1469-nsar-main.html

Ahmad A, Khan MU, Moorthy J, Jamshed SQ, Patel I (2015) Comparison of knowledge and attitudes about antibiotics and resistance, and antibiotics self-practicing between Bachelor of Pharmacy and Doctor of Pharmacy students in Southern India. Pharm Pract (Granada) 13:523

Al-Azzam S, Al-Husein B, Alzoubi F, Masadeh M (2007) Self-medication with antibiotics in Jordanian population. Int J Occup Med Environ Health 20:373–380. doi:10.2478/v10001-007-0038-9

Al-Haddad MS (2012) General public knowledge, source of knowledge and practice towards antibiotics in the State of Penang, Malaysia. Arch Pharm Practice 3:233–241

Alharbi SA, Wainwright M, Alahmadi TA, Salleeh HB, Faden AA, Chinnathambi A (2014) What if Fleming had not discovered penicillin? Saudi J Biol Sci 21:289–293. doi:10.1016/j.sjbs.2013.12.007

Alili-Idrizi E, Dauti M, Malaj L (2014) Validation of the parental knowledge and attitude towards antibiotic usage and resistance among children in Tetovo, the Republic of Macedonia. Pharm Pract (Granada) 12:467

Almaaytah A, Mukattash TL, Hajaj J (2015) Dispensing of non-prescribed antibiotics in Jordan. Patient Prefer Adher 9:1389–1395. doi:10.2147/PPA.S91649

Almeman AA, Ibrahim MIM, Rasool S (2014) Cost analysis of medications used in upper respiratory tract infections and prescribing patterns in University Sans Malaysia, Penang, Malaysia. Trop J Pharm Res 13:621–626. doi:10.4314/tjpr.v13i4.21

Al-Mohamadi A, Badr A, Bin Mahfouz L, Samargandi D, Al-Ahdal A (2013) Dispensing medications without prescription at Saudi community pharmacy: extent and perception. Saudi Pharm J 2:13–18. doi:10.1016/j.jsps.2011.11.003

Alsan M, Schoemaker L, Eggleston K, Kammili N, Kolli P, Bhattacharya J (2015) Out-of-pocket health expenditures and antimicrobial resistance in low-income and middle-income countries: an economic analysis. Lancet Infect Dis 15:1203–1210. doi:10.1016/s1473-3099(15)00149-8

André M, Vernby Å, Berg J, Lundborg CS (2010) A survey of public knowledge and awareness related to antibiotic use and resistance in Sweden. J Antimicrob Chemother 65:1292–1296. doi:10.1093/jac/dkq104

Antibiotic smart use, Thailand: involving community to curb antibiotic resistance 2016. http://www.reactgroup.org/wp-content/uploads/2016/10/Antibiotic-Smart-Use-project-case-study.pdf

Bahnassi A (2015) A qualitative analysis of pharmacists' attitudes and practices regarding the sale of antibiotics without prescription in Syria. J Taibah Univ Med Sci 10:227–233. doi:10.1016/j.jtumed.2014.09.001

Bajcetic M, Kalaba M, Godman B, Sipetic T, Coenen S, Versporten A, Goossens H (2012) Ongoing programs to reduce the high rate of antibiotic consumption in Serbia; current situation and future direction. Pharmacoepidemiol Drug Saf 21(Suppl 3):98

Balasegaram M, Clift C, Røttingen JA (2015) The global innovation model for antibiotics needs reinvention. J Law Med Ethics 43(S3):22–26. doi:10.1111/jlme.12270

Barlam TF, Gupta K (2015) Antibiotic resistance spreads internationally across borders. J Law Med Ethics 43(S3):12–16. doi:10.1111/jlme.12268

Barnett ML, Linder JA (2014) Antibiotic prescribing for adults with acute bronchitis in the United States, 1996-2010. JAMA 311:2020–2022. doi:10.1001/jama.2013.286141

Bejon P, Mwangi I, Ngetsa C, Mwarumba S, Berkley JA, Lowe BS, Maitland K, Marsh K, English M, Scott JA (2005) Invasive Gram-negative bacilli are frequently resistant to standard antibiotics for children admitted to hospital in Kilifi, Kenya. J Antimicrob Chemother 56:232–235. doi:10.1093/jac/dki145

Berkley JA, Lowe BS, Mwangi I, Williams T, Bauni E, Mwarumba S, Ngetsa C, Slack MP, Njenga S, Hart CA, Maitland K, English M, Marsh K, Scott JA (2005a) Bacteremia among children admitted to a rural hospital in Kenya. N Engl J Med 352:39–47. doi:10.1056/NEJMoa040275

Berkley JA, Maitland K, Mwangi I, Ngetsa C, Mwarumba S, Lowe BS, Newton CR, Marsh K, Scott JA, English M (2005b) Use of clinical syndromes to target antibiotic prescribing in seriously ill children in malaria endemic area: observational study. BMJ 330(7498):995. doi:10.1136/bmj.38408.471991.8F

Bigogo G, Audi A, Aura B, Aol G, Breiman RF, Feikin DR (2010) Health-seeking patterns among participants of population-based morbidity surveillance in rural western Kenya: implications for calculating disease rates. Int J Infect Dis 14:e967–e973. doi:10.1016/j.ijid.2010.05.016

Blomberg B, Manji KP, Urassa WK, Tamim BS, Mwakagile DS, Jureen R, Msangi V, Tellevik MG, Holberg-Petersen M, Harthug S, Maselle SY, Langeland N (2007) Antimicrobial resistance predicts death in Tanzanian children with bloodstream infections: a prospective cohort study. BMC Infect Dis 7:43. doi:10.1186/1471-2334-7-43

Bosch F, Roschi L (2008) The contributions of Paul Ehrlich to pharmacology: a tribute on the occasion of the centenary of his Nobel prize. Pharmacology 82:171–179

Braoios A, Pereira AC, Bizerra AA, Policarpo OF, Soares NC, Barbosa AS (2013) The use of antimicrobial drugs by the population in the city of Jataí, State of Goias, Brazil. Cien Saude Colet 18:3055–3060

Brogan DM, Mossialos E (2013) Incentives for new antibiotics: the Options Market for Antibiotics (OMA) model. Global Health 9:58. doi:10.1186/1744-8603-9-58

Brookes-Howell L, Elwyn G, Hood K, Wood F, Cooper L, Goossens H, Ieven M, Butler CC (2012) 'The body gets used to them': patients' interpretations of antibiotic resistance and the implications for containment strategies. J Gen Intern Med 27:766–772. doi:10.1007/s11606-011-1916-1

Brooks JT, Ochieng JB, Kumar L, Okoth G, Shapiro RL, Wells JG, Bird M, Bopp C, Chege W, Beatty ME, Chiller T, Vulule JM, Mintz E, Slutsker L (2006) Surveillance for bacterial diarrhea and antimicrobial resistance in rural western Kenya, 1997-2003. Clin Infect Dis 43:393–401. doi:10.1086/505866

Brooks L, Shaw A, Sharp D, Hay AD (2008) Towards a better understanding of patients' perspectives of antibiotic resistance and MRSA: a qualitative study. Fam Pract 25:341–348. doi:10.1093/fampra/cmn037

Chan YH, Fan MM, Fok CM, Lok ZL, Ni M, Sin CF, Wong KK, Wong SM, Yeung R, Yeung TT, Chow WC, Lam TH, Schooling CM (2012) Antibiotics nonadherence and knowledge in a community with the world's leading prevalence of antibiotics resistance: implications for public health intervention. Am J Infect Control 40:113–117. doi:10.1016/j.ajic.2011.03.017

Chandy SJ, Mathai E, Thomas K, Faruqul AR, Holloway K, Lundburg CS (2013) Antibiotic use and resistance: perceptions and ethical challenges among doctors, pharmacists and the public in Vellore, South India. Indian J Med Ethics 10:20–27

Chatterjee P, Fleck F (2011) Mobilizing political will to contain antimicrobial resistance. Bull World Health Organ 89:168–169. doi:10.2471/blt.11.030311

Chen C, Chen YM, Hwang KL, Lin SJ, Yang CC, Tsay RW, Liu CE, Young TG (2005) Behaviour, attitudes and knowledge about antibiotic usage among residents of Changhua, Taiwan. J Microbiol Immunol Infect 38:53–59

Cluzeau F, Manandhar A (2015) Towards an evidence-based National Action Plan on Antimicrobial Resistance in Vietnam. http://www.idsihealth.org/blog/towards-an-evidence-based-national-action-plan-on-antimicrobial-resistance-in-vietnam/

Coenen S, Ferech M, Haaijer-Ruskamp FM, Butler CC, Vander Stichele RH, Verheij TJ, Monnet DL, Little P, Goossens H, ESAC Project Group (2007) European Surveillance of Antimicrobial Consumption (ESAC): quality indicators for outpatient antibiotic use in Europe. Qual Saf Health Care 16:440–445. doi:10.1136/qshc.2006.021121

Colligan C, Sneddon J, Bayne G, Malcolm W, Walker G, Nathwani D (2015) Six years of a national antimicrobial stewardship program in Scotland: where are we now? Antimicrob Resist Infect Control 4:28. doi:10.1186/s13756-015-0068-1

Contopoulos-Ioannidis DG, Koliofoti ID, Koutroumpa IC, Giannakakis IA, Ioannidis JPA (2001) Pathways for inappropriate dispensing of antibiotics for rhinosinusitis: a randomised trial. Clin Infect Dis 33:76–82. doi:10.1086/320888

Costelloe C, Metcalfe C, Lovering A, Mant D (2010) Effect of antibiotic prescribing in primary care on antimicrobial resistance in individual patients: systematic review and meta-analysis. BMJ 340:c2096

Cotton MF, Wasserman E, Smit J, Whitelaw A, Zar HJ (2008) High incidence of antimicrobial resistant organisms including extended spectrum beta-lactamase producing Enterobacteriaceae and methicillin-resistant *Staphylococcus aureus* in nasopharyngeal and blood isolates of HIV-infected children from Cape Town, South Africa. BMC Infect Dis 8:40. doi:10.1186/1471-2334-8-40

Curry M, Sung L, Arroll B, Goodyear-Smith F, Kerse N, Norris P (2006) Public views and use of antibiotics for the common cold before and after an education campaign in New Zealand. NZ Med J 119:U1957

Dameh M, Norris P, Green J (2012) New Zealand pharmacists' experiences, practices and views regarding antibiotic use without prescription. J Prim Health Care 4:131–140

Dar OA, Hasan R, Schlundt J, Harbarth S, Caleo G, Dar FK, Littmann J, Rweyemamu M, Buckley EJ, Shahid M, Kock R, Li HL, Giha H, Khan M, So AD, Bindayna KM, Kessel A, Pedersen HB, Permanand G, Zumla A, Røttingen JA, Heymann DL (2016) Exploring the evidence base for national and regional policy interventions to combat resistance. Lancet 387:285–295. doi:10.1016/S0140-6736(15)00520-6

Davey P, Brown E, Charani E, Fenelon L, Gould IM, Holmes A, Ramsay CR, Wiffen PJ, Wilcox M (2013) Interventions to improve antibiotic prescribing practices for hospital inpatients. Cochrane Database Syst Rev 4:Cd003543. doi:10.1002/14651858.CD003543.pub3

Davies J, Davies D (2010) Origins and evolution of antibiotic resistance. Microbiol Mol Biol Rev 74:417–433

de Jong J, van den Berg PB, de Vries TW, de Jong-van den Berg LT (2008) Antibiotic drug use of children in the Netherlands from 1999 till 2005. Eur J Clin Pharmacol 64:913–919. doi:10.1007/s00228-008-0479-5

de Kraker ME, Jarlier V, Monen JC, Heuer OE, van de Sande N, Grundmann H (2013) The changing epidemiology of bacteraemias in Europe: trends from the European antimicrobial resistance surveillance system. Clin Microbiol Infect 19:860–868. doi:10.1111/1469-0691.12028

Del Fiol FDS, Lopes LC, Toledo MID, Barberato-Filho S (2010) Perfil de prescrições e uso de antibióticos em infecções comunitárias. Rev Soc Bras Med Trop 43:68–72. doi:10.1590/S0037-86822010000100015

Department of Health, Republic of South Africa (2014) Antimicrobial resistance National Strategy Framework 2014–2024. http://www.nda.agric.za/docs/media/A5%20Antimicrobial%20Resistance%20National%20Strategy%20Framework%202014-2024_final.pdf

Dibua UM, Onyemerela IS, Nweze EI (2014) Frequency, urinalysis and susceptibility profile of pathogens causing urinary tract infections in Enugu State, southeast Nigeria. Rev Inst Med Trop Sao Paulo 56:55–59. doi:10.1590/S0036-46652014000100008

Dooling K, Kandeel A, Hicks L, El-Shoubary W, Fawzi K, Kandeel Y, Etman A, Leena Lohiniva AL, Talaat M (2014) Understanding antibiotic use in Minya District, Egypt: physician and pharmacist prescribing and the factors influencing their practices. Antibiotics 3:233–243. doi:10.3390/antibiotics3020233

Dyar OJ, Beovic B, Vlahovic-Palcevski V, Verheij T, Pulcini C on behalf of ESGAP (the ESCMID [European Society of Clinical Microbiology and Infectious Diseases] Study Group for Antibiotic Policies) (2016) How can we improve antibiotic prescribing in primary care? Expert Rev Anti Infect Ther 14:403–413. doi:10.1586/14787210.2016.1151353

Dylst P, Vulto A, Godman B, Simoens S (2013) Generic medicines: solutions for a sustainable drug market? Appl Health Econ Health Policy 11:437–443. doi:10.1007/s40258-013-0043-z

Earnshaw S, Mendez A, Monnet DL, Hicks L, Cruickshank M, Weekes L, Njoo H, Ross S (2013) Global collaboration to encourage prudent antibiotic use. Lancet Infect Dis 13:1003–1004. doi:10.1016/S1473-3099(13)70315-3

Emeka PM, Al-Omar MJ, Khan TM (2012) A qualitative study exploring role of community pharmacy in the irrational use and purchase of nonprescription antibiotics in Al Ahsa. Eur J Gen Med 9:230–234

Enato EF, Sounyo AA, Einarson TR (2011) Medication utilization and illness management study in Nigeria. Ann Pharmacother 45:924–930. doi:10.1345/aph.1Q051

Eng JV, Marcus R, Hadler JL, Imhoff B, Vugia DJ, Cieslak PR, Zell E, Deneen V, McCombs KG, Zansky SM, Hawkins MA, Besser RE (2003) Consumer attitudes and use of antibiotics. Emerg Infect Dis 9:1128–1135. doi:10.3201/eid0909.020591

European Centre for Disease Prevention and Control (ECDPC) (2015) Annual epidemiological report. http://ecdc.europa.eu/en/publications/_layouts/forms/Publication_DispForm.aspx?List=4f55ad51-4aed-4d32-b960-af70113dbb90&ID=1292

Farah R, Lahoud N, Salameh P, Saleh N (2015) Antibiotic dispensation by Lebanese pharmacists: a comparison of higher and lower socio-economic levels. J Infect Public Health 8:37–46. doi:10.1016/j.jiph.2014.07.003

Fatokun O (2014) Exploring antibiotic use and practices in a Malaysian community. Int J Clin Pharm 36:564–569. doi:10.1007/s11096-014-9937-6

Finch RG, Metlay JP, Davey PG, Baker LJ (2004) Educational interventions to improve antibiotic use in the community: report from the International Forum on Antibiotic Resistance (IFAR) colloquium, 2002. Lancet Infect Dis 4:44–53. doi:10.1016/S1473-3099(03)00860-0

FIP statement of policy—control of antimicrobial medicines resistance (AMR). http://www.fip.org/www/uploads/database_file.php?id=289&table_id

Formoso G, Paltrinieri B, Marata AM, Gagliotti C, Pan A, Moro ML, Capelli O, Magrini N, LOCAAL Study Group (2013) Feasibility and effectiveness of a low cost campaign onantibiotic prescribing in Italy: community level, controlled, non-randomised trial. BMJ 347:f5391. doi:10.1136/bmj.f5391

Freimuth V, Linnan HW, Potter P (2000) Communicating the threat of emerging infections to the public. Emerg Infect Dis 6:337–347. doi:10.3201/eid0604.000403

Fürst J, Cizman M, Mrak J, Kos D, Campbell S, Coenen S, Gustafsson LL, Fürst L, Godman B (2015) The influence of a sustained multifaceted approach to improve antibiotic prescribing in Slovenia during the past decade: findings and implications. Expert Rev Anti Infect Ther 13:279–289. doi:10.1586/14787210.2015.990381

Gandra S, Barter DM, Laxminarayan R (2014) Economic burden of antibiotic resistance: how much do we really know? Clin Microbiol Infect 20:973–980. doi:10.1111/1469-0691.12798

Gebeyehu E, Bantie L, Azage M (2015) Inappropriate use of antibiotics and its associated factors among urban and rural communities of Bahir Dar city administration, Northwest Ethiopia. PLoS One 10:e0138179. doi:10.1371/journal.pone.0138179

Godman B, Abuelkhair M, Vitry A, Abdu S, Bennie M, Bishop I, Fahmy S, Garuoliene K, Herholz H, Martin A, Malmstrom R, Jan S, Schwabe U, Sermet C, Skiold P, Voncina L, Gustafsson LL (2012) Payers endorse generics to enhance prescribing efficiency; impact and future implications, a case history approach. GaBI Jn 1:69–83. doi:10.5639/gabij.2012.0102.017

Godman B, Campbell S, Suh HS, Finlayson A, Bennie M, Gustafsson L (2013) Ongoing measures to enhance prescribing efficiency across Europe: implications for other countries. J Health Tech Assess 1:27–42. doi:10.2217/cer.12.52

Godman B, Acurcio, Guerra Júnior AA, Alvarez-Madrazo S, Faridah Aryani MY, Bishop I, Campbell S, Eriksson I, Finlayson AE, Fürst J, Garuoliene K, Hussain S, Kalaba M, Jayathissa S, AbdolMalek AA, Małecka-Massalska T, Mansor Manan M, Martin A, Pedersen H, Sović Brkičić L, Smoleń A, Truter I, Wale J, Gustafsson LL (2014) Initiatives

among authorities to improve the quality and efficiency of prescribing and the implications. J Pharm Care Health Syst 1:15. doi: 10.4172/jpchs.1000113
Godman B, Malmstrom RE, Diogene E, Gray A, Jayathissa S, Timoney A, Acurcio F, Alkan A, Brzezinska A, Bucsics A, Campbell SM, Czeczot J, de Bruyn W, Eriksson I, Yusof FA, Finlayson AE, Fürst J, Garuoliene K, Guerra Júnior A, Gulbinovič J, Jan S, Joppi R, Kalaba M, Magnisson E, McCullagh L, Miikkulainen K, Ofierska-Sujkowska G, Pedersen HB, Selke G, Sermet C, Spillane S, Supian A, Truter I, Vlahović-Palčevski V, Vien LE, Vural EH, Wale J, Władysiuk M, Zeng W, Gustafsson LL (2015) Are new models needed to optimize the utilization of new medicines to sustain healthcare systems? Expert Rev Clin Pharmacol 8:77–94. doi:10.1586/17512433.2015.990380
Gonzales R (2005) Patients' preconceptions and convictions about antibiotics. Am Fam Phys 72:1900–1906
Gonzales R, Bartlett JG, Besser RE, Cooper RJ, Hickner JM, Hoffman JR, Sande MA (2001) Principles of appropriate antibiotic use for treatment of acute respiratory tract infections in adults: background, specific aims, and methods. Ann Intern Med 134:479–486. doi:10.7326/0003-4819-134-6-200103200-00013
Gonzales R, Corbett KK, Leeman-Castillo BA, Glazner J, Erbacher K, Darr CA, Wong S, Maselli JH, Sauaia A, Kafadar K (2005) The 'minimising antibiotic resistance in Colorado' project: impact of patient education in improving antibiotic use in private office practices. Health Serv Res 40:101–116. doi:10.1111/j.1475-6773.2005.00344.x
Goossens H (2009) Antibiotic consumption and link to resistance. Clin Microbiol Infect 15(Suppl 3):12–15. doi:10.1111/j.1469-0691.2009.02725.x
Goossens H, Ferech M, Vander Stichele R, Elseviers M (2005) Outpatient antibiotic use in Europe and association with resistance: a cross-national database study. Lancet 365:579–587. doi:10.1016/S0140-6736(05)17907-0
Goossens H, Guillemot D, Ferech M, Schlemmer B, Costers M, van Breda M, Baker LJ, Cars O, Davey PG (2006) National campaigns to improve antibiotic use. Eur J Clin Pharmacol 62:373–379. doi:10.1007/s00228-005-0094-7
Greene JB, Dolder C, Wallis ML (2011) The NC tars project: students leading the way to educate patients about proper use of antibiotics. J Am Pharm Assoc 51:539–543. doi:10.1331/JAPhA.2011.09159
Grigoryan L, Burgerhof JGM, Degener JE, Deschepper R, Lundborg CS, Monnet DL, Scicluna EA, Birkin J, Haaijer-Ruskamp FM (2008) Determinants of self-medication with antibiotics in Europe: the impact of beliefs, country wealth and the healthcare system. J Antimicrob Chemother 61:1172–1179. doi:10.1093/jac/dkn054
Gualano MR, Gili R, Scaioli G, Bert F, Siliquini R (2015) General population's knowledge and attitudes about antibiotics: a systematic review and meta-analysis. Pharmacoepidemiol Drug Saf 24:2–10. doi:10.1002/pds.3716
Hamm RM, Hicks RJ, Bemben D (1996) Antibiotics and respiratory infections: are patients more satisfied when expectations are met? J Fam Pract 43:56–62
Hassali MA, Kamil TK, Md Yusof FA, Alrasheedy AA, Yusoff ZM, Saleem F, Al-Tamimi SK, Wong ZY, Aljadhey H, Godman B (2015) General practitioners' knowledge, attitude and prescribing of antibiotics for upper respiratory tract infections in Selangor, Malaysia: findings and implications. Expert Rev Anti Infect Ther 13:511–520. doi:10.1586/14787210.2015.1012497
Hawkings NJ, Wood F, Butler CC (2007) Public attitudes towards bacterial resistance: a qualitative study. J Antimicrob Chemother 59:1155–1160. doi:10.1093/jac/dkm103
Hemo B, Shamir-Shtein NH, Silverman BG, Tsamir J, Heymann AD, Tsehori S, Friedman NL (2009) Can a nationwide media campaign affect antibiotic use? Am J Manag Care 15:529–534
Her Majesties Government (HMG) (2014) Antimicrobial resistance: tackling a crisis for the health and wealth of nations. http://amr-review.org/sites/default/files/AMR%20Review%20Paper%20-%20Tackling%20a%20crisis%20for%20the%20health%20and%20wealth%20of%20nations_1.pdf

Hoa NT, Chieu TT, Nghia HD, Mai NT, Anh PH, Wolbers M, Baker S, Campbell JI, Chau NV, Hien TT, Farrar J, Schultsz C (2011) The antimicrobial resistance patterns and associated determinants in *Streptococcus suis* isolated from humans in southern Vietnam, 1997-2008. BMC Infect Dis 11:6. doi:10.1186/1471-2334-11-6

Hoffman SJ, Ottersen T (2015) Addressing antibiotic resistance requires robust international accountability mechanisms. J Law Med Ethic 43(S3):53–64. doi:10.1111/jlme.12275

Hoffman SJ, Outterson K (2015) Introduction: What will it take to address the global threat of antibiotic resistance? J Law Med Ethics 43(S3):6–11. doi:10.1111/jlme.12267

Holloway KA (2011) Combating inappropriate use of medicines. Expert Rev Clin Pharmacol 4:335–348. doi:10.1586/ECP.11.14

Holloway KA, Henry D (2014) WHO essential medicines policies and use in developing and transitional countries: an analysis of reported policy implementation and medicines use surveys. PLoS Med 11, e1001724. doi:10.1371/journal.pmed.1001724

Holloway K, Mathai E, Gray A (2011) Surveillance of community antimicrobial use in resource-constrained settings-experience from five pilot projects. Trop Med Int Health 16:152–161. doi:10.1111/j.1365-3156.2010.02695.x

Holloway KA, Ivanovska V, Wagner AK, Vialle-Valentin C, Ross-Degnan D (2015) Prescribing for acute childhood infections in developing and transitional countries, 1990-2009. Paediatr Int Child Health 35:5–13. doi:10.1179/2046905514y.0000000115

Hoxha I, Malaj A, Tako R, Malaj L (2015) Survey on how antibiotics are dispensed in community pharmacies in Albania. Int J Pharm Pharm Sci 4:1545–1548

Huttner B, Goossens H, Verheij T, Harbarth S, CHAMP consortium (2010) Characteristics and outcomes of public campaigns aimed at improving the use of antibiotics in outpatients in high-income countries. Lancet Infect Dis 10:17–31. doi:10.1016/s1473-3099(09)70305-6

Hwang TJ, Powers JH, Carpenter D, Kesselheim AS (2015) Accelerating innovation in rapid diagnostics and targeted antibacterials. Nat Biotechnol 33:589–590. doi:10.1038/nbt.3251

IMCI (2004) Report of the IMCI health facility survey in Botswana. http://www.google.co.uk/url?sa=t&rct=j&q=&esrc=s&source=web&cd=1&ved=0CB0QFjAAahUKEwjRg-Ge8s7IAhXBdD4KHSBQBHY&url=http%3A%2F%2Fwww.afro.who.int%2Findex.php%3Foption%3Dcom_docman%26task%3Ddoc_download%26gid%3D648&usg=AFQjCNEOjGrBIDAqEfwnVFufFv9Cg69plw

Irawati L, Hassali MA, Saleem F (2015) Toppling the top-down approach: an example of managing antibiotic resistance in Malaysia. Aust Med J 8:100–102

Iroezindu MO, Chima EI, Isiguzo GC, Mbata GC, Onyedum CC, Onyedibe KI, Okoli LE (2014) Sputum bacteriology and antibiotic sensitivity patterns of community-acquired pneumonia in hospitalized adult patients in Nigeria: a 5-year multicentre retrospective study. Scand J Infect Dis 46:875–887. doi:10.3109/00365548.2014.954263

Islahudin F, Tamezi AMA, Shah NM (2014) Knowledge, attitudes and practices about antibiotic use among the general public in Malaysia. SE Asian J Trop Med Public Health 45:1474–1482

Jande M, Kongola G, Mwangi JW (2012) Drug prescribing pattern in two hospitals in Mwanza, Northwest Tanzania. East Cent J Pharm Sci 15:63–68

Jean SS, Hsueh PR (2011) High burden of antimicrobial resistance in Asia. Int J Antimicrob Agents 37:291–295. doi:10.1016/j.ijantimicag.2011.01.009

Jinks T, Lee N, Sharland M, Rex J, Gertler N, Diver M, Jones I, Jones K, Mathewson S, Chiara F, Farrar J (2016) A time for action: antimicrobial resistance needs global response. Bull World Health Organ 94(8):558–558A. doi:10.2471/BLT.16.181743

Kalungia AC, Burger J, Godman B, de Oliveira CJ, Simuwelu C (2016) Non-prescription sale and dispensing of antibiotics in community pharmacies in Zambia. Expert Rev Anti Infect Ther 14:1215–1223

Kang CI, Song JH (2013) Antimicrobial resistance in Asia: current epidemiology and clinical implications. Infect Chemother 45:22–31. doi:10.3947/ic.2013.45.1.22

Karambu I (2011) Abuse of 'prescription only' drugs on the rise. http://www.businessdailyafrica.com/Corporate-News/-/539550/1116580/-/vra01v/-/index.html. Accessed July 2015

Kariuki S, Muyodi J, Mirza B, Mwatu W, Daniels JJ (2003) Antimicrobial susceptibility in community-acquired bacterial pneumonia in adults. East Afr Med J 80:213–217

Kariuki S, Revathi G, Kariuki N, Muyodi J, Mwituria J, Munyalo A, Kagendo D, Murungi L, Hart CA (2005) Increasing prevalence of multidrug-resistant non-typhoidal salmonellae, Kenya, 1994-2003. Int J Antimicrob Agents 25:38–43. doi:10.1016/j.ijantimicag.2004.08.015

Kariuki S, Revathi G, Kariuki N, Kiiru J, Mwituria J, Muyodi J, Githinji JW, Kagendo D, Munyalo A, Hart CA (2006) Invasive multidrug-resistant non-typhoidal Salmonella infections in Africa: zoonotic or anthroponotic transmission? J Med Microbiol 55:585–591. doi:10.1099/jmm.0.46375-0

Kariuki S, Revathi G, Kiiru J, Mengo DM, Mwituria J, Muyodi J, Munyalo A, Teo YY, Holt KE, Kingsley RA, Dougan G (2010) Typhoid in Kenya is associated with a dominant multidrug-resistant *Salmonella enterica* serovar Typhi haplotype that is also widespread in Southeast Asia. J Clin Microbiol 48:2171–2176. doi:10.1128/jcm.01983-09

Kesselheim AS, Outterson K (2010) Fighting antibiotic resistance: marrying new financial incentives to meeting public health goals. Health Aff (Millwood) 29:1689–1696. doi:10.1377/hlthaff.2009.0439

Kho B, Ong C, Tan F, Wee C (2013) Antibiotic prescribing for upper respiratory tract infections in Sarawak district hospitals. Med J Malays 68:136–140

Kibuule D, Kagoya HR, Godman B (2016) Antibiotic use in acute respiratory infections in under-fives in Uganda: findings and implications. Expert Rev Anti Infect Ther 14:863–872. doi:10.1080/14787210.2016.1206468

Kiiru J, Kariuki S, Goddeeris BM, Butaye P (2012) Analysis of beta-lactamase phenotypes and carriage of selected beta-lactamase genes among *Escherichia coli* strains obtained from Kenyan patients during an 18-year period. BMC Microbiol 12:155. doi:10.1186/1471-2180-12-155

Kim SH, Song JH, Chung DR, Thamlikitkul V, Yang Y, Wang H, Lu M, So TM, Hsueh PR, Yasin RM, Carlos CC, Pham HV, Lalitha MK, Shimono N, Perera J, Shibl AM, Baek JY, Kang CI, Ko KS, Peck KR, ANSORP Study Group (2012) Changing trends in antimicrobial resistance and serotypes of *Streptococcus pneumoniae* isolates in Asian countries: an Asian Network for Surveillance of Resistant Pathogens (ANSORP) study. Antimicrob Agents Chemother 56:1418–1426. doi:10.1128/aac.05658-11

Klein EY, Makowsky M, Orlando M, Hatna E, Braykov NP, Laxminarayan R (2015) Influence of provider and urgent care density across different socioeconomic strata on outpatient antibiotic prescribing in the USA. J Antimicrob Chemother 70:1580–1587. doi:10.1093/jac/dku563

Kotwani A, Wattal C, Katewa S, Joshi PC, Holloway K (2010) Factors influencing primary care physicians to prescribe antibiotics in Delhi India. Fam Pract 27:684–690. doi:10.1093/fampra/cmq059

Kotwani A, Wattal C, Joshi PC, Holloway K (2012) Irrational use of antibiotics and role of the pharmacist: an insight from a qualitative study in New Delhi, India. J Clin Pharm Ther 37:308–312. doi:10.1111/j.1365-2710.2011.01293.x

Lai CC, Lee K, Xiao Y, Ahmad N, Veeraraghavan B, Thamlikitkul V, Tambuah P, Nelwan RHH, Shobl A, Wu J-J, Seto WH, Hsueh P-R, Nelwan (2014) High burden of antimicrobial drug resistance in Asia. J Glob Antimicrob Resist 2:141–147. doi:10.1016/j.ijantimicag.2011.01.009

Larsson M, Kronvall G, Chuc N, Karlsson I, Lager F, Hanh HD, Tomson G, Falkenberg T (2000) Antibiotic medication and bacterial resistance to antibiotics: a survey of children in a Vietnamese community. Trop Med Int Health 5:711–721

Laxminarayan R, Heymann DL (2012) Challenges of drug resistance in the developing world. BMJ 344:e1567

Laxminarayan R, Matsoso P, Pant S, Brower C, Røttingen J-A, Klugman K, Davies S (2016) Access to effective antimicrobials: a worldwide challenge. Lancet 387(10014):168–175, 179. doi:10.1016/S0140-6736(15)00474-2

Li Y, Xu J, Wang F, Wang B, Liu L, Hou W, Fan H, Tong Y, Zhang J, Lu Z (2012) Overprescribing in China, driven by financial incentives, results in very high use of antibiotics, injections, and corticosteroids. Health Aff 31:1075–1082. doi:10.1377/hlthaff.2010.0965

Lim KK, Teh CC (2012) A cross sectional study of public knowledge and attitude towards antibiotics in Putrajaya, Malaysia. South Med Rev 5:26–33

Little P, Stuart B, Francis N, Douglas E, Tonkin-Crine S, Anthierens S, Cals JW, Melbye H, Santer M, Moore M, Coenen S, Butler C, Hood K, Kelly M, Godycki-Cwirko M, Mierzecki A, Torres A, Llor C, Davies M, Mullee M, O'Reilly G, van der Velden A, Geraghty AW, Goossens H, Verheij T, Yardley L, GRACE Consortium (2013) Effects of internet-based training on antibiotic prescribing rates for acute respiratory-tract infections: a multinational, cluster, randomised, factorial, controlled trial. Lancet 382:1175–1182. doi:10.1016/s0140-6736(13)60994-0

Llor C, Bjerrum L (2014) Antimicrobial resistance: risk associated with antibiotic overuse and initiatives to reduce the problem. Ther Adv Drug Saf 5:229–241. doi:10.1177/2042098614554919

Loyola Filho AI, Uchoa E, Guerra HL, Firmo JO, Lima-Costa MF (2002) Prevalence and factors associated with self-medication: the Bambuí health survey. Rev Saude Publica 36:55–62

Macfarlane J, Holmes W, Macfarlane R, Britten N (1997) Influence of patients' expectations on antibiotic management of acute lower respiratory tract illness in general practice: questionnaire study. BMJ 315:1211–1214. doi:10.1136/bmj.315.7117.1211

Malcolm W, Nathwani D, Davey P, Cromwell T, Patton A, Reilly J, Cairns S, Bennie M (2013) From intermittent antibiotic point prevalence surveys to quality improvement: experience in Scottish hospitals. Antimicrob Resist Infect Control 2:3. doi:10.1186/2047-2994-2-3

Mangione-Smith R, McGlynn EA, Elliott MN, Krogstad P, Brook RH (1999) The relationship between perceived parental expectations and pediatrician antimicrobial prescribing behaviour. Pediatrics 103:711–718

Marwa KJ, Mushi MF, Konje E, Alele PE, Kidola J, Mirambo MM (2015) Resistance to cotrimoxazole and other antimicrobials among isolates from HIV/AIDS and Non-HIV/AIDS patients at Bugando medical centre, Mwanza, Tanzania. AIDS Res Treat 2015:103874. doi:10.1155/2015/103874

Massele A, Burger J, Katende-Kyenda NL, Kalemeera F, Kenaope T, Kibuule D, Mbachu O, Mubita M, Oluka M, Olusanya A, Paramadhas BD, van Zyl P, Godman B (2015) Outcome of the first medicines utilization research in Africa group meeting to promote sustainable and rational medicine use in Africa. Expert Rev Pharmacoecon Outcomes Res 15:885–888. doi:10.1586/14737167.2015.1088386

McKay RM, Vrbova L, Fuertes E, Chong M, David S, Dreher K, Purych D, Blondel-Hill E, Henry B, Marra F, Kendall PR, Patrick DM (2011) Evaluation of the do bugs need drugs? Program in British Columbia: can we curb antibiotic prescribing? Can J Infect Dis Med Microbiol 22:19–24

McNulty CA, Boyle P, Nichols T, Clappison P, Davey P (2007a) Don't wear me out: the public's knowledge of and attitudes to antibiotic use. J Antimicrob Chemother 59:727–738. doi:10.1093/jac/dkl558

McNulty CM, Boyle P, Nichols T, Clappison P, Davey P (2007b) The public's attitudes to and compliance with antibiotics. J Antimicrob Chemother 60S1:i63–i68. doi:10.1093/jac/dkm161

Md Rezal RS, Hassali MA, Alrasheedy AA, Saleem F, Md Yusof FA, Godman B (2015) Physicians' knowledge, perceptions and behaviour towards antibiotic prescribing: a systematic review of the literature. Expert Rev Anti Infect Ther 13:665–680. doi:10.1586/14787210.2015.1025057

Mehta SD, Maclean I, Ndinya-Achola JO, Moses S, Martin I, Ronald A, Agunda L, Murugu R, Bailey RC, Melendez J, Zenilman JM (2011) Emergence of quinolone resistance and cephalosporin MIC creep in *Neisseria gonorrhoeae* isolates from a cohort of young men in Kisumu, Kenya, 2002 to 2009. Antimicrob Agents Chemother 55:3882–38808. doi:10.1128/aac.00155-11

Mendelson M, Matoso M (2015) The South African Antimicrobial Resistance Strategy Framework. AMR Control 54–61

Meyer E, Gastmeier P, Deja M, Schwab F (2013) Antibiotic consumption and resistance: data from Europe and Germany. Int J Med Microbiol 303:388–395. doi:10.1016/j.ijmm.2013.04.004

Mikhael EM (2014) Evaluating the rationality of antibiotic dispensing in treating common cold infections among pharmacists in Baghdad, Iraq. Br J Pharm Res 4:2653–2661. doi:10.9734/BJPR/2014/13586

Mitsi G, Jelastopulu E, Basiaris H, Skoutelis A, Gogos C (2005) Patterns of antibiotic use among adults and parents in the community: a questionnaire-based survey in a Greek urban population. Int J Antimicrob Agents 25:439–443. doi:10.1016/j.ijantimicag.2005.02.009

Moura ML, Boszczowski I, Mortari N, Barrozo LV, Neto FC, Lobo RD, Pedroso de Lima AC, Levin AS (2015) The impact of restricting over-the-counter sales of antimicrobial drugs: preliminary analysis of national data. Medicine (Baltimore) 94:e1605. doi:10.1097/MD.0000000000001605

Mubita M, McCann M, Hachaambwa L, Lukwesa C (2013) An audit of compliance with the antimicrobial prescribing care bundle in internal medicine at University Teaching Hospital, Lusaka, Zambia. MSc (Clinical Pharmacy) thesis, Queen's University Belfast, Belfast, NI, UK

Muoneke V, Ibekwe M, Ibekwe R (2012) Childhood urinary tract infection in Abakaliki: etiological organisms and antibiotic sensitivity pattern. Ann Med Health Sci Res 2:29–32. doi:10.4103/2141-9248.96933

Mustafa M, Wood F, Butler CC, Elwyn G (2014) Managing expectations of antibiotics for upper respiratory tract infections: a qualitative study. Ann Fam Med 12:29–36. doi:10.1370/afm.1583

Napolitano F, Izzo MT, Di Giuseppe G, Angelillo IF (2013) Public knowledge, attitudes and experience regarding the use of antibiotics in Italy. PLoS One 4:e84177. doi:10.1371/journal.pone.0084177

Nathwani D, Sneddon J, Malcolm W, Wiuff C, Patton A, Hurding S, Eastaway A, Seaton RA, Watson E, Gillies E, Davey P, Bennie M, Scottish Antimicrobial Prescribing Group (2011) Scottish Antimicrobial Prescribing Group (SAPG): development and impact of the Scottish National Antimicrobial Stewardship Program. Int J Antimicrob Agents 38:16–26. doi:10.1016/j.ijantimicag.2011.02.005

National Collaborating Centre for Infectious Diseases—Community-acquired antimicrobial resistance: consultation notes, 2010. http://nccid.ca/publications/community-acquired-antimicrobial-resistance-consultation-notes/

National Information Programme on Antibiotics (2013) What do Canadians think? http://www.antibiotics-info.org/can_tnk.html

Ngumi ZW (2006) Nosocomial infections at Kenyatta National Hospital Intensive-Care Unit in Nairobi, Kenya. Dermatology 212(Suppl 1):4–7. doi:10.1159/000089192

Nicolini P, Nascimento JWL, Greco KV, Menezes FGD (2008) Fatoresrelacionados à prescriçãomédica de antibióticosemfarmáciapública da regiãoOeste da cidade de São Paulo. CienSaude Colet 13(sSupl). doi:10.1590/S1413-81232008000700018

O'Mara-Eves A, Brunton G, McDaid D, Oliver S, Kavanagh J, Jamal F (2013) Community engagement to reduce inequalities in health: a systematic review, meta-analysis and economic analysis. Public Health Res 1(4). doi:10.3310/phr01040

O'Neill (2015) Securing new drugs for future generations. https://amr-review.org/sites/default/files/SECURING%20NEW%20DRUGS%20FOR%20FUTURE%20GENERATIONS%20FINAL%20WEB_0.pdf

Ocan M, Obuku EA, Bwanga F, Akena D, Richard S, Ogwal-Okeng J, Obua C (2015) Household antimicrobial self-medication: a systematic review and meta-analysis of the burden, risk factors and outcomes in developing countries. BMC Public Health 15:742. doi:10.1186/s12889-015-2109-3

Oduyebo OO, Onwuezobe IA, Olayemi SO, Akintunde AS (2008) Correlating antibiotic consumption with antimicrobial resistance of uropathogens in a university teaching hospital in Lagos, Nigeria. Niger J Clin Pract 11:305–308

Oh AL, Hassali MA, Al-Haddad MS, Sulaiman SA, Shafie AA, Awaisu A (2011) Public knowledge and attitudes towards antibiotic usage: a cross-sectional study among the general public in the state of Penang, Malaysia. J Infect Dev Ctries 5:338–347

Okesola AO, Aroundegbe TI (2011) Antibiotic resistance pattern of uropathogenic *Escherichia coli* in South West Nigeria. Afr J Med Med Sci 40:235–238

Outterson K, Powers JH, Daniel GW, McClellan MB (2015) Repairing the broken market for antibiotic innovation. Health Aff 34:277–285. doi:10.1377/hlthaff.2014.1003

Pechere JC (2001) Patients' interviews and misuse of antibiotics. Clin Infect Dis 33(Suppl 3): S170–S173. doi:10.1086/321844

Petersen I, Johnson AM, Islam A, Duckworth G, Livermore DM, Hayward AC (2007) Protective effect of antibiotics against serious complications of common respiratory tract infections: retrospective cohort study with the UK General Practice Research Database. BMJ 335:982. doi:10.1136/bmj.39345.405243.BE

Plachouras D, Kavatha D, Antoniadou A, Giannitsioti E, Poulakou G, Kanellakopoulou K, Giamarellou H (2010) Dispensing of antibiotics without prescription in Greece, 2008: another link in the antibiotic resistance chain. Euro Surv 15:19488

Podolsky SH, Bud R, Gradmann C, Hobaek B, Kirchhelle C, Mitvedt T, Santesmases MJ, Thoms U, Berild D, Kveim Lie A (2015) History teaches us that confronting antibiotic resistance requires stronger global collective action. J Law Med Ethics 43(Suppl 3):27–32. doi:10.1111/jlme.12271

Pulcini C, Williams F, Molinari N, Davey P, Nathwani D (2011) Junior doctors' knowledge and perceptions of antibiotic resistance and prescribing: a survey in France and Scotland. Clin Microbiol Infect 17:80–87. doi:10.1111/j.1469-0691.2010.03179.x

Radyowijati A, Haak H (2003) Improving antibiotic use in low-income countries: an overview of evidence on determinants. Soc Sci Med 57:733–744

Rauber C, Feltrin MR, Piovezan AP (2009) Evaluation of antibiotics dispensing profile in Tubarao, Santa Catarina, Brazil. Braz J Pharm Sci 45:787–793

Raz R, Edelstein H, Grigoryan L, Haaijer-Ruskamp FM (2005) Self-medication with antibiotics by a population in northern Israel. Isr Med Assoc J 7:722–725

Reardon S (2014) Antibiotic resistance sweeping developing world. Nature 509:141–142. doi:10.1038/509141a

Reynolds L, McKee M (2009) Factors influencing antibiotic prescribing in China: an exploratory analysis. Health Policy 90:32–36. doi:10.1016/j.healthpol.2008.09.002

Reynolds L, McKee M (2011) Serve the people or close the sale? Profit-driven overuse of injections and infusions in China's market-based healthcare system. Int J Health Plann Manage 26:449–470. doi:10.1002/hpm.1112

Riaz H, Godman B, Hussain S, Malik F, Mahmood S, Shami A, Bashir S (2015) Prescribing of bisphosphonates and antibiotics in Pakistan: challenges and opportunities for the future. J Pharm Health Serv Res 6:111–121

Rodis JL, Green CG, Cook SC, Pedersen CA (2004) Effects of a pharmacist-initiated educational intervention on patient knowledge about the appropriate use of antibiotics. Am J Health Syst Pharm 6:1385–1389

Roque F, Soares S, Breitenfeld L, Lopez-Durán A, Figueiras A, Herdeiro MT (2013) Attitudes of community pharmacists to antibiotic dispensing and microbial resistance: a qualitative study in Portugal. Int J Clin Pharm 35:417–424. doi:10.1007/s11096-013-9753-4

Roque F, Soares S, Breitenfeld L, Gonzalez-Gonzalez C, Figueiras A, Herdeiro MT (2014) Portuguese community pharmacists' attitudes to and knowledge of antibiotic misuse: questionnaire development and reliability. PLoS One 9, e90470. doi:10.1371/journal.pone.0090470

Sabry NA, Farid SF, Dawoud DM (2014) Antibiotic dispensing in Egyptian community pharmacies: an observational study. Res Social Adm Pharm 10:168–184. doi:10.1016/j.sapharm.2013.03.004

Sabuncu E, David J, Bernede-Bauduin C, Pepin S, Leroy M, Boelle PY, Watier L, Guillemot D (2009) Significant reduction of antibiotic use in the community after a nationwide campaign in France, 2002-2007. PLoS Med 6:e1000084. doi:10.1371/journal.pmed.1000084

Saengcharoen W, Chongsuvivatwong V, Lerkiatbundit S, Wongpoowarak P (2008) Factors influencing dispensing of antibiotics for upper respiratory infections among Southern Thai community pharmacists. J Clin Pharm Ther 33:123–129. doi:10.1111/j.1365-2710.2007.00890.x

Salunkhe S, Pandit V, Dawane J, Sarda K, More C (2013) Study of over the counter sale of antimicrobials in pharmacy outlets in Pune, India: a cross sectional study. Int J Pharm Biol Sci 4:616–622

Santa-Ana-Tellez Y, Mantel-Teeuwisse AK, Leufkens HG, Wirtz VJ (2015) Seasonal variation in penicillin use in Mexico and Brazil: analysis of the impact of over-the-counter restrictions. Antimicrob Agents Chemother 59:105–110. doi:10.1128/AAC.03629-14

Sawair FA, Baqain ZH, Abu Karaky A, Abu Eid R (2009) Assessment of self-medication of antibiotics in a Jordanian population. Med Princ Pract 18:21–25. doi:10.1159/000163041

Seni J, Najjuka CF, Kateete DP, Makobore P, Joloba ML, Kajumbula H, Kapesa A, Bwanga F (2013) Antimicrobial resistance in hospitalized surgical patients: a silently emerging public health concern in Uganda. BMC Res Notes 6:298. doi:10.1186/1756-0500-6-298

Sharma A, Bukusi E, Gorbach P, Cohen CR, Muga C, Kwena Z, Holmes KK (2008) Sexual identity and risk of HIV/STI among men who have sex with men in Nairobi. Sex Transm Dis 35:352–354. doi:10.1097/OLQ.0b013e31815e6320

Shehadeh MB, Suaifan G, Hammad EA (2016) Active educational intervention as a tool to improve safe and appropriate use of antibiotics. Saudi Pharm J 24(5):611–615, Epub 2015 Mar 21

Siddiqi K, Newell J, Robinson M (2005) Getting evidence into practice: what works in developing countries? Int J Qual Health Care 17:447–454. doi:10.1093/intqhc/mzi051

So AD, Shah TA, Roach S, Ling Chee Y, NachmanK E (2015) An integrated systems approach is needed to ensure the sustainability of antibiotic effectiveness for both humans and animals. J Law Med Ethics 43(Suppl 3):38–45. doi:10.1111/jlme.12273

Song J, Jung S, Ko KS, Kim NY, Son JS, Chang HH, Ki HK, Oh WS, Suh JY, Peck KR, Lee NY, Yang Y, Lu Q, Chongthaleong A, Chiu CH, Lalitha MK, Perera J, Yee TT, Kumarasinghe G, Jamal F, Kamarulzaman A, Parasakthi N, Van PH, Carlos C, So T, Ng TK, Shibl A (2004) High prevalence of antimicrobial resistance among clinical *Streptococcus pneumoniae* isolates in Asia (an ANSORP study). Antimicrob Agents Chemother 48:2101–2107. doi:10.1128/AAC.48.6.2101-2107.2004

Sumpradit N, Chongtrakul P, Anuwong K, Pumtong S, Kongsomboon K, Butdeemee P, Khonglormyati J, Chomyong S, Tongyoung P, Losiriwat S, Seesuk P, Suwanwaree P, Tangcharoensathien V (2012) Antibiotics smart use: a workable model for promoting the rational use of medicines in Thailand. Bull WHO 90:905–913. doi:10.2471/BLT.12.105445

Sutradhar KB, Saha A, Huda NH, Uddin R (2014) Irrational use of antibiotics and antibiotic resistance in southern rural Bangladesh: perspectives from both the physicians and patients. Annu Res Rev Biol 4:1421–1430

Thoithi GN, Okalebo FA (2009) Country case study: Kenya. FIP Global Pharmacy workforce report. International Pharmaceutical Federation

Thriemer K, Katuala Y, Batoko B, Alworonga JP, Devlieger H, Van Geet C, Ngbonda D, Jacobs J (2013) Antibiotic prescribing in DR Congo: a knowledge, attitude and practice survey among medical doctors and students. PLoS One 8(2):e55495. doi:10.1371/journal.pone.0055495

Truter I (2015) Antimicrobial prescribing in South Africa using a large pharmacy database: a drug utilisation study. South Afr J Infect Dis 30:52–56

Van Boeckel TP, Gandra S, Ashok A, Caudron Q, Grenfell BT, Levin SA, Laxminarayan R (2014) Global antibiotic consumption 2000 to 2010: an analysis of national pharmaceutical sales data. Lancet Infect Dis 14:742–750. doi:10.1016/s1473-3099(14)70780-7

van de Sande-Bruinsma N, Grundmann H, Verloo D, Tiemersma E, Monen J, Goossens H, Ferech M, European Antimicrobial Resistance Surveillance System Group, European Surveillance of Antimicrobial Consumption Project Group (2008) Antimicrobial drug use and resistance in Europe. Emerg Infect Dis 14:1722–1730. doi:10.3201/eid1411.070467

Van Nguyen K, Do NTT, Chandna A, Nguyen TV, Van Pham C, Doan PM, Nguyen AQ, Thi Nguyen CK, Larsson M, Escalante S, Olowokure B, Laxminarayan R, Gelband H, Horby P, Thi Ngo HB, Hoang MT, Farrar J, Hien TT, Wertheim HF (2013) Antibiotic use and resistance in emerging economies: a situation analysis for Viet Nam. BMC Public Health 13:1158. doi:10.1186/1471-2458-13-1158

Van TD, Dinh QD, Vu PD, Nguyen TV, Pham CV, Dao TT, Phung CD, Hoang HT, Tang NT, Do NT, Nguyen KV, Wertheim H (2014) Antibiotic susceptibility and molecular epidemiology of *Acinetobacter calcoaceticus-baumannii* complex strains isolated from a referral hospital in northern Vietnam. J Glob Antimicrob Resist 2:318–321

Velickovic-Radovanovic R, Stefanovic N, Damnjanovic I, Kocic B, Antic S, Dinić M, Petrović J, Mitić R, Catić-Djordjević A (2015) Monitoring of antibiotic consumption and development of resistance by Enterobacteria in a tertiary care hospital. J Clin Pharm Ther 40:426–430. doi:10.1111/jcpt.12283

Versporten A, Bolokhovets G, Ghazaryan L, Abilova V, Pyshnik G, Spasojevic T, Korinteli I, Raka L, Kambaralieva B, Cizmovic L, Carp A, Radonjic V, Maqsudova N, Celik HD, Payerl-Pal M, Pedersen HB, Sautenkova N, Goossens H, WHO/Europe-ESAC Project Group (2014) Antibiotic use in Eastern Europe: a cross-national database study in coordination with the WHO Regional Office for Europe. Lancet Infect Dis 14:381–387. doi:10.1016/s1473-3099(14)70071-4

Volpato DE, Souza BVd, Dalla Rosa LG, Melo LH, Daudt CAS, Deboni L (2005) Use of antibiotics without medical prescription. Braz J Infect Dis 9:288–291

WHO (World Health Organization) Europe. The role of pharmacist in encouraging prudent use of antibiotics and averting antimicrobial resistance: a review of policy and experience. http://www.euro.who.int/__data/assets/pdf_file/0006/262815/The-role-of-pharmacist-in-encouraging-prudent-use-of-antibiotics-and-averting-antimicrobial-resistance-a-review-of-policy-and-experience-Eng.pdf

Wood SM, Shah SS, Bafana M, Ratner AJ, Meaney PA, Malefho KC, Steenhoff AP (2009) Epidemiology of methicillin-resistant *Staphylococcus aureus* bacteremia in Gaborone, Botswana. Infect Control Hosp Epidemiol 30:782–785. doi:10.1086/599003

World Health Organization (WHO) (2001) WHO global strategy for containment of antimicrobial resistance. http://www.who.int/csr/resources/publications/drugresist/en/EGlobal_Strat.pdf

World Health Organization (WHO) (2014) Antimicrobial resistance: global report on surveillance. http://www.who.int/drugresistance/documents/surveillancereport/en/

World Health Organization (2015) Worldwide country situation analysis: response to antimicrobial resistance. http://www.who.int/drugresistance/documents/situationanalysis/en/

World Health Organisation (WHO) Europe (2011) European strategic action plan on antibiotic resistance. http://www.euro.who.int/__data/assets/pdf_file/0008/147734/wd14E_AntibioticResistance_111380.pdf?ua=1

Wun YT, Lam TP, Lam KF, Ho PL, Yung WHR (2013) The public's perspectives on antibiotic resistance and abuse among Chinese in Hong Kong. Pharmacoepidemiol Drug Saf 22:241–249. doi:10.1002/pds.3339

You J, Yau B, Choi K, Chau C, Huang Q, Lee S (2008) Public knowledge, attitudes and behaviour on antibiotic use: a telephone survey in Hong Kong. Infection 36:153–157. doi:10.1007/s15010-007-7214-5

Zapata-Cachafeiro M, Gonzalez-Gonzalez C, Vaquez-Lago JM, López-Vázquez P, López-Durán A, Smyth E, Figueiras A (2014) Determinants of antibiotic dispensing without a medical prescription: a cross-sectional study in the north of Spain. J Antimicrob Chemother 6:1–5. doi:10.1093/jac/dku229

Zarb P, Goossens H (2012) Human use of antimicrobial agents. Revue Scientifique Et Technique 3:121–133

Zou XX, Fang Z, Min R, Bai X, Zhang Y, Xu D, Fang PQ (2014) Is nationwide special campaign on antibiotic stewardship program effective on ameliorating irrational antibiotic use in China? Study on the antibiotic use of specialized hospitals in China in 2011-2012. J Huazhong Univ Sci Technolog Med Sci 34:456–463. doi:10.1007/s11596-014-1300-6

In Silico Approaches Toward Combating Antibiotic Resistance

Rahul Shubhra Mandal and Santasabuj Das

Abstract Drug-resistant infections have become a major concern to human health worldwide, and the number of resistant bacteria is increasing each day. The conventional drug designing approaches are time-consuming and involve huge investment in addition to frequent failures at the clinical trial phase due to unwanted side effects. Because of these reasons, pharmaceutical companies are losing interest to invest in antibiotic research. Modern computational approaches have made the early process of drug target identification and lead compound optimization a lot easier. The anti-virulence strategy of target identification has proved to be safer as compared to the bactericidal or bacteriostatic drugs, since the chance of resistance development would be less due to non-interference with normal bacterial growth and survival. Identification of druggable targets and the use of chemical compound databases and computational tools made it possible to screen millions of molecules within a reasonably short time, taking care of individual ADMET properties. The early detection of potential drug targets and lead compounds is highly desirous in antibiotic research as it demands less time and cost. Therefore, a healthy collaboration between computational and experimental researchers is the future of novel antibiotic discovery.

1 Introduction

Combating antimicrobial drug resistance is a major challenge to the scientific community as it is spreading very fast. The emergence of drug resistances involves multiple factors but arises due to the selection pressure on microbial organisms imposed by the use of antimicrobial agents. This selection pressure alters microbial gene regulation, rendering the drugs ineffective or partially effective. An organism may become resistant to more than one drug, known as "multidrug resistance" or

R.S. Mandal (✉) • S. Das
Biomedical Informatics Center, National Institute of Cholera and Enteric Diseases, 57, Suresh Chandra Banerjee Road, Beliaghata, Kolkata 700010, India
e-mail: shubhra.rahul@gmail.com

G. Arora et al. (eds.), *Drug Resistance in Bacteria, Fungi, Malaria, and Cancer*,
DOI 10.1007/978-3-319-48683-3_25

may be "extensively drug resistant" when almost all drugs are ineffective. According to a WHO report (2014), multidrug-resistant tuberculosis (MDR-TB) affected 480,000 individuals globally in the year 2013, and extensively drug-resistant tuberculosis (XDR-TB) had spread to nearly 100 countries (http://www.weforum.org/reports/global-risks-2014-report). Antibiotic resistance has been declared as one of the greatest threats to human lives by the World Economic Forum Global Risks Report (Davies et al. 2013).

Emergence of resistant bacterial strains is an evolutionary phenomenon that occurs due to the genetic mutations or horizontal gene transfer from other microorganisms. Frequency of drug resistance increases with the overuse or misuse of antimicrobial drugs, as it exerts evolutionary pressure that provides survival advantage to the resistant microorganisms (Rice 2008). Overuse of antibiotics occurs where public awareness about judicial antibiotic use is lacking or the drugs are available over the counter. Inadequate infection control measures, poor sanitary conditions, and improper food-handling practices further complicate the situation by encouraging spread of antimicrobial resistance. In addition to the clinical settings, stockpiles of animal feed also consume large quantities of antibiotics, an increasing practice that contributes to the severity and spread of drug resistance. In dietary animals, antibiotics are frequently used for nontherapeutic purposes, such as prophylaxis and growth promotion (Palumbi 2001). Drug-resistant infections are a major concern globally due to the high mortality, the transmissible nature of the infections, and the huge economic burden they impose on the individuals and the society. Nearly 25,000 people die every year in the Europe because of multidrug-resistant bacterial infections, while two million people are getting infected in the United States (USA) with similar number of deaths (Hampton 2013). Antimicrobials currently account for more than 30 % sales from the hospital pharmacies in the USA (Sipahi 2008).

While resistance to the existing antibiotics is on the rise, development of new antibiotics has considerably slowed down. Salvarsan (Ehrlich and Hata 1910), penicillin (Fleming 2001), and Prontosil (Domagk 1935) were the first three antibiotics discovered during the first part of the nineteenth century. However, the mass production and distribution of penicillin started only in the year 1945. Paul Ehrlich was first to introduce the systematic screening approach during the discovery of Salvarsan (Ehrlich and Hata 1910). Soon this became a standard practice in the pharmaceutical industry for searching of new drug molecules. This strategy was successfully used during the period of 1950–1970, the so-called golden era for antibiotic discovery, leading to many new classes of antibiotics. In the succeeding years, the discovery of novel antibiotics was drastically slowed down, and no new drug with a completely different structure was developed.

Investment in antibiotic discovery research has significantly dropped in the recent years. In 2004, antibiotics constituted around 1.6 % of all drugs made by the world's 15 leading pharmaceutical companies that entered clinical trials. The number was reduced to only four multinational companies in 2013 who were still investing in antibiotic research (Boucher et al. 2013). On the other hand, government agencies failed miserably in discovering novel antibiotics. As a consequence, antibiotic discovery is currently almost entirely dependent on small pharmaceutical companies and academic institutions in the Western world (Sipahi 2008), while in Japan, large pharmaceutical companies continue to invest in antibiotic development (Jabes 2011; Moellering 2011; Boucher et al. 2013). Multiple factors contribute to the reduced antibiotics output. First, antibiotics are typically administrated for a very limited period of time, so it's less profitable than the drugs used for chronic diseases. Second, new antibiotics are usually kept in stock and only used when known drugs fail. In contrast, new drugs are prescribed immediately for other illnesses. This practice helps in delaying the emergence of resistant microbes but becomes a limiting factor for the initial investment returns. Third, the requirements of approval during clinical trials and other regulatory hurdles are raised in most cases (Spellberg et al. 2008). Because of these factors, many big pharmaceutical companies are dragging their feet from antibiotic research and development that involves an estimated investment of $1.7 billion per drug over a period of 10 years with less attractive return as profit (Projan 2003; Power 2006). The cost and duration of drug development process may be greatly reduced by adopting modern computational approaches, starting from determining targets to lead molecule identification. We will give an overview of the above computational tools and techniques and their application to drug discovery research in this chapter.

2 Computational Methods for Target Identification

2.1 The Antibacterial Drug Target

To become a potential drug target, a bacterial molecule should satisfy several criteria:

1. It should be essential for bacterial survival (i.e., responsible for cell division, metabolism, etc.).
2. The conservation of the target should be high enough, so that it can be used as a broad-spectrum therapeutic target.
3. There should not be any mammalian homologue with similar function.
4. The target should be "druggable," so that a ligand molecule could interact to the active site and is capable of altering its biological function (Silver 2011). The small molecule ligands should be highly selective toward its target to minimize off-target interaction and are expected to have efficacy against both Gram-positive and Gram-negative organisms.
5. The target molecule (protein/nucleic acid) should be evolutionarily stable (i.e., the occurrence of point mutation should be a rare event or should not interfere with the small molecule binding).
6. The target should be soluble and stable enough to produce high-quality and well-diffracting crystals, which are helpful for further investigation of the drug–target interaction. To remain aware of these factors while choosing an appropriate drug target is absolutely critical for a successful antibacterial drug discovery program (Silver 2011).

New antimicrobial drugs would either exploit alternative means to attack old targets or find out new "druggable" targets. It is now well understood that the bactericidal or bacteriostatic effects of available antibiotics enforce strong evolutionary pressure on the microorganisms. This leads to genetic mutations in bacteria making them drug resistant and giving significant survival advantage. Thus, scientists had felt the urge for novel strategies, such as the development of very narrow spectrum or even organism-specific antibiotics to avoid destruction of the beneficial flora or the use of adjuvants like β-lactamase inhibitors to enhance the efficacy of the currently available antibiotics against resistant organisms. Anti-virulence strategies may also be highly effective in the treatment of bacterial infections, while minimizing the chance of drug resistance. Drugs that target the virulence mechanisms, such as adhesion and invasion of the host cells, biofilm formation, toxin production, virulence gene expression, and secretion of virulence factors, will inhibit pathogenesis without compromising growth or survival of the organisms (Rasko and Sperandio 2010). Chemical inhibitors blocking toxins, pilins (Armstrong et al. 1995; Trachtman et al. 2003), quorum sensing molecules (Lesic et al. 2007), transcriptional regulators of virulence genes (Hung et al. 2005), type three secretion systems (Kauppi et al. 2003; Muschiol et al. 2006; Felise et al. 2008; Kline et al. 2008; Veenendaal et al. 2009), and histidine kinases (Rasko et al. 2008) have been reported in the literature.

2.2 Genomics-Based Approaches for Target Identification

Genomics-based approaches are currently playing a major role in antimicrobial drug discovery. Genome sequencing of microorganisms is essential to identify genes/proteins required for microbial survival. This powerful technique allows us to identify thousands of genes/proteins from humans and pathogenic microorganisms, among which many are potential therapeutic targets for drug development (Russ and Lampel 2005). Although genome sequence is an important resource for the identification of novel drug targets, it requires a methodical approach toward data analysis, and drug development against a potential target may not be possible in every case (Gao et al. 2008). Thus, FabH and FabI were new targets identified during the genomics era (Payne et al. 2007), but researchers have failed to establish new lead compounds for these targets. This results in a recent shift back to the classical methods, such as whole-cell-based phenotypic screening or reevaluation of the ligand molecules discarded earlier. However, whole-cell-based approach does not support the modern drug designing techniques, such as detection of novel drug targets for antimicrobial therapy (Chung et al. 2013).

Increasing number of microbial genome sequences is available these days due to the advancement of sequencing technology. However, it is difficult to identify a potential antimicrobial drug target only on the basis of genome sequences. A major bottleneck lies in the large number of genes/proteins for which no biological function is known. Some useful bioinformatic tools like NCBI BLAST may help to assign functions to the unknown genes by finding sequence homology to the known proteins. Development of computational databases and tools for prokaryotic genome comparison is reviewed elsewhere (Field et al. 2005). Comparative genomics-based approach is used to identify genes/proteins that codes for microbial virulence factors or necessary for microbial survival (Tang and Moxon 2001; Chan et al. 2002; Fritz and Raczniak 2002). This type of comparison is usually made between pathogenic and nonpathogenic microorganisms. The differential presence of genes in pathogenic strains may be related to pathogenicity and may be targeted, provided there are no homologues for them in the mammalian host cells. This method reduces the chance of undesirable side effects. Study of host–pathogen interactions have potential to deliver novel strategies in antimicrobial therapies, since genomics and its related technologies are not only used for target identification but also provide new insights into the antimicrobial mechanisms of action (Brazas and Hancock 2005). For example, cellular responses to antimicrobial therapy may be analyzed through whole-genome expression profiling (Fischer 2001) by microarray analysis or RNAseq technologies. Many tools are available for gene expression data analysis, which was earlier reviewed by Mandal et al. (2009).

The recent development and application of computational tools and techniques in biology has emerged as a new field called "bioinformatics," which can be used to search information contained in genomics for novel drug target identification. Network-based strategy is one of the frequently used techniques, which include endogenous metabolic, signaling, and regulatory information to reconstruct a highly interconnected network module. A gene or protein interaction network will allow us to map the known drug targets and their interactors, thus helping us to prioritize the known drug targets based on the associated regulatory information and also to explore novel drug targets. This approach will narrow down the selection of candidate drug targets which are not involved in multiple biological pathways, because inhibition of the target may otherwise result in undesirable effects due to blocking of the associated cellular activities.

With the development of the microarray technology, network-based strategy is largely supported by gene expression or protein expression data, which help to develop gene/protein network-based models for future predictions. In a study by P. Anitha et al., resistance-related genes were predicted through gene network-based study (Anitha et al. 2014). Different computational tools are available for the study of gene or protein networks. STRING is a database of known gene/protein interaction and is used for the prediction of novel interactions, including direct (physical) and indirect (functional) associations (Szklarczyk et al. 2015). DAVID, an integrated biological resource for functional enrichment of large gene/protein lists, helps in in-depth understanding of the biological systems. It is a precomputed database displayed in a tabular format containing similar annotation. It sorts the enriched gene sets into different, partially overlapping groups of all genes (Dennis et al. 2003; Huang et al. 2007, 2009). MCODE was the first algorithm used to screen protein complexes for the identification of highly interacting nodes present in any network. This was based on vertex weighting, complex prediction, and optimal post-processing by assigning weight to the vertex in local neighborhood density from the dense regions according to given parameters. The extracted sub networks were then ranked according to the individual size and density (Bader and Hogue 2003). Cytoscape software is used for modeling, analyzing, and visualizing molecular and genetic interaction network. It also provides graphical layouts that support three different algorithms, namely, hierarchical layout, spring-embedded layout, and circular layout (Shannon et al. 2003).

2.3 *Chemoinformatics-Based Approaches for Target Identification*

Different chemoinformatics approaches for target prediction are currently available. A ligand-based approach utilizes molecular three-dimensional architecture of the target, assuming that the chemical compounds will have similar biological activities if they have similar structures (Bender and Glen 2004). By this strategy, targets for small molecular compounds may be identified on the basis of similarities in shapes among known enzyme inhibitors or substrates present in different biological databases, such as PubChem (Wang et al. 2012), KiDB (Jensen and Roth 2008), and ChEMBl (Gaulton et al. 2012). Chemoinformatics is a rapid technique and can identify target proteins within a short period of time, but the major bottleneck is the requirement for prior knowledge on the structures of the protein and is ligand. A second approach is structure-based docking, in which a selected panel of target proteins or receptors with known ligand–protein interactions is docked by a "query" compound ("reverse" or "inverse" docking). The three-dimensional coordinates of the proteins are extracted from the PDB. To maintain uniformity of the chosen structures, several selection criteria need to be applied: (1) there should be no mutations or missing residues in and around the active site, (2) only a high-resolution structure (~1.5 Å) is acceptable, and (3) a ligand–protein complex is preferably selected. Finally, target proteins are predicted based on optimal interactions. However, traditional molecular docking approaches, which require significantly shorter computational time and often produce more reliable results, outperform this method (Koutsoukas et al. 2011).

Pharmacophore models, as per the International Union of Pure and Applied Chemistry, can be defined as "an ensemble of steric and electronic features that is necessary to ensure the optimal supramolecular interactions with a specific biological target and to trigger (or block) its biological response." In this method, query molecules were searched against a predefined pharmacophore model to find out the best-fitted one. PharmMapper (Liu et al. 2010) is a free web server that uses pharmacophore-based approaches to identify potential target proteins of a query molecule, such as any therapeutic agent or natural compounds for which binding targets are still unidentified. PharmMapper uses its own in-house collection of pharmacophore database created by the "target" molecules extracted from DrugBank (Law et al. 2014), BindingDB (Gilson et al. 2015), and PDTD (Gao et al. 2008) called PharmTargetDB. This database hosts more than 7000 pharmacophore models including 1627 drug targets. PharmMapper accepts query molecules in MDL SDF or Tripos Mol2 format and searches PharmTargetDB to predict probable drug targets.

Support vector machines are supervised learning methods containing learning algorithms for pattern recognition and data analysis. For the prediction of novel drug targets, this method uses the physicochemical properties of the amino acids of the primary protein sequences of known antibacterial targets (Han et al. 2007; Li and Lai 2007).

The UniDrug-Target server predicts drug targets in pathogenic bacteria for which FASTA sequences and sequence annotation are available. The server finds unique proteins of pathogenic bacteria by comparing the proteomes of pathogenic and nonpathogenic bacteria and the mammalian proteomes. A ranking system for the potential drug targets is developed based on the order of their functional importance for cell survival (Chanumolu et al. 2012).

In molecular docking, interactions between a query compound and target proteins are scored based on which, the likely identity of the query compound is determined. Popular examples of this method include INVDOCK (Chanumolu et al. 2012) and TarFisDock (Li et al. 2006). INVDOCK is a computer-based automated target prediction software for predicting nucleic acid or protein targets of any chemical compounds or therapeutic molecule. It accepts 3D structure of the query molecule and predicts its suitable target among 9000 protein and nucleic acid entries enlisted in their backend database. Potential applications of INVDOCK include early prediction of unwanted side effects (Ji et al. 2006) or toxic effects of any molecule under study and molecular target identification, such as biological target identification of genistein (Chen et al. 2006) and natural products (Chen et al. 2003). TarFisDock (target fishing docking) is a reverse docking server (Li et al. 2006), and it docked the query molecules such as drugs and drug candidates and natural products to the proteins listed in PDTD (potential drug target database) to predict their therapeutic target. PDTD contains 841 known drug targets with 3D structures available from PDB (Gao et al. 2008).

3 Computational Drug Designing Methods

In today's antimicrobial research and development, genomic, proteomic, and functional databases and related bioinformatic tools have become indispensible. Various computational approaches have been developed by merging independent subjects like mathematics, computer science, statistics, information technology, and molecular biology. Computer-aided drug design (CADD) method is elaborately discussed by Gregory Sliwoski et al. (2014), and the overall process is described in Fig. 1.

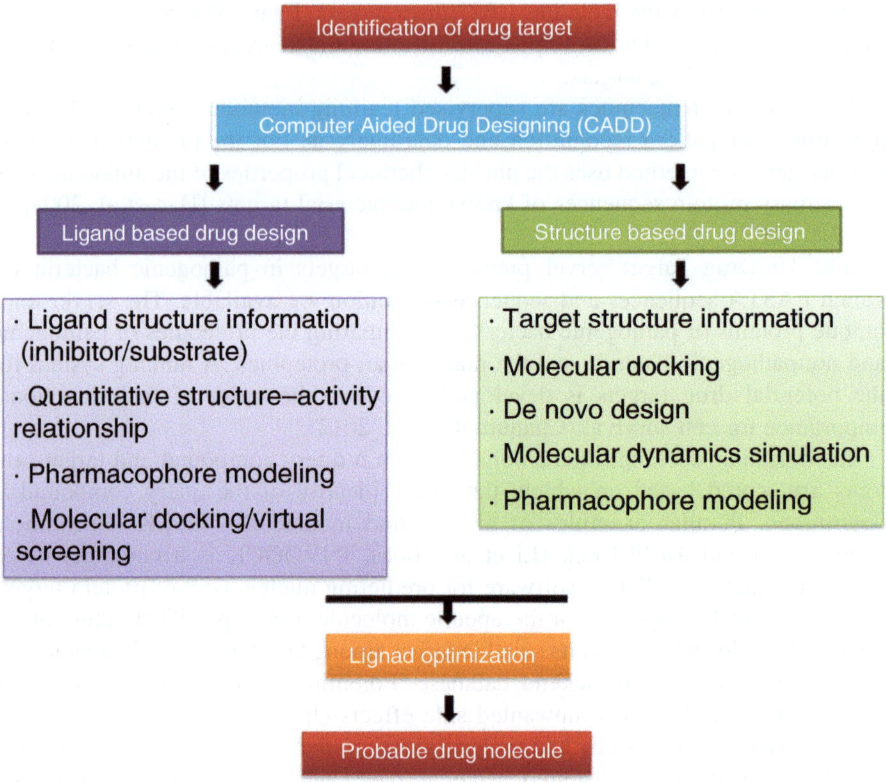

Fig. 1 The overall process of computer-aided drug designing

3.1 Structure-Based Methods (Docking, Homology Modeling)

Structure-based drug designing methods include protein structure modeling, molecular docking, and molecular dynamics simulation (MDS). After the identification of potential target protein for antibacterial drug development, the next step is to study the molecular architecture and structure–function relationship of the target. The X-ray crystallographic or NMR-based three-dimensional (3D) structures available in the PDB database (http://www.rcsb.org) are an authentic source for the above. When the target protein structure is not available, computational methods are employed to generate the structure. One widely used method for this purpose is homology modeling or comparative modeling. By this approach, an unknown protein structure is built based on its sequence similarity with a known structure called the template. MODELLER (Webb and Sali 2014) is an extensively used software for protein homology modeling. There are different homology modeling servers that are also available which are briefly described in Table 1.

Table 1 Frequently used homology modeling servers

Homology modeling servers	Description	URL	Reference
Phyre2	Web server for analyzing protein structure, function, and mutation. It uses advanced remote homology detection method for protein structure prediction	http://www.sbg.bio.ic.ac.uk/phyre2/	Kelley et al. (2015)
ModWeb	Web-accessible interface of MODELLER program for automated protein structure prediction	https://modbase.compbio.ucsf.edu/modweb/	Pieper et al. (2014)
LOMETS	It is a local meta-threading server. It predicts 3D structure of protein based on nine locally installed threading programs	http://zhanglab.ccmb.med.umich.edu/LOMETS/	Wu and Zhang (2007)
CPHmodels	Protein homology modeling server based on profile–profile alignment of secondary structure	http://www.cbs.dtu.dk/services/CPHmodels/	Nielsen et al. (2010)
SWISS-MODEL	Automated protein structure homology modeling server	http://swissmodel.expasy.org/	Guex et al. (2009)
I-TASSER	Protein structure and function prediction server. 3D structure of proteins are built based on multiple-threading alignments by LOMETS and iterative TASSER simulations	http://zhanglab.ccmb.med.umich.edu/I-TASSER/	Zhang (2008)

The knowledge of the 3D protein structures is indispensible for *in silico* inhibitor designing. By this method, inhibitors are searched out by screening small molecule databases like PubChem, ZINC, etc. (Table 2) using different docking and virtual screening software. AutoDock (Morris et al. 2009) is one of the most widely used freely available docking tools, and a second version of AutoDock is AutoDock Vina (Trott and Olson 2010), which is mostly used for virtual screening purpose. GOLD (Verdonk et al. 2003) and Glide (Friesner et al. 2004) are other frequently used software, and the predictive power of Glide is highly appreciated in the scientific community. Fragment-based designing of inhibitors is an approach where small molecular scaffolds are docked into protein active site and then potential hits are joined each other to build up a novel molecule. LUDI (Bohm 1992a, b), LigBuilder (Yuan et al. 2011), AutoGrow (Durrant et al. 2009, 2013), and eHiTS (Zsoldos et al. 2007) are tools for fragment-based ligand generation. This method often produces inhibitors with higher binding efficiency.

After getting the potential hits, binding efficiency is validated and measured *in silico* by MDS method. Both commercial software and freeware are widely available for MDS. CHARMM (Brooks et al. 2009) is a freeware to simulate inhibitor interactions within the protein active site. Another free package, GROMACS (Hess et al. 2008), is used in MD studies to generate "trajectories" which explains the behavior of the system under specific condition. Discovery Studio is an attractive package for QSAR studies, sequence analysis, protein structure modeling, and

Table 2 Different databases of small molecules for virtual screening

Databases	Summary	URL
PubChem	Database of chemical molecules containing three types of information, namely, substance, compound, and BioAssays	https://pubchem.ncbi.nlm.nih.gov/
ZINC	This database contains 21 million compounds and their physicochemical properties	http://zinc.docking.org/
ChEMBL	This database provides comprehensive information about 1 million bioactive (small drug-like molecules) compounds with 8200 drug targets	https://www.ebi.ac.uk/chembldb/
ChemDB	A repository of five million chemicals which physicochemical properties	http://cdb.ics.uci.edu/
ChemSpider	Contains more than 28 million unique chemical entities collected from more than 400 diverse data sources	http://www.chemspider.com/
BindingDB	Database of small molecules which contains 910,836 binding data, for 6263 protein targets and 378,980 small molecules	http://www.bindingdb.org/bind/index.jsp
PDB-Bind	It is a collection of 5671 protein–ligand complexes with their respective binding affinities and known three-dimensional structures	http://sw16.im.med.umich.edu/databases/pdbbind/index.jsp
PDBeChem	It provides comprehensive information of ligands, small molecules, and monomers. Presently it consists of 15,502 ligands	http://www.ebi.ac.uk/pdbe-srv/pdbechem/
HMDB	A database containing detailed information about small molecule metabolites found in the human body	http://www.hmdb.ca/
DrugBank	The database contains 8312 drug entries including 2036 FDA-approved small molecule drugs, 233 FDA-approved biotech (protein/peptide) drugs, 93 nutraceuticals, and over 6000 experimental drugs	http://www.drugbank.ca/
HIT	HIT is a comprehensive database for protein targets for FDA-approved drugs as well as the promising precursors. It currently contains about 1301 known protein targets (221 proteins are described as direct targets)	http://lifecenter.sgst.cn/hit/
SuperNatural	A database of natural products containing 50,000 natural compounds.	http://bioinfapplied.charite.de/supernatural_new/index.php
NPACT	It contains experimentally validated 1574 natural compound derived from plants exhibiting anti-cancerous activity	http://crdd.osdd.net/raghava/npact/
PharmGKB	Database containing clinical information of drug molecules	https://www.pharmgkb.org/
SuperDrug	This database contains approximately 2500 3D structures of active ingredients of essential marketed drugs	http://bioinf.charite.de/superdrug/

computational simulation, although not freely accessible (http://accelrys.com/products/discovery-studio/).

3.2 Ligand-Based QSAR Methods

Quantitative structure–activity relationship (QSAR) modeling is a chemoinformatics approach widely used for antimicrobial drug discovery. QSAR model quantitatively measures changes in the biological activities of a set of compounds with corresponding changes in their molecular structures. This method has been successfully applied for the analysis of antimicrobial peptide (AMP) data; the generated model was very useful for quantification of linear sequence patterns (Hilpert et al. 2006), amphipathicity (Frecer et al. 2004), and contact energy between neighboring amino acids (Jensen and Roth 2008). QSAR-based approaches significantly shorten the window for lead compound identification and cut the expenditure and time that would otherwise be spent for the synthesis of a large number of compounds and their evaluation by in vitro and in vivo experiments. QSAR may integrate very diverse chemical and biological data with the help of powerful statistical techniques like artificial neural networks (ANNs) (Speck-Planche and Cordeiro 2015). This is particularly useful, because no linear relationship often exists between the chemical structure and biological activities and/or toxicities of a compound (Prado-Prado et al. 2010; Tenorio-Borroto et al. 2012). The above technique was very successful for the discovery of antistreptococcal drugs (Speck-Planche et al. 2013). MATLAB software packages like Tsar or Neural Network Toolbox (http://www.mathworks.com/products/neuralnet/) are used for ANN analysis. Free software packages are also available for ANN analysis like Stuttgart Neural Network Simulator (http://www.ra.cs.uni-tuebingen.de/SNNS/).

4 Opinion

The conventional approaches yield significantly lesser number of new antibacterial drugs and fail to combat the menace of the rapid spread of drug-resistant pathogens. There is a strong urgency to adopt novel strategies in the field to avoid global catastrophe. Drug designing by anti-virulence strategy is a viable alternative, since it would not interfere with bacterial growth and survival, and thus minimize the chance of drug resistance. Computational methods are extremely useful for the initial phase of new drug development that involves target identification and lead optimization by shortening the time and reducing the cost. A healthy collaboration between bioinformaticians and experimental microbiologists is absolutely critical to achieve the above goal.

5 Concluding Remarks

Increasing prevalence of drug-resistant pathogenic bacteria in the food industry and in clinical settings has become a major threat worldwide, and there is an urgent need for novel therapeutic agents to combat them. The conventional drug design pipeline requires a huge investment and at least 12–15 years and despite that often fails to bring a product to the market. Thus the current scenario of antibiotic research and development is not very encouraging. The development and introduction of computational approaches toward target identification and drug designing has brought a dramatic change in the field and infused new hope. Integration of knowledge from diverse disciplines, such as statistics, mathematics, computer science, molecular biology, etc., made it possible to develop new computational tools and techniques for more accurate data analysis and predictions. Computer-based genomics and chemoinformatics approaches were able to significantly reduce the time and cost involved in this process. The whole-genome sequencing approach is particularly useful to understand the bacterial resistance mechanisms against antibiotics. Antimicrobial peptide-based therapy is a promising new approach toward combating drug resistance. Known structural and functional data of AMPs can be used for building computational models to predict novel candidate peptides as antimicrobial agents. A new trend has emerged that combines two complementary approaches, the experimental and computational biology, which has great potential in the future discovery of antimicrobial drugs.

Acknowledgments The authors wish to thank the Indian Council of Medical Research (ICMR) and the government of India for providing necessary funds and facilities. RSM thanks ICMR for funding support (IRIS ID: 2013–1551G).

References

Anitha P, Anbarasu A, Ramaiah S (2014) Computational gene network study on antibiotic resistance genes of *Acinetobacter baumannii*. Comput Biol Med 48:17–27. doi:10.1016/j.compbiomed.2014.02.009

Armstrong GD, Rowe PC, Goodyer P, Orrbine E, Klassen TP, Wells G, MacKenzie A, Lior H, Blanchard C, Auclair F, Thompson B, Rafter DJ, McLaine PN (1995) A phase I study of chemically synthesized verotoxin (Shiga-like toxin) Pk-trisaccharide receptors attached to chromosorb for preventing hemolytic-uremic syndrome. J Infect Dis 171:1042–1045. doi:10.1093/infdis/171.4.1042

Bader GD, Hogue CW (2003) An automated method for finding molecular complexes in large protein interaction networks. BMC Bioinformatics 4:2. doi:10.1186/1471-2105-4-2

Bender A, Glen RC (2004) Molecular similarity: a key technique in molecular informatics. Org Biomol Chem 2:3204–3218. doi:10.1039/B409813G

Bohm HJ (1992a) The computer program LUDI: a new method for the de novo design of enzyme inhibitors. J Comput Aided Mol Des 6:61–78

Bohm HJ (1992b) LUDI: rule-based automatic design of new substituents for enzyme inhibitor leads. J Comput Aided Mol Des 6:593–606

Boucher HW, Talbot GH, Benjamin DK Jr, Bradley J, Guidos RJ, Jones RN, Murray BE, Bonomo RA, Gilbert D (2013) 10 x '20 Progress--development of new drugs active against gram-negative bacilli: an update from the Infectious Diseases Society of America. Clin Infect Dis 56:1685–1694. doi:10.1093/cid/cit152

Brazas MD, Hancock RE (2005) Using microarray gene signatures to elucidate mechanisms of antibiotic action and resistance. Drug Discov Today 10:1245–1252. doi:10.1016/S1359-6446(05)03566-X

Brooks BR, Brooks CL 3rd, Mackerell AD Jr, Nilsson L, Petrella RJ, Roux B, Won Y, Archontis G, Bartels C, Boresch S, Caflisch A, Caves L, Cui Q, Dinner AR, Feig M, Fischer S, Gao J, Hodoscek M, Im W, Kuczera K, Lazaridis T, Ma J, Ovchinnikov V, Paci E, Pastor RW, Post CB, Pu JZ, Schaefer M, Tidor B, Venable RM, Woodcock HL, Wu X, Yang W, York DM, Karplus M (2009) CHARMM: the biomolecular simulation program. J Comput Chem 30:1545–1614. doi:10.1002/jcc.21287

Chan PF, Macarron R, Payne DJ, Zalacain M, Holmes DJ (2002) Novel antibacterials: a genomics approach to drug discovery. Curr Drug Targets Infect Dis 2:291–308

Chanumolu SK, Rout C, Chauhan RS (2012) UniDrug-target: a computational tool to identify unique drug targets in pathogenic bacteria. PloS One 7:e32833. doi:10.1371/journal.pone.0032833

Chen X, Ung CY, Chen Y (2003) Can an *in silico* drug-target search method be used to probe potential mechanisms of medicinal plant ingredients? Natl Prod Rep 20:432–444. doi:10.1039/B303745B

Chen X, Zhou H, Liu YB, Wang JF, Li H, Ung CY, Han LY, Cao ZW, Chen YZ (2006) Database of traditional Chinese medicine and its application to studies of mechanism and to prescription validation. Br J Pharmacol 149:1092–1103. doi:10.1038/sj.bjp.0706945

Chung BK, Dick T, Lee DY (2013) *In silico* analyses for the discovery of tuberculosis drug targets. J Antimicrob Chemother 68:2701–2709. doi:10.1093/jac/dkt273

Davies SC, Fowler T, Watson J, Livermore DM, Walker D (2013) Annual report of the Chief Medical Officer: infection and the rise of antimicrobial resistance. Lancet 381:1606–1609. doi:10.1016/S0140-6736(13)60604-2

Dennis G Jr, Sherman BT, Hosack DA, Yang J, Gao W, Lane HC, Lempicki RA (2003) DAVID: database for annotation, visualization, and integrated discovery. Genome Biol 4:P3. doi:10.1186/gb-2003-4-5-p3

Domagk G (1935) Ein Beitrag zur Chemotherapie der bakteriellen Infektionen. Dtsch Med Wochenschr 61:250. doi:10.1055/s-0028-1129486

Durrant JD, Amaro RE, McCammon JA (2009) AutoGrow: a novel algorithm for protein inhibitor design. Chem Biol Drug Des 73:168–178. doi:10.1111/j.1747-0285.2008.00761.x

Durrant JD, Lindert S, McCammon JA (2013) AutoGrow 3.0: an improved algorithm for chemically tractable, semi-automated protein inhibitor design. J Mol Graph Model 44:104–112. doi:10.1016/j.jmgm.2013.05.006

Ehrlich P, Hata S (1910) Die Experimentelle Chemotherapie der Spirilosen. Julius Springer, Berlin

Felise HB, Nguyen HV, Pfuetzner RA, Barry KC, Jackson SR, Blanc MP, Bronstein PA, Kline T, Miller SI (2008) An inhibitor of gram-negative bacterial virulence protein secretion. Cell Host Microbe 4:325–336. doi:10.1016/j.chom.2008.08.001

Field D, Feil EJ, Wilson GA (2005) Databases and software for the comparison of prokaryotic genomes. Microbiology 151:2125–2132. doi:10.1099/mic.0.28006-0

Fischer HP (2001) The impact of expression profiling technologies on antimicrobial target identification and validation. Drug Discov Today 6:1149–1150. doi:10.1016/S1359-6446(01)02047-5

Fleming A (2001) On the antibacterial action of cultures of a penicillium, with special reference to their use in the isolation of *B. influenzae*. 1929. Bull World Health Org 79:780–790. doi:10.1590/S0042-96862001000800017

Frecer V, Ho B, Ding JL (2004) De novo design of potent antimicrobial peptides. Antimicrob Agents Chemother 48:3349–3357. doi:10.1128/AAC.48.9.3349-3357.2004

Friesner RA, Banks JL, Murphy RB, Halgren TA, Klicic JJ, Mainz DT, Repasky MP, Knoll EH, Shelley M, Perry JK, Shaw DE, Francis P, Shenkin PS (2004) Glide: a new approach for rapid, accurate docking and scoring. 1. Method and assessment of docking accuracy. J Med Chem 47:1739–1749. doi:10.1021/jm0306430

Fritz B, Raczniak GA (2002) Bacterial genomics: potential for antimicrobial drug discovery. BioDrugs 16:331–337

Gao Z, Li H, Zhang H, Liu X, Kang L, Luo X, Zhu W, Chen K, Wang X, Jiang H (2008) PDTD: a web-accessible protein database for drug target identification. BMC Bioinformatics 9:104. doi:10.1186/1471-2105-9-104

Gaulton A, Bellis LJ, Bento AP, Chambers J, Davies M, Hersey A, Light Y, McGlinchey S, Michalovich D, Al-Lazikani B, Overington JP (2012) ChEMBL: a large-scale bioactivity database for drug discovery. Nucleic Acids Res 40:D1100–D1107. doi:10.1093/nar/gkr777

Gilson MK, Liu T, Baitaluk M, Nicola G, Hwang L, Chong J (2015) BindingDB in 2015: a public database for medicinal chemistry, computational chemistry and systems pharmacology. Nucleic Acids Res. doi:10.1093/nar/gkv1072

Guex N, Peitsch MC, Schwede T (2009) Automated comparative protein structure modeling with SWISS-MODEL and Swiss-PdbViewer: a historical perspective. Electrophoresis 30(Suppl 1): S162–S173. doi:10.1002/elps.200900140

Hampton T (2013) Report reveals scope of US antibiotic resistance threat. JAMA 310 (16):1661–1663. doi:10.1001/jama.2013.280695

Han LY, Zheng CJ, Xie B, Jia J, Ma XH, Zhu F, Lin HH, Chen X, Chen YZ (2007) Support vector machines approach for predicting druggable proteins: recent progress in its exploration and investigation of its usefulness. Drug Discov Today 12:304–313. doi:10.1016/j.drudis.2007.02.015

Hess B, Kutzner C, van der Spoel D, Lindahl E (2008) GROMACS 4: algorithms for highly efficient, load-balanced, and scalable molecular simulation. J Chem Theory Comput 4:435–447. doi:10.1021/ct700301q

Hilpert K, Elliott MR, Volkmer-Engert R, Henklein P, Donini O, Zhou Q, Winkler DF, Hancock RE (2006) Sequence requirements and an optimization strategy for short antimicrobial peptides. Chem Biol 13:1101–1107. doi:10.1016/j.chembiol.2006.08.014

Huang DW, Sherman BT, Tan Q, Collins JR, Alvord WG, Roayaei J, Stephens R, Baseler MW, Lane HC, Lempicki RA (2007) The DAVID Gene Functional Classification Tool: a novel biological module-centric algorithm to functionally analyze large gene lists. Genome Biol 8: R183. doi:10.1186/gb-2007-8-9-r183

Huang DW, Sherman BT, Lempicki RA (2009) Systematic and integrative analysis of large gene lists using DAVID bioinformatics resources. Nat Prot 4:44–57. doi:10.1038/nprot.2008.211

Hung DT, Shakhnovich EA, Pierson E, Mekalanos JJ (2005) Small-molecule inhibitor of *Vibrio cholerae* virulence and intestinal colonization. Science 310:670–674. doi:10.1126/science.1116739

Jabes D (2011) The antibiotic R&D pipeline: an update. Curr Opin Microbiol 14:564–569. doi:10.1016/j.mib.2011.08.002

Jensen NH, Roth BL (2008) Massively parallel screening of the receptorome. Comb Chem High Throughput Screen 11:420–426. doi:10.2174/138620708784911483

Ji ZL, Wang Y, Yu L, Han LY, Zheng CJ, Chen YZ (2006) *In silico* search of putative adverse drug reaction related proteins as a potential tool for facilitating drug adverse effect prediction. Toxicol Lett 164:104–112. doi:10.1016/j.toxlet.2005.11.017

Kauppi AM, Nordfelth R, Uvell H, Wolf-Watz H, Elofsson M (2003) Targeting bacterial virulence: inhibitors of type III secretion in *Yersinia*. Chem Biol 10:241–249. doi:10.1016/S1074-5521(03)00046-2

Kelley LA, Mezulis S, Yates CM, Wass MN, Sternberg MJ (2015) The Phyre2 web portal for protein modeling, prediction and analysis. Nat Prot 10:845–858. doi:10.1038/nprot.2015.053

Kline T, Felise HB, Barry KC, Jackson SR, Nguyen HV, Miller SI (2008) Substituted 2-imino-5-arylidenethiazolidin-4-one inhibitors of bacterial type III secretion. J Med Chem 51:7065–7074. doi:10.1021/jm8004515

Koutsoukas A, Simms B, Kirchmair J, Bond PJ, Whitmore AV, Zimmer S, Young MP, Jenkins JL, Glick M, Glen RC, Bender A (2011) From *in silico* target prediction to multi-target drug design: current databases, methods and applications. J Proteomics 74:2554–2574. doi:10.1016/j.jprot.2011.05.011

Law V, Knox C, Djoumbou Y, Jewison T, Guo AC, Liu Y, Maciejewski A, Arndt D, Wilson M, Neveu V, Tang A, Gabriel G, Ly C, Adamjee S, Dame ZT, Han B, Zhou Y, Wishart DS (2014) DrugBank 4.0: shedding new light on drug metabolism. Nucleic Acid Res 42:D1091–D1097. doi:10.1093/nar/gkt1068

Lesic B, Lepine F, Deziel E, Zhang J, Zhang Q, Padfield K, Castonguay MH, Milot S, Stachel S, Tzika AA, Tompkins RG, Rahme LG (2007) Inhibitors of pathogen intercellular signals as selective anti-infective compounds. PLoS Pathog 3:1229–1239. doi:10.1371/journal.ppat.0030126

Li Q, Lai L (2007) Prediction of potential drug targets based on simple sequence properties. BMC Bioinformatics 8:353. doi:10.1186/1471-2105-8-353

Li H, Gao Z, Kang L, Zhang H, Yang K, Yu K, Luo X, Zhu W, Chen K, Shen J, Wang X, Jiang H (2006) TarFisDock: a web server for identifying drug targets with docking approach. Nucleic Acid Res 34(Web Server issue):W219–W224. doi:10.1093/nar/gkl114

Liu X, Ouyang S, Yu B, Liu Y, Huang K, Gong J, Zheng S, Li Z, Li H, Jiang H (2010) PharmMapper server: a web server for potential drug target identification using pharmacophore mapping approach. Nucleic Acid Res 38(Web Server issue):W609–W614. doi:10.1093/nar/gkq300

Mandal S, Moudgil M, Mandal SK (2009) Rational drug design. Eur J Pharmacol 625:90–100. doi:10.1016/j.ejphar.2009.06.065

Moellering RC Jr (2011) Discovering new antimicrobial agents. Int J Antimicrob Agents 37:2–9. doi:10.1016/j.ijantimicag.2010.08.018

Morris GM, Huey R, Lindstrom W, Sanner MF, Belew RK, Goodsell DS, Olson AJ (2009) AutoDock4 and AutoDockTools4: automated docking with selective receptor flexibility. J Comput Chem 3:2785–2791. doi:10.1002/jcc.21256

Muschiol S, Bailey L, Gylfe A, Sundin C, Hultenby K, Bergstrom S, Elofsson M, Wolf-Watz H, Normark S, Henriques-Normark B (2006) A small-molecule inhibitor of type III secretion inhibits different stages of the infectious cycle of *Chlamydia trachomatis*. Proc Natl Acad Sci USA 103:14566–14571. doi:10.1073/pnas.0606412103

Nielsen M, Lundegaard C, Lund O, Petersen TN (2010) CPHmodels-3.0--remote homology modeling using structure-guided sequence profiles. Nucleic Acid Res 38(Web Server issue):W576–W581. doi:10.1093/nar/gkq535

Palumbi SR (2001) Humans as the world's greatest evolutionary force. Science 293:1786–1790. doi:10.1126/science.293.5536.1786

Payne DJ, Gwynn MN, Holmes DJ, Pompliano DL (2007) Drugs for bad bugs: confronting the challenges of antibacterial discovery. Nat Rev Drug Discov 6:29–40. doi:10.1038/nrd2201

Pieper U, Webb BM, Dong GQ, Schneidman-Duhovny D, Fan H, Kim SJ, Khuri N, Spill YG, Weinkam P, Hammel M, Tainer JA, Nilges M, Sali A (2014) ModBase, a database of annotated comparative protein structure models and associated resources. Nucleic Acid Res 42(Database issue):D336–D346. doi:10.1093/nar/gkh095

Power E (2006) Impact of antibiotic restrictions: the pharmaceutical perspective. Clin Microbiol Infect 12(Suppl 5):25–34. doi:10.1111/j.1469-0691.2006.01528.x

Prado-Prado FJ, Garcia-Mera X, Gonzalez-Diaz H (2010) Multi-target spectral moment QSAR versus ANN for antiparasitic drugs against different parasite species. Bioorg Med Chem 18:2225–2231. doi:10.1016/j.bmc.2010.01.068

Projan SJ (2003) Why is big Pharma getting out of antibacterial drug discovery? Curr Opin Microbiol 6:427–430. doi:10.1016/j.mib.2003.08.003

Rasko DA, Sperandio V (2010) Anti-virulence strategies to combat bacteria-mediated disease. Nat Rev Drug Discov 9:117–128. doi:10.1038/nrd3013

Rasko DA, Moreira CG, de Li R, Reading NC, Ritchie JM, Waldor MK, Williams N, Taussig R, Wei S, Roth M, Hughes DT, Huntley JF, Fina MW, Falck JR, Sperandio V (2008) Targeting QseC signaling and virulence for antibiotic development. Science 321:1078–1080. doi:10.1126/science.1160354

Rice LB (2008) The Maxwell Finland Lecture: for the duration-rational antibiotic administration in an era of antimicrobial resistance and clostridium difficile. Clin Infect Dis 46:491–496. doi:10.1086/526535

Russ AP, Lampel S (2005) The druggable genome: an update. Drug Discov Today 10:1607–1610. doi:10.1016/S1359-6446(05)03666-4

Shannon P, Markiel A, Ozier O, Baliga NS, Wang JT, Ramage D, Amin N, Schwikowski B, Ideker T (2003) Cytoscape: a software environment for integrated models of biomolecular interaction networks. Genome Res 13:2498–2504. doi:10.1101/gr.1239303

Silver LL (2011) Challenges of antibacterial discovery. Clin Microbiol Rev 24:71–109. doi:10.1128/CMR.00030-10

Sipahi OR (2008) Effects of antibiotic resistance on industrial antibiotic R&D. Expert Rev Anti Infect Ther 6:523–539

Sliwoski G, Kothiwale S, Meiler J, Lowe EW Jr (2014) Computational methods in drug discovery. Pharmacol Rev 66:334–395. doi:10.1124/pr.112.007336

Speck-Planche A, Cordeiro MN (2015) A general ANN-based multitasking model for the discovery of potent and safer antibacterial agents. Methods Mol Biol 1260:45–64. doi:10.1007/978-1-4939-2239-0_4

Speck-Planche A, Kleandrova VV, Cordeiro MN (2013) Chemoinformatics for rational discovery of safe antibacterial drugs: simultaneous predictions of biological activity against streptococci and toxicological profiles in laboratory animals. Bioorg Med Chem 21:2727–2732. doi:10.1016/j.bmc.2013.03.015

Spellberg B, Guidos R, Gilbert D, Bradley J, Boucher HW, Scheld WM, Bartlett JG, Edwards J Jr, Infectious Diseases Society of America (2008) The epidemic of antibiotic-resistant infections: a call to action for the medical community from the Infectious Diseases Society of America. Clin Infect Dis 46(2):155–164. doi:10.1086/524891

Szklarczyk D, Franceschini A, Wyder S, Forslund K, Heller D, Huerta-Cepas J, Simonovic M, Roth A, Santos A, Tsafou KP, Kuhn M, Bork P, Jensen LJ, von Mering C (2015) STRING v10: protein-protein interaction networks, integrated over the tree of life. Nucleic Acid Res 43: D447–D452. doi:10.1093/nar/gku1003

Tang CM, Moxon ER (2001) The impact of microbial genomics on antimicrobial drug development. Annu Rev Genomics Hum Genet 2:259–269. doi:10.1146/annurev.genom.2.1.259

Tenorio-Borroto E, Penuelas Rivas CG, Vasquez Chagoyan JC, Castanedo N, Prado-Prado FJ, Garcia-Mera X, Gonzalez-Diaz H (2012) ANN multiplexing model of drugs effect on macrophages; theoretical and flow cytometry study on the cytotoxicity of the anti-microbial drug G1 in spleen. Bioorg Med Chem 20:6181–6194. doi:10.1016/j.bmc.2012.07.020

Trachtman H, Cnaan A, Christen E, Gibbs K, Zhao S, Acheson DW, Weiss R, Kaskel FJ, Spitzer A, Hirschman GH, Investigators of the HUSSPMCT (2003) Effect of an oral Shiga toxin-binding agent on diarrhea-associated hemolytic uremic syndrome in children: a randomized controlled trial. JAMA 290:1337–1344. doi:10.1001/jama.290

Trott O, Olson AJ (2010) AutoDock Vina: improving the speed and accuracy of docking with a new scoring function, efficient optimization, and multithreading. J Comput Chem 31:455–461. doi:10.1002/jcc.21334

Veenendaal AK, Sundin C, Blocker AJ (2009) Small-molecule type III secretion system inhibitors block assembly of the Shigella type III secreton. J Bacteriol 191:563–570. doi:10.1128/JB.01004-08

Verdonk ML, Cole JC, Hartshorn MJ, Murray CW, Taylor RD (2003) Improved protein-ligand docking using GOLD. Proteins 52:609–623. doi:10.1002/prot.10465

Wang Y, Xiao J, Suzek TO, Zhang J, Wang J, Zhou Z, Han L, Karapetyan K, Dracheva S, Shoemaker BA, Bolton E, Gindulyte A, Bryant SH (2012) PubChem's BioAssay Database. Nucleic Acid Res 40(Database issue):D400–D412. doi:10.1093/nar/gkr1132

Webb B, Sali A (2014) Comparative protein structure modeling using MODELLER. Curr Prot Bioinfo 47:5.6.1–5.6.32. doi:10.1002/0471250953.bi0506s47

Wu S, Zhang Y (2007) LOMETS: a local meta-threading-server for protein structure prediction. Nucleic Acid Res 35:3375–3382. doi:10.1093/nar/gkm251

Yuan Y, Pei J, Lai L (2011) LigBuilder 2: a practical de novo drug design approach. J Chem Info Model 51:1083–1091. doi:10.1021/ci100350u

Zhang Y (2008) I-TASSER server for protein 3D structure prediction. BMC Bioinformatics 9:40. doi:10.1186/1471-2105-9-40

Zsoldos Z, Reid D, Simon A, Sadjad SB, Johnson AP (2007) eHiTS: a new fast, exhaustive flexible ligand docking system. J Mol Graph Model 26:198–212. doi:10.1016/j.jmgm.2006.06.002

Synthetic Solutions to Drug Resistance

Gunjan Arora, Richa Misra, and Andaleeb Sajid

Abstract Synthetic biology has the potential to revolutionize human health standards. Advances in molecular and cellular biology, computational science, and systems biology have contributed to this new field that is trying to apply engineering principles to answer biological questions. These applications of synthetic biology may bring hope to our battle against the ever-emerging drug-resistant forms of disease. Advances in medical science helped us to achieve early victory against various common diseases in the last century. However, newer forms of drug-resistant diseases have emerged and are threatening the human race immensely. To fight this challenge, several strategies have been adopted to various degrees of success. For some diseases such as cancer, tuberculosis, HIV/AIDS, malaria, and diabetes, the threat is still imminent. This chapter discusses the early efforts of synthetic biologists toward better management, wherein we summarize some of the very recent work in the field. We hope the arsenal of synthetic solutions will provide superior solutions toward early diagnosis and treatment of drug-resistant disease forms.

1 Introduction

One of the main features of any living cell is adaptability to survive. Every living organism on the earth adapts strategies to grow and dominate the environmental challenges. Therefore, it is no surprise that the acquired drug resistance is possibly a

G. Arora (✉)
National Institute of Allergy and Infectious Diseases, National Institutes of Health, Rockville, MD, USA
e-mail: arorag193@gmail.com

R. Misra
Department of Zoology, Miranda House, University of Delhi, Delhi, India

A. Sajid (✉)
Tuberculosis Research Section, National Institute of Allergy and Infectious Diseases, National Institutes of Health, Bethesda, MD, USA
e-mail: sajid.andaleeb@gmail.com

G. Arora et al. (eds.), *Drug Resistance in Bacteria, Fungi, Malaria, and Cancer*,
DOI 10.1007/978-3-319-48683-3_26

universal truth for all cells. From early days, humans are using various compounds to combat diseases. The first evidence dates back to 3300 BC where humans carried an antibacterial fungus (McFarlane 1991). There is additional evidence that humans use various drugs to enhance their own strength. Therefore, drugs have provided the economic, social, and physical strength to human society (Guerra-Doce 2015). It is because of advances in medical science that there is radical shift in life expectancy. In 1900, average human life expectancy was only 31 years, which is now 67.2 years worldwide (http://www.who.int/kms/initiatives/indiana.pdf). So, to last these triumphs and to provide our future generation a longer quality life, we need drugs. The problem is the usage of every drug or medicine has a life, after which acquired resistance makes the drug less useful. Therefore, we need new therapeutic solutions against diseases and syndromes. Lots of efforts are already being made in this direction, and although a lot has been covered on this topic, this chapter, in particular, is dedicated on new emerging themes in drug discovery by the study of synthetic biology.

2 Synthetic Biology for Novel Molecules

Is synthetic biology really just a beta version of bioengineering/genetic engineering? Is it fair to say that synthetic biology is launched on advances in genetic engineering, computational biology, and systems biology? It is probably not an exaggeration if we say that being an application-oriented theme, synthetic biology has the potential to change the course of human history. In the late seventeenth century, industrial revolution changed the course of world history. It created sustained growth of economy by which humans achieved better developmental index that measures life expectancy, economic growth, and knowledge. In the next two centuries, humans dominated land, water, and space. In the current era, we can not only appreciate biological complexity on molecular level but also tinker with it to create more complex systems. This knowledge gifted a toolbox to our generation that every human being in the history aspired for. We can decide the course of our history by designing a life that nurtures our environment and us both.

In the last 70 years, antibiotics or related antimicrobials have been used extensively for human and animal welfare, and we have exhausted antibacterial chemical scaffolds (Wright 2012). This is one of the reasons for over-relying on certain drugs and faster rates of emergence of drug resistance. In the present scenario with rampant drug resistance, synthetic biologists are trying to solve molecular mechanisms of drug action, drug delivery, diagnostics, and novel therapeutic discoveries (Trosset and Carbonell 2015). Synthetic biology can help us not only engineer the known biological modules but also design nonnatural products made by alien enzymes. Engineered cells expressing artificial enzymes and expressing synthetic pathways will synthesize new complex molecules that will provide new therapeutics, e.g., peptide antibiotics.

The following sections will shed light on how synthetic biology approaches will be useful in tackling drug resistance.

2.1 Engineered Therapeutics

One of the earliest applications in this regard is rewiring the cellular pathways and modeling signaling response. For instance, immune cell signaling is a stochastic and efficient means to rewire signaling pathways, and activating these cells is the key to immunotherapy (Lim 2010; Chakravarti and Wong 2015). T-cell immunotherapies have been shown to hold great potential toward drug-resistant forms of cancer (Chakravarti and Wong 2015). However, heterogeneity in cytolytic lymphocyte activation and cytotoxicity does hinder in effectiveness of this response. To address this situation, specific receptors known as chimeric antigen receptors were expressed on T-cell surface (Kalos et al. 2011). These receptors allow T cells to identify and destroy tumor cell. These engineered T cells are infused in the body by the adoptive transfer method, which can now specifically lyse tumor cells (Kalos et al. 2011). However, these activated T cells can have serious side effects too, and there are associated complications that limit their use (Stauss and Morris 2013). An engineered T cell with a regulatable switch could potentially limit the T-cell activation and thus the related side effects and can help in more effective response. Using synthetic biology approach in a recent study, T cells were engineered to express CAR (cancer antigen receptors), but they do not show cytotoxic response unless they are switched on (Wu et al. 2015). The switch further allows controlling the response strength and time. The early progress in T-cell engineering and their importance in immunotherapies indicate the potential of these cells as arguably best mammalian model to apply synthetic biology tools. Further research is needed to apply Boolean logic gates on the immune cells by which we can have multiple input/output controls, each regulated independently. This is how T-cell developmental plasticity controls its regulatory response and our strength to fight infections. This real-time control will allow us to realize potential of "a living drug."

2.2 Bacteriophages

Bacteriophages (or phages) are the viruses that infect and thrive within bacteria as host (Seed 2015). They are very specific in selecting their host. Like other viruses, they utilize the bacterial cell machinery, multiply, and then lyse the bacteria to dissipate. Phages are possibly one of the most omnipresent group (Clokie et al. 2011). Bacteriophages were discovered by a British scientist, Frederick William Twort, in 1915. In his landmark paper in Lancet, he described these "ultramicroscopic viruses" that can infect bacteria (Twort 1915). This study was soon followed with similar report by French-Canadian scientist, Félix d'Herelle, who is considered a co-discoverer of bacteriophages and also the pioneer of phage therapy (Chanishvili 2012). In that era, serum therapy was the only effective treatment known to scientists since antibiotics were still not discovered. Phages, being very specific, were very attractive arsenal against bacterial pathogens (Chanishvili 2012; Salmond and Fineran 2015). However, after the discovery of more effective antibiotic therapy that was effective against broad spectrum of bacterial infections, the phage therapy lost its value (Wei 2015). Also, compared to phages, the antibiotics could be produced easily, and it was easier to scale up their production. The phage therapy was used intermittently in some countries in the last century; however, there is increased interest in phage therapy with the emergence of multidrug-resistant bacteria (Salmond and Fineran 2015; Wei 2015).

However, the conventional phage therapy has certain specific limitations such as:

- Many antibiotics are wide spectrum while phages are specific.
- Phage therapy requires skilled workforce and more organized supply chain that can coordinate and respond according to diagnostics.
- Phage preparations are not yet "one-size-fits-all" approach and need constant attention. There is not enough data known about the possibility of bacteria developing resistance to phages.
- For drug regulatory authorities, there is further need to find gold standard that can be used as reference.
- Generation of antibodies against phages in humans.

2.2.1 Use of D29 Mycobacteriophages Against Tuberculosis

Tuberculosis is one of the most deadly diseases in the world, and the major hurdle in treatment is emergence of drug-resistant *Mycobacterium tuberculosis* strains (Arora et al. 2010; Sajid et al. 2015; Singhal et al. 2015; Marais 2016; Yates et al. 2016). Multidrug-resistant TB (MDR-TB) is described as infection from *M. tuberculosis* strains that are resistant to isoniazid and rifampicin (the two most important first-line drugs) (Altimari et al. 2015). Extensively drug-resistant TB (XDR-TB) is infection from *M. tuberculosis* strain that is resistant to first-line drugs in addition to at least one fluoroquinolone and one injectable drug (Wilson and Tsukayama 2016). Due to poor medical facilities and lengthy diagnostic tests regime, the rate of death among patients with XDR-TB is very high (Velasquez et al. 2014). Despite the success in eradicating polio and treating leprosy and controlling AIDS, most developing countries still face continuing threats from infectious diseases like tuberculosis and malaria. The situation could be evaluated from the recent WHO tuberculosis report which mentions that at least 300,000 Indians die each year from tuberculosis (WHO global tuberculosis report 2015).

Mycobacteriophages are the viruses that specifically infect mycobacteria (Hatfull 2014a, b). They were discovered in mid-1900s, and currently their number has crossed 600 (Pope and Hatfull 2015). Among others, D29 is one of the most studied mycobacteriophages. D29 is a lytic phage, which can thrive in *M. tuberculosis* and *M. ulcerans*, the slow-growing pathogens (Rybniker et al. 2006; Samaddar et al. 2015). D29 infects mycobacteria and replicates within it, followed by lysis of cell wall and release of amplified generation of phages. Thus, when TB samples are inoculated with D29, an increase in the number of phages indicates the presence of viable mycobacteria (Rybniker et al. 2006).

Since phages utilize host cell machinery for their division, they respond differently in the presence of drugs. For example, phages cannot replicate in the presence of rifampicin, which disrupts host transcription machinery (Pai et al. 2005). On the other hand, phages infecting drug-resistant strains can undergo replication normally even in the presence of drugs (Pai et al. 2005; Schofield et al. 2012; Smartt et al. 2012; https://www.google.com/patents/US8501400). Other first-line drugs such as isoniazid and ethambutol provided in the DOTS program do not inhibit phage replication. Thus, these parameters may be used to assess the susceptibility of mycobacteria to these drugs by incubating the bacteria with critical concentrations of the drug that kills the bacteria. This way when the phages will be added to the system, they will not be able to replicate in those bacteria that were killed by the drug. As phages are known to be quickly removed from the body, they can be of great clinical value. Additionally, to neutralize the bacteriophage inside the body, antibodies or simple elements such as ferrous (iron II) compounds can be administered which will inactivate mycobacteriophages such as D29 (Toussaint and Muschel 1959; Park et al. 2003). With these advances, phages can prove to be useful for the development of detection methods for mycobacterial presence. For example, mycobacteriophages are administered to the patients to allow infection along with ferrous ammonium sulfate (Park et al. 2003). Following treatment with FAS, only the phages within the host mycobacteria that have resulted from successful infection and replication will remain (McNerney et al. 1998; Park et al. 2003).

With the introduction of new drugs, the ratio of drug-resistant to drug-sensitive patients is speedily going upward. Rifampicin, which is a first-line agent in the treatment of tuberculosis and whose resistance is responsible for majority of deaths in combination with isoniazid resistance, the problem is engraved with the lengthy diagnostic time to detect rifampicin resistance (Drobniewski et al. 2015; Rifat et al. 2015). The use of D29 mycobacteriophages that can kill the nonresistant cells and also can be used for detection of resistance is novel solution to detect and/or treat rifampicin resistance. Therefore, the challenge is to design synthetic bacteriophages for therapeutic use and diagnostics that can be easily monitored and neutralized/controlled by genetic switches and are less antigenic.

2.2.2 Synthetic Bacteriophages

Engineered bacteriophages that are more potent, safer, and have higher efficacy can be lucrative solutions to tackle antibiotic-resistant bacteria (Jassim and Limoges 2014). Since the inception of synthetic biology, the idea is to develop therapeutic candidates that are safer, more sensitive, and specific. Lu et al. have reported synthetic phages toward gene networks that are not currently targeted by antibiotics (Lu and Collins 2009). Successful pathogens are probably nature's best-designed machines that try to circumvent the antibiotic pressure by phenotypic and genotypic countermeasures. These measures are the key to their survival and evolution of drug resistance. The engineered phages target the bacteria by not letting them to adapt for survival measures thus making them more susceptible to antibiotics. Clinical efficacy analyses reveal that the antibiotic-enhancing phages make bacterial killing more potent compared to the antibiotic administered alone (Lu and Collins 2009). The first successful example of synthetic phages revalidates hopes from elegant engineered biological machines of future (Smith et al. 2003).

3 Synthetic Biosensors to Tackle Drug Resistance

Recent surge in drug resistance can be attributed to multiple reasons including the most prominent theme of direct correlation between diagnostic methods and therapeutic management. To a great extent, therapeutic misuse, in particular, of antibiotics is responsible for evolution of drug resistance (Laxminarayan et al. 2013). In resource-poor settings, physicians often rely on symptoms and conventional wisdom and not on accurate identification tools. To safeguard patients' life, often physicians take a decision to include certain antimicrobials/antibiotics in their prescription (Leekha et al. 2011). This abundant use of antibiotics has led to acquired drug resistance in pathogens much faster than otherwise. Additionally, it is even more difficult to accurately identify drug-resistant forms that lead to further administration of multiple antibiotics. Accurate diagnostic tools that can identify the disease in resource-poor settings will be the key to help the physicians' conundrum (Hedt et al. 2011). Thus, accurate, inexpensive, and diagnostic methods that can easily identify the drug-resistant forms should be a part of long-term solution for most drug resistance-prone diseases. Unless we provide therapeutic treatment based on sound scientific methods and not just a guess on symptoms, the cells will keep acquiring resistance to newer drugs leaving us with little maneuvering (Hedt et al. 2011).

Synthetic biology is trying to work on this aspect in many novel ways. The three major components of any diagnostic method are sensory input signal, processing component, and output method. The synthetic tools could help us identify novel components in all three aspects. For example, for the detection of Lyme disease-causing bacteria *Borrelia burgdorferi*, Burbelo et al. (2010) used a synthetic protein consisting of a repeated antigenic peptide sequence (Burbelo et al. 2010). The use of synthetic protein leads to a highly sensitive sero-analysis test. Similarly, Pardee et al. devised a new diagnostic method for the deadly pathogen Ebola virus by combining cell-free extracts with synthetic gene circuit on paper (Pardee et al. 2014). This study used RNA sensors created on toehold module that can also be used to identify drug-resistant forms in the future for other diseases. Toehold switches are de novo created riboregulators that act as transcriptional activators in reaction to related arbitrary RNA sequences (Green et al. 2014). In another study, Weber et al. identified drug-resistant forms of *M. tuberculosis* by using an engineered gene circuit (Weber et al. 2008). Such synthetic gene circuits will allow discovery of new antibiotic compounds against drug-resistant forms.

4 Novel Therapeutics

Immune system is our primary defense mechanism; it not only restrains us from unknown infectious microbes but also from the "self" sick cells. However, sometimes the immune system can't distinguish between self and nonself, and this misjudgment leads to tissue destruction and other related autoimmune diseases (Her and Kavanaugh 2016). Many autoimmune disorders are related to pro-inflammatory cytokine release like psoriasis, a chronic inflammatory skin disease (Burfield and Burden 2013). As with most autoimmune disorders, there is no cure for psoriasis, and use of anti-inflammatory medications can only provide temporary relief from symptoms (Gupta et al. 2015). However, after some time, the patient develops treatment-resistant psoriasis (Burfield and Burden 2013; Busard et al. 2014). This is a typical case where not only we need effective drugs but also have to tackle drug resistance. To solve this problem, synthetic biologists are offering a novel solution (Schukur et al. 2015). Psoriasis pathology is related to increase in inflammatory cytokines TNF-α and IL-22 (Lowes et al. 2014). By employing mammalian engineered circuits that can detect increases in these inflammatory cytokines, disease manifestations can be ceased. The gene circuit further responds by producing anti-inflammatory cytokines IL-4 and IL-10 (Schukur et al. 2015). This results in modulation of immune regulation and suppression of inflammatory cytokines that result in induction of psoriasis-related skin flares. This AND gate logic circuit will be the key in the future to the personalized and cell-based therapies to treat complex disorders.

5 Synthetically Engineered Protein Machinery: An Alternate to Conventional Drug Therapy

Cellular phenotypes are an outcome of protein assemblies working together. For every cellular function, be it replication, transcription, translation, signaling, sensing, transport, metabolism, stress management, division, or even cell senescence, many proteins come together and work in concert to produce the desired effect. Manipulation of these machineries can significantly alter the cell behavior that can be utilized for the benefit of science and technology as in whole.

Synthetic biology approaches have provided efficient means for such manipulations, for example, incorporation of nonnatural amino acids and chemicals to produce synthetic proteins that do not occur in nature and have better potential to carry out their functions (Lang and Chin 2014; Schutz and Mootz 2014; Lang et al. 2015). Another utilization of synthetic proteins is in production of chemicals and drugs. Many chemicals that are synthesized in chemistry take a lot of space, instrumentation, and time of scientists to achieve the desired purity and quality. Synthetic protein assemblies take advantage of enzymatic properties to synthesize the chemical of interest in a self-maintained system that does not require constant monitoring (Keasling 2010; Moura et al. 2016) (Fig. 1). One of the famous examples of protein engineering is manufacturing of synthetic glycoproteins. Glycoproteins have a carbohydrate moiety attached to the polypeptide chain, co-translationally or posttranslationally. Most of these proteins are surface

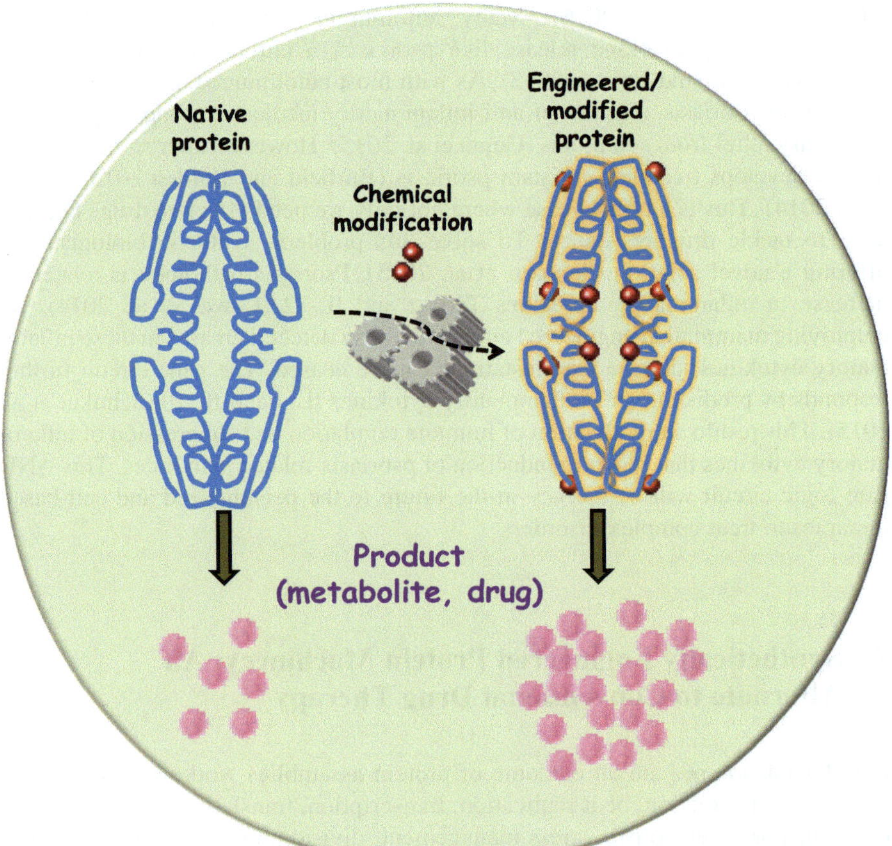

Fig. 1 Conceptual representation of engineered proteins: Synthetic or engineered proteins are generated by chemical modifications of amino acid derivatives. These proteins are used to produce substantial quantities of desired products such as metabolites and drugs, within a cell system

associated or secreted from the cell (Funakoshi and Suzuki 2009). Glycosylation affects protein functions directly as well as their distribution and metabolism; hence, glycosylation defects are associated with variety of disorders including cancer (Varki et al. 2009; Christiansen et al. 2014), neurological disorders (Freeze et al. 2015), diabetic neuropathy (Sugimoto et al. 2008), autoimmune diseases like rheumatoid arthritis (Delves 1998; Goulabchand et al. 2014), etc. Rectification of glycoprotein disorders is extremely difficult, if achievable at all. But recent progress in synthetic protein biology has made this arduous task not only possible but also motivating (Chen 2015).

Besides being useful in medical treatments, synthetic protein assemblies are an important tool for biotechnologists and industrialists. There is strong potential for their use in generating tools for DNA/RNA polymer modifications and technologies, unnatural protein and peptide synthesis, and synthesis of engineered bacteria for protein of molecules of pharmaceutical interests (Glasscock et al. 2016).

5.1 Nonconventional Drug Therapy

The next step is to design an artificial cellular system that can perform according to the desired function and produce the compounds of interest (Lucks et al. 2008). These so-called designer cells have been synthesized, for example, for the production of lipid biofuels (d'Espaux et al. 2015), small molecule drugs (Moura et al. 2016), or even treatment of diseases like cancer and psoriasis (Schukur et al. 2015, 2016; http://www.bbc.com/news/health-34731498). The designer cells have prominent expression of artificially designed protein assemblies that can alter the overall behavior of cells. Alternatively, engineered protein antibodies have been utilized as an efficient means for drug delivery. The advantages of synthetic proteins are as follows: they can be used as delivery vehicles without being recognized by immune cells (have self-recognizable epitopes), are stable in the gastrointestinal tract and are resistant to proteolytic enzymes. Such engineered proteins can also be utilized in treatment of diseases that target only the defective or infected cells. In a similar approach for protein engineering, prominent use of cyclotides has been discussed. Cyclotides are cysteine-rich macrocyclic peptides that form multiple disulfide bonds, thus providing extra strength to the peptides (Craik et al. 2006). These cyclotides, which originate from plants, have been used to manipulate the protein structures in order to make them resistant to unnatural conditions such as high pressure, temperature, enzymatic digestion, or other stresses (Colgrave and Craik 2004). Additionally, these cysteine-rich cyclotides also provide an interesting scaffold for peptide-based drug designing, having better stability and pharmacokinetic properties (Poth et al. 2013; Ackerman et al. 2014; Henriques et al. 2015). Thus, these approaches promise to provide a hope for disease treatment strategies without development of resistance.

6 Conclusion

A major reason why cancer kills approximately 8.2 million people, AIDS 1.2 million, tuberculosis 1.5 million, and malaria 483,000 people every year, worldwide, is that emerging drug resistance is making most of the diseases untreatable. This is despite the fact that potent therapeutics were developed against all of these and many other disease forms. To stop the spread of drug-resistant disease forms, there is an urgent need to employ interdisciplinary approaches. Synthetic biologists who are advocating such synthetic solutions for societal benefits face a lot of early challenges. Initial apprehension toward synthetic biology is fading way, and soon the early synthetic therapeutic solutions will reach the clinics to benefit human life. It will take some time before synthetic solutions against drug-resistant forms will become a reality for patients. Further, for research purposes, it will be interesting to find the response of drug-resistant disease agents against these novel solutions. Combination of traditional knowledge, modern medicine, and synthetic solutions will help in finding better solutions against cancer, infectious disease, and other deadly disorders.

References

Ackerman SE, Currier NV, Bergen JM, Cochran JR (2014) Cystine-knot peptides: emerging tools for cancer imaging and therapy. Expert Rev Proteomics 11:561–572. doi:10.1586/14789450.2014.932251

Altimari JM, Hockey SC, Boshoff HI, Sajid A, Henderson LC (2015) Novel 1,4-substituted-1,2,3-triazoles as antitubercular agents. ChemMedChem 10:787–791. doi:10.1002/cmdc.201500051

Arora G, Sajid A, Gupta M, Bhaduri A, Kumar P, Basu-Modak S, Singh Y (2010) Understanding the role of PknJ in *Mycobacterium tuberculosis*: biochemical characterization and identification of novel substrate pyruvate kinase A. PLoS One 5:e10772. doi:10.1371/journal.pone.0010772

Burbelo PD, Issa AT, Ching KH, Cohen JI, Iadarola MJ, Marques A (2010) Rapid, simple, quantitative, and highly sensitive antibody detection for lyme disease. Clin Vaccine Immunol 17:904–909. doi:10.1128/CVI.00476-09

Burfield L, Burden AD (2013) Psoriasis. J R Coll Phys Edinb 43:334–338. doi:10.4997/JRCPE.2013.414

Busard C, Zweegers J, Limpens J, Langendam M, Spuls PI (2014) Combined use of systemic agents for psoriasis: a systematic review. JAMA Dermatol 150:1213–1220. doi:10.1001/jamadermatol.2014.1111

Chakravarti D, Wong WW (2015) Synthetic biology in cell-based cancer immunotherapy. Trends Biotechnol 33:449–461. doi:10.1016/j.tibtech.2015.05.001

Chanishvili N (2012) Phage therapy--history from Twort and d'Herelle through Soviet experience to current approaches. Adv Virus Res 83:3–40. doi:10.1016/B978-0-12-394438-2.00001-3

Chen R (2015) The sweet branch of metabolic engineering: cherry-picking the low-hanging sugary fruits. Microb Cell Fact 14:197. doi:10.1186/s12934-015-0389-z

Christiansen MN, Chik J, Lee L, Anugraham M, Abrahams JL, Packer NH (2014) Cell surface protein glycosylation in cancer. Proteomics 14:525–546. doi:10.1002/pmic.201300387

Clokie MR, Millard AD, Letarov AV, Heaphy S (2011) Phages in nature. Bacteriophage 1:31–45. doi:10.4161/bact.1.1.14942

Colgrave ML, Craik DJ (2004) Thermal, chemical, and enzymatic stability of the cyclotide kalata B1: the importance of the cyclic cystine knot. Biochemistry 43:5965–5975. doi:10.1021/bi049711q

Craik DJ, Cemazar M, Wang CK, Daly NL (2006) The cyclotide family of circular miniproteins: nature's combinatorial peptide template. Biopolymers 84:250–266. doi:10.1002/bip.20451

Delves PJ (1998) The role of glycosylation in autoimmune disease. Autoimmunity 27:239–253. doi:10.3109/08916939808993836

d'Espaux L, Mendez-Perez D, Li R, Keasling JD (2015) Synthetic biology for microbial production of lipid-based biofuels. Curr Opin Chem Biol 29:58–65. doi:10.1016/j.cbpa.2015.09.009

Drobniewski F, Cooke M, Jordan J, Casali N, Mugwagwa T, Broda A, Townsend C, Sivaramakrishnan A, Green N, Jit M, Lipman M, Lord J, White PJ, Abubakar I (2015) Systematic review, meta-analysis and economic modelling of molecular diagnostic tests for antibiotic resistance in tuberculosis. Health Technol Assess 19:1–188, vii–viii. doi:10.3310/hta19340

Freeze HH, Eklund EA, Ng BG, Patterson MC (2015) Neurological aspects of human glycosylation disorders. Annu Rev Neurosci 38:105–125. doi:10.1146/annurev-neuro-071714-034019

Funakoshi Y, Suzuki T (2009) Glycobiology in the cytosol: the bitter side of a sweet world. Biochim Biophys Acta 1790:81–94. doi:10.1016/j.bbagen.2008.09.009

Glasscock CJ, Lucks JB, DeLisa MP (2016) Engineered protein machines: emergent tools for synthetic biology. Cell Chem Biol 23:45–56. doi:10.1016/j.chembiol.2015.12.004

Goulabchand R, Vincent T, Batteux F, Eliaou JF, Guilpain P (2014) Impact of autoantibody glycosylation in autoimmune diseases. Autoimmun Rev 13:742–750. doi:10.1016/j.autrev.2014.02.005

Green AA, Silver PA, Collins JJ, Yin P (2014) Toehold switches: de-novo-designed regulators of gene expression. Cell 159:925–939. doi:10.1016/j.cell.2014.10.002

Guerra-Doce E (2015) Psychoactive substances in prehistoric times: examining the archaeological evidence. Time Mind 8:91–112. doi:10.1080/1751696X.2014.993244

Gupta MA, Simpson FC, Gupta AK (2015) Psoriasis and sleep disorders: a systematic review. Sleep Med Rev 29:63–75. doi:10.1016/j.smrv.2015.09.003

Hatfull GF (2014a) Molecular genetics of mycobacteriophages. Microbiol Spectr 2:1–36. doi:10.1128/microbiolspec.MGM2-0032-2013

Hatfull GF (2014b) Mycobacteriophages: windows into tuberculosis. PLoS Pathog 10:e1003953. doi:10.1371/journal.ppat.1003953

Hedt BL, Laufer MK, Cohen T (2011) Drug resistance surveillance in resource-poor settings: current methods and considerations for TB, HIV, and malaria. Am J Trop Med Hyg 84:192–199. doi:10.4269/ajtmh.2011.10-0363

Henriques ST, Huang YH, Chaousis S, Sani MA, Poth AG, Separovic F, Craik DJ (2015) The prototypic cyclotide Kalata B1 has a unique mechanism of entering cells. Chem Biol 22:1087–1097. doi:10.1016/j.chembiol.2015.07.012

Her M, Kavanaugh A (2016) Alterations in immune function with biologic therapies for autoimmune disease. J Allergy Clin Immunol 137:19–27. doi:10.1016/j.jaci.2015.10.023

Jassim SA, Limoges RG (2014) Natural solution to antibiotic resistance: bacteriophages 'The Living Drugs'. World J Microbiol Biotechnol 30:2153–2170. doi:10.1007/s11274-014-1655-7

Kalos M, Levine BL, Porter DL, Katz S, Grupp SA, Bagg A, June CH (2011) T cells with chimeric antigen receptors have potent antitumor effects and can establish memory in patients with advanced leukemia. Sci Transl Med 3:95ra73. doi:10.1126/scitranslmed.3002842

Keasling JD (2010) Manufacturing molecules through metabolic engineering. Science 330:1355–1358. doi:10.1126/science.1193990

Lang K, Chin JW (2014) Cellular incorporation of unnatural amino acids and bioorthogonal labeling of proteins. Chem Rev 114:4764–4806. doi:10.1021/cr400355w

Lang K, Davis L, Chin JW (2015) Genetic encoding of unnatural amino acids for labeling proteins. Methods Mol Biol 1266:217–228. doi:10.1007/978-1-4939-2272-7_15

Laxminarayan R, Duse A, Wattal C, Zaidi AK, Wertheim HF, Sumpradit N, Vlieghe E, Hara GL, Gould IM, Goossens H, Greko C, So AD, Bigdeli M, Tomson G, Woodhouse W, Ombaka E, Peralta AQ, Qamar FN, Mir F, Kariuki S, Bhutta ZA, Coates A, Bergstrom R, Wright GD, Brown ED, Cars O (2013) Antibiotic resistance-the need for global solutions. Lancet Infect Dis 13:1057–1098. doi:10.1016/S1473-3099(13)70318-9

Leekha S, Terrell CL, Edson RS (2011) General principles of antimicrobial therapy. Mayo Clin Proc 86:156–167. doi:10.4065/mcp.2010.0639

Lim WA (2010) Designing customized cell signalling circuits. Nat Rev Mol Cell Biol 11:393–403. doi:10.1038/nrm2904

Lowes MA, Suarez-Farinas M, Krueger JG (2014) Immunology of psoriasis. Annu Rev Immunol 32:227–255. doi:10.1146/annurev-immunol-032713-120225

Lu TK, Collins JJ (2009) Engineered bacteriophage targeting gene networks as adjuvants for antibiotic therapy. Proc Natl Acad Sci USA 106:4629–4634. doi:10.1073/pnas.0800442106

Lucks JB, Qi L, Whitaker WR, Arkin AP (2008) Toward scalable parts families for predictable design of biological circuits. Curr Opin Microbiol 11:567–573. doi:10.1016/j.mib.2008.10.002

Marais BJ (2016) The global tuberculosis situation and the inexorable rise of drug-resistant disease. Adv Drug Deliv Rev. doi:10.1016/j.addr.2016.01.021

McFarlane M (1991) It's time to manage your time. Dermatol Nurs 3(172):182

McNerney R, Wilson SM, Sidhu AM, Harley VS, al Suwaidi Z, Nye PM, Parish T, Stoker NG (1998) Inactivation of mycobacteriophage D29 using ferrous ammonium sulphate as a tool for the detection of viable *Mycobacterium smegmatis* and *M. tuberculosis*. Res Microbiol 149:487–495. doi:10.1016/S0923-2508(98)80003-X

Moura M, Finkle J, Stainbrook S, Greene J, Broadbelt LJ, Tyo KE (2016) Evaluating enzymatic synthesis of small molecule drugs. Metab Eng 33:138–147. doi:10.1016/j.ymben.2015.11.006

Pai M, Kalantri S, Pascopella L, Riley LW, Reingold AL (2005) Bacteriophage-based assays for the rapid detection of rifampicin resistance in *Mycobacterium tuberculosis*: a meta-analysis. J Infect 51:175–187. doi:10.1016/j.jinf.2005.05.017

Pardee K, Green AA, Ferrante T, Cameron DE, DaleyKeyser A, Yin P, Collins JJ (2014) Paper-based synthetic gene networks. Cell 159:940–954. doi:10.1016/j.cell.2014.10.004

Park DJ, Drobniewski FA, Meyer A, Wilson SM (2003) Use of a phage-based assay for phenotypic detection of mycobacteria directly from sputum. J Clin Microbiol 41:680–688. doi:10.1128/JCM.41.2.680-688.2003

Pope WH, Hatfull GF (2015) Adding pieces to the puzzle: new insights into bacteriophage diversity from integrated research-education programs. Bacteriophage 5:e1084073. doi:10.1080/21597081.2015.1084073

Poth AG, Chan LY, Craik DJ (2013) Cyclotides as grafting frameworks for protein engineering and drug design applications. Biopolymers 100:480–491. doi:10.1002/bip.22284

Rifat M, Hall J, Oldmeadow C, Husain A, Milton AH (2015) Health system delay in treatment of multidrug resistant tuberculosis patients in Bangladesh. BMC Infect Dis 15:526. doi:10.1186/s12879-015-1253-9

Rybniker J, Kramme S, Small PL (2006) Host range of 14 mycobacteriophages in *Mycobacterium ulcerans* and seven other mycobacteria including *Mycobacterium tuberculosis*--application for identification and susceptibility testing. J Med Microbiol 55:37–42. doi:10.1099/jmm.0.46238-0

Sajid A, Arora G, Singhal A, Kalia VC, Singh Y (2015) Protein phosphatases of pathogenic bacteria: role in physiology and virulence. Annu Rev Microbiol 69:527–547. doi:10.1146/annurev-micro-020415-111342

Salmond GP, Fineran PC (2015) A century of the phage: past, present and future. Nat Rev Microbiol 13:777–786. doi:10.1038/nrmicro3564

Samaddar S, Grewal RK, Sinha S, Ghosh S, Roy S, Das Gupta SK (2015) Dynamics of mycobacteriophage-mycobacterial host interaction: evidence for secondary mechanisms for host lethality. Appl Environ Microbiol 82:124–133. doi:10.1128/AEM.02700-15

Schofield DA, Sharp NJ, Westwater C (2012) Phage-based platforms for the clinical detection of human bacterial pathogens. Bacteriophage 2:105–283. doi:10.4161/bact.19274

Schukur L, Geering B, Charpin-El Hamri G, Fussenegger M (2015) Implantable synthetic cytokine converter cells with AND-gate logic treat experimental psoriasis. Sci Transl Med 7:318ra201. doi:10.1126/scitranslmed.aac4964

Schukur L, Geering B, Fussenegger M (2016) Human whole-blood culture system for ex vivo characterization of designer-cell function. Biotechnol Bioeng 113:588–597. doi:10.1002/bit.25828

Schutz V, Mootz HD (2014) Click-tag and amine-tag: chemical tag approaches for efficient protein labeling in vitro and on live cells using the naturally split Npu DnaE intein. Angew Chem Int Ed Engl 53:4113–4117. doi:10.1002/anie.201309396

Seed KD (2015) Battling phages: how bacteria defend against viral attack. PLoS Pathog 11: e1004847. doi:10.1371/journal.ppat.1004847

Singhal A, Arora G, Virmani R, Kundu P, Khanna T, Sajid A, Misra R, Joshi J, Yadav V, Samanta S, Saini N, Pandey AK, Visweswariah SS, Hentschker C, Becher D, Gerth U, Singh Y (2015) Systematic analysis of mycobacterial acylation reveals first example of acylation-mediated regulation of enzyme activity of a bacterial phosphatase. J Biol Chem 290:26218–26234. doi:10.1074/jbc.M115.687269

Smartt AE, Xu T, Jegier P, Carswell JJ, Blount SA, Sayler GS, Ripp S (2012) Pathogen detection using engineered bacteriophages. Anal Bioanal Chem 402:3127–3146. doi:10.1007/s00216-011-5555-5

Smith HO, Hutchison CA 3rd, Pfannkoch C, Venter JC (2003) Generating a synthetic genome by whole genome assembly: phiX174 bacteriophage from synthetic oligonucleotides. Proc Natl Acad Sci USA 100:15440–15445. doi:10.1073/pnas.2237126100

Stauss HJ, Morris EC (2013) Immunotherapy with gene-modified T cells: limiting side effects provides new challenges. Gene Ther 20:1029–1032. doi:10.1038/gt.2013.34

Sugimoto K, Yasujima M, Yagihashi S (2008) Role of advanced glycation end products in diabetic neuropathy. Curr Pharm Des 14:953–961. doi:10.2174/138161208784139774

Toussaint AJ, Muschel LH (1959) Neutralization of bacteriophage by normal serum. Nature 183:1825–1827. doi:10.1038/1831825a0

Trosset JY, Carbonell P (2015) Synthetic biology for pharmaceutical drug discovery. Drug Des Devel Ther 9:6285–6302. doi:10.2147/DDDT.S58049

Twort FW (1915) An investigation on the nature of ultra-microscopic viruses. Lancet 2:1241–1243. doi:10.1016/S0140-6736(01)20383-3

Varki A, Kannagi R, Toole BP (2009) Glycosylation changes in cancer. In: Varki A, Cummings RD, Esko JD, et al (eds) Essentials of glycobiology, Chap 44, 2nd edn. Cold Spring Harbor Laboratory Press, New York. http://www.ncbi.nlm.nih.gov/books/NBK1963/

Velasquez GE, Becerra MC, Gelmanova IY, Pasechnikov AD, Yedilbayev A, Shin SS, Andreev YG, Yanova G, Atwood SS, Mitnick CD, Franke MF, Rich ML, Keshavjee S (2014) Improving outcomes for multidrug-resistant tuberculosis: aggressive regimens prevent treatment failure and death. Clin Infect Dis 59:9–15. doi:10.1093/cid/ciu209

Weber W, Schoenmakers R, Keller B, Gitzinger M, Grau T, Daoud-El Baba M, Sander P, Fussenegger M (2008) A synthetic mammalian gene circuit reveals antituberculosis compounds. Proc Natl Acad Sci USA 105:9994–9998. doi:10.1073/pnas.0800663105

Wei H (2015) Bacteriophages, revitalized after 100 years in the shadow of antibiotics. Virol Sin 30:1–2. doi:10.1007/s12250-014-3562-y

WHO global tuberculosis report (2015) http://apps.who.int/iris/bitstream/10665/191102/1/9789241565059_eng.pdf?ua=1

Wilson JW, Tsukayama DT (2016) Extensively drug-resistant tuberculosis: principles of resistance, diagnosis, and management. Mayo Clin Proc. doi:10.1016/j.mayocp.2016.01.014

Wright GD (2012) Antibiotics: a new hope. Chem Biol 19:3–10. doi:10.1016/j.chembiol.2011.10.019

Wu CY, Roybal KT, Puchner EM, Onuffer J, Lim WA (2015) Remote control of therapeutic T cells through a small molecule-gated chimeric receptor. Science 350:aab4077. doi:10.1126/science.aab4077

Yates TA, Khan PY, Knight GM, Taylor JG, McHugh TD, Lipman M, White RG, Cohen T, Cobelens FG, Wood R, Moore DA, Abubakar I (2016) The transmission of *Mycobacterium tuberculosis* in high burden settings. Lancet Infect Dis 16:227–238. doi:10.1016/S1473-3099(15)00499-5

Future of Drug Discovery

Ahmed Kamal, Shalini Nekkanti, Nagula Shankaraiah, and Manda Sathish

Abstract Over the past few decades, tremendous effort has been put into developing agents to address the multidrug resistance (MDR) mechanisms. Lately, pharmaceutical industry is under pressure owing to the reduced output of new drugs, rising research and development costs, current economic recession, and stricter regulatory guidelines. This chapter highlights the future of drug discovery particularly with reference to overcoming drug resistance and other related challenges by improving the efficiency of this process.

1 Introduction

The decade of 1990s has yielded several blockbuster drugs which resulted in the advancement in healthcare, quality of life, and life expectancy (Munos 2009). Lately, the pharmaceutical industry is undergoing unprecedented transformation, mostly due to the reduced output of new drugs, rising research and development costs, current economic recession, drug pricing pressures, and stricter regulatory guidelines. This puts burden on all pharmaceutical companies for innovation to increase their output of new drugs, improve efficiency, and boost R&D productivity in a cost-effective manner. Another critical problem faced by the pharmaceutical

A. Kamal (✉)
Medicinal Chemistry and Pharmacology, CSIR-Indian Institute of Chemical Technology, Hyderabad 500 007, India

Department of Medicinal Chemistry, National Institute of Pharmaceutical Education and Research (NIPER), Hyderabad 500 037, India
e-mail: ahmedkamal915@gmail.com

S. Nekkanti • N. Shankaraiah
Department of Medicinal Chemistry, National Institute of Pharmaceutical Education and Research (NIPER), Hyderabad 500 037, India

M. Sathish
Medicinal Chemistry and Pharmacology, CSIR-Indian Institute of Chemical Technology, Hyderabad 500 007, India

G. Arora et al. (eds.), *Drug Resistance in Bacteria, Fungi, Malaria, and Cancer*,
DOI 10.1007/978-3-319-48683-3_27

industry is the emergence of multidrug resistance (MDR) in tumor cells and microorganisms. It involves the simultaneous development of resistance to the drug originally used and also to other related chemicals, resulting in lower intra-cellular concentrations. New molecules that can inhibit MDR are expected from diverse sources, and the future will witness the increased screening of molecules from a vast range of settings for this purpose.

2 The Emerging Horizons of Drug Discovery

Contrast in the trends of investment and productivity has made the major pharma sectors to reconsider their objectives and their interactions with one another (Cressey 2011; Maxmen 2011). A changing paradigm for the drug discovery is emerging which involves a highly efficient strategy compared to the traditional model (Khanna 2012; Kamal et al. 2014).

2.1 Partnering Initiatives/Collaborations

Traditionally, the academia, small biotech, and large pharma companies worked independent of each other. However, the continual financial strain in the academic and biotech sectors as well as low output from large pharma industries has led to their collaboration to overcome innovation hurdles and promote success in drug development. In such collaborations, the academia and small biotech companies generate hits, whereas pharma industries develop these hits into drug candidates (Mullard 2012). For example, the Manchester Collaborative Centre for Inflammation Research (MCCIR) was formed by GlaxoSmithKline (GSK), AstraZeneca, and the University of Manchester with the aim to develop new drugs for inflammatory diseases. Further, the pharma industry is venturing into open innovation or "crowdsourcing" model which aims to utilize the joint expertise of a "crowd" of external scientists. Lilly has initiated some open innovation projects, such as the web-based solution providers InnoCentive and YourEncore. It is predicted that variants of these partnering models will be adopted in pharma industry in the near future.

2.2 New Molecular Entities (NMEs) and Novel Biological Entities (NBEs)

A ratio of 30:70 for the first-time to follow-on approaches is considered to be the best in the drug discovery field. Moreover, when compared to blockbuster drugs, the addition of smaller market size prospects can decrease risk and develop sustainability. The life cycle management prospects consisting of drug combinations, stereoisomers, polymorphs, and drug delivery systems are becoming important strategies to maximize profit.

The last decade has seen huge R&D investments by large pharma companies in the field of biologics. The drug approval rates of biologics are also higher compared to small molecules due to higher target specificity and efficacy in unmet medical needs. The proprietary protection is ending for the biologics in the market, and they shall experience severe competition from biosimilars. Unlike the highly competitive generic market for small molecules, the competition is lower for biologics as they are considered as specialty products and NBEs can claim a price of almost 60–80 % of the existing product.

2.3 Generics and Supergenerics

Currently, the market quota for generics is estimated to rise from 10 % of the global pharmaceutical sales due to the projected patent expiry of proprietary R&D products. Numerous generic companies are entering into partnerships with R&D labs to create "supergenerics" as alternates to marketed generics. They can be fixed dose combinations, dosage variations, polymorphs, stereoisomers, new formulations, and delivery systems and offer improved efficacy, reduced side effects, and challenge to life cycle management approaches of proprietary products. In 2010, Sandoz has launched enoxaparin, a generic version of Sanofi's Lovenox, which surfaced as the first "generic blockbuster" with $531 million sales in the first half of 2011 (Thayer 2011).

2.4 Drug Repositioning

During the past decade, the drug repositioning methods for finding innovative uses of current or discontinued drugs have garnered popularity as they offer diminished development time, cost, and safety profile uncertainty. Researchers look for potential targets for repurposing of existing combinatorial libraries originally intended for different targets (Reymond et al. 2010). Consequently, there is increased interest to systematically search targets for potent herbal remedies employed in traditional Chinese medicine (TCM) using computational approaches (Chen 2011). The research area of "foodinformatics" involves searching for prospective applications of food chemicals via computational methods (Medina-Franco et al. 2012). "Nutritional epigenomics" has emerged as a consequence of amplified interest in studying the role of diet in controlling epigenetic events (Szic et al. 2010). Flavoring substances in the "generally recognized as safe" (GRAS) list of food materials are beginning to appear as a source of drug candidates (Burdock et al. 2006).

2.5 Mechanistic Systems Modeling

The two mechanistic systems modeling methods employed in a number of therapeutic areas include phenotype-driven models of disease and genome-scale metabolic models (GEMs). The phenotype-driven models are developed to ascertain the dependence of clinical outputs on disease mechanisms, to replicate the effects of current therapies and drug candidates, and to facilitate in silico target screening. One of the most successful investigations that could be carried out with phenotype-driven models was that of rheumatoid arthritis (Rullmann et al. 2005). Alternatively, GEMs represent the most comprehensive database accessible in a computable configuration for immediate study, especially in cancer (Duarte et al. 2007). System models need to be practical across the research pipeline to help overcome the productivity challenges faced by the pharmaceutical industry.

2.6 Molecular Networks

Integration of specific network models may significantly enhance the outcome of the drug development process by identifying numerous novel drug targets (Csermely et al. 2013). Network robustness has been identified as the main culprit in the development of drug resistance, where unconventional cellular pathways are triggered, to counter the effects of drug action (Kitano 2007). Co-targeting of another critical level of drug-affected networks is an effective means to overcome resistance (Zimmermann et al. 2007). Comparison of molecular network pathways of resistant pathogens with those of normal pathogens aids in the identification of important drug-resistant targets and co-targets (Kim et al. 2010).

2.7 Chemogenomics

The mounting proof for polypharmacology (i.e., therapeutic effects are due to the interaction of drugs with multiple targets) is enabling the shift toward multitarget approaches (Jacoby 2011). Ideally, polypharmacology can be completely understood if there was readily accessible data interlinking the whole ligand and target spaces. Thus, chemogenomics strives to determine a number of probable ligands for most of the probable targets. Polypharmacological activity of bioactive compounds can be predicted through either the ligand- or the structure-based methods (Koutsoukas et al. 2011). Moreover, computational advances are being exploited to study the SAR of chemogenomic data sets (Lounkine et al. 2012) and investigate prospective targets for probes employed in chemical biology (Gregori-Puigjané et al. 2012).

2.8 *Phenotypic Screening*

In the phenotypic screening approach or "forward pharmacology," a characteristic related to the disease is exploited to develop a cell-based assay (Vogt and Lazo 2005). Compounds are then screened in the assay to identify active lead compounds even without the knowledge of the molecular mechanism and protein target. This approach can have a practical role in drug discovery for numerous rare diseases in which a drug target has not been validated and can also be useful in the discovery of novel drug targets (Swinney and Anthony 2011). The development of robotic screening techniques and highly sensitive detection systems allows phenotypic assays to be miniaturized and facilitates rapid screening of large compound libraries.

2.9 *Peptide Therapeutics*

At present, there are approximately 500 and 140 peptide drugs in preclinical development and clinical trials, respectively, with more than 60 US Food and Drug Administration (FDA)-approved ones on the market (Kaspar and Reichert 2013). Multifunctional peptides represent more than one therapeutic effect (polypharmacology), which enables the possibility of more personalized treatment of diverse patient groups. Recently, the GLP-1 (glucagon-like peptide-1) agonist, Tanzeum™ (albiglutide), was launched in the multifunctional peptide field with great commercial success. "Cell-penetrating peptides," such as "penetratin," enable peptides previously limited to extracellular targets to reach intracellular targets. Finally, the conjugation of small molecules, oligoribonucleotides, and antibodies to peptides facilitates the development of peptide therapeutics with enhanced safety and efficacy. This principle has been applied for the conjugation of a radioactive ligand to neurotensin 1 (NT1) receptor peptide agonist for the treatment of pancreatic cancer (Okarvi 2008).

3 Novel Approaches to Battle Multidrug Resistance (MDR) in Cancer

The side effects associated with traditional MDR inhibitors have resulted in the development of numerous alternatives. Modern advances in antisense oligonucleotide technologies have led to a more specific manner to handle MDR through downregulation of ABC transporter proteins. MDR mechanisms are an essential part of normal physiology of living cells; hence, scientists prefer circumventing rather than inhibiting them. It is anticipated that the elucidation of resistance mechanisms arising in the "targetable" tumors will aid in the development of rational therapeutic combinations.

3.1 Targeting Glc-Cer Synthase (GCS)

Ceramide is a secondary messenger in apoptotic signaling pathways, where a reduction in its synthesis enhances cellular resistance to apoptosis (Liu et al. 2000). It was demonstrated that elevated Glc-Cer synthase (GCS) activity results in glucosylceramide (GC), which accumulates in multidrug-resistant tumors that are less receptive to chemotherapy. Stimulation of ceramide glycosylation leads to drug resistance, whereas its blockade enhances the sensitivity of cancer cells to cytotoxics (Lucci et al. 1999). In cancer cell models, drug combinations that stimulate ceramide production and block glycosylation were found to augment the efficacy of chemotherapy (Lavie et al. 1997). In adriamycin-resistant breast cancer cells, downregulation of ceramide glycosylation reinstated cell sensitivity to adriamycin (Liscovitch and Lavie 2002). Thus, GCS leads to drug resistance by diminishing drug-induced formation of apoptotic ceramide and implies a new drug target in cancer MDR (Senchenkov et al. 2001).

3.2 Multidrug-Resistant Bone Marrow Stem Cells

Cancer stem cells are innately resistant to cancer drugs owing to their dormancy, ability to repair DNA, and expression of ABC transporters. The use of anticancer drugs is highly influenced by a dose-limiting toxicity factor, i.e., avoiding suppression of bone marrow stem cells. Recently, multiple drug-resistant stem cells from bone marrow have been developed (that carry vectors with the MDR1 cDNA), which permitted treatment at otherwise lethal doses (Gottesman et al. 2002).

Studies have shown that stem cells overexpress ATP-binding cassette subfamily G member 2 (ABCG2), rather than the most clinically targeted ATP-binding cassette subfamily B member 1 (ABCB1) (Hirschmann-Jax et al. 2004). During chemotherapy, administration of ABCG2 inhibitors might aid in eliminating tumor stem cells. Dual ABCG2 and ABCB1 inhibitors, elacridar and tariquidar (1 and 2, respectively, Fig. 1), are approved for clinical trials, and more ABCG2 inhibitors are under development (Sparreboom et al. 1999). Antibodies against ABCG2 might be valuable in killing cancer stem cells by delivering toxins. The antibodies may also be used diagnostically in detecting tumors, visualizing metastasis, and monitoring therapeutic response.

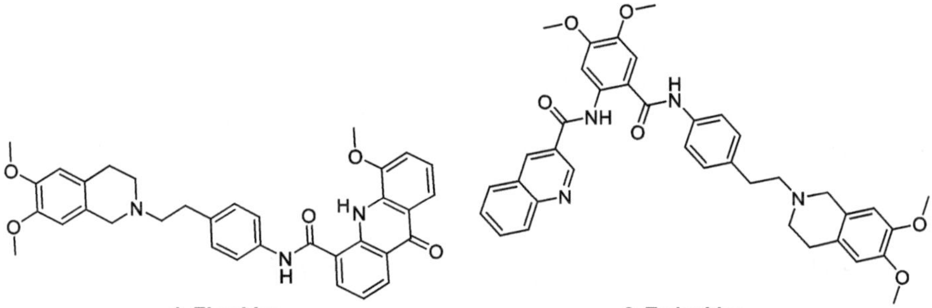

1, Elacridar **2, Tariquidar**

Fig. 1 Dual ABCG1 and ABCB2 inhibitors

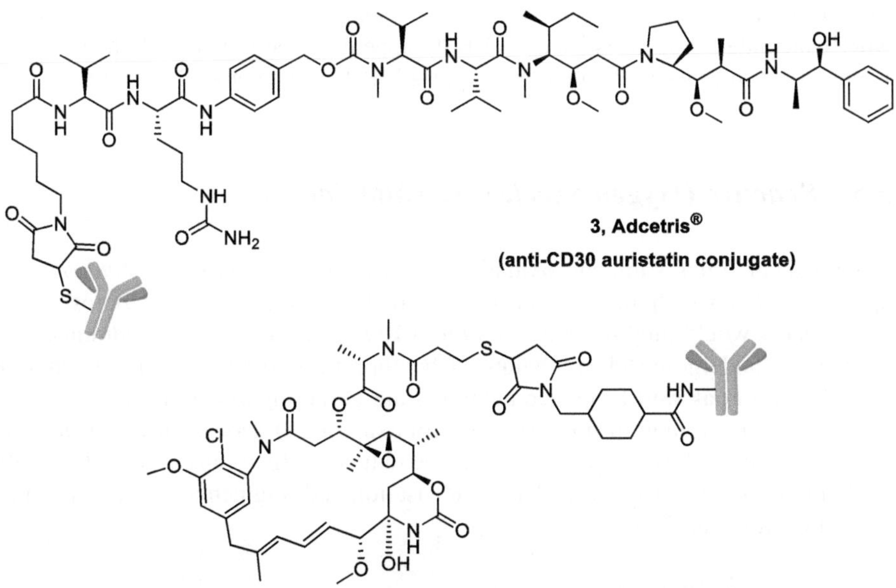

3, Adcetris®
(anti-CD30 auristatin conjugate)

4, Kadcycla®
(anti-Her2 maytansine conjugate)

Fig. 2 Representative antibody-drug conjugate (ADC) structures

3.3 Antibody-Drug Conjugates (ADCs)

Antibody-drug conjugates (ADCs) exploit the specificity of monoclonal antibodies (mAbs) to deliver drugs to antigen-expressing cancer cells in a selective manner. Recent advances in ADC research led to the faster FDA approval of Adcetris® (brentuximab vedotin, 3, Fig. 2) that can treat Hodgkin's lymphoma as well as anaplastic large-cell lymphomas (Younes et al. 2010; Senter and Sievers 2012).

In 2013, Kadcyla® (adotrastuzumab emtansine, T-DM1, 4, Fig. 2), which links trastuzumab and maytansine (a potent antimicrotubule agent), was approved to treat Her2-positive breast cancer (Verma et al. 2012). Labetuzumab-SN-38, which combines cathepsin B-cleavable linker and SN-38 (the active metabolite of irinotecan), has proceeded to phase II trials for colorectal cancer (Segal et al. 2013).

3.4 Inhibition of Glutathione (GSH) Synthesis

The precise mechanism of MRP1 in MDR has not been proved yet, though glutathione (GSH) is expected to be involved. N-acetylcysteine (NAC) and DL-buthionine (*S*, *R*)-sulfoximine (BSO) are a pro-glutathione drug and an inhibitor of GSH synthesis, respectively. They display mutually reverse response in MRP1-mediated drug resistance, and BSO appears to be a potential chemotherapeutic agent in cancer cells that overexpress MRP1 (Akan et al. 2004, 2005).

3.5 Reactive Oxygen Species vs. Antioxidants

Many cancer drugs stimulate oxidative stress by production of reactive oxygen species (ROS) which might contribute to their cytotoxicity via apoptosis. So, antioxidants would inhibit the generation of ROS and should not be administered in patients undergoing chemotherapy. Chemopreventive agents that can inhibit ROS formation as well as induce apoptosis are garnering attention. Interestingly, novel catechins from grape procyanidins, for example, 4b-(S-cysteinyl)epicatechin 3-O-gallate, have the ability to scavenge free radicals, which can cause S-phase cell cycle inhibition, and trigger nuclear condensation and fragmentation in melanoma cells (Lozano et al. 2005).

3.6 Plant Flavonoids

Large amount of data exists on the cytotoxic activity of plant flavonoids involving inhibition of cell growth and activity of kinase, induction of apoptosis, and inhibition of the secretion of matrix metalloproteinases, angiogenesis, and tumor invasive nature (Kanadaswami et al. 2005). Furthermore, their anticancer activities depend upon the hydroxylation of the flavones B ring, such as luteolin and quercetin. Further elucidation of the mechanisms underlying their activity and in vivo studies are essential to develop anticancer therapeutics based on flavonoids. Antioxidants can exhibit diverse effects on the efficacy of anticancer drugs depending upon their additional pharmacological properties that even dominate their antioxidant effects.

3.7 *Alternative Kinase Inhibitors*

Knowledge of the drug resistance mutational mechanisms has aided in the discovery and study of alternative kinase inhibitors. For instance, dasatinib was determined to be effective against imatinib-resistant mutations and linked to a considerably higher rate of cytogenetic and molecular responses when compared with imatinib (Bradeen et al. 2006; Kantarjian et al. 2010). An alternative strategy involves targeting the (onco) protein through the next-generation selective inhibitors such as nilotinib. Nilotinib not only shows activity against many imatinib resistance mutations but also is associated with a significantly higher rate of major molecular response (Saglio et al. 2010). Structural insights into gatekeeper residue (e.g., T315I) mutation have helped in the development of ABL kinase inhibitors AP24534 and HG-7-85-01, which are currently undergoing clinical development and do not lead to resistance (O'Hare et al. 2009; Weisberg et al. 2010). The design of potent kinase inhibitors that can bind to the bioactive conformation of the gatekeeper mutation should be considered during the drug discovery process.

3.8 *Epigenetic Modifications*

Epigenetic modifications that cause resistance mainly include DNA methylation and histone modification, which lead to tumor progenitor cell formation. Epigenetic switches are considered as ideal targets to overcome drug resistance, as they can simultaneously improve differentiation, suppress growth, and regulate the stage-specific development of metastatic cancer. They also make resistant cancer cells susceptible to other drugs; for example, lung cancer patients pretreated with the epigenetic drugs DAC and HDACi had lower incidences of relapse to conventional chemotherapy (Juergens et al. 2011; Sarkar et al. 2013a, b; Byler et al. 2014; Byler and Sarkar 2014). On the whole, these results indicate that epigenetic drugs in combination with conventional therapies may be valuable in the treatment of drug-resistant cancer.

3.9 *Selective CDK4 and CDK6 Inhibitors*

Cyclin-dependent kinases (CDKs) are strictly controlled enzymes that regulate all cell cycle transitions (Lim and Kaldis 2013). Recently, it has become apparent that deregulation of CDK4 and CDK6 are the major oncogenic triggers in many cancers (Scaltriti et al. 2011). Pyrido[2,3-*d*]pyrimidin-7-ones possessing a 2-aminopyridine side chain at the C2 position were identified to exhibit higher selectivity for CDK4 and CDK6 compared to other CDKs (VanderWel et al. 2005). Pfizer has developed

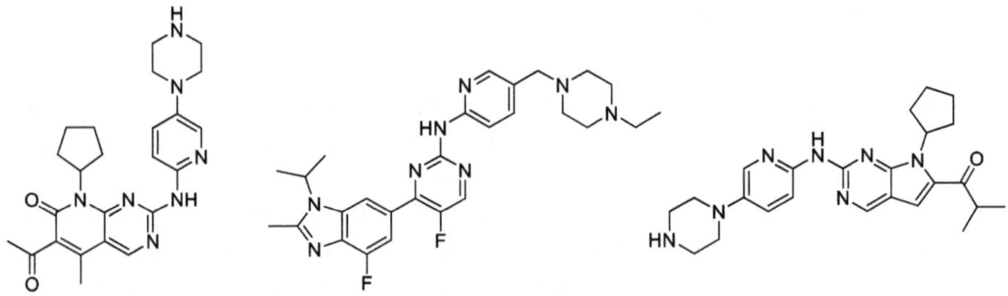

5, PD-0332991 **6, LY-2835219** **7, LEE011**

Fig. 3 Selective CDK4 and CDK6 inhibitors

one of such compound, PD-0332991 (palbociclib, 5, Fig. 3) that exhibited strong G1 arrest in preliminary studies (Fry et al. 2004; Toogood et al. 2005). Eli Lilly and Novartis have developed the drugs LY-2835219 (abemaciclib, 6) and LEE011 (7), respectively, which were presumed to bind to the ATP-binding pocket of CDK4 and CDK6 (Gelbert et al. 2014; Vora et al. 2014).

3.10 Targeting the Mitochondria

Many anticancer drugs affect pathways which occur earlier than mitochondria, which subsequently merge onto the intrinsic senescence pathway. The disruption in these upstream pathways leads to drug resistance, which can be overcome by directly targeting the mitochondria (Fulda et al. 2010). The cancer cells have elevated levels of lactate due to high metabolic activity and the activation of the membrane Na^+/H^+-antiporter (NHE). The activation of NHE results in acidic extracellular pH, which alters the cellular uptake of many chemotherapeutic drugs (Raghunand et al. 1999a, b). Hence, proton pump inhibitors (PPIs) have been shown to sensitize multidrug-resistant cells to cancer drugs (Uwagawa et al. 2010). Owing to the high energy demand of cancer cells, drugs that cause depletion of ATP may sensitize resistant cells. It was demonstrated that 2-deoxyglucose (2DG), metformin, and lonidamine (LND) in combination therapy with conventional chemotherapeutic drugs caused depletion of cellular ATP and sensitized resistant cancer cells (Floridi et al. 1981; Ben Sahra et al. 2010).

4 Antibiotic Resistance Breakers (ARBs)

Resistance to the existing antibiotics is fast escalating; therefore, strategies of breaking this resistance should be established as soon as possible. One such strategy is to coadminister suitable nonantibiotic drugs along with the antibiotic, to restore adequate therapeutic activity, for example, the successful coadministration of clavulanic acid (a β-lactamase inhibitor) with β-lactam antibiotics (Prabhudesai et al. 2011). ARBs can directly inhibit bacteria, reduce the minimum inhibitory concentration of the antibiotic, and/or modulate host defense by altering inflammation and autophagy (elimination of unwanted constituents from cells) (Brown 2015). The first priority is the development of ARBs against the four Gram-negative organisms, *Escherichia coli*, *Pseudomonas aeruginosa*, *Acinetobacter baumannii*, and *Klebsiella pneumoniae*, followed by ARBs against Gram-positive bacteria, in particular the highly lethal methicillin-resistant *Staphylococcus aureus* (MRSA) and *Clostridium difficile*. Consequently, these antibiotics need ARBs: cephalosporins and carbapenems, polymyxins, fluoroquinolones, tetracyclines and aminoglycosides, and macrolides. The drugs with the strongest evidence in breaking resistance are grouped into three classes: ARBs for Gram-positive bacteria, ARBs for Gram-negative bacteria, and ARBs for both classes.

4.1 ARBs for Gram-Negative Bacteria

Ciclopirox (8, Fig. 4) is effective against a broad range of MDR Gram-negative and Gram-positive species (Dittmar et al. 1981). Its activity against Gram-negative bacteria may be attributed to the inhibition of the surface coat lipopolysaccharide (LPS) synthesis. Ciclopirox also chelates intracellular iron (inhibition of metal-dependent enzymes) and downregulates nucleotide-binding proteins and mammalian target of rapamycin (mTOR) signaling (Niewerth et al. 2003; Dihazi et al. 2013; Zhou et al. 2014). Inhibition of LPS has the advantage of making Gram-negative species susceptible to antibiotics that are usually effective only for Gram-positive species.

Loperamide (9, Fig. 4), a μ-opioid receptor agonist, has no innate antibacterial activity but exhibits synergism with a wide range of antibiotics such as tetracyclines, cephalosporins, novobiocin, and polymyxin B (Ejim et al. 2011). Upon intravenous administration, it does not show opiate-like effects, as it cannot cross the blood-brain barrier due to efflux by P-glycoprotein. Other effects may include the modification of the small-molecule permeability of Gram-negative bacteria, causing dysregulation of the influx and efflux mechanisms and thus resulting in the accumulation of otherwise effluxed antibiotics (Tascini et al. 2013).

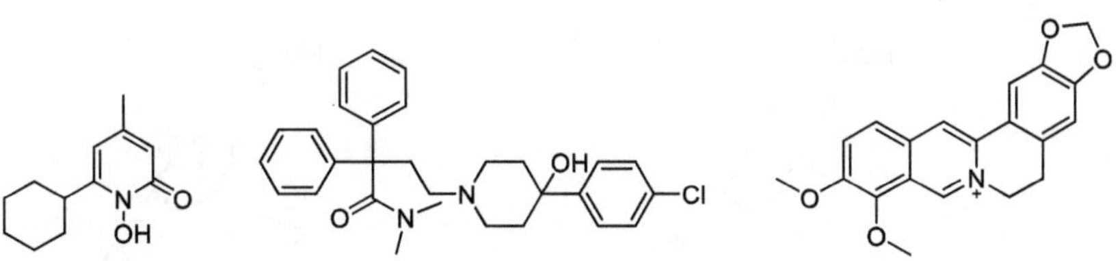

8, Ciclopirox **9, Loperamide** **10, Berberine**

Fig. 4 ARBs for Gram-negative (8, 9) and Gram-positive bacteria (10)

4.2 ARBs for Gram-Positive Bacteria

The alkaloid berberine (10, Fig. 4) is highly effective against streptococcal, staphylococcal, and enterococcal species, including MDR strains of *Mycobacterium tuberculosis* and *S. aureus*. It is also effective against *E. coli*, *Giardia lamblia*, *Vibrio cholerae*, and *Klebsiella spp*. This activity may be attributed to the inhibition of the sortase enzyme in Gram-positive bacteria (Kim et al. 2004) and cell division protein FtsZ in the Gram-negative *E. coli* (Domadia et al. 2008). Studies have suggested that berberine elevates the host defense response by suppression of pro-inflammatory responses, and bacteria are weak at developing resistance to berberine (Jeong et al. 2009).

4.3 ARBs for Both Bacterial Classes

Curcumin (11, Fig. 5), a constituent of turmeric, has direct antibiotic activity against many Gram-negative and Gram-positive species (Moghadamtousi et al. 2014). It also exhibits synergism with antibiotics of broad range against MRSA (Mun et al. 2013). The potential mechanisms of action include inhibition of sortase, autophagy by inhibition of the AKT-mTOR pathway, and modulation of host gut cytokine response (Park et al. 2005; Aoki et al. 2007).

Epigallocatechin-3-gallate (EGCG, 12, Fig. 5) is a polyphenol found in green tea and exhibits synergism with antibiotics against both Gram-positive and Gram-negative bacteria (Steinmann et al. 2013). The mechanisms through which EGCG

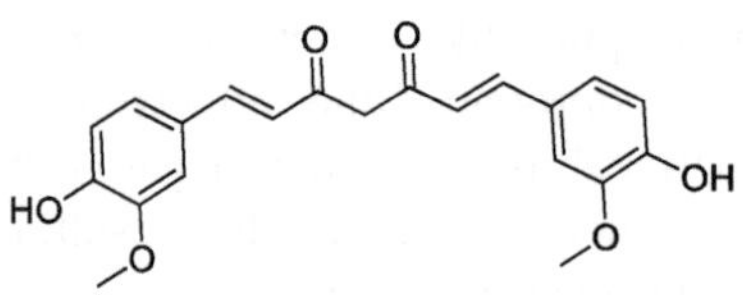

11, Curcumin

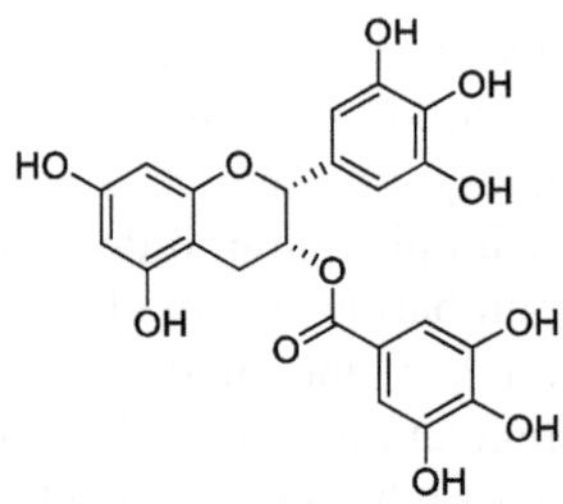

12, Epigallocatechin-3-gallate (EGCG)

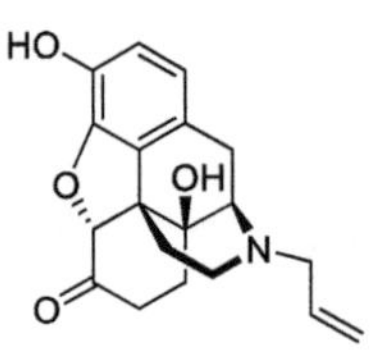

13, (+)-Naloxone

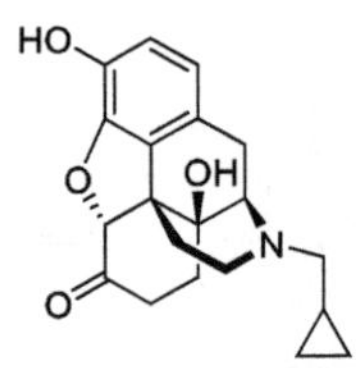

14, (+)-Naltrexone

Fig. 5 ARBs for both Gram-negative and Gram-positive bacteria

exerts its activity are very diverse owing to the presence of phenolic groups in its chemical structure capable of interacting with multiple proteins. EGCG inhibits DNA gyrase and dihydrofolate reductase in the synthesis of bacterial type II fatty acids (Zhang and Rock 2004). It protects penicillins from bacterial inactivation by inhibiting penicillinase activity (Zhao et al. 2002).

(+)-Naloxone and (+)-naltrexone (13 and 14, Fig. 5) could be co-administered with antibiotics for treating intestinal infections caused by *E. coli* or *Shigella* spp. Their ability to inhibit the release of pro-inflammatory cytokines has been exploited in the treatment of LPS-driven systemic endotoxic shock (Hutchinson et al. 2010).

No effective ARBs were identified for the fluoroquinolones and the aminoglycosides. Future research toward ARBs for these two useful classes would be greatly valuable.

5 Future Trends in Anti-HIV Drug Discovery

In spite of the availability of nearly 30 licensed anti-HIV drugs, the rapid emergence of MDR mutations requires the development of new, safe, and effective antiviral agents. The triumph of combinatorial antiretroviral therapy (cART) in the treatment of HIV infection has been short-lived due to the quick emergence of HIV strains that are multidrug resistant, poor bioavailability to host, and cumulative toxicities. Hence, there is a necessity for alternative strategies for discovery of antiretroviral drugs and additional therapeutic agents with novel action modes or targets. In recent years, several innovative strategies have been employed to discover anti-HIV agents with novel scaffolds and better aqueous solubility and resistance profiles, including fragment-based screening, privileged fragment-based reconstruction, dynamic ligation screening (DLS)-based drug discovery, rapid diversity-oriented synthesis combined with in situ screening, and hierarchical multiple-filter database searching. These strategies boost lead identification and optimization with huge potential for yielding new antiviral drugs.

5.1 Substrate Envelope Hypothesis to Reduce Drug Resistance

In the drug design for targeting rapidly evolving proteins like HIV-1 protease (HIV-1 PR), the principal concern revolves around drug resistance. The analysis of the crystal structure of HIV-1 PR in complex with its substrates suggested that the specificity for substrates depends on matching a defined shape within the binding pocket termed as the "substrate envelope." Hence, inhibitors that remain within the substrate envelope and are less likely to induce resistant mutants need to be identified (Logsdon et al. 2004; Prabu-Jeyabalan et al. 2006; Altman et al. 2008).

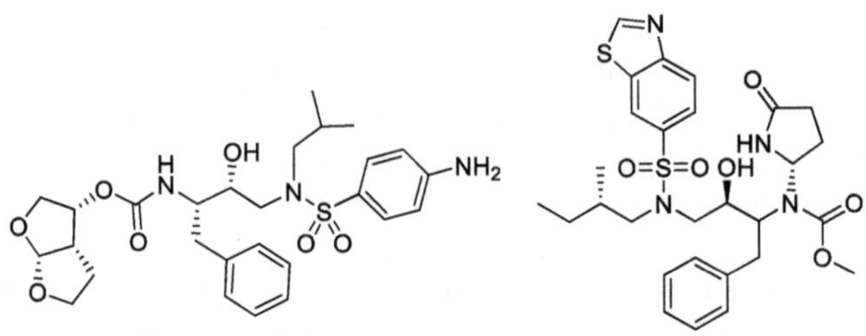

Fig. 6 Discovery of HIV-1 PR inhibitors based on the substrate envelope hypothesis

In recent years, a combination of computational structure-based design with substrate envelope strategies has been utilized to develop subnanomolar HIV-1 PR inhibitors (15 and 16, Fig. 6) toward patient-derived MDR viruses (Nalam and Schiffer 2008). Interestingly, resistance to nucleoside RT inhibitor (NRTI) drugs zidovudine and lamivudine has been attributed to their protrusion beyond the substrate envelope, creating an opportunity for HIV-1 reverse transcriptase (HIV-1 RT) to develop resistance (Shen et al. 2013). In fact, the substrate envelope hypothesis has already been applied in the discovery of tenofovir which might elude drug resistance through residing within the substrate envelope, but might still be vulnerable to resistance involving loss of key interactions (Tuske et al. 2004). Application of this hypothesis may revolutionize the efforts toward future drug design (Xue et al. 2014).

5.2 *Clinically Validated and Promising Antiretroviral Targets*

New antiretroviral drugs need to target emerging sites on clinically validated targets, alternative mechanisms, or newly emerging targets in order to circumvent MDR. There is still tremendous scope for targeting the clinically validated HIV targets (RT, IN, PR, and CCR5) through diverse mechanisms of action, such as RNase H inhibitors (Cao et al. 2014), nucleotide-competing RT inhibitors (NcRTIs) (Maga et al. 2010), allosteric IN inhibitors, and PR dimerization inhibitors (Kang et al. 2014). Further, other targets have been deemed as potentially druggable alternatives as they are essential for the key steps of viral replication such as protein-protein and protein-nucleic acid interactions (Mori et al. 2011), DNA G-quadruplex formation (Amrane et al. 2014), protein conformational transitions (Herschhorn et al. 2014), and HIV assembly (Nguyen et al. 2011). These targets include the dimerization initiation site (DIS) of the HIV-1 genomic RNA (Ennifar et al. 2013), HIV-1 matrix protein (Zentner et al. 2013), HIV-1 capsid (Dewan et al. 2012), nuclear import of pre-integration complex (Zhan et al. 2010), retroviral

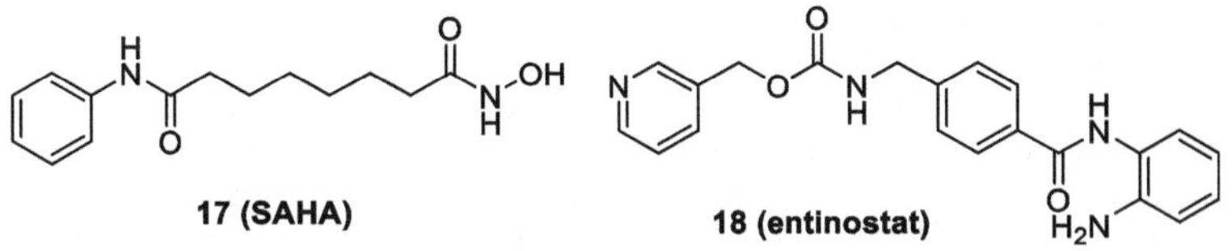

17 (SAHA) 18 (entinostat)

Fig. 7 Histone deacetylase (HDAC) inhibitors

nucleocapsid zinc fingers (Vercruysse et al. 2012), and many host antiviral restriction factors (Sloan and Wainberg 2013).

One of the several strategies that are being evaluated to eliminate latent viral reservoirs is "kick (shock)-and-kill" strategy which employs compounds that stimulate viral replication, allowing them to be eliminated by the current cART (Marsden and Zack 2013; Mbonye and Karn 2014). Histone deacetylase (HDAC) inhibitors such as vorinostat/SAHA and entinostat (17 and 18, respectively) (Fig. 7), can reactivate latent HIV in some patients, providing support for the feasibility of shock-and-kill strategy (Huber et al. 2011).

6 Perspectives

In spite of advances in drug discovery and the development of a few good MDR modulators, we are far behind in clinical application of these agents. It is clear that a combination of multiple targeted therapies will be indispensable to effectively prevent and/or treat drug-resistant cancers and microbes. Such efforts will require collaboration between academia and industry to bring together the appropriate resources and innovation. Finally, the knowledge of drug resistance thus gained and holds great promise to improve the lives of scores of patients with debilitating diseases.

Acknowledgments SN is thankful to DoP, Ministry of Chemicals and Fertilizers Govt. of India, New Delhi, for the award of Research Fellowship. NS gratefully acknowledges SERB, DST, Govt. of India for research grant (YSS-2015-001709).

References

Akan I, Akan S, Akca H, Savas B, Ozben T (2004) N-acetylcysteine enhances multidrug resistance-associated protein mediated doxorubicin resistance. Eur J Clin Invest 34:683–689. doi:10.1111/j.1365-2362.2004.01411.x

Akan I, Akan S, Akca H, Savas B, Ozben T (2005) Multidrug resistance-associated protein 1 (MRP1) mediated vincristine resistance: effects of N-acetylcysteine and buthionine sulfoximine. Cancer Cell Int 5:22. doi:10.1186/1475-2867-5-22

Altman MD, Ali A, Reddy GS, Nalam MN, Anjum SG, Cao H, Chellappan S, Kairys V, Fernandes MX, Gilson MK, Schiffer CA, Rana TM, Tidor B (2008) HIV-1 protease inhibitors from inverse design in the substrate envelope exhibit subnanomolar binding to drug-resistant variants. J Am Chem Soc 130:6099–6113. doi:10.1021/ja076558p

Amrane S, Kerkour A, Bedrat A, Vialet B, Andreola ML, Mergny JL (2014) Topology of a DNA G-quadruplex structure formed in the HIV-1 promoter: a potential target for anti-HIV drug development. J Am Chem Soc 136:5249–5252. doi:10.1021/ja501500c

Aoki H, Takada Y, Kondo S, Sawaya R, Aggarwal BB, Kondo Y (2007) Evidence that curcumin suppresses the growth of malignant gliomas in vitro and in vivo through induction of autophagy: role of Akt and extracellular signal-regulated kinase signaling pathways. Mol Pharmacol 72:29–39. doi:10.1124/mol.106.033167

Ben Sahra I, Laurent K, Giuliano S, Larbret F, Ponzio G, Gounon P, Le Marchand-Brustel Y, Giorgetti-Peraldi S, Cormont M, Bertolotto C, Deckert M, Auberger P, Tanti JF, Bost F (2010) Targeting cancer cell metabolism: the combination of metformin and 2-deoxyglucose induces p53-dependent apoptosis in prostate cancer cells. Cancer Res 70:2465–2475. doi:10.1158/0008-5472.CAN-09-2782

Bradeen HA, Eide CA, O'Hare T, Johnson KJ, Willis SG, Lee FY, Druker BJ, Deininger MW (2006) Comparison of imatinib mesylate, dasatinib (BMS-354825), and nilotinib (AMN107) in an N-ethyl-N-nitrosourea (ENU)-based mutagenesis screen: high efficacy of drug combinations. Blood 108:2332–2338. doi:10.1182/blood-2006-02-004580

Brown D (2015) Antibiotic resistance breakers: can repurposed drugs fill the antibiotic discovery void? Nat Rev Drug Discov 14:821–832. doi:10.1038/nrd4675

Burdock GA, Carabin IG, Griffiths JC (2006) The importance of GRAS to the functional food and nutraceutical industries. Toxicology 221:17–27. doi:10.1016/j.tox.2006.01.012

Byler S, Sarkar S (2014) Do epigenetic drug treatments hold the key to killing cancer progenitor cells? Epigenomics 6:161–165. doi:10.2217/epi.14.4

Byler S, Goldgar S, Heerboth S, Leary M, Housman G, Moulton K, Sarkar S (2014) Genetic and epigenetic aspects of breast cancer progression and therapy. Anticancer Res 34:1071–1077

Cao L, Song W, De Clercq E, Zhan P, Liu X (2014) Recent progress in the research of small molecule HIV-1 RNase H inhibitors. Curr Med Chem 21:1956–1967. doi:10.2174/0929867321666140120121158

Chen CYC (2011) TCM Database@Taiwan: the world's largest traditional Chinese medicine database for drug screening in silico. PLoS One 6:e15939. doi:10.1371/journal.pone.0015939

Cressey D (2011) Traditional drug-discovery model ripe for reform. Nature 471:17–18. doi:10.1038/471017a

Csermely P, Korcsmáros T, Kiss HJ, London G, Nussinov R (2013) Structure and dynamics of molecular networks: a novel paradigm of drug discovery: a comprehensive review. Pharmacol Ther 138:333–408. doi:10.1016/j.pharmthera.2013.01.016

Dewan V, Liu T, Chen KM, Qian Z, Xiao Y, Kleiman L, Mahasenan KV, Li C, Matsuo H, Pei D, Musier-Forsyth K (2012) Cyclic peptide inhibitors of HIV-1 capsid-human lysyl-tRNA synthetase interaction. ACS Chem Biol 7:761–769. doi:10.1021/cb200450w

Dihazi GH, Bibi A, Jahn O, Nolte J, Mueller GA, Engel W, Dihazi H (2013) Impact of the antiproliferative agent ciclopirox olamine treatment on stem cells proteome. World J Stem Cells 5:9–25. doi:10.4252/wjsc.v5.i1.9

Dittmar W, Grau W, Raether W, Schrinner E, Wagner WH (1981) Microbiological laboratory studies with ciclopiroxolamine. Drug Res 31:1317–1322

Domadia PN, Bhunia A, Sivaraman J, Swarup S, Dasgupta D (2008) Berberine targets assembly of *Escherichia coli* cell division protein FtsZ. Biochemistry 47:3225–3234. doi:10.1021/bi7018546

Duarte NC, Becker SA, Jamshidi N, Thiele I, Mo ML, Vo TD, Srivas R, Palsson BØ (2007) Global reconstruction of the human metabolic network based on genomic and bibliomic data. Proc Natl Acad Sci USA 104:1777–1782. doi:10.1073/pnas.0610772104

Ejim L, Farha MA, Falconer SB, Wildenhain J, Coombes BK, Tyers M, Brown ED, Wright GD (2011) Combinations of antibiotics and nonantibiotic drugs enhance antimicrobial efficacy. Nat Chem Biol 7:348–350. doi:10.1038/nchembio.559

Ennifar E, Aslam MW, Strasser P, Hoffmann G, Dumas P, van Delft FL (2013) Structure-guided discovery of a novel aminoglycoside conjugate targeting HIV-1 RNA viral genome. ACS Chem Biol 8:2509–2517. doi:10.1021/cb400498n

Floridi A, Paggi MG, Marcante ML, Silvestrini B, Caputo A, De Martino C (1981) Lonidamine, a selective inhibitor of aerobic glycolysis of murine tumor cells. J Natl Cancer Inst 66:497–499. doi:10.1093/jnci/66.3.497

Fry DW, Harvey PJ, Keller PR, Elliott WL, Meade M, Trachet E, Albassam M, Zheng X, Leopold WR, Pryer NK, Toogood PL (2004) Specific inhibition of cyclin-dependent kinase 4/6 by PD 0332991 and associated antitumor activity in human tumor xenografts. Mol Cancer Ther 3:1427–1438

Fulda S, Galluzzi L, Kroemer G (2010) Targeting mitochondria for cancer therapy. Nat Rev Drug Discov 9:447–464. doi:10.1038/nrd3137

Gelbert LM, Cai S, Lin X, Sanchez-Martinez C, Del Prado M, Lallena MJ, Torres R, Ajamie RT, Wishart GN, Flack RS, Neubauer BL, Young J, Chan EM, Iversen P, Cronier D, Kreklau E, de Dios A (2014) Preclinical characterization of the CDK4/6 inhibitor LY2835219: in-vivo cell cycle-dependent/independent anti-tumor activities alone/in combination with gemcitabine. Invest New Drugs 32:825–837. doi:10.1007/s10637-014-0120-7

Gottesman MM, Fojo T, Bates SE (2002) Multidrug resistance in cancer: role of ATP-dependent transporters. Nat Rev Cancer 2:48–58. doi:10.1038/nrc706

Gregori-Puigjané E, Setola V, Hert J, Crews BA, Irwin JJ, Lounkine E, Marnett L, Roth BL, Shoichet BK (2012) Identifying mechanism-of-action targets for drugs and probes. Proc Natl Acad Sci USA 109:11178–11183. doi:10.1073/pnas.1204524109

Herschhorn A, Gu C, Espy N, Richard J, Finzi A, Sodroski JG (2014) A broad HIV-1 inhibitor blocks envelope glycoprotein transitions critical for entry. Nat Chem Biol 10:845–852. doi:10.1038/nchembio.1623

Hirschmann-Jax C, Foster AE, Wulf GG, Nuchtern JG, Jax TW, Gobel U, Goodell MA, Brenner MK (2004) A distinct 'side population' of cells with high drug efflux capacity in human tumor cells. Proc Natl Acad Sci USA 101:14228–14233. doi:10.1073/pnas.0400067101

Huber K, Doyon G, Plaks J, Fyne E, Mellors JW, Sluis-Cremer N (2011) Inhibitors of histone deacetylases: correlation between isoform specificity and reactivation of HIV type 1 (HIV-1) from latently infected cells. J Biol Chem 286:22211–22218. doi:10.1074/jbc.M110.180224

Hutchinson MR, Zhang Y, Shridhar M, Evans JH, Buchanan MM, Zhao TX, Slivka PF, Coats BD, Rezvani N, Wieseler J, Hughes TS, Landgraf KE, Chan S, Fong S, Phipps S, Falke JJ, Leinwand LA, Maier SF, Yin H, Rice KC, Watkins LR (2010) Evidence that opioids may have toll like receptor 4 and MD-2 effects. Brain Behav Immun 24:83–95. doi:10.1016/j.bbi.2009.08.004

Jacoby E (2011) Computational chemogenomics. Wiley Interdiscip Rev Comput Mol Sci 1:57–67. doi:10.1002/wcms.11

Jeong HW, Hsu KC, Lee JW, Ham M, Huh JY, Shin HJ, Kim WS, Kim JB (2009) Berberine suppresses proinflammatory responses through AMPK activation in macrophages. Am J Physiol Endocrinol Metab 296:955–964. doi:10.1152/ajpendo.90599.2008

Juergens R, Wrangle J, Vendetti F, Murphy SC, Zhao M, Coleman B, Sebree R, Rodgers K, Hooker CM, Franco N, Lee B, Tsai S, Delgado IE, Rudek MA, Belinsky SA, Herman JG, Baylin SB, Brock MV, Rudin CM (2011) Combination epigenetic therapy has efficacy in patients with refractory advanced non-small cell lung cancer. Cancer Discov 1:598–607. doi:10.1158/2159-8290.CD-11-0214

Kamal A, Shankaraiah N, Shalini N (2014) Recent horizons in drug discovery and development. Curr Trends Pharm Sci 2:6–11

Kanadaswami C, Lee LT, Lee PP, Hwang JJ, Ke FC, Huang YT, Lee MT (2005) The antitumor activities of flavonoids. In Vivo 19:895–909

Kang D, Song Y, Chen W, Zhan P, Liu X (2014) "Old Dogs with New Tricks": exploiting alternative mechanisms of action and new drug design strategies for clinically validated HIV targets. Mol Biosyst 10:1998–2022. doi:10.1039/c4mb00147h

Kantarjian H, Shah NP, Hochhaus A, Cortes J, Shah S, Ayala M, Moiraghi B, Shen Z, Mayer J, Pasquini R, Nakamae H, Huguet F, Boqué C, Chuah C, Bleickardt E, Bradley-Garelik MB, Zhu C, Szatrowski T, Shapiro D, Baccarani M (2010) Dasatinib versus imatinib in newly diagnosed chronic-phase chronic myeloid leukemia. N Engl J Med 362:2260–2270. doi:10.1056/NEJMoa1002315

Kaspar AA, Reichert JM (2013) Future directions for peptide therapeutics development. Drug Discov Today 18:807–817. doi:10.1016/j.drudis.2013.05.011

Khanna I (2012) Drug discovery in pharmaceutical industry: productivity challenges and trends. Drug Discov Today 17:1088–1102. doi:10.1016/j.drudis.2012.05.007

Kim SH, Shin DS, Oh MN, Chung SC, Lee JS, Oh KB (2004) Inhibition of the bacterial surface protein anchoring transpeptidase sortase by isoquinoline alkaloids. Biosci Biotechnol Biochem 68:421–424. doi:10.1271/bbb.68.421

Kim HU, Kim TY, Lee SY (2010) Genome-scale metabolic network analysis and drug targeting of multi-drug resistant pathogen *Acinetobacter baumannii* AYE. Mol Biosyst 6:339–348. doi:10.1039/b916446d

Kitano H (2007) A robustness-based approach to systems-oriented drug design. Nat Rev Drug Discov 6:202–210. doi:10.1038/nrd2195

Koutsoukas A, Simms B, Kirchmair J, Bond PJ, Whitmore AV, Zimmer S, Young MP, Jenkins JL, Glick M, Glen RC, Bender A (2011) From in silico target prediction to multi-target drug design: current databases, methods and applications. J Proteomics 74:2554–2574. doi:10.1016/j.jprot.2011.05.011

Lavie Y, Cao H, Volner A, Lucci A, Han TY, Geffen V, Giuliano AE, Cabot MC (1997) Agents that reverse multidrug resistance, tamoxifen, verapamil, and cyclosporin A, block glycosphingolipid metabolism by inhibiting ceramide glycosylation in human cancer cells. J Biol Chem 272:1682–1687. doi:10.1074/jbc.272.3.1682

Lim S, Kaldis P (2013) Cdks, cyclins and CKIs: roles beyond cell cycle regulation. Development 140:3079–3093. doi:10.1242/dev.091744

Liscovitch M, Lavie Y (2002) Cancer multidrug resistance: a review of recent drug discovery research. IDrugs 5:349–355

Liu YY, Han TY, Giuliano AE, Hansen N, Cabot MC (2000) Uncoupling ceramide glycosylation by transfection of glucosylceramide synthase antisense reverses adriamycin resistance. J Biol Chem 275:7138–7143. doi:10.1074/jbc.275.10.7138

Logsdon BC, Vickrey JF, Martin P, Proteasa G, Koepke JI, Terlecky SR, Wawrzak Z, Winters MA, Merigan TC, Kovari LC (2004) Crystal structures of a multidrug-resistant human immunodeficiency virus type 1 protease reveal an expanded active-site cavity. J Virol 78:3123–3132. doi:10.1128/JVI.78.6.3123-3132.2004

Lounkine E, Kutchukian P, Petrone P, Davies JW, Glick M (2012) Chemotography for multi-target SAR analysis in the context of biological pathways. Bioorg Med Chem 20:5416–5427. doi:10.1016/j.bmc.2012.02.034

Lozano C, Torres JL, Julia L, Jimenez A, Centelles JJ, Cascante M (2005) Effect of new antioxidant cysteinyl-flavanol conjugates on skin cancer cells. FEBS Lett 579:4219–4225. doi:10.1016/j.febslet.2005.06.051

Lucci A, Giuliano AE, Han TY, Dinur T, Liu YY, Senchenkov A, Cabot MC (1999) Ceramide toxicity and metabolism differ in wild-type and multidrug-resistant cancer cells. Int J Oncol 15:535–540. doi:10.3892/ijo.15.3.535

Maga G, Radi M, Gerard MA, Botta M, Ennifar E (2010) HIV-1 RT inhibitors with a novel mechanism of action: NNRTIs that compete with the nucleotide substrate. Viruses 2:880–899. doi:10.3390/v2040880

Marsden MD, Zack JA (2013) HIV/AIDS eradication. Bioorg Med Chem Lett 23:4003–4010. doi:10.1016/j.bmcl.2013.05.032

Maxmen A (2011) Translational research: the American way. Nature 478:S16–S18. doi:10.1038/478S16a

Mbonye U, Karn J (2014) Transcriptional control of HIV latency: cellular signaling pathways, epigenetics, happenstance and the hope for a cure. Virology 454–455:328–339. doi:10.1016/j.virol.2014.02.008

Medina-Franco JL, Martínez-Mayorga K, Peppard TL, Del Rio A (2012) Chemoinformatic analysis of GRAS (Generally Recognized as Safe) flavor chemicals and natural products. PLoS One 7:e50798. doi:10.1371/journal.pone.0050798

Moghadamtousi SZ, Kadir HA, Hassandarvish P, Tajik H, Abubakar S, Zandi K (2014) A review on antibacterial, antiviral, and antifungal activity of curcumin. Biomed Res Int 2014:186864. doi:10.1155/2014/186864

Mori M, Manetti F, Botta M (2011) Targeting protein-protein and protein-nucleic acid interactions for anti-HIV therapy. Curr Pharm Des 17:3713–3728. doi:10.2174/138161211798220972

Mullard A (2012) 2011 in reflection. Nat Rev Drug Discov 11:6–8. doi:10.1038/nrd3643

Mun SH, Joung DK, Kim YS, Kang OH, Kim SB, Seo YS, Kim YC, Lee DS, Shin DW, Kweon KT, Kwon DY (2013) Synergistic antibacterial effect of curcumin against methicillin-resistant *Staphylococcus aureus*. Phytomedicine 20:714–718. doi:10.1016/j.phymed.2013.02.006

Munos B (2009) Lessons from 60 years of pharmaceutical innovation. Nat Rev Drug Discov 8:959–968. doi:10.1038/nrd2961

Nalam MN, Schiffer CA (2008) New approaches to HIV protease inhibitor drug design II: testing the substrate envelope hypothesis to avoid drug resistance and discover robust inhibitors. Curr Opin HIV AIDS 3:642–646. doi:10.1097/COH.0b013e3283136cee

Nguyen AT, Feasley CL, Jackson KW, Nitz TJ, Salzwedel K, Air GM, Sakalian M (2011) The prototype HIV-1 maturation inhibitor, bevirimat, binds to the CA-SP1 cleavage site in immature Gag particles. Retrovirology 8:101. doi:10.1186/1742-4690-8-101

Niewerth M, Kunze D, Seibold M, Schaller M, Korting HC, Hube B (2003) Ciclopirox olamine treatment affects the expression pattern of *Candida albicans* genes encoding virulence factors, iron metabolism proteins, and drug resistance factors. Antimicrob Agents Chemother 47:1805–1817. doi:10.1128/AAC.47.6.1805-1817.2003

O'Hare T, Shakespeare WC, Zhu X, Eide CA, Rivera VM, Wang F, Adrian LT, Zhou T, Huang WS, Xu Q, Metcalf CA 3rd, Tyner JW, Loriaux MM, Corbin AS, Wardwell S, Ning Y, Keats JA, Wang Y, Sundaramoorthi R, Thomas M, Zhou D, Snodgrass J, Commodore L, Sawyer TK, Dalgarno DC, Deininger MW, Druker BJ, Clackson T (2009) AP24534, a pan-BCR-ABL inhibitor for chronic myeloid leukemia, potently inhibits the T315I mutant and overcomes mutation-based resistance. Cancer Cell 16:401–412. doi:10.1016/j.ccr.2009.09.028

Okarvi SM (2008) Peptide-based radiopharmaceuticals and cytotoxic conjugates: potential tools against cancer. Cancer Treat Rev 34:13–26. doi:10.1016/j.ctrv.2007.07.017

Park BS, Kim JG, Kim MR, Lee SE, Takeoka GR, Oh KB, Kim JH (2005) *Curcuma longa* L. constituents inhibit sortase A and *Staphylococcus aureus* cell adhesion to fibronectin. J Agric Food Chem 53:9005–9009. doi:10.1021/jf051765z

Prabhudesai PP, Jain S, Keshvani A, Kulkarni KP (2011) The efficacy and safety of amoxicillin-clavulanic acid 1000/125mg twice daily extended release (XR) tablet for the treatment of bacterial community-acquired pneumonia in adults. J Indian Med Assoc 109:124–127

Prabu-Jeyabalan M, King NM, Nalivaika EA, Heilek-Snyder G, Cammack N, Schiffer CA (2006) Substrate envelope and drug resistance: crystal structure of RO1 in complex with wild-type human immunodeficiency virus type 1 protease. Antimicrob Agents Chemother 50:1518–1521. doi:10.1128/AAC.50.4.1518-1521.2006

Raghunand N, He X, van Sluis R, Mahoney B, Baggett B, Taylor CW, Paine-Murrieta G, Roe D, Bhujwalla ZM, Gillies RJ (1999a) Enhancement of chemotherapy by manipulation of tumour pH. Br J Cancer 80:1005–1011. doi:10.1038/sj.bjc.6690455

Raghunand N, Martínez-Zaguilán R, Wright SH, Gillies RJ (1999b) pH and drug resistance. II. Turnover of acidic vesicles and resistance to weakly basic chemotherapeutic drugs. Biochem Pharmacol 57:1047–1058. doi:10.1016/S0006-2952(99)00021-0

Reymond JL, van Deursen R, Blum LC, Ruddigkeit L (2010) Chemical space as a source for new drugs. MedChemComm 1:30–38. doi:10.1039/C0MD00020E

Rullmann JA, Struemper H, Defranoux NA, Ramanujan S, Meeuwisse CM, van Elsas A (2005) Systems biology for battling rheumatoid arthritis: application of the Entelos PhysioLab platform. IEEE Proc Syst Biol 152:256–262. doi:10.1049/ip-syb:20050053

Saglio G, Kim DW, Issaragrisil S, le Coutre P, Etienne G, Lobo C, Pasquini R, Clark RE, Hochhaus A, Hughes TP, Gallagher N, Hoenekopp A, Dong M, Haque A, Larson RA, Kantarjian HM (2010) Nilotinib versus imatinib for newly diagnosed chronic myeloid leukemia. N Engl J Med 362:2251–2259. doi:10.1056/NEJMoa0912614

Sarkar S, Goldgar S, Byler S, Rosenthal S, Heerboth S (2013a) Demethylation and re-expression of epigenetically silenced tumor suppressor genes: sensitization of cancer cells by combination therapy. Epigenomics 5:87–94. doi:10.2217/epi.12.68

Sarkar S, Horn G, Moulton K, Oza A, Byler S, Kokolus S, Longacre M (2013b) Cancer development, progression and therapy: an epigenetic overview. Int J Mol Sci 14:21087–21113. doi:10.3390/ijms141021087

Scaltriti M, Eichhorn PJ, Cortés J, Prudkin L, Aura C, Jiménez J, Chandarlapaty S, Serra V, Prat A, Ibrahim YH, Guzmán M, Gili M, Rodríguez O, Rodríguez S, Pérez J, Green SR, Mai S, Rosen N, Hudis C, Baselga J (2011) Cyclin E amplification/ overexpression is a mechanism of trastuzumab resistance in HER2+ breast cancer patients. Proc Natl Acad Sci USA 108:3761–3766. doi:10.1073/pnas.1014835108

Segal NH, Verghis J, Govindan S, Maliakal P, Sharkey RM, Wegener WA, Goldenberg DM, Saltz LB (2013) A phase I study of IMMU-130 (labetuzumab-SN38) anti-CEACAM5 antibody–drug conjugate (ADC) in patients with metastatic colorectal cancer (mCRC). Cancer Res 73: LB-159. doi:10.1158/1538-7445.AM2013-LB-159

Senchenkov A, Litvak DA, Cabot MC (2001) Targeting ceramide metabolism–a strategy for overcoming drug resistance. J Natl Cancer Inst 93:347–357. doi:10.1093/jnci/93.5.347

Senter P, Sievers EL (2012) The discovery and development of brentuximab vedotin for use in relapsed Hodgkin lymphoma and systemic anaplastic large cell lymphoma. Nat Biotechnol 30:631–637. doi:10.1038/nbt.2289

Shen Y, Altman MD, Ali A, Nalam MN, Cao H, Rana TM, Schiffer CA, Tidor B (2013) Testing the substrate-envelope hypothesis with designed pairs of compounds. ACS Chem Biol 8:2433–2441. doi:10.1021/cb400468c

Sloan RD, Wainberg MA (2013) Harnessing the therapeutic potential of host antiviral restriction factors that target HIV. Expert Rev Anti Infect Ther 11:1–4. doi:10.1586/eri.12.146

Sparreboom A, Planting AS, Jewell RC, van der Burg ME, van der Gaast A, de Bruijn P, Loos WJ, Nooter K, Chandler LH, Paul EM, Wissel PS, Verweij J (1999) Clinical pharmacokinetics of doxorubicin in combination with GF120918, a potent inhibitor of MDR1 P-glycoprotein. Anticancer Drugs 10:719–728. doi:10.1097/00001813-199909000-00005

Steinmann J, Buer J, Pietschmann T, Steinmann E (2013) Anti-infective properties of epigallocatechin-3-gallate (EGCG), a component of green tea. Br J Pharmacol 168:1059–1073. doi:10.1111/bph.12009

Swinney DC, Anthony J (2011) How were new medicines discovered? Nat Rev Drug Discov 10:507–519. doi:10.1038/nrd3480

Szarc vel Szic K, Ndlovu MN, Haegeman G, Vanden Berghe W (2010) Nature or nurture: let food be your epigenetic medicine in chronic inflammatory disorders. Biochem Pharmacol 80:1816–1832. doi:10.1016/j.bcp.2010.07.029

Tascini C, Tagliaferri E, Giani T, Leonildi A, Flammini S, Casini B, Lewis R, Ferranti S, Rossolini GM, Menichetti F (2013) Synergistic activity of colistin plus rifampin against colistin-resistant KPC-producing *Klebsiella pneumoniae*. Antimicrob Agents Chemother 57:3990–3993. doi:10.1128/AAC.00179-13

Thayer AM (2011) Shifting generics to the super side. Chem Eng News 22–26

Toogood PL, Harvey PJ, Repine JT, Sheehan DJ, VanderWel SN, Zhou H, Keller PR, McNamara DJ, Sherry D, Zhu T, Brodfuehrer J, Choi C, Barvian MR, Fry DW (2005) Discovery of a potent and selective inhibitor of cyclin-dependent kinase 4/6. J Med Chem 48:2388–2406. doi:10.1021/jm049354h

Tuske S, Sarafianos SG, Clark AD Jr, Ding J, Naeger LK, White KL, Miller MD, Gibbs CS, Boyer PL, Clark P, Wang G, Gaffney BL, Jones RA, Jerina DM, Hughes SH, Arnold E (2004) Structures of HIV-1 RT-DNA complexes before and after incorporation of the anti-AIDS drug tenofovir. Nat Struct Mol Biol 11:469–474. doi:10.1038/nsmb760

Uwagawa T, Misawa T, Iida T, Sakamoto T, Gocho T, Wakiyama S, Hirohara S, Yanaga K (2010) Proton-pump inhibitor as palliative care for chemotherapy-induced gastroesophageal reflux disease in pancreatic cancer patients. J Palliat Med 13:815–818. doi:10.1089/jpm.2009.0404

VanderWel SN, Harvey PJ, McNamara DJ, Repine JT, Keller PR, Quin J 3rd, Booth RJ, Elliott WL, Dobrusin EM, Fry DW, Toogood PL (2005) Pyrido[2,3-d]pyrimidin-7-ones as specific inhibitors of cyclin-dependent kinase 4. J Med Chem 48:2371–2387. doi:10.1021/jm049355+

Vercruysse T, Basta B, Dehaen W, Humbert N, Balzarini J, Debaene F, Sanglier-Cianferani S, Pannecouque C, Mely Y, Daelemans D (2012) A phenyl-thiadiazolylidene-amine derivative ejects zinc from retroviral nucleocapsid zinc fingers and inactivates HIV virions. Retrovirology 9:95. doi:10.1186/1742-4690-9-95

Verma S, Miles D, Gianni L, Krop IE, Welslau M, Baselga J, Pegram M, Oh D-Y, Diéras V, Guardino E, Fang L, Lu MW, Olsen S, Blackwell K (2012) Trastuzumab emtansine for HER2-positive advanced breast cancer. N Engl J Med 367:1783–1791. doi:10.1056/NEJMoa1209124

Vogt A, Lazo JS (2005) Chemical complementation: a definitive phenotypic strategy for identifying small molecule inhibitors of elusive cellular targets. Pharmacol Ther 107:212–221. doi:10.1016/j.pharmthera.2005.03.002

Vora SR, Juric D, Kim N, Mino-Kenudson M, Huynh T, Costa C, Lockerman EL, Pollack SF, Liu M, Li X, Lehar J, Wiesmann M, Wartmann M, Chen Y, Cao ZA, Pinzon-Ortiz M, Kim S, Schlegel R, Huang A, Engelman JA (2014) CDK 4/6 inhibitors sensitize PIK3CA mutant breast cancer to PI3K inhibitors. Cancer Cell 26:136–149. doi:10.1016/j.ccr.2014.05.020

Weisberg E, Choi HG, Ray A, Barrett R, Zhang J, Sim T, Zhou W, Seeliger M, Cameron M, Azam M, Fletcher JA, Debiec-Rychter M, Mayeda M, Moreno D, Kung AL, Janne PA, Khosravi-Far R, Melo JV, Manley PW, Adamia S, Wu C, Gray N, Griffin JD (2010) Discovery of a small-molecule type II inhibitor of wild-type and gatekeeper mutants of BCR-ABL, PDGFRalpha, Kit, and Src kinases: novel type II inhibitor of gatekeeper mutants. Blood 115:4206–4216. doi:10.1182/blood-2009-11-251751

Xue W, Ban Y, Liu H, Yao X (2014) Computational study on the drug resistance mechanism against HCV NS3/4A protease inhibitors vaniprevir and MK-5172 by the combination use of molecular dynamics simulation, residue interaction network, and substrate envelope analysis. J Chem Inf Model 54:621–633. doi:10.1021/ci400060j

Younes A, Bartlett NL, Leonard JP, Kennedy DA, Lynch CM, Sievers EL, Forero-Torres A (2010) Brentuximab vedotin (SGN-35) for relapsed CD30-positive lymphomas. N Engl J Med 363:1812–1821. doi:10.1056/NEJMoa1002965

Zentner I, Sierra LJ, Maciunas L, Vinnik A, Fedichev P, Mankowski MK, Ptak RG, Martin-Garcia J, Cocklin S (2013) Discovery of a small-molecule antiviral targeting the HIV-1 matrix protein. Bioorg Med Chem Lett 23:1132–1135. doi:10.1016/j.bmcl.2012.11.041

Zhan P, Liu X, De Clercq E (2010) Blocking nuclear import of pre-integration complex: an emerging anti-HIV-1 drug discovery paradigm. Curr Med Chem 17:495–503. doi:10.2174/092986710790416335

Zhang YM, Rock CO (2004) Evaluation of epigallocatechin gallate and related plant polyphenols as inhibitors of the FabG and FabI reductases of bacterial type II fatty-acid synthesis. J Biol Chem 279:30994–31001. doi:10.1074/jbc.m403697200

Zhao WH, Hu ZQ, Hara Y, Shimamura T (2002) Inhibition of penicillinase by epigallocatechin gallate resulting in restoration of antibacterial activity of penicillin against penicillinase-producing *Staphylococcus aureus*. Antimicrob Agents Chemother 46:2266–2268. doi:10.1128/AAC.46.7.2266-2268.2002

Zhou H, Shen T, Shang C, Luo Y, Liu L, Yan J, Li Y, Huang S (2014) Ciclopirox induces autophagy through reactive oxygen species-mediated activation of JNK signaling pathway. Oncotarget 5:10140–10150. doi:10.18632/oncotarget.2471

Zimmermann GR, Lehar J, Keith CT (2007) Multi-target therapeutics: when the whole is greater than the sum of the parts. Drug Discov Today 12:34–42. doi:10.1016/j.drudis.2006.11.008

MIX
Papier aus verantwortungsvollen Quellen
Paper from responsible sources
FSC® C105338

If you have any concerns about our products,
you can contact us on
ProductSafety@springernature.com

In case Publisher is established outside the EU,
the EU authorized representative is:
Springer Nature Customer Service Center GmbH
Europaplatz 3, 69115 Heidelberg, Germany

Printed by Libri Plureos GmbH
in Hamburg, Germany